METAL, CERAMIC AND POLYMERIC COMPOSITES FOR VARIOUS USES

Edited by **John Cuppoletti**

Metal, Ceramic and Polymeric Composites for Various Uses
Edited by John Cuppoletti

Published by InTech

Janeza Trdine 9, 51000 Rijeka, Croatia

Publishing Process Manager Romina Krebel

Technical Editor Teodora Smiljanic

Cover Designer InTech Design Team

Additional hard copies can be obtained from orders@intechopen.com

Metal, Ceramic and Polymeric Composites for Various Uses, Edited by John Cuppoletti
p. cm.
ISBN 978-953-307-353-8

Meet the editor

Dr. Cuppoletti is Professor of Molecular and Cellular Physiology at the University of Cincinnati. (Ph.D. 1977 University of California, Davis). His research interests include pumps, carriers, and ion channels of the epithelia. In recent years, potassium and chloride channel regulation, pharmacology, and molecular engineering have become increasingly important tools with which to study the functional mechanisms of these proteins. More recently, as the X-ray crystal structures of these proteins have become available, it has become possible to combine these types of studies with computational chemistry approaches in attempts to understand how the structure relates to the function of these proteins. These transport proteins have been placed into synthetic membranes in order to obtain more detailed information regarding protein function that cannot be obtained from purely biological membranes. The natural and engineered proteins in these membranes are functional, and have led to the development of many interesting devices. Examples include drug screening devices, devices for identification of toxic bacterial products, and membranes for use in fuel cells.

Contents

Preface XIII

Part 1 New Materials and Analytic Methods 1

Chapter 1 **Growth Reinforcing Composite Materials from Liquidus Phase: Mechanical and Microstructural Parameters Relationship Essentially Evincing the Predominance of an Akin Mass Composition over the Domain of Compositions 3**
B.L. Sharma and Parshotam Lal

Chapter 2 **In-Situ Formation of TiC Using Laser Cladding 33**
Ali Emamian, Stephen F. Corbin and Amir Khajepour

Chapter 3 **New Superhard Ternary Borides in Composite Materials 61**
Zachary Zachariev

Chapter 4 **Thermo-Mechanical Treatment of Glass-Balloon-Dispersed Metal Matrix Composite 79**
Osamu Umezawa

Chapter 5 **Inversion of Physical Properties for Determining the Microstructure of Natural Composites 97**
Kazatchenko E., Markov M., Mousatov A. and Pervago E.

Chapter 6 **Hybrid Inorganic–organic Composites of Layered Double Hydroxides Intercalated with Organic Acid Anions for the Uptake of Hazardous Substances from Aqueous Solution 123**
Tomohito Kameda and Toshiaki Yoshioka

Chapter 7 **Hybrid Materials Based on Zn-Al Alloys 149**
G. Torres-Villaseñor and E. Martínez-Flores

Chapter 8 **Mechanical Behavior of Filled Thermoplastic Polymers 171**
S. Bazhenov

Chapter 9 **Porous Ceramic Matrix Al_2O_3/Al Composites as Supports and Precursors for Catalysts and Permeable Materials 195**
S.F.Tikhov, N.A.Pakhomov, E.I.Nemykina, A.N.Salanov, V.A.Sadykov, V.E.Romanenkov and Ya.Ya. Piatsiushyk

Chapter 10 **On Thermal Conductivity of an In-Situ Metal Matrix Composite - Cast Iron 211**
J.K.Chen and S.F.Chen

Chapter 11 **Synthesis and Characterization of Ordered Mesoporous Silica Pillared Clay with HPW Heteropoly Acid Encapsulated into the Framework and Its Catalytic Performance for Deep Oxidative Desulfurization of Fuels 225**
Baoshan Li, Zhenxing Liu, Jianjun Liu, Zhiyuan Zhou, Xiaohui Gao, Xinmei Pang and Huiting Sheng

Chapter 12 **Biocomposites Based on New Monomer Systems Reinforced with Micro / Nanoparticles and Glass Fibers 239**
Cristina Prejmerean, Tinca Buruiana, Teresa Nunes, Marioara Moldovan and Loredana Colceriu

Chapter 13 **Wollastonite-Based Chemically Bonded Phosphate Ceramic Composites 265**
Henry A. Colorado, Clem Hiel, Thomas Hahn and Jenn Ming Yang

Chapter 14 **Microstructure and High-Temperature Strength of 9CrODS Ferritic Steel 283**
Shigeharu Ukai

Chapter 15 **Spider Silk Composites and Applications 303**
Yang Hsia, Eric Gnesa, Felicia Jeffery, Simon Tang and Craig Vierra

Chapter 16 **Natural Silkworm Silk-Epoxy Resin Composite for High Performance Application 325**
K.M. Kelvin Loh and C.K. Willy Tan

Chapter 17 **Noncrystalline L-Phenylalanine-Silica Hybrid Composite Materials for High Selective Reversed Phase Liquid Chromatography 341**
M. Mizanur Rahman, Makoto Takafuji and Hirotaka Ihara

Part 2 **New Materials with Unique Properties 355**

Chapter 18 **High Dispersion Power of Cardo-Typed Fluorene Moieties on Carbon Fillers 357**
Shinichi Kawasaki, Fengzhe Jin and Toshikazu Takata

Chapter 19 **Optical Properties and Some Applications of Plasmonic Heterogeneous Materials 377**
Sergey Moiseev

Chapter 20 **Polymer-Bioglass Composite Coatings: A Promising Alternative for Advanced Biomedical Implants 393**
Floroian Laura, Popescu Andrei, Serban Natalia and Mihailescu Ion N.

Chapter 21 **Effects of Filler Content on Mechanical and Optical Properties of Dental Composite Resins 421**
Seyed Mostafa Mousavinasab

Part 3 **Applications of New Materials 429**

Chapter 22 **Tuneable Composites Containing Magnetic Microwires 431**
L. Panina, M. Ipatov, V. Zhukova, J. Gonzalez and A. Zhukov

Chapter 23 **3D Network SiC-Metals Composites for Heavy Duty Brake Applications 461**
Hongqiang Ru, Jingyang Li and Wei Wang

Chapter 24 **Smart Magnetic Composites 475**
Jerzy Kaleta, Daniel Lewandowski, Rafał Mech and Piotr Zając

Chapter 25 **Magnetic and Dynamic Mechanical Properties of Nd-Fe-B Composite Materials with Polymer Matrix 505**
Aleksandar Grujić, Jasna Stajić-Trošić, Mirko Stijepović, Jasmina Stevanović and Radoslav Aleksić

Chapter 26 **Composite Materials for Some Radiophysics Applications 525**
S.B.Bibikov and M.V. Prokof'ev

Chapter 27 **The Composite Materials for Localization of Volatile Radioactive Iodine Forms from Steam-Air Phase during Severe Accidents at NPPs 545**
Sergey A. Kulyukhin, Lubov' V. Mizina, Igor' A. Rumer and Nikolay B. Mikheev

Chapter 28 **Composite Materials under Extreme Radiation and Temperature Environments of the Next Generation Nuclear Reactors 571**
Nicholas Simos

Chapter 29 **Graphite-Composites Alternatives for Electrochemical Biosensor 597**
Ninoska Bojorge and Eliana Alhadeff

Chapter 30 **Composite Cathode Material for Li-Ion Batteries Based on $LiFePO_4$ System 621**
Janina Molenda and Marcin Molenda

Chapter 31 **Multilayer Fresnel Zone Plate with High- Diffraction Efficiency: Application of Composite Layer to X-ray Optics 637**
Shigeharu Tamura

Chapter 32 **New Composite Materials for Decreasing of Radioactive Molecular Iodine in the Water Coolant in Working and New Developed NPPs 655**
Sergey A. Kulyukhin and Lubov' V. Mizina

Chapter 33 **The Glass Transition Temperature in Dental Composites 669**
J.C.S. Moraes, M.M.D.S. Sostena and C.R. Grandini

Preface

This book contains chapters on composites for engineering lighter and stronger materials for structural engineering. In addition to passive structural engineering materials, there are also materials for applications such as lithium ion batteries and improved materials for "smart" magnetic materials which can be tuned or change shape. The book contains chapters on materials that are in use, as well as chapters on the development and testing of materials that could be used in the future.

Human health in the developed and developing world can benefit from advances in composite materials. Examples in this book include materials for monitoring the state of human physiology, the development of materials for repair of teeth and materials that will stimulate the growth of bone.

With the large impact of 7 billion people on earth and rapid development throughout the world, some of these new composites will have a positive impact on the environment. Examples include the development of approaches to improve natural silk materials without using technology that might harm the environment, development of materials to reduce radioactive materials in the case of a nuclear accident, and an environmentally friendly Portland cement replacement.

The work shows that some of the work of the investigators is driven by concern for the environment and health of society. The work could have an impact in improving the quality of life.

The chapters in this book on composites are diverse. The 33 chapters comprising this book have been grouped into three main parts:

1) New materials and analytic methods: This part focuses on research on composites of metals and ceramics as well as natural silk and silk epoxy composites.

2) New Materials with unique properties: This part presents studies in which composite materials have been developed for applications of plasmons and for composites which will stimulate bone growth, rather than simply substitute the bone.

3) Applications of new materials: This part presents studies on the development of new materials with unique magnetic properties, materials for use in extreme environ-

ments such as nuclear reactors, and materials for use in dental restoration and bone repair. In addition, there are several chapters on composites for improvement in lithium ion batteries which show promise over previous materials.

I am pleased to have had the opportunity to work with the authors and to have served as editor of this book which expands composite materials research into so many exciting areas of development of materials, engineering, medicine and dental restoration.

The book contains a wide variety of studies from authors from all around the world. I would like to thank all the authors for their efforts in sending their best papers to the attention of audiences including students, scientists and engineers throughout the world. The world will benefit from their studies and insights. The new possibilities of the open access press bringing together such a diverse group and to disseminate widely on the web is revolutionary, and without the contributions of the group and the mechanism of InTech Open Access Publisher, this Book titled "Metal, Ceramic and Polymeric Composites for Various Uses" would not be possible.

I also wish to acknowledge the help given by InTech Open Access Publisher, in particular Ms. Romina Krebel, for her assistance, guidance, patience and support.

John Cuppoletti, Ph.D.
Department of Molecular and Cellular Physiology
University of Cincinnati
Cincinnati OH
USA

Part 1

New Materials and Analytic Methods

1

Growth Reinforcing Composite Materials from Liquidus Phase: Mechanical and Microstructural Parameters Relationship Essentially Evincing the Predominance of an Akin Mass Composition over the Domain of Compositions

B.L. Sharma and Parshotam Lal
Department of Chemistry, University of Jammu
India

1. Introduction

Composite materials are heterogeneous systems consisting of two or more physically distinct and mechanically separable materials belonging to different space groups, and are in equilibrium with a single liquidus phase. The essence of composite materials lies in the concept that their properties must essentially be superior, and possibly unique in some specific respects, to those of the constituent phases, particularly if these are geometrically oriented to one another in some periodic or anisotropic manner (Sharma et al., 2008, 2009). Composite materials designed by anisotropic growth process, can offer advanced possibilities as reinforced materials for the constructions and further development of supersonic aircraft, space vehicles, high pressure tanks, for which strength properties are required that cannot be provided by the existing homogeneous materials (Ashbee & Woishnis, 1993). Besides mechanical fields, these materials are equally promising in a variety of other fields (Hull & Clyne, 2006). For example, most metallic permanent magnets are made of alnico alloys, obtained by unidirectional solid-state decomposition under cooling conditions in an applied magnetic field. The rod composite structure of quasi binary NiSb-InSb (Caram et al., 1990) obtained by unidirectional solidification from its homogeneous eutectic melt, finds a representative example in electronics. Currently, the femoral stem is constructed from Co-Cr-Mo, Co-Ni-Cr-Mo and Ti-Al-V alloys (Callister, 2006; Mallick, 1993 & Pillar, 1984) in preference to stainless steels. The advantages of composite materials appear when the specific modulus (modulus per unit weight) and specific strength (strength per unit weight) are considered. The higher specific modulus and specific strength of composite materials means that the weight of constituent phases can be reduced (Callister, 2006 & Courtney, 2000). This is a factor of greater importance in moving constituent phases in all forms of transport where reductions in weight result in greater efficiency and energy savings (Callister & Rethwisch, 2008).

The essence of anisotropic growth applied to composite materials is the ability to put strong stiff fibers in the right place, in the right orientation with right volume fraction. Implicit in the present searching approach is the concept that while in designing the composite material by variable anisotropic growth, the solidification aspect is also related to discover the final product as an acceptable analog for manufacturing the engineering product. Because of the complex nature of the phenomena involved in an anisotropic growth process from the melt, a considerable judgment is required to assess both theoretical and experimental observations, since the composite materials that can be developed by this process exhibit a large diversity of micro-morphologies. However, a physical understanding of the mechanical properties of composite materials as a basis for the improvement of the properties by controlling the growth rates from the melt can readily be distinguished when examined in an optical or electronic microscope. In view of watching the growth process visually, binary organic eutectic systems being transparent are a most suitable option for investigation, since metals are opaque and the observations had to be made on external surfaces. A very interesting finding of the current investigation is an akin mass composition (the composition whereat masses of the constituent members would almost overlap each other) that predominates the domain of compositions, since anisotropic mechanical parameters obeying the Weibull probability distribution curve evince the predominance, particularly at moderate anisotropic growth rate. Likewise, of greater interest is the moderate anisotropic growth in the domain of solidification modes as the modal products structured by the growth process comprise microstructures which are three-to-four fold superior in mechanical strength to their isotropic growth performed in an ice bath (~273 K) and many fold to their respective individual components irrespective of the growth pattern. Vickers microhardness offers supporting evidence to the essence of an identical form of the Weibull distribution curve demonstrating the strength-growth relationship. The liquidus structure encompassing the entire composition of binary organic eutectic systems is characterized by rheological view point comprising of viscosity and surface tension measurements at different temperatures, which manifests specific interactions occurring in the eutectic compositional melt that are complimentary supported by thermodynamic analysis in terms of excess functions. The experimental data on kinetics and anisotropic crystal growth from the melt verify nucleation theory and evince the essence of dislocation mechanism. Thermal and X-ray diffraction studies of binary composite materials are accomplished in order to ascertain thermal stability, purity, composition, enthalpies of fusion and mechanical combination of homogeneous materials.

A series of binary organic eutectic systems, viz., naphthalene-o-nitrophenol; α-naphthol-naphthalene; diphenylamine-α-naphthol; benzil-diphenyl; benzoic acid-benzamide; benzoic acid-naphthalene; naphthalene-phenanthrene; α-naphthylamine-diphenylamine and binary metallic eutectic systems, namely, Sn-Cd; Cd-Bi; Pb-Sn and Pb-Bi, were selected for the investigation. The choice of homogeneous materials in aforementioned systems is restricted to the methodical considerations: low temperatures of melting, purity and the possibility of precisely controlling and measuring temperatures.

2. Experimental details

2.1 Materials and methodology

A great precautious care was taken to the selection of utmost pure homogeneous materials, since organic materials are less pure and even small portion of soluble impurities can essentially influence the crystal growth. Consequently, preference was given to analytical reagent, especially of The British Drug House Limited and Merck for 99.9% pure organic materials (naphthalene, o-nitrophenol, α-naphthol, diphenylamine, benzil, diphenyl, benzoic acid, benzamide, phenanthrene and α-naphthylamine) and Alfa Aesar for 99.999%

pure metals (Pb, Sn, Bi and Cd). The melting temperatures and enthalpies of fusion of the experimental homogeneous materials were determined by thaw-melt technique and thermal analysis (DSC Linseis STA PT-1000) and found consistent with the literal values (Lide, 2009).

2.2 Diagrams of state

Mixtures of the constituent materials were prepared covering the whole of the mole fraction composition range (binary organic systems) and weight percent composition converted into mole fraction composition range (binary metallic systems). The homogeneity in mixtures was ascertained by heat-chill method. The solidus-liquidus equilibrium curves of binary organic systems were studied by thaw-melt method, while those of binary metallic systems by thermal analysis.

2.3 Heterogeneous nucleation

Nucleation studies of binary organic systems being transparent in nature were performed following the procedure referred earlier (Sharma et al., 2004a). A nearly constant amount (4.0 g) of each composition of the systems was carefully taken into different pyrex tubes of uniform internal diameters. The sample tubes were sealed and successively undertaken for spontaneous nucleation observations in a paraffin bath manipulated at the temperature ~30K higher than the melting temperature of each specimen. The bath temperature attained by the tube was ensured by thermometric technique before cooling process for spontaneous crystallization. The melt-nucleation process for each sample tube was recycled for several observations (minimum six runs) necessary in comparative estimate of undercooling. Maximum care was taken to perform undercooling experiments for different compositional melts under almost identical conditions by keeping the volume of the melt and the rate of cooling nearly the same in all the cases.

2.4 Anisotropic growth velocity

The experimental technique for determining the anisotropic growth velocity from different compositional melts of binary organic systems was similar to that adopted earlier (Sharma, 2003). The velocity of anisotropic growth from a definite compositional melt was measured in a transverse direction along the length 10×10^{-2} m and internal diameter of 6×10^{-2} m of the pyrex tube with terminal right-angled bends. The experimental tube was placed in a paraffin thermostat kept constant to ±0.01 K for experimental observations at several degrees of undercooling. The rate of advance of the crystal boundary was recorded by stop watch. Several observations (minimum six runs) of anisotropic growth velocity were recorded for each compositional melt at the selected experimental supercooling. The results are found less accurate from the eutectic melts of each system, since within the observation time the growing crystal front moved only a very small distance which could not be ascertained very accurately.

2.5 Rheological measurements

Viscosities of different compositional melts representing the complete mole fraction composition range of each binary organic system, were measured relative to the literal viscosity values of reference liquid p-xylene (Lide, 2009) in a paraffin bath with an accuracy of ±0.05 K in the temperature range 391-408 K. The viscometer employed for the purpose was modified as shown in Fig.A. Likewise, an attempt was made for the first time to measure surface tension of binary α-napthylamine-diphenylamine eutectic melt system related to the reference liquid p-xylene (Lide, 2009 & Edward, 1930) in the aforementioned bath in the temperature range 323-373 K using modified stalagmometer (Fig.B) compatible with the suitability of the experiment. The viscometer and stalagmometer are essentially modified by the side tube 'S' connecting the bulb 'B' right below the capillary of the viscometer to its upper mark 'X', and the bulb 'B' of the stalagmometer to the upper mark 'X' in the light of the following:

i. the excess pressure of the melt increasing the kinetic energy of the flow is made to escape through the side tube 'S' leaving the experimental volume of the melt to flow freely down the capillary under its own gravity. This minimizes kinetic energy correction.
ii. the bulb 'B' is an important component of the viscometer in connection with the elimination of upward thrust due to surface tension which could resist the free flow of the melt. The bulb 'B_1' of the stalagmometer would also overcome in nipping the resistance.
iii. the advantage of the modified stalagmometer over other stalagmometer lies in the concept that the relative surface tension measurements of low temperature solid materials in general and binary or ternary solid mixtures in particular can easily be undertaken at different temperatures in an air oven or a suitable bath to assess the specific interactions in their respective melts' surface.

2.6 Growth technique

2.6.1 Anisotropic crystal form

Pyrex tubes of nearly the same dimensions were selected as dies to achieve anisotropic rod form of crystals from different compositional melts in the following experimental setup. An experimental pyrex tube containing half-full melt of freshly prepared eutectic or non-eutectic phase, was vertically clamped to the centre of an empty graduated beaker (volume capacity $1dm^3$) arranged in an oven set at a temperature ~30 K higher than the melting temperature of the experimental specimen. The molten mass was nucleated by circulating water in case of organic melt, and silicone oil in case of metallic melt from their respective reservoirs manipulated to perforate at ~300 K. The melt in the tube started nucleating when the rising level of water or silicone oil as the case may be, just touched the bottom of the tube creating sharp temperature gradient at the solidus-liquidus interface that resulted in the occurrence of melt-growth process anisotropically towards longitudinal direction along the length of the tube. In this manner, each compositional melt in a separate pyrex tube was grown to crystal form at 18 different intervals in the experimental time range 5-45 min. The crystal sample grown anisotropically was removed from an experimental pyrex tube by warming its section in both longitudinal and transverse directions just to glide the solidus mass to its easier exit.

2.6.2 Isotropic crystal form

Nuclei of critical size almost grew out instantaneously in all directions when a pyrex tube containing the melt at ~30 K higher than its freezing temperature was dropped vertically into an ice bath maintained at ~273 K. This particular mode of solidification is presumed to be time independent and thereby zero order reference growth. The solidified crystal from the tube was removed following the aforementioned procedure. A good many crystal samples of eutectic and akin mass eutectic compositions, and the constituent materials were grown isotropically and anisotropically from their respective melts at different but nearly constant growth rates determined by circulating approximately the same volume of water or silicone oil for each experimental interval. An interesting feature of time variable growth is the yield of moderate anisotropic growth velocity (~2.90 X 10^{-7} m^3s^{-1}) determined by setting the flow-interval of reservoir liquid at ~5 X 10^{-4} m^3 for 30 min. At this moderate solidification rate, the growth from an akin mass compositional melt of each experimental binary organic system is found to be more efficient in promoting its mechanical properties which are superior, as revealed by the experimental evidences in the present work, to those of the eutectic composite and constituent materials grown under the same environmental conditions.

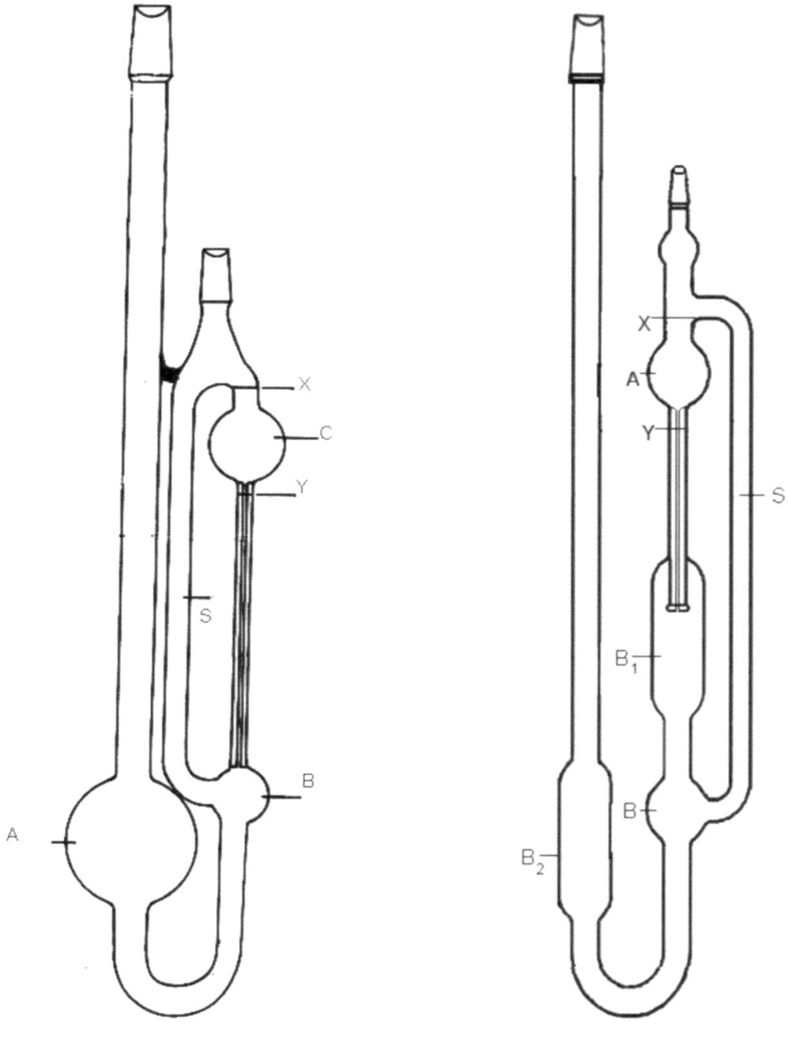

A) Viscometer B) Stalagmometer

2.7 Mechanical test

2.7.1 Macrorupture test

The experimental samples after dimensions' ascertainment were subsequently subjected to tensile, modulus of rupture (flexural) and compressive tests in a VEB Thuringer Industrie Werk Rauenstein tensometer where a steadily increasing load would determine the rupture force of a crystal specimen until it shows least resistance. These tests were performed visually watching the fracture process. Though the tests are of conventional nature, yet can be related also to the macro scale, particularly relevant to engineering products.

2.7.2 Microdeformation test

Metallic systems are suitable materials for micro hardness testing because of atomically rough surfaces. Indentations were induced on selected points chosen diagonally on an ansiotropically or isotropically grown specimen at room temperature (~300 K) using Vickers microhardness tester attached to an incident-light metallurgical research microscope (Neophot-2) in the applied load range 10 X 10^{-2} - 100 X 10^{-2} N. For each test, a very small diamond indenter having pyramidal geometry was forced into the surface of the specimen at room temperature and the size of the indent was found growing with rising applied load.

2.8 Microscopic studies

2.8.1 Optical microscopic examination

The micro-slides of binary organic eutectic and akin mass composite phases grown anisotropically and isotropically were examined in a polarizing microscope to view the growth habits of crystallites which were later photographed.

2.8.2 Electron microscope examination

The metallic specimens grown isotropically and anisotropically were polished at room temperature by following the procedure referred in connection with the analogous problem (Caram et al., 1991). A thin layer of a specimen etched in ferric chloride was mounted on stub with gold coated holder and examined in a scanning electron microscope (Jeol T-330) for microgrowth observations. Several grown samples of eutectic and akin mass eutectic composites were viewed in this manner and the growth habits acquired by the growing eutectic phases during solidification process at different growth rates were photographed.

2.9 X-ray diffraction (XRD) studies

X-ray diffraction patterns of eutectics and non-eutectics were scanned in a Rigaku X-ray diffractometer with Cu K_α radiation of wave length 1.540 A^0 at room temperature to reveal their nature of combination.

3. Observations

The experimental evidences offer convincing similarity among binary organic systems in physical behavior and characterization. Metallic systems Sn-Cd and Pb-Sn also exhibit alike behavior with distinct morphology (regular) from that (complex regular) of Cd-Bi and Pb-Bi systems. In view of physical habits' resemblance of the systems, two representative examples, namely, naphthalene–o-nitrophenol from organic systems and Sn-Cd from metallic systems shall be the main components revealing the objectivity of the work. The melting temperatures of their constituent materials are recorded in Table1, also containing their entropies of fusion and α-values (where $\alpha = \xi \Delta S_f / R$, is called dimensionless entropy, since R is the molar gas constant) as well computed thereby.

material	melting temperature(K)	enthalpy of fusion ($kJmol^{-1}$)	entropy of fusion ($Jmol^{-1}K^{-1}$)	α-value
naphthalene	353.00	19.00	53.81	6.47
o-nitrophenol	318.70	17.50	54.91	6.60
Sn	509.00	7.00	13.70	1.65
Cd	597.00	6.10	10.22	1.23

Table 1. Melting temperatures, enthalpies and entropies of fusion, and α-values

Diagrams of state representing the experimentally determined liquidus temperatures and theoretically computed ideal temperatures of naphthalene–o-nitrophenol and Sn-Cd systems are represented in Fig.1 (a and b). The undercooling curve resulting in by experimentally determined temperatures of spontaneous crystallization for naphthalene–o-nitrophenol system is also included in its phase diagram (Fig.1a). Thermodynamic analysis of the systems reveals specific molecular interactions in their molten states and hence their non-ideal nature. The non-ideal character of the system is authenticated by excess functions (E), namely, Gibbs free energy (G^E), entropy (S^E) and enthalpy (H^E), computed by the following standard relations (Sharma et al.,2004b):

$$G^E = RT[x_1^l \ln\gamma_1^l + x_2^l \ln\gamma_2^l] \quad (1)$$

$$S^E = -R[x_1^l \ln\gamma_1^l + x_2^l \ln\gamma_2^l + Tx_1^l (\partial \ln\gamma_1^l / \partial T)_P + Tx_2^l (\partial \ln\gamma_2^l / \partial T)_P] \quad (2)$$

$$H^E = -RT^2[x_1^l \ln\gamma_1^l + x_2^l \ln\gamma_2^l] \quad (3)$$

where x_i^l and γ_i^l respectively, denote the mole fraction and activity coefficient of the eutectic phase i (i = 1, 2) in binary mixture of liquidus temperature T, with the enthalpy of fusion $\Delta_f H_i^\circ$, obtained by thermal analysis, at its melting point T_i°. The superscript l refers to the condensed phase of the systems. The differential quantities $(\partial \ln\gamma_1^l / \partial T)_P$ and $(\partial \ln\gamma_2^l / \partial T)_P$ are the slopes of the liquidus lines obtained by plotting activity coefficients of the eutectic phases in eutectic mixtures versus their liquidus temperatures.
The quantitative magnitudes of G^E, S^E and H^E, obtained by Eqs. (1, 2 and 3) at different mole fraction compositions with their respective liquidus temperatures, are plotted in Fig.2. The minimal and maximal inflections respectively of G^E, and S^E and H^E at the eutectic compositions evidentially evince their obedience to the spontaneity criteria and Planck probability distribution (S = k ln w, k and w respectively being the Boltzmann constant and the configurational weight of the phase molecules). Microstructural parameters structuring the modal microstructures of the systems drive movement strength from excess functions (Sharma et al., 2008,2009) to overcome their faults called lamellar faults (Hunt & Jackson, 1966). The following symmetry relations obtained from the limit of undercooling (Table 2) are found consistent over the whole of the compositional range expressing their obedience to the criteria of nucleation theory:

$$\xi = \frac{T_C}{T_m} \tag{4}$$

$$\frac{\Delta T}{T_m} = \frac{T_m - T_c}{T_m} \tag{5}$$

at which ΔT, T_m and T_c respectively, are the undercooling, liquidus and spontaneous crystallization temperatures. ξ is a crystallographic factor which is less than and almost one, and represents fraction of total number of neighbors situated in the newly formed crystal layer.

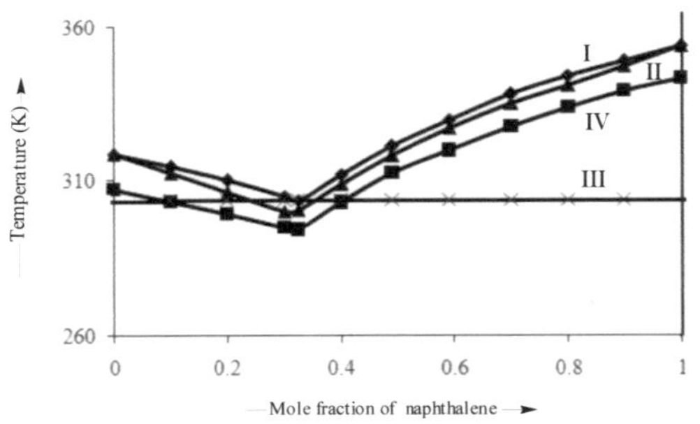

(a) Naphthalene- o-nitrophenol system
(I) liquidus temperatures curve
(II) ideal temperatures curve
(III) solidus temperatures curve
(IV) under cooling curve

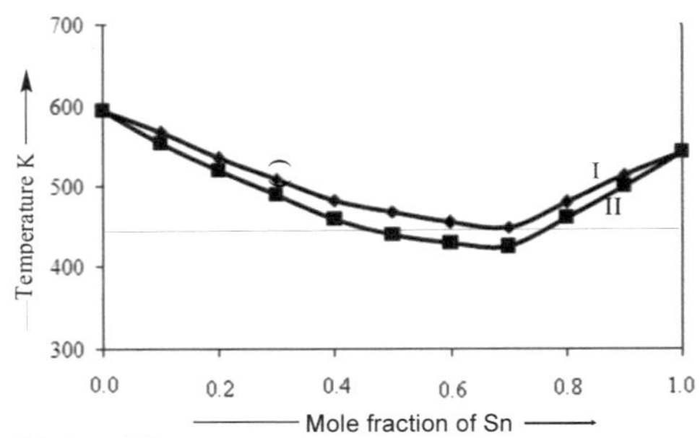

(b) Sn-Cd system
(I) liquidus temperatures curve
(II) ideal temperatures curve

Fig. 1. Diagrams of State

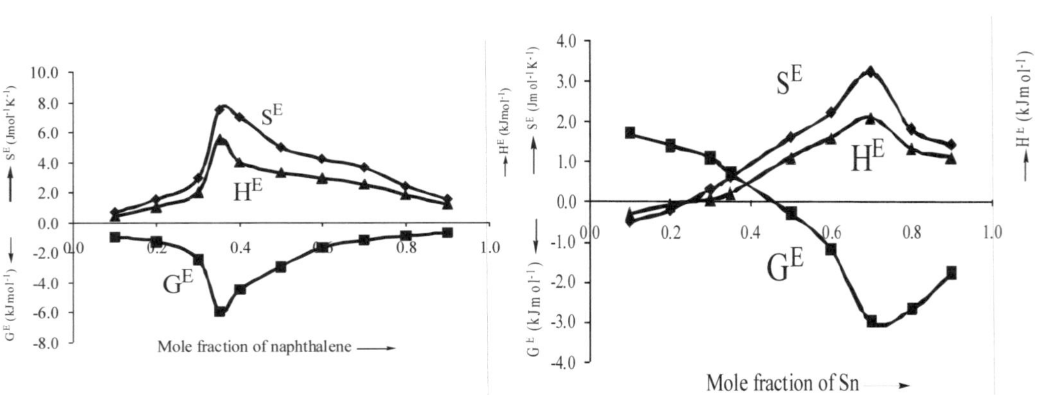

a) b)

Fig. 2. Excess functions G^E, H^E and S^E, a) Naphthalene- o-nitrophenol system, b) Sn-Cd system

mole fraction of naphthalene	liquidus temperature T_m (K)	temperature of spontaneous crystallization T_c (K)	maximum undercooling $\Delta T=(T_m- T_c)$ (K)	$\xi = \frac{T_c}{T_m}$	$\frac{\Delta T}{T_m}$ x10-2
0.0	318.70	307.20	11.50	0.96	3.61
0.1	315.20	303.10	12.10	0.97	3.84
0.2	311.70	299.00	12.65	0.96	4.06
0.3	304.70	294.70	10.00	0.97	3.28
0.33(e*):	303.70	293.20	10.50	0.97	3.46
0.4	313.70	304.70	9.00	0.97	2.87
0.5	321.20	312.40	8.80	0.97	2.74
0.6	329.70	319.80	9.90	0.97	3.00
0.7	339.90	327.70	12.20	0.97	3.59
0.8	343.20	333.70	9.50	0.97	2.77
0.9	348.70	339.10	9.60	0.97	2.75
1.0	353.50	343.70	9.80	0.97	2.77

Table 2. Heterogeneous nucleation data for the naphthalene-o-nitrophenol eutectic system. (e*: eutectic)

The kinetics of anisotropic crystallization velocity, V, from the molten states of the eutectic and non-eutectic phases follows the parabolic law:

$$\log V = \log k + n \log \Delta T \qquad (6)$$

since k and n respectively, being the intercept and the slope are determined from the straight lines obtained by plotting log V versus log ΔT (Fig.3) wherein n equals or almost equals 2 for the system. The observed anisotropic velocity of crystallization estimated from the eutectic melt is relatively low in comparison to its values from the non-eutectic melts (ΔT = 10 K).

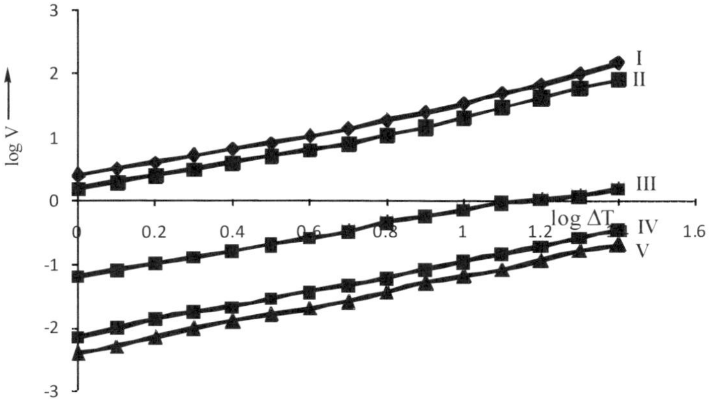

Fig. 3. Anisotropic velocity of crystallization at various degrees of undercooling for naphthalene- o-nitrophenol system: (I) pure naphthalene; (II) pure o-nitrophenol; (III)-(V) 0.2, 0.4 and 0.3250 (eutectic) mole fractions of naphthalene.

The viscosities for different compositional melts of the naphthalene–o-nitrophenol system measured in the temperature range 337–402 K are represented in Fig.4, which evidentially predicts the anomalous viscous behavior by the eutectic compositional melt, particularly in the vicinal temperatures to its liquidus temperature and this unusual behavior starts declining with rising temperature and completely evanesces at 402 K.

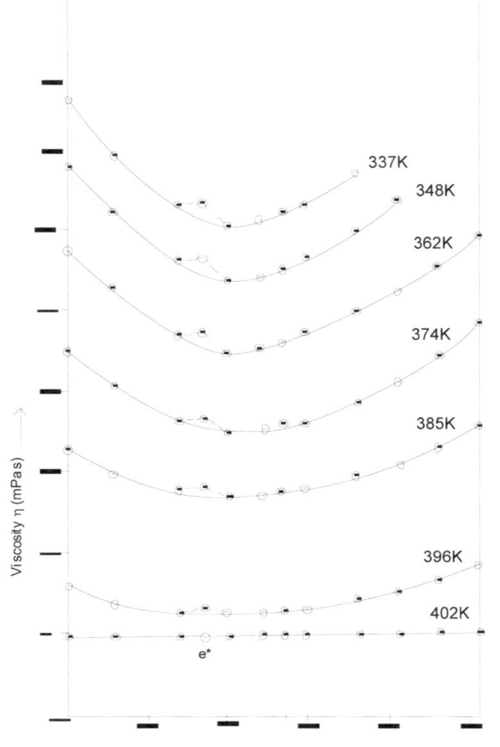

Fig. 4. Viscosity-composition plot for naphthalene-o-nitrophenol system at different temperatures (e* eutectic)

The viscosity data of the system in the experimental temperature range follow the Arrhenius equation:

$$\log \eta = \log \eta_0 + \frac{Evis}{2.303RT} \tag{7}$$

the parameters η and E_{vis} respectively, being the viscosity and activation energy for viscous flow of any compositional melt of the system at different rising temperatures T. The log η values are plotted against 1/T for any compositional melt at different temperatures and the slope of the best straight line obtained thereby, determines the activation energy and the intercept of the line yields the constant η_0 for that composition. The activation energy data (Table 3) for experimental compositional melts of the system obtained likewise reveal the predominance of activation energy for the eutectic melt over its values for the pre- and post-eutectic melts, which finally, consociates with a normal value of the system at 402 K (Fig.5).

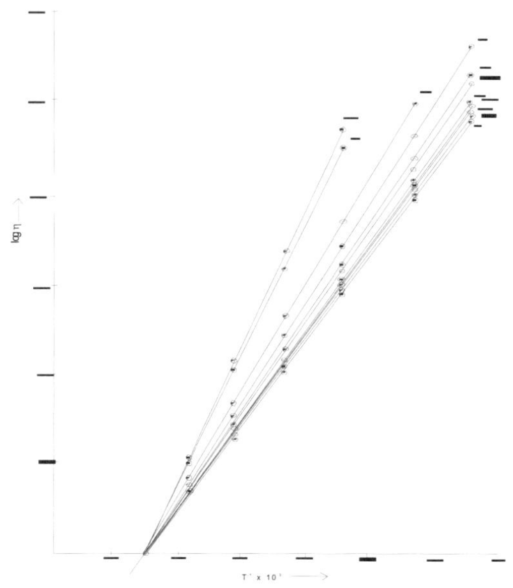

I o-nitrophenol
XI naphthalene
II-X mixtures with 0.1, 0.26, 0.32 (eutectic), 0.38, 0.46, 0.52, 0.7, 0.8 and 0.9 mole fractions od naphthalene respectively

Fig. 5. A plot of log η vs T for naphthalene-o-nitrophenol system

mole fraction of naphthalene	activation energy E_{VIS}(X 10^{-2} kJmol^{-1})	mole fraction of naphthalene	activation energy E_{VIS}(X 10^{-2} kJmol^{-1})
0.00	3.96	0.50	3.92
0.10	3.72	0.57	3.86
0.26	3.62	0.80	3.78
0.33 (e*)	3.88	0.90	3.64
0.38	3.64	1.00	3.56
0.46	3.82		

Table 3. Activation energy data in the temperature range 337-402 K for naphthalene- o-nitrophenol system in the paraffin thermostat with an accuracy of ±0.05 K

Alike, the surface tension, γ, measurements for different compositional melts of the system α-naphthylamine-diphenylamine in the temperature range 323–373 K (Fig.6) offers supporting evidence to the abnormal tendency of the phase molecules in the eutectic compositional melt.

The surface tension data of the system in the experimental temperature range obey Eotvos empirical equation:

$$\gamma = k(T_c - T) [d/M]^{3/2} \quad (8)$$

where k is the constant with numerical value 2.1 x 10^{-7} $JK^{-1}mol^{-2/3}$ valid for almost all materials and the parameters T_c, T, d and M respectively, are the critical temperature at which internal pressure equals 1 atmospheric pressure, experimental temperature, density and molar mass. The situation at which T_c equals T, the surface tension approaches zero. Figure 7 is a plot showing the variation of surface tension over its entire compositional

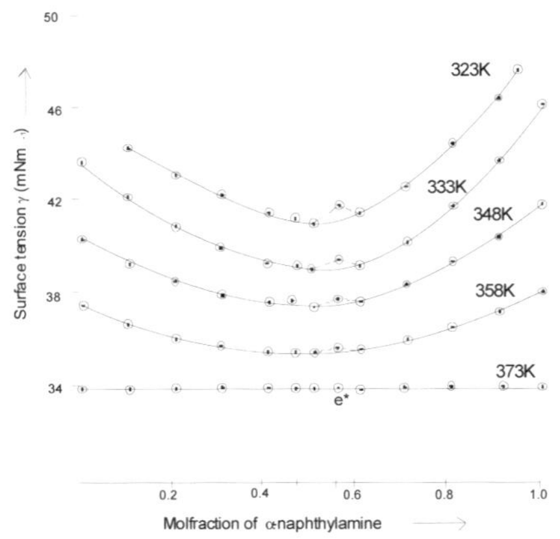

Fig. 6. Surface tension-composition plot for α-naphthalene-diphenylamine system at different temperatures, e* eutectic

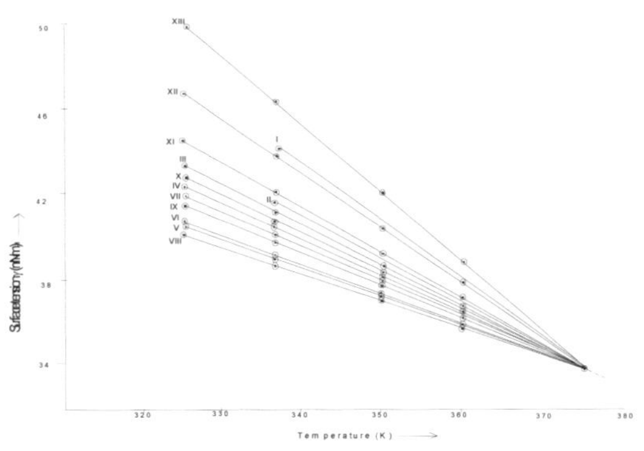

I diphenylamine
XIII α-naphthalene
II-XII mixtures with 0.1, 0.2, 0.3, 0.4, 0.46, 0.5, 0.55,(eutectic) 0.6, 0.7, 0.8, 0.9 molfractions α-naphthalene respectively

Fig. 7. A plot of γvs temperature for naphthalene-diphenylamine system range with increasing temperature. The temperature greatly effects the surface tension reducing it to its limiting value at T = T_c, an experimental observation possible only when there is no meniscus or barrier between the melt and the vapors (1 atm pressure). This implies that experimental observations of both viscosity and surface tension strongly support the excess functions predicting prominent molecular interactions and consequently, the structural changes in the eutectic melt evincing the endothermic nature of the systems.

The germinating crystallites from organic eutectic phases melt in an ice bath environment (~273 K) are of short size, aggressive, non-attaching and crossing each other showing no crystallite-matrix relationship resulting in overall morphology amply rich in small spherulites (Fig. 8a). The spherulitic density (spherulites per unit area) decreases with

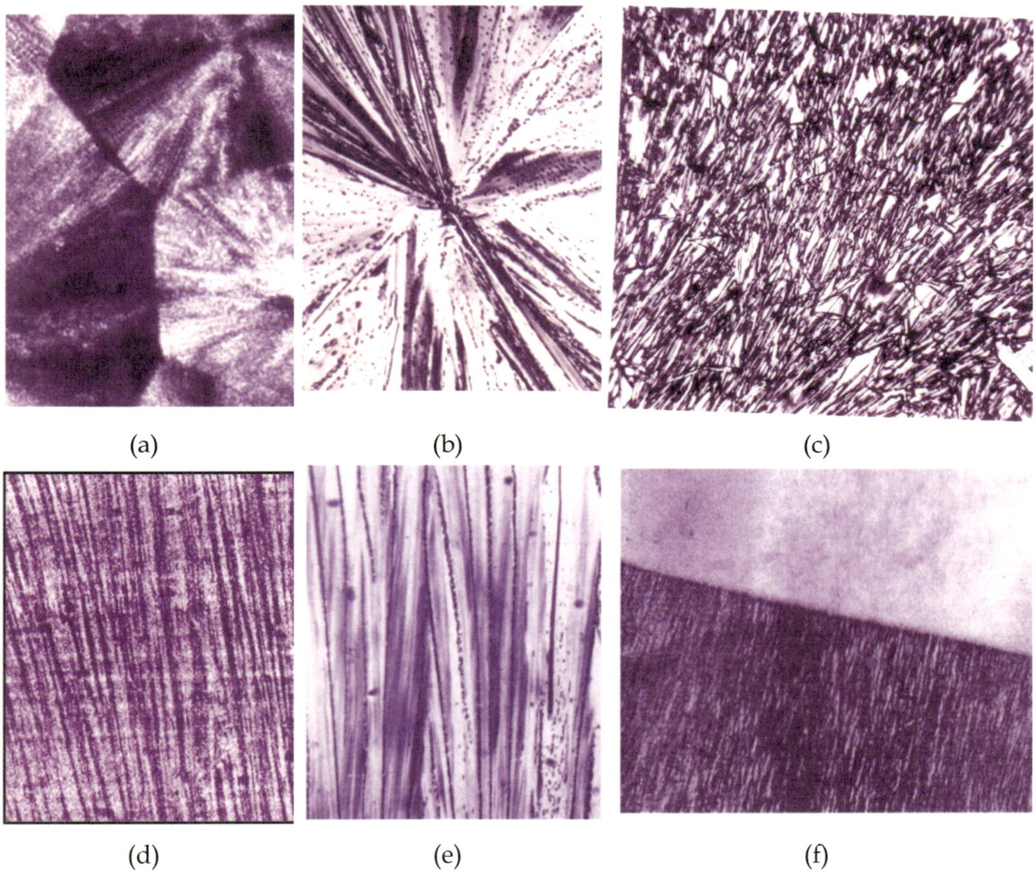

Fig. 8. Morphological diversity of naphthalene-o-nitrophenol eutectic composite materials by different modes of solidification (100 x). Microstructure of eutectic composite material in (a) ice bath (~273 K), isotropic growth, (b) at room temperature (~300 K), random growth, (c) at 315 K, (d) at moderate anisotropic growth from bottom to top (~2.90 X 10^{-7} m^3s^{-1}), (e) akin mass eutectic composite material at moderate anisotropic growth from bottom to top (~2.90 X 10^{-7} m^3s^{-1}), (f) eutectic composite material at anisotropic growth from bottom to top (~2.14 X 10^{-7} m^3s^{-1})

decreasing undercooling (Fig. 8b) and spherulitic size increases (~300 K) showing its centre as the seat of nucleation which finally evanesces (Fig.8c) with further undercooling decrease (~315 K). Nevertheless, the crystallites' growth habits of eutectic and akin mass eutectic compositions can be transformed from spheric or irregular morphology to unique fibrous and lamellar structures consisting of crystallites which are non-aggressive, attaching and embedded parallel to each other reinforcing the matrix by anisotropic growth modes of solidification. In the domain of solidification rates, microstructures represented by Figs. 8d and 8e are respectively discovered for eutectic and akin mass eutectic composites of naphthalene–o-nitrophenol system at moderate anisotropic growth rate (~2.90 X 10^{-7} m^3s^{-1}). Both the microstructures indispensably exhibit long and continuous crystallites with damaged free surfaces embedded parallel to each other along the growth direction. The akin mass eutectic composite structure is annexed superior to that of eutectic composite, vis-à-vis strength point of view. Figure 8f is a microstructure of the eutectic phases developed at anisotropic growth velocity ~2.14 X 10^{-7} m^3s^{-1} showing crystallites of short size and disconnected with aligning tendency to the longitudinal growth direction. On the contrary, pure eutectic components grew out from their respective melts as lamellar cells that crystallized either from the bulk of the melt or through secondary nucleation during the entire anisotropic solidification process. In view of anisotropic growth ability to organize the growing eutectic phases from the molten state, microstructures represented by Fig.9 (a, b and c) comprising of aligning lamellae, respectively are the specimens of Sn-Cd eutectic phases grown at moderate (~2.90 X 10^{-7} m^3s^{-1}), post-moderate (~3.50 X 10^{-7} m^3s^{-1}) and pre-moderate (~2.30 X 10^{-7} m^3s^{-1}) anisotropic growth rates. The basic distinction in morphologies of non-metallic and metallic eutectic systems lies in the thermodynamic concept (Ovsienko et al., 1980) that coupled growth occurs when both phases have low entropies of fusion ($\alpha<2$); i.e.; the lamellae or crystallites of the phases grow in contact resulting in regular structure with their round growth fronts and the overall growth is termed non-faceted–non-faceted (nf-nf). A coupled growth doesn't occur when both the eutectic phases have high entropies of fusion ($\alpha>2$) and consequently a regular structure result in consisting of crystallites or lamellae of the phases that grow side by side near each other with sharp growth fronts and the type of the growth is differentiated faceted-faceted (f-f). A complex regular structure comprising of either nf- f lamellae or nf-f crystallites is produced when the entropy of fusion is low ($\alpha<2$) in one phase but high ($\alpha>2$) in the other and the growth being the intermediate case, is distinguished non-faceted–faceted (nf-f). In the naphthalene-o-nitrophenol system, both the phases have high entropy of fusion ($\alpha>2$) furnishing an irregular structure, whereas in Sn-Cd system, both the phases have low entropy of fusion producing regular morphology. Regular morphology implies that the spacing among lamellae or crystallites must appear constant throughout the structure; otherwise irregular in nature. For this very reason, rod-type growth has a perfect ordered structure merely because of periodic constant inter-rod spacing, whereas lamellar structure proves to be less regular sheerly because of the movement of lamellae or crystallites termed lamellar or fiber faults that would become the cause of slight uncertainty in their spacing consistency.

Macro-rupture of naphthalene–o-nitrophenol system is virtually a collection of the following stress-strain fractures (Boyer, 1999; Callister,2006 & Kollmann, 1975):

(i) Modulus of rupture, $$Y_{rup} = \frac{PL}{\pi r^3} \text{ (circular cross section)} \quad (9)$$

(ii) Tensile strength, $$T_{rup} = \frac{P}{\pi r^2} \quad (10)$$

(iii) Compressive strength, $$\sigma_{rup} = \frac{P}{\pi r^2} \quad (11)$$

where the parameters P, L and r respectively, are the applied load in Newton force (N), span, and radius of an experimental specimen in metres, herewith P and r being dependent on the dimensions of the specimen and nature of the fracture, have variable magnitudes. In each fracture, the specimens yielding strength deviations by >15% from the averaged result were not included in the computed mechanical property data presented for eutectic and akin mass eutectic composite phases grown at both moderate anisotropic (2.90 X 10^{-7} m^3s^{-1}) and isotropic (~273 K) solidification rates, in Table 4.

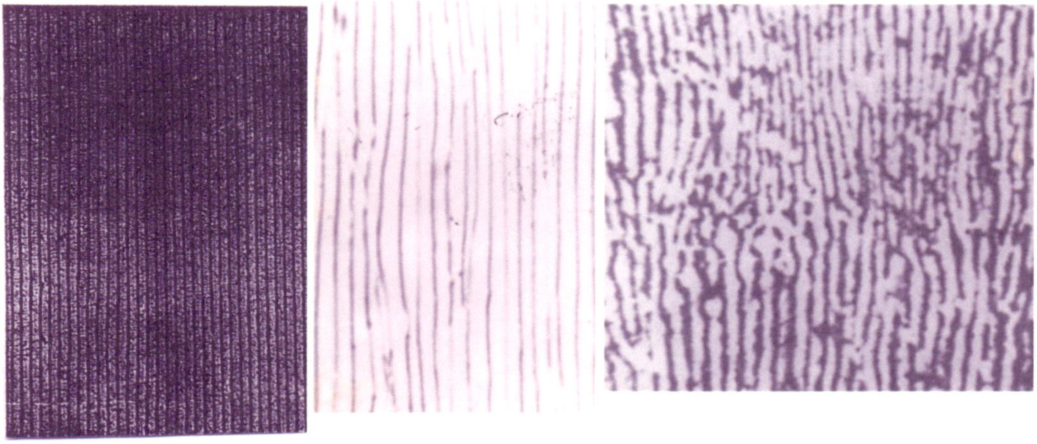

(a) (b) (c)

Fig. 9. Lamellar growth of Sn-Cd eutectic alloy in the moderate growth region (1500x): (a) lamellar Sn-Cd eutectic at moderate anisotropic growth from bottom to top (~2.90 X 10-7 m3s-1), (b) disconnected lamellar Sn-Cd eutectic at anisotropic growth from bottom to top (~3.50 X 10-7 m3s-1),(c) disconnected and short lamellae structure of Sn-Cd eutectic at anisotropic growth from bottom to top (~ 2.30 X 10-7 m3s-1).

The mechanical observations data , obtained by Eqs. (8, 9 and 10) for the entire composition range of the naphthalene-o-nitrophenol system at constant moderate anisotropic growth (~2.90 X 10^{-7} m^3s^{-1}), are plotted in Fig.10, implicitly indicating the predominance of the akin mass eutectic composition. In search of authenticity onto this experimental evidence, both eutectic and akin mass eutectic composite materials grown at constant but different anisotropic growth rates, were subsequently subjected to tensile, flexural (modulus of rupture), and compressive tests and the computed tensile strength data are represented in Fig.11, while data for modulus of rupture and compressive modes are recorded in Table 5, supporting the predominance of the akin mass eutectic composite over eutectic composite, both expressing obedience to the Weibull distribution. Evidently, this aspect generates the strength-growth relationship which follows an identical form of the Weibull probability distribution curve inculcating the obedience of the microstructural paramaters, namely, crystallite diameter, crystallite length, crystallite length distribution, volume fraction of crystallites, and the alignment and packing arrangements of crystallites, to the distribution.

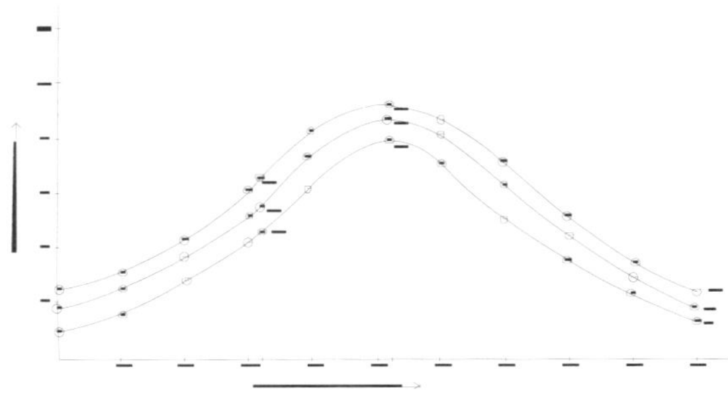

Fig. 10. Variation of mechanical properties of naphthalene-o-nitrophenol system at moderate growth (~2.90 x $10^{-7}m^3s^{-1}$) (i) modulus of rupture (ii) tensile strength (iii) compressive strength a* akin mass composition, e* eutectic

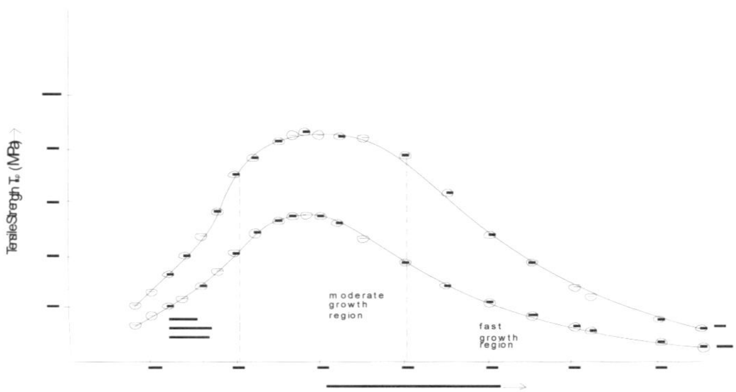

Fig. 11. Plots of tensile strength at variable growth (i) akin mass composition and (ii) eutectic composition of the naphthalene-o-nitrophenol system

mechanical property	akin mass composition		eutectic composition		Noneutectic materials			
					naphthalene		o-nitrophenol	
	moderate anisotropic growth	isotropic growth	moderate anisotropic growth	isotropic growth	moderate anisotropic growth	isotropic growth	moderate anisotropic growth	isotropic growth
(i) Modulus of rupture Y_{rup} (Mpa)	7.90	2.96	5.10	2.50	1.30	0.60	1.00	0.50
(ii) Tensile Strength T_{rup} (Mpa)	8.60	3.40	6.00	2.90	1.80	1.0	1.80	1.00
(iii) Compressive Strength σ_{rup} (Mpa)	9.20	3.70	6.70	3.30	2.30	1.10	2.40	1.30

Table 4. Mechanical properties of akin mass and eutectic compositions of naphthalene-o-nitrophenol system at moderate anisotropic (~ 2.90 x$10^{-7}m^3s^{-1}$) and isotropic (~273 K) growth rates.

growth velocity (V X 10-7m3s-1)	akin mass composition		eutectic composition	
	modulus of rupture Y_{rup} (MPa)	compressive strength σ_{rup} (MPa)	modulus of rupture Y_{rup} (MPa)	compressive strength σ_{rup} (MPa)
0.8	1.10	2.80	0.70	1.50
1.0	1.80	3.50	1.00	1.90
1.2	2.80	4.40	1.30	2.40
1.4	4.20	5.10	1.70	2.70
1.6	5.40	6.00	2.30	3.70
1.8	6.20	6.80	2.80	4.30
2.0	6.80	7.60	3.30	5.00
2.2	7.10	8.20	3.80	5.70
2.5	7.70	8.90	4.20	6.30
2.7	7.80	9.10	4.40	6.50
2.8	7.90	9.20	4.50	6.60
3.0	7.80	9.10	4.40	6.50
3.2	7.60	9.00	4.30	6.40
3.5	7.10	8.80	3.80	6.00
4.0	6.10	7.80	2.70	5.20
4.5	4.80	6.20	2.00	3.80
5.0	3.70	4.70	1.50	2.90
5.5	2.60	3.80	1.10	2.30
6.0	2.10	3.00	0.80	1.80
6.2	1.70	2.70	0.60	1.60
7.0	1.20	1.80	0.30	1.50
7.5	0.80	1.30	0.20	0.80

Table 5. Variation of modulus of rupture and compressive strength with growth velocity of naphthalene- o-nitrophenol binary system.

Micro deformation test virtually represents the Vickers microhardness, H_v, which is a measure of a materials resistance to localized plastic deformation (e.g., a small dent or a scratch) by the stress-strain relation (Beigh et al., 1995 & Hayden, 1965):

$$H_v = 1.8544 P/d^2 \tag{12}$$

herewith, P signifies the same as mentioned above, whereas the parameter d denotes the average diagonal length of the indentation mark in meter and varies with the size of the cracks.

The variation of Vickers microhardness for specimens of the eutectic alloy Sn-Cd grown by different modes of experimental anisotropic solidification rates at constant applied load of 50 X 10^{-2} N for a specified interval of 20s, offers supporting evidence to the essence of an identical form of the Weibull distribution curve (Fig.12) exhorting the strength-growth relationship. The specimens of the eutectic alloy Sn-Cd obtained by moderate anisotropic (~2.90 X 10^{-7} m^3s^{-1}) growth process inflicted with indentation impressions by a constant load (60 X 10^{-2} N) at different indentation times (Fig.13 a, b and c), measurement of crack length and growing indents' dimensions by a variable load are illustrated in terms of photomicrographs in Fig.13 (d, e and f). A maximum care was exercised to consider only well defined cracks developed during indentation process. The average crack length of all such cracks was estimated for a particular indentation impression. The crack length was measured from the centre of the indentation mark to the tip of the crack (Fig.13d). The size growth with a variable load is represented in Fig.13 (e and f).

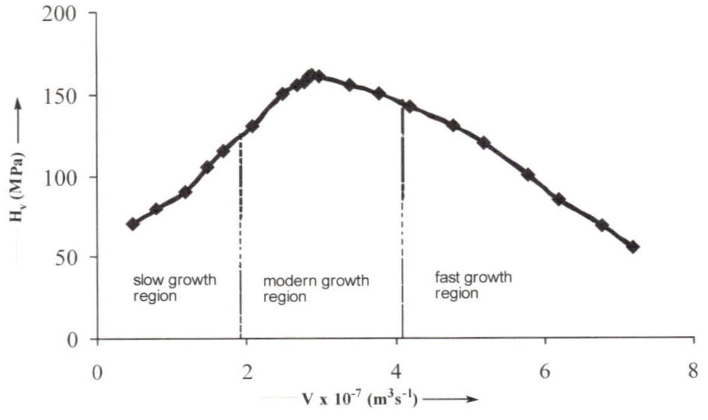

Fig. 12. Variation of micro hardness with growth velocity for the eutectic alloy Sn-Cd at constant applied load of 50x10^{-2} N for a specified interval of 20 s

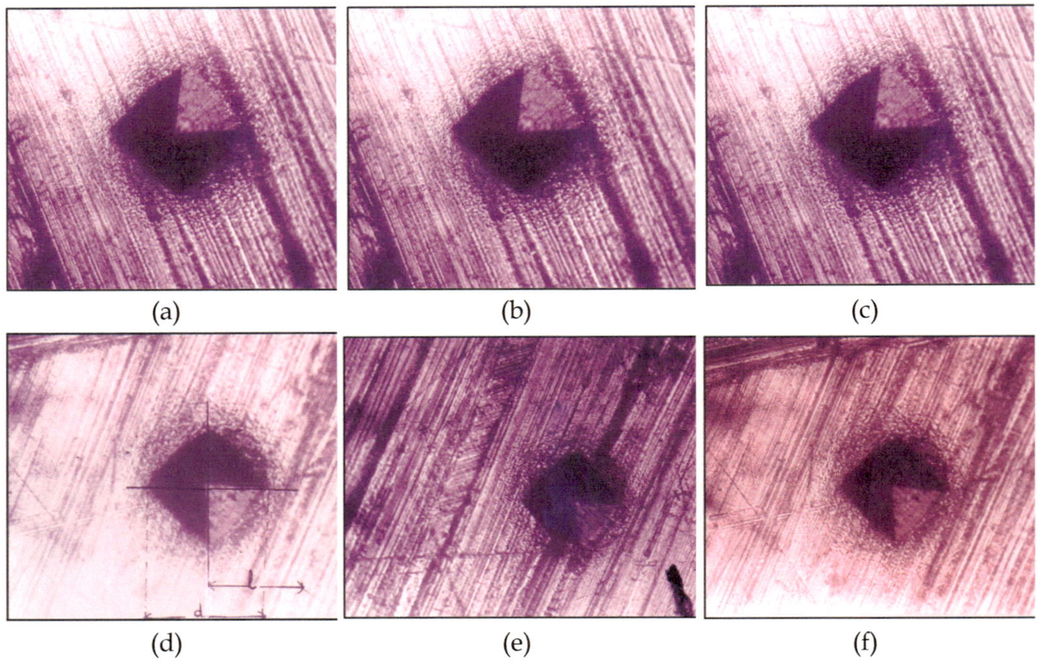

Fig. 13. Photomicrographs showing indentation impressions inflicted on the anisotropic specimens of the eutectic alloy Sn-Cd (625x): (a) size of indentation with a load of 60X10^{-2}N at 50s; (b) size of indentation with a load of 60X10^{-2}N at 100s;(c) size of indentation with a load of 60X10^{-2}N at 150s; (d) measurement of crack length;(e) size of indentation with a load of 20X10^{-2}N; (f) size of indentation with a load of 40X10^{-2}N.

Figures 14 and 15 respectively predict the variation of microhardness with increasing indentation time and applied load. The analysis of microhardness is summarized in Figs. 16-19.

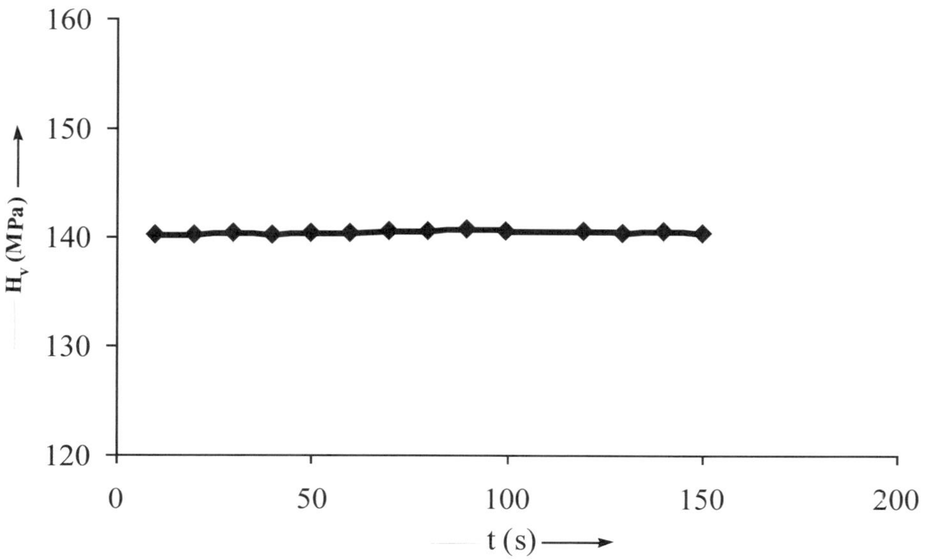

Fig. 14. A plot showing invariability of micro-hardness with variable indetation time for eutectic alloy Sn-Cd for constant applied load. P (30 x 10^{-2} N)

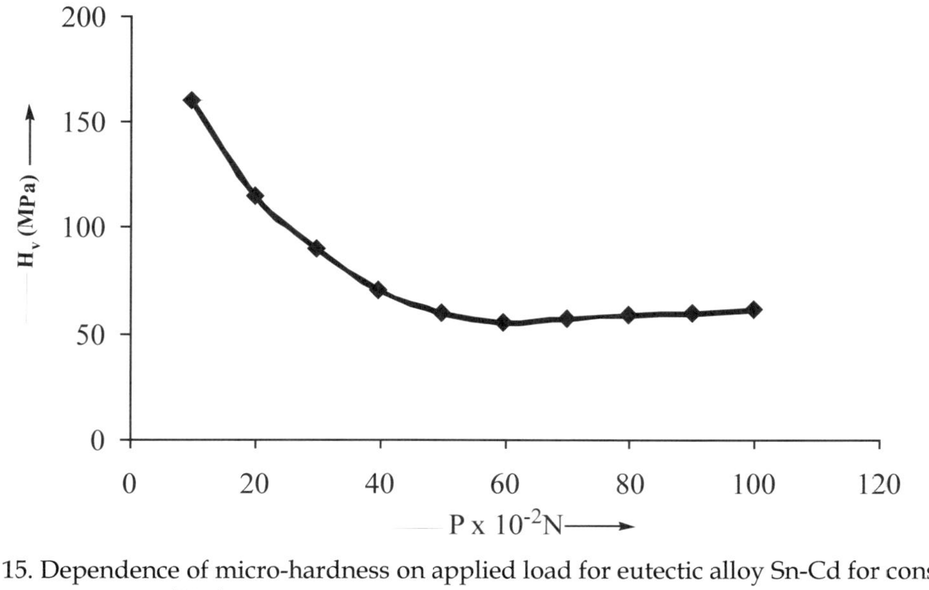

Fig. 15. Dependence of micro-hardness on applied load for eutectic alloy Sn-Cd for constant indentation time, t (20 s)

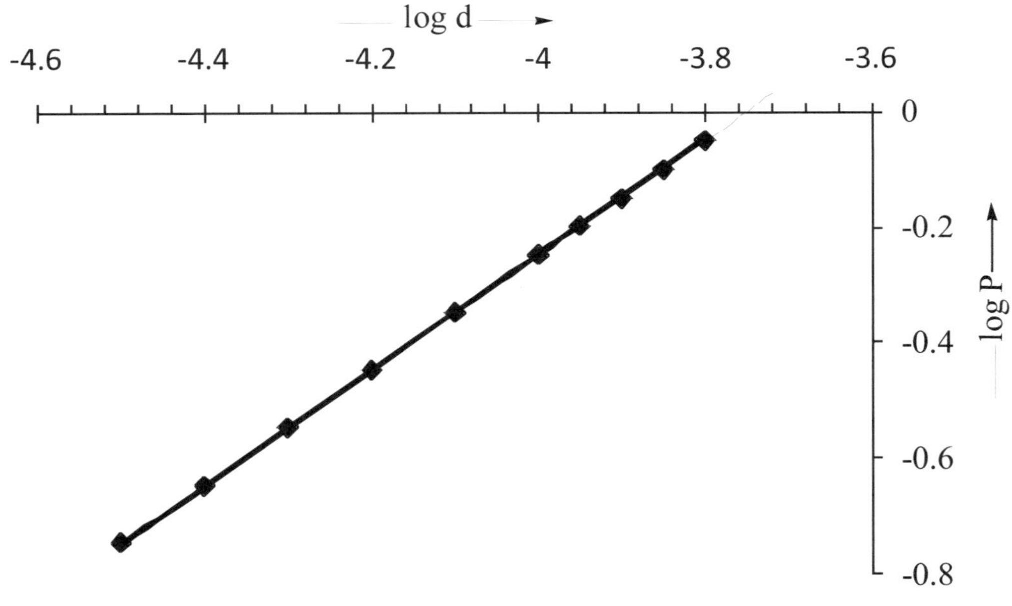

Fig. 16. Relationship between log P and log d for the eutectic alloy Sn-Cd

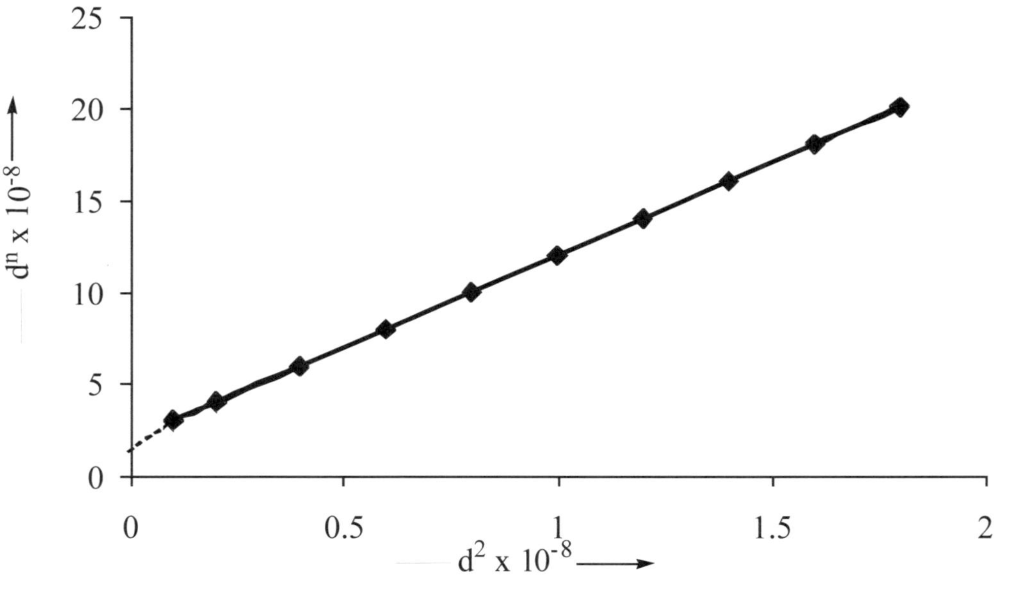

Fig. 17. Variation of d^{π} and d^2 for the eutectic alloy Sn-Cd

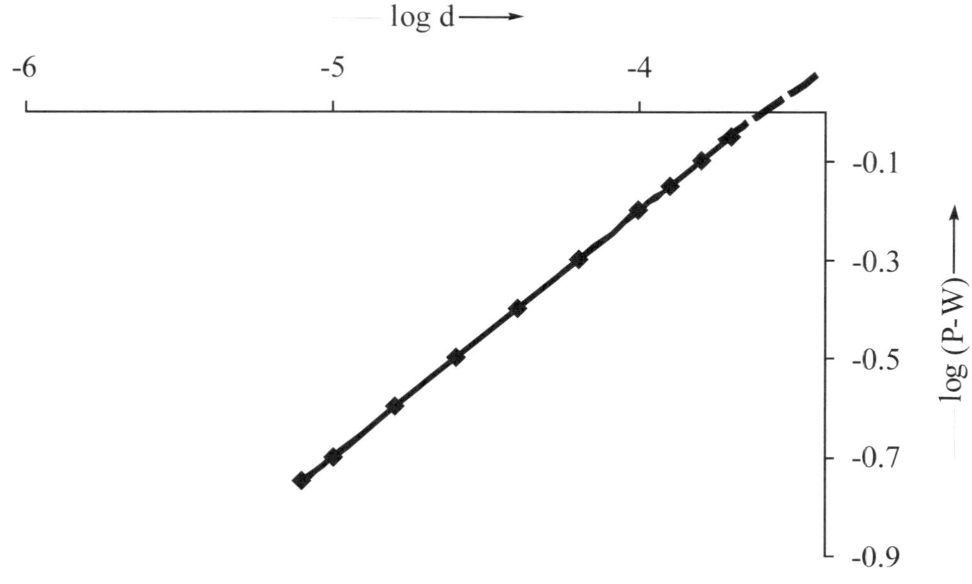

Fig. 18. Relationship between log (P-W) and log d for eutectic Sn-Cd

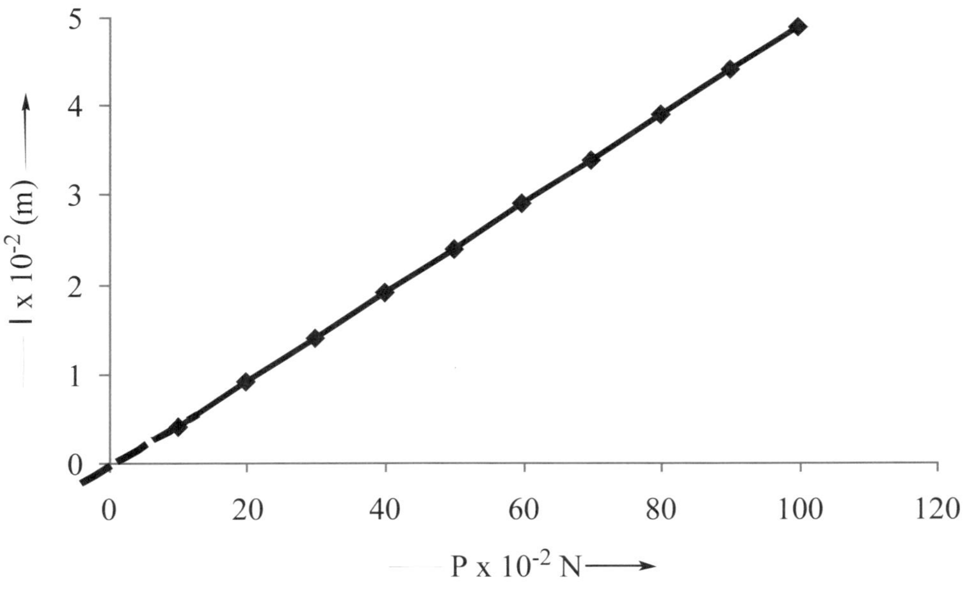

Fig. 19. A plot showing variation of cracklength (l) with applied load (P) for eutectic alloy Sn-Cd

The parameters extracted from the analysis are recorded in Tables 6-7.

eutectic alloy	n	$10^6 Nm^{-2}$ k_1	$10^6 Nm^{-2}$ k_2	$k_2/k1$	W (10^{-3}N)	W/k_1 ($10^{-8}m^2$)
Sn-Cd	1.5672	1.3824	7.4155	5.3642	8.7868	6.3562

Table 6. Hardness analysis constants for the eutectic alloy Sn-Cd

applied load P x 10^{-2}N	crack length (l x 10^{-2}m)	fracture toughness K_c ($Nm^{-3/2}$)	brittleness B_i ($10^6 m^{-1/2}$)	hardness H_v (MPa)	yield strength σ_y (MPa)
10	0.3	89.29	1.79	160	53.33
20	0.7	84.03	1.37	115	38.33
30	0.9	50.19	1.79	90	30.00
40	1.2	43.62	1.60	70	23.33
50	1.6	35.36	1.69	60	20.00
60	1.9	32.72	1.68	55	18.33
70	2.1	32.89	1.73	57	19.00
80	2.7	25.74	2.29	59	19.67
90	3.2	22.48	2.67	60	20.00
100	3.4	22.78	2.68	61	20.33

Table 7. Hardness parameters for the eutectic alloy Sn-Cd

The XRD data summarized in Table 8 and X-ray patterns represented by Fig. 20 for the eutectic alloy Pb-Bi, selected as representative example in the current work, specify that the number of peaks and intensities exhibited by the eutectic alloy are complementary to its constituent phases.

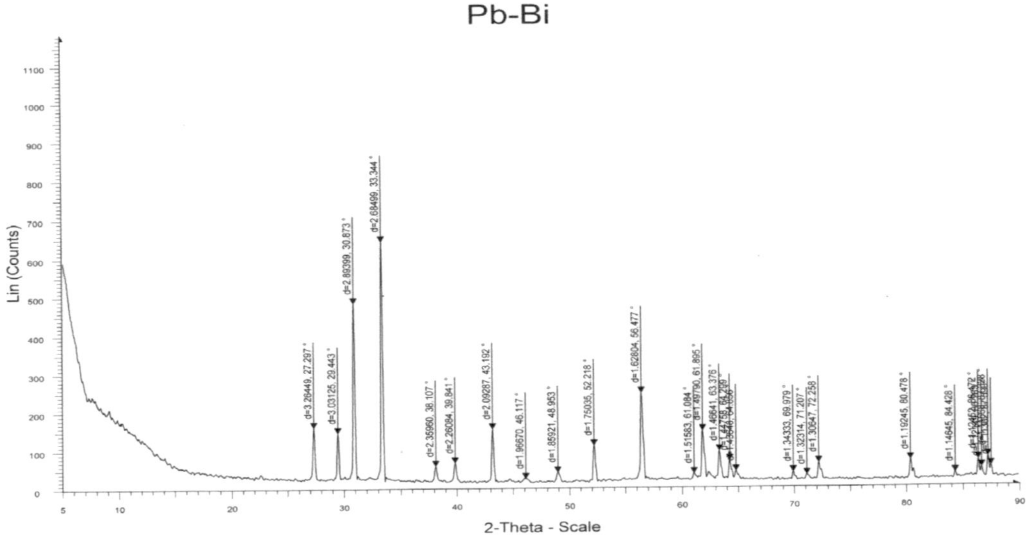

Fig. 20. X-ray diffraction patterns of the eutectic alloy Pb-Bi

	Pb		Bi		Pb-Bi eutectic		Corresponding Planes (hkl)
S. No	Pos [2°TH]	d-spacing [Å]	Pos [2°TH]	d-spacing [A°]	Pos [2°TH]	d-spacing [A°]	
1	6.487	13.6113	6.224	14.189	22.469	3.95389	Bi (003)
2	28.918	3.0850	22.220	3.997	27.297	3.26449	Bi (012)
3	31.448	2.8423	22.805	3.8962	29.443	3.03125	Pb (110)
4	36.440	2.4636	24.091	3.6911	30.873	2.89399	Pb (111)
5	43.381	2.0842	26.821	3.3213	33.344	2.68499	Pb (111)
6	49.973	1.8236	27.427	3.2492	38.107	2.35960	Bi (104)
7	52.407	1.7444	31.502	2.8376	39.841	2.26084	Bi (110)
8	56.271	1.6335	37.740	2.3817	43.192	2.09287	Pb (200)
9	62.314	1.4888	38.211	2.3524	46.117	1.96670	Bi (113)
10	65.414	1.4255	39.872	2.2589	48.953	1.85921	Bi (202)
11	77.135	1.2353	44.826	2.0202	52.218	1.75035	Pb(220)
12	77.397	1.2321	45.684	1.9843	56.477	1.62804	Bi (024)
13	85.579	1.1339	46.199	1.9634	61.084	1.51583	Bi (200)
14	85.852	1.1310	48.511	1.8751	61.895	1.49790	Pb (311)
15	88.363	1.1052	48.954	1.8591	63.376	1.46641	Bi (211)
16	88.646	1.1024	49.561	1.8378	64.299	1.44758	Bi (122)
17	89.714	1.0921	56.286	1.6331	64.856	1.43648	Pb (222)
18			59.566	1.5507	69.979	1.34333	Bi (214)
19			61.325	1.5140	71.207	1.32314	Bi (009)
20			62.434	1.4862	72.258	1.30647	Bi (300)
21			64.761	1.4383	80.478	1.19245	Bi (208)
22			67.682	1.3832	84.428	1.14645	Bi (119)
23			71.013	1.3262	86.472	1.12452	Bi (101)
24			72.110	1.3087	86.763	1.12149	Bi (217)
25			73.917	1.2812	87.328	1.11568	Bi (217)
26			75.542	1.2576	87.590	1.11302	Pb (420)

Table 8. X-ray analysis for the eutectic alloy Pb-Bi

4. Analysis and discussion

The deviations of the systems from ideal regions covered by their respective diagrams of state (Fig.1) consisting of solidus-liquidus equilibrium curves, predict specific interactions between unlike phase molecules or unlike atoms in the miscibility phenomenon. The thermodynamic analysis of the systems confirms the deviations (Fig.2) and comprehensively yields quantitative idea of the interactions, inasmuch the excess functions are being computed from experimentally determined parameters, are reliable and discovered expedient for predicting the phase equilibrium curves of binary systems (Sharma et al.,2004).

The heterogeneous nucleation data obtained from the limit of undercooling (Table 2) provide a lucid consistency in the symmetry relations, $\xi = T_c/T_m$ and $\Delta T/T_m$, which certainly authenticates the nucleation theory for the system entirety. This essential constancy in symmetry relations implicitly explores a definite relationship existing between the nucleating solidus form obtained from the limit of undercooling and the liquid structure conceived by the entropy of fusion.

The critical analysis of the plot (Fig. 3) shows that the anisotropic velocity of crystallization from binary compositional melts essentially decreases significantly with the gradual addition of one eutectic phase to the other. Consequently, the inferential interference of the eutectic phases decreases the crystallization velocity from the eutectic melt ($10 \times 10^{-4} ms^{-1}$) which is much lower than its value from the non eutectic melts; naphthalene ($158 \times 10^{-4} ms^{-1}$) and o-nitrophenol ($92 \times 10^{-4} ms^{-1}$) for fixed undercooling $\Delta T = 10K$. The kinetics of the anisotropic crystal growth being dependent on parabolic form, follow the dislocation mechanism wherein eutectic phases would diffuse and grow at the sites at which surface with the stages is formed owing to lattice imperfections and screw dislocations. These imperfections do intersect the surfaces and produce steps of one or more molecular diameters in height. These steps are the center of lattice disturbance and wind themselves in spirals during crystal growth. The logical growth concept extracted from the theoretical interpretation of the experimental observations is consistent with the diffusion growth of the eutectic phases from their binary melt that the growing phases mutually sustain the diffusion process through melt layer becoming rich in their concentration at different times of successive nucleation.

The anomalous behavior of viscosity (Fig.4) and surface tension (Fig.6) for the eutectic melt furnishes supporting evidence to the essence of molecular interactions conceiving magnitudes already delivered by the excess functions. The most probable structural change (configurational weight) in phase molecules occurs in the eutectic compositional melt owing to three phases' coexistence, resulting in significant molecular clusters' formation rich in one of the eutectic phases. The degree of molecular clustering tendency decreases as the temperature is gradually raised that would finally become the cause of limiting values of rheological parameters. The explanation for the higher value of the activation energy for the eutectic melt may be attributed to the following contributions: (i) the activation energy for viscous flow; (ii) the energy essentially required overcoming the molecular clusters' tendency. The molecular clustered array of atoms would inevitably result in an increase both in viscosity and surface tension, and a reduction in surface area. Consequent thereupon, the thermal energy expended to break the clusters would certainly exceed activation energy. An increasing temperature vitiates clustering phenomenon and the particular temperature whereat it completely evanesces only the first contribution will be the activation energy. In the current work, the cluster-free flow is evidentially illustrated for viscosity at 402K in Fig.4 and for surface tension at 373 K (Fig.6). Likewise, the activation energy for eutectic compositional melts also consociates with normal value of the systems as indicated by dotted lines in Fig.5 for viscous flow of naphthalene–o-nitrophenol system at 402K, and in Fig 7 for loose surface flow of α-naphthylamine-diphenylamine system at 373K.

An irregular morphology (Fig 8 a –c) by f-f growth of eutectic phases from molten state results in because of their high entropies of fusion, since the formation of packs of their crystallites from the eutectic melt would arise under the influence of large thermal and mechanical stresses during the growth process. These stresses seem to be the cause of splitting the main single crystallite into separate single crystallites or groups of single crystallites in the absence of orientation relationship among faceted crystallites, eventually accomplishing an irregular morphology. The crystallite spherulitic form occurs at a large undercooling (~273 K) merely because the rich eutectic phase first nucleates and grows depleting the vicinal melt to be rich in other phase that also nucleates at a certain supersaturation. Alike, both the phases grow side by side. The large thermal and mechanical stresses influence the splitting of crystallites to occur along the direction and successive splitting lead to the formation of spherulites. As mentioned in the observation section, the moderate anisotropic growth is predominant finding in the domain of anisotropic solidification rates because of its aesthetics ability to produce in- situ oriented composite materials consisting of crystallites (Fig.8 d – e) and lamellae (Fig.9a) which are attaching , nonaggressive and embedded parallel to each other exhibiting nearly consistent periodic orientation relationship. Microstructure (Fig.9b) consisting of long and continuous parallel lamellae indicates lamellar faults (Hunt &Jackson, 1966). Microstructures (Figs.8f and 9c) consisting of crystallites or lamellae of short size, exhibit their disconnected but non aggressive alignment along the growth direction. The supercrescent crystallites or lamellae growth virtually produces the complete lamellar or rod –type microstructures .The eutectic phases with $\alpha > 2$ grow side by side near each other from binary melt resulting in lamellar structures of long and continuous non aggressive crystallites lamina (Fig.8d - e), whereas eutectic phases with $\alpha < 2$ grow supercrescently (edge-wise) from the melt to produce lamellar or rod-type lamina consisting of long and continuous nonaggressive lamellae which are embedded parallel to each other showing orientation relationship almost consistent throughout the structure (Fig. 9a).

An insight analysis of Table 4 reveals approximately one and a half to two fold enhancement in each mode of mechanical property of the akin mass eutectic composite material at the moderate anisotropic growth velocity (~2.90 X 10^{-7} m^3s^{-1}) in comparison to the eutectic composite material and many fold superiority over its isotropic growth (~273 K) and constituent materials whether grown anisotropically or isotropically. The plot of Fig.10 drawn between different modes of macrohardness and the entire range of composition of the naphthalene-o-nitrophenol system at nearly constant moderate anisotropic growth velocity (~2.90 X 10^{-7} m^3s^{-1}) strengthens the quantitative analysis by showing the predominance of the akin mass eutectic composition. Consequently, the curve (Fig.11) divides the experimental range of anisotropic growth velocity into three regions, namely, (i) slow growth region; (ii) moderate growth region and (iii) fast growth region .Among which, the moderate growth region, as is evident from the plot (Fig.11), would seem to be the most probable for producing lamellar or rod-type structures of both eutectic (Fig.8d) and akin mass eutectic (Fig.8e) composite materials, since microstructural parameters appear to be nearly obeying the Gauss distribution. Consequentiality, the microstructural crystallites of the akin mass composite material nearly follow the rod- type growth habits (Fig.8e) reinforcing the matrix where there is a perfect crystallite- matrix bond. On the contrary, the crystallites of the eutectic microstructure (Fig.8d) emerge with lamellar growth habits reinforcing the matrix but the crystallite- matrix bond exist with less perfection .The mechanical parameters of both the composite materials in the slow and fast growth regions are found comparable (Table 5 and Fig.11) because microstructural parameters express their obedience to the Weibull distribution in these regions showing no crystallite matrix relationship (Fig.8 a- c).The analysis necessarily involves a physical understanding of the relationship between the microstructures of materials and their mechanical properties .Evidentially, the curve (Fig.11) has two cut-off points corresponding to a lower strength limit in the slow and fast growth regions, and an upper strength limit in the moderate growth region. The later is equivalent to the theoretical strength of the matrix reinforced crystallites with damaged free surfaces, practically having no density of dislocations; i.e.; internal defects or surface flaws responsible for reduced strength. Likewise, the microhardness of the eutectic alloy Sn-Cd (Fig. 12) can be explained essentially involving its microstructures presented in Fig.9 on the similar dependence pattern with variable anisotropic growth velocity as tensile strength for the eutectic and akin mass eutectic composite materials of the naphthalene-o-nitrophenol system, follows in Fig.11. Implicit in the plots (Figs.11 and 12) is the concept that the variation of an anisotropic mechanical property over the entire range of growth velocity furnishes an evidence of its dependence as linear, optimum, and linear respectively in slow, moderate, and fast growth regions of solidification. An interesting finding to be noted from Tables 4 -5 is that, both eutectic and akin mass eutectic composite materials with isotropic growth (~273 K) are respectively found nearly twice and thrice stronger than that of their constituent materials merely because of their high specific modulus and high specific strength due to alignment of the crystallites although having dislocations. Likewise, moderate anisotropic growth of pure constituent materials is superior to their isotropic growth. Besides, the quantitative perusal of mechanical parameters presented in Tables 4 and 5,explores that the compressive mode slightly exceeds the tensile mode, which in turn shows an edge on modulus of rupture ($\sigma_{rup} > T_{rup} > Y_{rup}$).The inequality existing between mechanical modes essentially involves crystallite's length, which in the current work, accomplishes with the attachment of alike crystallites at different times and their efficiency in gripping and the reinforcing the matrix decreases with the reduction of crystallite's length. Crystallite-ends play important role in the fracture of short crystallite composite materials (Fig.8f and Fig.9c) and also in continuous crystallite composite materials (Fig.8d and Fig.9b), since the long crystallites may break down into discrete lengths.

Figure 13a-c reveals that the indentation size by a constant load ($60x10^{-2}$N) is independent of variable indentation times inferring the microhardness (H_v) to be so. The observation is strongly supported by Fig.14 indicating practically no change in the microhardness with the variable indentation time for a constant applied load of ($30 X10^{-2}$N) at room temperature, which is consistent with the plastic deformation of the eutectic alloy that remains unaffected with variable indentation times. The nonlinear variation of microhardness (Fig.15) with variable applied load in the range $10 X 10^{-2}$ - $100 X10^{-2}$ N at constant indentation time, t (20s) shows that H_v decreases with increasing load until about $60 X10^{-2}$N and thereafter , H_v tends to saturation which is full at $70 X10^{-2}$N.The qualitative explanation of the variation follows that the indenter penetrates only surface layers at small loads, consequently, the effect is more pronounced at these loads. However, the increasing load increases the penetration depth that damages the inner layers more effectively to shattering the ability of hardness and ultimately, H_v appears to be independent of further applied load .The variation is in good agreement with the microhardness increase during early stages of plastic deformation (Brookes, 1986) but is contrary to Kick's law that H_v remains constant irrespective of the magnitude of applied load, P (Ascheron et al.,1989),qualitatively:

$$P = k_1 d^n \qquad (13)$$

the Meyer's index, n = 2, accounts for constant H_v, and k_1 is a constant. A linear plot between log P and log d (Fig.16) of Eq. (13) yields slope, n and intercept, k_1 at the particular applied load P that exists at $d = 10^{-3}$ m for any set of discrete data. The literature (Beigh et al., 1995) speaks that for any material with rising applied load H_v decreases when $n<2$ and increases when $n>2$. Both n and k_1 computed for the eutectic alloy Sn- Cd are recorded in Table 6.In fact, hardness is a measure of the resistance to localized plastic deformation, defined as the ratio W/A where W, the load in N and A, the area of indentation in m^2.This implies that a load applied to a specimen is partially affected, since a portion of it being utilized to overcome the Newtonian resultant pressure, W of the specimen itself. Consequently, the H_v data of the specimen necessarily bound to be analyzed in terms of the actual load (P- W) acting on the specimen. The inference from this concept to be drawn is that a load less than W will not result in plastic deformation. In view of the resultant pressure, Hays and Kendall (Hays & Kendall, 1973) modified Eq. (13) in a manner:

$$(P-W) = k_2 d^2 \qquad (14)$$

where k_2 is a constant and n=2, in the Eq. (14) called the logarithmic index, since W allows the limiting case to prevail where hardness is not markedly dependent on the load and the index can be evaluated by subtracting Eq. (13) from Eq. (14), which results:

$$d^n = (k_2 / k_1)\, d^2 + W/ k_1 \qquad (15)$$

again, the plot of Eq. (15) for d^n versus d^2 (Fig. 17) yields the slope k_2 / k_1 and the intercept W/k_1. From the known value of k_1, obtained by the plot (Fig. 16), k_2 and W computed thereby are recorded in Table 6 accomplishing the key data on microhardness. Furthermore, the plot (Fig .18) of log (P-W) versus log d of Eq. (14) yields the value of the logarithmic index, $n < 2$, virtually evincing the validity of the Newtonian resistant pressure concept of the eutectic alloy Sn- Cd.

The fracture toughness, K_c is a measure of resistance to brittle fracture when a crack is developed on the alloy specimen and follows the relation:

$$K_c = P/\beta l^{3/2} \qquad (16)$$

where l is the crack length measured from the center of the indentation mark to the crack tip. β is a numerical constant that depends on indenter's geometry, β = 7 for Vickers indenter. The satisfactory values of K_c can be obtained from Eq.(16) only if $l/a \geq 3$ or $l/a < 3$ respectively for median or radial crack system (Fig. 13 d) where $a = d / 2$. Table 7 records the parameter K_c and factor l/a at variable applied load P. The plot in (Fig. 19) shows the linear dependence of the crack length l with increasing applied load P. The brittleness is usually expressed as the brittleness index B_i and can be estimated from K_c values by the relationship (Lawn & Fuller, 1975):

$$B_i = H_v / K_c \tag{17}$$

B_i values calculated by Eq.(17) for the eutectic alloy Sn- Cd are presented in Table 7.The yields strength of a material can be estimated from the microhardness values using the relation (Beigh et al., 1995) valid for n<2:

$$\sigma_V = H_v /3 \tag{18}$$

the yields strength of the eutectic alloy in the load range 10 X 10^{-2} - 100 X 10^{-2} N for constant indentation interval for 20 s is recorded in Table 7. All these parameters authenticate the plastic deformation of the eutectic alloy Sn- Cd and comprehensively accomplish the key data on hardness.
Figure 20 shows the XRD patterns for the eutectic alloy Pb-Bi containing a number of peaks and intensities of pure Pb and Bi atoms and expressing their immiscibility in the solidus eutectic, since the eutectic alloy does not indicate any peak of its own. The analysis implicitly, confirms the eutectic alloy to be a mechanical mixture of its constituent phases simulating weak interactions at their atomic levels. This implies that the eutectic composite is a terminal solidus solution.

5. Conclusions

Thermodynamic analysis of binary solidus-liquidus equilibrium data besides predicting the non-ideal nature of the systems, also offers an alternative thermal device for studying the phase equilibrium curve of binary systems, particularly capable of forming eutectic mixtures, in terms of excess functions G^E, S^E and H^E by variation in a mole fraction composition. The heterogeneous nucleation data verify the nucleation theory revealing the consistency in symmetry relations which predict that there is a definite relationship between the nucleation determined from the limit of undercooling and the liquidus structure of eutectic and non-eutectic melts associated with the entropy of fusion. The analysis of anisotropic growth kinetics emphasizes that the nucleation of eutectic phases from a binary melt follows the diffusion growth process. The rheological properties evince the essence of prominent molecular interactions in the binary eutectic melt and their emulation as a function of temperature. The excess functions and rheological properties are found complementary in predicting the non-ideal nature of binary eutectic systems and hence their liquidus structures. Microscopic studies confirm the f-f growth of the eutectic phases with $\alpha > 2$ and nf-nf growth of the eutectic phases with $\alpha < 2$, and also reveal their dependence on both temperature and modes of solidifications. The strength-growth relationship follows an identical form of the Weibull probability distribution curve that acquires two cut-off points corresponding to a lower strength limit in the slow and fast growth regions, and an upper strength limit in the moderate growth region. The moderate growth is an experimental evidence for being the most probable one in the domain of solidification modes in structuring the modal microstructure consisting of crystallites or lamellae reinforcing the matrix where there is a perfect crystallite-matrix bond. Of greater interest is the akin mass composition that predominates the domain of compositions in strength view point because of its superior microstructure. The chapter also discusses the micohardness of binary metallic eutectic system Sn-Cd and the experimental evidences confirm the plastic deformation of the system. The chapter work authenticates with experimentally investigated strength data that the physical properties, and in particular, the mechanical behavior of a material depend on the growth habits to produce the modal microstructure of the material. X-ray studies affirm the mechanical combination of the eutectic phases resulting in terminal solidus solution and hence the eutectic solidus structure.

6. Acknowledgement

The authors gladly express thanks and acknowledge the instinct cooperative spirit of Ms. Surby Gupta, Mr. Arun Kumar, Ms. Savita Gupta, Mr. Sahil Sharma (Computer Assistant), Mr. Rajesh and Mr. Asim Sharma in making the chapter work more effective as a teaching and learning, and research tool, and all those who have shared their input and contributions in compiling the chapter.

7. References

Ashbee, K.H. (1993) *Fundamental Principles of Fiber Reinforced Composites*, 2nd ed., Technomic Publishing Company, Lanchaster, PA (1993).

Ascheron C., Hasse C., Kuhn G., & Neumann H., (1989). Microhardness of Sn-doped InP. *Crystal research and technology*. Vol. 24, Issue 2, (February 1989) , pp. (K33-K35).

Beigh, S., Kotru, P. N., & Wanklyn, B.N. (1995). Indentation induced microhardness and fracture studies on (110) and (001) planes of flux-grown dysprosium orthoferrite single crystals. *Materials chemistry and physics*, Vol.40 (1995), pp (.99-104).

Boyer, H.E. (1999). *Hardness Testing*, 2nd ed., ASM International Materials Park, OH (1999).

Brookes.C.A.(1986). *Inst. Phys. Ser*, Vol. 75, Chapter 3.

Callister, W.D.(2006). & Rethwisch D.G. (2008). *Composites: Polymer-Matrix Composites. Fundamentals of Materials Science and Engineering*, 3rd ed., Hoboken, New York (2008).

Callister, W.D. *Materials Science and Engineering. An introduction*. 6th Ed., John Wiley and Sons Inc., Canada (2006).

Caram, R., Banan, M.,& Wilcox, W. R. (1991). Directional solidification of Pb-Sn eutectic with vibration. *Journal of crystal growth*, Vol.114 (July 1991), pp.(249-254).

Caram, R., Chandrashekher, S.,& Wilcox, W.R. (1990). Influence of convection on rod spacing of eutectics. *Journal of crystal growth*, Vol.106 (march1990), pp.(294-302).

Courtney, T.H. (2000). *Mechanical Behavior of Materials*, 2nd Ed., Mc Graw- Hill Higher Education, Burr Ridge, IL, (2000.)

Edward, W. W .(1930). *International Critical Tables of Numerical Data*, Vol.7,McGraw-Hill,New York (1930).

Hayden, H.W., Moffatt, W.G., & Wulff, J. (1965), *The Structure and Properties of Materials*, Vol. III, *Mechanical Behaviour*, John Wiley & Sons, New York (1965).

Hays C.,& Kendall E.G. (1973). Analysis of knoop microhardness. *Metallography*. Vol. 6 (August 1973), pp. (275-282).

Hull, D., & Clyne, T.W. (2006). *An introduction to composite ma*terials, 2nd Ed., Cambridge University Press, Newyork, 2006.

Hunt,D., & Jackson K. A.,(1966). Binary Eutectic Solidification. *Trans. AMIE*, Vol.236 (June 1966), pp.(843-852).

Kollman, F. P., Ed. F. P. Kollman, Berlin, (1975). *Principles of Wood Science and Technology*, Heildberg, New York (1975).

Lawn B. R.,*& Fuller E. R.* (1975). Equilibrium penny - like cracks in indentation fracture. *Journal of Material science.* Volume 10, (1975), pp. (2016-2024).

Lide,D.R.(2009). *CRC Handbook of Chemistry and Physics, A Ready Preference Book of Chemical and Physical Data:*90thEd.,CRC Press,London 2009.

Mallick, P.K. (1993). *Fiber-Reinforced Composites, Materials, Manufacturing, and Design*, 2nd ed., Marcel Dekker, New York (1993).

Ovsienko, D.E., Alfinstev, G. A.,& Arizumi,T. Ed. 2nd (1980) *Crystal Growth, properties and Application*, vol. 2, Springer-Verlag, Berlin Heidelberg New York 1980.]

Pillar, R.M. (1984). *"Manufacturing Processes of Metals: The Processing and Properties of Metal Implants," Metal and Ceramic Biomaterials*, Ducheyne, P & Hastings, G (Editors), Franklin Book Company, Elkins Park, PA (1984).

Sharma, B.L. (2003). Structural models of faceted -faceted eutectic system vanillin - acenaphthene *Mater. Chem. Phys.* 78 (2003), pp. (691-701).

Sharma, B.L., Jamwal, R. & Kant, R. (2004a). Thermodynamic and lamella models relationship for the eutectic system benzoic acid -cinnamic acid *Cryst. Res. Technol.* 39, No. 5, (2004), pp.(454-464).

Sharma, B.L., Tandon, S. Kant, R , & Sharma, R. (2004b). Quantitative essence of molecular interactions in binary organic eutectic melt systems *Thermochimica Acta* 421 (2004), pp. (161-169).

Sharma, B.L., Gupta, S.,Tandon,S , & Kant, R.(2008). Physico-mechanical properties of naphthalene- acenaphthene eutectic system by different modes of solidification. *Mater.Chem.,Phys.*Vol. 111, (2008) pp. (423-430)

Sharma, B.L., Tandon, S., & Gupta, S. (2009). Characteristics of the binary faceted eutectic : benzoic acid -salicylic acid system, *Cryst. Res. Technol.* 44, No.3, (2009). pp. (258-268)

Woishnis W. A. (1993). *Engineering Plastics and Composites*, 2nd ed., ASM International Materials Park, OH (1993).

2

In-Situ Formation of TiC Using Laser Cladding

Ali Emamian, Stephen F. Corbin and Amir Khajepour
University of Waterloo
Canada

1. Introduction

Composite materials result from the combination of two or more dissimilar materials with different physical and mechanical properties. The final product has superior properties compared to the individual components. Particles or reinforcements in different geometries, including particulates, fibres and whiskers are used in various types of matrices such as polymers, ceramics or metals. The toughness and strength of composites are thus functions of the matrix and reinforcements properties and phase morphology.

Most composites are made from a ductile matrix in which hard particles are distributed. The matrix of a composite can be a polymer, ceramic, or metal. Metal matrix composites (MMCs) are a type of composite in which ceramics, such as TiC, WC and TiB_2 with a high melting point and high hardness, are distributed in a metal matrix like Fe, Co or Ni. Fe and its alloys have attracted substantial attention of late due to their advantages over other alloys. These advantages include availability, low cost, and chemical compatibility with a wide range of steels. Al_2O_3, ZrO_2, TiN, TiC and VC are examples of hard particles used as reinforcements in metal matrices. Among these particles, TiC has sparked considerable interest because of its good wettability with Fe, high hardness and low density compared to the other components. In metal matrix composites, hard particles distributed in the matrix cause an increase in strength, stiffness, wear resistance and decreased density (Cui et al., 2007; Emamian et al., 2010c; Pu, 2008).

Different methods are used to form Fe-TiC composites. For example, Das et al. (2002) reviewed various synthesis routes of TiC-reinforced Fe-based composites such as powder metallurgy, conventional melting, casting, combustion synthesis and laser surface melting. Additionally, the authors formed TiC by a combustion synthesis method known as aluminothermy. Similarly, Jiang et al. (2000, 2007) fabricated a metal matrix composite by a casting process. Using WC and TiC as reinforcing particles, they determined that these particles are evenly distributed in the metal matrix. Unfortunately, these methods are not well suited to produce an MMC coating.

Laser cladding is an effective method which is extensively used in surface modifications of metallic materials. The highly-focused generated heat in this process results in a minimal heat-affected zone. Therefore, the process has a minimum impact on a substrate's mechanical properties. In conventional laser cladding, the majority of reinforcing phases are directly added into the coating materials (i.e., an ex-situ process). Ariely et al. (1991) carried out laser alloying of AISI 1045 steel with TiC powder fed by the dynamic blowing method. They studied the effect of changing the laser power, scan speed and feed rate on the process.

While optimum parameters significantly increase the surface hardness, some solubility of TiC in the molten Fe produces a small fraction of TiC dendrites upon re-solidification of the coating. Tassin et al. (1995) used a laser process to enhance the surface hardness of AISI 316L by adding TiC particles. As well, they added chromium carbide (Cr_2C_3) instead of TiC to augment the hardness by as much as 450 to 900 HVN.

In the ex-situ method, chemical incompatibility between the reinforcement and matrix can occur, leading to poor interfacial bonding. Different thermal properties between the composite phases (e.g., coefficient of thermal expansion, CTE) increase the risk of cracks at the poorly bonded ceramic/matrix interface in ex-situ synthesis. In contrast, in-situ laser cladding can eliminate the interface problem. The in-situ process is a method where the hard ceramic phase is formed during coating by the reaction between powder ingredients. This creates a thermodynamically stable ceramic and matrix phase with sufficient strength to transfer stresses. Therefore, the probability of crack formation and failure at the composite interface decreases. Moreover, the high solidification rate in laser cladding opens the option of controlling particle size during the evolution of reinforcement particles.

The degree of uniformity of the particle distribution throughout the matrix directly affects the degree of uniformity of the composite properties. It is therefore important to control particle distribution during a in-situ laser cladding process. The distribution of phases and the overall clad microstructure is sensitive to the relatively large number of laser cladding operating parameters. These parameters, as well as the physical phenomena which govern the in-situ process and determine the final quality of the fabricated parts, make this layer-based technique a complex process.

Du et al. (2007a, 2007b) enhanced the wear resistance and hardness of low carbon steel through the formation of TiC-VC particles in an iron matrix by laser cladding. They also achieved a similar result for TiC-TiB_2. The increase in hardness was due to carbides being distributed uniformly in the metal matrix. Wu et al (2001) increased the surface hardness with TiC_P/Ni using laser cladding.

Wang et al. (2008, 2009) increased the wear resistance of AISI 1045 steel by the in-situ synthesis of a composite coating by preplacing FeCrBSi alloy and graphite powders on the substrate surface prior to the laser surface treatment. They proposed that increasing the wear resistance might be due to the formation of a variety of in-situ carbides including TiC particles. The in-situ process was also selected by Yan et al. (1996), who formed TiC in a nickel-based alloy coating on mild steel. During the cladding process, other types of particles such as $Cr_{23}C_6$, Ni_5Si_2 and Cr_2B were formed and uniformly increased the hardness.

From the above literature review, it is clear that the in-situ formation of TiC particle metal matrix composites through laser cladding is a feasible process. Furthermore, it is widely reported that the hardness and wear resistance of the substrate can be significantly enhanced by the formation of a TiC-based composite coating. It is also clear that the TiC morphology can vary throughout the clad thickness, but no detailed or satisfactory explanation on why this happens, or what the mechanism of TiC formation is during laser cladding, has been published. This is partly due to the lack of knowledge and understanding about the relationship between clad microstructure and laser processing conditions. In addition, most researchers have focused on the use of rather complex, multi-component metal matrix powders containing various combinations of Ni, Fe, Co, Cr, B or Si. This can produce a variety of carbides other than TiC and creates complex solidification behaviour during cooling.

This chapter discusses the development of an in-situ laser cladding technique to deposit a TiC-based MMC coating on a steel substrate with no cracks and excellent bonding to the substrate and a high hardness. The approach involves feeding graphite, titanium and iron powders into a melt pool to form an in-situ clad consisting of TiC in a Fe-rich metal matrix. To study the effects of laser parameters on the quality of the clad, they are combined in two general parameters (effective energy and powder deposition density) to gain a better understanding of their roles. These parameters can help to predict the clad quality prior to conducting experiments (Emamian et al., 2009).
In order to study the effect of chemical compositions on TiC morphology, micro hardness and wear resistance of the clad, diverse chemical compositions resulting from different C:Ti ratios (i.e., 45:55 and 55:45 at%) and Fe content (i.e., 70, 60, 50 and 10 wt%) were selected for the powder composition. Additionally, several laser parameters were selected to study their effects on the clad microstructure.

1.1 Phase diagram
In this research, Fe, Ti and C are the primary components. The binary and ternary diagrams help to find the phases which are expected to appear during the process. Although the laser cladding process is a non-equilibrium process, phase diagrams are useful guides to interpret and estimate the result-ant phase formation.

1.1.1 Ti-C phase diagram
The selection of the atomic ratio forming a TiC phase is important in the formation of the TiC particles (see Fig. 1).

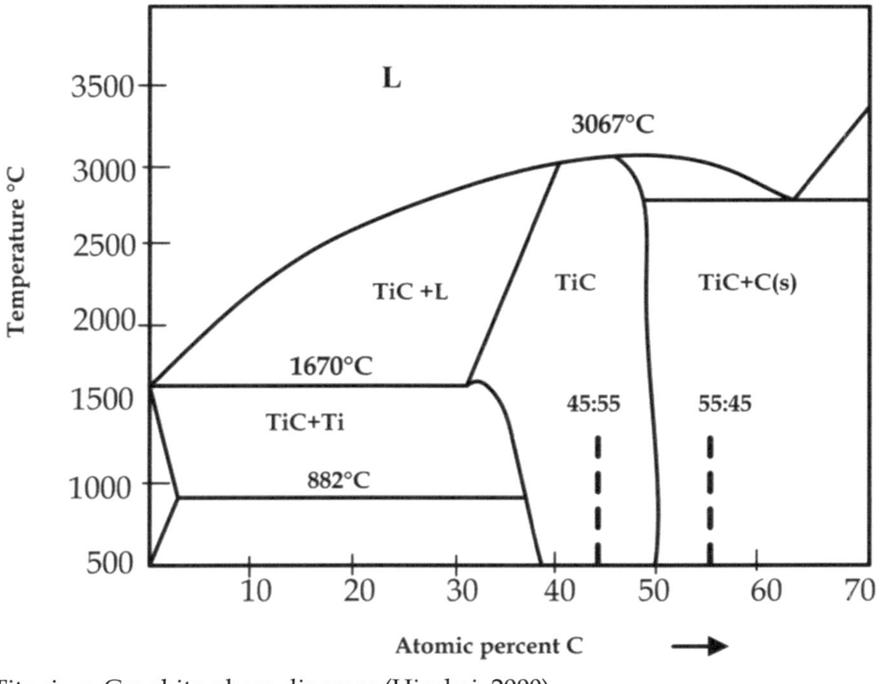

Fig. 1. Titanium-Graphite phase diagram (Hirokai, 2000)

It can be seen that TiC is stable over a wide range of composition as a intermediate phase. To avoid the formation of secondary phases such as Ti (α), Ti (ß) or C, a composition close to Ti-45%at C should be selected. For the composition study section C:Ti, the composition was changed to 55:45. Obviously, in the non-equilibrium condition phase formation can deviate from the predictions of the equilibrium phase diagram. Nonetheless, these diagrams are very useful in predicting, at least qualitatively, phase formation under non-equilibrium states.

1.1.2 Fe-Ti phase diagram

Since Fe and Ti are in contact during the melting of non-equilibrium solidification, the phase diagram below should be considered. Titanium is a ferrite stabilizer and austenite is stable in a narrow zone.

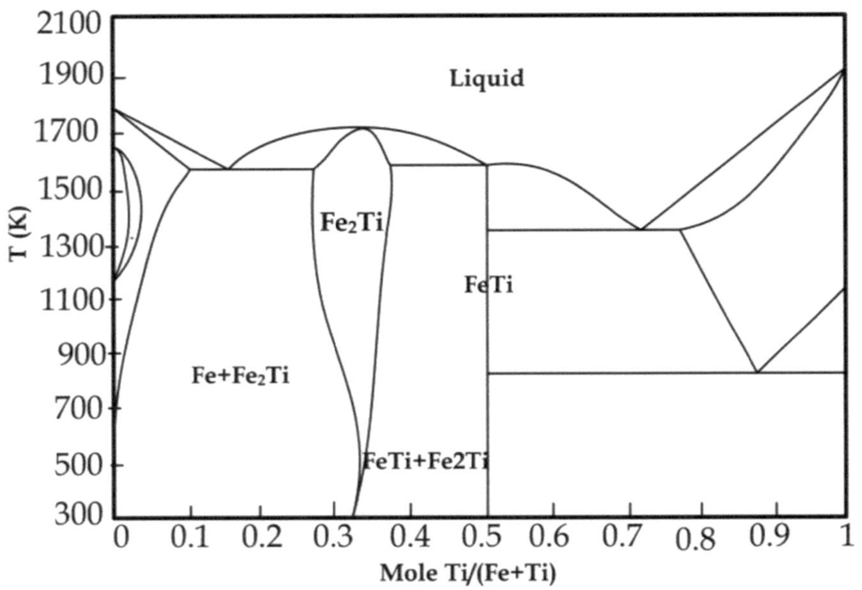

Fig. 2. Iron-Titanium Phase diagram (Hirokai, 2000)

According to Fig. 2, there are two types of intermetallics: Fe_2Ti and FeTi. Neither shows acceptable mechanical properties and hence are undesirable phases. The phase diagram of Fe-Ti shows that, by moving to the right of the phase diagram, the likelihood of both intermetallic phases being present increases. This is particularly so between the 0.3-0.5 mole fractions of titanium, where both intermetallics are in equilibrium. The goal is to form a metal matrix composite and decrease the chance of an intermetallic formation.

1.1.3 Thermodynamics of formation

Phase transformation calculations indicate that it is highly probable to have TiC particles instead of any other carbide such as Fe_3C in an in-situ TiC-Fe cladding. Fe_3C is one of several compositions which can be formed in the melt pool. The following thermodynamic equations provide information about the reactions (Dubourg & St-Georges, 2006):

$$[Ti] + [C] = <TiC> \ \Delta G = -139{,}0461.31 + 98.96\ T \ \ j/mole \tag{1}$$

$$3Fe_{(l)} + <C> = <Fe_3C>, \ \ \Delta G = -53416 + 55.68\ T \ \ j/mole \tag{2}$$

The equations prove that the Gibbs free energy of TiC formation is more negative than that of Fe_3C. Therefore, the probability of Fe_3C formation is lower than TiC. Although the equations are useful in equilibrium conditions, they are still useful to predict the formed phases. Moreover, it can be seen in the next chapter that there are no results to confirm the presence of Fe_3C.

2. Experimental set-up

In the experiments, AISI 1030 medium carbon steel is chosen as the substrate, with 0.28-0.34 C, 0.6-0.9 Mn, P less than 0.04, and S less than 0.05 in wt%. The substrate dimensions on which the claddings are deposited are 100×30×6.35 mm. Prior to laser cladding, the substrates are shot-blasted and rinsed with ethanol, followed by acetone.

Mixtures of Ti, graphite, and Fe powder are fed into a melt pool formed by the laser using a Sulzer Metco (TWIN 10-C) powder feeder. The ratio of C: Ti is 45:55 at %C and Ti-55 at %C. A three-digit code is considered for powder compositions. The first digit stands for Fe percentages (7 for 70) and the next two digits are the C:Ti ratios in atomic percentages (i.e., 45 or 55%). In the first two sections presented below, the powder composition was set at Ti-45% at C (745). In the composition study section, the remainder of the samples are studied (i.e. 755, 655, 555, 155). Table 1 presents the chemical compositions of powders which were studied in this chapter.

Mixtures of the powders are pre-cursors, with 99.5% purity; a 40 μm maximum powder size is further mixed with 98% purity Fe powder. To obtain a homogeneous distribution, all powders are blended for four hours at 300 rpm in a quarter-filled jar of 165×60 mm with a milling media of glassy balls where the ball-to-powder-size ratio is 20:1.

Sample group code	Fewt %	Ti wt%	C wt%	Fe at%	Ti at%	C at%
Group 745	70	25	5	57	24	19
Group 755	70	23	7	55	20	25
Group 655	60	31	9	44	26	30
Group 555	50	38	12	33	30	37
Group 155	10	69	21	5	43	52

Table 1. Chemical composition of investigated powders

An IPG fiber laser model YLR-1000-IC operating in continuous mode with a maximum power of 1 kW is used to produce a series of single-clad tracks. The diameter of the laser beam spot size on the workpiece (WP) surface is fixed at 2.5 mm. The laser machine is integrated with a five-axis CNC vertical machining centre to control the velocity of the WP. To protect the melt pool from oxidation, Argon shielding gas is supplied through a nozzle at 10 $Lmin^{-1}$.

The specimens are sectioned for micro-structural examination in a longitudinal direction. The samples are prepared using SiC grit paper with grit sizes ranging from 240 to 2400 and polished with diamonds from 6 microns to 1 micron. Afterwards, the samples are placed in an ultrasonic bath to remove any surface contaminants and then rinsed with alcohol and air dried.

A LEO SEM with 20keV is used to examine the microstructure and morphology of the phases formed during cladding. SEM is selected to determine the quality of the clad deposit (i.e., bonding to the substrate) and to determine the in-situ TiC particle morphology. The working distance is 9-14 mm and the vacuum system is set at 1.53e-0.005 mBar. Different magnifications are used to capture the microstructure. The mounting material used was conductive in order to prevent charging during sample observation.

The micro-hardness at different depths from the substrate-clad interface is measured using a Vickers micro-hardness tester, and the average of three hardness measurements is reported.

A low magnification EDS analysis of the whole longitudinal section of the clads helps to measure the Fe percentage to calculate the dilution. Further, an XRD analysis is performed with a Rigaku AFC-8 diffractometer.

Clad heights were measured by a caliper vernier. Specifically, measurements were made at the middle of the clad which normally were uniform in height. Clad height was also measured using scale bars in low magnification SEM images.

3. Laser processing conditions

The laser process parameters to determine the optimum laser cladding conditions are given in Table 2. Initial experiments were held at a constant powder feed rate of 8 g/min while systematically increasing the work piece (WP) scan speed from 2,4 to 6 mm/s. Each scan speed was explored at laser powers from 250 to 650 W (i.e., samples 1 to 9). Visual inspection of the formed clad tracks revealed that, for the power setting, a scan speed of 6 mm/s was required to produce a clad deposit.

In the second series of experiments (samples 10 to 13), the scan speed was increased from 6 to 16 mm/s while fixing the laser power and powder feed rate at 650 W and 8 g/min., respectively. Again, a clad deposit was produced in each case (except at the highest scan speed) but no bonding with the substrate occurred.

In an effort to create a clad/substrate bond, the laser power was further increased to 700 and 800 W while fixing the scan speed at 6 mm/sand using powder feed rates of 4 and 8 g/min (samples 14 to 17). Partial bonding of the clad to the substrate occurred in these cases. For example, at the higher powder feed rate, the clad adhered to the substrate immediately after cladding but detached during complete cooling. For the 4 g/min feed rate, clads remained attached to the substrate but were easily removed upon application of a small force by hand. Lowering the WP scan speed to 4, 3 or 2 mm/s (i.e., samples 18-20) did not rectify this situation.

In the final series of experiments, the laser power was increased to 900 W while the powder feed rate was at 4 or 8 g/min and the scan speed was at 6 or 8 mm/s. A laser power of 1000 W was also used with a scan and powder feed rate of 4 mm/s and 4 g/min, respectively. Visual examination of these clad tracks revealed that in all cases a clad deposit was produced and was well bonded to the substrate.

3.1 SEM results

SEM is selected to determine the quality of the clad deposit and bonding with the substrate and to find out if in-situ TiC particles are formed. For this investigation, samples which indicated a good clad deposit but no bonding with the substrate (i.e., samples 3 and 9) and a successful clad deposit (i.e., samples 23 and 24) were chosen.

However, none of these processing conditions were able to create a bond between the clad deposit and substrate.

Fig. 3 illustrates an SEM back scatter image of a typical region from the un-bonded clad deposit sample 9. It is clear that the clad microstructure consists of a matrix in which a relatively high volume fraction of uniformly dispersed particles are present. The size of these particles ranges from approximately 1 to 7 μm.

No	Power (W)	Scan Speed (mm/s)	Feed Rate (g/min)	Effective Energy (J/mm^2)	Powder Deposition Density (g/mm^2)	Observation
1	250	2	8	50	0.027	No Bond-No Clad
2	250	4	8	25	0.013	No Bond-No Clad
3	250	6	8	17	0.009	No Bonding-Formed clad
4	400	2	8	80	0.027	No Bond-No Clad
5	400	4	8	40	0.013	No Bond-No Clad
6	400	6	8	27	0.009	No bonding-Formed clad
7	650	2	8	130	0.027	No Bond-No Clad
8	650	4	8	65	0.013	No Bond-No Clad
9	650	6	8	43	0.009	No Bonding-Formed Clad
10	650	8	8	32	0.007	No Bonding-Formed Clad
11	650	10	8	26	0.005	No Bonding-Formed Clad
12	650	12	8	22	0.004	No Bonding-Formed Clad
13	650	16	8	16	0.003	No Bond-No Clad
14	700	6	8	47	0.009	Partial bonding with substrate melting
15	700	6	4	47	0.004	Weak bond-Removable Clad
16	800	6	8	53	0.009	Partial bonding with substrate
17	800	6	4	63	0.004	Weak bond-Removable Clad
18	800	2	8	160	0.027	No bonding-Formed Clad
19	800	3	8	106	0.018	No bonding-Formed Clad
20	800	4	4	80	0.007	Partial bonding with substrate
21	900	6	8	60	0.009	Successful Clad
22	900	8	8	45	0.007	Successful Clad
23	900	6	4	60	0.004	Successful Clad
24	900	8	4	45	0.003	Successful Clad
25	900	4	4	90	0.007	Successful Clad
26	1000	4	4	100	0.007	Successful Clad
27	1000	4	4	100	0.007	Successful Clad

Table 2. Results and Observation

Fig. 3 also illustrates a higher magnification image of a eutectic structure in grain boundaries of the matrix of sample 9 (i.e. region 2). Table 3 reports a typical EDS analysis performed on the dark grey dispersed particles and regions 1 and 2 of the matrix. The high Ti content of the particles is consistent with TiC particles. The majority of the matrix is a very Fe-rich phase with slightly elevated Ti concentrations in region 2. The very small black particles in Fig. 3 were too small to obtain reliable EDS analysis.
Carbon was not included in the analysis due to inaccuracies in its inclusion in a quantitative analysis.

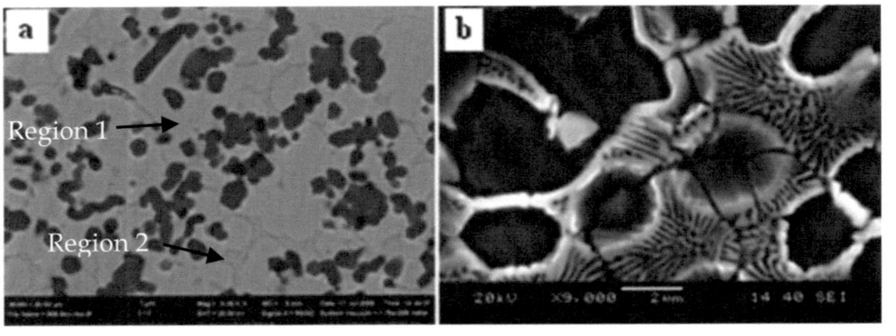

Fig. 3. BSE image of sample 9 a) distributed black particles in individual and longitudinal shape in matrix of un-bonded clad b) High magnification of Region 2 Eutectic structure of Fe-Fe_2Ti

Region	Ti conc.(wt%)	Fe conc.(wt%)
Dark grey particles	95.2	4.8
Region 1	8.7	91.3
Region 2	16.5	83.5

Table 3. EDS analysis of phases in sample 9

Fig. 4 shows a high magnification image of the two phase microstructure of sample 23. The microstructure consists of a high volume fraction of fine (i.e., < 2 μm) particles dispersed in a matrix. As with Table 3, EDS analysis confirmed that the dark grey particles are Ti-rich, whereas the matrix is Fe-rich.

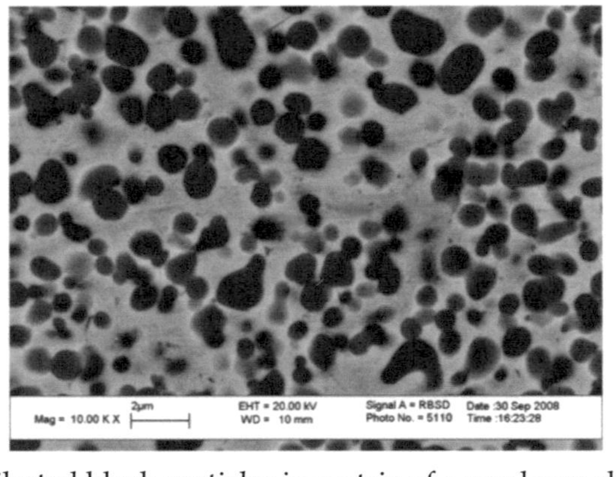

Fig. 4. BSE image-distributed black particles in matrix of a one layer clad (sample 23)

Fig. 5 shows a longitudinal section of sample 24 which confirms a uniform distribution of Ti-rich particles in an iron-rich matrix. However, in some areas, such as those labeled region A, larger black particles exist. Qualitative EDS analysis of the particles, which included carbon in the analysis, indicated a 95 wt% carbon composition with the balance as Ti and Fe. Note that the black particle in the centre of the image is surrounded by the Ti-rich grey phase. An EDS analysis of this region reveals an approximately 95 wt% Ti: 5 wt% Fe composition.

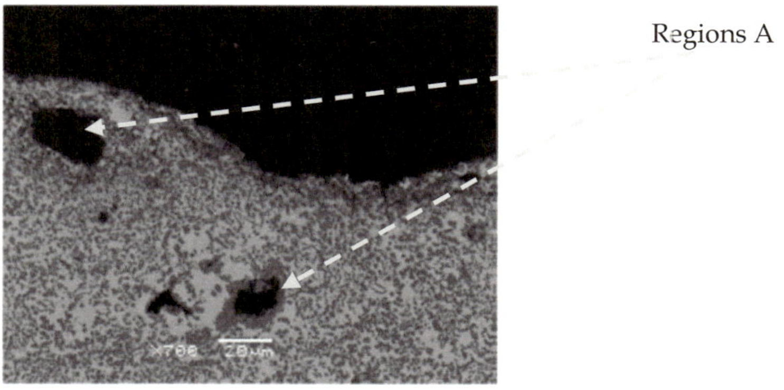

Fig. 5. Sample 24 TiC distributions

3.2 XRD results

XRD analysis was performed on samples 9 (un-bonded clad) and 23 (one layer clad). The XRD spectrums for these samples are shown in Fig. 6.

Fig. 6 depicts the XRD result for the un-bonded clad. According to this graph, TiC, Fe_2Ti and iron-titanium solid solution exist in the clad. Since there is no bonding with the substrate, dilution does not contribute to the results.

The X-ray diffraction pattern for a one layer clad is plotted in Fig. 6 and confirms the presence of the iron-titanium solid solution and TiC in the clad zone. It is worth noting that the intermetallic phase (Fe_2Ti) did not appear in the diffraction peaks of the clad coating, in contrast to the result for the un-bonded case.

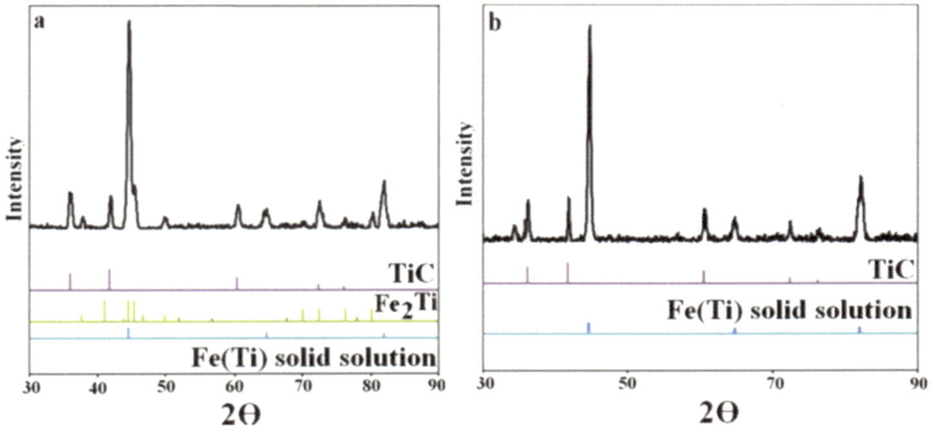

Fig. 6. XRD result of a) un-bonded clad (sample 9) and b) one layer clad (sample 23)

3.3 Further laser process/clad development and analysis

From the data in Table 2, it is clear that laser power scan speed and powder feed rate all play a vital role in clad formation, bonding, and clad quality. From the table, we can see that increasing the power increases the probability of forming a high quality clad. However, it is not the only effective parameter. In any period of time, the laser energy must provide sufficient heat to melt a given volume of the substrate as well as the incoming powder stream. This will be determined not only by the laser power but also the work piece (WP) scan speed, and powder feed rate. Two combined parameters (i.e., the effective energy per unit area (E_{eff}) and the powder deposition density (*PDD*)) are needed in order to analyze the influence of these three parameters.

Effective energy is defined as the parameter which provides a measure of the delivered energy to the process by the laser. This energy is principally responsible for melting the substrate surface and powder and can be defined by computing (Emamian et al., 2009, 2010a).

$$\text{Energy per unit area } (\text{J/mm}^2) = P/(V.D) \quad (3)$$

where *P* is the laser power, *V* is the scan speed of WP, and *D* is the laser spot size in *mm*. Therefore the units of effective energy are W ($mm^{-2}s^{-1}$) or Jmm^{-2}.

The powder deposition density (*PDD*) is introduced as (Emamian et al., 2010a, 2010b):

$$\text{Powder deposition density } (\text{g/mm}^2) = R/(V.D) \quad (4)$$

where *V* and D have the previous meaning, *R* is the powder feed rate in g/min, such that the powder deposition density (*PDD*) has units of g/mm². Table 2 reports these calculated parameters for each of the laser conditions. Fig. 7 plots the data where *PDD* is the x-axis and (E_{eff}) is the y-axis.

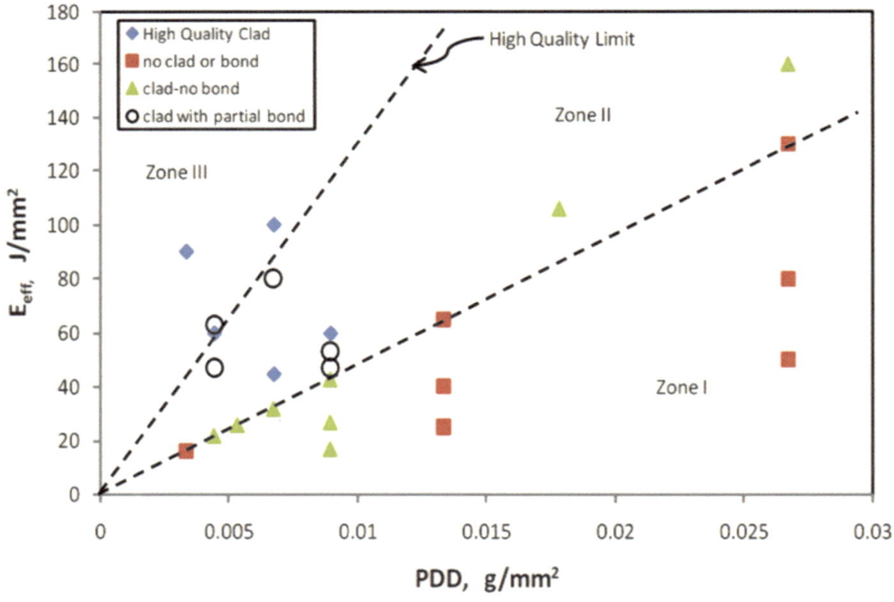

Fig. 7. Effective energy versus powder deposition density

As seen in Fig. 7, low E_{eff} and high *PDD* values leads to no clad deposit formation or bonding with the substrate. (i.e., Zone I). While increasing the E_{eff} and reducing *PDD* leads to the formation of a clad deposit, it is not bonded to the substrate (i.e., Zone II). Finally, high effective energy along with lower powder deposition density leads to the formation of a dense clad deposit which is well bonded to the substrate (i.e., Zone III). Therefore, to achieve an acceptable clad quality, effective energy and powder deposition density should be adjusted for Zone III. This graph helps us to predict the proper parameters to create an acceptable clad prior to conducting any experiment.

To further explore the predictability of this figure, a range of clad conditions which lie close to the boundary line between Zones III and II were chosen and the clad experiments completed. These conditions are discussed in the next section. All of the conditions produced a clad deposit that could not be removed from the substrate, indicating good bonding. Metallographic examination of these samples was completed to further analyze clad quality and microstructure.

3.4 TiC formation and morphology

In this section, the TiC formation mechanism is explained based on observations and phase diagrams. This explanation can be followed in two steps–namely, melting and solidification during laser cladding.

In order to explain the TiC morphology formed during laser cladding, the melting stage of the process must first be considered. When the powder mixture enters the laser beam, the powder will rapidly heat up (Fig. 8a). The melting points of pure Fe, Ti and C (or graphite) are 1538, 1668 and 3400 °C, respectively. Given the relative melting temperatures of the pure powders, it is reasonable to assume that Fe melts first, creating a liquid phase that spreads and surrounds the Ti and graphite powders (e.g., Fig. 8b). The interaction of the elements of the powder stream will cause the formation of a ternary mixture. Fig. 9 presents the liquidus surface projections for the entire ternary C-Fe-Ti composition. The actual ternary composition of the powder mixture deposited on the 1030 steel substrate is indicated by the black dot (745 PC) in Fig. 9. From these indicators, it is clear that both Ti and C have significant solubilities in the Fe-rich liquid phase and that this solubility increases with an increase in temperature. It is also clear from Fig. 9 that the Fe-rich liquid phase will be in equilibrium with solid TiC. Therefore, alloying Fe with Ti and C actually creates a ternary liquid through the gradual dissolution of the Ti and graphite powders.

In addition, a TiC layer will form between this ternary liquid and the original graphite particles in order to establish the required phase equilibrium. As the clad temperature increases, dissolution of the solid phases will continue as the solubility of Ti and C in the liquid increases. Also the growth of the TiC phase toward the graphite core will increase (e.g., Fig. 8-c). Evidence for the microstructure depicted in Fig. 8-c is presented in Fig. 5 , which illustrates graphite particles surrounded by a TiC layer.

The next stage in clad microstructural development largely depends on the peak temperature induced by the laser. If the peak melt pool temperature exceeds the liquidus temperature of the ternary composition, all of the TiC/C particles will dissolve into the melt, creating a completely liquid melt pool. If no dilution by the substrate is assumed, then it can be stated that the liquidus temperature of the clad melt pool for composition of 745 is 2345 °C. If the peak melt temperature is less than the liquidus temperature of the melt pool clad (e.g., in the presence of Fe substrate dissolution), then a semi-solid mixture will persist where the solid graphite particle core will eventually convert through the further development of the diffusion couple to TiC.

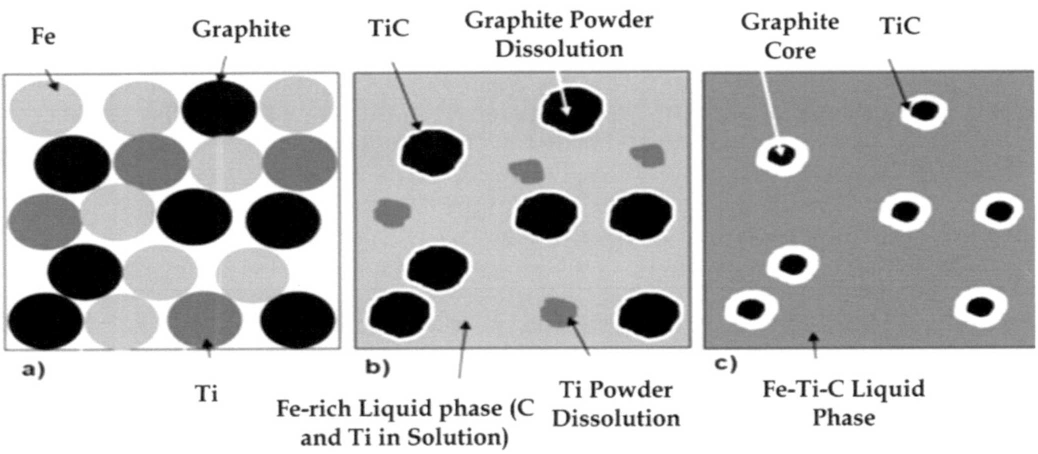

Fig. 8. TiC formation mechanism

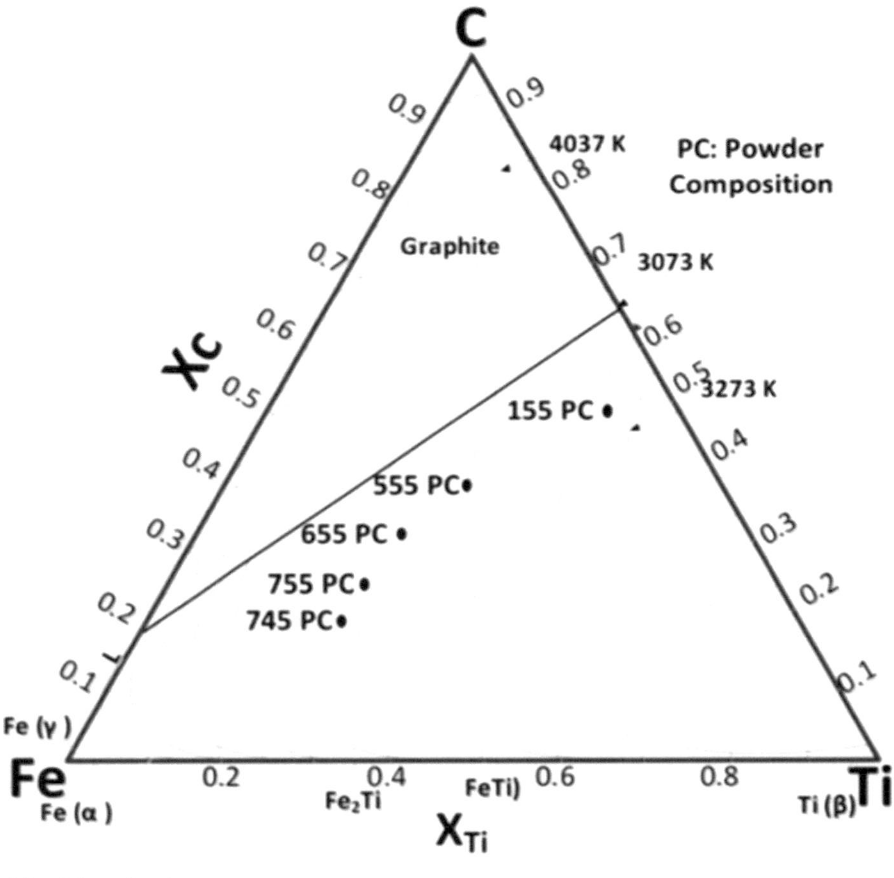

Fig. 9. Liquid projection

Fig. 10 confirms that there is a large phase field where both liquid and solid TiC will exist in a semi-solid state. It also confirms that Fe-rich mixtures (e.g., the left-hand side of the diagram) have lower liquidus temperatures. Dilution with the substrate increases the Fe content in the clad, thereby decreasing the liquidus temperature. Dilution thus plays a crucial role in establishing the clad microstructure and TiC morphology.

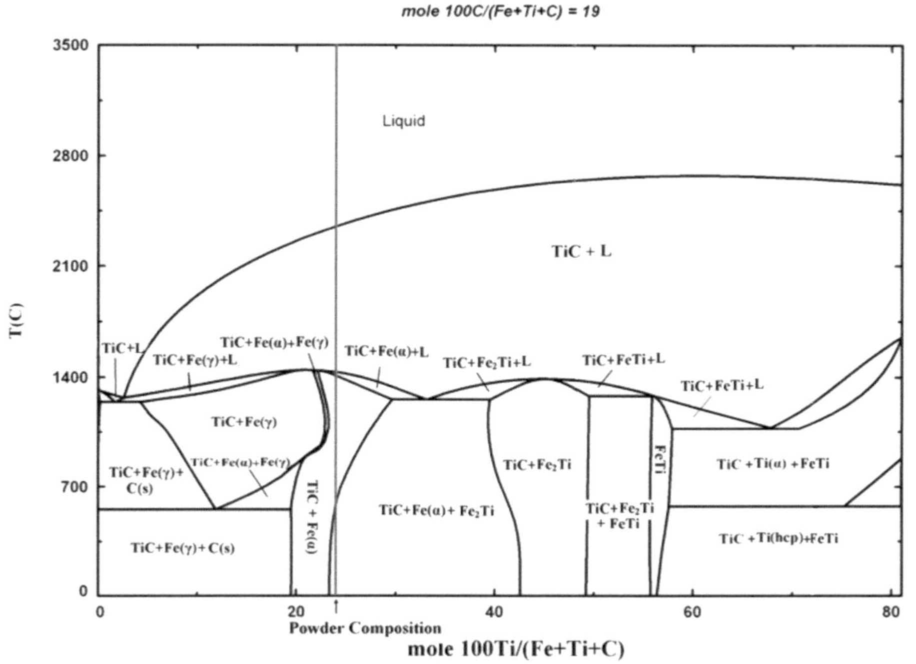

Fig. 10. Cross section of Fe-Ti-C ternary diagram in 19%at C

4. Effect of combined parameters on TiC morphology and clad microstructure

In this section, the effect of laser parameters on TiC morphology is studied. Results show that laser parameters play a crucial role in the morphology of TiC. Dendritic or spherical TiC particles with different distributions are observed, depending on the applied laser parameters. Two combined parameters, effective energy and powder deposition density, are used in order to study the effect of laser process parameters on TiC morphology.

A series of experiments are conducted in constant laser power and scan speed, constant effective energy and constant powder deposition density in order to study the TiC morphologies. Laser parameters are selected in order to be in a high quality zone as well as to result in particular values for combined parameters (namely, effective energy and powder deposition density). These series of samples help to confirm the high quality zone (e.g., zone III) found in the previous section. Fig. 11 depicts the process map for the experiments.

Samples with powder deposition density (*PDD*) greater than or equal to 0.010 g/mm^2 were selected for detailed analysis. Low clad height was the main reason for the lack of ability to analyze the samples with low *PDD* (less than 0.010 g/mm^2).

4.1 Constant laser power with different scan speeds (type A experiments)

Two groups of samples were investigated in this section: samples A1, A2, and A3, with 961 W laser powers at a scan speed of 120, 210, and 330 mm/min (Fig. 11). The second group has 907 W laser power explored with scan speeds similar to the first group.

As observed in Fig. 12, sample A1 shows dendrites of TiC particles in the clad. Table 4 shows that sample A1 has the maximum value for E_{eff} and *PDD* among the A series of experiment. As well, it exhibits the lowest value of dilution and highest clad height amongst the three samples. In samples A2 and A3, there is no evidence of dendrites, and the TiC particles are very fine. TiC particle size in sample A2 is in the range of 1-4 μm, while sample A3 has a smaller size range.

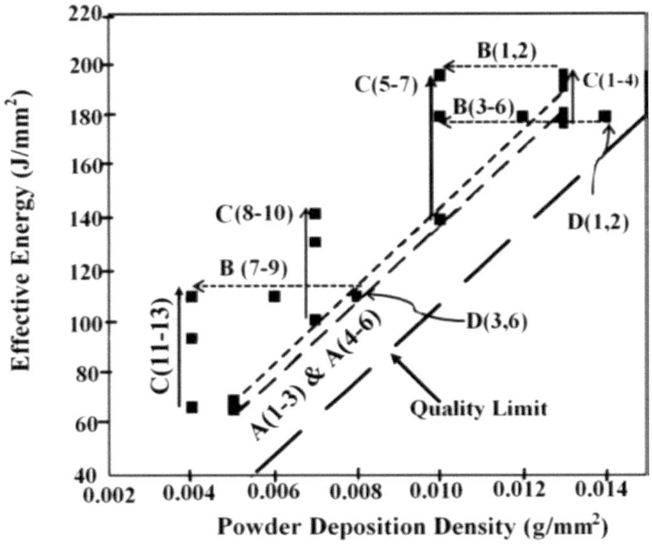

Fig. 11. Map study for groups A to D

The same trends were observed in samples A4, A5 and A6. In both groups of samples, it is noted that, by increasing the scan speed, dilution is increased, clad height is decreased, and the TiC microstructure is finer. This can be attributed to the relationship between scan speed and *PDD*. Specifically, as the scan speed increases, the *PDD* decreases. Since the laser power is constant, a larger portion of energy can be assigned for melting the substrate by decreasing the *PDD*, which then increases dilution.

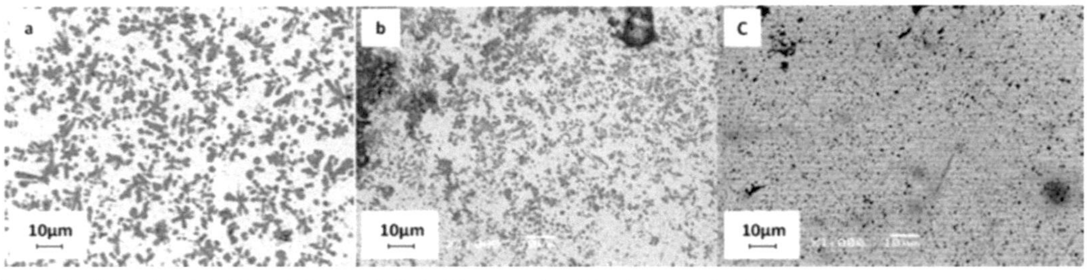

Fig. 12. Sample A1 (dendrites of TiC), b) sample A2, c) sample A3

Sample Category	Sample NO.	Laser Power (W)	Scan speed (mm/min)	Powder feed rate (g/min)	Effective Energy (J/mm²)	Powder deposition density(g/mm²)	Hardness (HVN)	Dilution %	Clad Height (mm)
A	1	961	120	4	192	0.013	575	72.8	1.40
	2	961	209.7	4	110	0.008	317	89.0	0.40
	3	961	330	4	70	0.005	750	94.9	0.14
	4	907	120	4	181	0.013	600	69.6	1.00
	5	907	210	4	104	0.008	400	88.0	0.10
	6	907	330	4	66	0.005	570	93.0	0.07
B	1	982	120	4	196	0.013	350	73.0	1.10
	2	982	120	3	196	0.010	650	77.0	0.60
	3	884	117	4	180	0.014	650	69.1	1.60
	4	907	120	4	181	0.013	600	69.6	1.00
	5	758.2	101.1	3	180	0.012	900	64.0	0.60
	6	589.7	78.6	2	180	0.010	2400	66.7	0.50
	7	961	209.7	4	110	0.008	317	89	0.4
	8	926.7	202.2	3	110	0.006	550	95	0.2
	9	864.9	188.7	2	110	0.004	600	--	---
C	1	884	120	4	177	0.013	650	70.3	1.60
	2	907	120	4	181	0.013	600	69.6	1.00
	3	961	120	4	192	0.013	575	72.8	1.40
	4	982	120	4	196	0.013	350	73.0	1.10
	5	700	120	3	140	0.010	850	70.0	0.90
	6	589.7	78.6	2	180	0.010	2400	66.7	0.50
	7	982	120	3	196	0.010	650	77.0	0.60
	8	506	120	2	101	0.007	644	79.0	0.07
	9	982	180	3	131	0.007	593	63.0	0.04
	10	708	120	2	142	0.007	584	75.5	0.1
	11	506	180	2	67	0.004	240	79.0	Low
	12	708	180	2	94	0.004	1000	80.0	0.14
	13	864.9	188.7	2	110	0.004	600	--	---
D	1	884	117	4	180	0.014	650	69.1	1.60
	2	663.4	88.5	3	180	0.014	1800	65.0	0.70
	3	961	209.7	4	110	0.008	317	89.0	0.4
	4	720.7	157.3	3	110	0.008	932	63.0	0.25
	5	480.5	104.8	2	110	0.008	450	88.8	0.10
	6	961	210	4	110	0.008	317	88.0	0. 40

Table 4. Results and data for groups A to D

4.2 Constant effective energy with variable powder deposition density (Type B Experiments)

The results of samples (e.g., sample B1 and B2 with 196 J/mm^2 and sample B3, B4, B5 and B6 with 180 J/mm^2) are presented in this section. Fig. 13 a-b illustrate the TiC morphology in samples B1 and B2, respectively. These samples have identical E_{eff} (i.e., 196), laser power and scan speeds, but their respective *PDD* values are 0.013 and 0.010, which result from different powder feed rates (4 and 3 g/min). This shows that sample B1 has a dendritic structure in almost the entirety of the clad, whereas sample B2 has very fine TiC particles distributed in the clad, as illustrated at Fig. 13-b.

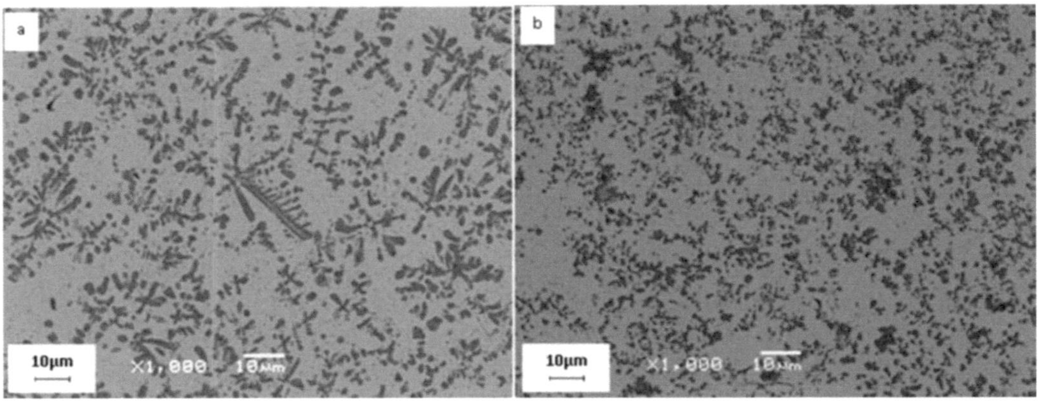

Fig. 13. BSE images for a) sample B1 and b) sample B2

Fig. 14 a-b show the clad in samples B3 and B4. These samples have different *PDD*, 0.014 and 0.013. Sample B3 with *PDD* of 0.014 shows the dendritic structure of the TiC in the clad. At the bottom of the clad, very fine TiC particles can be seen approximately 100 μm from the clad/substrate interface. Fig. 14-b also shows a dendritic structure, specifically at the middle and top of the clad in sample B4. Sample B3 shows well developed TiC dendrites compared to that of sample B4.

Fig. 14-c shows that sample B5 has fine TiC particles at the bottom of the clad (which gradually become coarser) and dendritic TiC in the middle. B5 microstructure shows both the dendritic and spherical morphology of TiC. The observed microstructure in B5 is not fully dendritic, similar to B3. However, dendrites of TiC can be observed in some locations in the middle and at top of the clad.

Fig. 14-d depicts sample B6 with a *PDD* of 0.010. TiC particles are distributed uniformly in the clad. There is no evidence for the presence of a dendritic structure of TiC particles. TiC morphology in sample B6 is spherical and the TiC particle aspect ratio can be found between 1:1 and 1:5. As Table 4 indicates, hardness values for this sample are high (e.g., 1300-3500 HVN).

This portion of the study on TiC morphology proves that *PDD* plays a crucial role in TiC formation and on the observed morphology of TiC particles. A considerable difference in TiC morphology can be detected for samples B1 and B2 which have identical laser parameters and E_{eff}, but different *PDD*, resulting from varying powder feed rate. The same can be said for the B3 to B6 series of data.

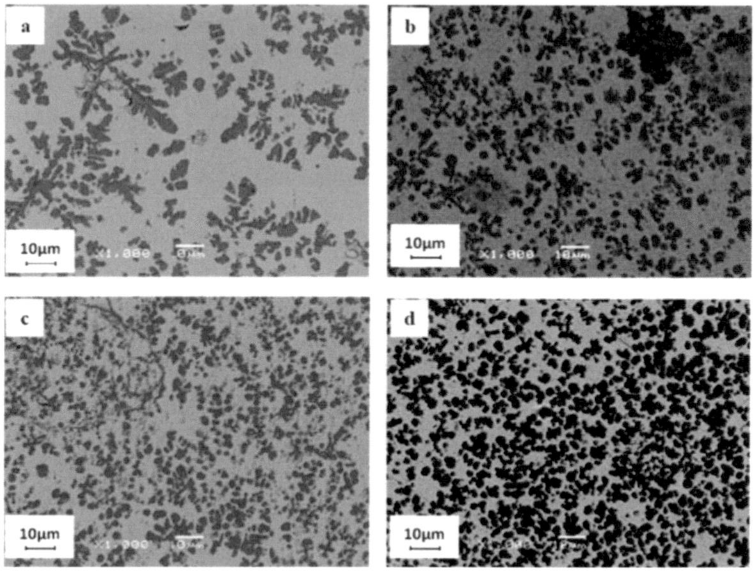

Fig. 14. a) Sample B3 b) sample B4 c) sample B5 d) sample B6

4.3 Constant powder deposition density with variable effective energy (experiments series C)

In this section, TiC particle morphology resulting from a constant powder deposition density with varying E_{eff} was studied. For this purpose, a series of samples which are identical in *PDD* and different in E_{eff} are selected. Samples C1 to C4 in the first group and C5 to C7 in the second group were investigated.

In the first group (C1 to C4), the scan speed and powder feed rate are constant. Therefore, laser power is the only parameter that is variable to create a different E_{eff}. All the samples show a dendritic structure for the TiC particles. An E_{eff} range of 177 to 196 does not have a considerable effect on TiC morphology with a constant *PDD* (0.013). Table 4 shows little change in dilution for this series of samples. Since all the parameters (except laser power) are the same, it can be seen that in the range of 884 to 982 W, these parameters form a dendritic structure.

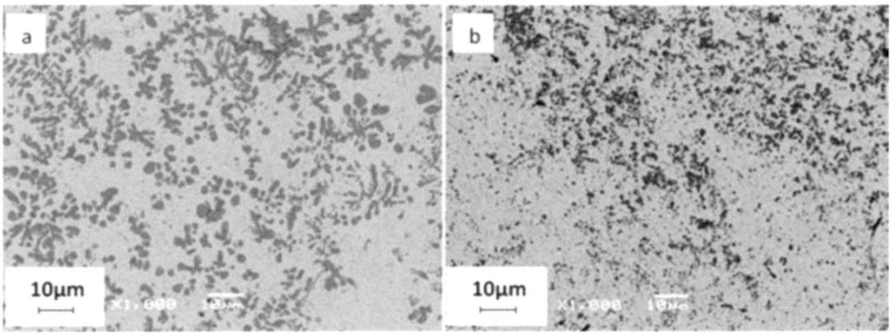

Fig. 15. BSE image of TiC morphology resulting from different *PDD* a) sample C4 b) sample C7

Thus, it may be concluded that with constant powder deposition density and laser parameters (such as scan speed, powder feed rate), increasing the E_{eff} (i.e., increasing the laser power) likewise increases the probability of creating a liquid melt pool. As a result, the TiC particle structure is mostly dendritic. This interpretation is applicable to the first group of samples (C1 to C4).

Observing samples C5 and C7 provides valuable information. Other than for the laser power, they have the same laser parameters. Sample C5 has a lower laser power (700) than sample C7 (982). From the clad microstructures, it can be seen that dendritic TiC can be detected in sample C5 whereas sample C7 shows very fine TiC particles but no evidence of a dendritic structure. Higher laser power can cause increased dilution, which changes the clad composition to the Fe-rich side and precludes dendrite formation due to the small freezing range, as shown at the left side of the phase diagram.

4.4 Identical effective energy and powder deposition density (experiment series D)

The purpose of this section is to clarify the effect of identical effective energy and powder deposition density on the morphology of the TiC particles formed during in-situ laser cladding. As mentioned earlier, the process parameters have been calculated to be in the high quality clad zone and also resulted in a specific effective energy and powder deposition density. Table 4 shows that sample D1 and D2 have the same effective energies (E_{eff}) and powder deposition densities (*PDD*), but dissimilar process parameters. The same situation can be seen in samples D3, D4, D5 and D6. Laser conditions of D1 and D2 have been used later in powder composition study (i.e. conditions A and B, respectively).

Fig. 16-a illustrates back scatter images of sample D1 where dendritic TiC can clearly be seen. Conversely, Fig. 16-b illustrates the TiC morphology for sample D2, and no evidence of a dendritic structure for TiC can be found. Moreover, the distribution of TiC particles in the clad is very uniform compared to sample D1. A simple comparison between Fig. 16-a and 17-b shows there is a considerable difference between TiC morphology, although both have the same E_{eff} and *PDD* values. Both samples (D1, D2) show fine TiC at the clad bottoms which gradually becomes coarser at the top of the clad.

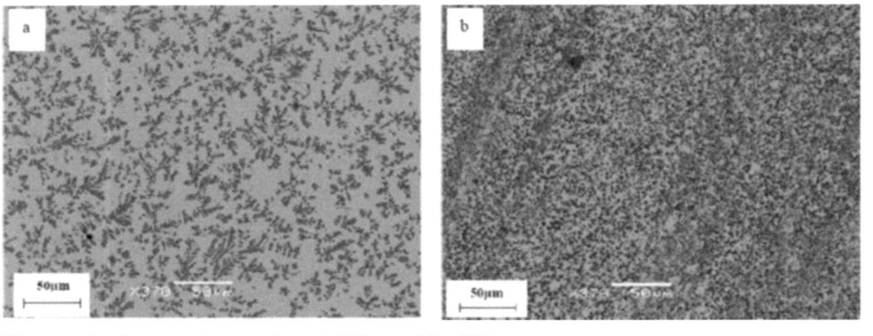

Fig. 16. TiC morphology of samples a) D1 and b) D2

4.5 Discussion

In order to create a metallurgical bond between the clad and the substrate, dilution must occur. Fig. 9 shows that dilution by Fe decreases the liquidus temperature of the melt pool (labeled dot). Fig. 10 illustrates a vertical section through the ternary Fe-Ti-C phase diagram at a constant composition of 19%at C using the Factsage software. According to the Fe-Ti-C ternary diagram, depending on the melt pool temperature (conditional on laser process parameters), ternary liquid can be present either in a single phase or in equilibrium with TiC/C to form a semi-solid melt pool. In a semi-solid melt pool, TiC and liquid are in equilibrium, as is shown in Fig. 10. Further, a semi-solid melt pool resulted from a peak temperature less than the liquidus temperature, and spherical TiC particles which are distributed uniformly or in clusters can form from this mechanism.

In contrast, if the peak temperature exceeds the liquidus temperature, a liquid melt pool can be formed. Therefore, according to Figure 13a, TiC dendrites are the first phase to solidify during cooling and create dendritic TiC.

Fig. 16 depicted different microstructures for samples D1 and D2. Sample D1 contains dendritic TiC, whereas sample D2 shows spherical TiC particles which are distributed in the matrix. Since the E_{eff} and *PDD* are identical for both, the only reason for the dissimilar microstructure could be the laser parameters, specifically the laser power. According to Fig. 9, dilution with Fe decreases the liquidus temperature. Peak power (i.e., 884) increases the melt pool temperature and thus increases the probability of having a liquid melt pool. Hence, the clad microstructure exhibits a well-developed dendritic structure of TiC.

In sample D6, although a higher laser power has been applied compared to sample D1 (i.e., 961), a dendritic TiC could not be detected. The diagram confirms that there is a large region in which liquid and solid TiC can be in equilibrium. This region decreases by increasing the Fe content. Therefore, in samples D4 and D6 (which have lower powder deposition density [i.e., 0.008]), dilution plays a vital role. The clad composition increases in Fe, and dendritic structures of TiC did not develop during solidification. On the other hand, Fig. 10 shows that by increasing the Fe content by dilution, the freezing range decreases. By decreasing the freezing range, the chance of dendrite formation decreases. This phenomenon can explain the TiC morphology in C5 and C7. Dendrites of TiC were observed in C5 with a laser power of 700 W, whereas no evidence of dendritic TiC was observed in sample C7 with a laser power of 900 W.

In laser cladding, powder and substrate absorbed the energy in order to melt and create the clad. By increasing the powder feed rate, the energy portion absorbed by the powder increases, allowing for a decrease in laser energy absorbed by the substrate. As a result, dilution with the substrate is decreased. Increasing the laser power, or increasing the scan speed, increases the contribution of dilution (as seen in Table 4). By comparing samples D1 and D2, it can be seen that dilution in sample D1 is higher than sample D2.

By increasing the dilution amount, the liquidus temperature decreases and shifts the clad composition to the Fe-rich side of the phase diagram. This results in a change in microstructure of the clad and TiC morphology.

Several modelling techniques have been used to model laser cladding by powder injection. These help to illustrate the physics of the process. A review of the attenuation of laser power in different powder feed rates and nozzle angles were presented, and the temperature distribution in the substrate and the powder particles on the substrate's surface was calculated (Huang, 2005).

5. Powder composition study

In this section, the effect of powder composition on TiC morphology and clad hardness using the in-situ laser cladding process was studied. Two atomic ratios, 45:55 and 55:45, were selected for C:Ti (Fig. 1), the first one of which has the potential to form near stociometric TiC, while the second potentially forming residual graphite which may be beneficial from a wear perspective.. Fe percentages were explored with 70, 60, 50 and 10wt% to increase the volume fraction of TiC in the clad (Table 1). Table 5 shows the selected laser parameters for deposition as a function of composition. Note that, for samples 155, the powder feed rates were decreased, compared to the more Fe rich compositions, in order to achieve a bonded clad. Any feed rate higher than that reported in Table 5 did not create metallurgical bonding between the clad and substrate.

Parameters condition	Sample No	Laser power W	Scan speed (mm/sec)	Powder feed rate (g/min)	Effective Energy (J/mm^2)	Powder deposition density (g/mm^2)
A	745, 755,655,555	884.2	117	4	180	0.014
B	745, 755,655,555	663.4	88.5	3	180	0.014
A	155	884.2	117	2.12	180	0.007
B	155	663.4	88.5	0.95	180	0.004

Table 5. Applied laser parameters

5.1 SEM, EDS and image analysis results

In this section, clad microstructures of samples with different clad compositions are studied. Fig. 17 shows the SEM images of samples with identical laser parameters but different chemical compositions. Results show that clad composition affects the clad microstructure and morphology. Fig. 17-a and 18-b are related to the samples which are identical in Fe weight percentage but different in C:Ti ratio. Fig. 17-a shows the dendiritic microstructure of TiC, while Fig. 17-b depicts the spherical TiC distributed uniformly in the matrix. Decreasing the Fe percentage in the powder composition and increasing the C:Ti ratio increases the probability of a reaction between Ti and C and thus increases the volume fraction of TiC.

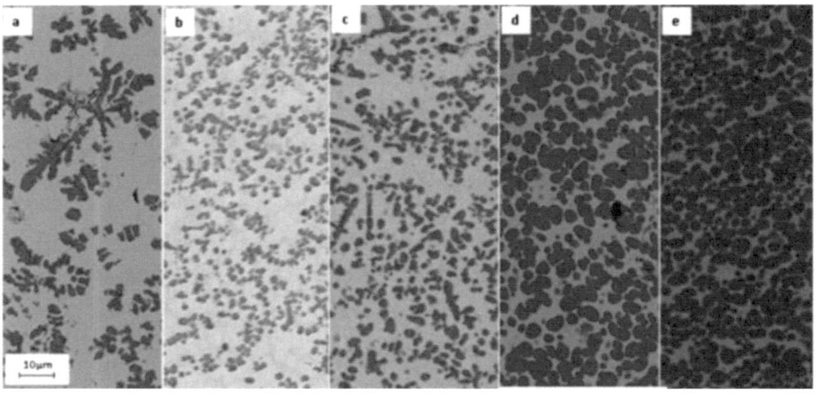

Fig. 17. Developed TiC morphologies using laser condition A for compositions of: a) 745 b) 755 c) 655 d) 555 e) 155

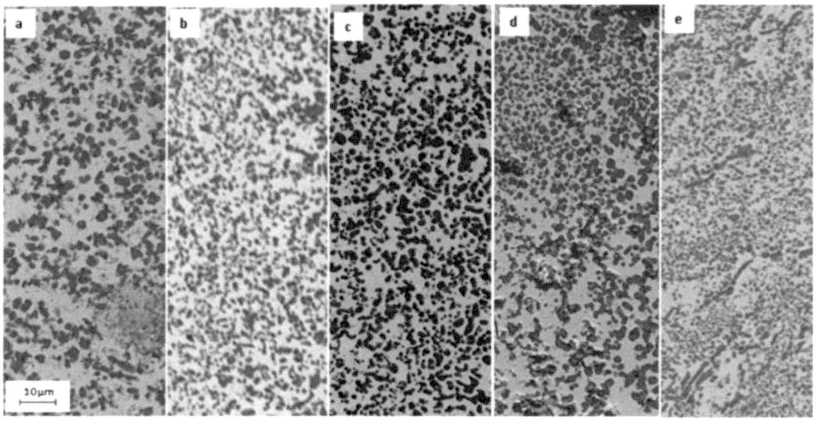

Fig. 18. Developed TiC morphologies using laser condition B for compositions of: a) 745 b) 755 c) 655 d) 555 e) 155

Fig. 17-b to 18-d show a higher volume fraction of TiC distributed in the clad. Except for sample 745-A, dendritic TiC cannot be observed in the rest of the samples. This trend also can be observed in samples deposition using condition B, as indicated in
Fig. 18. Fig. 17 and
Fig. 18 show that by decreasing the Fe percentage and increasing the C:Ti ratio, the volume fraction of TiC increases. Moreover, in condition B, dendritic TiC cannot be observed in sample 745-B and the size of the TiC particles are generally smaller for condition B compared to condition A. Image analysis helps to measure the trend of volume fraction qualitatively. Table 6 depicts the image analysis results. It can be seen that in similar applied laser conditions, the volume fraction of TiC increases by decreasing the Fe percentages. It is clear that decreasing the Fe content increases the TiC volume fraction but that this feature is more significant in laser condition A.

Sample	%Matrix	%TiC	Sample	%Matrix	%TiC
745-A	81	19	745-B	56	44
755-A	69	31	755-B	63	37
655-A	66	34	655-B	50	50
555-A	43	57	555-B	54	46
155-A	30	70	155-B	45	55

Table 6. Image analysis results

5.2 Hardness results

Hardness results are depicted in
Fig. 19, which provides information about samples with different chemical compositions deposited by laser condition A. Each hardness value is an average of at least three measurements at an identical position from the clad/substrate interface. It can be seen that, by decreasing the Fe percentage, general hardness values are increased. Sample 155-A with a minimum Fe percentage shows the highest clad hardness. This trend can also be seen in samples deposited by condition B.

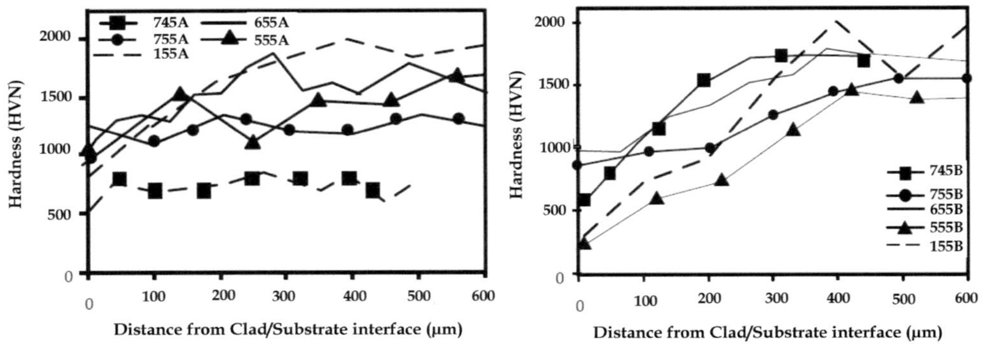

Fig. 19. Hardness profile for samples deposited with laser condition A, B

Fluctuation in hardness results normally happens in composites because of different hardness values of matrix and reinforcement particles. Moreover, an average of three measurements can be a source of fluctuation depending on which phase is punched by the indenter. Sample 745-A with dendritic TiC has the lowest fraction of TiC particles (as illustrated in Fig. 17-a and Table 6), causing a generally lower hardness compared to the other samples. Deposition of the same composition (sample 745-A) under different laser conditions (i.e., sample 745-B) increases the hardness profile dramatically. The main reason for hardness enhancement is the different distribution and morphology of TiC particles in 745-A compared to those in 745-B, which also significantly increases the TiC volume fraction from 19 to 44%.

5.3 Discussion

The temperature and composition of a clad melt pool during deposition are two main factors that affect final clad microstructure and TiC morphology. Laser power establishes the peak power that determines the clad temperature. Depending on the clad position, different cooling and solidification rates are developed which affect the clad microstructure and carbide morphologies. Composition is affected by initial fed powder composition and the extent of dilution by the Fe-rich matrix. Fig. 9 shows the liquidus projection in a ternary phase diagram of Fe-Ti-C. If no dilution by the substrate is assumed, then it is possible to state that the liquidus temperature of the powder mixture is about 2618 K (2345 °C) in a powder composition of 745 and the liquidus temperature of composition 155 is about 3384 K (3111°C). Thermal modelling of A and B laser conditions shows peak temperatures of 2200°C and 1831°C, respectively (Emamian, 2010d). Only sample 745A exhibits a dendritic TiC morphology arising from solidification of a completely liquid melt pool. This indicates that the laser peak temperature exceeded the liquidus temperature of the clad composition. The main reason for the formation of a complete liquid melt pool in this sample is dilution by the substrate which shifts the clad composition to the Fe-rich side of Fig. 10, resulting in a decrease in the liquidus temperature. In the rest of the samples, as Fig. 9 shows, increasing the C:Ti ratio and decreasing the Fe percentage in the powder mixture, causes an increase in the liquidus temperature. Therefore, the peak temperature developed by the laser heating does not exceed the liquidus temperature for the clad composition. Therefore a full liquid melt pool does not develop and spherical TiC particles are formed from the semi solid melt pool. While samples 745-A and 755-A have very close chemical compositions, differences in liquidus temperatures result in a considerable change in TiC morphology.

6. Wear

Wear resistance is an important function in the balance of properties of metal matrix composites. Wear starts from softer components which are in contact with counterfaces (i.e., pin, abrasive material or steel ball). Therefore, in metal matrix composites with hard particles, the matrix is worn in the preliminary stages of the wear process. Different mechanisms of matrix degrading such as plastic deformation, micromachining, crack propagation and brittle chipping, which are controlled by the matrix mechanical properties (i.e., hardness or fracture toughness) can increase the wear rate.

Wear can usually be classified on the basis of the nature of interactions, such as two-body or three-body. For two-body abrasive wear, a delamination model is feasible. In this model, wear resistance strongly depends on fracture toughness and not on composite strength, because crack propagation is the controlling factor. In three-body abrasive wear, abrasive particles tend to break down and bury themselves in alloys with relatively soft matrices. However, hard particles in composites resist scratching because of their high hardness under low loads and crack into small pieces under heavy loads. The broken particles remain embedded in the matrix during wear. In this case, composite strength and fracture toughness both play important roles (Jahanmir et al., 1976; Saka et al. 1985).

In this section, the wear resistances of six samples with different chemical composition deposited using two laser parameters (i.e., A, B) are studied. The objectives of this section are:

1. to find the optimal micro-structure (resulting from the laser process parameters and powder composition) of TiC in an Fe matrix with respect to the highest wear resistance ;
2. to measure the wear properties of in-situ deposited Fe-TiC by laser cladding; and
3. to find the optimal operating parameters of laser cladding that develop the best wear properties.

The wear resistance of samples is measured based on standard ASTM G65-04 using an adjusted volume loss (mm^3), as below:

$$\textit{Volume loss } (mm^3) = \textit{weight loss } (g)/\textit{density of coating } (g.cm^{-3})*1000 \qquad (5)$$

$$\textit{Adjusted volume loss } (mm^3) = \textit{measured volume loss } (mm^3)*228.6\ (mm)/\textit{wheel diameter after use } (mm) \qquad (6)$$

6.1 Results

In the previous section, Fig. 17 and
Fig. 18 showed the TiC morphology resulting from different powder compositions and laser process parameters. In this section, wear properties of Fe-TiC composite coatings with different morphology are studied. Fig. 17 and
Fig. 18 show the distribution of TiC; however, the size of the particles and volume fraction are different.

The micro-hardness profile of conditions A and B show that samples 155A and 155B have a higher hardness value, which primarily have results from a higher volume fraction of the clad's TiC particles. In addition, sample 745A shows the lowest profile among the other samples. In considering Fig. 17, it can be concluded that a larger area of the matrix remained without the support of TiC particles. Fig. 20 illustrates the wear test results based on ASTM G65-04.

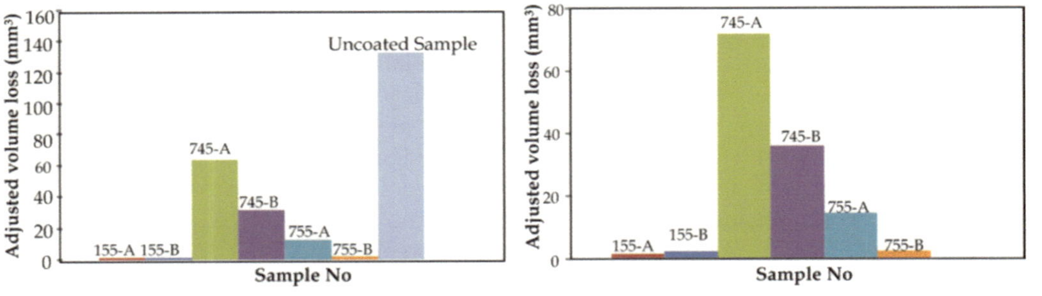

Fig. 20. Wear test results

The maximum volume loss after the uncoated sample (AISI 1030) belongs to sample 745A, while samples 155A and B have a minimum volume loss. Fig. 20 shows that by increasing the C:Ti ratio in samples, volume loss decreases. By comparing samples 745 and 755 in both conditions A and B, group 755 shows better wear resistance. Moreover, although 755B has 70%wt Fe, the results show that it has good wear resistance compared to group 155. By comparing hardness and wear test results, it can be seen that samples with higher hardness show higher wear resistance.

6.2 Discussion

By increasing the C:Ti ratio from 45:55 at% and 55:45 at% with the same laser condition, the TiC particles' morphology was formed into spherical shapes with a uniform distribution. Group 155A has a higher volume fraction of TiC compared to the 155B conditions (Table 6). By considering the melting points of ingredients (3400, 1670, and 1535° C for C, Ti and Fe, respectively), higher amounts of C and Ti in group 155 absorb a considerable portion of the laser energy compared to 745 or 755 group composition. Hence, the higher laser power in sample 155A compared to 155B encourages a reaction between C and molten Ti by the quick melting of Ti and Fe. However, in 155B, the laser energy is not sufficient to promote TiC formation in the clad. By comparing Fig. 17-c and
Fig. 18-c, it can be seen that the volume fraction and size of the TiC particles are larger in 155A than in 155B, resulting in less volume loss in 155A. Also, the larger TiC particle size in 155A compared to 155B can be another reason for better wear resistance in 155A than in 155B.

In sample 755A, a higher laser beam energy melts the substrate and increases the dilution. Hence, the volume fraction of the TiC decreases compared to 755B. Therefore, laser parameters can be selected based on powder composition in order to have a maximum TiC volume fraction.

Fig. 20 depicts how the abrasive wear performance of samples is affected by the volume fraction of TiC particles. The volume fraction of TiC particles in reverse is related to the distance between carbides occupied by the Fe matrix. It is obvious that, in abrasive conditions, the softer part is removed by grits. Therefore, it is expected that by increasing the volume fraction of the coating, the wear resistance increases (He et al., 2009; Nurminen et al., 2009; Pu, 2008). The volume fraction in this research is increased firstly by increasing the C:Ti ratio in the constant Fe percentage and secondly by decreasing the Fe percentage in the powder composition. Optimal results for wear resistance of in-situ Fe-TiC can be found in group 155A. Since Ti can dissolve in Fe during the laser cladding process, nominal 90%

TiC (93%vol TiC) is not achieved during the process; however, maximum 70% volume fraction of TiC shows considerable wear resistance.
The usual mechanism of abrasives in wear tests are grooving, plastic deformation and fracture of hard phases. Table 7 provides information about size and hardness values of hard-phase particles and abrasives.

Particle	Size range, μm	Hardness (HVN)
TiC	1-7	3200
Abrasive	212-300	1160

Table 7. Particle size information

Abrasive grits first start deforming the Fe matrix, which is the softer component in the composite coating. If the carbide interlock is sufficiently high, the exposure of the soft matrix to the abrasive will be decreased. Since the hardness of the TiC is higher than abrasives (Table 7), carbide damage is less than for grits. Moreover, in-situ TiC particles are developed during the laser cladding process, and hence the TiC/Fe interface has a strong bonding. Therefore, the probability of deboning and crack formation in the particle/binder interface decreases. Pulling out and crushing the WC particles are the main problems that have been noted by researchers (Stachowiak, G. B. & Stachowiak G.W, 2010). Furthermore, identical wear test conditions were conducted on sample WC-12%Ni (with 80%vol WC) to compare it with 155A (with 80% vol TiC). Results show a similar adjusted volume loss for both (i.e., 1.9744 and 1.7731 mm^3 volume loss for WC and 155A, respectively).

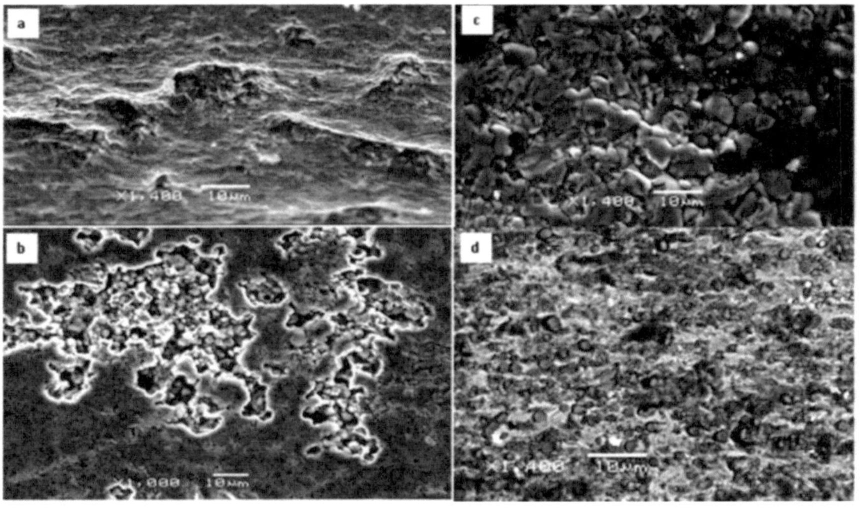

Fig. 21. (a) Plastic deformations in worn surface of 745A (b) Pulling out of carbides in worn surface of 755A (c) Worn matrix and remaining TiC particles in sample 155A (d) Worn matrix and remaining TiC particles in sample 755B

It is normal that the matrix of sample 745A, which has minimum support of the TiC, contains more plastic deformation than the others (Fig. 21 a). Matrix plastic deformation increases the stress on TiC particles, pulling them off the surface. It is worth noting that, during the wear test, increasing the temperature in the matrix around the TiC particles helps to drop the yield stress of the matrix, resulting in easier plastic deformation around the carbides. By increasing the volume fraction of the TiC in samples 755 and 155, plastic deformation could not be observed. However, in some areas of 755A, a pulling out of the TiC is detected. Fig. 21 b shows a pulling out of the carbides in sample 755A. Group 155 shows a uniform worn surface without carbide removal. The higher volume fraction of the TiC results in the minimizing of the plastic deformation and removing of the matrix. However, volume loss still belongs to the matrix area. Fig. 21-c shows the interlocked TiC and a small area of worn matrix in 155A. Fig. 21-d depicts the worn matrix for sample 755B with the remains of the TiC. Although sample 755B contains 70% Fe, the uniform distributed particles and high volume fraction of the TiC resulted in higher wear resistance than that of 755A and 745A and B.

7. Conclusion

From the above experiments, we can conclude that two important factors - temperature and chemical composition of the clad - establish clad microstructures. We can also conclude that laser process parameters play a crucial role in the quality and microstructure of the clad, the ternary phase diagram of Fe-Ti-C can be used to predict the final microstructure of the clad. Furthermore, the melt pool chemical composition defines the liquidus temperature. Hence, the clad temperature can be higher or less than the liquidus temperature in different locations of the clad. Therefore, depending on clad and liquidus temperatures, the melt pool would be semi-solid or liquid, which defines the TiC formation mechanism and morphology.

The experiments also showed that, depending on powder deposition density (resulting from the powder feed rate and scan speed) and peak power of the clad temperature, different TiC morphologies and distributions could result. Moreover, identical combined parameters do not guarantee identical TiC morphology, as laser parameters still play a role in TiC morphology.

Furthermore, a uniform distribution of spherical TiC particles indicates that increasing the fraction of carbides in a matrix results in higher values for hardness (sample D2). From this, it can be concluded that, in order to enhance the clad hardness and wear resistance and achieve uniform carbide distribution, laser parameters must be adjusted to form a semi-solid melt pool instead of a liquid molten pool. A liquid melt pool forms dendritic TiC particles, which are not well distributed through the clad. Hence, a considerable region of clad cannot be supported by TiC particles. This conclusion also was confirmed by wear resistance results in a series of chemical composition and two laser conditions.

8. References

Ariely, S., Shen, J., Bamberger, M., Dausiger, F. & Hugel, H. (1991). Laser surface alloying of steel with TiC. *Surface and Coatings Technology*, 45 403-408.

Cui, C., Guo, Z., Wang, H. & Hu, J. (2007). In situ TiC particles reinforced grey cast iron composite fabricated by laser cladding of Ni–Ti–C system. Journal of Materials Processing Technology, 183, 380-385.

Das, K., Bandyopadhyay, T. K. & Das, S. (2002). A review on the various synthesis routes of TiC reinforced ferrous based composites. Journal of Materials Science 37, 3881-3892.

Dubourg, L. & St-Georges, L., Optimization of laser cladding process using Taguchi & EM methods for MMC coating production (2006). Journal of Thermal Spray Technology 15, 790-795

Du, B., Li, Q., Wang, X. & Zou, Z. (2007). In situ synthesis of TiC-VC particles reinforced Fe-based metal matrix composite coating by laser cladding. Hanjie Xuebao/Transactions of the China Welding Institution 28, 65-68.

Du, B., Zou, Z., Wang, X. & Li, Q. (2007). In situ synthesis of TiC-TiB2 reinforced FeCrSiB composite coating by laser cladding. Surface Review and Letters 14, 315-319.

Emamian A., Corbin, S. F. & Khajepour,A. (2010). Effect of laser cladding process parameters on clad quality and in-situ formed microstructure of Fe–TiC composite coatings. Surface and Coatings Technology 205, 2007-2015.

Emamian A., Corbin, S.F, & Khajepour A., (2010). "Study on Laser Parameters Effect on Morphology of In-Situ Fe-TiC particles Deposition on Mild Steel Using Laser Cladding Process" ICALEO Conference 2010, Anheim, California, USA.

Emamian, A., Corbin, S.F, & Khajepour A., (2010). "Correlation between temperature distribution and formed microstructure of in-situ laser cladding of Fe-TiC on carbon steel" ICALEO Conference 2010, Anheim, California, USA.

Emamian A., Corbin, S.F., & Khajepour A. (2009). "In-situ TiC particles reinforced carbon steel composite Fabricated by laser cladding of Fe-Ti-C system", 21st Canadian Material Science Conference, Kingston, Canada.

He, Q., Wang, Y., Zhao, W. & Cheng, Y. (2009). Interface microstructure and wear properties of TiC-Ni-Mo coatings prepared by in-situ fabrication of laser cladding. Hanjie Xuebao/Transactions of the China Welding Institution 30.

Hirokai Okamoto, (2000). Phase Diagrams for Binary Alloys (ASM)

Huang Y.L (2005). Interaction between laser beam and powder stream in the process of laser cladding with powder feeding. Journal of Modeling and simulation in material science and engineering 13, 47-56.

Jahanmir S., Abrahamson E.P., and Suh N.P., (1976). Wear, Vol 40, p. 75.

Jiang, W. H., Fei, J. & Han, X. L. (2000). In situ synthesis of (TiW)C/Fe composites. Materials Letters, 46, 222-224.

Jiang, W. H. & Kovacevic, R. (2007). Laser deposited TiC/H13 tool steel composite coatings and their erosion resistance. Journal of Materials Processing Technology, 186, 331-338.

Nurminen, J., Näkki, J. & Vuoristo, P. (2009). Microstructure and properties of hard and wear resistant MMC coatings deposited by laser cladding. International Journal of Refractory Metals and Hard Materials 27, 472-478.

Pu, Y. (2008). Microstructure and tribological properties of in situ synthesized TiC, TiN, and SiC reinforced Ti_3Al intermetallic matrix composite coatings on pure Ti by laser cladding. Appl. Surf. Sci. 255, 2697-2703.

Saka N. and Karalekas D.P., (1985). Friction and Wear of Particle Reinforced Metal-Ceramic Composites, Wear of Materials, p 784.

Stachowiak, G. B. & Stachowiak, G. W. (2010). Tribological characteristics of WC-based claddings using a ball-cratering method. International Journal of Refractory Metals and Hard Materials 28 95-105.

Tassin, C., Laroudie, F., Pons, M. & Lelait, L. (1995). Carbide-reinforced coatings on AISI 316 L stainless steel by laser surface alloying. Surface and Coatings Technology, 76-77, 450-455.

Wang, X. H., Song, S. L.& Qu, S. Y. & Zou, Z. D. (2007). Characterization of in situ synthesized TiC particle reinforced Fe-based composite coatings produced by multi-pass overlapping GTAW melting process. Surface & Coatings Technology ,201, 5899-5905.

Wang, X. H., Zhang, M., Liu, X. M.& Qu, S. Y. & Zou, Z. D. (2008). Microstructure and wear properties of TiC/FeCrBSi surface composite coating prepared by laser cladding. Surface and Coatings Technology, 202, 3600-3606.

Wang, X. H., Qu, S. Y., Du, B. S. & Zou, Z. D. (2009). In situ synthesised TiC particles reinforced Fe based composite coating produced by laser cladding. Materials Science and Technology 25, 388-392.

Wu, X. & Hong, Y. (2001). Microstructure and mechanical properties at TiCp/Ni-alloy interfaces in laser-synthesized coatings. Materials Science and Engineering A 318, 15-21.

Yan, M. & Hanqi, H. (1996). In situ laser surface coating of TiC metal-matrix composite layer. J. Mater. Sci. 31, 4303-4306.

3

New Superhard Ternary Borides in Composite Materials

Zachary Zachariev
Institute of Polymers, Bulgarian Academy of Sciences
Bulgaria

1. Introduction

Superhard substances are those a hardness above 20 GPa, i.e. higher than that of corundum (Kislii et al.,1988) or with a Vickers hardness Hv exceeding 40 GPa (Sologenko&Gregorianz, 2005). The ten utmost non-metal substances and refractory compounds form a "hardness pyramid" (Kislii et al.,1988) diamond being on top, followed by cubic-boron nitride and boron carbide.

Metal-like ones form a similar pyramid, the transition metal borides occupying its top. However, the maximum hardness found for them is inferior to that of the non-metal substances. The interest in borides is due to their extraordinary hardness (up to 1873 ^{0}C) as compared to other refractory compounds.

The hardest boride ($B_{12}C_3$) is used as a wear resistant polycrystalline material, armor tiles, nuclear industry, etc. (Anderson, 2002).However, applications are restricted by its high brittleness due to the strong covalent bonds in its crystal lattice. It has been established (Zakhariev&Radev, 1988) that superhard ternary compounds based on boron carbide with dissolved IV-VI group metals ($B_{12-n}C_3Me_n$) can be obtained by sintering CMC composite materials without pressure.

Sintering of TiB_2-MeC-Me systems with no pressure applied is a new trend in the field of superhard boride composite materials. The binding metal (e.g. Ni or Co) in TiB_2-Me system reacts with the boride and forms a low-melting phase (Lecrivan&Provost, 1968); (Fitzer, 1973). The growing interest in borides comes from their high-temperature behavior: high melting point, hardness, wear resistance and chemical inertness. These allows using them to produce cathodes for electrolytic aluminium, first wall coatings and neutron absorbers for nuclear technology, valve components in cool liquefaction plants and crucible materials for metal evaporation.

Studies on the TiB_2-TiC and TiB_2-TaC systems show that the carbides inhibits the grain growth of the boride phase (Murata et al., 1967).The TiB_2-MeC-Me materials have the following advantages as compared with the binary systems mentioned (Petzow&Telle, 1984): application of hot pressing is not necessary, the grain growth is completely inhibited, the mechanical properties are improved due to the small grain size (about 1μm) (Zakhariev et al., 1993).

WC-Co is the source material in the field of the metal-working whereas superhard ternary boride (WCoB) coating entails a sharp rise in wear resistance of the composite materials (Zakhariev et al., 1987). The micro hardness of synthetic polycrystalline WCoB amounts to

4650 κgf/mm^2 (Zakhariev et al., 1986) which explains the increased service tool - life of the WC-Co instruments. Similar ternary compounds are MoCoB and WFeB with the same orthorhombic unit cells (Jeitschko, 1968).
Combining borides (B_4C, TiB_2 et al) with carbides (MeC, WC-Co et al) allows their chemical interactions upon heating, the building up of eutectics etc. This can result in densification of the composite material without applying of any additional pressure as well as in enhanced physico-mechanical properties due to the arising of superhard ternary compounds ($B_{12-n}C_3Me_n$, $(Ti,W)_2B$, WCoB) (Zachariev, 2001).
Under thermochemical treatment of steels (Fe-C) with a transition metal borides (ZrB_2, TiB_2, CrB2) (Zakhariev et all., 1970) it appears a ternary compound $(Fe,Me)_2B$ of considerable hardness 20-23 GPa. The layer it has built up combines the advantages of consecutive metalizing and boronizing layers thus bringing about the enhanced resistance to wear and corrosion of the metal-matrix composite (MMC) system.
Similar properties with respect to corrosion and wear have another complex boride (synthesized Mo_2FeB_2 with ferrous binder), known as "new hard alloy" (Komai et al., 1989). The ternary boride $Cr_xFe_yB_z$ ($Cr_3Fe_{80}B_{17}$) also shows unique properties which have been utilized in several amorphous materials for corrosion applications.
Fukatsu et al. (Fukatsu, 1967) have shown that the wear resistance of hard alloys increases with the increasing of their hardness provided all other conditions are the same.
The present paper, based mainly on research done by the author and associated colleagues, aims at a concise unification of the results on new superhard boride composites in view of their practical applications.

2. Experimental methods

The carbides ($B_{12}C_3$, TiB_2 and $Me^{IV-VI}C$ produced by "ESK-Kempten", "Merck" and "Ventron Alfa Products", respectively) were homogenized in a Frisch planetary mill, pressed at 200 MPa and then sintered at 1700 - 2250°C in a Degussa furnace with graphite heater in argon atmosphere.
The hardness was determined using a "Leitz-Durimet" hardness tester with loads of $HV_{0.5}$ to HV_1 and a "Carl Zeiss" micrometer with indentation load 30-100 g. Compound hardness as a function of indentation load has been traced.
An automated DRON-3 diffractometer, with CuKα radiation was used to investigate the structure of the materials under study. Their morphology was characterized by scanning electron spectroscopy (SEM) on a JSM 840 apparatus equipped with a Link QK 200 dispersive X-ray analyzer.
The initial TiB_2 powder (type PIII; Ti-68.6, B-27.3,TiO_2-3.9, Co-0.2 wt.%) was prepared under industrial conditions using a technology developed in the Institute of problems in Materials Science, Kiev. The chemical analysis of another initial TiB_2 Koch-Light powder showed Ti-68.8 and B-31.2 wt.%. The powders TiB_2 were milled 75-120 min using hard-alloys bodies and vessels of WC-Co (K10) in the Fritsch planetary mill with acetone serving as a medium.
The polycrystalline samples of the ternary boride WCoB were obtained by crystallization from a cobalt-rich melts of the corresponding powdery components at 1600 °C according to (Petrov&Will, 1981).
Standard WC-Co cutting plates K10 (92 wt.%WC, 6%Co and 2 wt.% TaNbC) were packed separately in Borozar-HM (powdery product, 325 mesh) and B_4C (F220 technical grade, ESK-Kempten) and heated at 1000-1400 °C for 30-120 min in an inert medium (argon). The

heating was carried out in a large-scale Bor 6-CM-3 installation for deposition of boride coating on K10 plates.

For simultaneous boron-zirkonizing (or boron-chromizing) of steel samples by powdery borides ZrB_2 (CrB_2) and some activators were used the heating occurring at 950-1050 °C in argon atmosphere.

Obtaining the superhard boron-metallizing layer on steel tools or parts of them requires their coating in a patented paste (commercial paste "Zahobor-P"). The technology is very simple: a coat of the paste is applied over the working surfaces of machine tools and parts. The metal surfaces thus coated are dried and then subjected to heat-treatment. Additional procedures, staff and equipment are not required.

3. Results and discussion

3.1 Ternary $B_{12-n}C_3Me_n$ borides in CMC-composite $B_{12}C_3$ + Me_xB_y

It is obvious that a sintering of $B_{12}C_3$ is only possible at temperatures above 2100°C, i.e. close to the melting point of boron carbide (2447°C). SEM of a fracture surface of $B_{12}C_3$ (10 wt %) + W_2B_5 material sintered at 2150°C and a grain size of 2-7 μm is shown in Fig.1. With larger magnifications (2000x), a thin eutectic binder is visible at the grain boundaries.

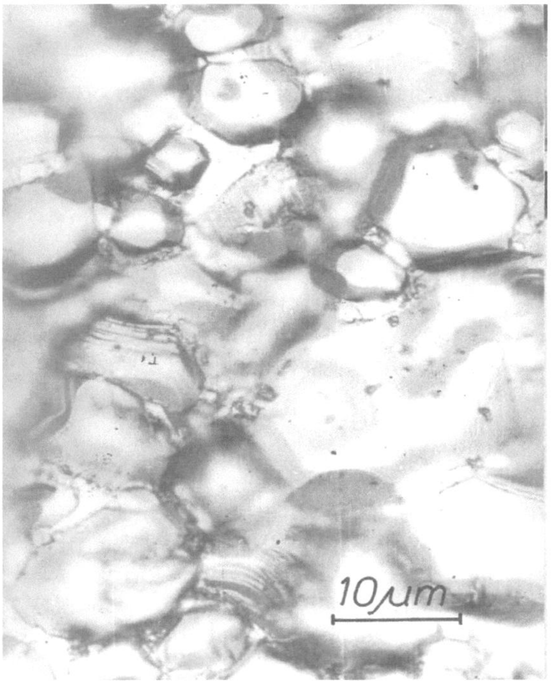

Fig. 1. SEM of the fracture surface of $B_{12}C_3$ + 10 wt%W_2B_5 composite material (sintered at 2150°C; 20 min)

An eutectic resulting from interaction between the two carbides (B_4C + WC → W_2B_5 + C) does not seem unexpected when taking into account the eutectic character of the ternary diagrams B-C-Me $^{IV-VI}$ (Schouler et al.,1983) and the quasi-binary systems B_4C-MeB_2 (Portnoi& Samsonov,1960).

Data in Tabl.1 show that the systems containing 4th Period metals, such as $B_{12}C_3$-TiB_2 (VB_2,CrB_2) have the lowest T_{eut} (2150-2200 °C) as compared with the other ones ($B_{12}C_3$-ZrB_2 /NbB_2 and $B_{12}C_3$-HfB_2/ TaC), which is of importance for the assessment of the sintering temperature of composites.

			MeB₂ in eutectic, %		
Compounds	T_m,°C	ΔH,KJ/mol	vol.	Mol	T_{eut}, °C
$B_{12}C_3$	2447	70,0	-	-	-
TiB_2	3217	280,0	20,0	26	2197
ZrB_2	3247	314,0	20,0	24	2277
HfB_2	3347	335,0	23,0	22	2377
VB_2	2747	142,4	35,0	46	2167
NbB_2	2997	174,6	35,6	36	2247
TaB_2	3097	217,7	27,0	32	2367
CrB2	2217	125,6	63,0	70	2147

Table 1. Melting point and Eutectic temperature of composite system $B_{12}C_3$-MeB_2

Due to the Me_xB_y particles it creates, appearance of the eutectic leads to brittle boron carbide becoming stronger. This results in a significant decrease in brittleness of the recrystallized composite and an increase in its crack-resistance. Measured by the indentation method, the critical coefficient of stress intensity (KIc) is in differently oriented eutectics ($B_{12}C_3$-MeB_2) ranges from 6-12 MPa to 2.5-3.5 MPa depending on their components. This result indicates that the use of eutectics as a new class of CMC composite materials resistant under extreme conditions is promising.

The microhardness values of a ternary phase $B_{12-n}C_3Me_n$ borides in composite $B_{12}C_3$ + Me_xB_y, as compared to hot-pressed pure boron carbide, and the values obtained by other authors are presented in Table 2. The data show that the boron carbide hardness in the ternary borides phases is much higher than that of pure boron carbide and other composites based on $B_{12}C_3$, *c*-BN and c-BC_2N.

This sharp increase in hardness seems to be due to dissolution of the transition metals in the crystal lattice of boron carbide. During the chemical interaction of the two carbides the boron needed for the reaction comes from the boron carbide lattice.

Since from the investigations of Lipp et al. (Lipp&Schwetz, 1975) it is known that the homogeneity region of $B_{4+x}C$ has no substantial effect on its hardness, the only reason for the sharp increase in boron carbide hardness should be the formation of new superhard ternary compounds with the formula $B_{12-n}C_3Me_n$ (Fig. 2).

These compounds do not differ essentially from the ternary compounds $B_{12}C_{3-n}Me_n$ (predicted by Lipp &Roder, 1966 and proved to Okada et al., 1990). The difference is that with the latter compounds, the replacement of C*o* (at. Radius 0.91 Å) by the larger element Al (at. radius 1.43 Å) leads to a considerable increase in the size of the unit cell in the direction of the hexagonal C- axis and of the lattice volume (Table 3). In our case, just the opposite happens. The volume shows no substantial change, whereas the lattice parameter *c* decreases (see Table 3).

This is attributed to an increase in carbon content in the homogeneity region of the boron carbide, which is due to elimination of boron from the compound. The decrease in the parameter *co*, while the crystal lattice volume remains unchanged, indicates that the metal atoms are incorporated most probably in the icosahedron B2 and B1 sites.

Initial Material (weight %)	HV1	HV05	Reference
$B_{12}C_3$+W_2B_5 (10 ; 50)	50; 58	53* ; 77**	(Zakhariev, 1988)
$B_{12}C_3$+CrB_2 (10)	56	77**	(Zachariev, 2001)
$B_{12}C_3$ + VB_2 (10)	43	52	(Zachariev, 2001)
$B_{12}C_3$+Ti (7)	-	63	(Makarenko et al. ,1977)
$B_{12}C_3$+Zr (10)	-	60	(Makarenko et al.,1977)
$B_{12}C_3$+ Al (50)	25	-	(Lipp & Roder ,1966)
$B_{48}C_2Al_3$ - crystal	30.5	-	(Okada et al., 1990)
$B_{12}C_3$ + TiB_2 (10;20)	-	45 ; 55	(Nishiyama , 1985)
$B_{12}C_3$+TiB_2 (10;25)	-	30 ; 34	(Telle & Petzow ,1987)
$B_{12}C_3$+TiB_2 (10)+W_2B_5 (5)	-	37 ; 42	(Telle & Petzow ,1987)
c-BC_2N	-	76	(Solozhenko , 2001)
B12C3-HP "pure"	34	30 ; 38 ; 30	(Lipp &Schwetz , 1975)

* - Length of pyramid diagonal indentation [μm]: * 4.4; ** 3.15

Table 2. Microhardness (GPa) of $B_{12-n}C_3$ Me_n from the initial materials $B_{12}C_3$ + Me_xB_y and other composites

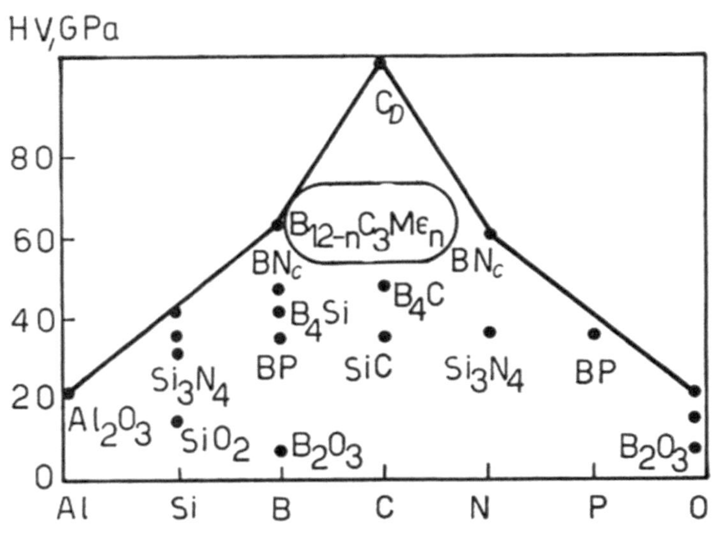

Fig. 2. Non-metal hardness pyramid with the new ternary boride phase $B_{12-n}C_3Me_n$

Compounds	C_o[Å]	a_o[Å]	Volume [nm³]	Reference
$B_{11}C_3W$	12,078	5,601	0,32757	(Zachariev , 2001)
$B_{11}C_3W$	12,024	5.604	0,32697	(Zachariev , 2001)
$B_{11}C_3W$	12,054	5,606	0,32802	(Zachariev , 2001)
$B_{11}C_3Ti$ (10 - 20%)	12,06-12,07	5,607	-	(Zachariev , 2001)
$B_{11}C_3V$ (10- 20%)	12,02-12,05	5,605	-	(Zachariev , 2001)
$B_{12}C_2Al$	12,39	5,65	0,34252	(Lipp & Roder , 1966)
$B_{12}(C,Si)_3$	12,1-12,4	5,60	0,3286-	(Kislii at all. , 1988)
$B_{13}C_2$	12,19	5,67	0,33938	(Thevenot & Bouchacour, 1979)
$B_{12}C_3$ (initial ESK)	12,12	5,60	0,32915	(Lipp & Schvetz, 1975)

Table 3. Parameters of the crystal lattice of the ternary metal borides compounds C_o[Å] , a_o[Å]; V[nm³]

Due to the strongly extended lattice during the incorporation of Al and Si into the C-C-C axis, the phases $B_{12}C_2A1$, $B_{48}C_2Al_3$ and $B_{12}C_2Si$ have a microhardness (HV1 = 25 - 30.5 GPa) even lower that that of pure boron carbide (Table 2).

A novel superhard phase c-BC_2N was synthesized using the laser-heated diamond anvil cell with a hardness Hv=76 GPa [Solozhenko, 2001] (Tabl.2).

The hardness (50-77GPa) of the new ternary borides is much higher than that of "pure" boron carbide sintered by hot pressing (Tabl.2). It is equal to that of cubic boron nitride (i.e. next to diamond) and even of some polycrystalline diamonds of the type "Carbonado" and "CB". However, the price of the new boride is several orders of magnitude lower.

We have developed an original method to obtain articles of boron carbide composite, in which the hot-pressing stage is avoided. The new superhard boron carbide has already found several applications: in the production of armor plates to protect people and machines from bullets and shrapnel (Fig. 3 and Fig. 4), for protection against neutron radiation in nuclear power plants as well as in nuclear therapy of tumours.

Other possible applications will use the large capacity for neutron absorption by the investigated CMC- composites. The cross-sections of neutron absorption are as follows: boron carbide - $1,98.10^{-2}$ barn, tungsten boride (W_2B_5) - $2,14.10^{-1}$ barn, the proposed CMC material B_4C + W_2B_5 - $3,18.10^{-1}$ barn. The latter has been successfully tested in the nuclear power plant in Kozloduy, Bulgaria for 14 months.

3.2 Ternary (Ti,W)B_2 boride in CMC-composite materials TiB_2-WC-Co

The milled TiB_2-WC powders containing up to 1 wt. % cobalt are sintered in order to obtain two-phase cermets, i.e. to avoid crystallization of the brittle ternary phases WCoB and W_2CoB_2. The ternary phases were found in sintered samples which are obtained from initial

mixtures containing 2 to 10 wt. % Co. The densification of the alloys during the sintering began above 1100 °C. Increasing the temperature, the decomposition of WC causes formation of facets on the grain surfaces. During the isothermal sintering above 1700 °C to 1850 °C, β-WB and (Ti,W)B_2 solid solutions were precipitated on the TiB_2 nuclei.
The microstructure of the initial powder TiB_2 +TiO_2+ 18.64WC + 1Co is characterized by homogeneous distribution of fine TiB_2 grains (4.4 µm). The presence of the (Ti,W)B_2 -phase seems to be more pronounced now as the latter is not restricted to the periphery of TiB_2 (the Koch-Light sample) entire grains of it being observed (Fig. 5). Some of the boride crystals have partially lost their boron through the eutectic Co-WCoB film (TiB_2: a=3.018; c=3.209Å). The eutectic built up seems uniformly distributed remaining all but invisible with a thickness of less than 0.4 µm. The size of the pores is 1-3 µm.

Fig. 3. Protective ceramic equipment

Fig. 4. Composite boroncarbide ceramic plates on armored vehicle (Company Rafael - Israel)

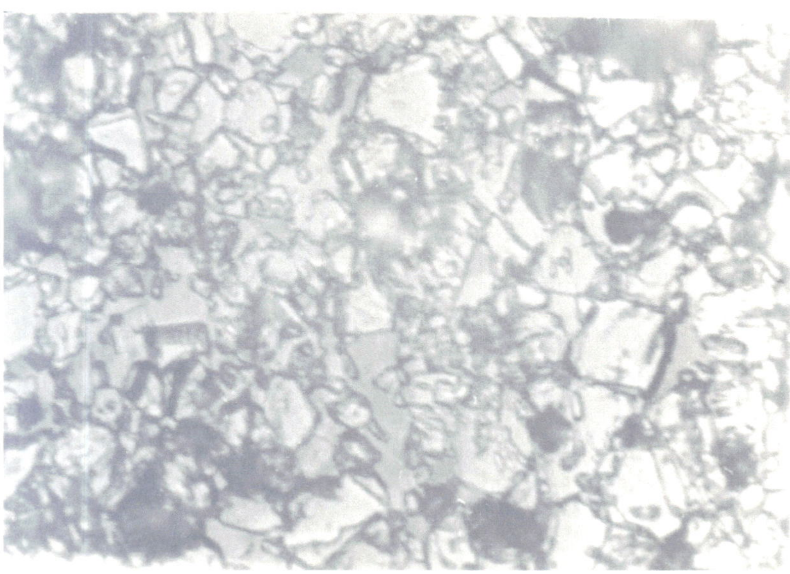

Fig. 5. The microstructure of TiB_2- WB – $(Ti,W)B_2$ composite; x 1600

The mechanical properties of the TiB_2-WB-Co composite materials under study have exhibited a very high hardness combined with a high strength.
Small amounts of the binding metal (cobalt - 1%) considerably decrease the sintering temperature. This is due to reactions which yield eutectics. The same binding metal allows densification of the alloys during sintering without pressure application.
As a general rule, introducing small amounts of the binding metal (1 wt-% with colalt) in TiB_2-WB system considerably decreases the sintering temperature considerably without any resort to pressure while yielding a very high hardness 50.6 GPa (92.3 HRA) (Fig.6 and Tabl.4) of the ternary phase $(Ti,W)B_2$.

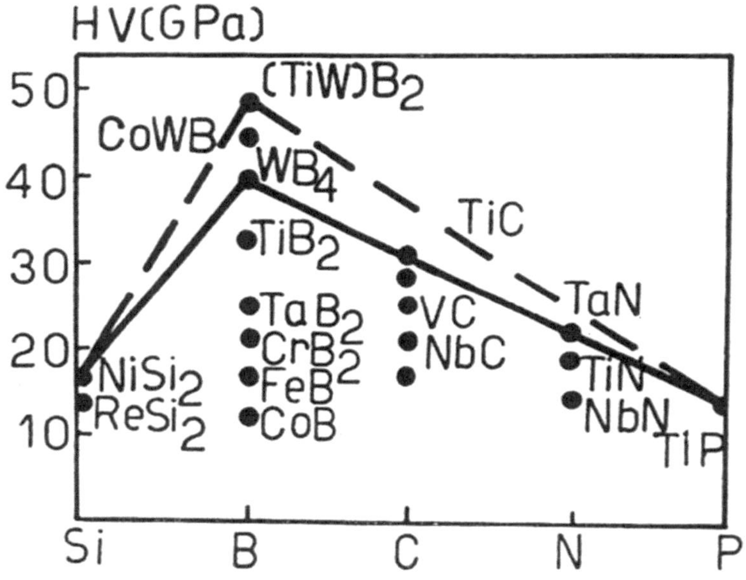

Fig. 6. Metal-like hardness pyramid with the new superhard ternary phases CoWB and $(Ti,W)B_2$

The difference in hardness between of the two composite materials (Tabl.4) seems to come from more ternary $(Ti,W)B_2$ being present in the initial powder TiB_2 (PIII), due to contamination of the latter with TiO_2.

Initial milled powder[wt%]	Specific surface area [m^2/g]	Experimental density[g/cm^3]	Relative density[%]	Hardness RockwellA	Microhardness Hμ0.1 [GPa]
TiB_2(Koch-Light)	2.58	5.22	98.30	91.3	33.83
+ 20.25WC+0.75Co					
TiB_2(PIII)+18.64WC	3.53	5.09	98.80	92.3	50.6
+ 1 Co+ 3.9TiO2					

Table 4. Properties of TiB_2-WB-Co sintered in Ar at 1850 ^oC, 120 min

3.3 Ternary WCoB boride in MMC-composite materials WC(TiC)-Co

For the sintered ternary polycrystalline tungsten-cobalt boride WCoB chemical analysis gave: W- 66.3±6 + Co- 24.7±2 + B- 6±3 wt.% while X-ray studies showed that it is orthorhombic WCoB (ordered $PbCl_2$ structure, Type E-TiNiSi), a=5.724Å; b=3.240Å; c=6.632Å.

Several large WCoB crystals (dimensions of 1x1.5x3 mm) are shown in Fig.7-1.

The dependence of the compound hardness on the indentation load was plotted. In this way, the straight line of Meyer and the line of microhardness (Fig.7-2) were obtained. Obviously, within the range of indentor loads used (below 100 gf), the microhardness gradually decreased with increasing loading. Hence, the microhardness value for WCoB depends on the loading. This correlation is due mainly to plastic deformation, which was observed in our case at a low indentation load (less than 50 gf).

The {001} and {100} faces of WCoB showed a reticular anisotropy of microhardness in directions *c* and *b*.

The hardness of the ternary compound, Hμ50 = 4650 ± 230 kgf.mm^{-2}, is associated mainly with the type and the distribution of bonds in it and corresponds to the usual high hardness of the transition metal borides. Gilman (Gilman, 1970) is of the opinion that this hardness is due for the most part to overlapping of themetal-nonmetal bonds during the shearing of the dislocations. Within the framework of this model $H\mu=-2\Delta H_f/V_m$, where ΔH_f denotes the heat of formation of borides (kcal mol^{-1}) and V_m is the molecular volume (cm^3).Thereupon, after converting 4650 kgf.mm^{-2} into 5.45 kcal.cm^{-3} (the value of Hμ) and 30.758 Å^3 into 18.5 cm^3.mol $^{-1}$ (the value of V_m), one obtains for WCoB ΔH_f = 100.5 kcal mol^{-1}. Comparison of the heat of formation of WCoB with that of TiB_2 (77.4 kcal mol^{-1}) confirms the increase in hardness of the compound with rising heat of formation.

Fig. 7-1. Ternary polycristals WCoB

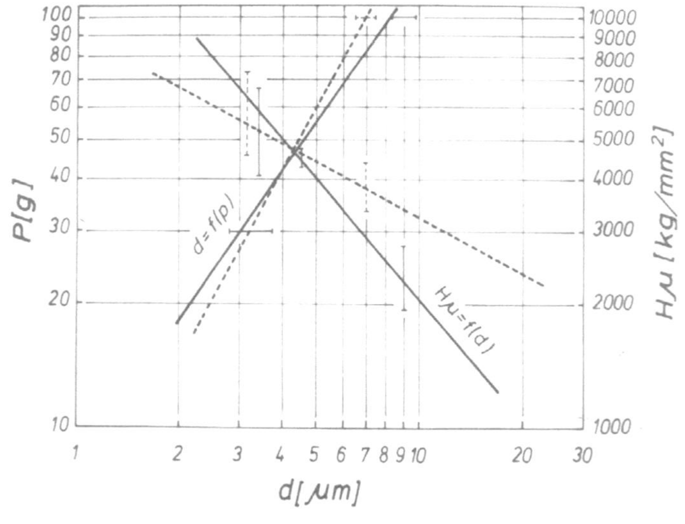

Fig. 7-2. Dependence of Hμ on the indentor loading for WCoB

Formation of WCoB upon the WC(TiC)-Co matrix during thermo-chemical treatment could result from interaction of the type $WC + TiB_2 + Co \rightarrow WCoB + TiC + CoB$.

TiB_2 most probably participates in the coating formation as a donor of boron which diffuses into the samples and interacts with the cobalt and the tungsten carbide. A similar mechanism is proposed in order to explain the formation of boride coatings on iron and steel during thermo-chemical treatment with other boronizing agents (boron and B_4C). The concentration of WCoB in the diffusion layer depends on the composition of the initial alloys and the experimental conditions of thermal treatment 950-1100 °C (Fig.8).

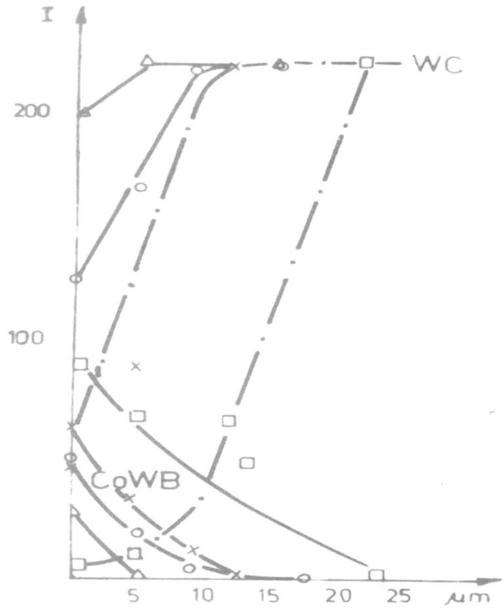

Fig. 8. The phase composition us depth through the diffusion layer formed on a WC-Co: Δ- 950 °C; O- 1000 °C; x- 1050 °C; □- 1100 °C; ▬ WCoB; -.- WC

The presence of the ternary orthorhombic compound WCoB in the surface layer of carbide alloys enhances their wear resistance in metal-cutting. The enhancement seems to follow from the increase in their hardness (Fig. 9. The difference in phase composition of the diffusion layers obtained using the two powders affects the layer hardness. The use of Borozar-HM (base TiB_2) leads to the formation of WCoB only in the diffusion layer, whereas in the case of B_4C the ternary boride W_2CoB_2, which is richer in boron, prevails. This can be explained on the basis of the Co-W-B phase diagram (Stadelmaier, 1967) taking into account the boron content (i.e. the transfer of boron from boron carbide). The termo-chemical treatment with Borozar-HM of the cutting alloy results in the formation on its surface of superhard layers whose hardness exceeds that of the layers obtained with B_4C. The maximum hardness value 23.4 GPa was found for layer with Borozar-HM at 1200 °C, which is assigned to the formation at this temperature of a single-phase ternary WCoB layer.

Ternary borides are useful in drawing plain wire and metalworking where a superhard layer of them is formed by diffusion on the main material consisting of carbide - cobalt alloys WC (TiC) - Co.

Layers of this kind improve the performance of nozzles, turning-lathes and other devices used in drawing and cutting of metal articles such as wire, rods, pipes, plates (Fig.10).

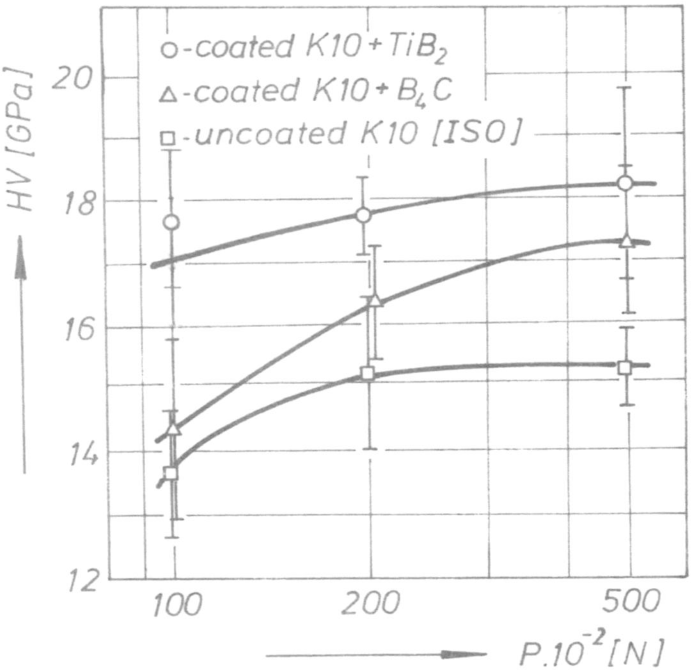

Fig. 9. Dependence of the boride layer hardness on the temperature of matrix (K10, WC-Co-10%) treatment with Borozar-HM (TiB_2) and B_4C

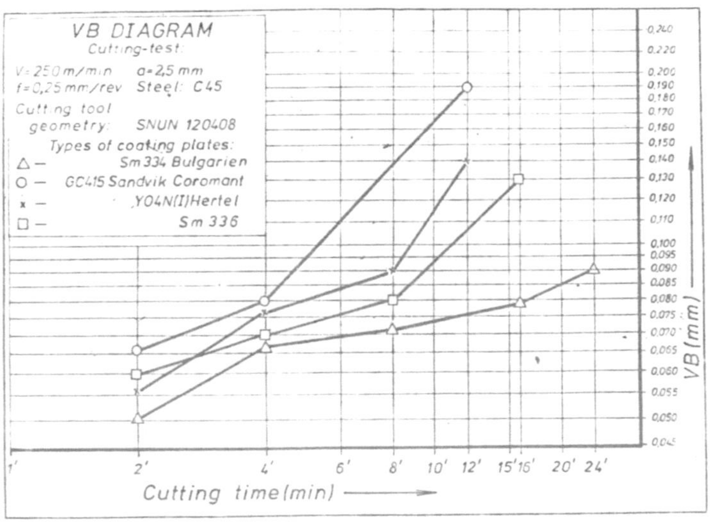

Fig. 10. Comparative diagram (VB) of the wear in steel working of various types of coated cutting tools: □ - WCoB+ TiC-Al_2O_3 deposited on Sm 336 plate (Gabrovo,Bulgaria); X- TiC-Al_2O_3 deposited on Cp 1 -V04 (Hertel, Germany); O- TiC-Al_2O_3-TiN deposited on GC 415 plate (Coromant, Sweden) ; Δ- WCoB + TiC-Al_2O_3 on Sm334 plate (Gabrovo, Bulgaria)

3.4 Ternary $(Fe,Zr)_2B$ boride in MMC- composite material (Fe-C matrix)

The thermochemical treatment with ZrB_2 (CrB_2, ,TiB_2) and activators on steels yields a diffusion layer with a thickness between 50 µm and 1 mm . Our investigation of ZrB_2 (KBF_4 having been used as a activator) were carried out over the temperature range 1000-1100 °C for 4 h. The diffusion layer obtained at 1050 °C was of a two-zone nature: zone "A" and the underlying (in the direction of the sample centre) zone "B" (Fig .11(A) and Fig.11 (B), respectively). The photograph of the polished section in Fig.11(A) shows that zone "b" consist of needle-like crystals characteristic for the gas transport of boron. Zone "B", which is situated above layer b-d, has a different structure with small grain size and a very high microhardness (3575-2438 $kg.mm^{-2}$ (Fig. 11(B)).

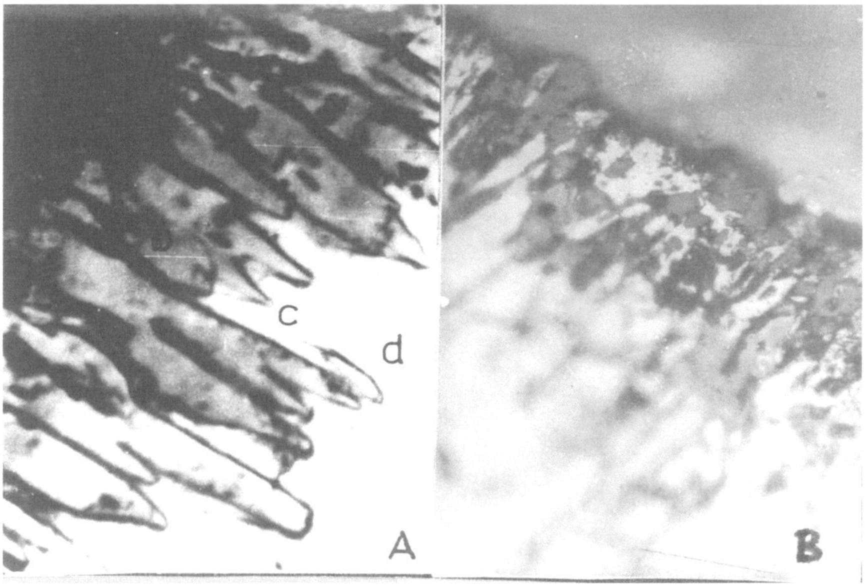

Fig. 11. The two-zone diffusion layer on Armco-iron treated with ZrB_2 and KBF_4(3wt.%): A - boronized zone (Fe_2B) ; B - upper superhard $(Fe,Zr)_2B$ - ZrO_2 zone; x 600

For the zones below the zirconium a greater microhardness (as compared with that of "pure" Fe_2B) has been detected by microprobe analysis. It might be due to replacement of some iron atoms of the Fe_2B-phase by zirconium ones. This is also indicated by the change in lattice parameters of the underlying iron boride.

The ternary boride $(Fe,Zr)_2B$ enhances the wear- and heat resistance of the steels coated.

With boron-chromizing, the ternary boride $(Fe,Cr)_2B$ imparts additionally augmented resistance to corrosion (Fig.12).

The microstructure of the boron-chromium layer obtained on steel C45 with Zahobor paste is presented in Fig.13. X-ray microanalysis has shown that black grains of chromium- iron boride $(Fe,Cr)_2B$ contain 14-50 wt.%Cr.

Boron-metalizing with paste during hardening of steel is a new process resulting in a surface layer with a high wear resistance and stability with respect to oxidation and corrosion, minimum time and cost losses needed. The proces carried out with pastes for boronization and boronmetalization leads to products of higher hardness, i.e. durability (twice as long as in cases of nitration and cementation), a higher stability towards high temperatures and a higher corrosion resistance.

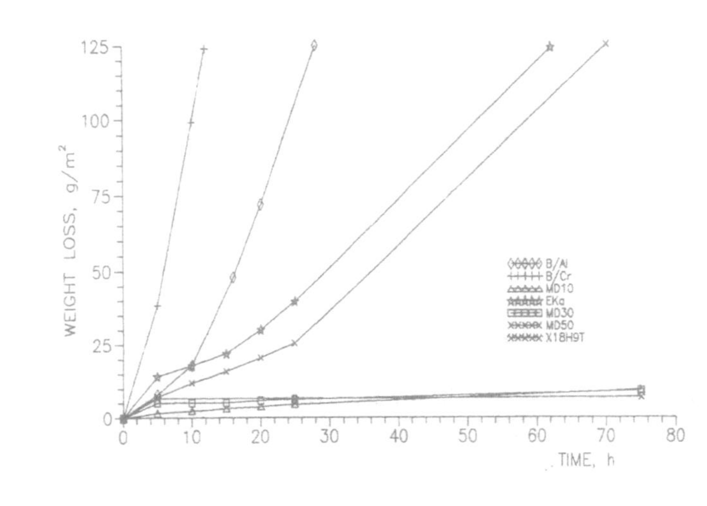

Fig. 12. Corrosion resistance in 10%H_2SO_4 of boron-chromium (Zahobor paste MD10, MD30, MD50); boronizing with Ekabor-Paste (Eka-Germany); subsequently boronized electrolyte chromium (B/Cr); boro-aluminizing (B/Al) layers deposited on C45 and chrom-nickel steel (X18H9T)

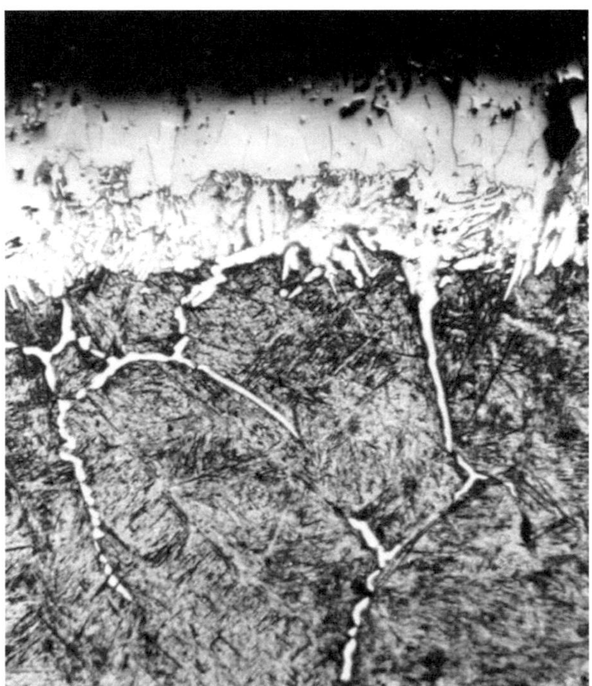

Fig. 13. Microstructure of boron-chromium layer deposited on C45 steel by Zahobor paste (MD30); X600

Table 5 shows the results of x-ray phase analysis of the layers as well as data on their thickness and microhardness. Obviously, the highest hardness corresponds to borchromium layer obtained using Zahobor (MD30).

Paste	Phase composition	Thickness, μm	Hμ 30, GPa	Hμ50, GPa
Zahobor - Bulgaria	CrB, $(Fe,Cr)_2B$	150	20.4	18.6
Ekabor-Germany	Fe_2B	160	15.6	15

Table 5. X-ray phase analysis, thickness and microhardness of the diffusion layers with pastes on C45

On the basis of these results it may be inferred that doping of the layers with chromium, which leads to the appearance of $(Fe,Cr)_2B$ phase increases significantly their microhardness. The phase composition of the diffusion layers determines their microhardness, i.e. their wear resistance. Hence, we may predict an even higher wear resistance of our boron-chromium layers. Indeed, the results on their wear resistance correlate with those on their microhardness (Fig.14). The most stable Zahobor (MD30) layer is more than twice as stable as the boronized one according to the fifty-hours-test.

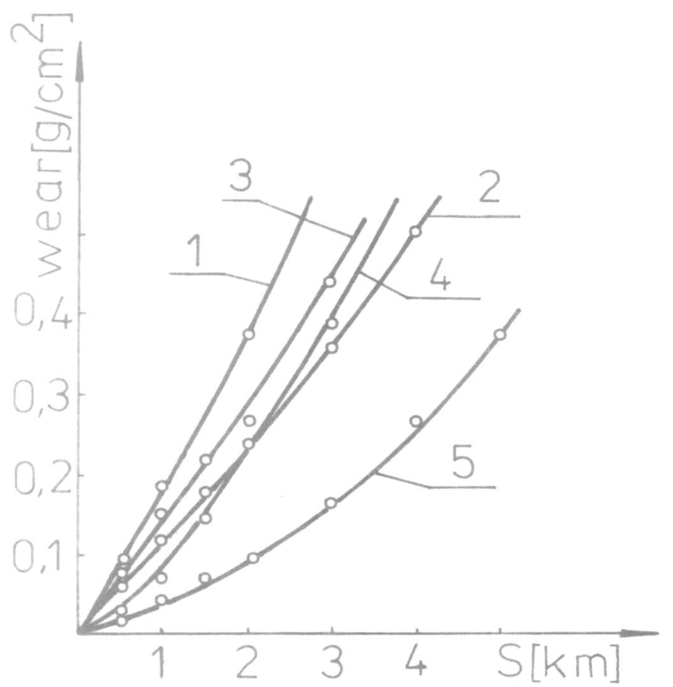

Fig. 14. Wear resistance of the boron-chrominized (5-MD30), boronized (4-Ekabor), boron-aluminized (2- B/Al) , electrodeposited chromium (3) layers and base chromium-nickel steel

The positive effect from boron-chromizing is illustrated on Fig.15 and Fig.16 for landholder's steel instruments treated with the "Zahobor"-paste and used in the Netherlands.
The paste is suitable for treatment of steel machine tools and parts of large dimensions, e.g. metal stamps, hammering press matrices, guides, rolls for wiredrawing, steel pulleys, steel belt conveyor rolls, ploughshares, tracks, extruder screws and other similar machine parts, subjected to wear and corrosion. Machine tools, instruments and parts with larger design tolerances as regards cross-section dimensions are especially suitable for boron-metalizing.

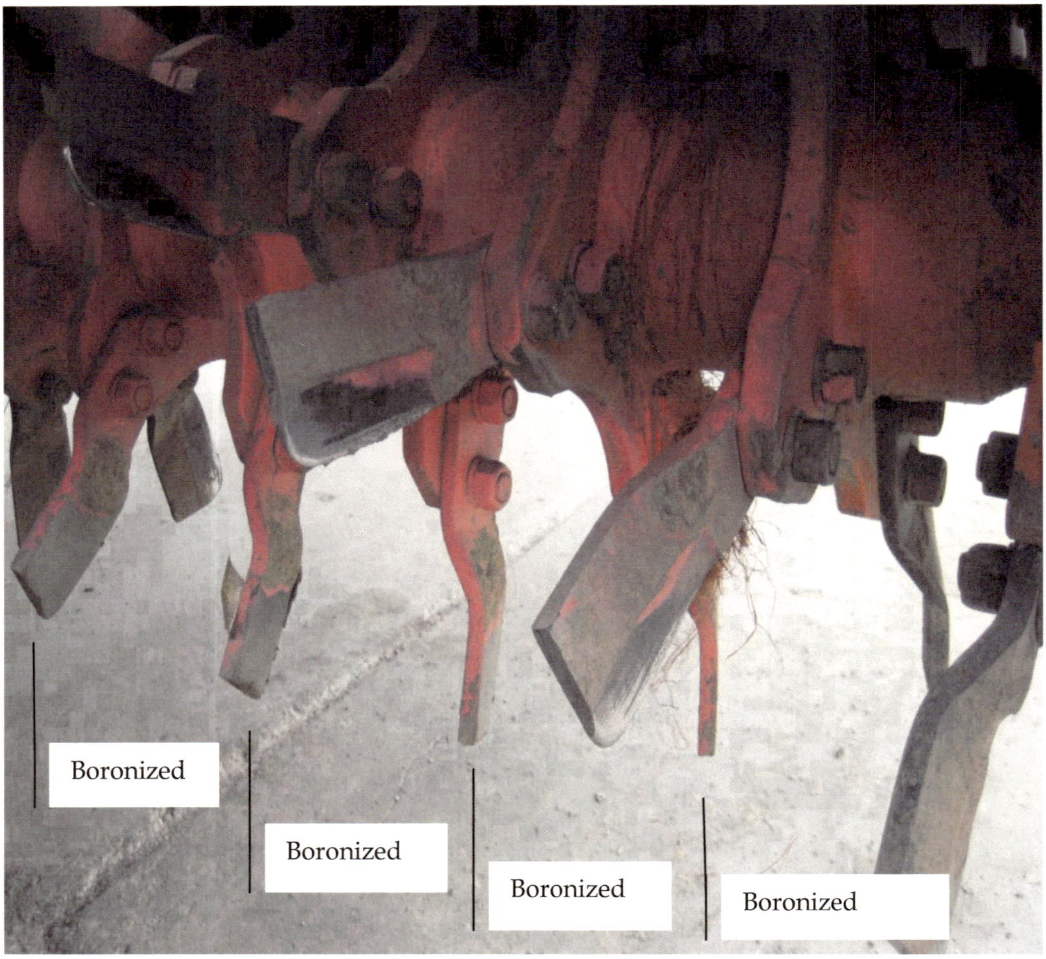

Fig. 15. Boronized parts have been used for 6 ha

Fig. 16. The knifes "Tortella" - Italy boronized with paste "ZAHOBOR"

4. Conclusion

Sintering without applying a pressure is a new trend in the field of superhard boride materials. The present paper deals with the microhardness of some boride CMC and MMC composite materials ($B_{12}C_3$ + Me_xB_y, TiB_2 + WB + Co, WCoB + WC–Co, $(Fe,Me)_2B$ + Fe-C) obtained in this way.

It is shown that the transition metal dissolves in the crystal lattice of $B_{12}C_3$ building up new superhard ternary borides $B_{12-n}C_3Me_n$ with a hardness of 50 - 77 GPa. The latter values exceed considerably the hardness of pure $B_{12}C_3$ and coincide with those for cubic-BN and some synthetic diamonds of the type "CB" or "Carbonado ACPK".

Another example is the composite material TiB_2 - WB, where the surface of its grains proves enriched of tungsten to $(Ti,W)B_2$, this leading to an extremely high value of 50.6 GPa.

The hardness of the ternary boride WCoB amounts to 38 - 43 GPa depending on the indentor loading. Presipitate in the form of a boronizing coating upon the carbide cutting tools, WCoB- phase increase their tool-life.

Thermochemical treatment of steels with ZrB_2 (TiB_2 or CrB_2) leeds to form a diffusion layer with superhard ternary phases $(Fe,Me)_2B$. This phases improve the wear- and corrosion resistance of the steels.

In comparison with other metal- like refractory compounds, the superhardness of the materials studied points to new applications in industry. In this field of view, include my own researches, is make an attempt to unification of the scientific results and to show the perspectives about using of the obtaining superhard ternary composite (CMC and MMC) materials.

5. References

Anderson Ch., (2002). Ceramic armor materials by design, *The Amer.Ceram.Soc.*,pp.487-489

Fitzer E. (1973). *Arch.Eisenhuttenwes* , v.44, pp.703-709

Fukatsu T.; Yuhara K.& Kobori K. (1967). *Nippon Kinzoku Gakkhaishi*, N3, 1127-1131

Gilman J. (1970), *J.Appl.Phys.*, N41, pp.1664-1669

Jeitschko W. (1968). *Acta Cryst.*, B24, pp. 930-934

Kislii P,; Kuzenkova M.& N.Bondarchuk (1988). Carbide bora, *Naukova dumka*, Kiev

Komai M. et al.(1989). *MRS Int'lMtg.on Adv.Mats.*, v.4, Materials Research Society, pp.475-480

Lecrivian L.&.Provost G. (1968). *Berichte der Deutschen Keramischen Gesellschaft*, v.45, 7, pp.347-351

Lipp A.& Schwetz K. (1975). *Berichte Dt. Keram.* Ges, N52, pp.335-340

Lipp A.& Roder M. (1966). *Z.Anorg. Algemeine Chem.*, v.343, pp.1-9

Makarenko G.;Kosolapova T.& Marek E. (1977). Tugoplavkie boridi I silizidi AN USSR, *Naukova Dumka*, Kiev, pp.66-77

Murata Y,;Julien H.& Whitney E. (1967). *Ceramic Bulleting 46*, N7, pp.643-648

Nishiyama K. (1985). *JSCM N11*, pp.53-61

Okada Sh.;Kudou K.;Hiyoshi H.;Higashi I.;& Lundstrom T. (1990). *J.of the Ceram.Society of Japan, Int.Edition*, v.98-1342, pp.42-47

Petrov K.&Will G. (1981). *J.Materials Science*, v,16, pp.3218-3223

Petzow G,&Telle R. (1984).New Development in the Field of Refractory Hard Metals Based on Cemented Borides, in *Lectures on Advanced Ceramics*, Uchida Rokakuho, Tokyo

Portnoi K.&Samsonov G.(1960). *Gurnal Prikladnoi Chimii*,v.33, pp.577-584

Schouler M.;Ducarroir M.& Bernard C. (1983). *Rev.Int.Hautes Temp.Refract.* v.20, pp.26-30

Solozhenko V. (2001). *Diamond Relat.Mater.* N10, pp.2228-2234

Solozhenko V. & Gregoryanz E. (2005). Synthesis superhard materials, *Materials Today, Elsevier Ltd.*,pp.76-85

Stadelmaier H.& Lowder J. (1967). *Metall (Berlin)*,N21 (10),pp.1023-102

Zachariev Z. (2001). "Neue Superharte Kompositionswerkstoffe", *Metall, (Internationale Fachzeitschrift fur Metallurgie), Giesel Verlag GmbH, Isernhagen*, pp. 23-87

Zakhariev Z. & Radev D. (1988). Properties of polycristaline boron carbide sintered in the presence of W2B5 with out pressing, *J.Materials Science Letters*, v.7, pp. 695-697

Zakhariev Z.; Ivanova M.& Serebriakova T.(1993). Hard Materials Based on Cemented TiB2-WB-Co Alloys, *XI Inter.Symp.Boron, Borides and Related Compounds, Tsukuba*, Japan

Zakhariev Z. ; Marinov M.; Zlateva R.& Chistov Ch.(1987). A new combination of coatings on carbide cutting tools, *Surface and Coatings Technology*, v.31, pp.265-273

Zakhariev Z.;Ziateva R.& Petrov K, (1986). Microhardness and high-temperature oxidation stability of CoWB, *J. Less-Common Metals*, v.117, pp.129-133

Zakhariev Z.;Belopitov N.& Razkazov N. (Jul 1970). *Pat.Bulg.* N16115

4

Thermo-Mechanical Treatment of Glass-Balloon-Dispersed Metal Matrix Composite

Osamu Umezawa
Yokohama National University
Japan

1. Introduction

Mechanical aspects of recycled composites are demonstrated in this chapter. Aluminium and its alloys are probably the most widely used matrix materials for metal matrix composites (MMCs) (Rodriguez-Castro, 2002). Although reinforcements in the form of continuous and discontinuous fibers have already been investigated in detail (Yu & Lee, 2000), discontinuous reinforcements such as those of dispersoids have also become more popular. Subsequent working of such dispersoid-reinforced MMCs can enhance their mechanical properties (Seah & Hemanth, 2007). Aluminium alloy/glass composites cast with glass particles of sizes in the range of several tens of micrometers have been fabricated and have exhibited superior mechanical properties with regard to strength and wear resistance (Nicom & Nomura, 2006; Gibson & Ashby, 1999).

Since magnesium is the lightest metal in the structural materials, magnesium alloys also have great potential for applications in automotive and aerospace industries. Furthermore, solid-state recycling processes of magnesium alloy chips and/or fragments such as hot extrusion, mechanical alloying, equal channel angular extrusion (ECAE) and cyclically repeated plastic working (RPW) have been developed (Mabuchi et al., 1995; Clark et al., 1997; Kondoh et al., 2003a, 2003b; Kondoh & Luagnvaranaunt, 2003c). The recycled materials show superior mechanical properties such as high strength and high strain-rate super-plasticity. Kondoh et al. (Kondoh & Luagnvaranaunt, 2003c) has established a process to fabricate high performance magnesium composite, in which Mg_2Si and MgO particles were synthesized due to deoxidization of SiO_2 glass by magnesium, and a refinement of both those dispersoids and magnesium matrix grains was achieved during RPW. Under only 2 mass% SiO_2 addition in AZ31 alloy, the Mg_2Si compound was employed as a reinforcement of the composite, because of its high hardness, high Young's modulus and superior corrosion resistance. In order to fabricate the composites of glass balloons and magnesium alloy chips, however, their mixture should be consolidated with the condition for synthesis of Mg_2Si and MgO phases.

On the other hand, various porous metals have been fabricated and their specific properties such as ultra-lightweight and low strength have been considered with more attention (Gibson & Ashby, 1999). Further, porous ceramics (e.g., Al_2O_3, ZrO_2, TiO_2, SiC, TiC, ZrC, etc.) have been used to develop ceramic matrix composites by various processing routes such as the melt stirring process and pressure infiltration technique (Chou et al., 2000;

Wilkes et al., 2006; Cao et al., 2004a, 2004b; Rambo et al., 2004, 2005). Therefore, the dispersion of porous particles in a metal matrix is one of the methods employed to control the size and distribution of the cell structure in the composite.

Recently, new recyclable materials composed of recycled aluminium and glass balloons have been developed (Kashiwaya et al., 1999) and fabricated for building panels (Naigai Building Materials Corporation). However, in order to take the maximum advantage of their specific properties such as ultra lightweight and good electromagnetic shielding for mechanical or electronic parts, a workability that can deform them into any form is required. Since the balloon is a brittle second-phase particle, it is necessary to include either a thermomechanical treatment to plasticize the composite or perform near-net-shape casting.

Umezawa et al. (Umezawa & Nagai, 1999) developed a repeated thermomechanical treatment (RTMT) to refine the microstructure of Al-Si cast materials; the materials successfully obtained good plasticity by this treatment. The RTMT was adopted to control the microstructure of the glass-balloon-dispersed aluminium alloy (AC3A) matrix composite in order to achieve a satisfactory stress-strain relationship and ductility for the composite (Shiga & Umezawa, 2007). Furthermore, the mixture of glass balloons and AZ31 magnesium alloy chips was thermomechanically treated to form their composite with Mg_2Si and MgO phases.

2. Tensile and compressive properties of glass-balloon-dispersed aluminium alloy composite

A repeated thermomechanical treatment (RTMT) was adopted to control the microstructure of a glass-balloon-dispersed aluminium alloy (AC3A) composite. The RTMT, which involves the repetition of a multi-step process and followed by heat treatment, was applied to a cast plate of the composite. The composite was successfully worked into a rod or sheet by either swaging or flat rolling. The porous glass balloons were deformed or cracked so that the composite could be worked easily. The swaged material exhibited higher tensile strength, Young's modulus, and elongation as compared to the cast material. The composite materials also exhibited excellent energy absorption properties.

2.1 Microstructural modification by repeated thermomechanical treatment

2.1.1 Test material

The test material (AC) was a 10-mm-thick cast plate (Alcelite™ (Naigai Building Materials Corporation)), which consisted of a recycled glass balloon and aluminium alloy AC3A. The major compositions of the glass balloon were 68 SiO_2, 6.3 Al_2O_3, 0.6 Fe_2O_3, 0.6 MgO, 9.5 CaO, 11.7 Na_2O, and 1.3 K_2O in mass%. The major properties of the AC and AC3A are listed in Table 1. Figure 1 shows the microstructure of the AC material and glass balloon. The porous

Material	Tensile strength, σ_B /MPa	Compressive strength, σ_C /MPa	Bending strength, σ_D /MPa	Bending elastic modulus, G /MPa	Thermal conductivity, κ /W·m^{-1}K^{-1}	Thermal expansion coefficient, α / × 10^{-6}K^{-1}
AC	15.7	49	41.2	144	54.4	16.5
AC3A	170	170	196	686	217.7	23.6

Table 1. Mechanical and thermal properties of the test material (AC) and AC3A.

glass balloon with a diameter of approximately 1 mm is dispersed in the AC3A matrix. The balloon contains numerous closed bubbles. The specific gravities of the glass balloon and AC3A are 1.58×10^{-4} and 2.7 g/cm^3, respectively. The specific gravity of the AC material, which was measured with a rectangular bar of dimensions 9.57 x 9.47 x 32 mm, was obtained as 1.5 g/cm^3. The volume fraction of the glass balloon was estimated to be approximately half in the AC. The material annealed at 793 K for 1.8 ks was designated as AN.

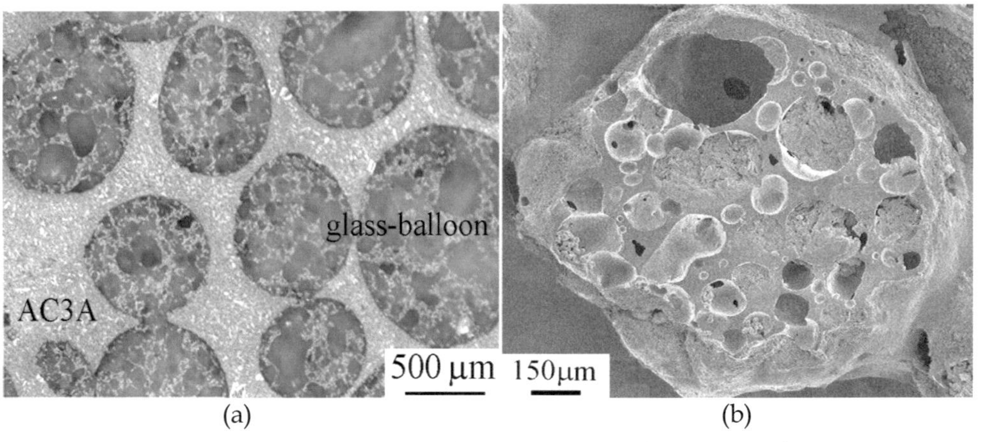

Fig. 1. Microstructure of the (a) AC material dispersed glass balloons in the AC3A matrix and (b) whole view of a glass balloon.

2.1.2 Repeated thermomechanical treatment and modified microstructure

In order to avoid fracture, the AN samples were worked in multiple steps with intermediate annealing at 793 K for 1.8 ks by using the RTMT process (Umezawa & Nagai, 1999). The samples were machined either as round bars (10 mm in diameter × 150 mm in length) for swaging or as plates (10 mm in thickness × 20 mm in width × 200 mm in length) for flat-rolling. After the samples were worked by either swaging or flat-rolling in multiple steps at room temperature, they were formed into a rod (CS) or sheet (CR). The reduction in section area per working step was approximately 10%, and the working-annealing cycle was repeated for a total section area reduction of approximately 90%. For the resulting RTMT materials in the final state, a multiple-step process involving either swaging or flat-rolling was performed at a different value of total section area reduction. The working strains were defined as $\eta = \ln A_0/A$, where η denotes the working strain; A_0, the initial section area of the sample; and A, the section area of the worked sample. The specific gravities of the worked CR and the selected CS samples (η = 0.46 and 1.38) were measured and their microstructures were observed. In the final step, the samples with or without annealing were indicated as -H or -A, respectively.

The specific gravity of the materials increases from 1.5 to 2.3 g/cm^3 with the working strain up to η= 2.6 (Fig. 2). This indicates that the relative density of the materials increases. In the case of rolling, a linear relationship is obtained between the working strain and specific gravity. However, the CS material exhibits a higher value of specific gravity for a lower working strain. Although the data plots for the CS material are limited, the glass balloons in the CS after swaging may be divided and aligned along the working direction with a lower working strain due to a type of hydrostatic stress effect.

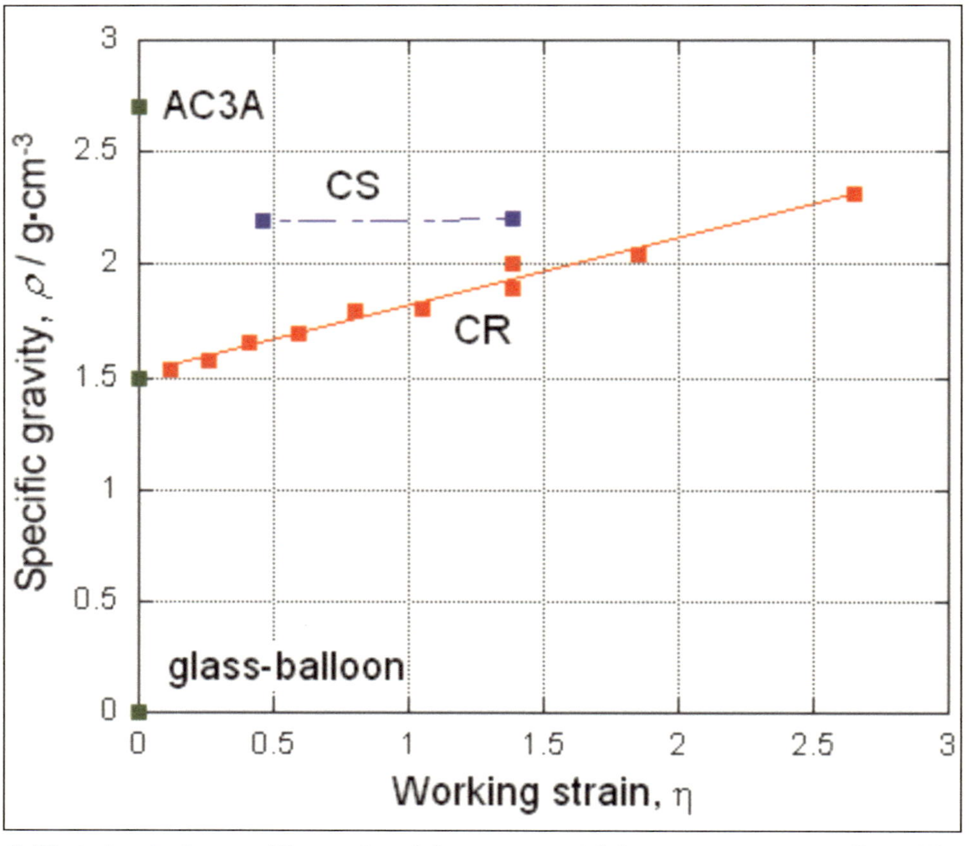

Fig. 2. Variation in the specific gravity of the test material during swaging or rolling. (Shiga & Umezawa, 2007)

Then microstructures of the composites after swaging or rolling was observed and compared, as shown in Figs. 3 and 4. Rod-like voids in the CS (Figs. 3c and 3d) or pancake-like voids in the CR (Figs. 4c and 4d) are observed at higher working strains. The porous balloons may be cracked and dropped from the sample surface. Comparing the microstructure in the transverse section of the CS and CR materials that were worked at lower working strains, it is observed that the radius of the voids in the CS (Fig. 3b) is lower than that in the CR (Fig. 4b). This observation agrees with the higher value of specific gravity at lower working strains.

2.2 Tensile and compressive properties

2.2.1 Tensile properties

A tensile test was performed for the AC, AN, and CS (η = 1.38) materials. Smooth-type specimens with a gauge diameter of 3.5 mm and length of 30 mm were used for the AC and AN materials. For the CS, the specimens were obtained from rods parallel to the longitudinal direction, and the gauge geometry was mechanically polished to obtain a diameter of 2.4 mm. The elongation was monitored by a clip gauge with the knife-edges set onto the tensile specimen. Its gauge length was 25 mm. A crosshead speed of 0.5 mm/min was selected in a motor-driven testing machine at 293 K (in air) under displacement control.

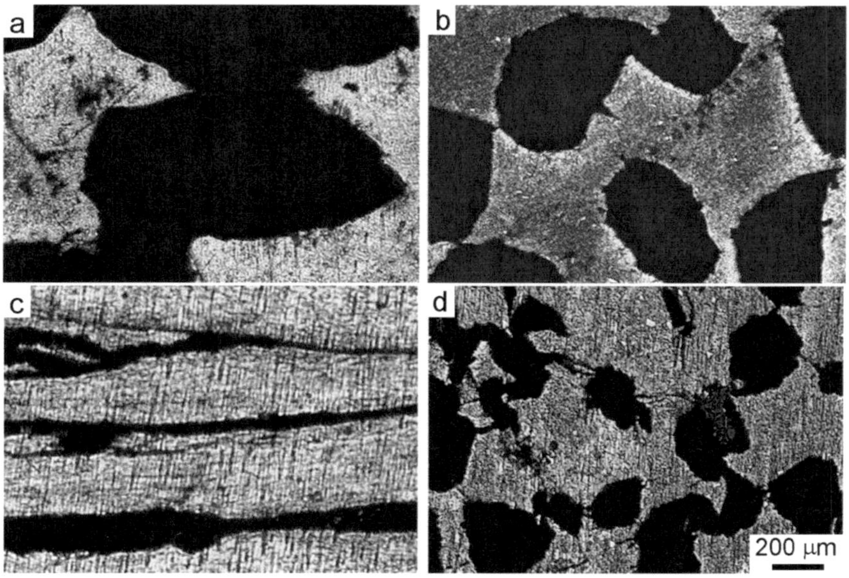

Fig. 3. Microstructure of the CS materials in the longitudinal section (a) and (c) and transverse section (b) and (d): (a),(b) $\eta = 0.46$ and (c),(d) $\eta = 1.38$. (Shiga & Umezawa, 2007)

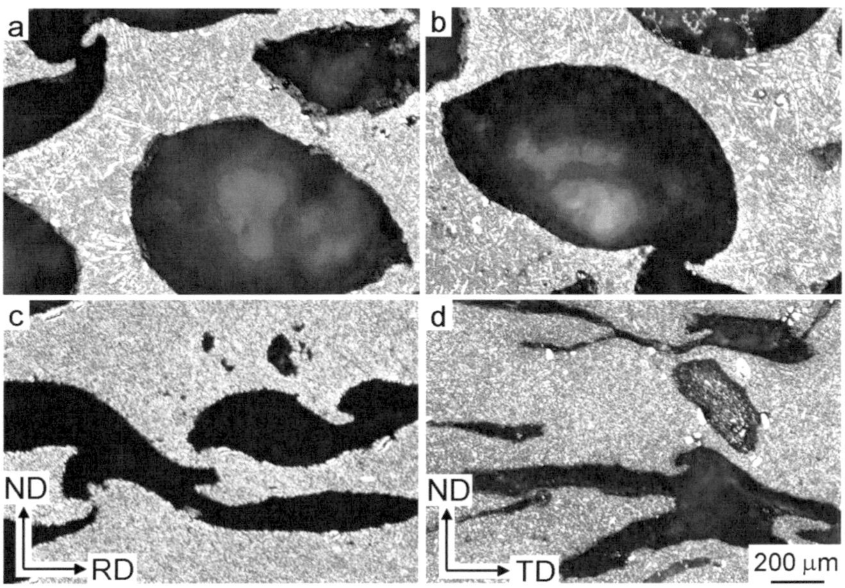

Fig. 4. Microstructure of the CR materials with working strain (a),(b) $\eta = 0.41$ and (c),(d) $\eta = 1.38$. (Shiga & Umezawa, 2007)

Figure 5 shows the tensile stress-strain curves and Table 2 summarizes the tensile properties. The swaged materials (CS-H and CS-A) exhibit higher values of Young's modulus as compared to the cast materials (AC and AN). This may be related to the increase in specific gravity due to deformation of the balloons. The tensile strength of the CS materials is approximately thrice that of the cast materials. The RTMT not only increases the specific gravity but also the uniform elongation. The CS materials exhibit necking instability; however, the cast materials are fractured before approaching it. Hence, a microstructural modification by the RTMT could result in the avoidance of early fracture due to tension, which is characteristic of the cast materials. The load drops that appear in the curves of the swaged materials and the early fracture in the cast materials may result from the instantaneous separation between the balloons and the matrix. In fact dimple in the matrix traced the interface with dropped glass-balloon or brittle fractured glass-balloon mostly covers the fracture surface of the AC, as shown in Fig. 6a. The manner in which these fracture occurred was not detected in the CS. The annealing at 793 K for 1.8 ks results in higher uniform elongation in both the AC and CS materials. In particular, the annealed CS material (CS-A) showed an excellent elongation up to a strain of approximately 20%. This manner is typical of materials that undergo the RTMT (Umezawa & Nagai, 1999). Although delamination is detected in the tensile fracture in the CS materials, as shown in Fig. 6b, the RTMT may allow greater flexibility in the working of the composite.

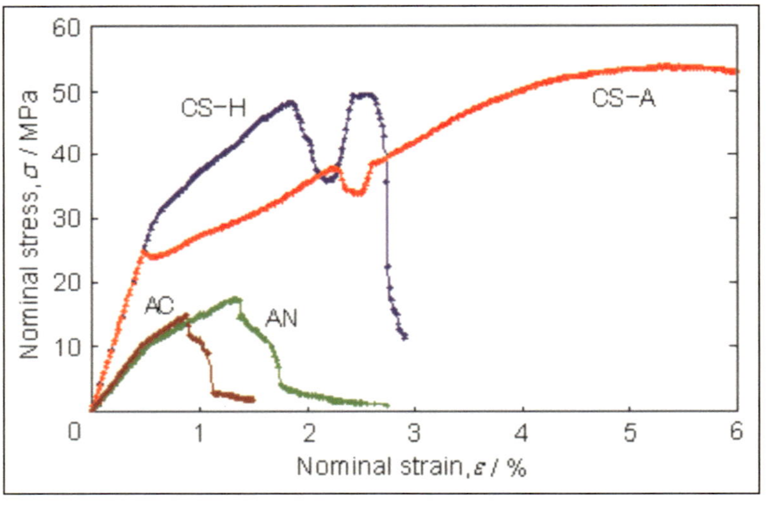

Fig. 5. Stress-strain curves in the tensile test. (Shiga & Umezawa, 2007)

Material	0.2% proof stress, $\sigma_{0.2}$ /MPa	Tensile strength, σ_B /MPa	Elongation, *El* / %	Young's modulus, μ /GPa
AC	15	15	1.4	2.2
AN	14	17	2.7	1.9
CS-H	35	49	2.7	5.2
CS-A	27	54	19	5.3

Table 2. Tensile properties of the CS materials. (Shiga & Umezawa, 2007)

The area fraction of the matrix in the transverse section of the CS is higher than that of the AC, and the matrix in the CS continuously exists along the tensile direction in the form of a fiber. Therefore, the CS may exhibit the tensile performance of the aluminium matrix rather than the AC; such behaviour results from the reduction in the stress concentration, which induces fracture of the matrix.

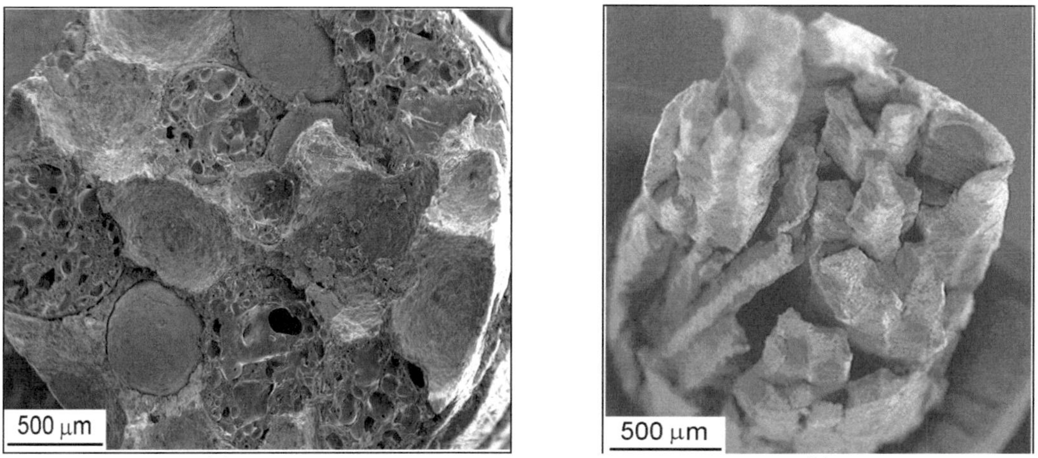

(a) (b)

Fig. 6. Fracture surfaces of the (a) AC and (b) CS-A. (Shiga & Umezawa, 2007)

2.2.2 Finite element analysis

Interactive algorithms (Yamada, 2006) used for computing the averaged response of non-linear composites have been applied to microscopically analyse the tensile deformation in the elastic regime. For the AC, spherical glass balloons were installed into the aluminium matrix as a closed-cell foam. Pores were distributed in the balloon. For the CS, ellipsoidal (rod-like) pores were installed into the matrix since the balloon was collapsed into pieces and could be plastically deformed within the matrix.

A simple stress analysis was applied to the materials in order to discuss the fracture manner and stress-strain relationship mentioned in above. Figure 7 shows an analysis map for the principal stress distribution under a microstrain in the elastic regime where the applied stress is 1 MPa in tension along the y axis. The stress concentration appears around the glass balloons. When a glass balloon represents as a sphere, its diameter along the y-axis is taken for the earth's axis (NS). Then the equator is perpendicular to NS. The glass balloon at the lower left in Fig. 7 represents a reference sphere. The compressive stress with the σ_{xx} component is relatively high at both N and S poles (indicated by arrows in Fig. 7a). The principal stress, σ_{yy}, acts near the interface, especially around the equator. These behaviours agree with the stretching of the faces of the closed-cell foam under tension [6]. Thus, the tensile stress around the equator on the sphere occurs due to tension and may lead to fracture. The stress concentration of the σ_{yy} is estimated about three times of the applied stress. On the contrary, the compressive stress around the equator on the neighbouring balloon occurs in Fig. 7b. The maximum magnitude of the σ_{yy} in compression is almost the same in tension. These stress concentration phenomena agrees with the fracture surface shown in Fig. 6a. The brittle fractured glass-balloon may result from the concentrated tensile stress, and the glass-balloon concentrated compressive stress may be separated from the matrix. Even though the interface strength between the glass-balloon and the matrix is very low, it reveals that the stress concentration around the equator on the glass-balloon may predominantly affect the tensile fracture behaviour.

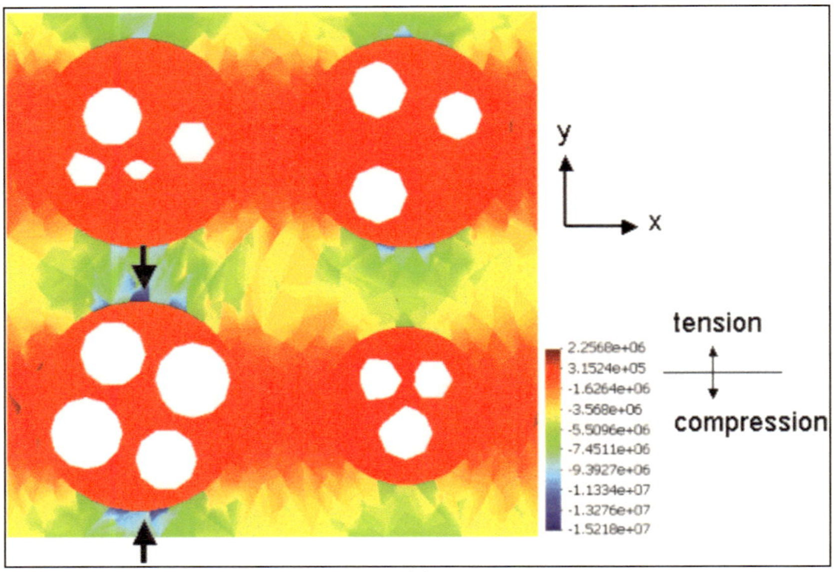

(a)

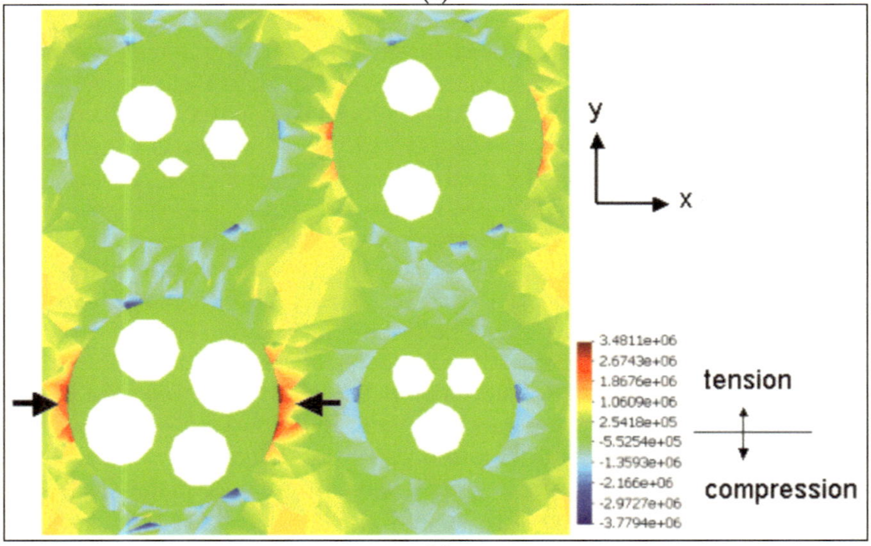

(b)

Fig. 7. Principal stresses in the AC material: (a) σ_{xx} and (b) σ_{yy}. Applied tensile stress is 1 MPa and its direction is parallel to the y-axis. The area in white colour shows the pores in the glass balloon. Arrows in (a) indicate a compressive stress concentration part. Arrows in (b) indicate a tensile stress concentration part. (Shiga & Umezawa, 2007)

In the case of the CS material, the stress concentration mostly appears near at both the top and bottom parts of the elongated pore, as shown in Fig. 8. The influence of tensile stress concentration at the ellipsoidal closed-cell foam due to fracture may be lower than that at the spherical foam, although the maximum magnitude of tensile stress concentration in the CS is higher than that in the AC. The tensile stress of σ_{xx} between elongated pores occurs. It may give an origin of delamination of the composite.

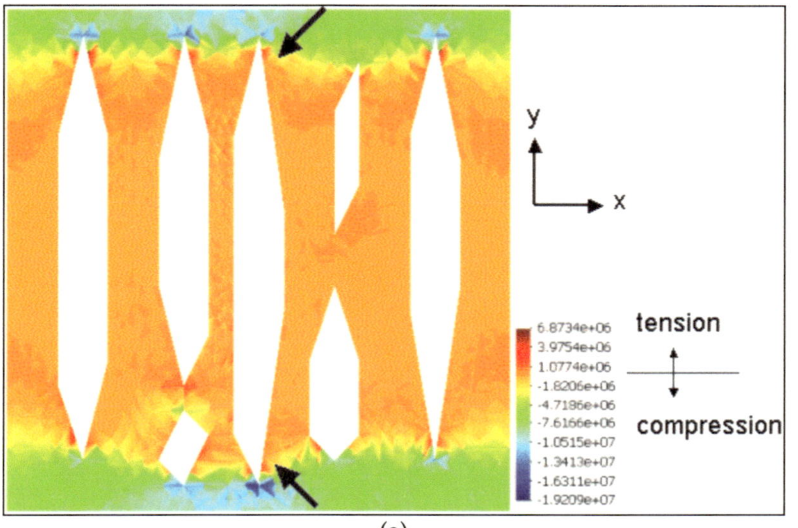

(a)

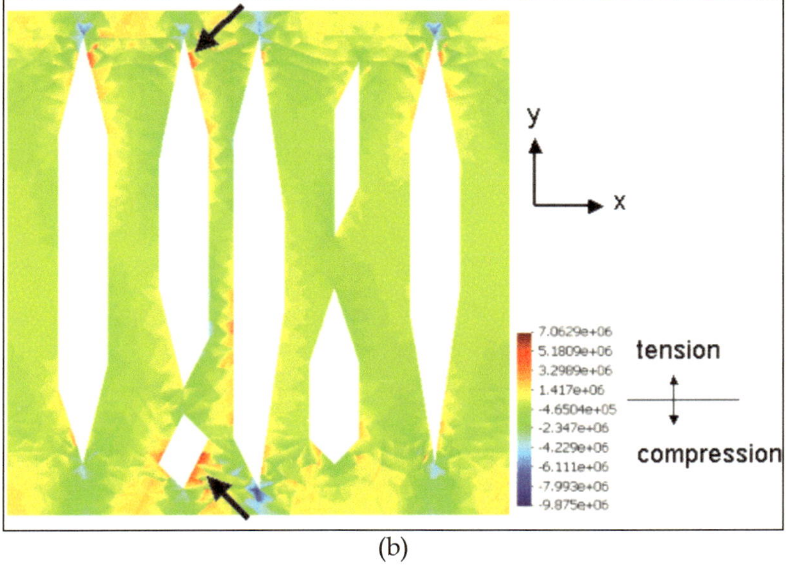

(b)

Fig. 8. Principal stresses in the CS-A material: (a) σ_{xx} and (b) σ_{yy}. Applied tensile stress is 1 MPa and its direction is parallel to the y-axis. The elongated pores are indicated in white colour. Arrows in both (a) and (b) indicate a tensile stress concentration part. (Shiga & Umezawa, 2007)

2.2.3 Compressive properties and energy absorption

A compressive test was also performed for the AC, AN, and CS (η = 0.46) materials. The cylindrical specimens were machined from rods parallel to the longitudinal direction. For the AC and AN materials, the diameter and height of the specimens were 9 mm. For the CS material, the diameter and height of the specimens were 6 mm. In the compressive test, the diameter at the centre of the specimen was measured by using a micrometer. In both the tests, a crosshead speed of 0.5 mm/min was selected in a motor-driven testing machine at 293 K (in air) under displacement control.

Figure 9 schematically illustrates the compressive stress–strain curve of the porous material. The curve generally shows a very low increase in stress (plateau regime); this is followed by an acceleration in the stress (densification). The energy absorbed per unit volume up to a strain at the transition point from the plateau to the accelerated level, ε_1, is defined as (Gibson & Ashby, 1999)

$$W = \int_0^{\varepsilon_1} \sigma d\varepsilon \qquad (1)$$

The absorbed energy is schematically shown by the hatched area in Fig. 9. The energy absorption efficiency, E, is also evaluated as

$$E = \frac{\int_0^{\varepsilon_1} \sigma d\varepsilon}{\sigma_0 \cdot \varepsilon_1} \qquad (2)$$

Where, σ_0 is the stress at strain ε_1.

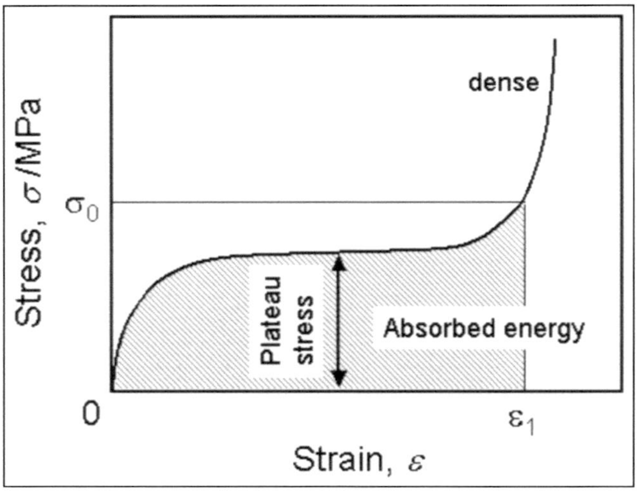

Fig. 9. Schematic illustration of compressive stress–strain curve in a porous material.

The plateau regime after linear elasticity appears in all the compressive stress-strain curves, as shown in Fig. 10. The arrows in Fig. 10 show the strain of the transition point from the plateau to the accelerated stress, ε_1 for the present evaluation. The higher the strain range of the plateau, the larger is the magnitude of energy absorption. When the material is compressed, the glass balloon within the AC3A matrix is compressed. The working during the plateau range may result in the plastic deformation of the matrix and fracture of the glass balloon. However, the balloon is brittle and collapses easily. It may then be harmonized with plastic deformation of the matrix. In fact, the energy absorption efficiency, E, of the AC and AN is approximately 0.8–0.9, as listed in Table 3; these values are excellent as they are similar to those of an open-cell foam material (nearly E = 1). There is a possibility to use this composite material instead of cell foam materials.

The CS materials show lower inclination of linear elasticity and higher yield stress as compared to the AC and AN materials. Since the CS materials exhibited higher Young's modulus and relative density, the linear elasticity may involve the effect of bending of the closed-cell. The glass-balloons were broken into pieces and could be easily deformed so that the matrix containing elongated pores might result in the lower modulus. The absorbed energy, W, of the CS material is considerably higher than that of the cast material, although the energy absorption efficiency is lower. This may be related to the lower inclination of linear elasticity of the CS.

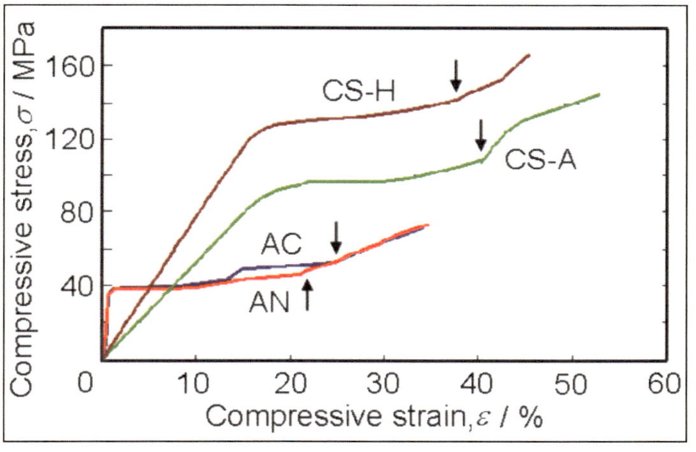

Fig. 10. Stress-strain curves in the compressive test. The arrows indicate the strain, ε_1, for each curve. (Shiga & Umezawa, 2007)

Material	ε_1	σ_0 /MPa	W /MPa	E
AC	0.22	52	9.5	0.8
AN	0.20	46	8.0	0.9
CS-H	0.41	109	31	0.7
CS-A	0.38	143	39	0.7

Table 3. Energy absorption properties. Parameters W and E are the energy absorbed per unit volume as shown in Fig. 9 and the energy absorption efficiency, respectively.

3. Fabrication of glass-balloon-dispersed magnesium alloy composite

The mixture of glass-balloon and AZ31 magnesium alloy chips with 1:1 in volume was heated and extruded to form a brick of composites. The bricks were heat-treated in α phase region, and their microstructure was characterized. As the heating time increased, oxidation of magnesium was accelerated. As heat-treatment temperature was higher the oxidation was also promoted. The heat-treatment condition at 873 K - 1h was chosen to form Mg_2Si. As press load increased, the volume fraction of Mg_2Si was increased.

3.1 Extrusion of magnesium alloy chip and glass balloon mixture

3.1.1 Test materials

Machine chips of AZ31 magnesium alloy and 50 mol% porous glass-balloons were mixed by rocking-mill with 60 rpm for 600 sec. The major compositions of the glass balloon were 68 SiO_2, 6.3 Al_2O_3, 0.6 Fe_2O_3, 0.6 MgO, 9.5 CaO, 11.7 Na_2O, and 1.3 K_2O in mass%, and its diameter was approximately 1 mm. The balloon contains numerous closed bubbles. The specific gravities of the glass balloon and AZ31 are 1.58×10^{-4} and 1.78 g/cm^3, respectively. The mixture was heated in the container, and extruded into cylindrical form with 20 mm in diameter by 100 ton hydraulic press machine under the conditions listed in Table 4.

Sample	Weight /g	Heating temperature /K	Start temperature /K	Finish temperature /K	Pass load /ton
No. 1	15	300	300	300	4
No. 2	15	418	403	393	6
No. 3	15	468	456	451	14
No. 4	18	468	456	448	14
No. 5	18	488	481	472	15
No. 6	18	538	526	522	15
No. 7	18	568	553	544	20
No. 8	18	588	577	567	20
No. 9	18	638	618	608	20
No. 10	18	588	577	567	10
No. 11	18	588	577	567	15

Table 4. The extruded samples from the mixture of AZ31 alloy chip and glass balloon

3.1.2 Macrostructure and microstructure of preform

In appearance a brick of composite was formed into a cylinder at the temperature over 400 K as shown in Fig. 11, but the heating over 573 K was needed to be almost no chipping at the edge of cylinder. Figure 12 shows the macrostructure of the composite in the longitudinal cross section of No.8 sample. The glass-balloon was broken into pieces and dispersed along plastic flow. Density of glass pieces was relatively higher in the vicinity of sample surface, as the sample was extruded at lower temperature.

Figure 13 shows microstructure of the section for No. 9 sample. Grain size was finer in the vicinity of the sample surface and boundary, and pieces of glass balloon were dispersed in the matrix. Elemental analysis by EPMA demonstrated that fragmented glass-balloons were also distributed not only at the particle boundaries but also in the inside of garins as shown in Fig. 14, and that enriched oxygen was detected at the boundaries. Therefore the boundaries resulted from the interface of AZ31 magnesium alloy chips, and fragmented glass-balloons were partially contaminated there.

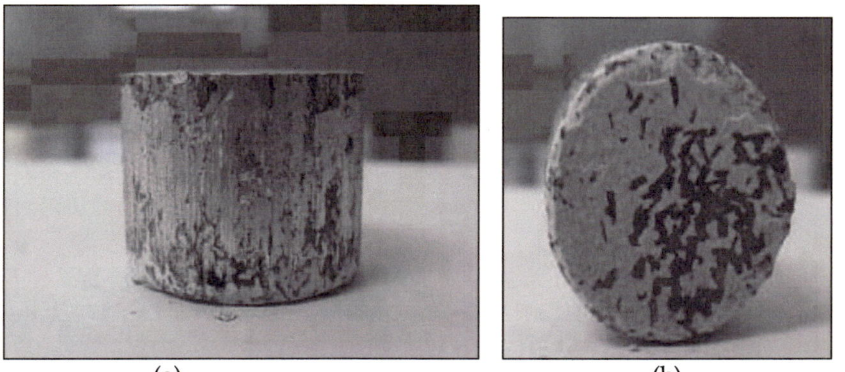

(a) (b)

Fig. 11. Sample side view (22 mm in height) (a) and bottom view (b) of No. 8

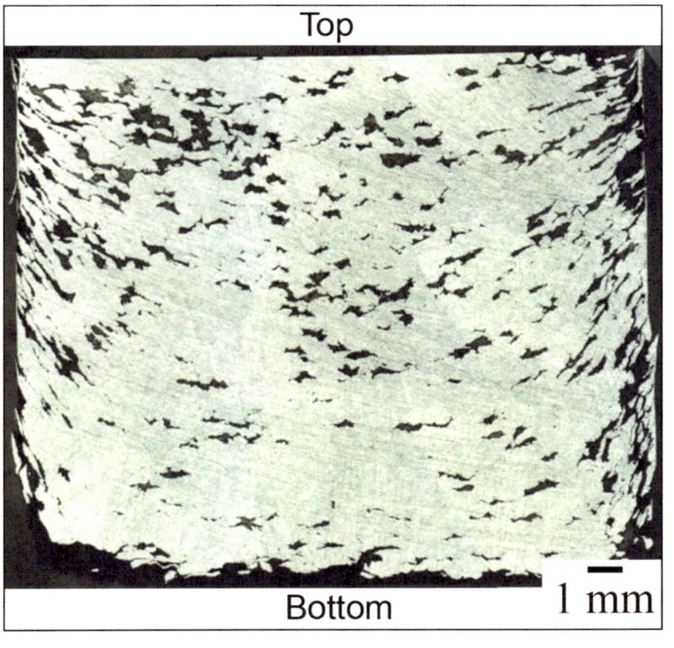

Fig. 12. Macrostructure of No. 8 in the longitudinal section

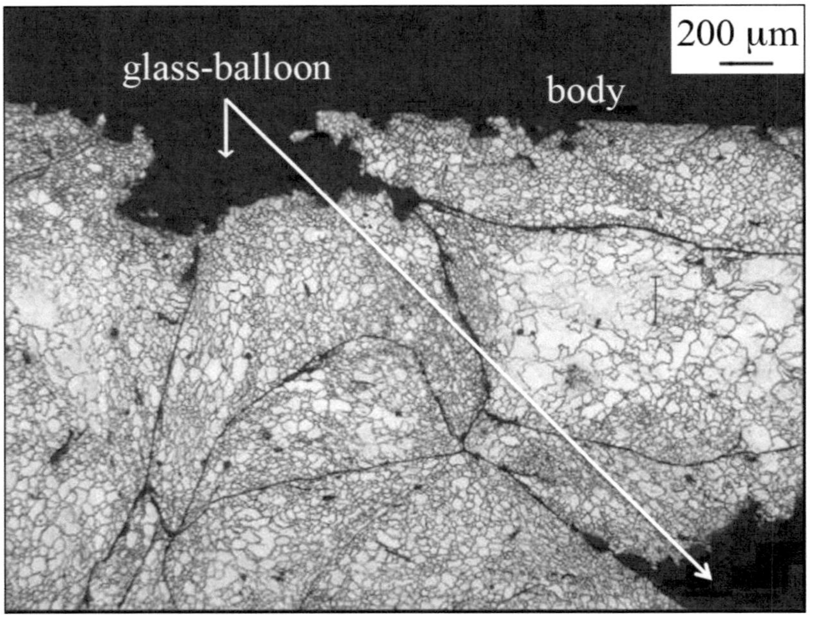

Fig. 13. Microstructure in the cross section of No. 9.

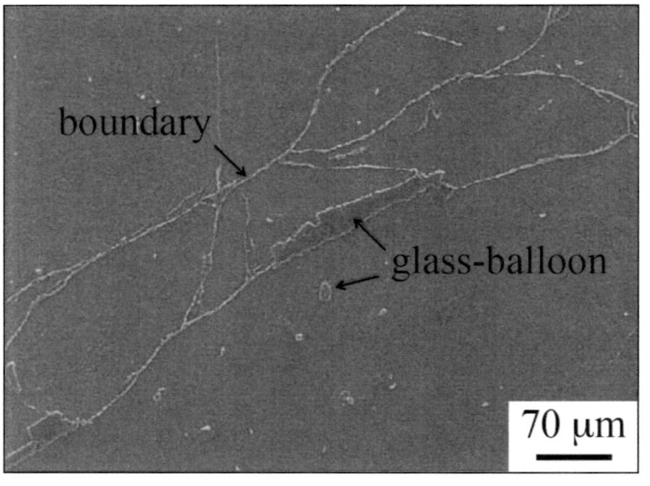

Fig. 14. Secondary electron image in the cross section of No. 8.

3.2 Solid-state synthesis of Mg_2Si and MgO

The DSC (differential scanning calorimeter) curves for the compacts in employing the elemental AZ31 and SiO_2 glass powder mixture expect the exothermic peaks at 923 K due to the latent heat of magnesium melting and at 844 K (SiO_2 glass transition temperature) due to the phase transformation from α to β (Reynard et al., 1996). However, they showed the starting temperature of the exothermic at 730 K (Kondoh & Luagnvaranaunt, 2003c). Therefore, the deoxidizing SiO_2 by magnesium has a large possibility to form Mg_2Si and

MgO by the reaction in eq.(3), although the ignition temperature in DSC curve may depend on the compositions and crystalline in the glass-balloon.

$$4Mg + SiO_2 \rightarrow Mg_2Si + 2MgO \tag{3}$$

3.2.1 Heat treatment of preform

The samples were heat-treated in vacuum at the temperature of 773 K, 823 K, 873 K and 898 K for 1 hr. The heat-treated materials were characterized by X-ray diffraction (XRD) with Cu-Kα_1 radiation operated at 40 kV - 200 mA (range: 20-90 deg, step: 0.02, scan: 2 deg/min) and micro-Vickers hardness (load: 1 kg, holding time: 15 s).

In the sample No. 2, a solid-state synthesis was available at the temperature heating over 873 K, and not at lower temperatures due to oxidation of magnesium chips. The isothermal treatment at higher temperature may cause a partial remelting of magnesium α phase.

Then the No. 2 samples were heated at the temperatures of 873 K and 898 K for longer time to identify the reacted phases. Figure 15 summarized the estimated volume fractions of crystalline phases formed by isothermal heating, which were calculated from the integrated intensity of peaks in XRD profiles. As the heating time increased, the oxidation of magnesium was accelerated as well as deoxidization of SiO_2. As heat-treatment temperature was higher, the oxidation was also promoted. As a result, the heat-treatment condition at 873 K for 1 h was chosen to form Mg_2Si with less MgO.

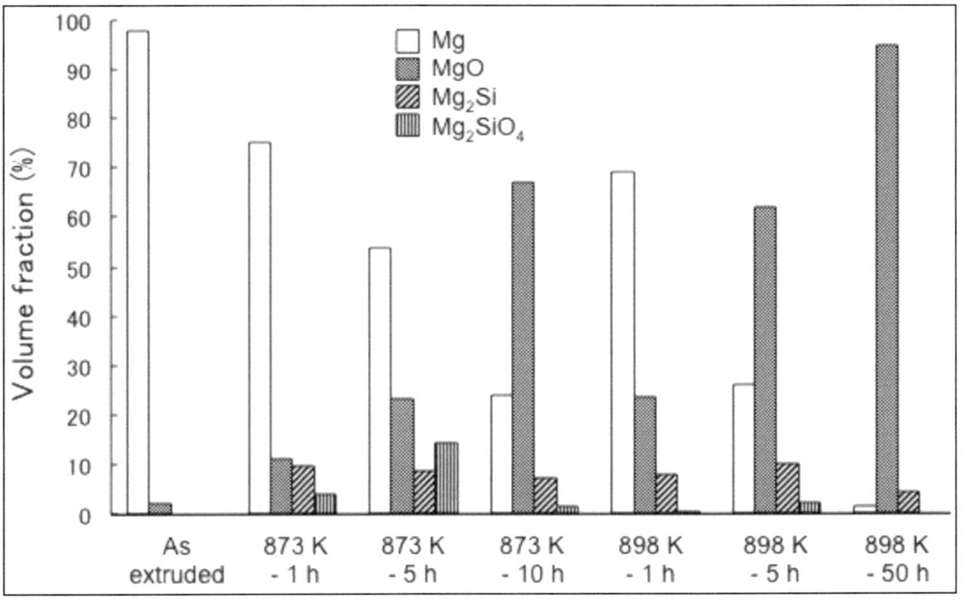

Fig. 15. Effect of heat treatment on estimated volume fractions of crystalline phases for No.2 sample

3.2.2 Effect of press load on preform

The samples extruded at about 573 K, i.e. No. 8, 10 and 11, were pressed with different load from 10 to 20 ton, so that the higher press load resulted in finer dispersion of glass balloons and higher adhesion of matrix. Figure 16 showed the estimated volume fractions of crystalline phases after heating at 873 K for 1 h. The less deoxidization of SiO_2 may be due to smaller faction of interface between SiO_2 and magnesium, as the lower press load is. The No. 8 sample of pressed with load 20 ton exhibited much higher volume fractions of Mg_2Si and MgO and showed 400 HV after the heat treatment.

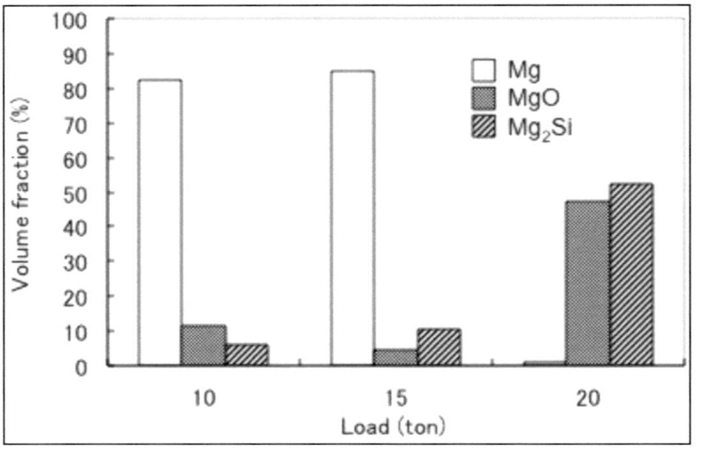

Fig. 16. Effect of press load on the volume fraction of crystalline phases after heating at 873 K for the extruded samples at about 573 K

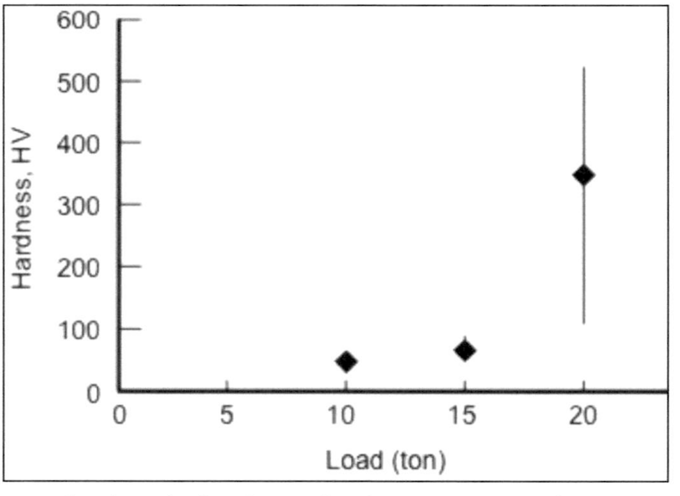

Fig. 17. Effect of press load on the hardness after heating at 873 K for the extruded samples at about 573 K

4. Conclusion

Thermomechanical treatments were adopted to control the microstructure of porous glass-balloon-dispersed aluminium alloy and magnesium alloy composites. The mechanical properties and the microstructure of the composite were characterized.

The porous glass-balloon-dispersed AC3A aluminium alloy composite, which was subjected to the repeated thermomechanical treatment along with the working by either swaging or flat rolling, was successfully formed into a rod or sheet. Microstructural modification by the treatment avoided early fracture due to tension and resulted in good ductility. The swaged material exhibited higher tensile strength and Young's modulus as compared to the cast material. A stress concentration around the glass balloons was confirmed. The stress may lead to fracture. The composite materials also exhibited excellent energy absorption properties. The swaged material showed lower inclination of linear elasticity and higher yielding stress as compared to the cast material.

A new process for fabricating magnesium composites with a solid-state reaction by glass balloons has been proposed. The mixture of glass balloons and AZ31 magnesium alloy chips was thermomechanically treated to form their composite and applied to solid-state synthesis of Mg_2Si and MgO phases. It demonstrated a possibility to employ SiO_2 glass balloons to fabricate a lightweight magnesium alloy composite.

5. Acknowledgment

The author thanks S. Shiga, Y. Kosuge and Dr. S. Morooka for their help on the experimental works.

6. References

Cao, J., Rambo, C.R. & Sieber, H. (2004a), Preparation of porous Al_2O_3-Ceramics by biotemplating of wood, *Journal of Porous Material*, Vol. 11, No. 3, (July 2004), pp. 163-172, ISSN 1380-2224

Cao, J., Rusina, O. & Sieber, H. (2004b), Processing of porous TiO_2-ceramics from biological preforms, *Ceramics International*, Vol. 30, No. 7, (2004), pp. 1971-1974, ISSN 0272-8842

Chou, S.-N., Huang, J.-L., Li, D.F. & Lu, H.-H. (2007), The mechanical properties and microstructure of Al2O3/aluminum alloy composites fabricated by squeeze casting, *Journal of Alloys and Compounds*, Vol. 436, No. 1-2, (June 2007), pp. 124-130, ISSN 0925-8388

Clark, C.R., Wright, C., Suyanarayana, C., Baburaj, E.G. & Froes, F.H., Synthesis of Mg_2X (X = Si, Ge, or Sn) intermetallics by mechanical alloying, *Materials Letters*, Vol. 33, No. 1-2, (November 1997), pp. 71-75, ISSN 0167-577X

Gibson, L.J. & Ashby, M.F. (1999), *Cellular Solids: Structure and Properties*, 2nd ed., Cambridge University Press, Cambridge, UK, (July 1999), ISBN 9780521499118

Kashiwaya, K., Kasetani, N., Nagao, K. & Mitani, H. (1999), New recyclable material composed of aluminium & glass-balloon: part 1 general properties, *Summaries of technical papers of Annual Meeting Architectural Institute of Japan A-1 Materials & construction*, (July 1999), pp. 545-546, ISSN 1341-4437

Kondoh, K., Oginuma, H., Tsuzuki, R. & Aizawa, S. (2003a), Magnesium matrix composite with solid-state synthesized Mg_2Si dispersoids, *Materials Transactions*, Vol. 44, No. 4, (April 2003), pp. 611-618, ISSN 1345-9678

Kondoh, K., Oginuma, H., Kimura, A., Matsukawa, S. & Aizawa, S. (2003b), *In-situ* synthesis of Mg_2Si intermetallics via powder metallurgy process, *Materials Transactions*, Vol. 44, No. 5, (May 2003), pp. 981-985, ISSN 1345-9678

Kondoh, K. & Luagnvaranaunt, T. (2003c), New process to fabricate magnesium composites using SiO_2 glass scraps, *Materials Transactions*, Vol. 44, No. 12, (December 2003), pp. 2468-2474, ISSN 1345-9678

Mabuchi, M., Kubota, K. & Higashi, K. (1995), New recycling process by extrusion for machined chips of AZ91 magnesium and mechanical properties of extruded bars, Materials Transactions JIM, Vol. 36, No. 10, (October 1995), pp. 1249-1254, ISSN 1345-9678

Naigai Building Materials Corporation, Alcelite, Available from http://www.naigai-technos.co.jp/kenzai/english/k_products.html

Nicom, N. & Nomura, H. (2006), Melt infiltration of SiCp reinforced Al matrix composite by newly designed pressure infiltration technique, *Materials Science & Engnieering A*, Vol. 441, No. 1-2, (December 2006), pp. 97-105, ISSN 0921-5090

Rambo, C.R., Cao, J. & Sieber, H. (2004), Preparation and properties of highly porous, biomorphic YSZ ceramics, *Material Chemistry Physics*, Vol. 87, No. 2-3, (August 2004), pp. 345-352, ISSN 0254-0584

Rambo, C.R., Cao, J., Rusina, O. & Sieber, H. (2005), Manufacturing of biomorphic (Si, Ti, Zr)-carbide ceramics by sol–gel processing, *Carbon*, Vol. 43, No. 6, (May 2005), pp. 1174-1183, ISSN 0008-6223

Reynard, B., Takir, F., Guyot, F., Gawanmesia, G.D., Liebermann, R.C. & Gillet, P. (1996), High-temperature Raman spectroscopic and X-ray diffraction study of β-Mg_2SiO_4: Insights into its high-temperature thermodynamic properties and the β to α phase-transformation mechanism and kinetics, *American Mineralogist*, Vo. 81, No. 5-6, (June 1996), pp. 585-594, ISSN 0003-004X

Rodriguez-Castro, R. (2002), Microstructure and mechanical behavior of functionally graded Al A359/SiCp composite, *Materials Science & Engineering A*, Vol. A323, No. 1-2, (January 2002), pp. 445-456, ISSN 0921-5090

Seah, K.H.W. and Hemanth, J. (2007), Cryogenic effects during casting on the wear behavior of aluminum-alloy/glass MMCs, *Composites: Part A*, Vol. 38, No. 5, (May 2007), pp. 1395-1402, ISSN 1359-835X

Shiga, A. & Umezawa, O. (2007), Effects of thermo-mechanical treatment on the tensile and compressive properties of a glass-balloon-dispersed aluminum alloy composite, Materials Transactions, Vol. 48, No. 12, (December 2007), pp. 3088-3094, ISSN 1345-9678

Shorowordi, K.M., Laoui, T., Haseeb, A.S.M.A., Celis, J.P. & Froyen, L. (2003), Microstructure and interface characteristics of B_4C, SiC and Al_2O_3 reinforced Al matrix composites: a comparative, *Journal of Material Process & Technology*, Vol. 142, No. 3, (December 2003), pp. 738-743, ISSN 0924-0136

Umezawa, O. & Nagai, K. (1999), Microstructural refinement of as cast Al-12.6wt%Si alloy by repeated thermomechanical treatment to produce a heavily deformable material, *Metallurgical Materials Transactions A*, Vol. 30A, No. August, (August 1999), pp. 2221-2228

Wilkes, T.E., Young, M.L., Sepulveda, R.E., Dunand, D.C. & Faber, K.T. (2006), Composites by aluminum infiltration of porous silicon carbide derived from wood precursors, *Scripta Materialia*, Vol. 55, No. 12, (December 2006), pp. 1083-1086, ISSN 1359-6462

Yamada, T. (2006), Iterative algorithms for computing the averaged response of nonlinear composites under stress-controlled loadings, *International Journal for Multiscale Computational Engineering*, Vol. 4, No. 4, (2006), pp. 475-486, ISSN 1543-1649

Yu, X.X. & Lee, W.B. (2000), The design and fabrication of an alumina reinforced aluminum composite material, *Composites: Part A*, Vol. 31, No. 3, (March 2000), pp. 245-258, ISSN 1359-835X

5

Inversion of Physical Properties for Determining the Microstructure of Natural Composites

Kazatchenko E., Markov M., Mousatov A. and Pervago E.
Instituto Mexicano del Petróleo
Mexico

1. Introduction

The microstructure of sedimentary rocks depends on their generation processes. Ninety five percent of the sedimentary rocks are clastic and carbonate formations. Clastic rocks are generated during weathering of crystalline rocks, transportation, sorting by size, etc. In contrast to the clastic formations, carbonate rocks are composed during a sedimentary deposition of organic origin or chemical precipitations. As a result of different grain shapes and posterior processes of diagenesis, carbonate formations have a complicated double-porosity microstructure. Because these sedimentary rocks are potentially hydrocarbon deposits, the determination of their porosity, saturation and permeability is required for the oil and gas field exploration and correct reserve assessment. To evaluate formation petrophysical characteristics, the interpretation of the well logging data that contain the measurements of different physical properties in boreholes is fundamental.

Geological formations, especially sedimentary rocks, can be treated as a natural composite material that consists of solid grains of different minerals and system of pores saturated with fluids (water, oil, and gas). In comparison with the synthetic-material characterization, the study of such a natural composite includes one specific additional problem that is recognition of the composite microstructure (component concentrations, shapes, spatial distributions, connectivity, size, etc) from its effective physical properties. In this chapter we present a method for characterizing the pore system structure of double porosity carbonates by using the joint inversion of well logging data. This inversion, which we have called petrophysical inversion, is based on the concept of unified microstructure model and micromechanical self-consistent method for simulating the natural-composite effective physical properties such as: elastic module, electrical conductivity, dielectric constant and thermal conductivity. We present and discuss an example of the joint nonlinear multimethod inversion of well log data for petrophysical characterization of double porosity carbonate formations.

2. Unified model for double – porosity medium

Generally, carbonate formations comprise two pore systems with pores of different scales, spatial distribution and shapes. According to numerous investigations (Choquette and Pray, 1970; Lucia, 1999) the carbonate porosity can be treated as a double porosity medium comprising two pore components: primary small-scale pores and secondary large-scale pores. Some examples of the carbonate rocks (Fig. 1) show existence of the complicated fracture systems generated during tectonic processes and presence of strong vug porosity. The saturating fluid flow provokes vug generation through the material dissolution in the fracture intersection or organic forms. The typical size of primary (intergranular, intercrystalline) pores does not exceed 30-50 μm while the characteristic length of secondary pores such as vugs and cracks (microfractures) are in the range of 0.5 - 50 mm (Bagrintseva, 1999). Currently the double porosity concept for modeling of rock physical properties is widely used (Barrenblatt et al., 1960; Wilson and Aifantis, 1982; Auriault and Boutin, 1994; Berryman and Wang, 2000).

The scale of the macroscopic medium description (determination of the effective macroscopic properties) corresponds to the characteristic length of well logging investigations, (0.1-0.15 m) and it is less than wavelengths of the acoustic log (0.4-0.5 m).

Since the scale of measurements on rock samples or in a borehole is much larger than the size of rock components (grains and all type pores) such a medium can be treated as a microheterogeneous material characterized by the effective physical properties. Because of the ratios between different pore scales and macroscopic scale we apply the hierarchical two step homogenization approach proposed by Auriault and Boutin (1994), and Auriault and Royer (2002).

Fig. 1. Examples of the carbonate rocks.

We considered the solid frame composed by rock grains and primary pores as a matrix with effective physical parameters (non -zero conductivity and non-zero shear modulus). These matrix properties correspond to the effective properties obtained after homogenization of the heterogeneous porous frame. Then we have used this homogeneous matrix as a host in which the inclusions of different sizes and shapes representing the secondary pores are embedded. In this model we assume that all pores are completely saturated with water and are randomly distributed and oriented. Therefore the matrix and the effective double-porosity medium are treated as isotropic media. The shapes of grains, matrix and secondary pores are approximated by three-axial ellipsoids. By varying the ellipsoid aspect ratios we have modeled different secondary porosity types as cracks, vugs, and channels.

3. Effective medium methods

The micromechanical effective medium methods developed for composite materials have a wide application for simulation of rock properties. These methods are based on the solution of one-particle problem for a single inclusion placed in some effective host and take into account the inclusion interactions for high concentrations. Generally two basic approaches are used for the modeling of rock geophysical parameters: the effective media approximation (EMA) and differential effective medium (DEM) methods (Budiansky and O'Connell, 1976; Berryman, 1980; Sen et al., 1981; Feng and Sen, 1985; Norris, 1985; Berge et al., 1993; Zimmerman, 1991; Hornby et al., 1994; Pribnow and Sass, 1995; Le Ravalec and Gueguen, 1996; Kazatchenko et al., 2004a, b).

The DEM method based on the sequentially repeated iterative solution of the homogenization problem works well for two-component materials in which the first constituent is accepted as a host and the second one represents inclusions placed in this host (Berryman, 1980; Norris, 1985).

However, in the case of simulating different physical properties from the unified microstructure model, the introduction of any model component as the initial host leads to serious contradictions. For example, the selection of the conductive fluid (water) saturating pore components as the host guarantees the existence of electrical conductivity for small porosities but does not allow the effective shear modulus to be calculated (Sen et al., 1981; Mendelson and Cohen, 1982; Norris et al., 1985). The host corresponding to high-resistive solid-grains inclusions (second model component) does not provide the correct simulation of the effective electrical conductivity for the low porosity range. That is why for calculating effective physical properties of geological formations we have used the symmetrical EMA method that treats equally all N components with no one distinguished as a host (Berryman, 1980, 1992; Mavko et al., 1998).

This method allows us to calculate the effective physical properties from the unified multicomponent model composed of solid grains and fluid-filled primary and secondary pores. The elements of each component are approximated by three-axis ellipsoids with different aspect ratios.

Equations for the elastic properties were obtained by Korringa et al. (1979), Berryman (1980), and Norris (1985)

$$\sum_{i=1}^{N} C_i (L_i - L) T^{(i)} = 0 \qquad (1)$$

where C_i is a volumetric concentration of the i-th component ($\Sigma C_i=1$), L is the elastic tensor of the effective medium, L_i is the elastic tensor of i-th component, and $T^{(i)}$ is Wu's tensor that relates the strain tensor inside an individual element of i-th component with the uniform stain field far from it (Wu, 1966).

For a medium composed of arbitrary distributed isotropic elements, the components of the tensor L_i are defined as $L_{jklm} = K_i \delta_{jk} \delta_{lm} + \mu_i (\delta_{jl} \delta_{km} + \delta_{jm} \delta_{kl} - 2\delta_{jk} \delta_{lm} / 3)$, where K_i and μ_i are the bulk and shear moduli of i-th component, δ_{ij} is Kronecker delta. In this case the system (1) can be written as:

$$\sum_{i=1}^{N} C_i (\mu_i - \mu) Q_i = 0, \ \sum_{i=1}^{N} C_i (K_i - K) P_i = 0 \qquad (2)$$

where K and μ are the bulk and shear moduli of effective medium respectively, $P_i = T^{(i)}_{jjll} / 3$ and $Q_i = \left(T^{(i)}_{jjll} - P_i \right) / 5$ in these equations, repeated subscripts are summed.

For the ellipsoidal inclusion Wu´s tensor is

$$T^{(i)} = \left[I + S^{(i)} L^{-1} (L_i + L)\right]^{-1} \tag{3}$$

where I is the fourth-order isotropic identity tensor, and $S^{(i)}$ is Eshelby's tensor (Eshelby, 1957). The components of Eshelby's tensor $S^{(i)}$ for the ellipsoidal inclusions are given by the series of expressions (Eshelby, 1957). When the ellipsoid semi-axes satisfy the inequality $a_1>a_2>a_3$ these expressions are:

$$S_{1111} = \frac{3}{8\pi(1-\nu)} a_1^2 J_{11} + \frac{1-2\nu}{8\pi(1-\nu)} J_1 ,$$

$$S_{1122} = \frac{1}{8\pi(1-\nu)} a_2^2 J_{12} + \frac{1-2\nu}{8\pi(1-\nu)} J_1 , \tag{4}$$

$$S_{1212} = \frac{1}{16\pi(1-\nu)} (a_1^2 + a_2^2) J_{12} + \frac{1-2\nu}{16\pi(1-\nu)} (J_1 + J_2)$$

where υ is Poisson's coefficient of effective medium, J_j , J_{ik} are integrals given by

$$J_1 = \frac{4\pi a_1 a_2 a_3}{(a_1^2 - a_2^2)\sqrt{(a_1^2 - a_3^2)}} \{F(\theta,k) - E(\theta,k)\} ,$$

$$J_2 = \frac{4\pi a_1 a_2 a_3}{(a_2^2 - a_3^2)\sqrt{(a_1^2 - a_3^2)}} \left\{ \frac{a_2\sqrt{(a_1^2 - a_3^2)}}{a_1 a_3} - E(\theta,k) \right\} ,$$

$$J_1 + J_2 + J_3 = 4\pi , \quad 3J_{11} + J_{12} + J_{13} = 4\pi / a_1^2 ,$$

$$3a_1^2 J_{11} + a_2^2 J_{12} + a_3^2 J_{13} = 3J_1 , \quad J_{21} = (J_2 - J_1) / (a_1^2 - a_2^2) .$$

In these expressions $F(\theta,k)$ and $E(\theta,k)$ are the elliptic integrals of the first and second kind where $\theta = \arcsin\left\{\left[(a_1^2 - a_3^2)/a_1^2\right]^{1/2}\right\}$ and $k = \left[(a_1^2 - a_2^2)/(a_1^2 - a_3^2)\right]^{1/2}$.

We used Carlson's algorithm (Carlson, 1979) to calculate the elliptical integrals $F(\theta,k)$ and $E(\theta,k)$. The components of the Wu's tensor $T^{(i)}$ (Wu, 1966) were calculated numerically using equations (4).

Electrical conductivity equations for isotropic medium with ellipsoidal components were obtained by Sen et al. (1981) and Norris et al. (1985)

$$\sum_{i=1}^{N} C_i(\sigma_i - \sigma) R_i = 0 \tag{5}$$

where $R_i = D_{kk} / 3$, σ_i and σ are electrical conductivities of the i-th component and effective medium, respectively. The tensor $D^{(i)}$ is analog of the Wu's tensor for the electromagnetic field.

$$D_i = \left[I + n^{(i)}\sigma^{-1}(\sigma_i - \sigma)\right]^{-1} \tag{6}$$

where $n^{(i)}$ is depolarization factor of the i-component (Stratton, 1941). The depolarization factor has only three none zero components $n_{kk}^{(i)}$. Using the appropriate ellipsoidal coordinate system (Landau and Lifshitz, 1960) these components can be determined as following (index i of the constituents is omitted)

$$n_{kk} = \frac{a_1 a_2 a_3}{2}\int_0^{\infty} \frac{ds}{(s + a_k^2)\sqrt{(s + a_1^2)(s + a_2^2)(s + a_3^2)}} \tag{7}$$

and calculated using the elliptic integrals $F(\theta,k)$ and $E(\theta,k)$

$$n_{11} = \frac{a_1 a_2 a_3}{(a_1^2 - a_2^2)\sqrt{(a_1^2 - a_3^2)}}\{F(\theta,k) - E(\theta,k)\},$$

$$n_{22} = \frac{a_1 a_2 a_3}{(a_2^2 - a_3^2)\sqrt{(a_1^2 - a_3^2)}}\left\{\frac{a_2\sqrt{(a_1^2 - a_3^2)}}{a_1 a_3} - E(\theta,k)\right\}, \tag{8}$$

$$n_{11} + n_{22} + n_{33} = 1,$$

here a_1, a_2, and a_3 are the ellipsoid semi-axis.
The thermal conductivity and dielectrical permittivity for composite medium are described by equation (5) where σ_i and σ are substituted by λ_i, ε_i and by λ, ε, respectively where λ_i, ε_i are thermal conductivity and dielectrical permittivity of i-th component and λ, ε are thermal conductivity and dielectrical permittivity of effective medium

$$\sum_{i=1}^{N} C_i(\lambda_i - \lambda)R_i = 0, \quad \sum_{i=1}^{N} C_i(\varepsilon_i - \varepsilon)R_i = 0. \tag{9}$$

Equations (2), (5), and (9) are solved numerically by iterations.

4. Simulation of physical properties

By using the unified model and EM method we have modeled the following effective properties of a double porosity medium: acoustic P- and S-wave velocities, electrical conductivity, dielectric permittivity, and thermal conductivity as functions of the primary and secondary porosity, and shapes of secondary pores (aspect ratios). In this section we consider a matrix model and determination of its parameters, and then modeling of the physical properties of a double-porosity medium.

4.1 Matrix model

As it was mentioned, for simulating physical properties of the double porosity formation we apply the hierarchical two step homogenization. At the first step the matrix effective properties are obtained. The matrix is presented by a two-component composite medium of solid grains and pores. The grains of high resistivity form the solid frame. The pores are saturated by a conductive fluid. The elements of each component are approximated by three-axis ellipsoids with different aspect ratios which are introduced as functions of porosity.

The introduction of the varied aspect ratios is based on the supposition that the microstructures of porous rocks (pore-size distribution and its connection network) for high and low porosities are different. The variations of the aspect ratios can be obtained from two classes of experimental data. The first includes data about the porosity and microstructural statistical characteristics measured on cores. The second class (used in this work) presents the acoustic velocities and electrical conductivity from cores or well log data. In this case the aspect ratios are determined by adjusting the calculated parameters (acoustic velocities and electrical conductivity) with experimental data. For acoustic-wave velocities (V_P and V_S) we applied the generalized regression equation for a limestone formation saturated with water for porosities from 0.01 to 0.32 (Mavko et al., 1998).

$$V_{Pm} = 5.62 - 6.65\phi_m\,,\quad V_{Sm} = 3.05 - 3.87\phi_m\,,\quad \rho_m = 2.72(1-\phi_m) + \rho_f\phi_m \tag{10}$$

where V_{Pm} and V_{Sm} are matrix P- and S-wave velocities (km/s), ρ_m is matrix bulk density (g/cm^3), ϕ_m is the matrix porosity, and ρ_f is the bulk density of a saturating fluid (that in the considered case is equal to 1 g/cm^3).

The relationship of the electrical conductivity to a primary porosity for the carbonate formations that has no secondary porosity as microfractures and vugs, was obtained by Kazatchenko and Mousatov (2002) based on the statistical analysis of numerous experimental data. The formation factor for this type of rock is described by Archie's law (Archie, 1942)

$$\sigma^* / \sigma_f = \phi^m \tag{11}$$

with the cementation exponent m=2, here σ^* and σ_f are electrical conductivities of a medium and a saturating fluid respectively.

Using equations (2) - (8) for the components of corresponding tensors obtained for ellipsoids, we calculated the P-wave velocity and conductivity as functions of porosity. The minimization of a fitting error between experimental and synthetic data allows determination of the pore and grain geometry (i.e., values of the aspect ratios) over the porosity range of 2-31%. These aspect ratios were used for calculating S-wave velocities and dielectric permittivity to compare them with the regression equation (10) and (12) to further confirm the validity of the model structure obtained. Semi-empirical equation (12) (CRIM) was obtained for homogeneous rocks with one-pore system (Guéguen and Palciauskas 1994)

$$\varepsilon_m = \left[\phi_m\varepsilon_f^{1/2} + (1-\phi_m)\varepsilon_g^{1/2}\right]^2. \tag{12}$$

Saturating water parameters are: ρ_f=1 g/cm^3, K_f=2.25 GPa, electrical conductivity σ_f=1 Ωm, and the ratio of the water and grain conductivities is taken to be 105. The limestone and water dielectrical permittivity are ε_l =7.5 and ε_w=80+i0.05.

First, we assumed constant aspect ratios of pore and grain elements on the all porosity interval (2-31%). For this situation, the best coincidence of theoretical and experimental data is achieved when the aspect ratios of grains and pores are α_{1g}=0.3048, α_{2g}=0.0903 and α_{1p}=0.0803, α_{2p}=0.0252. The matching of the acoustic data is satisfactory, with a relative error less than 2%. However, the conductivity relative error is rather high and corresponds to 30%

in the porosity interval 5-10%. For porosities lower than 5% the predicted conductivity is reduced considerably and tends to the percolation threshold.

The variation of the aspect ratios allows the predicted and experimental data in the porosity interval 3-31% to be adjusted with a high degree of accuracy (Fig. 2 A and B). The relative error for the P-wave velocity is less than 0.05%, and for the conductivity does not exceed 5% for porosities over the range of 4-31%.

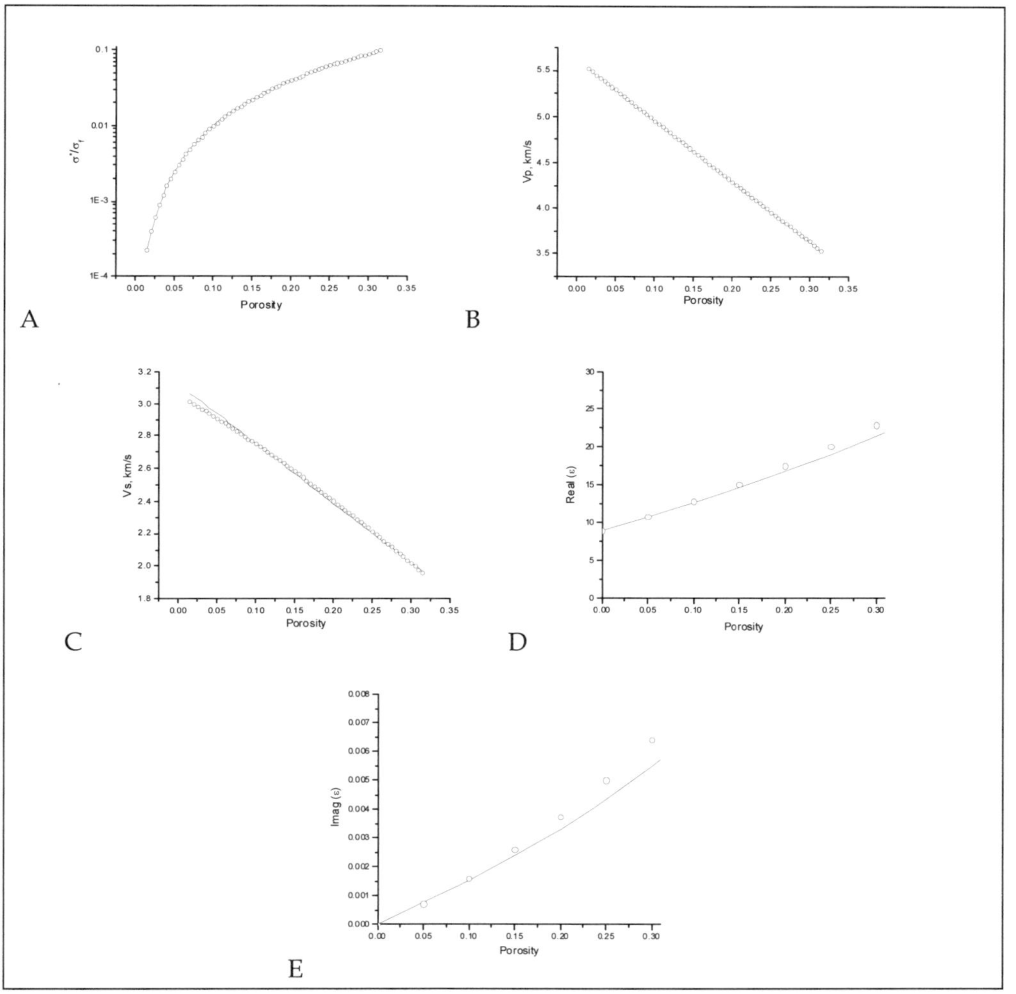

Fig. 2. Results of the matching regression curves described experimental data (solid lines) and simulated properties (circles) for the matrix model. (A) - electrical conductivity; (B) - P-wave velocity; (C) - S-wave velocity ; (D) - real part of the effective dielectric permeability at the frequency of 1 MHz; (E) - imaginary part of the effective dielectric permeability.

The EMA using spheroids with non variable aspect ratios was rarely applied to model electrical properties, because it predicted vanishing conductivity (percolation threshold) at relatively high porosity values. Allowing variation of the pore and grain aspect ratios moves the percolation threshold to lower porosities near 2% in agreement with measured data for carbonate collections.

The calculated S-wave velocities coincide with experimental equation both for constant and for variable aspect ratios (Fig. 2C). The maximum relative errors are 2% and 3% respectively. The dielectrical permittivities obtained with variable shapes of pores and grains are in a good agreement with CRIM equation. Small differences between them increases with a porosity value however it does not exceed 7% for porosity 30% (Fig. 2 D and E).

For high porosities (28-31%) the pores have ellipsoidal form with aspect ratios α_{1p}=0.300 and α_{2p}=0.097. When the porosity decreases from 31% to 2% the ellipsoids transform into the needle-form pores with the small aspect ratios α_{1p}=0.065 and α_{2p}=0.022. This pore geometry maintains the electrical conductivity for low concentrations of conductive component. The solid grain geometry changes from the ellipsoids with α_{1g}=0.408, α_{2g}=0.269, to penny-shapes characterized by α_{1g}=0.984, α_{2g}=0.098. The variations of aspect ratios with porosity are approximated by polynomials of the second power with the correlation coefficient $R^2 > 0.96$

$$\alpha_{1g} = 3.81\phi^2 - 3.70\phi + 1.11\,, \quad \alpha_{2g} = 4.27\phi^2 - 0.86\phi + 1.19\,,$$

$$\alpha_{1p} = 5.40\phi^2 - 0.67\phi + 0.09\,, \quad \alpha_{2p} = 1.56\phi^2 - 0.17\phi + 0.03\,, \tag{13}$$

The predicted variations of shapes for skeleton grains and saturated pores correspond to the recently reported experimental data for carbonate formations on pore microstructure changes depending on porosity (Song et al., 2002).

For the matrix thermal conductivity we have applied the regression equation from Lichtenecker's model (Pribnow and Sass, 1995; Schoen, 1996)

$$\lambda_m = 2.71^{(1-\phi_m)}\lambda_f^{\phi_m} \tag{14}$$

where λ_m and λ_f are thermal conductivities of matrix and saturating water, λ_f =0.63 W/m*K.

4.2 Model of double-porosity rock

On the second step of the hierarchical homogenization we considered the solid frame composed by rock grains and primary pores as a matrix with effective physical parameters (non -zero conductivity and non-zero shear modulus). The secondary pores correspond to inclusions placed into the matrix and they are approximated by three-axial ellipsoids with aspect ratios α_1 and α_2. Selection of the aspect ratios allows us to model different type of secondary porosity such as vugs (close to sphere inclusions), vugs connected by channels (quasi needle shapes), and cracks (flattened ellipsoids). The shape of the matrix component (taking into account that the matrix is an isotropic homogeneous medium) is given by spheres.

We have simulated the compressional and shear wave velocities, electrical and thermal conductivities for two cases: (1) the secondary porosity comprised by one type of pore-shapes (unimodal secondary-porosity shapes) and (2) the secondary porosity consisted of two components with different aspect ratios (bimodal secondary-porosity shapes). The effective physical properties were calculated for different inclusion aspect ratios $\alpha_{1,2}$=1-0.001, and values of the matrix and secondary porosities: ϕ_m=0.02-0.24 and ϕ_s=0.0025-0.08. The dielectrical permittivity was simulated only for the first case.

Because the experimental measurements in double porosity rocks provide the value of total porosity, for quantitative evaluation of the secondary-pore influence it is convenient to calculate deviations between the effective physical properties for two media with the same total porosity $\phi_{t1}=\phi_{t2}$ but different concentrations of the primary and secondary pores. The first medium has only matrix porosity ϕ_m ($\phi_{t1}=\phi_{m1}$), while another one has both the matrix (ϕ_{m2}) and secondary (ϕ_s) porosities. The total porosity of the second medium is $\phi_{t2}=\phi_{m2}+\phi_s-\phi_{m2}\phi_s$. The deviations of the acoustic velocities ΔV_P and ΔV_S, electrical $\Delta\sigma$, thermal conductivities $\Delta\lambda$, and dielectrical permittivity $\Delta\varepsilon$ are $\Delta V_P=V_P-V_{Pm}$, $\Delta V_S=V_S-V_{Sm}$, $\Delta\sigma=\ln(\sigma)-\ln(\sigma_m)$, $\Delta\lambda=\lambda-\lambda_m$, $\Delta\varepsilon=\varepsilon-\varepsilon_m$, where V_P, V_S, σ, λ, and ε are the effective parameters of double porosity medium with the total porosity ϕ_{t2}, V_{Pm}, V_{Sm}, σ_m, λ_m, and ε_m are parameters of a medium with the total porosity ϕ_{t1}.

It should be noted that instead of the fracture density ($\tau = 3\phi / 4\pi\alpha$) widely used in micromechanics (Budiansky B. and O'Connell R. J., 1976), we consider the fracture porosity, because the well logging measurements provide the porosity values. The pore volume is one of the most important characteristics of rocks that determines their capacity of oil and gas storing.

4.3 Secondary-porosity with one type of pore-shapes

We have calculated the differences of the effective elastic, electrical, thermal and dielectrical parameters for secondary porosity ϕ_s=0.02 and two matrix porosities ϕ_m=0.08, 0.18 (ϕ_{t2}=0.1, 0.2). Depending on the aspect ratios, the parameters studied lie below (negative deviations) or above (positive deviations) the corresponding matrix regression equations (Fig. 3). The electrical conductivity has considerable sensitivity to the secondary pore shape and value. For the matrix porosity ϕ_m=0.08 the absolute deviation $\Delta\sigma$ changes from +0.5 for cracks (α_1=1, α_2=0.001) to -0.3 for vugs ($\alpha_1=\alpha_2$=1) that corresponds to the relative sensitivity of +12% and -8% (Fig. 3A). Both cracks and vugs affect the effective electrical conductivity approximately in the same way. However, the relative sensitivity of the electrical conductivity drops up to ±5% when the matrix porosity increases.

Cracks increase and vugs decrease the dielectric permittivity (Fig. 3 B). Its influence depends directly on the value of the secondary pores. $\Delta\varepsilon$ may achieve 5-6% for ϕ_s=0.02 and matrix porosity ϕ_m=0.08. However, for ϕ_m=0.18 the difference $\Delta\varepsilon$ is less than 2%.

The thermal conductivity demonstrates a very low relation with the secondary-pore shapes (Fig. 3 C). The difference $\Delta\lambda$ does not exceed 1% even for model with the ϕ_s=0.02 and ϕ_m=0.08 and the parameter $\Delta\lambda$ decreases with increasing matrix porosity.

Presence of the secondary pores with $\alpha_1<\alpha_2$ results in a decrease of the acoustic velocities with respect to the matrix velocities (Fig. 3 D and E). Strongly flattened ellipsoids with $\alpha_1<<\alpha_2$ (cracks) cause considerable relative deviations of velocities: 30% and 85% for P-wave and S-wave respectively. The sensitivity of the S-waves to cracks is about three times bigger than that of the P waves. The secondary inclusions with $\alpha_1\approx\alpha_2$ slightly increase the effective velocities (in the range of 1%) in comparison with the matrix values. If the secondary porosity remains the same, the increment of the matrix porosity leads to smoothing the secondary porosity effect.

Variations of the ΔV_P, $\Delta\sigma$, and their zero-isolines versus the aspect ratios of the secondary pore are plotted in Fig. 4. We confined our examination of velocity variation to ΔV_P

parameter because ΔV_S versus aspect ratios has the same character as ΔV_P. The $\Delta\lambda$ and $\Delta\varepsilon$ have a small sensitivities to the secondary-pore shapes. When the secondary ellipsoids have aspect ratios α_1 and α_2 bigger than 0.03 the effective electrical conductivity is lower than the matrix conductivity ($\Delta\sigma<0$). Outside this region (Fig. 4), when aspect ratios decrease, i.e. the secondary pores become elongated in one or two dimensions, the electrical conductivity exceeds the matrix one ($\Delta\sigma>0$).

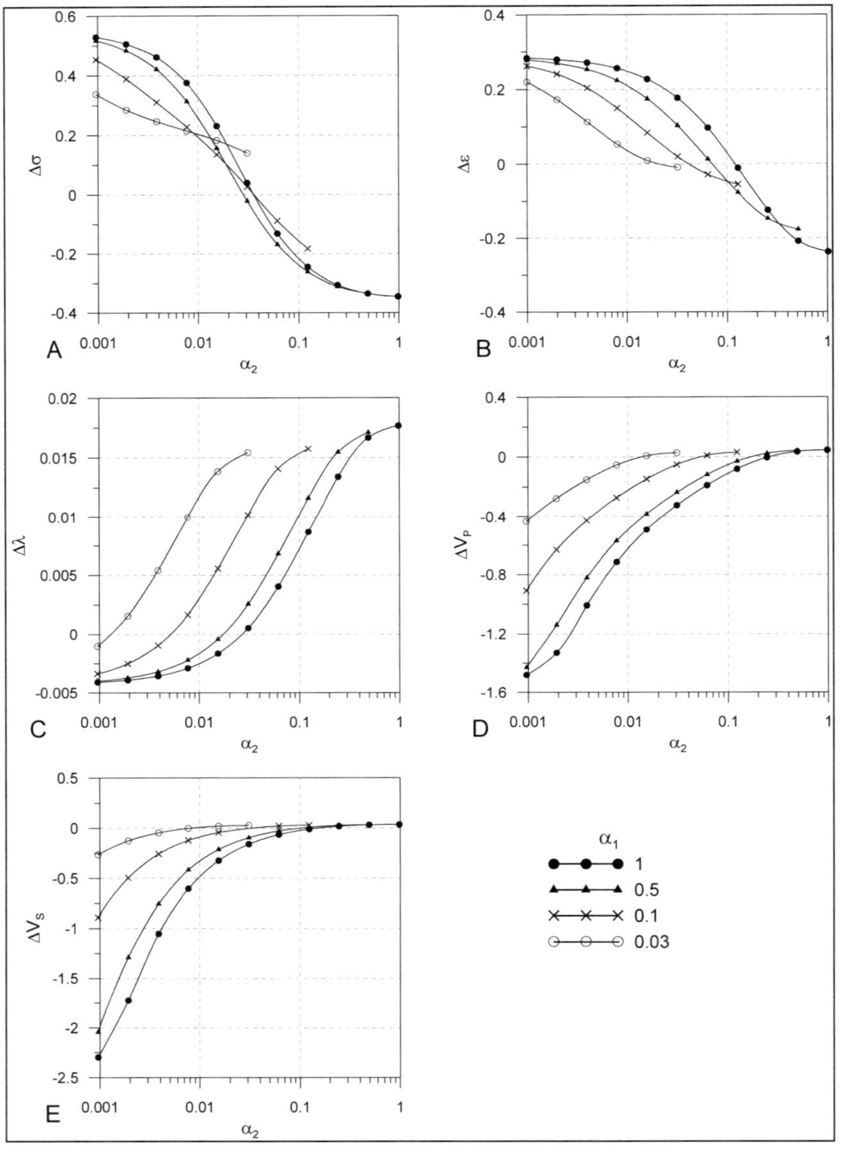

Fig. 3. Deviations of the electric conductivity (A), dielectric permittivity (B), thermal conductivity (C), and P- (D) and S-wave (E) velocities versus inclusion aspect ratios. $\phi_m=0.08$, $\phi_s=0.02$, $\sigma_m=0.006\ \Omega$m. Curve parameter is aspect ratio α_1.

The central area on the plots for P-wave velocity (Fig. 4), which correspond to nearly sphere or prolate spheroids, is characterized by velocities bigger than the matrix one. A medium with such inclusion shapes is more rigid than the matrix with the porosity $\phi_{t1}=\phi_{m1}$ ($\Delta V_P>0$). In two periphery zones, where the secondary-pore shapes approach flattened ellipsoids, there is a reduction of the effective compressional wave velocities with respect to the matrix. Presence of the secondary pores corresponded to the central zone in Fig 4 generally increases the thermal conductivity. However, strongly flattened ellipsoids (α_1=0.03-1, α_2=0.001-0.03 and α_1=0.001-0.03, α_2=0.03-1) make the thermal conductivity lower than that of the matrix.

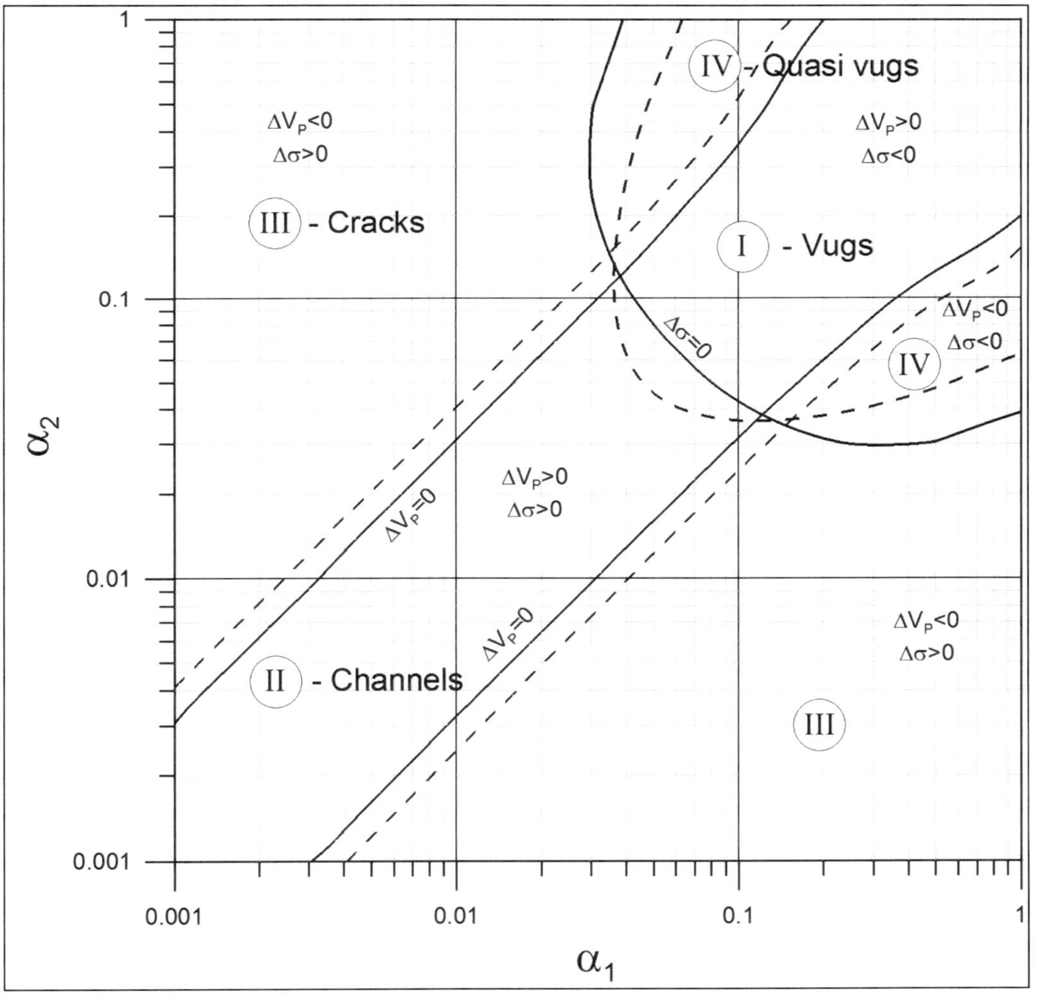

Fig. 4. Classification of the secondary porosity types. Lines correspond to the zeroth isolines of the ΔVP and $\Delta\sigma$ variations for models: solid lines - ϕ_m=0.1, ϕ_s=0.01; dash lines - ϕ_m=0.2, ϕ_s=0.01.

It should be noted that for certain combinations of the α_1, α_2, ϕ_m, and ϕ_s the effective parameters of double porosity medium is equal to the matrix parameters (the zero-isolines in Fig. 4). The distribution of these isolines has a quasi orthogonal behavior and, consequently, the responses of the acoustic and electrical parameters can be considered independent from each one. The elongation of inclusions determines the electrical conductivity changes. Elongated inclusions improve the electrical current propagation but do not affect the elastic module. On the other hand, the elastic velocities are sensitive to the flatness of inclusions (two dimensional elongations). This secondary-porosity type corresponds to cracks or microfractures that make the medium softer and result in significant decrease of the P- and S-wave velocities.

4.4 Classification of the secondary-porosity types

Based on the ΔV_P and $\Delta\sigma$ responses we have divided the inclusion shapes into four general groups that are associated with the specific types of secondary porosity. Due to the weak inclusion effect on the thermal conductivity and dielectric permittivity we do not consider these parameters in the secondary-pore classification. We used the zero-isolines (boundaries where the signs of the ΔV_P and $\Delta\sigma$ deviations are changed) for delimiting the pore-shape groups (Fig 4). Such a pore-shape separation is convenient because the position of sign-change boundaries weakly depends on the values of matrix and secondary porosities.

The first group (I) of inclusions is announced to $\Delta\sigma<0$ and $\Delta V_P>0$. It corresponds to the vuggy porosity. Inclusions from the second group (II), where $\Delta\sigma>0$ and $\Delta V_P>0$ have needle shapes and can be considered as channel porosity (Fig. 4). Flattened ellipsoids with aspect ratios in the area III ($\Delta\sigma>0$ and $\Delta V_P<0$) describe the crack porosity type. Fourth group (IV) is characterized by $\Delta\sigma<0$ and $\Delta V_P<0$. This pore–shape group forms a transition porosity type between vuggy and crack porosities. We call this quasi-vuggy porosity type.

The deviations $\Delta\sigma$, $\Delta\lambda$, ΔV_P, and $\Delta\varepsilon$ as functions of the total and secondary porosity for each porosity types are shown in Fig. 5-9. These graphs were calculated for the following aspect ratios corresponding to vugs ($\alpha_1=\alpha_2=1$), channels ($\alpha_1=\alpha_2=0.01$), cracks ($\alpha_1=1$, $\alpha_2=0.001$), and quasi-vugs ($\alpha_1=1$, $\alpha_2=0.1$). For all pore-shapes considered, the deviations of the physical properties are proportional to the secondary porosity and inversely depend on the matrix porosity. In addition to the general tendency, each parameter has specific features:

a. Electrical conductivity (Fig. 5). As noted earlier, electrical conductivity has a high sensitivity to the presence of secondary pores for all porosity types. However, the sensitivity strongly decreases with increasing matrix porosity. For matrix-porosity value above 0.1-0.15, the effective conductivity rapidly approaches the matrix conductivity and practically coincides with it for $\phi_m>0.2$.
b. Thermal conductivity (Fig. 6). For all porosity types the deviations $\Delta\lambda$ are low and may be neglected in the range of matrix and secondary porosity values considered here.
c. Dielectric permittivity has complicated relations with total and secondary porosities for models with channels and quasi-vugs. These relations result in the increase of the dielectric permittivity for high total porosity (effect of cracks) and the decrease of the dielectric permittivity for low total porosity (effect of vugs) (Fig. 7 B and D).
d. Acoustic wave velocities (Fig. 8 and 9). The acoustic wave velocities have variable sensitivities to the types of secondary porosity. In the cases of vug and channel porosities (Fig. 8 A, B and Fig. 9 A, B) the acoustic velocities are weakly depended on the matrix porosity.

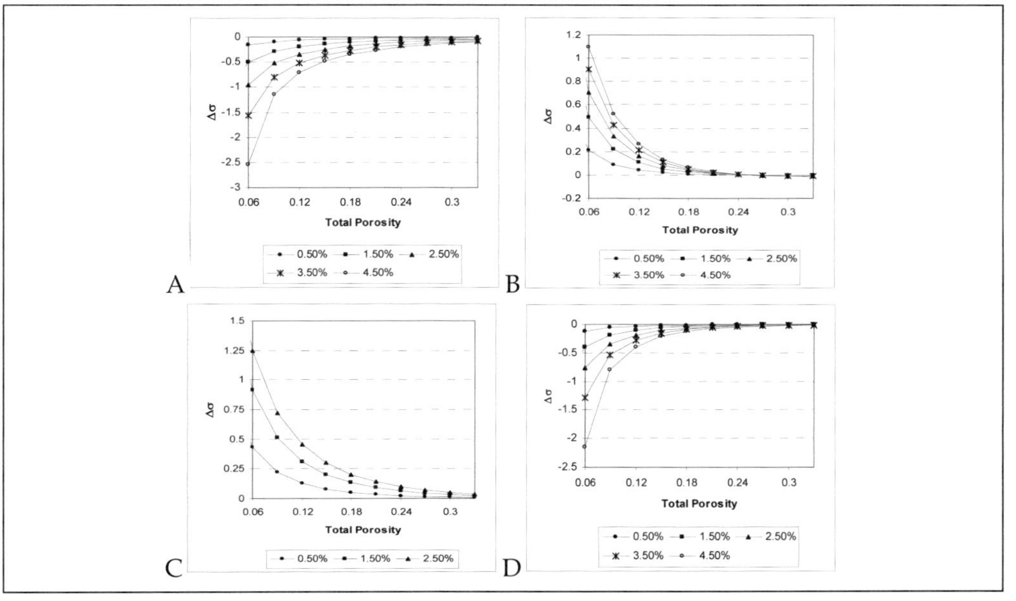

Fig. 5. Influence of secondary- and matrix-porosity values on the electrical conductivity for the media with: A - vugs ($\alpha_1=\alpha_2=1$); B - channels ($\alpha_1=\alpha_2=0.01$); C - cracks ($\alpha_1=1$, $\alpha_2=0.001$); D - quasi-vugs ($\alpha_1=1$, $\alpha_2=0.1$). Curve parameter is value of secondary porosity.

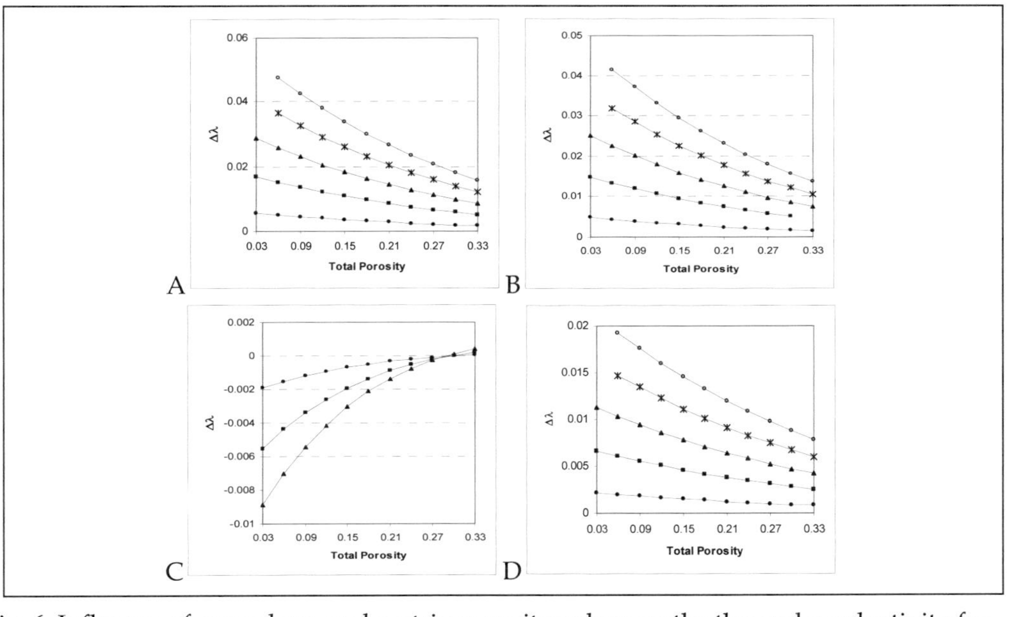

Fig. 6. Influence of secondary- and matrix-porosity values on the thermal conductivity for the media with: A - vugs ($\alpha_1=\alpha_2=1$); B - channels ($\alpha_1=\alpha_2=0.01$); C - cracks ($\alpha_1=1$, $\alpha_2=0.001$); D - quasi-vugs ($\alpha_1=1,\alpha_2=0.1$). Curve parameter is value of secondary porosity.

It should be noted that a direct relation between the ΔV_P and the matrix porosity occur for channel porosity. This fact can be explained by decreasing the effective V_P more slowly than the matrix V_{Pm} with the growth of the matrix porosity. Cracks and quasi-vugs strongly affect the acoustic wave velocities. Small crack concentration (ϕ_s=0.005) leads to significant deviations of the effective velocities from the matrix. The sensitivities of velocities to cracks and quasi-vugs are maintained high in the whole matrix-porosity interval despite of weakly decreasing with matrix porosity increase.

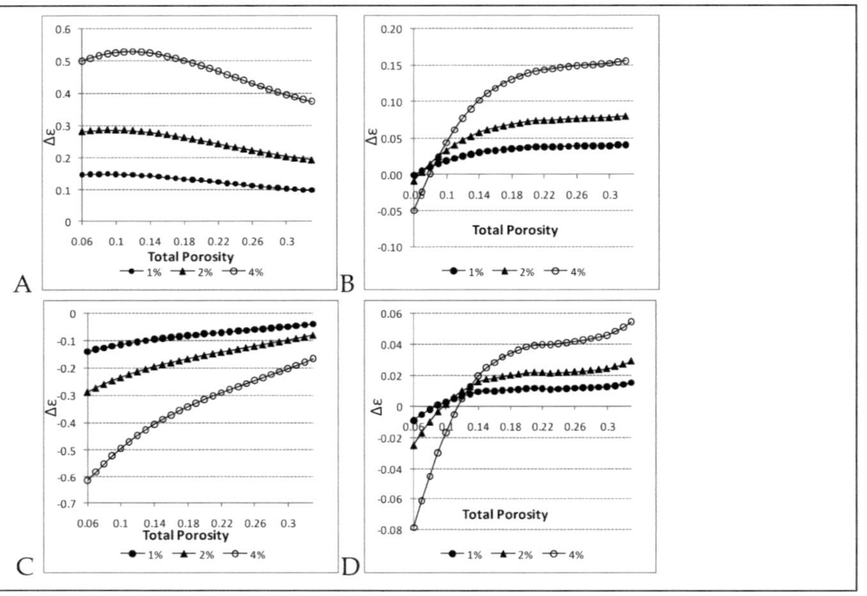

Fig. 7. Influence of the secondary- and matrix-porosity values on the dielectric permittivity for media with: A. - cracks (α_1=1, α_2=0.001); B. - quasi-vugs (α_1=α_2=0.1); C. - vugs (α_1=α_2=1); D. - channels (α_1=α_2=0.01). Curve parameter is value of secondary porosity.

4.5 Secondary porosity composed by the mixture of two pore-shape types

To calculate the effective properties of media with two types of secondary pores we applied the EMA scheme for three-component: matrix represented by elements of spherical shape and secondary pores formed two different groups of ellipsoids.

Small concentrations of vugs in a medium with channel porosity (Fig. 10) or channels in a rock with vuggy porosity (Fig. 11) keep $\Delta\sigma$, $\Delta\lambda$, and ΔV_P close to the deviations corresponding to the dominant porosity type. Presence of vugs reduces the electrical conductivity of rocks with the channel-porosity type and, conversely, introducing channels in vuggy-porosity media enhances electrical conductivity. When the values of channel and vuggy component are the same, the electrical conductivity coincides with the matrix regression equation ($\Delta\sigma = 0$, Fig. 10 A and 11 A). For such porosity-type combinations the variations of the acoustic velocities and thermal conductivity are very small. For acoustic and thermal parameters adding the channels or vugs results in an increase of their deviations (as for single-type porosity) proportionally to the secondary-porosity value.

Addition of a low concentration of cracks in media with dominant vuggy or channel porosity leads to significant changes of the acoustic velocities (Fig. 12 B and 13 B). For example, a crack fraction of 0.005 added to a rock with the 0.08 porosity of vugs or channels is enough to reduce the effective velocities to values lower than the matrix velocities. In this situation the velocity response corresponds to the cracked medium and does not reflect the vuggy or channel porosities. Presence of crack weakly increases the electrical conductivity of the channel or vuggy porosity media (Fig. 12 A and 13 A) and it does not affect the general behavior of $\Delta\sigma$-curves defined by their dominant porosity types. Furthermore, the influence of crack on the conductivity decreases for increasing vuggy or channel porosities. The thermal conductivity response depends mostly on the value of secondary porosity (vuggy or channel) and does not vary when cracks are added.

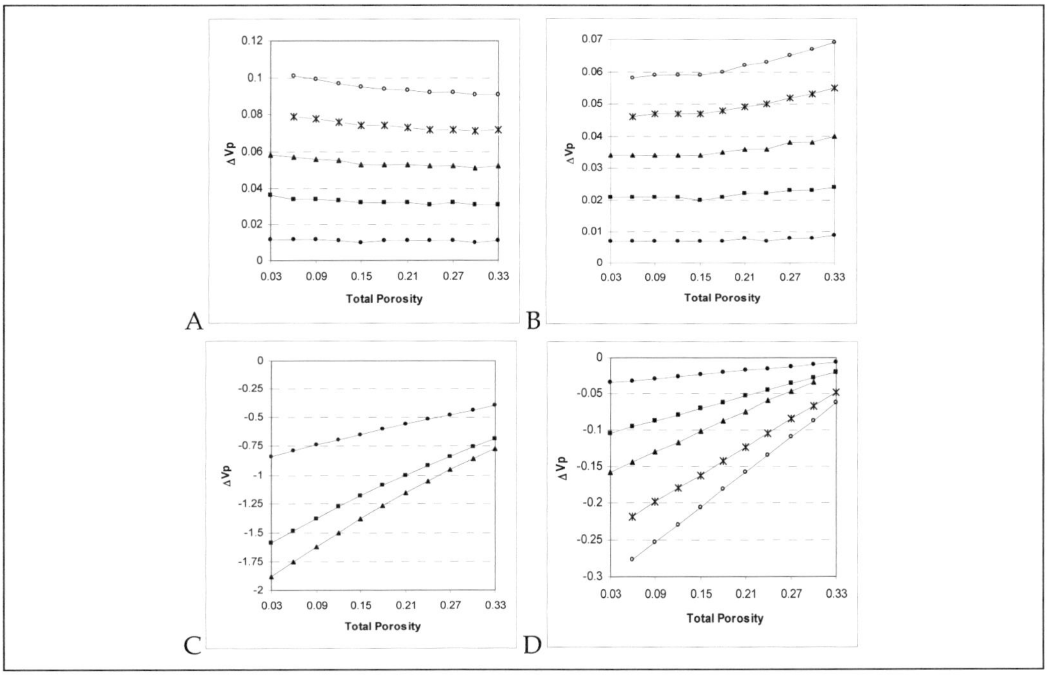

Fig. 8. Influence of secondary- and matrix-porosity values on the P-wave velocity for the media with: A - vugs ($\alpha_1=\alpha_2=1$); B - channels ($\alpha_1=\alpha_2=0.01$); C - cracks ($\alpha_1=1$, $\alpha_2=0.001$); D - quasi-vugs ($\alpha_1=1$, $\alpha_2=0.1$). Curve parameter is value of secondary porosity presented in Fig. 5.

Thus a secondary porosity with two pore-shape components that may occur in real carbonate formations produces specific variations in the effective physical properties and significantly changes the relationship between the electrical and acoustic parameters obtained. When vug porosity is equal to the channel porosity, effective electrical conductivity coincides with the matrix conductivity. In this case, only acoustic velocities indicate the presence of secondary pores ($\Delta V_P>0$). Addition of a low crack concentration in vuggy- or channel-porosity rocks changes completely the velocity response, resulting in characterization of such media as fractured only. In this case electrical conductivity response will indicate the dominant type of secondary pores. A specific combination of secondary-porosity types can lead to the formal equality of one effective parameter to the matrix value, while the response of the other physical property will show the dominant porosity type. Therefore the joint analysis of the acoustic wave velocities and electrical conductivity allow us to estimate correctly the secondary- porosity with one or two types of pore-shapes.

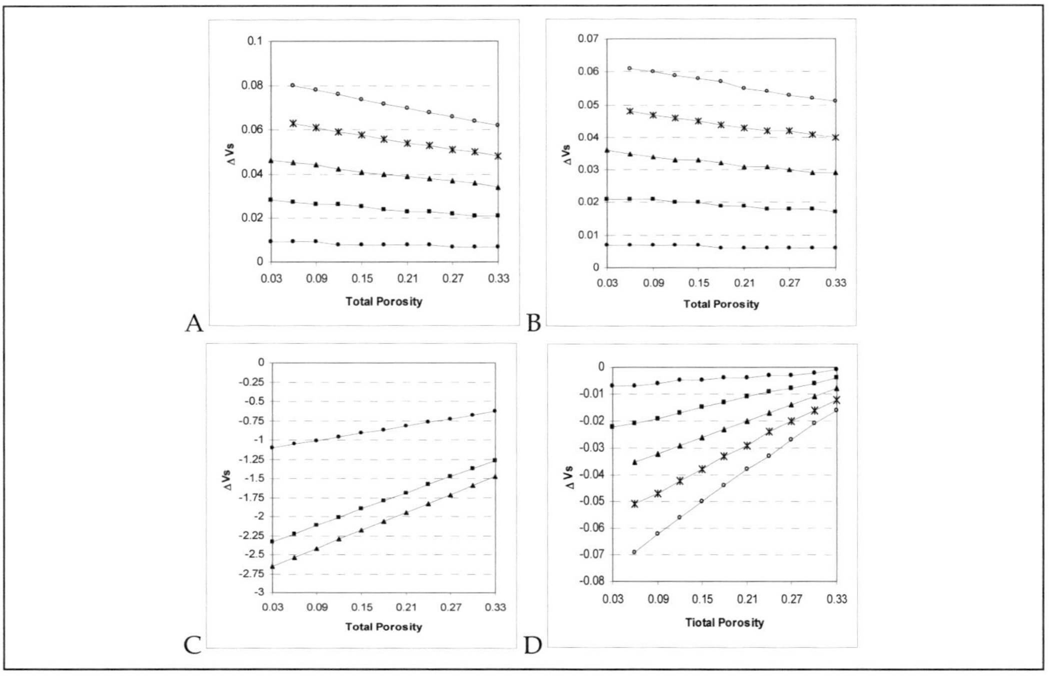

Fig. 9. Influence of secondary- and matrix-porosity values on the S-wave velocity for the media with: A - vugs ($\alpha_1=\alpha_2=1$); B - channels ($\alpha_1=\alpha_2=0.01$); C - cracks ($\alpha_1=1$, $\alpha_2=0.001$); D - quasi-vugs ($\alpha_1=1$, $\alpha_2=0.1$). Curve parameter is value of secondary porosity and it is presented in Fig. 5.

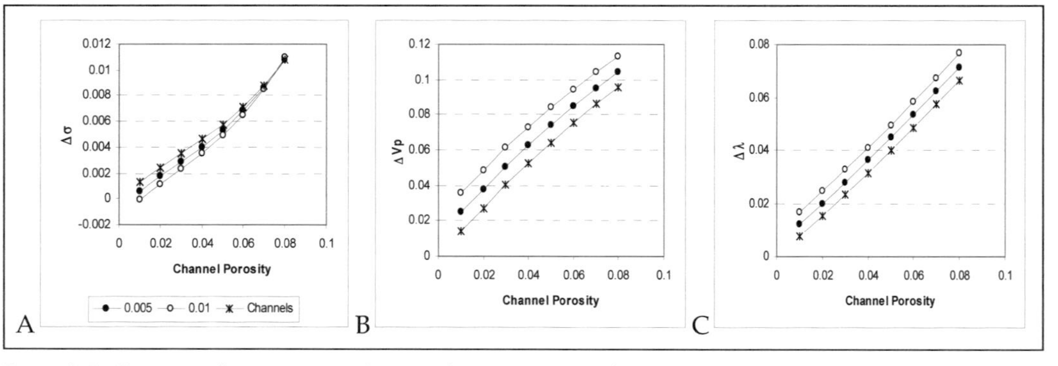

Fig. 10. Influence of two types of secondary pores on the P-wave velocity (B), electrical (A) and thermal (C) conductivities. Stars - channel porosity; solid circles - channel porosity with 0.005 of vugs, open circles - channel porosity with 0.01 of vugs. Channel aspect ratios $\alpha_1=1$, $\alpha_2=100$; vug aspect ratios $\alpha_1=\alpha_2=1$; total porosity $\phi_t=0.1$.

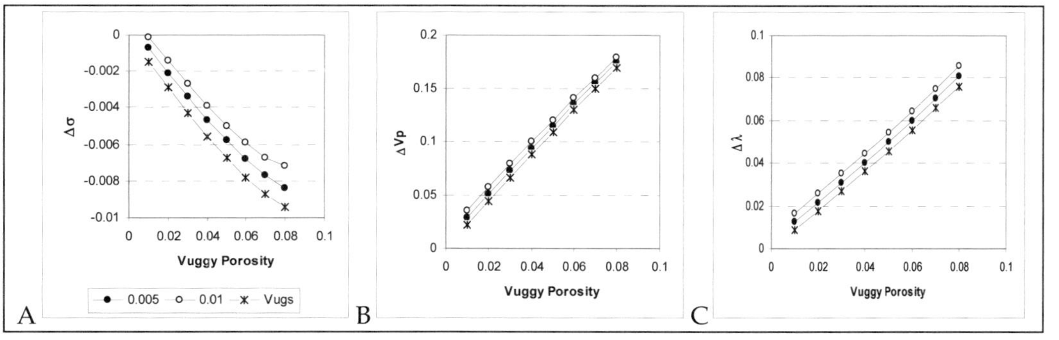

Fig. 11. Influence of two types of secondary pores on the P-wave velocity (B), electrical (A) and thermal (C) conductivities. Stars - vuggy porosity; solid circles - vuggy porosity with 0.005 of channels, open circles - vuggy porosity with 0.01 of channels. Vug aspect ratios $\alpha_1=\alpha_2=1$; channel aspect ratios $\alpha_1=1$, $\alpha_2=100$; total porosity $\phi_t=0.1$.

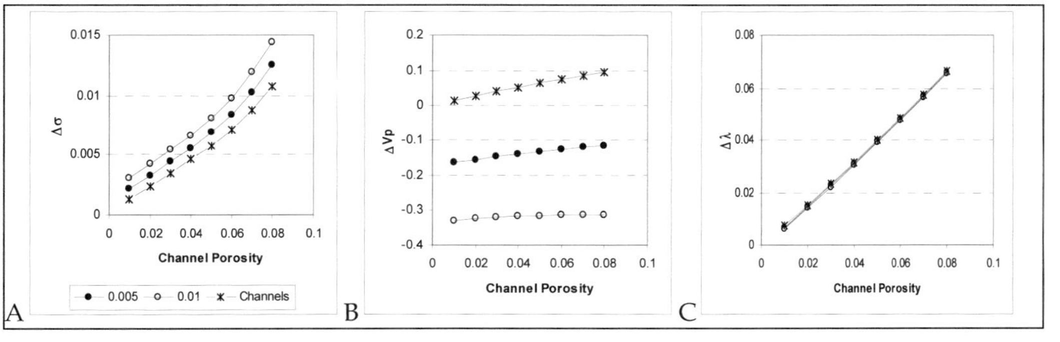

Fig. 12. Influence of two types of secondary pores on the P-wave velocity (B), electrical (A) and thermal (C) conductivities. Stars - channel porosity; solid circles - channel porosity with 0.005 of cracks, open circles - channel porosity with 0.01 of cracks. Channel aspect ratios $\alpha_1=1$, $\alpha_2=100$; crack aspect ratios $\alpha_1=1$, $\alpha_2=0.01$; total porosity $\phi_t=0.1$.

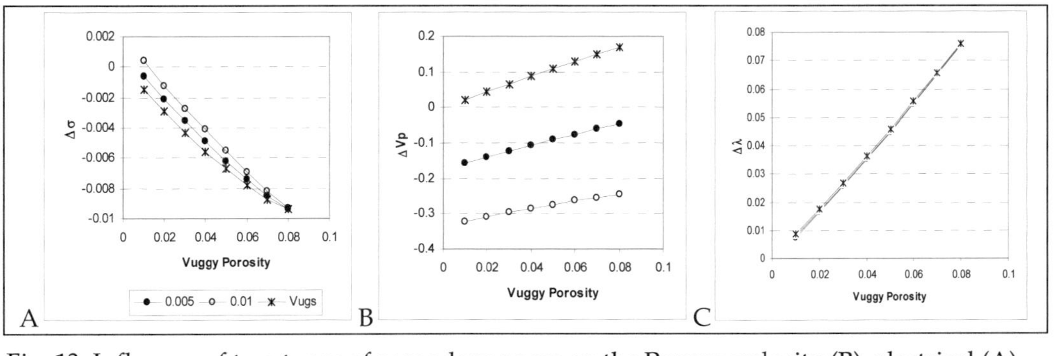

Fig. 13. Influence of two types of secondary pores on the P-wave velocity (B), electrical (A) and thermal (C) conductivities. Stars - only vuggy porosity; solid circles - vuggy porosity with 0.005 of cracks, open circles - vuggy porosity with 0.01 of cracks. Vug aspect ratios $\alpha_1=\alpha_2=1$; crack aspect ratios$\alpha_1=1$, $\alpha_2=0.01$; total porosity $\phi_t=0.1$.

The calculation results for the models completely saturated by water, show different sensitivities of the acoustic-wave velocities and electrical conductivity to the inclusion shapes, matrix and secondary porosities. The electrical conductivity demonstrates the similar sensitivity to both the vuggy and crack porosity types. While the acoustic wave velocities are significantly affected by flattened inclusions and have very low sensitivity to nearly spherical inclusions. The thermal conductivity is characterized by a low sensitivity to the water-saturated secondary pores for all considered shapes.

Based on the results obtained we propose a new secondary-porosity classification: vugs (close to sphere inclusions), quasi-vugs (weakly oblate inclusions), channels (needle-shape inclusions), and cracks (strongly flattened inclusions).

The variations of effective characteristics when the secondary porosity was constituted by mixtures of two pore-shape types (such as vugs and cracks, channels and cracks, vugs and channels) are also assessed. The joint analysis of the velocities and conductivities allows us to determine correctly the composition of the secondary porosity constituted by two pore-shape components.

5. Joint inversion for the rock microstructure determination

In this work, based on the unified microstructure model, we perform the joint nonlinear multimethod inversion of well log data. A set of experimental data includes measurements of different physical properties of geological formations (total porosity, density, P- and S-wave velocities, electrical conductivity, and natural radioactivity) carried out along a borehole with high density of observations.

The inversion procedure consists in minimizing a cost function that includes the sum of weighted square differences between the measured well-logging data and theoretical logs calculated using unified model and EMA method. Taking into account that this optimization procedure is an ill-posed problem because the experimental data contain noise related to equipment accuracy and environmental conditions, we have included additional regularization functional containing complementary information (initial model parameters and ranges of their variations) in the cost function. The petrophysical parameters determined by the inversion are the mineral concentrations, matrix porosity, secondary porosity and secondary porosity type (aspect ratios of spheroids approximating secondary pores). The physical properties of mineral components and fluid are assumed as known and can be adjusted by the posterior analysis of fitting-error distributions for each log.

To determine the pore microstructure parameters we have formulated the joint inversion procedure of well log data as solving an optimization problem where the following quadratic cost function is minimized for each point of measurements

$$F(z_i) = \left\| W_d\left(d(m) - d_{obs}\right)\right\|^2 + \lambda \left\| W_m(m - m_0)\right\|^2 , \tag{15}$$

here the vectors d_{obs} and d(m) are the measured and theoretically calculated well log data, respectively. These vectors can be presented in the form

$$\mathrm{d_{obs}} = \left[\log t_P, \log t_S, \log r_{MSFL}, \log\phi, \rho, \log r_{LLD}, \log\gamma, PEF\right]^T , \tag{16}$$

$$\mathrm{d(m)} = \left[\log t_P(m), \log t_S(m), \log r_{MSFL}(m), \log\phi(m), \rho(m), \log r_{LLD}(m), \log\gamma(m), PEF(m)\right]^T \tag{17}$$

where their components correspond to the logarithmically scaled conventional well logs used in the inversion: P-wave transit time (the parameter is inverse to velocity) – t_P, S-wave transit time – t_S, resistivity measured by MSFL tool - r_{MSFL}, resistivity measured by LLD tool - r_{LLD}, neutron porosity - ϕ_t, density - ρ, gamma ray - γ, and the log of photoelectric effect – PEF. We have found that the use of the logarithmic scale is more suitable for the joint inversion of different logs because they are positive and some of them are characterized by large variations (resistivity, porosity, gamma values) and close to lognormal distributions (or at least non-Gaussian distributions). In this case the datum residual term of equation (16) represents normalized deviations between calculated and measured data and allows us to avoid the problem related to scales and units of different measured quantities.

The variable m is the vector of unknown pore microstructure parameters which have to be found by minimizing the cost function $F(z_i)$

$$m = \left[\log\phi_m, \log\phi_{s1}, \log\phi_{s2}, \log\alpha_{s1}, \log\alpha_{s2}, \log V_{sh}, \log V_d,\right] \tag{18}$$

where ϕ_m is the matrix porosity, ϕ_{s1} is the secondary porosity of the first type (microfractures), ϕ_{s2} is the secondary porosity of the second type (vugs and channels), α_{s1} and α_{s2} are the aspect ratios of spheroids (pore shape) of the first and second types of the secondary porosity respectively, V_{sh} and V_d are the shale and dolomite volumes respectively. The vector m_0 represents a reference model that can be given using a priori information

$$m_0 = \left[\log\phi_m^0, \log\phi_{s1}^0, \log\phi_{s2}^0, \log\alpha_{s1}^0, \log\alpha_{s2}^0, \log V_{sh}^0, \log V_d^0,\right] \tag{19}$$

W_d is a diagonal matrix of weight coefficients which allow us to account different error scales and distributions for each log. These coefficients are a nonlinear functions of the relationships between the physical characteristics (given logs) and model parameters m, and accuracy of measurements for each log. A diagonal matrix W_m assigns the weights for model parameters based on a priori information. We adjust the component values of both matrices using analysis of the posterior error distributions for each log.

The scalar factor λ is a regularization parameter that introduces a relative weight between the misfit term (first term) and Tikhonov's stabilizer (second term) of the cost function (Tikhonov and Arsenin, 1977). The optimal value of this parameter has to provide the minimal deviation from the reference model m_0 (maximal influence of a priori information) and keeps the misfit between the calculated and measured logs within a prescribed error values (Zhdanov, 2002).

To solve the optimization problem for the cost function (equation 16) we have applied the Levenberg-Marquardt method (Levenberg, 1944) that provides the stable global minimum search without calculating functional derivatives. This method is especially effective when the minimized functional has a complex surface as in the case of joint inversion of different physical properties.

We have tested the presented joint inversion technique on the theoretically simulated well logs with prescribed levels of noise for synthetic models (matrix and secondary porosities, secondary pore shapes) and on experimental data by comparing inversion results with geological and petrophysical information (core data, NMR, FMI).

5.1 Synthetic model

Here we present a verification of the inversion convergence on a theoretical model that consists of grains, matrix pores, cracks and vugs approximated by spheroidal inclusions. The parameters of the solid grains composing matrix (density, P- and S-wave velocities) correspond to limestone. The matrix and secondary pores are saturated with water. The matrix, crack, and vuggy porosities were changed in the ranges of 3 - 5%, 0.1 - 1.5%, and 3-6%, respectively. The crack and vug aspect ratios were varied from 10^{-3} up to $5*10^{-3}$, and 0.1-0.7, respectively. The shapes of matrix pores and grains are functions of the matrix porosity (equation 13). The values of all parameters were chosen in the mentioned intervals using uniform random distribution (the Monte-Carlo method). This model have been generated by 1000 realizations. For each realization the P- and S-wave transit times, electrical resistivity, density, and total porosity were calculated adding a random and normally distributed noise of 3% for probability 0.97 which correspond to well logging measuring errors. Then we have applied the joint inversion of the simulated logs to reconstruct the model parameters. The comparison of the matrix, vuggy and fracture porosity, and aspect ratios of vugs and fractures obtained is presented in figures 14. The dotted lines mark the prescribe error-intervals: absolute error ±0.003 for porosities and relative error ±10% for aspect ratios. More than 50% of the matrix and secondary porosity determined by inversion are in the range of ±0.002 and the fracture and vug aspect ratios are in the range of ±10%.

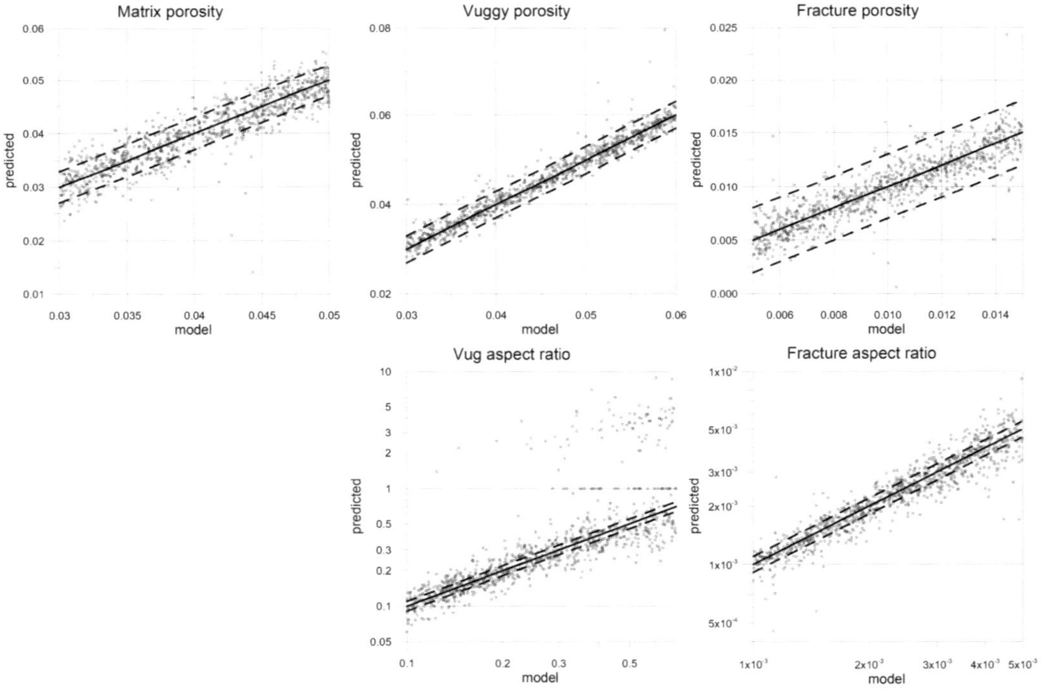

Fig. 14. Comparison of the model parameters: matrix, vuggy and fracture porosities, fracture and vug aspect ratios with parameters obtained by joint inversion.

5.2 Model of carbonate formation with thin shale layers

The volume of shale in the carbonate formations is not high generally and it does not make difficult the potential zone detection using logging data. On the other hand the presence of shale can lead to erroneous results in the determination of secondary pore types if the model of pure carbonate is applied for the joint inversion of logs. In carbonate sediments the shale component are mostly presented as thin layers and in some special facies, it can be arbitrary distributed as inclusions in the matrix or in the secondary pores (Scholle and Ulmer-Scholle, 2003).

Based on this geological data we have introduced a model for carbonate formations containing shale that corresponds to a transversely isotropic medium composed of intercalated layers of pure carbonates and shale. In this model the axis of macroscopic anisotropy (appeared due to formation texture) is perpendicular to the stratification plan and the effective physical properties are described by tensors. For the vertical borehole and horizontal layers the tensors of effective parameters have only two principal components which we will name as vertical (perpendicular to layers) and horizontal (parallel to layers) components.

Taking into account that in the acoustic logs the vertical velocities of elastic wave are measured, we can apply the average time equation for the P- and S – waves' travel times

$$t_{\upsilon}^{P} = t_{c}^{P}(1 - V_{sh}) + t_{sh}^{P} V_{sh}, \quad t_{\upsilon}^{P} = t_{c}^{P}(1 - V_{sh}) + t_{sh}^{P} V_{sh} \tag{20}$$

where t_{υ}^{P}, t_{c}^{P}, t_{sh}^{P}, t_{υ}^{S}, t_{c}^{S}, and t_{sh}^{S} are the travel times of vertical effective velocity component, travel times in carbonates and shales for P - and S – waves, respectively, and V_{sh} is the shale volume. The resistivity well logs in thick layers due to anisotropy paradox measure the horizontal component of conductivity tensor

$$r_{h}^{-1} = r_{c}^{-1}(1 - V_{sh}) + r_{sh}^{-1} V_{sh}, \tag{21}$$

where r_h, r_c and r_{sh} are the horizontal component of the effective resistivity tensor, carbonate and shale resistivities, respectively.

The effective density ρ_{υ} and total porosity ϕ_{υ} in this model can be given as follows

$$\rho_{\upsilon} = \rho_{c}(1 - V_{sh}) + \rho_{sh} V_{sh}, \quad \phi_{\upsilon} = \phi_{c}(1 - V_{sh}) + \phi_{sh} V_{sh} \tag{22}$$

where ρ_c, ρ_{sh}, ϕ_c and ϕ_{sh} are the densities and total porosities of carbonates and shales, respectively. The relationships of the gamma ray and photoelectric (PEF) logs as functions of the shale volume was approximated by a linear commonly used equations (Bassiouni, 1994)

$$\gamma_{\upsilon} = (\gamma_{sh} - \gamma_{c})V_{sh} + \gamma_{c}, \quad PEF_{\upsilon} = (PEF_{c} V_{c}\rho_{c} + PEF_{sh} V_{sh}\rho_{sh}) / (V_{c}\rho_{c} + V_{sh}\rho_{sh}) \tag{23}$$

where γ_{υ}, γ_{sh}, and γ_c, are the effective gamma, and gamma values for the shale and clean carbonates, respectively; PEF_{υ}, PEF_{sh}, and PEF_c are the effective PEF parameter, and PEF parametres for the shale and clean carbonates, respectively.

The effective elastic moduli (P – and S – waves' velocities) and electrical resistivity for the pure carbonate component are found by using the modeling approach described above for the simulation of physical properties starting from the unified pore-space model of double porosity medium. The shale physical properties as well mineral and fluid characteristics can be considered as known and then they are adjusted by the posterior analysis of inversion error distributions for each log.

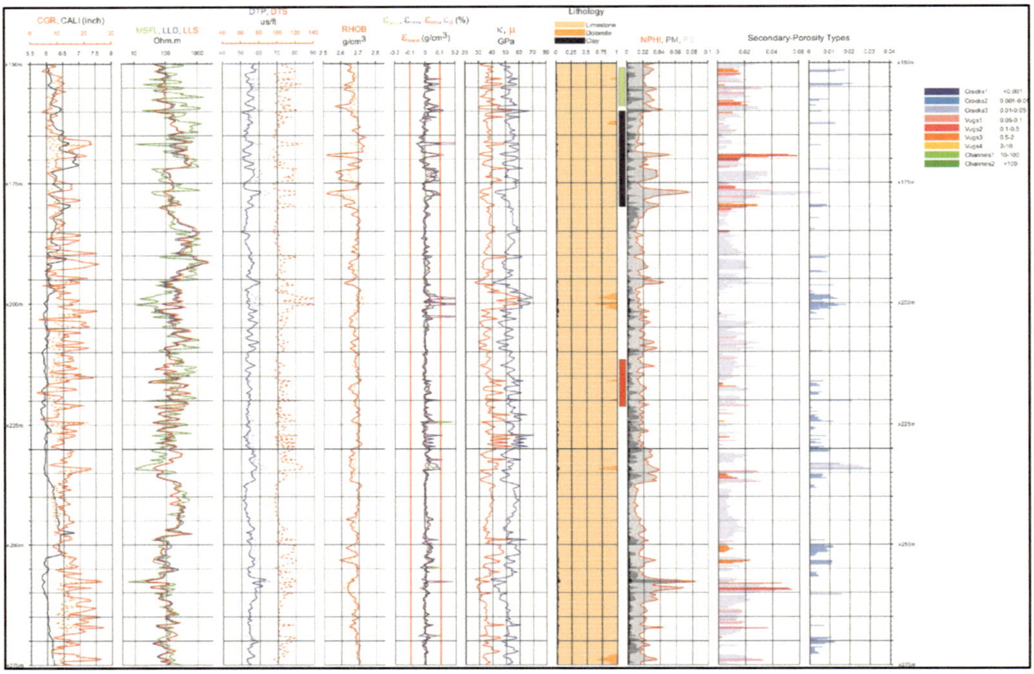

Fig. 15. Results of the joint inversion for the borehole W1. Track 1: CGR - corrected gamma log, CALI – caliper; Track 2: MSFL, LLD, and LLS - resistivity logs; Track 3: DTP and DTS - transit times of the P- and S-wave; Track 4: RHOB - formation density; all measured logs are presented by the solid curves and predicted logs are presented by the dotted curves; Track 5: inversion errors for the resistivity (εMSFL), P-wave transit time (εDTP), total porosity (εϕ), and density (εRHOB), the red lines mark ±10% interval; Track 6: K and μ - bulk module; Track 7: lithology; Track 8: NPHI - neutron porosity, PM - matrix porosity (dark grey area), PS - secondary porosity (grey area), shale porosity (black area); Track 8 and 9: types of two secondary porosity systems. Color legend for the secondary porosity types: deep blue – cracks with $\alpha<0.001$, blue – cracks with $\alpha=0.001-0.01$; light blue - cracks with $\alpha=0.01-0.05$; pink – vugs with $\alpha=0.05-0.1$; crimson – vugs with $\alpha=0.1-0.5$; red - vugs with $\alpha=0.5-2$; yellow - vugs with $\alpha=2-10$; chlorine green - channels with $\alpha=10-100$; deep-green - channels with $\alpha>100$.

6. Examples of the well log inversion

Below we present the results of joint inversion for two boreholes in carbonate formations characterized by different types of secondary porosity.

The first formation is described as compact limestone that are characterized by the low neutron porosity and composed of matrix porosity and microfractures with porosity in the range of 0.01-0.03 (Fig. 15). The matrix porosity does not exceed 0.01-0.02. The average fitting errors between each calculated and measured log is about 5%. The interval x150-x160 m contains microfractures with the porosity about 0.015-0.02 and the relativity high aspect ratios 0.01-0.05 that can be associated with open microfractures affected by dissolution. In this interval the presence of gas was detected. The zone x160-x180 m corresponds to the approved oil production interval with the secondary porosity up to 0.04-0.06 formed by open microfractures and vugs. The vug shapes represents oblate spheroids with the aspect ratios in the range from 0.01 up to 0.1. The results of inversion indicate that the production interval is associated with secondary-porosity increasing and can be extended downwards up to depth of x185 m. The additional perspective zone can be marked in the interval of x245-x265 m. In the core sampling interval x210-x218 m the pore microstructure characteristics obtained by the inversion correspond to compact formations (the matrix porosity is lower than 0.01) with the secondary porosity of 0.01 represented mostly by cracks. This result coincides with the core description as compact fractured mudstone. The second inversion example represents high porosity carbonate formations with the vuggy porosity type (Fig. 16). This formation is composed of dolomite and limestone with low enough shale volume. At the depth of core sampling (x370 m) the dolomite concentration determined by inversion coincides with one obtained from core. The matrix porosity varies in the range of 0.02-0.04 while the secondary vuggy porosity changes from 0.02 up to 0.16.

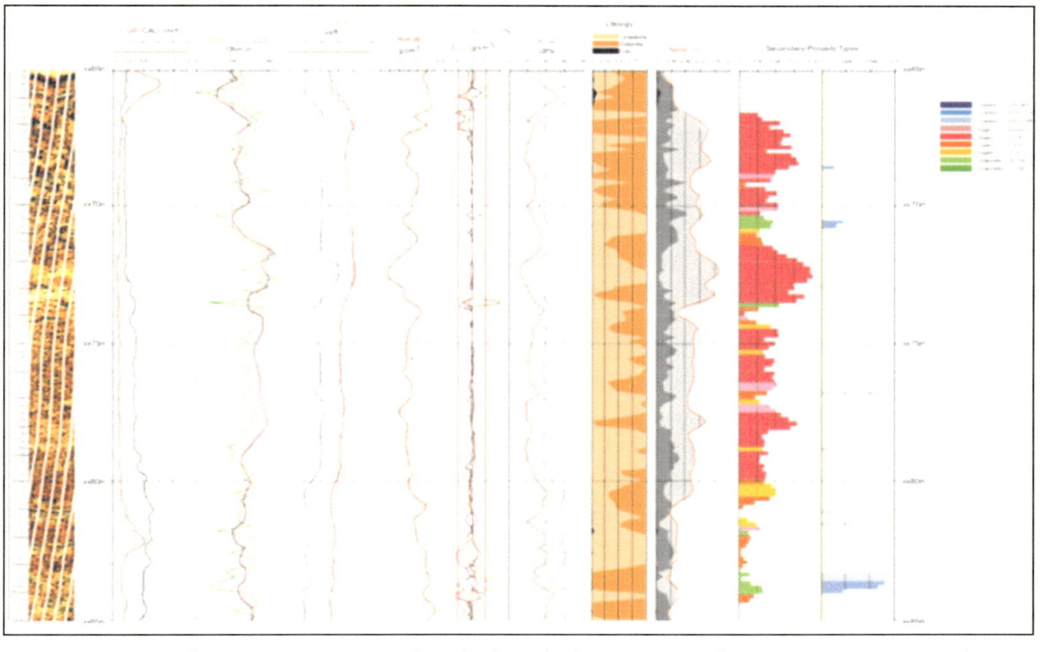

Fig. 16. Results of the joint inversion for the borehole W2. The left track presents FMI data. Description of tracks is given in figure 15.

The misfit distributions demonstrate that the average errors between the measured and calculated logs (except the gamma log) are about 5%. High dispersion of the corrected gamma log is related to the incomplete correction of uranium concentration and probable presence of disperse clay that is not took into account in the applied model. The inversion results show a good agreement with the images obtained by FMI tool (Fig. 16). The results of joint inversion obtained for various boreholes from vuggy and fractured carbonate reservoirs show a good correspondence with core data, image log, and geological descriptions. The output inversion parameters as the matrix porosity, secondary porosity, and aspect ratios can be considered as important quantitative characteristics of pore microstructure to improve classification of carbonate lithotypes, processes of secondary-pore generation (fracturing, dolomitization, dissolution), and permeability prediction.

7. Conclusions

We have presented a technique for joint simulating the acoustic velocities, electrical and thermal conductivities, and dielectric permittivity of double porosity rocks treated as a natural composite material. The technique is based on the unified microstructure model and hierarchical two step homogenization using the EMA symmetrical approach. The peculiarity of the model consists in calculating the effective properties of the porous matrix with the variable aspect ratios for ellipsoids approximated by grains and pores which depend on the porosity value (first step of homogenization). This approach provides simultaneously a non-zero shear modulus for high porosity and electrical conductivity for low porosity and allows us to simulate the effective matrix properties with a high accuracy.
In the second homogenization step we calculate the effective properties for the double porosity medium where the secondary pores of different shapes are embedded into the effective matrix. The presence of secondary pores changes significantly the acoustic and electrical properties of the double porosity medium in comparison with composites containing just one pore system despite the same total porosity values. For correct estimation of the secondary-porosity with one or two types of pore-shapes, both acoustic wave velocities and electrical conductivity have to be used.
Based on the modeling results we have developed the method of nonlinear joint inversion of measured physical properties for determining the microstructure parameters of natural composite porous material. The method has been successfully applied for the integral interpretation of experimental well log data for petrophysical characterization of double porosity carbonate formations containing cracks and vugs.
The matrix and secondary porosities, and shapes of secondary pores obtained as the results of joint inversion of well logs can be considered as important quantitative characteristics of the pore microstructure to improve evaluation of carbonate formations and allow us (1) to characterize carbonate lithotypes and processes of secondary-pore generation (fracturing, dolomitization, dissolution), (2) to determine saturation of double-porosity rocks, (3) to determine geomechanical parameters and calibrate seismic data, (4) to assess probability of the secondary pore percolation for predicting formation permeability.

8. References

Archie, G.E., 1942. The electrical resistivity log as an aid in determining some reservoir characteristics, Trans. AIME, 146, 54-62.

Auriault, J. L., Boutin, C., 1994. Deformable porous media with double porosity, III: Acoustics. Transport in Porous Media 14, 143-162.

Auriault, J. L., Royer, P., 2002. Seismic waves in fractured porous media. Geophysics 67, 259-263.

Bagrintseva, K. I., 1999. Conditions of generation and properties of carbonate reservoirs of oil and gas. RGGU, Moscow. (in Russian).

Barrenblatt, G. I., Zheltov, I. P., Kochina, I. N., 1960. Basic concepts in the theory of seepage of homogeneous liquids in fissured rocks. Prikladnaya Matematica y Mechanica 24,

852–864 (in Russian), (English translation: 1960, Soviet Applied Mathematic & Mechanic 24, 1286–1303).

Bassiouni, Z., 1994. Theory, Measurement, and Interpretation of Well Logs. SPE Textbook Series, v. 4, Richardson, TX.

Berge, P.A., Berryman, J.G., Bonner, B.P., 1993. Influence of microstructure on rock elastic properties. Geophysical Research Letters 20 (23), 2619-2622.

Berryman, J., 1980. Long wavelength propagation in composite elastic media. Journal of Acoustic Society of America 68, 1809–1831.

Berryman, J.G., 1992. Single-scattering approximations for coefficients in Biot's equations of poroelasticity. Journal of Acoustic Society of America 91, 551-571

Berryman, J.G., Wang, H.F., 2000. Elastic wave propagation and attenuation in a double-porosity dual-permeability medium. International Journal of Rock Mechanics and Mining Sciences 37, 63-78.

Budiansky, B., O'Connell, R.J., 1976. Elastic module of a cracked solid. International Journal of Solid and Structures 12, 81-97.

Carlson, B.C., 1979. Computing Elliptic Integrals by Duplication, Numerische Matematic, 33, 1 - 16.

Choquette, P. W., Pray, L. C., 1970. Geologic nomenclature and classification of porosity in sedimentary carbonates. AAPG Bulletin 54, 207-250.

Eshelby, J.D., 1957. The determination of the elastic field of an ellipsoidal inclusion and related problems. Proc. R. Soc. Lond. A. 241, 376-396.

Guéguen Y. and Palciauskas V. 1994. Introduction to the Physics of Rocks. Princeton University Press.

Feng, S., Sen, P.N., 1985. Geometrical model of conductive and dielectric properties of partially saturated rocks. Journal of Applied Physics 58 (8), 3236-3243.

Hornby, B. E., Shwartz, L. M., Hudson, J. A., 1994. Anisotropic effective medium modeling of elastic properties of shales. Geophysics 59, 1570-1583.

Kazatchenko, E., Mousatov, A., 2002. Estimation of secondary porosity for carbonate formation using core data: total porosity and formation factor. SPE 77787, SPE Annual Technical Conference and Exhibition, San Antonio, Texas.

Kazatchenko, E., Markov, M., Mousatov, A., 2004a. Joint modeling of acoustic velocities and electrical conductivity from unified microstructure of rocks. Journal of Geophysical Research, 109, B01202, doi: 10-1029/2003JB002443.

Kazatchenko, E., Markov, M., Mousatov, A., 2004b. Joint inversion of acoustic and resistivity data for carbonate microstructure evaluation. Petrophysics 45 (2), 130-140.

Korringa, J., Brown, R.J.S., Thompson, D.D., Runge, R.J., 1979. Self-consistent imbedding and the ellipsoidal model for porous rocks. Journal of Geophysical Research 84, 5591-5598.

Landau, L.D., Lifshitz, E., 1960. Electrodynamics of continuous media. N.-Y. Pergamon press, New York.

Le Ravalec, M., Gueguen, Y., 1996. High and low frequency elastic moduli for a saturated porous cracked rock. Geophysics 61, 1080-1094.

Levenberg K., 1944. A Method for the Solution of Certain Non-Linear Problems in Least Squares. The Quarterly of Applied Mathematics, 2, 164–168.

Lucia, F. J., 1999. Carbonate reservoir characterization. Springer.

Mavko, G., Mukerji, T., Dvorkin, J., 1998. The rock physics handbook. Tools for seismic analysis in porous media. Cambridge University Press.

Mendelson, K.S., Cohen, M.N., 1982. The effect of grain anisotropy on the electrical properties of isotropic sedimentary rocks. Geophysics 47, 257–263.

Norris, A.N., 1985. A differential scheme for the effective moduli of composites. Mechanics of Materials 4, 1-16.

Norris, A.N., Sheng, P., Callegari, A.J., 1985. Effective–medium theories for two-phase dielectric media. Journal of Applied Physics 57, 1990–1996.

Pribnow, D., Sass, J., 1995. Determination of thermal conductivity from deep boreholes. Journal Geophysical Research 100, 9981-9994.

Schoen, J. H., 1996. Physical properties of rocks: Fundamentals and principles of Petrophysics. In Handbook of Geophysical Exploration. Section I, Seismic exploration 18, Redwood Books, Trowbridge.

Scholle, P. A., and D. S. Ulmer-Scholle, 2003. A color guide to the petrography of carbonate rocks: grains, textures, porosity, diagenesis. Tulsa, OK, American Association of Petroleum Geologists Memoir 77.

Sen, P., Scala, C., Cohen, M.H., 1981. A self - similar model for sedimentary rocks with application to the dielectric constant of fused glass beard. Geophysics 46, 781–796.

Song, Y.-Q., N.V. Lisitza, D.F. Allen, and W.E. Kenyon, 2002. Pore geometry and its geological evolution in carbonate rocks, Petrophysics, 43(5), 420-424.

Stratton, J.A., 1941. Electromagnetic theory. Mc Graw-Hill, New York.

Tikhonov, A. N., and Arsenin, V.Y., 1977. Solution of Ill - Posed Problems, Winston, Washington, D.C Willis.

Wilson, R.K., Aifantis, E.C., 1982. On the theory of consolidation with double porosity. International Journal of Engineering Science 20, 1009-1035.

Wu, T.T., 1966. The effect on inclusion shape on the elastic moduli of a two – phase material. International Journal of Solids Structures 2, 1–8.

Zhdanov, M.S., 2002. Geophysical Inverse Theory and Regularization Problems. Elsevier, Amsterdam - New York – Tokyo.

Zimmerman, R.W, 1991. Compressibility of Sandstones, Elsevier, Amsterdam - New York – Tokyo.

6

Hybrid Inorganic–organic Composites of Layered Double Hydroxides Intercalated with Organic Acid Anions for the Uptake of Hazardous Substances from Aqueous Solution

Tomohito Kameda and Toshiaki Yoshioka
Graduate School of Environmental Studies, Tohoku University
Japan

1. Introduction

Layered double hydroxide (LDH) is an inorganic compound with a layered structure and exhibits anion–exchange properties. Its basic chemical composition is $[M^{2+}{}_{1-x}M^{3+}{}_{x}(OH)_2](A^{n-})_{x/n}\bullet mH_2O$, where x is the $M^{3+}/(M^{2+} + M^{3+})$ molar ratio and A^{n-} indicates an anion with a valency of n (Ingram and Taylor, 1967; Allmann, 1968; Taylor, 1973). Its crystal structure consists of positively charged brucite-like octahedral hydroxide layers containing M^{3+} at M^{2+} sites, which are electrically neutralized by the interlayer anions, and water molecules in the hydration shell of the intercalated anions occupying the remaining interlayer space. Among various types of LDHs composed of different compositions of M^{2+} and M^{3+}, Mg–Al LDH has been used practically as an antacid and a flame retardant (Cavani et al., 1991). Mg–Al LDH intercalated with CO_3^{2-} ($CO_3\bullet$Mg–Al LDH) can be transformed into Mg–Al oxide by calcination at 450–800 °C according to the following reaction:

$$Mg_{1-x}Al_x(OH)_2(CO_3)_{x/2} \rightarrow Mg_{1-x}Al_xO_{1+x/2} + x/2CO_2 + H_2O \quad (1)$$

Mg–Al oxide can rehydrate and combine with anions to reconstruct the LDH structure in the following manner:

$$Mg_{1-x}Al_xO_{1+x/2} + x/n\ A^{n-} + (1 + x/2)H_2O \rightarrow Mg_{1-x}Al_x(OH)_2A_{x/n} + xOH^- \quad (2)$$

LDH can intercalate various anions. Recently, modification of LDH with organic anions has received considerable attention, as summarized in a review by Newman and Jones (1998), with possible applications in catalysis, photochemistry, and electrochemistry.

We carried out a series of studies to develop hybrid inorganic–organic composite materials by intercalation of organic anions having functional groups in their structures into the interlayer of LDH. Due to the action of the functional groups of the organic anions, the composite materials are expected to selectively take up and collect both inorganic and organic contaminants present at low concentrations in aqueous solutions.

The aims and concerns presented in this chapter are the preparation of organically modified LDH by intercalation of organic anions into the interlayer of LDH and the uptake of heavy metal ions and nonionic organic compounds by the organically modified LDH.

2. Uptake of heavy metal ions by organically modified LDH

Due to slow reactivity and inefficiency, the application of chelate resins in the treatment of wastewater with a high concentration of heavy metal ions is difficult. Generally, a two-step process is considered acceptable for wastewater treatment. First, the heavy metal ions are removed from the wastewater by a chemical precipitation technique, and then the remaining heavy metal ions are treated using chelate resins. A one-step method capable of achieving highly effective wastewater treatment is desired. A one-step method is thought to be possible by applying a scavenger that can function as chelate resins and can rapidly remove heavy metal ions from wastewater, decreasing the heavy metal ion concentration to the level of the effluent standards in Japan. The aim of the present study was to develop an organic–inorganic composite material to scavenge heavy metal ions from wastewaters. Chelate resin contains functional groups such as aminocarboxylate and polyamine that form chelate bonds with heavy metal ions. Therefore, we developed a scavenger by incorporating an organic acid anion having functional groups in its structure into the inorganic compound. Aminocarboxylates such as an ethylenediaminetetraacetate (EDTA) anion were examined as functional groups. Then, LDH was examined as the inorganic compound.

Several studies have examined the use of LDH and calcined LDH for the preservation of aqueous environments. LDH and calcined LDH can take up oxometalates such as $Cr_2O_7^{2-}$, SeO_3^{2-}, SeO_4^{2-}, AsO_3^{3-}, AsO_4^{3-}, TcO_4^-, and ReO_4^- from aqueous solution (Dousova et al., 2003; Gillman, 2006; Goswamee et al., 1998; Kang et al., 1996; Lazaridis and Asouhidou, 2003; Rhee et al., 1997; Tsuji et al., 1999; Wang et al., 2009; Yang et al., 2005; You et al., 2001a; You et al., 2001b). Goh et al. (2008) reviewed the application of LDH for removal of oxometalates. LDH, when acting as a hydroxide, has also acted as a precipitation reagent for heavy metal ions such as Cu^{2+}, Pb^{2+}, and Zn^{2+} in wastewater via hydroxide formation (Fujii et al., 1992). However, heavy metal cations cannot be intercalated in the interlayer of LDH. In order to take up heavy metal cations using inorganic materials from aqueous solution, cationic clay such as montmorillonite is typically used due to its cation-exchange properties. On the other hand, EDTA anions can be intercalated in the interlayer of LDH (Narita et al., 1995). Similarly, stable anionic chelates formed between heavy metal ions and aminocarboxylates such as EDTA can be also intercalated in the interlayer of LDH. For example, the intercalations of $Ni(EDTA)^{2-}$, $Co(EDTA)^{2-}$, $Cu(EDTA)^{2-}$, $Zn(EDTA)^{2-}$, and $Cd(EDTA)^{2-}$ have been examined (Sato et al., 1990; Narita et al., 1995; Tsyganok et al., 2001a; Tsyganok et al., 2001b; Lukashin et al., 2003; Tsyganok and Sayari, 2006). Accordingly, we have developed a scavenger by intercalating an EDTA anion in the interlayer of LDH and applied the scavenger to decrease the concentrations of heavy metal ions in aqueous solutions for the first time. In order for the scavenger to take up heavy metal ions from an aqueous solution, the EDTA anion should function as a chelating agent in the interlayer of LDH. In the following section, we discuss our work along with related studies.

2.1 LDH intercalated with EDTA

The EDTA•Mg–Al LDH, which had an EDTA anion intercalated in the interlayer of Mg–Al LDH, was synthesized by adding a solution of $Mg(NO_3)_2$ and $Al(NO_3)_3$ to a solution of NaOH and EDTA-2Na (disodium dihydrogen EDTA dihydrate) according to the coprecipitation reaction expressed by Eq. (3) and by mixing at 30 °C for 1 h (Kameda et al., 2005a).

$$0.75Mg^{2+} + 0.25Al^{3+} + 2OH^- + 0.0625EDTA^{4-} \rightarrow Mg_{0.75}Al_{0.25}(OH)_2(EDTA)_{0.0625} \quad (3)$$

Since the EDTA•Mg–Al LDH was prepared in the pH range where EDTA anions exist stably as tetravalent anions by setting the NaOH concentration in the EDTA solution to 1.2 times the stoichiometric quantity defined by Eq. (3), the EDTA•Mg–Al LDH was considered to be a Mg–Al LDH having $EDTA^{4-}$ in the interlayer.

EDTA•Mg–Al LDH was added to 1.0 mM MCl_2 (M = Cu, Cd) solution or a 1:1 mixed chloride solution of Cu^{2+} and Cd^{2+}, and then the resultant suspension was kept standing at 20 °C under mild agitation for 60 min (Kameda et al., 2005a). The solution pH was adjusted to 5.0 by addition of 0.1 M hydrochloric acid, and Cu^{2+} and Cd^{2+} did not precipitate as their hydroxides. In this case, the molar ratios of $EDTA^{4-}$ in the EDTA•Mg–Al LDH added to M^{2+} in the MCl_2 solution were set at 1.0–2.0 ($EDTA^{4-}$ /M^{2+} = 1.0–2.0). For the Cu–Cd mixed chloride solution, the $EDTA^{4-}$/(Cu^{2+} + Cd^{2+}) molar ratio was set at 0.5. The capacity of CO_3•Mg–Al LDH, the interlayer of which was not chemically modified, was also examined to estimate the function of $EDTA^{4-}$ in the interlayer of EDTA•Mg–Al LDH. Figure 1 shows the variation in the concentration of Cu^{2+} with time during the stirring of EDTA•Mg–Al LDH or CO_3•Mg–Al LDH in the $CuCl_2$ solution. In this case, the molar ratio of $EDTA^{4-}$ in the EDTA•Mg–Al LDH added to Cu^{2+} in the $CuCl_2$ solution was set at 1.0 ($EDTA^{4-}$/Cu^{2+} = 1.0). The amount of CO_3•Mg–Al LDH added was equal to that of EDTA•Mg–Al LDH.

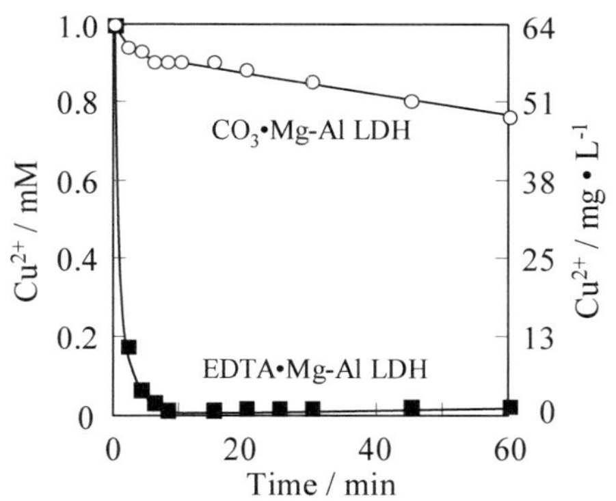

Fig. 1. Variation in the concentration of Cu^{2+} with time during the stirring of EDTA•Mg–Al LDH or CO_3•Mg–Al LDH in the $CuCl_2$ solution.

For EDTA•Mg–Al LDH, the concentration of Cu^{2+} decreased very rapidly to 0.02 mM (1 mg/L) in 10 min. The uptake of Cu^{2+} was 20 mg-Cu^{2+}/g-EDTA•Mg–Al LDH. On the other hand, for CO_3•Mg–Al LDH, the concentration decreased in a substantially slower manner, to 0.9 mM (57 mg/L) in 10 min (2 mg-Cu^{2+}/g-CO_3•Mg–Al LDH). This difference in the uptake of Cu^{2+} was attributed to the function of $EDTA^{4-}$ in the interlayer of EDTA•Mg–Al LDH. In particular, the $Cu(EDTA)^{2-}$ complex is thought to form in the interlayer of EDTA•Mg–Al LDH. Along with formation of the chelate complexes, Cl^- is presumed to intercalate in the interlayer of EDTA•Mg–Al LDH to compensate for the change in negative charge resulting from the change in anionic species from $EDTA^{4-}$ to $Cu(EDTA)^{2-}$ in the interlayer. On the other hand, a slight decrease in the concentration of Cu^{2+} was observed for CO_3•Mg–Al LDH (Fig. 1). In this experiment, 0.1 M HCl was added to the solution to maintain the pH at 5.0. Due to the decrease in the local pH by the addition of HCl, Mg^{2+} and Al^{3+} dissolved from CO_3•Mg–Al LDH. Mg^{2+} was observed to be dissolved from the

CO_3•Mg-Al LDH. On the contrary, the concentration of Al^{3+} in the solution was less than 1% at any time. Therefore, the slight removal of Cu^{2+} by CO_3•Mg-Al LDH may be attributable to the coprecipitation of Cu^{2+} with the dissolved Al^{3+} to form Cu-Al LDH. Mg^{2+} dissolved from EDTA•Mg-Al LDH as well as CO_3•Mg-Al LDH, while Al^{3+} was detected to a substantially lower concentration than Mg^{2+} in the solution. Consequently, the uptake of Cu^{2+} by EDTA•Mg-Al LDH was attributed to both the formation of chelate complexes with $EDTA^{4-}$ in the interlayer and coprecipitation with the dissolved Al^{3+}. Note that EDTA•Mg-Al LDH removed Cu^{2+} from an aqueous solution with an initial concentration of 64 mg/L to less than the effluent standards in Japan (3 mg/L) in a short span of 10 min. For EDTA•Mg-Al LDH, the concentration of Cd^{2+} decreased with time, and the concentration decreased quickly with increasing $EDTA^{4-}/Cd^{2+}$ molar ratio. The concentration decreased to 0.02 mM (2 mg/L) in 10 min at an $EDTA^{4-}/Cd^{2+}$ molar ratio of 2.0 (17 mg-Cd^{2+}/g-EDTA•Mg-Al LDH). However, for CO_3•Mg-Al LDH, the decrease in the concentration was extremely slow, and the concentration was still 0.93 mM (104 mg/L) after 10 min (1 mg-Cd^{2+}/g-CO_3•Mg-Al LDH). The differences in the removal of Cd^{2+} were also attributed to the function of $EDTA^{4-}$ in the interlayer of EDTA•Mg-Al LDH. Note that the concentration of Cd^{2+} decreased rapidly from 112 mg/L to less than 1 mg/L at an $EDTA^{4-}/Cd^{2+}$ molar ratio of 2.0 in 15 min. For the Cu-Cd mixed chloride solution, the concentration of Cu^{2+} decreased with time more rapidly than that of Cd^{2+}, confirming the preferential uptake of Cu^{2+} over Cd^{2+} from the mixed solution by EDTA•Mg-Al LDH. This preferential uptake was probably due to the difference between the stabilities of $Cu(EDTA)^{2-}$ and $Cd(EDTA)^{2-}$. The chelate formation constants for $Cu(EDTA)^{2-}$ and $Cd(EDTA)^{2-}$ are reported to be 18.8 and 16.5; i.e., $Cu(EDTA)^{2-}$ is more stable than $Cd(EDTA)^{2-}$. Thus, it is believed that $EDTA^{4-}$ in the interlayer formed the chelate complex with Cu^{2+} preferentially over Cd^{2+} in the mixed solution. From these results, it can be deduced that $EDTA^{4-}$ maintains its chelating function even when intercalated in the interlayer of Mg-Al LDH.

Perez et al. (2006) examined the uptake of Cu^{2+}, Cd^{2+}, and Pb^{2+} on Zn-Al LDH intercalated with EDTA. EDTA•Zn-Al LDH was obtained from NO_3•Zn-Al LDH using the anion exchange method. The uptake of Cu^{2+}, Cd^{2+}, and Pb^{2+} from the aqueous solutions was due to chelation between EDTA and metal cations. The shape of the adsorption isotherms suggests specific interaction and high host-guest affinity. At pH 5.5 and an initial concentration of 10 mM, the amounts adsorbed were 1117, 375, and 871 μmol/g for Cu^{2+}, Cd^{2+}, and Pb^{2+}, respectively. Kulyukhin et al. (2008) examined the adsorption of Sr radionuclides from the aqueous phase on EDTA•Mg-Al LDH. The adsorption was apparently associated with the complexation of Sr^{2+} with $[H_2(EDTA)]^{2-}$ incorporated in the EDTA•Mg-Al LDH. Rojas et al. (2009) examined EDTA-modified Zn-Al LDH as a Cu^{2+} scavenger based on removal kinetics and sorbent stability. They noted that the exchange process between NO_3•Zn-Al LDH and $[H_2(EDTA)]^{2-}$ anions takes place with partial erosion of the layers, causing the intercalation of $[Zn(EDTA)]^{2-}$ anions instead of the ligand. They insisted that the uptake of $[Cu(H_2O)_6]^{2+}$ by $[Zn(EDTA)]^{2-}$ intercalated in sorbent-LDH results from an exchange reaction between chelated zinc cations in the interlayer and copper cations in solution. Gasser and Aly (2009) examined the kinetic and adsorption mechanisms of Cu^{2+} and Pb^{2+} using EDTA•Mg-Fe LDH. Cu^{2+} and Pb^{2+} were adsorbed by EDTA•Mg-Fe LDH due to chelation with EDTA from the LDH interlayer. The adsorption of Cu^{2+} was higher than that of Pb^{2+}; this was attributed to the stability constant of $[Cu(EDTA)]^{2-}$, which is higher than that of $[Pb(EDTA)]^{2-}$. The shape of the adsorption isotherm suggests specific interaction and high affinity. The kinetics of the adsorption of Cu^{2+} onto EDTA•Mg–Fe LDH follows first-order reversible kinetics, and the values of the first-order rate constants increased with increasing temperature. The chemisorption step may be the rate-limiting step of adsorption from solution.

In our previous research (Kameda et al., 2005a), the pH of the solution during uptake of Cu^{2+} and Cd^{2+} by EDTA•Mg-Al LDH was adjusted to 5.0 by the addition of 0.1 mol/L acid; this pH value was maintained to prevent precipitation of the hydroxides of Cu^{2+} and Cd^{2+}. However, this method resulted in the dissolution of Mg^{2+} from EDTA•Mg–Al LDH because of the increasing acidity as Mg^{2+} precipitated as $Mg(OH)_2$ at pH >10. Therefore, Cu^{2+} was selected as the divalent metal ion of the LDH host layer because the hydrolysis product of Cu^{2+} (pK_{sp} = 17.3) is less soluble than that of Mg^{2+} (pK_{sp} = 11.2). Recently, we prepared Cu–Al LDH intercalated with EDTA (EDTA•Cu–Al LDH) by suspending Cu–Al oxide, obtained by calcination of CO_3•Cu-Al LDH, in EDTA solution (Kameda et al., 2011). The reconstruction of Cu–Al oxide to Cu–Al LDH was promoted with increasing temperature and time. The reaction at a pH of ~8 suggests that $[H(EDTA)]^{3-}$ was intercalated in the interlayer of Cu–Al LDH. EDTA•Cu–Al LDH was found to take up rare metal ions such as Sc^{3+} and La^{3+} in an aqueous solution at a pH of ~6–6.5. The uptake of Sc^{3+} was caused not only by the chelating function of $[H(EDTA)]^{3-}$ in the interlayer but also by the chemical behavior of Cu–Al LDH itself. On the other hand, the uptake of La^{3+} was caused only by the chelating function of $[H(EDTA)]^{3-}$ in the interlayer. $[H(EDTA)]^{3-}$ in the interlayer of EDTA•Cu–Al LDH formed a chelate complex with Sc^{3+} more preferentially than with La^{3+}.

2.2 LDH intercalated with carboxylates

We examined organic acids for leaching heavy metals from fly ash generated by gasification and melting furnace treatments for municipal solid wastes (Saito et al., 2007). Organic acids such as citric, malic, and tartaric acids were found to be effective for leaching heavy metals due to the formation of chelate complexes between the metals and acids on the surface of the fly ash. Since citric, malic, and tartaric acids have the capacity for heavy metal extraction, Mg–Al LDHs intercalated with citrate ($C_6H_5O_7^{3-}$), malate ($C_4H_4O_5^{2-}$), and tartrate ($C_4H_4O_6^{2-}$) anions can be expected to take up heavy metal ions from aqueous solution into the interlayers. Citric, malic, and tartaric acids exist abundantly in nature, and, therefore, the corresponding organic acid anion-intercalated Mg–Al LDHs are more appropriate for commercial-scale applications than EDTA•Mg–Al LDH.

Meyn et al. (1990) used the anion exchange method to prepare Mg–Al LDH composite materials with intercalated $C_6H_5O_7^{3-}$ and $C_4H_4O_6^{2-}$, whereas Zhang et al. (2004) prepared Mg–Al LDH intercalated with $C_6H_5O_7^{3-}$, $C_4H_4O_5^{2-}$, and $C_4H_4O_6^{2-}$ through a unique method involving the dissolution of CO_3•Mg–Al LDH to form the corresponding carboxylic acid followed by precipitation in basic solution. Tronto et al. (2001, 2004) used coprecipitation to form Mg–Al LDH intercalated with $C_6H_5O_7^{3-}$.

We prepared Mg–Al LDHs intercalated with $C_6H_5O_7^{3-}$, $C_4H_4O_5^{2-}$, and $C_4H_4O_6^{2-}$ by coprecipitation via dropwise addition of a Mg–Al nitrate solution to citrate, malate, and tartrate solutions, respectively, at a constant pH of 10.5 and investigated the effects of the organic anion concentrations on the formation of the composites (Kameda et al., 2009a). For Mg–Al LDH intercalated with citrate (citrate •Mg–Al LDH), low solution concentrations of $C_6H_5O_7^{3-}$ resulted in intercalation of $C_6H_5O_7^{3-}$ into the interlayer at a 26° angle relative to the brucite-like layers of Mg-Al LDH. At higher $C_6H_5O_7^{3-}$ concentrations, both $Mg(C_6H_5O_7)^-$ and $C_6H_5O_7^{3-}$ were intercalated into the interlayer with an irregular orientation, and $C_6H_5O_7^{3-}$ adsorbed to the surface of the Mg-Al LDH. For Mg-Al LDHs intercalated with malate and tartrate (malate•Mg-Al LDH and tartrate•Mg-Al LDH), the intercalated $C_4H_4O_5^{2-}$ and $C_4H_4O_6^{2-}$ were also oriented at an angle of 26° relative to the brucite-like layers of the Mg-Al LDH, regardless of solution concentration. The amount of intercalated $C_4H_4O_5^{2-}$ for malate•Mg-Al LDH was governed solely by the electric charge balance in Mg-Al LDH and was not affected by increased solution concentration. However, increasing the solution concentration of $C_4H_4O_6^{2-}$ during the formation of tartrate•Mg-Al LDH resulted in both interlayer intercalation and adsorption of $C_4H_4O_6^{2-}$ to the surface of the composite. These Mg-Al LDHs were found to take up Cu^{2+} and Cd^{2+} rapidly from an aqueous solution at a constant pH of 5.0 (Kameda et al., 2008a). This capacity was mainly attributed to the formation of citrate-metal, malate-metal, and tartrate-metal complexes in the interlayers of the Mg-Al LDHs. The uptake of Cu^{2+} increased in the order malate•Mg-Al LDH < tartrate•Mg-Al LDH < citrate•Mg-Al LDH. The uptake of Cd^{2+} increased in the order malate•Mg-Al LDH < tartrate•Mg-Al LDH = citrate•Mg-Al LDH. These differences in Cu^{2+} and Cd^{2+} uptake were attributable to differences in the stabilities of the citrate-metal, malate-metal, and tartrate-metal complexes. These results indicate that $C_6H_5O_7^{3-}$, $C_4H_4O_5^{2-}$, and $C_4H_4O_6^{2-}$ were adequately active as chelating agents in the interlayers of Mg-Al LDHs.

We also examined the kinetics of the uptake of Cu^{2+} and Cd^{2+} by citrate •Mg-Al LDH, malate•Mg-Al LDH, and tartrate•Mg-Al LDH (Kameda et al., 2010a). The Mg-Al LDHs were added to 500 mL of 1.0 mmol/L $Cu(NO_3)_2$ or $Cd(NO_3)_2$ solution. The molar ratio of citrate, malate, or tartrate in Mg-Al LDH to Cu^{2+} or Cd^{2+} in the nitrate solution was set to 1.0. The resultant suspension was left to stand at 10, 20, and 30 °C for 60 min with stirring at a constant pH of 5.0. The concentrations of Cu^{2+} and Cd^{2+} decreased with time and increasing temperature. The accelerated rate of Cu^{2+} and Cd^{2+} uptake by the Mg-Al LDHs at high temperatures implies that the uptake reactions were governed by chemical reactions rather than mass transfer. The reactions were examined kinetically based on the chemical equation. The uptake of Cu^{2+} by citrate•Mg-Al LDH was attributed to the 1:1 chelation between Cu^{2+} and citrate as follows:

$$C_6H_5O_7^{3-} + Cu^{2+} \rightarrow Cu(C_6H_5O_7)^- \tag{4}$$

This reaction gives the following rate equation:

$$d[Cu(C_6H_5O_7)^-]/dt = k[C_6H_5O_7^{3-}][Cu^{2+}] \tag{5}$$

where t is the reaction time, and k is the apparent rate constant. Because

$$d[Cu(C_6H_5O_7)^-]/dt = d[Cu^{2+}]/dt \text{ and } [C_6H_5O_7^{3-}]/[Cu^{2+}] = 1, -d[Cu^{2+}]/dt = k[Cu^{2+}]^2 \tag{6}$$

The integration of Eq. (6) gives

$$1/[Cu^{2+}] = kt + 1 \tag{7}$$

Assuming that x is the degree of Cu^{2+} uptake, $[Cu^{2+}]$ can be expressed as $1 - x$ because the initial Cu^{2+} concentration was 1.0 mmol/L. Equation 7 can be converted to Eq. (8):

$$1/(1 - x) = kt + 1 \tag{8}$$

The uptake rate of Cu^{2+} by citrate•Mg–Al LDH was determined from Eq. (8), and the plot of $1/(1 - x)$ versus t is shown in Fig. 2.

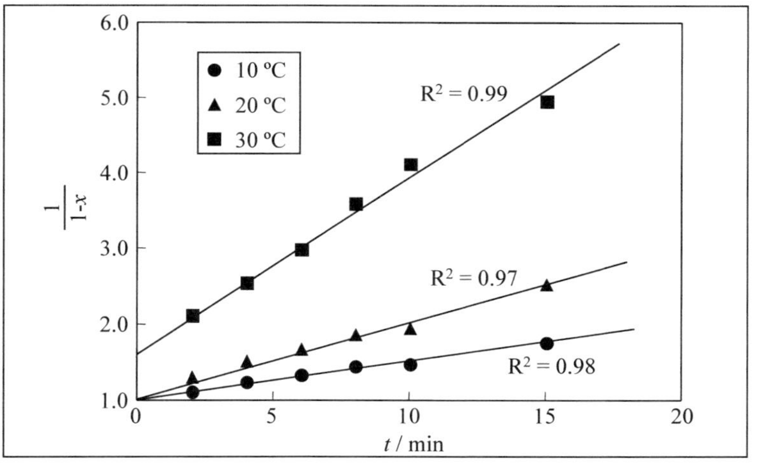

Fig. 2. Plot of $1/(1 - x)$ vs. t for uptake of Cu^{2+} by citrate•Mg–Al LDH.

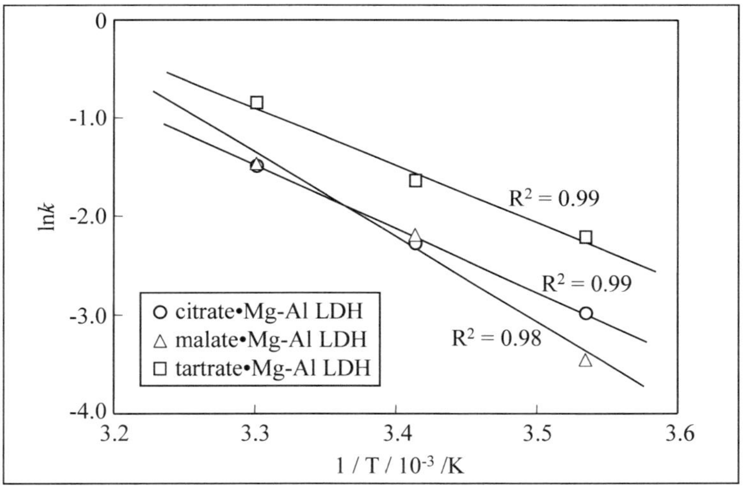

Fig. 3. Arrhenius plot of the apparent rate constant for uptake of Cu^{2+} by Mg–Al LDHs.

Good linearity was obtained at each temperature, indicating that the reaction rate could be expressed by Eq. (5). At 30 °C, the value of the intercept was not 1.0, suggesting that the precipitation rate of Cu–Al LDH, which formed between a few Cu^{2+} cations and dissolved Al^{3+}, was very fast. The apparent rate constants at 10, 20, and 30 °C were 5.1×10^{-2}, 1.0×10^{-1}, and 2.3×10^{-1}, respectively. Thus, the apparent rate constant increased with increasing temperature. An Arrhenius plot of k, determined from the slopes of the straight lines in Fig. 2, is shown in Fig. 3; this plot yields an apparent activation energy of 53.1 kJ·mol^{-1}. This value confirms that the uptake of Cu^{2+} by citrate•Mg–Al LDH proceeded under chemical reaction control because the apparent activation energy for the chemical reaction was larger than 30 kJ·mol^{-1}. The uptake of Cu^{2+} by this LDH occurred via chelation between Cu^{2+} and citrate. The uptake of Cu^{2+} by malate•Mg–Al LDH was attributed to the 2:2 chelation between Cu^{2+} and malate and is expressed as

$$2C_4H_4O_5^{2-} + 2Cu^{2+} \rightarrow [Cu_2(C_4H_3O_5)(C_4H_4O_5)]^- + H^+ \quad (9)$$

This reaction gives the following rate equation:

$$d[[Cu_2(C_4H_3O_5)(C_4H_4O_5)]^-]/dt = k[C_4H_4O_5^{2-}]^2[Cu^{2+}]^2 \quad (10)$$

where t is the reaction time, and k is the apparent rate constant. Because

$$d[[Cu_2(C_4H_3O_5)(C_4H_4O_5)]^-]/dt = -(1/2)\bullet d[Cu^{2+}]/dt \text{ and } [C_4H_4O_5^{2-}]/[Cu^{2+}] = 1,$$

$$-(1/2)\bullet d[Cu2+]/dt = {}^{k}[Cu2+]4 \quad (11)$$

Integration of Eq. (11) gives

$$(1/6)\bullet[Cu^{2+}]^{-3} = kt + 1/6 \quad (12)$$

Equation 12 can be converted to Eq. (13):

$$(1/6)\bullet(1 - x)^{-3} = kt + 1/6 \quad (13)$$

The uptake rate of Cu^{2+} by malate•Mg–Al LDH was determined from Eq. (13). The plot of $(1/6)\bullet(1 - x)^{-3}$ versus t showed good linearity at each temperature, indicating that the reaction rate could be expressed by Eq. (10). At 30 °C, the value of the intercept was not 1/6. This was due to the very fast precipitation rate of Cu-Al LDH. The apparent rate constants at 10, 20, and 30 °C were 3.1×10^{-2}, 1.1×10^{-1}, and 2.3×10^{-1}, respectively. The apparent rate constant increased with increasing temperature. An Arrhenius plot of k, determined from the slopes of the straight lines, is also shown in Fig. 3. The apparent activation energy in this case was 71.2 $kJ\cdot mol^{-1}$, confirming that the uptake of Cu^{2+} by malate•Mg–Al LDH proceeded under chemical reaction control. Furthermore, the uptake of Cu^{2+} by this LDH occurred via chelation between Cu^{2+} and malate. The uptake of Cu^{2+} by tartrate•Mg–Al LDH was attributed to the 2:2 chelation between Cu^{2+} and tartrate, and is expressed as

$$2C_4H_4O_6^{2-} + 2Cu^{2+} \rightarrow [Cu_2(C_4H_3O_6)(C_4H_4O_6)]^- + H^+ \quad (14)$$

This reaction gives the following rate equation:

$$d[[Cu_2(C_4H_3O_6)(C_4H_4O_6)]^-]/dt = k[C_4H_4O_6^{2-}]^2[Cu^{2+}]^2 \quad (15)$$

Because Eq. (15) is similar to Eq. (10), it can be converted to Eq. (13). The uptake rate of Cu^{2+} by tartrate•Mg–Al LDH was determined from Eq. (13). The reaction rate in this case could be expressed by Eq. (15) because the plot of $(1/6)\bullet(1 - x)^{-3}$ versus t gave good linearity at each temperature. The apparent rate constants at 10, 20, and 30 °C were 1.1×10^{-1}, 1.9×10^{-1}, and 4.3×10^{-1}, respectively. An Arrhenius plot of k, determined from the slopes of the straight lines, is also shown in Fig. 3; in this case, the apparent activation energy is 48.9 $kJ\cdot mol^{-1}$. Thus, it was confirmed that the uptake of Cu^{2+} by tartrate•Mg–Al LDH proceeded under chemical reaction control. Furthermore, the uptake of Cu^{2+} by this LDH occurred via chelation between Cu^{2+} and tartrate. Similarly, the uptake of Cd^{2+} by three types of Mg–Al LDHs was also examined kinetically based on the chemical equation. For citrate•Mg–Al LDH, the apparent rate constants at 10, 20, and 30 °C were 5.9×10^{-3}, 1.0×10^{-2}, and 2.7×10^{-2}, respectively; thus, it is clear that the rate constant increased with increasing temperature. The apparent activation energy was calculated to be 53.6 kJ·mol^{-1}. This value confirms that the uptake of Cd^{2+} by citrate•Mg–Al LDH proceeded under chemical reaction control. The uptake of Cd^{2+} by this LDH occurred via 1:1 chelation between Cd^{2+} and citrate. For malate•Mg–Al LDH, the apparent rate constants at 10, 20, and 30 °C were 1.3×10^{-2}, 2.3×10^{-2}, and 5.5×10^{-2}, respectively. The apparent activation energy was calculated to be 50.8 kJ·mol^{-1}, confirming that the uptake of Cd^{2+} by malate•Mg–Al LDH proceeded under chemical reaction control. Furthermore, the uptake of Cd^{2+} by this LDH occurred via 1:2 chelation between Cd^{2+} and malate. For tartrate•Mg–Al LDH, the apparent rate constants at 10, 20, and 30 °C were 5.8×10^{-3}, 2.0×10^{-2}, and 1.2×10^{-1}, respectively. An Arrhenius plot of k provides an apparent activation energy of 106.5 kJ·mol^{-1}. This value confirms that the uptake of Cd^{2+} by tartrate•Mg–Al LDH proceeded under chemical reaction control. Furthermore, the uptake of Cd^{2+} by tartrate•Mg–Al LDH occurred via 1:2 chelation between Cd^{2+} and tartrate. To summarize, the reaction simply proceeded via chelation at a certain rate, which was different for each combination.

The citrate-modified Mg–Al LDH was suggested to have citrate anions not only intercalated in the interlayer but also adsorbed on the surface (Kameda et al., 2008a, 2009a). Because Mg–Al LDH has the basic structure of $Mg(OH)_2$, organic acid anions such as citrate are likely to be adsorbed on the surface of $Mg(OH)_2$. The organic acid anion-adsorbed $Mg(OH)_2$ is also expected to take up heavy metal ions from aqueous solution. Furthermore, by coprecipitation as well as Mg–Al LDHs intercalated with organic anions, new layered magnesium hydroxides modified with organic acid anions may be produced. Narita et al. and Tagaya et al. have previously reported the formation of layered zinc hydroxides modified with aliphatic carboxylates, aromatic carboxylates, anionic surfactants, and organic oxychlorides (Tagaya et al., 1996; Takahashi et al., 1997; Ogata et al., 1998; Ogata et al., 2000). Hence, layered magnesium hydroxides modified with organic acid anions are also expected to take up heavy metal ions from aqueous solutions. $Mg(NO_3)_2$ solution was added dropwise to citrate, malate, and tartrate solutions at 30 °C under mild agitation (Kameda et al., 2009b). Solution pH was adjusted to 10.5 by addition of NaOH solution. After addition of $Mg(NO_3)_2$ solution, the resultant suspensions were incubated at 30 °C for 1 h. For malate and tartrate, new layered magnesium hydroxides in which brucite layers were bridged by $C_4H_4O_5^{2-}$ and $C_4H_4O_6^{2-}$ were prepared. The $C_4H_4O_5^{2-}$ and $C_4H_4O_6^{2-}$ may also have been absorbed on the surfaces of the hydroxides. For citrate, $Mg(OH)_2$ with absorbed $C_6H_5O_7^{3-}$ was produced. These materials were found to take up Cu^{2+} rapidly from an aqueous solution at a constant pH of 5.0. The Cu^{2+} uptake was attributed to the formation of chelate complexes of Cu^{2+} with $C_6H_5O_7^{3-}$, $C_4H_4O_5^{2-}$, and $C_4H_4O_6^{2-}$.

Nakayama et al. (2007) examined mercaptocarboxylic acid as a chelating agent. Mercaptocarboxylic acids were intercalated into Mg–Al LDH accompanied by oxidation of mercapto groups to form disulfide. The intercalation compound selectively adsorbed Hg^{2+} and Ag^{+}, whereas there was almost no adsorption of Pb^{2+} and Cu^{2+}. Quantitative adsorption of Hg^{2+} was observed for mercaptosuccinic acid and 3,3′-dithiodipropionic acid intercalated by LDH. The adsorption sites were verified to be disulfide bonds. The selective adsorption of heavy metal ions was due to the effect of a confined field in the interlayer of LDH. Pavlovic et al. (2009) investigated the adsorption of Cu^{2+}, Cd^{2+}, and Pb^{2+} by LDHs intercalated with the chelating agents diethylenetriaminepentaacetate (dtpa) and meso-2,3-dimercaptosuccinate (dmsa). NO_3•Zn–Al LDH was intercalated with the chelating agents dtpa and dmsa by anion exchange, and the adsorption of these metal ions occurred mainly through chelation by interlayer ligands. The adsorption isotherms indicated a high host-guest affinity. The presence of dmsa in the Zn–Al LDH interlayer increased the affinity for the metal ions relative to dtpa•Zn–Al LDH. Although Cu^{2+}, Cd^{2+}, and Pb^{2+} were mainly adsorbed by chelation, a certain amount may have been precipitated due to the higher surface alkalinity of the LDH. $PbCO_3$ was also detected. Precipitation of amorphous metal and isomorphous substitution of Zn^{2+} by another metal ion in the brucite-like layer might also have been involved in the adsorption processes. Liang et al. (2010) also examined the sorption of Pb^{2+} by LDH intercalated with dtpa. Mg–Al LDH intercalated with dtpa was prepared by coprecipitation. The maximum adsorption of Pb^{2+} was about 170 and 40 mg/g for dtpa•Mg–Al LDH and Cl•Mg–Al LDH, respectively. Langmuir isotherms described the sorption data well, and a pseudo-second-order kinetic model fit the sorption kinetic processes better for both the LDHs. The mechanism of Pb^{2+} adsorption on dtpa•Mg–Al LDH can be explained by Pb–dtpa chelation, while that for Cl•Mg–Al LDH was primarily surface-induced precipitation.

3. Uptake of nonionic organic compounds by organically modified LDH

Recently, LDH and its oxide have been investigated as scavengers to treat wastewater containing anionic organic contaminants. The materials are known to function as effective anion adsorbents in the uptake of phenols, terephthalate, anionic surfactants, ionizable pesticides and herbicides, humic and fulvic acid, anionic dyes, and colored organics from aqueous solutions (Ulibarri et al., 1995; Hermosin et al., 1996; Pavlovic et al., 1997; Barriga et al., 2002; Crepaldi et al., 2002; You et al., 2002a; Orthman et al., 2003; Pavlovic et al., 2005; Li et al., 2005; Cardoso and Valim, 2006; Vreysen and Maes, 2008; Chao et al., 2009; Gaini et al., 2009; Valente et al., 2009; Setti et al., 2010; Sun et al., 2010). Based on the ability of LDH to intercalate various types of anions in interlayers, organic–inorganic composite materials could be produced by intercalation of organic anions having functional groups in their structure into the interlayers of LDH. LDHs modified with organic anions are expected to take up nonionic organic compounds from aqueous solution depending on the functional groups in the structure of the intercalated organic anions. For example, Mg–Al LDH intercalated with dodecylsulfate (DS) could adsorb hydrophobic pesticides such as linuron, atrazine, acephate, and diazinon from aqueous solution (Villa et al., 1999). This is due to the modification of the interlayer surface of the LDH from hydrophilic to hydrophobic. Modification of Mg–Al LDHs with DS and dodecylbenzenesulfonate resulted in very high adsorption of the uncharged pesticide triadimefon (Celis et al., 2000). The high adsorption of triadimefon was due to the paraffin-like sorbents. Dissolution of the hydroxide structure of the LDHs at low pH (< 4) limited their use as sorbents in acidic conditions. For interlayer adsorption of triadimefon in LDHs, X-ray diffraction (XRD) and Fourier transform infrared (FT-IR) studies indicated weak interactions between the pesticide molecules and the sorbents, in agreement with the high reversibility observed in the adsorption–desorption isotherms. You et al. (2002b) examined the surfactant-enhanced adsorption of organic compounds by Mg–Al LDHs. Organo-Mg–Al LDHs were prepared by incorporating anionic surfactants, octylsulfate, DS, 4-octylbenzenesulfonate, and dodecylbenzenesulfonate, into Mg–Al LDH via ion exchange. The anionic surfactants were intercalated into Mg–Al LDH with the surfactants oriented perpendicular in the interlayer. The octylsulfate formed bimolecular films, and other surfactants resulted in monolayer structures. Intercalation of surfactants into Mg–Al LDH decreased the surface area, whereas surfactants dramatically enhanced the LDH affinity for 1,2,4-trichlorobenzene and 1,1,1-trichloroethane in aqueous solutions. Adsorption potential depended on the type of surfactant as well as the configuration of surfactant molecules within Mg–Al LDH interlayers. The adsorption characteristics indicated that the retention of organic compounds by organo-Mg–Al LDHs was due to a partitioning mechanism. Zhao and Nagy (2004) investigated DS•Mg–Al LDHs with a Mg/Al molar ratio of 2–5 for trapping chlorinated organic pollutants in water. Adsorption of trichloroethylene and tetrachloroethylene by DS•Mg–Al LDH did not follow a simple trend with increasing organic content but rather varied with probable DS configuration and LDH composition. DS•Mg–Al LDH with a Mg/Al molar ratio of 3 had the highest affinity for both hydrophobic organic compounds in water because it has the optimal charge density for forming an effective partition medium with the LDH interlayer. DS•Mg–Al LDH with a Mg/Al molar ratio of 2 had the lowest sorption affinity, although it contained the highest amount of DS, and adsorption is thought to occur on edge/external surface area. Wang et al. (2005) examined the surface modification of Mg–Al LDH and incorporation of hydrophobic organic compounds. Surface properties of Mg–Al LDHs intercalated with DS and dodecylbenzenesulfonate were modified from hydrophilic to hydrophobic. Both Mg–Al LDHs adsorbed the nonionic hydrophobic organic pesticide, chlorpyrifos, which was successfully retained in the hydrophobic region of both Mg–Al LDHs. Chuang et al. (2008) examined the removal of 2-chlorophenol from aqueous solution by NO_3•Mg–Al LDH and DS•Mg–Al LDH. The dissociation of 2-chlorophenol was 8.56. A high concentration of 2-chlorophenol was adsorbed by the hydrophilic–hydrophilic physical interaction of NO_3•Mg–Al LDH when the pH was >8.56; however, the hydrophobic-hydrophobic partition interaction of DS•Mg–Al LDH was more important when the pH was <8.56. The efficiencies of both LDHs with respect to the enhancement of 2-chlorophenol adsorption are strongly dependent on pH, speciation, ionic strength, and metal dissolution.

As mentioned above, organo LDH has the potential to take up nonionic organic contaminants from an aqueous solution. We studied the preparation of organo-Mg–Al LDH and its uptake of bisphenol A. Furthermore, we studied the selective uptake of nonionic organic contaminants from aqueous solution. These studies are discussed in the following sections.

3.1 Nonselective uptake of aromatic compounds

Recently, the contamination of effluents and water in our environment with hazardous organic materials has become more serious, and a great deal of effort has been made to develop water-cleaning technology and materials usable for this purpose. It is generally known that toxic organic contaminants, including PCB (polychlorinated biphenyls), dioxin, fran, and other so-called endocrine disruptors and organic halides have low solubility in water. For effective elimination of toxic organic materials from aqueous environments, the development of an adsorbent or absorbent for these materials at fairly low concentrations is an urgent requisite. Some types of oils are candidates for the role of scavengers. Sufficiently frequent contact of the contaminants with these oils will require an energy-intensive process and dispersion of oil droplets of adequately small size coupled with oil–water phase separation. To overcome these challenges, the primary objective of this series of studies was to investigate the use of solid particles modified with organic materials that perform the function of picking up the target organic contaminant. Solids of this type are expected to enable handling of the solid particles housing the oily and viscous organic materials and even the water-soluble materials in their structure. The first attempt was the formation of LDH particles holding an organic substance with the desired function in their interlayers. When anionic organic acid species are intercalated into the interlayers of LDH, they remain functional; the chemically modified LDH can be an effective scavenger for organic contaminants in aqueous effluents.

We prepared DS•Mg–Al LDH with a Mg/Al molar ratio of 3 by adding a solution of $Mg(NO_3)_2$ and $Al(NO_3)_3$ to an aqueous solution of DS at a constant pH of 10.0 (Kameda et al., 2005b). The intercalation of DS^- into Mg–Al LDH was controlled primarily by the charge balance between the positive electric charge of Al-bearing brucite-like octahedral layers and the anion in the interlayer. The capacity of DS•Mg–Al LDHs and CO_3•Mg–Al LDH to take up bisphenol A from an aqueous solution was examined by suspending 100-mg samples in 20 mL of 100 mg/L bisphenol A solution under shaking at 20 °C for 10~360 min. For DS•Mg–Al LDHs with a DS/Al molar ratio of 1.0, the concentration of bisphenol A decreased very rapidly to 15 mg/L within 30 min. For DS•Mg–Al LDH with a DS/Al molar ratio of 0.5, the concentration of bisphenol A decreased to 50 mg/L within 30 min. However, for CO_3•Mg–Al LDH with no DS content, only a slight decrease in the concentration of bisphenol A was observed. These results indicate that DS^- in the interlayer gives rise to the uptake of bisphenol A from the aqueous solution into the chemically modified Mg–Al LDH, probably caused by the hydrophobic interactions between the DS^- and bisphenol A. The organic material DS maintained adequate functionality in the interlayer of DS•Mg–Al LDH and can be handled as a fine powder with hydrophobic properties. In the XRD patterns of DS•Mg–Al LDH before and after the uptake of bisphenol A, no detectable change was observed. This implies that the basal spacing of DS•Mg–Al LDH was not affected by the uptake of bisphenol A, i.e., bisphenol A did not cause the interlayer to expand by being incorporated into the space between the intercalated DS^- and $Mg(OH)_2$-type host lattice. Bisphenol A is thought to be incorporated into the openings between the alkyl groups of DS^- in the interlayer of DS•Mg–Al LDH; this space is approximately 22 Å. It is generally accepted that the large interlayer space, occupied by organic anions, accommodates organic compounds of long molecular lengths, such as bisphenol A. The extraction of bisphenol A taken-up by DS•Mg–Al LDH with a DS/Al molar ratio of 1.0 has been attempted using ethanol. In the extraction with deionized water free of ethanol, the concentration of bisphenol A was 16.8 mg/L, corresponding to extraction of less than 5% of the bisphenol A uptaken by DS•Mg–Al LDH; extraction of bisphenol A with water was minimal. The concentration of bisphenol A in the extraction media increased sharply with increasing ethanol concentration and was as high as 320 mg/L for 99.5% ethanol solution. This observed value is equivalent to approximately 90% of the bisphenol A taken-up by DS•Mg–Al LDH. The bisphenol A taken-up by DS•Mg–Al LDH was easily extracted with ethanol. This extraction was attributed to the hydrophobic interactions between ethyl groups of ethanol and methyl groups of bisphenol A as well as hydrogen bonds formed by hydroxyl groups of the two compounds. The DS^- intercalated in the interlayer of DS•Mg–Al LDH was not dissolved by ethanol solution. To confirm the stable intercalation of DS^- in the interlayer after extraction, the uptake capacity of ethanol-treated DS•Mg–Al LDH for bisphenol A in an aqueous solution was compared with that of fresh DS•Mg–Al LDH. In the presence of a suspension of ethanol-treated DS•Mg–Al LDH, the concentration of bisphenol A decreased from 100 mg/L to 11.9 mg/L, similar to that for fresh DS•Mg–Al LDH (10.9 mg/L). This result proves that DS•Mg–Al LDH was regenerated by the extraction of bisphenol A with ethanol, i.e., the intercalated DS^- remained in the interlayer. DS^- was rigidly fixed in the interlayer, in contact with ethanol as well as bisphenol A solution. To summarize, it was confirmed that DS•Mg–Al LDH retains its activity during repeated cycles of uptake and extraction of bisphenol A. The DS^- in the solid particles was in a stable arrangement in the solid phase while simultaneously reacting rapidly with bisphenol A outside the particles.

We examined the preparation and characterization of Mg–Al LDHs intercalated with 2-naphthalene sulphonate (2-NS^-) and 2,6-naphthalene disulphonate (2,6-NDS^{2-}) (Kameda et al., 2006). 2-NS•Mg–Al LDH and 2,6-NDS•Mg–Al LDH, with 2-NS^- or 2,6-NDS^{2-} intercalated in the interlayer of Mg–Al LDH, were prepared by dropwise addition of a mixed aqueous solution of $Mg(NO_3)_2$ and $Al(NO_3)_3$ at a Mg/Al molar ratio of 3.0 to an aqueous ethanol solution of 2-NS-Na or to an aqueous solution of 2,6-NDS-Na_2 at a constant pH of 10.0 and left standing at 30 °C for 1 h. The intercalated 2-NS^- and 2,6-NDS^{2-} maintained their intrinsic molecular structure in the interlayer of Mg–Al LDH. XRD analysis confirmed that 2-NS•Mg–Al LDH had two basal spacings of approximately 19 and 9 Å. It is likely that 2-NS^- was intercalated in the interlayer of Mg–Al LDH in two orientations. It is believed that the naphthalene ring of 2-NS^- was oriented parallel and perpendicular to the brucite ($Mg(OH)_2$)-like layers of Mg–Al LDH. The 2-NS/Al molar ratio in the 2-NS•Mg–Al LDH increased with increasing 2-NS/Al molar ratio in solution, and the interlayer of Mg–Al LDH was intercalated with 2-NS^- to near full capacity for neutralization of the positive charge of the host lattice at a 2-NS/Al molar ratio of 2.0 in solution. The XRD patterns of 2,6-NDS•Mg–Al LDH exhibited very weak, broad peaks corresponding to the basal spacings at 2θ ranging from 6–11°, suggesting an irregular arrangement of 2,6-NDS^{2-} in the Mg–Al LDH interlayers. This was attributed to the limited mobility of the 2,6-NDS^{2-} resulting from the pin-up of the anions at the two $-SO_3^-$ groups on either side of the interlayer. In contrast to 2-NS^-, the intercalation with 2,6-NDS^{2-} was approximately 80% of the value expected due to the charge balance on the Mg–Al LDH. The disordered configuration of 2,6-NDS^{2-} in the interlayer of Mg–Al LDH was presumed to restrict the intercalation.

We also investigated the properties of 2-NS•Mg–Al LDH and 2,6-NDS•Mg–Al LDH as scavengers for organic contaminants having aromatic rings in their molecular structures (Kameda et al., 2007). The ability of 2-NS•Mg–Al LDH and 2,6-NDS•Mg–Al LDH to take up bisphenol A from aqueous solution was examined by suspending 0.1 g of the samples in 20 mL of 100 mg/L bisphenol A solution while shaking at 20 °C for 10–360 min. Figure 4 shows the variation in the concentration of bisphenol A with time for 2-NS•Mg–Al LDHs and for CO_3•Mg–Al LDH as a reference material.

For 2-NS•Mg–Al LDHs, the concentration of residual bisphenol A decreased very rapidly from 100 mg/L to 10 mg/L or less in the first 30 min and then decreased fairly slowly. In particular, at a 2-NS/Al of 0.43, the lowest molar ratio, the concentration decreased to less than 3 mg/L within 30 min. In contrast, for CO_3•Mg–Al LDH without chemical modification of the interlayer, only a slight decrease in the bisphenol A concentration was observed. These results demonstrate that the adsorption of bisphenol A on the LDH particles plays a minor role and that the intercalated organic entity may be largely responsible for the uptake of bisphenol A from aqueous solution. The uptake of bisphenol A from aqueous solution was attributed to the association of the naphthalene rings of the 2-

NS^- with the benzene rings of the incorporated bisphenol A. 2,6-NDS•Mg–Al LDH also had the potential to take up bisphenol A from aqueous solution. The intrinsic functions of the naphthalene ring of 2,6-NDS^{2-} were unchanged in the interlayer and positively affected the uptake of bisphenol A due to the association between the aromatic rings. Mg–Al LDH with lower contents of 2-NS^- or 2,6-NDS^{2-} in the interlayer had larger potential to take up bisphenol A from aqueous solution than that with more intercalated organic acid. The former is thought to have larger spaces to accommodate more bisphenol A in the interlayer than the latter, which had limited space due to the higher content of the intercalated entity. It is expected that practical water-cleaning processes should be able to continuously treat wastewater contaminated with hazardous organic compounds at an acceptable flow rate. This can be achieved by passing wastewater through a column containing absorbents. In our study, the test solutions containing bisphenol A and ethanol were alternately passed through a DS•Mg–Al LDH particle column (length = 48 mm, inner diameter = 4.6 mm) to examine the cycles of uptake of bisphenol A and elution of uptaken bisphenol A by ethanol (Kameda et al., 2009c). The test solution containing bisphenol A, ethanol, or deionized water was passed through the column at a rate of 1.5 mL/min at 80 psi using a gradient pump according to the designed pattern of flow. The column packed with fine particles of the adsorbent produced a high pressure drop, and a high operation pressure was applied to maintain the desired flow rates. The volume of the column was 0.8 mL, and the space velocity was calculated to be 112.5 h^{-1} under the conditions used in this study. The column was filled with 0.5 g of DS•Mg–Al LDH particles and maintained at 20 °C using a water bath. Metal filters with pore diameters of 2.0 and 0.5 μm were placed at the top and bottom of the column to hold the particles. Residual bisphenol A in the effluent from the column was continuously monitored at a wavelength of 280 nm using a photomultiplier detector. First, deionized water was passed through the column for 10 min to clean the column and to monitor the system. Then, bisphenol A solution was fed to the column. Figure 5 presents the variation in the absorbance (λ = 280 nm) observed using the photomultiplier detector for a flow of 100 mg/L bisphenol A solution without the column and that for the effluent from the column containing DS•Mg–Al LDH particles.

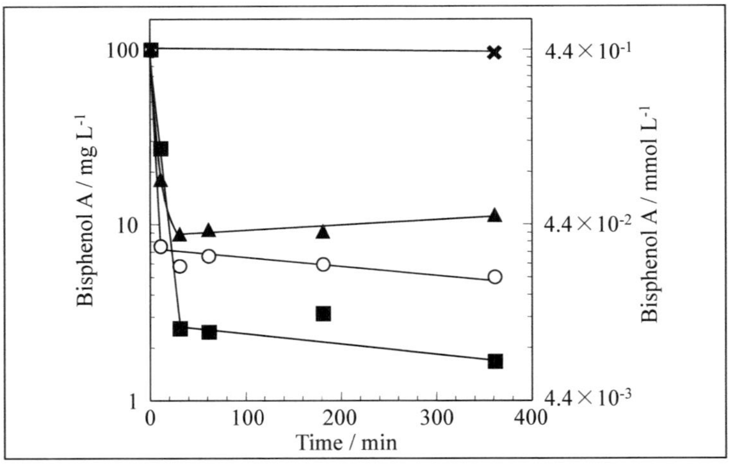

Fig. 4. Variation in the concentration of bisphenol A with time for suspensions of 2-NS•Mg–Al LDHs and CO_3•Mg–Al LDH in bisphenol A solution. 2-NS•Mg–Al LDH: the 2-NS/Al molar ratio was 0.43 (■), 0.82 (○), and 0.96 (▲) CO_3•Mg–Al LDH: (×)

In the absence of the column (Fig. 5(a)), no absorbance was detected during the first 10 min of deionized water flow. The line corresponding to the first 10 min in Fig. 5(a) is hereafter referred to as the baseline. Immediately after the flow of bisphenol A solution, the absorbance quickly increased to a value corresponding to the concentration of bisphenol A and remained constant. This proves that the bisphenol A in aqueous solution was continuously monitored with high sensitivity by the detector. When deionized water was passed through the column containing DS•Mg–Al LDH particles (Fig. 5(b)) for 10 min, the absorbance rapidly increased and immediately decreased to the baseline. The rapid increase in the absorbance may be attributable to the dissociation of molecular sodium dodecylsulfate (SDS) dissolved in the intercalated DS^- medium in DS•Mg–Al LDH due to hydrophobic interactions. As a result, the DS•Mg–Al LDH particles in the column required an initial washing with water to remove SDS from the LDH. After the first 10 min (Fig. 5(b)), when bisphenol A was passed through the column, the measured absorbance remained on the baseline for 200 min, i.e., the concentration of bisphenol A in the effluent was less than the detection limit of the detector used in this study (0.1 mg/L). This elimination of bisphenol A was observed for the effluent with a volume 356 times that of the column. This suggests that bisphenol A was taken up by the DS•Mg–Al LDH particles during its flow through the column. However, the absorbance gradually increased after 200 min, indicating the presence of bisphenol A in the effluent. This reflects the gradual decrease in the accommodation space for bisphenol A in the interlayers of DS•Mg–Al LDH with time, approaching a break-through point, 260 min, at which the concentration reached 10 mg/L. The column has the potential to be used for the treatment of bisphenol A solution for 260 min under the conditions used in this study. Due to the flow of bisphenol A solution through the column, the pH of the effluent increased immediately from an initial value of 5.2 to near 7.0 and then remained in the range of 6.0–8.0. The Mg^{2+} concentrations were very low at each stage. These results are acceptable according to the water quality standard in Japan ($5 < pH < 9$; Mg^{2+} is not listed there). From the XRD analysis, it can be deduced that the configuration of the intercalated DS^- in DS•Mg–Al LDH was not affected by the flow of bisphenol A solution.

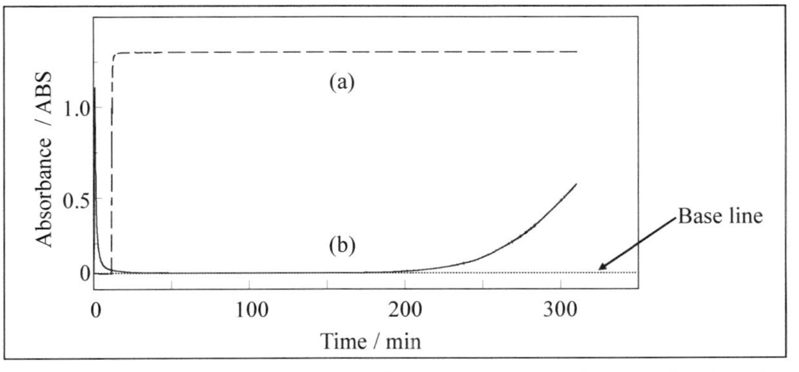

Fig. 5. Variation in absorbance (λ = 280 nm) observed using a photomultiplier detector for (a) flow of 100 mg/L bisphenol A solution without a column and (b) effluent from a column containing DS•Mg–Al LDH particles.

Furthermore, bisphenol A solution and ethanol were alternately passed through a column containing DS•Mg–Al LDH particles to investigate the possibility of continuous treatment, consisting of repeated cycles of bisphenol A uptake from the aqueous solution and ethanol elution of the uptaken bisphenol A. The designed pattern of the alternate flow of bisphenol A solution and ethanol is presented in Fig. 6 (A). This figure represents a flow series of deionized water for 0–10 min, 100 mg/L bisphenol A solution for 10–40 min, deionized water for 40–45 min, and ethanol solution for 45–60 min to elute bisphenol A from the DS-layer. In the ethanol elution stage, the ethanol concentration increased to 50 vol% at 10 vol%/min during the first 5 min and remained constant at 50 vol% for the following 5 min.

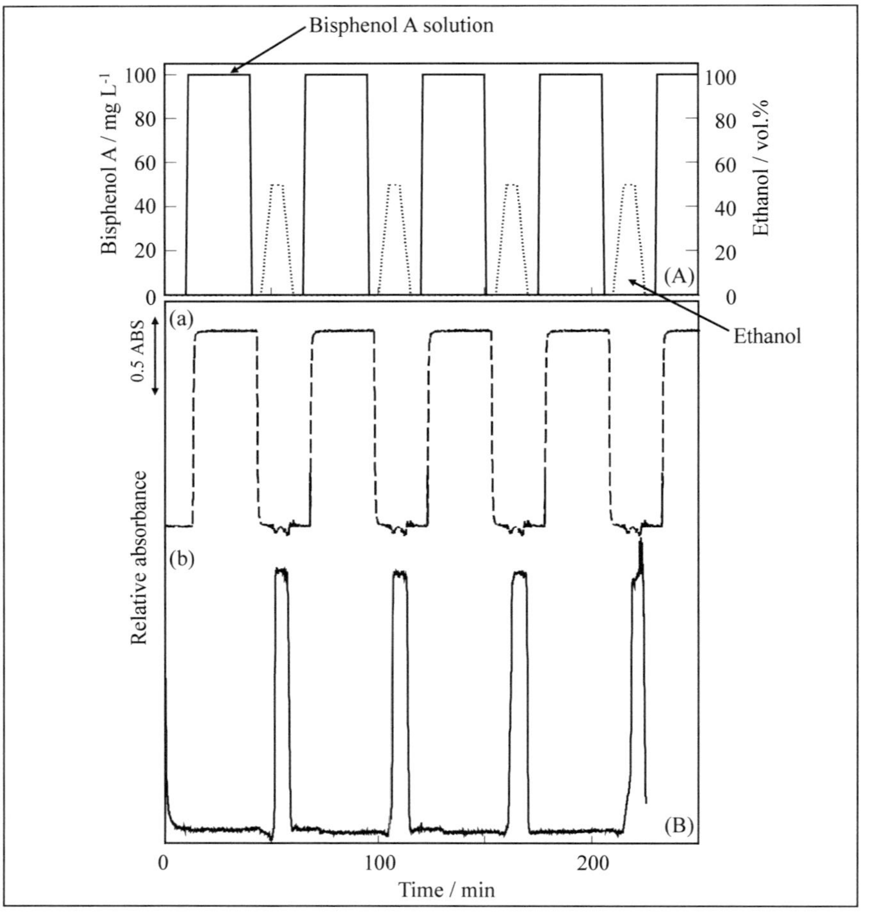

Fig. 6. (A) Designed pattern for alternate flow of bisphenol A solution and ethanol with a flow rate of 1.5 mL/min. (B) Variation in the absorbance (λ = 280 nm) of the effluent obtained by alternate flow of bisphenol A solution and ethanol (a) without a column and (b) through a column containing DS•Mg-Al LDH particles.

Then, for the last 5 min, the concentration decreased to 0 vol% at a constant rate of 10 vol%/min. After the flow of deionized water for 60–65 min, 100 mg/L bisphenol A solution was passed through the column again, and this cycle was repeated. Figure 6 (B) presents the variation in the absorbance (λ = 280 nm) for the effluent from the column by the alternate flow of bisphenol A and ethanol solutions and the corresponding curve for the system without the column. For the system without the column (Fig. 6 (B)(a)), a constant absorbance was obtained for the flow of bisphenol A solution, while no absorbance was detected for the flow of deionized water and ethanol. The variation in the absorbance corresponds well to the flow conditions presented in Fig. 6 (A). For the system with a column containing DS•Mg–Al LDH particles (Fig. 6 (B)(b)), no absorbance was detected during the flow of bisphenol A solution through the column, indicating that bisphenol A was almost completely eliminated by the DS•Mg–Al LDH particles. In contrast, the absorbance increased with the flow of ethanol solution, confirming that bisphenol A taken-up by the DS•Mg–Al LDH particles was eluted by ethanol. The maximum concentration of bisphenol A in ethanol was 130 mg/L, which was higher than the concentration (100 mg/L) of bisphenol A solution, indicating that the bisphenol A was concentrated to 130% under the conditions in this study. These results show that bisphenol A could be continuously eliminated by DS•Mg–Al LDH, which can be regenerated through ethanol elution. This suggests that the uptake of bisphenol A from the aqueous phase and its elution from the DS^- layer in the LDH are qualitative and fairly quick processes. The absorbance measurement was performed over 224 min of flow, as presented in Fig. 6 (B)(b), and ceased due to the increase in pressure inside the column. This may be attributable to the swelling of the DS•Mg–Al LDH particles caused by the incorporation of ethanol in the interlayers of the LDH. To summarize, the DS•Mg–Al LDH particles packed in a column were found to continuously take up bisphenol A from the aqueous solution. Furthermore, they were regenerated by elution of the uptaken bisphenol A by ethanol. The column containing DS•Mg–Al LDH particles is expected to be applicable for the treatment of wastewater containing organic substances. According to the results of this study, the composite materials coupling LDH with organic anions are expected to be effective absorbents for organic as well as inorganic contaminants in aqueous solution. By such a technique, water-soluble or intrinsically liquid materials can be applied to water-cleaning processes based on column-style operations.

3.2 Selective uptake of aromatic compounds

In environmental systems, hazardous organic contaminants can coexist with natural organic materials in wastewater effluents. If Mg–Al LDHs intercalated with organic anions selectively take up target organic materials from wastewater streams, they may be useful as effective scavengers for water treatment. To take up target organic materials selectively, the electronic state of the intercalated aromatic anions can be manipulated by the addition of electron-donating or -withdrawing substituents, thereby activating or deactivating the aromatic rings, respectively. Activated and deactivated systems interact strongly, generating a π–π stacked geometry (Gung and Amicangelo, 2006). Mg–Al LDHs intercalated with aromatic anions are thus expected to selectively take up hazardous aromatic compounds from wastewater due to the strong electrostatic attraction between the aromatic rings of Mg–Al LDHs and those of the hazardous aromatics.

We investigated the selective uptake of six aromatic compounds with different functional groups from aqueous solutions by Mg–Al LDH intercalated with 2,7-naphthalene disulfonate ($2,7\text{-}NDS^{2-}$) (Kameda et al., 2009d, 2009e). The six examined compounds were 1,3-dinitrobenzene (DNB), nitrobenzene (NB), benzaldehyde (BA), anisole (AS), *N,N*-dimethylaniline (DMA), and 1,2-dimethoxybenzene (DMB). 2,7-NDS•Mg–Al LDH (Mg/Al = 2.9; 2,7-NDS/Al = 0.31) was added to 20 mL of a single solution consisting of each aromatic compound at a concentration of 0.5 mM, and it was shaken at 20 °C for 120 min. In this case, the amount of 2,7-NDS•Mg–Al LDH added was such that the molar ratio of intercalated 2,7-NDS^{2-} to each compound was 50. To demonstrate the effect of the interlayer anions on the uptake of aromatics, DS•Mg–Al LDH (Mg/Al = 3.2; DS/Al = 0.98) and CO_3•Mg–Al LDH were also used as reference materials. When DS•Mg–Al LDH was added to the single solution, the molar ratio of intercalated DS^- to each compound was 50. The amount of CO_3•Mg–Al LDH was 0.5 g. Table 1 shows the uptake of each aromatic compound from the single solution by 2,7-NDS•Mg–Al LDH, DS•Mg–Al LDH, and CO_3•Mg–Al LDH.

Mg-Al LDH	Uptake (%)					
	DNB	NB	BA	AS	DMA	DMB
2,7-NDS^{2-}	84.7	56.5	46.4	28.3	20.4	14.2
DS^-	56.0	61.9	41.5	66.9	75.0	62.8
CO_3^{2-}	0	5.7	5.1	10.5	13.3	0

Table 1. Uptake of each aromatic compound from a single solution by 2,7-NDS•Mg–Al LDHs, DS•Mg–Al LDH, and CO_3•Mg–Al LDH.

The uptake of each aromatic compound by 2,7-NDS•Mg–Al LDHs and DS•Mg–Al LDH was greater than that by CO_3•Mg–Al LDH. In the case of CO_3•Mg–Al LDH, in which no chemical modification of the interlayer occurred, the uptake was attributed to the adsorption of the aromatic compounds on the surface of the CO_3•Mg–Al LDH particles. The high uptake by 2,7-NDS•Mg–Al LDHs and DS•Mg–Al LDH was attributed to the interaction between the aromatic compounds and the 2,7-NDS^{2-} and DS^- intercalated in the interlayer of Mg–Al LDH, respectively. In the case of DS•Mg–Al LDH, the maximum and minimum uptakes were 75.0% for DMA and 41.5% for BA, and the dependence of the uptake on the type of aromatic compound was low. This nonselective uptake was likely due to a hydrophobic interaction between the alkyl groups of DS^- intercalated in the interlayer and the benzene ring of the compound in the aqueous solution. For 2,7-NDS•Mg–Al LDH, the uptake decreased in the following order: DNB > NB > BA > AS > DMA > DMB. The maximum and minimum uptakes were 84.7% for DNB and 14.2% for DMB, and the dependence of the uptake on the type of aromatic compound was high. This selective uptake was attributed to π-π stacking interactions between the benzene ring of the compound and the naphthalene ring of 2,7-NDS^{2-} intercalated in the interlayer spaces of Mg–Al LDH. The uptake is thought to depend on the electronic state of the aromatic compound, which, in turn, depends on the functional group. The electronic states of the benzene rings of DNB, NB, and BA are compared as follows. All these compounds contain electron-withdrawing groups, namely, $-NO_2$ and -CHO groups. The strength of $-NO_2$ is greater than that of -CHO. DNB has two NO_2 groups, and, therefore, the benzene ring of DNB has the lowest electron density among the aromatics. Additionally, the benzene ring of NB has a lower electron density than that of BA. Next, the electronic states of the benzene rings of AS, DMA, and DMB are compared. These compounds are grouped for comparison because they all contain electron-donating groups, namely, $-OCH_3$ and $-N(CH_3)_2$ groups. DMB has two $-OCH_3$ groups, and, thus, the benzene ring of DMB has the largest electron density among these aromatics. The strength of the $-N(CH_3)_2$ group is greater than that of the $-OCH_3$ group. This suggests that the benzene ring of DMA has a larger density of electrons than that of AS. In other words, the density of electrons in the benzene ring in the aromatics is expected to decrease in the following order: DMB > DMA > AS > BA > NB > DNB. The electronic state of the naphthalene ring of 2,7-NDS^{2-} intercalated into the interlayer of 2,7-NDS•Mg–Al LDH can be described as follows. 2,7-NDS^{2-} has two -SO_3^- groups in its structure that are weak electron-withdrawing groups. The naphthalene ring is richer in electrons than the benzene ring is, and, thus, the naphthalene ring of 2,7-NDS^{2-} probably has a larger density of electrons due to the weak electron-withdrawing ability of the two -SO_3^- groups. The high uptake of aromatic compounds in the decreasing order of DNB > NB > BA > AS > DMA > DMB by 2,7-NDS•Mg–Al LDH was caused by π–π stacking interactions between the electron-rich naphthalene ring of 2,7-NDS^{2-} intercalated into the interlayer of 2,7-NDS•Mg–Al LDH and the benzene ring of the aromatics, which has lower electron density, in the decreasing order of DMB > DMA > AS > BA > NB > DNB. That is, the electron-poor benzene ring of aromatics showed strong interactions with the electron-rich naphthalene ring of intercalated 2,7-NDS^{2-}, and these interactions resulted in the large uptake of aromatics by 2,7-NDS•Mg–Al LDH. Table 2 shows the uptake of multiple aromatic compounds from mixed solutions I and II by 2,7-NDS•Mg–Al LDH. In this case, the mixed solution I contained 0.5 mM DNB, BA, and DMB. The mixed solution II contained 0.5 mM DNB, NB, BA, AS, and DMB.

	Uptake (%)				
	DNB	NB	BA	AS	DMB
Mixed solution I	82.8	—	30.2	—	3.7
Mixed solution II	79.3	47.5	32.2	12.0	6.6

Table 2. Uptake of multiple aromatic compounds from mixed solutions I and II by 2,7-NDS•Mg–Al LDH.

Even for the mixtures, uptake was high; uptake decreased in the order DNB > (NB) > BA > (AS) > DMB. This order matched that of the uptake from the single solution (Table 1). 2,7-NDS•Mg–Al LDH was found to selectively take up aromatic compounds with electron-poor benzene rings from the mixtures in aqueous solution.

NTS•Mg–Al LDH and ANDS•Mg–Al LDH, which respectively contained 1,3,6-naphthalenetrisulfonate (NTS^{3-}) and 3-amino-2,7-naphthalenedisulfonate ($ANDS^{2-}$) intercalated in the interlayer spaces, were also prepared by a coprecipitation technique (Kameda et al., 2010b). The naphthalene core of NTS^{3-} was likely oriented parallel to the brucite-like host layers of Mg–Al LDH, lying at the midpoint of the interlayer space. The naphthalene core of $ANDS^{2-}$ was oriented parallel to the brucite-like layers, at the midpoint of the interlayer space, when ANDS•Mg–Al LDH was prepared from a solution with a low concentration of $ANDS^{2-}$. When a solution with a high concentration of $ANDS^{2-}$ was used in the preparation, the naphthalene core of $ANDS^{2-}$ was thought to be oriented in an irregular fashion with respect to the brucite-like layers. NTS•Mg–Al LDH and ANDS•Mg–Al LDH were both capable of absorbing large quantities of aromatics from aqueous solution in the decreasing order of DNB > NB > BA > DMA > AS > DMB. The difference in absorption was attributed to differences in the extent of π–π stacking interactions occurring between the benzene rings of the aromatics and the naphthalene core of the intercalated NTS^{3-} and $ANDS^{2-}$. The uptake of aromatics in these cases probably occurred via insertion of the benzene rings of the aromatics between the naphthalene cores of the intercalated NTS^{3-} and $ANDS^{2-}$ and the brucite-like layers.

We also examined the effect of intercalated aromatic sulfonates on uptake of aromatic compounds from aqueous solutions by modified Mg–Al LDH (Kameda et al., 2010c). We utilized Mg–Al LDHs modified by intercalation with three aromatic sulfonates: 2,7-NDS^{2-}, benzenesulfonate (BS^-), and benzenedisulfonate (BDS^{2-}). BS•Mg–Al LDH and BDS•Mg–Al LDH, which respectively contained BS^- and BDS^{2-} intercalated in the interlayer space, were prepared by a coprecipitation technique (Kameda et al., 2008b). The intercalated BS^- and BDS^{2-} maintained their intrinsic molecular structures within the Mg–Al LDH interlayers. At low intercalation levels, the benzene ring of BS^- in BS•Mg–Al LDH was inclined at 30° relative to the plane of the brucite-like layers of Mg–Al LDH. With increasing BS^- content, the benzene ring adopted an additional configuration perpendicular to the Mg–Al LDH layers. In BDS-intercalated Mg–Al LDH, the benzene ring of BDS^{2-} was tilted at 26° relative to the plane of the Mg–Al LDH layers. Intercalation levels of BDS^{2-} were smaller than those of BS^- despite the greater charge density of BDS^{2-}, which was likely attributable to a greater degree of electrostatic repulsion between intercalated anions. The uptake of DNB was notably high and in the order 2,7-NDS•Mg–Al LDH > BS•Mg–Al LDH > BDS•Mg–Al LDH. This was definitively attributed to the π–π stacking interactions between the electron-poor benzene ring of DNB and the electron-rich aromatic rings of the intercalated aromatic sulfonates, which have electron density in the order 2,7-NDS^{2-} > BS^- > BDS^{2-}. To summarize, the electron-rich aromatic rings of intercalated aromatic sulfonates undergo stronger π–π stacking interactions with the electron-poor benzene ring of DNB. The uptake of AS by 2,7-NDS•Mg–Al LDH and BS•Mg–Al LDH was considerably lower than that of DNB by the same LDHs. The more electron-rich benzene ring of AS was thought to undergo weak π–π stacking interactions with the electron-rich aromatic rings of intercalated 2,7-NDS^{2-} and BS^-.

Finally, we examined the effect of the interlayer spacing of Mg–Al LDH on the ability to take up a nonionic organic material (Kameda et al., 2009f). Mg–Al LDHs, intercalated with 1-propanesulfonate (PS^-), 1-hexanesulfonate (HS^-), and 1-dodecanesulfonate (DDS^-), were prepared by coprecipitation, yielding PS•Mg–Al LDH, HS•Mg–Al LDH, and DDS•Mg–Al LDH, respectively. The increase in the alkyl chain lengths of the Mg–Al LDHs (PS^- < HS^- < DDS^-) resulted in the perpendicular orientation of the organic acid anions in the interlayer of Mg–Al LDH, which in turn resulted in more organic acid anions being accommodated in the interlayer space. An organic acid anion with a long molecular length was more easily intercalated in the interlayer of Mg–Al LDH than one with a short molecular length. This was attributed to the hydrophobic interaction between the alkyl chains, affecting the intercalation of the organic acid anions. The uptake of *N,N*-dimethylaniline (DMA) by Mg–Al LDHs increased in the order PS•Mg–Al LDH < HS•Mg–Al LDH < DDS•Mg–Al LDH. The uptake was attributed to the hydrophobic interactions between DMA and the intercalated PS^-, HS^-, and DDS^-. Thus, Mg–Al LDH, which has many large interlayer spacings when intercalated with organic acid anions, can take up a large number of DMA molecules from an aqueous solution.

4. Conclusion

This chapter summarizes the preparation of organically modified LDH by intercalation of organic anions into the interlayer of LDH and the uptake of heavy metal ions and nonionic organic compounds by the organically modified LDH. In particular, we developed hybrid inorganic–organic composite materials capable of selective uptake of heavy metal ions and nonionic organic compounds from aqueous solution. It is significant that we developed a relatively simple method for the preparation of the composite materials. Moreover, it must be emphasized that the utilization of a water-soluble reagent was possible by the incorporation of the reagent into an inorganic compound. The composite materials could remove heavy metal ions and bisphenol A from aqueous solutions with high concentrations to very low concentrations in a short time. These materials are expected to be applied for actual wastewater treatment.

5. Acknowledgment

This research was partially supported by the Ministry of Education, Science, Sports, and Culture, Grand-in-Aid for Young Scientists (A), 18681010, 2006-2008.

6. References

Allmann, R. (1968). The crystal structure of pyroaurite. *Acta Crystallographica*, Vol. B24, pp. 972–977.

Barriga, C.; Gaitan, M.; Pavlovic, I.; Ulibarri, M. A.; Hermosin, M. C.; Cornejo, J. (2002). Hydrotalcites as sorbent for 2,4,6-trinitrophenol: influence of the layer composition and interlayer anion. *Journal of Materials Chemistry*, Vol. 12, pp. 1027-1034.

Cardoso, L. P.; Valim, J. B. (2006). Study of acids herbicides removal by calcined Mg-Al-CO_3-LDH. *Journal of Physics and Chemistry of Solids*, Vol. 67, pp. 987-993.

Cavani, F.; Trifiro, F.; Vaccari, A. (1991). Hydrotalcite-type anionic clays: preparation, properties, and applications. *Catalysis Today*, Vol. 11, pp. 173-301.

Celis, R.; Koskinen, W. C.; Hermosin, M. C.; Ulibarri, M. A.; Cornejo, J. (2000). Triadimefon interactions with organoclays and organohydrotalcites. *Soil Science Society of America journal*, Vol. 64, pp. 36-43.

Chaara, D.; Pavlovic, I.; Bruna, F.; Ulibarri, M. A.; Draoui, K.; Barriga, C. (2010). Removal of nitrophenol pesticides from aqueous solutions by layered double hydroxides and their calcined products. *Applied Clay Science*, Vol. 50, pp. 292-298.

Chao, Y. F.; Lee, J. J.; Wang, S. L. (2009). Preferential adsorption of 2,4-dichlorophenoxyacetate from associated binary-solute aqueous systems by Mg/Al-NO_3 layered double hydroxides with different nitrate orientations. *Journal of Hazardous Materials*, Vol. 165, pp. 846-852.

Chuang, Y. H.; Tzou, Y. M.; Wang, M. K.; Liu, C. H.; Chiang, P. N. (2008). Removal of 2-chlorophenol from aqueous solution by Mg/Al layered double hydroxide (LDH) and modified LDH. *Industrial & Engineering Chemistry Research*, Vol. 47, pp. 3813-3819.

Crepaldi, E. L.; Tronto, J.; Cardoso, L. P.; Valim, J. B. (2002). Sorption of terephthalate anions by calcined and uncalcined hydrotalcite-like compounds. *Colloids and Surfaces A: Physicochemical and Engineering Aspects*, Vol. 211, pp. 103-114.

Dousova, B.; Machovic, V.; Kolousek, D.; Kovanda, F.; Dornicak, V. (2003). Sorption of As(V) species from aqueous systems. *Water, Air, and Soil Pollution*, Vol. 149, pp. 251-267.

Fujii, S.; Sugie, Y.; Kobune, M.; Touno, A.; Touji, J. (1992). Uptakes of Cu^{2+}, Pb^{2+} and Zn^{2+} on synthetic hydrotalcite in aqueous solution. *Nippon Kagaku Kaishi*, Vol. 1992, pp. 1504-1507.

Gaini, L. E.; Lakraimi, M.; Sebbar, E.; Meghea, A.; Bakasse, M. (2009). Removal of indigo carmine dye from water to Mg-Al-CO_3-calcined layered double hydroxides. *Journal of Hazardous Materials*, Vol. 161, pp.627-632.

Gasser, M. S.; Aly, H. F. (2009). Kinetic and adsorption mechanism of Cu(II) and Pb(II) on prepared nanoparticle layered double hydroxide intercalated with EDTA. *Colloids and Surfaces A: Physicochemical and Engineering Aspects*, Vol. 336, pp. 167-173.

Gillman, G. P. (2006). A simple technology for arsenic removal from drinking water using hydrotalcite. *Science of the Total Environment,* Vol. 366, pp. 926-931.

Goh, K. H.; Lim, T. T.; Dong, Z. (2008). Application of layered double hydroxides for removal of oxyanions: A review. *Water Research,* Vol. 42, pp. 1343-1368.

Goswamee, R. L.; Sengupta, P.; Bhattacharyya, K. G.; Dutta, D. K. (1998). Adsorption of Cr(VI) in layered double hydroxides. *Applied Clay Science,* Vol. 13, pp. 21-34.

Gung, B. W.; Amicangelo, J. C. (2006). Substituent effects in C_6F_6-C_6H_5X stacking interactions. *Journal of organic chemistry*, Vol. 71, pp. 9261-9270.

Hermosin, M. C.; Pavlovic, I.; Ulibarri, M. A.; Cornejo, J. (1996). Hydrotalcite as sorbent for trinitrophenol: Sorption capacity and mechanism. *Water Research,* Vol.30, pp. 171-177.

Ingram, L.; Taylor, H. F. W. (1967). The crystal structures of sjogrenite and pyroaurite. *Mineralogical Magazine*, Vol. 36, pp. 465-479.

Kameda, T.; Saito, S.; Umetsu, Y. (2005a). Mg–Al layered double hydroxide intercalated with ethylenediaminetetraacetate anion: Synthesis and application to the uptake of heavy metal ions from an aqueous solution. *Separation and Purification Technology,* Vol. 47, pp. 20-26.

Kameda, T.; Saito, M.; Umetsu, Y. (2005b). Preparation of a composite material for the uptake of bisphenol A from aqueous solutions, the dodecylsulfate ion-intercalated Mg-Al layer-structured double hydroxide particles. *Journal of Alloys and Compounds,* Vol. 402, pp. 46-52.

Kameda, T.; Saito, M.; Umetsu, Y. (2006). Preparation and characterisation of Mg-Al layered double hydroxides intercalated with 2-naphthalene sulphonate and 2,6-naphthalene disulphonate. *Materials Transactions*, Vol. 47, pp. 923-930.

Kameda, T.; Saito, M.; Umetsu, Y. (2007). Uptake of bisphenol A from aqueous solution by Mg–Al-layered double dydroxides intercalated with 2-naphthalene sulfonate and 2,6-naphthalene disulfonate. *Materials Transactions*, Vol. 48, pp. 2225-2229.

Kameda, T.; Takeuchi, H.; Yoshioka, T. (2008a). Uptake of heavy metal ions from aqueous solution using Mg–Al layered double hydroxides intercalated with citrate, malate, and tartrate. *Separation and Purification Technology*, Vol. 62, pp. 330-336.

Kameda, T.; Yamazaki, T.; Yoshioka, T. (2008b). Preparation and characterization of Mg–Al layered double hydroxides intercalated with benzenesulfonate and benzenedisulfonate. *Microporous and Mesoporous Materials*, Vol. 114, pp. 410-415.

Kameda, T.; Takeuchi, H.; Yoshioka, T. (2009a). Hybrid inorganic/organic composites of Mg–Al layered double hydroxides intercalated with citrate, malate, and tartrate prepared by co-precipitation. *Materials Research Bulletin*, Vol. 44, pp. 840-845.

Kameda, T.; Takeuchi, H.; Yoshioka, T. (2009b). Preparation of organic acid anion-modified magnesium hydroxides by coprecipitation: A novel material for the uptake of heavy metal ions from aqueous solutions. *Journal of Physics and Chemistry of Solids*, Vol. 70, pp. 1104-1108.

Kameda, T.; Saito, M.; Umetsu, Y. (2009c). Continuous treatment of bisphenol A and diethyl phthalate solutions using dodecylsulfate-intercalated Mg-Al layered double hydroxide particles packed in column. *Materials Transactions*, Vol. 50, pp. 1541-1547.

Kameda, T.; Yamazaki, T.; Yoshioka, T. (2009d). Selective uptake of aromatic compounds from aqueous solutions by Mg–Al layered double hydroxide intercalated with 2,7-naphthalene disulfonate. *Chemistry Letters*, Vol. 38, pp. 522-523.

Kameda, T.; Yamazaki, T.; Yoshioka, T. (2009e). Preparation of Mg–Al layered double hydroxide intercalated with 2,7-naphthalene disulfonate and its selective uptake of aromatic compounds from aqueous solutions. *Bulletin of the Chemical Society of Japan*, Vol. 82, pp. 1436-1440.

Kameda, T.; Tsuchiya, Y.; Yamazaki, T.; Yoshioka, T. (2009f). Preparation of Mg–Al layered double hydroxides intercalated with alkyl sulfates and investigation of their capacity to take up *N,N*-dimethylaniline from aqueous solutions. *Solid State Sciences*, Vol. 11, pp. 2060-2064.

Kameda, T.; Takeuchi, H., Yoshioka, T. (2010a). Kinetics of uptake of Cu^{2+} and Cd^{2+} by Mg–Al layered double hydroxides intercalated with citrate, malate, and tartrate. *Colloids and Surfaces A: Physicochemical and Engineering Aspects*, Vol. 355, pp. 172-177.

Kameda, T.; Yamazaki, T., Yoshioka, T. (2010b). Preparation of Mg-Al layered double hydroxides intercalated with 1,3,6-naphthalenetrisulfonate and 3-amino-2,7-naphthalenedisulfonate and assessment of their selective uptake of aromatic compounds from aqueous solutions. *Solid State Sciences*, Vol. 12, pp. 946-951.

Kameda, T.; Yamazaki, T., Yoshioka, T. (2010c). Effect of intercalated aromatic sulfonates on uptake of aromatic compounds from aqueous solutions by modified Mg–Al layered double hydroxide. *Materials Research Bulletin*, Vol. 45, pp. 751-753.

Kameda, T.; Hoshi, K.; Yoshioka, T. (2011). Uptake of Sc^{3+} and La^{3+} from aqueous solution using ethylenediaminetetraacetate-intercalated Cu–Al layered double hydroxide reconstructed from Cu–Al oxide. *Solid State Sciences*, Vol. 13, pp. 366-371.

Kang, M. J.; Rhee, S. W.; Moon, H.; Neck, V.; Fanghanel, Th. (1996). Sorption of MO_4^- (M = Tc, Re) on Mg/Al layered double hydroxide by anion exchange. *Radiochimica Acta*, Vol. 75, pp. 169-173.

Kulyukhin, S. A.; Krasavina, E. P.; Gredina, I. V.; Rumer, I. A.; Mizina, L. V. (2008). Sorption of cesium, strontium, and yttrium radionuclides from the aqueous phase on layered double hydroxides. *Radiochemistry*, Vol. 50, pp. 493-501.

Lazaridis, N. K.; Asouhidou, D. D. (2003). Kinetics of sorptive removal of chromium(VI) from aqueous solutions by calcined Mg–Al–CO_3 hydrotalcite. *Water Research*, Vol. 37, pp. 2875-2882.

Li, F.; Wang, Y.; Yang, Q.; Evans, D. G.; Forano, C.; Duan, X. (2005). Study on adsorption of glyphosate (*N*-phosphonomethyl glycine) pesticide on MgAl-layered double hydroxides in aqueous solution. *Journal of Hazardous Materials*, Vol. B125, pp. 89-95.

Liang, X.; Hou, W.; Xu, Y.; Sun, G.; Wang, L.; Sun, Y.; Qin, X. (2010). Sorption of lead ion by layered double hydroxide intercalated with diethylenetriaminepentaacetic acid. *Colloids and Surfaces A: Physicochemical and Engineering Aspects*, Vol. 366, pp. 50-57.

Lukashin, A. V.; Vertegel, A. A.; Eliseev, A. A.; Nikiforov, M. P.; Gornert, P.; Tretyakov, Y. D. (2003). Chemical design of magnetic nanocomposites based on layered double hydroxides. *Journal of Nanoparticle Research*, Vol. 5, pp. 455-464.

Meyn, M.; Beneke, K.; Lagaly, G. (1990). Anion-exchange reactions of layered double hydroxides. *Inorganic Chemistry*, Vol. 29, pp. 5201-5207.

Nakayama, H.; Hirami, S.; Tsuhako, M. (2007). Selective adsorption of mercury ion by mercaptocarboxylic acid intercalated Mg-Al layered double hydroxide. *Journal of Colloid and Interface Science*, Vol. 315, pp. 177-183.

Narita, E.; Yamagishi, T.; Tazawa, K.; Ichijo, O.; Umetsu, Y. (1995). Uptake behavior of chelating agents by magnesium-aluminum oxide precursor with reconstruction of hydrotalcite-like layer structure. *Clay Science,* Vol. 9, pp. 187-197.

Newman, S. P.; Jones, W. (1998). Synthesis, characterization and applications of layered double hydroxides containing organic guests. *New journal of chemistry*, Vol. 22, pp. 105-115.

Ogata, S.; Tasaka, Y.; Tagaya, H.; Kadokawa, J.; Chiba, K. (1998). Preparation of new fibrous layered compounds by the reaction of zinc hydroxide with organic compounds. *Chemistry Letters*, Vol. 27, pp. 237-238.

Ogata, S.; Tagaya, H.; Kadokawa, J.; Chiba, K. (2000). Preparation of new fibrous organic–inorganic layered compounds derived from zinc hydroxide. *Molecular Crystals and Liquid Crystals*, Vol. 341, pp. 419-424.

Orthman, J.; Zhu, H. Y.; Lu, G. Q. (2003). Use of anion clay hydrotalcite to remove coloured organics from aqueous solutions. *Separation and Purification Technology*, Vol. 31, pp. 53-59.

Pavlovic, I.; Perez, M. R.; Barriga, C.; Ulibarri, M. A. (2009). Adsorption of Cu^{2+}, Cd^{2+} and Pb^{2+} ions by layered double hydroxides intercalated with the chelating agents diethylenetriaminepentaacetate and meso-2,3-dimercaptosuccinate. *Applied Clay Science*, Vol. 43, pp. 125-129.

Pavlovic, I.; Ulibarri, M. A.; Hermosin, M. C.; Cornejo, J. (1997). Sorption of an anionic surfactant from water by a calcined hydrotalcite-like sorbent. *Fresenius Environmental Bulletin*, Vol. 6, 266-271.

Pavlovic, I.; Barriga, C.; Hermosin, M. C.; Cornejo, J.; Ulibarri, M.A. (2005). Adsorption of acidic pesticides 2,4-D, Clopyralid and Picloram on calcined hydrotalcite. *Applied Clay Science*, Vol. 30, pp. 125-133.

Perez, M. R.; Pavlovic, I.; Barriga, C.; Cornejo, J.; Hermosin, M. C.; Ulibarri, M.A. (2006). Uptake of Cu^{2+}, Cd^{2+} and Pb^{2+} on Zn–Al layered double hydroxide intercalated with EDTA. *Applied Clay Science*, Vol. 32, pp. 245-251.

Rhee, S. W.; Kang, M. J.; Kim, H.; Moon, C. H. (1997). Removal of aquatic chromate ion involving rehydration reaction of calcined layered double hydroxide (Mg−Al−CO_3). *Environmental Technology,* Vol. 18, pp. 231-236.

Rojas, R.; Perez, M. R.; Erro, E. M.; Ortiz, P. I.; Ulibarri, M. A.; Giacomelli, C. E. (2009). EDTA modified LDHs as Cu^{2+} scavengers: Removal kinetics and sorbent stability. *Journal of Colloid and Interface Science*, Vol. 331, pp. 425-431.

Saito, C.; Okada, H.; Titus, M. J.; Yoshioka, T.; Mizoguchi, T. (2007). Leaching of heavy metals from fly ash generated from gasification and melting furnace for municipal solid wastes by organic acids. *Haikibutsu Gakkai Ronbunshi*, Vol. 18, pp. 157-166.

Sato, T.; Okuyama, H.; Endo, T.; Shimada, M. (1990). Preparation and photochemical properties of cadmium sulphide-zinc sulphide incorporated into the interlayer of hydrotalcite. *Reactivity of Solids,* Vol. 8, pp. 63-72.

Setti, N. D.; Jouini, N.; Derriche, Z. (2010). Sorption study of an anionic dye - benzopurpurine 4B - on calcined and uncalcined Mg-Al layered double hydroxides. *Journal of Physics and Chemistry of Solids*, Vol. 71, pp. 556-559.

Sun, Z.; Jin, L.; Shi, W.; Wei, M.; Duan, X. (2010). Preparation of an anion dye intercalated into layered double hydroxides and its controllable luminescence properties. *Chemical Engineering Journal*, Vol. 161, pp. 293-300.

Tagaya, H.; Ogata, S.; Morioka, H.; Kadokawa, J.; Chiba, K. (1996). New preparation method for surface-modified inorganic layered compounds. *Journal of Materials Chemistry*, Vol. 6, pp. 1235-1237.

Takahashi, S.; Iwasa, T.; Kanazawa, Y.; Umetsu, Y.; Narita, E. (1997). Formation of organic anion-pillared layered zinc hydroxides by coprecipitation. *Nippon Kagaku Kaishi*, Vol. 1997, pp. 502-507.

Taylor, H. F. W. (1973). Crystal structures of some double hydroxide minerals. *Mineralogical Magazine*, Vol. 39, pp. 377-389.

Tronto, J.; Crepaldi, E. L.; Pavan, P. C.; Paula, C. C.; Valim, J. B. (2001). Organic anions of pharmaceutical interest intercalated in magnesium aluminum LDHs by two different methods. *Molecular Crystals and Liquid Crystals*, Vol. 356, pp. 227-237.

Tronto, J.; Reis, M. J.; Silverio, F.; Balbo, V. R.; Marchetti, J. M.; Valim, J. B. (2004) In vitro release of citrate anions intercalated in magnesium aluminium layered double hydroxides. *Journal of Physics and Chemistry of Solids*, Vol. 65, pp. 475-480.

Tsuji, M.; Matsunami, J.; Tamamura, Y. (1999). Selective removal of selenium in wastewater by hydrotalcite with high layer charge. *Transactions of the Materials Research Society of Japan*, Vol. 24, pp. 357-360.

Tsyganok, A. I.; Suzuki, K.; Hamakawa, S.; Takehira, K.; Hayakawa, T. (2001a). Alternative approach to incorporation of nickel into layered structure of Mg−Al double hydroxides: intercalation with $[Ni(EDTA)]^{2-}$ species. *Chemistry Letters*, Vol. 30, pp. 24-25.

Tsyganok, A. I.; Suzuki, K.; Hamakawa, S.; Takehira, K.; Hayakawa, T. (2001b). Mg−Al layered double hydroxide intercalated with $[Ni(EDTA)]^{2-}$ chelate as a precursor for an efficient catalyst of methane reforming with carbon dioxide. *Catalysis Letters*, Vol. 77, pp. 75-86.

Tsyganok, A.; Sayari, A. (2006). Incorporation of transition metals into Mg−Al layered double hydroxides: Coprecipitation of cations vs. their pre-complexation with an anionic chelator. *Journal of Solid State Chemistry*, Vol. 179, pp. 1829-1840.

Ulibarri, M. A.; Pavlovic, I.; Hermosin, M. C.; Cornejo, J. (1995). Hydrotalcite-like compounds as potential sorbents of phenols from water. *Applied Clay Science*, Vol. 10, pp. 131-145.

Valente, J. S.; Tzompantzi, F.; Prince, J.; Cortez, J. G. H.; Gomez, R. (2009). Adsorption and photocatalytic degradation of phenol and 2,4 dichlorophenoxiacetic acid by Mg-Zn-Al layered double hydroxides. *Applied Catalysis B: Environmental*, Vol. 90, pp. 330-338.

Villa, M. V.; Sanchez-Martin, M. J.; Sanchez-Camazano, M. (1999). Hydrotalcites and organo-hydrotalcites as sorbents for removing pesticides from water. *Journal of Environmental Science & Health*, Vol. B34, pp. 509-525.

Vreysen, S.; Maes, A. (2008). Adsorption mechanism of humic and fulvic acid onto Mg/Al layered double hydroxides. *Applied Clay Science*, Vol. 38, pp. 237-249.

Wang, B.; Zhang, H.; Evans, D. G.; Duan, X. (2005). Surface modification of layered double hydroxides and incorporation of hydrophobic organic compounds. *Materials Chemistry and Physics*, Vol. 92, pp. 190-196.

Wang, S. L.; Liu, C. H.; Wang, M. K.; Chuang, Y. H.; Chiang, P. N. (2009). Arsenate adsorption by Mg/Al−NO_3 layered double hydroxides with varying the Mg/Al ratio. *Applied Clay Science*, Vol. 43, pp. 79-85.

Yang, L.; Shahrivari, Z.; Liu, P. K. T.; Sahimi, M.; Tsotsis, T.T. (2005). Removal of trace levels of arsenic and selenium from aqueous solutions by calcined and uncalcined layered double hydroxides (LDH). *Industrial & Engineering Chemistry Research*, Vol. 44, pp. 6804-6815.

You, Y. W.; Zhao, H. T.; Vance, G. F. (2001a). Removal of arsenite from aqueous solutions by anionic clays. *Environmental Technology*, Vol. 22, pp. 1447-1457.

You, Y.; Vance, G. F.; Zhao, H. (2001b). Selenium adsorption on Mg−Al and Zn−Al layered double hydroxides. *Applied Clay Science*, Vol. 20, pp. 13-25.

You, Y.; Zhao, H.; Vance, G. F. (2002a). Adsorption of dicamba (3,6-dichloro-2-methoxy benzoic acid) in aqueous solution by calcined-layered double hydroxide. *Applied Clay Science*, Vol. 21, pp. 217-226.

You, Y.; Zhao, H.; Vance, G. F. (2002b). Surfactant-enhanced adsorption of organic compounds by layered double hydroxides. *Colloids and Surfaces A: Physicochemical and Engineering Aspects*, Vol. 205, pp. 161-172.

Zhang, J.; Zhang, F.; Ren, L.; Evans, D. G.; Duan, X. (2004). Synthesis of layered double hydroxide anionic clays intercalated by carboxylate anions. *Materials Chemistry and Physics*, Vol. 85, pp. 207-214.

Zhao, H.; Nagy, K. L. (2004). Dodecyl sulfate-hydrotalcite nanocomposites for trapping chlorinated organic pollutants in water. *Journal of Colloid and Interface Science*, Vol. 274, pp. 613-624.

7

Hybrid Materials Based on Zn-Al Alloys

G. Torres-Villaseñor and E. Martínez-Flores

Instituto de Investigaciones en Materiales, Universidad Nacional Autónoma de México

Facultad de Ingeniería, Universidad Autónoma de San Luís Potosí

México

1. Introduction

In this research we investigate hybrid materials based on Zn-Al alloys near eutectic composition. Two kinds of composites materials were studied bimetallic sheet and particle reinforced Zn-Al-Cu alloys. The particles used were ceramics and intermetallics. The bimetallic is produced by a process comprising providing a strip or sheet of a zinc alloy core and creating a strip or sheet of composite material by metallurgically bonding commercially pure aluminum cladding layers to the zinc alloy. Particle reinforced metal matrix composites were prepared using zinalco as metal matrix by means of powder metallurgy techniques. Alumina, graphite and intermetallic particles were used as reinforcement. The composites prepared using this technique exhibit good consolidation even before sintering. Samples of the base alloy and of the different composites were sintered in air at 473 K in periods of 10, 20, 40 and 80 h. Measurements of density, hardness and yield strength in compression were performed on green and sintered materials. A decrease of 30% in density is achieved for the 27 vol.% alumina composite as compared with the unreinforced base alloy. An improvement of 13% in the values of conventional yield strength and hardness is shown for the composite with 7 vol.% alumina. Metal matrix composites have now been extensively studied, and much of the research concerns the way in which the reinforcing element is added. In some cases ceramic or intermetallic particles are used as a reinforcement material. In other cases, cladding techniques produce a bimetallic with better corrosion resistance properties in extreme corrosion environments, and good mechanical properties. In general the composites have excellent specific mechanical properties, such as specific modulus, specific strength, fatigue resistance, high-temperature performance and very good wear resistance, and so on. Discontinuous reinforcement (particulate, chopped and short fiber, whisker, etc.) composites are currently attracting considerable attention in research and application development fields due to their low cost and simple manufacture. On the other hand, most of the reinforced metallic materials show lower plasticity and toughness than those of the metals matrix. This is one of the main limitations to their use in many industrial fields. Thus, more and more researchers and producers begin to be concerned with the plasticity and toughness of these materials. Special efforts have been made to determine the main factors, which influence the materials plasticity, and to estimate the fracture strain of the reinforced materials.From the viewpoint of running safety and reliability, the quantitative evaluation of their plasticity is an important topic for materials workers.

The primary aim of the present study was to investigate the feasibility of the fabrication of foams, composite and bimetallic materials based on the Zn–Al eutectoid alloy. Dense Zn–

22Al alloy is one of the most commonly used materials for studying superplastic behavior. The reason for this lies in the easy production method of fine equiaxial crystal grains by solution treatment in the single-phase region followed by quenching. The equilibrium diagram Al-Zn has been one of the most investigated over a period of more than 80 years. Two types of phase diagrams have been published (Fig.1) for the Al-Zn system; they differ in the presence or absence of the ß high temperature phase and two reactions, a monotectoid one at 340 °C and a peritectic one at 443 °C. For many years the possibility of a peritectic reaction in the zinc-aluminum system has been a subject of discussion. Nayak(1973), studied a small portion of the diagram in great detail and discovered anomalies in high temperature lattice parameters. This led to the reinstatement of a very narrow two-phase region ($\alpha + \beta$) at 72 mass % Zn. The cell parameter of ß (ZnAl) phase at 340°C decrease from 0.404 nm at 70 mass% Zn to 0.403 nm at 76mass% Zn. Our recent high-temperature X-ray investigation carried out on ß phase, at the composition 78,2 mass%Zn, shows on the diffraction peaks the presence of extra reflections. The analysis shows that the structure corresponds to a triclinic structure. The structure transition from f.c.c (aluminum solid solution at high temperature) to triclinic could be originated by a weak distortion of the cubic phase when the Zn content reaches 70,6 mass%Zn. When this occurs the primitive rhombohedral unit cell (a = 285,671 pm; α = 60°) of the f.c.c structure is also distorted, originating a triclinic cell. Our measured triclinic cell parameters are: a= 405,057 pm, b= 403,466 pm, c= 403,437 pm and α = 90,22, β = 89,99°, γ = 90,23°. The new triclinic unit cell originates on the distortion of the rhombohedral primitive cell contained in this distorted high temperature f.c.c structure. If the ß phase is accepted as an intermetalic, the diagram of Fig. 1b accounts for all the experimental facts. In this diagram the invariant transformations in solidification are : a peritectic, liq + Al$\rightarrow$ZnAl(ß), at aproximate 70%Zn, 443°C; and a eutectic , liq. $\rightarrow$ Al(α) + ZnAl(ß) at 94.9%Zn, 381°C. At 275°C it decomposes by an eutectoid reaction into Al(α) + Zn(η) with the eutectoid point at 78%Zn. It is important to realize that the properties of Zn-Al eutectoid alloy, like those of steel, vary with heat treatment. The alloy is put in the superplastic condition, by heating between 280 and 350˚C and water quenching. This treatment produces fine grains of α and η phases. Under this condition, it is capable of severe elongations in excess of 2000% at 250˚C.

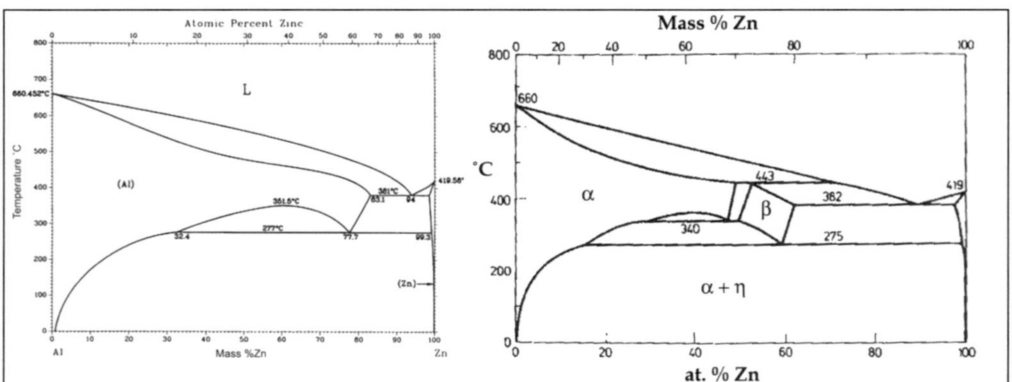

Fig. 1. Zn-Al phase diagrams

The microstructure consists of extremely fine grains (2-5 μm), which are produced by a cellular transformation, which are almost unresolvable except by electron microscopy. Slowly cooling to room temperature, a pearlitic structure, similar to that of pearlite in steel (Fig.2) is produced

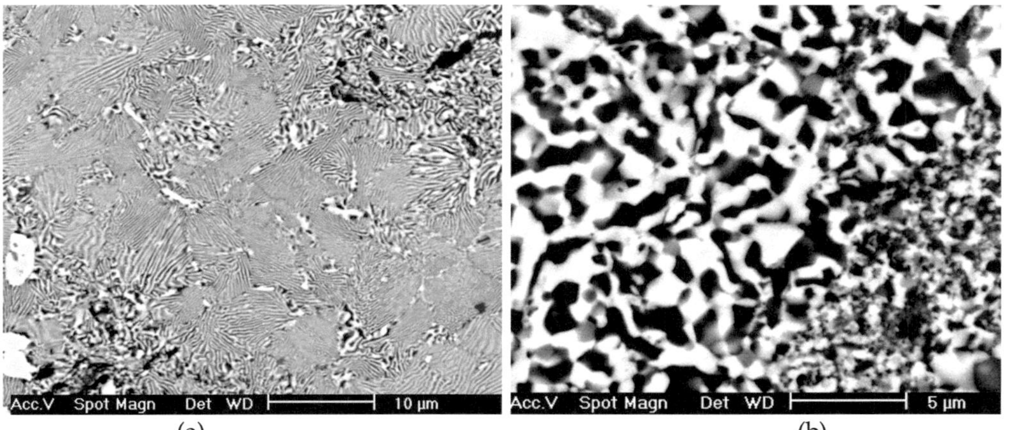

Fig. 2. Microstructures in Zn-Al-Cu. (a) Slowly cooled alloy. (b) Quenched alloy

The addition of copper to the Zn-Al alloy increases the strength and the creep resistance without seriously affecting the superplasticity. Several gravity casting alloys with different cooper contents was developed by the international Lead and Zinc Research Organization (ILZRO); they were introduced into the market with the names ZA-8, ZA-12 and ZA-27, the number indicates the amount of aluminum of the alloy. In all cases Magnesium is part of the composition and because of that, the superplasticity is avoided in these alloys. The alloy Zn-20 mass %Al- 2 mass %Cu (zinalco) is an alloy with good properties for gravity casting and in addition it can be rolled and extruded. Copper originates new intermetallic phases at high and low temperatures (ε and t´), which undergo time dependent transformations into equilibrium phases. The zinalco alloy is also unstable at room temperature and transforms by a solid state reaction into a room temperature stable compound containing α (Al solid solution), η (Zn solid solution) and t' (Al_4Cu_3Zn). However, the reactions take place extremely slowly at room temperature and at a rapidly increased rate as the temperature is raised or by thermo-mechanical treatments i.e. rolling or extrusion. The tensile properties, impact resistance and hardness are all affected by these changes. Furthermore, since some reactions are accompanied by volumetric changes, castings will undergo slight but measurable dimensional changes. The physical and mechanical properties of the zinalco alloy are shown in Tables 1 and 2, it is a high-strength material with engineering properties comparable, in most instances, to those of structural steel and heat-treated aluminum alloys.

Density (g/cm^3)	Melting range (°C)	Elastic Modulus (GPa)
5,3 g/cm^3	420-480	90

Table 1. Physical properties of Zn-20mass%Al-2mass%Cu.

	Yield Strength (MPa)	UTS (MPa)	Elongation (%)	Brinell Hardness
As cast	280	300	3-5	83-90
Die Cast	310	320	8-10	107-116
Extruded	320	410	30-35	40-55
Rolled (ssuperplastic)	290	310	80-100	25-30

Table 2. Mechanical Properties (Room Temperature) of Zn-20mass%Al-2mass%Cu.

2. Foams

Metallic foams (porous metals with high porosity ranging from 40 to 98 vol%) are constantly developed and growing as new engineering materials. These exceptionally light weighted materials possess unique combinations of properties, such as impact energy absorption capacity, air and water permeability, unusual acoustic properties, low thermal conductivity and good electrical insulating properties. Several of the engineering properties are superior to those of polymeric foams: they are stiffer by an order of magnitude, they have a higher melting point, they dissipate heat efficiently, they possess superior fire resistance and they do not involve toxic fumes in a fire. Their applications include shock and impact absorbers, dust and fluid filters, engine exhaust mufflers, porous electrodes, high-temperature gaskets, heaters and heat exchangers, flame arresters, catalyst supporters, etc. The field of applications of metallic foams is growing steadily. There are many methods available to produce metallic foams (Davies & Zhen 1983). Kitazono & Takiguchi (2007) have succeeded in developing close cell Zn-22mass%Al eutectoid alloy foams with superplastic structure through the powder metallurgy process. The strain rate sensitivity exponent (m> 0,3), classical of a fine grain structure and absorbed energy of Zn-22Al foams are much larger than those of conventional aluminum foams. This may be due to superplastic deformation of the cell walls. The relative low melting point (480°C) of the eutectoid Zn-Al alloy gave advantages either in compacting or in foaming stage. Daodud (2008) succeeded in preparing a foam composite comprising Zn-22 mass%Al eutectoid alloy and Ni-coated fly ash micro-balloons through the stir casting method. The plateau stress of this foam reach as 100 MPa over a region of a 10-60 % strain and with a density of 3,3 g/cm^3. The damping property of the eutectoid Zn-Al alloy was studied by Sirong et al. (2007) in composite foams reinforced by 10 vol.% SiC, fabricated with the melt foaming route using $CaCo_3$ blowing agent. The damping properties of this compound were higher than those of the eutectoid alloy. The addition of Al_2O_3 short fibers to the Zn-22Al foams increases the compressive yield stress and energy absorption capacity (Jiaan 2008). The behavior of internal friction (IF) and relative dynamic modulus (RDM) in a foamed Zn-Al eutectoid alloy was studied by Wei et al (2002). The specimens with macroscopic pores (0,5-1 mm) were prepared in an air pressure infiltration process. The IF peak is of a grain boundary, which is associated with the diffusive flux on a crystalline boundary between the like-phases of Al/Al. Zn-22Al foams, with an open-cell structure fabricated by the replication process using NaCl preform, with equivalent cell size to salt particle (840- 3900 μm in size), shows a relation between the relative plastic collapse stress and relative density that can be described with Gibson and Ashby´s model (Yu et al. 2009). The relative low melting point (480°C) of the eutectoid Zn-Al alloy give advantages either in compacting or in foaming stage, besides it requires less energy for production. Foams based on Zn-20mass%Al-2mass%Cu (zinalco) alloys have been produced by using granules which can be incorporated into the melt. This method produces an interconnected cellular structure or sponge metal by casting metal around granules introduced into the casting mould. These granules can be soluble (but heat-resistant), such as sodium chloride (ordinary table salt), which is later leached out to leave a porous metal. The eutectoid Zn-Al alloy is well known because of its superplastic properties. Copper additions up to 3% maintains the superplastic properties of the alloy and has obvious strain rate sensitivity, i.e. the stress rises rapidly with increasing strain rate deformation. In addition, the alloy with copper and without copper has excellent damping properties, as we can see from figure 3. Zn-20 mass%Al-2 mass%Cu (zinalco) foams containing up to 50 vol% NaCl grains have been successfully produced using the stir casting technique. Salt grains were pre-heated to 450°C in a graphite mould and molten zinalco alloy (650°C) was poured on the preheated salt grains and then stirred for one minute.

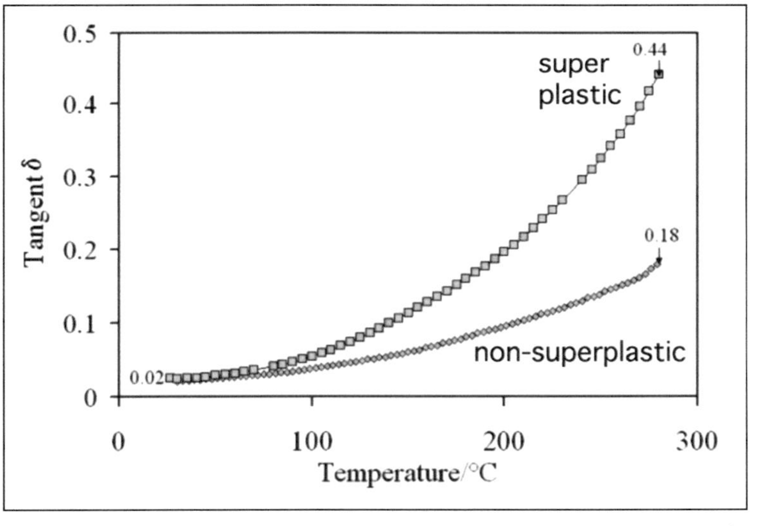

Fig. 3. Dynamical mechanical analysis of the damping behavior (tan δ), carried out in both superplastic and non-superplastic zinalco.

The mixture was annealed at 450°C for 5 minutes and then furnance cooled. A good wetting between the salt grains and molten zinalco alloy was observed. Figure 4a shows the distribution of the salt grains after solidification, Figure 4b shows the microstructure of the as cast zinalco. It reveals Zn-rich phase (η) surrounding the dendrites of the aluminum-rich phase (α). The light and dark phases in Figure 4b are α and η-phases, respectively.

The NaCl grains embedded in the metallic matrix were removed washing the specimen with hot water, leaving macroscopic pores in the material and forming foamed specimen. Figure 5. The mean diameter of the macroscopic pores ranges from 2 to 4 mm and a pore volume fraction calculated from:

$$V_f = 1 - (\rho_f / \rho) \times 100$$

where V_f is the volume fraction of the macroscopic pores, ρ is the density of dense zinalco (ρ = 5,4g/cm^3) and ρ_f the measured density of the foam. The calculated pore volume fraction is about 62%. The macroscopic pores are open and with irregular shapes. Two different NaCl crystals sizes (average 4 mm and 1,5 mm) were used which produce two different porous alloys. Type A, with a porous size of 3 ± 0,2 mm with a measured density of 2 ± 0,3 g/cm^2 and Type B with a porous size of 1,3 ± 0,3 mm and a density of 2,5 ± 0,2 g/cm^3.

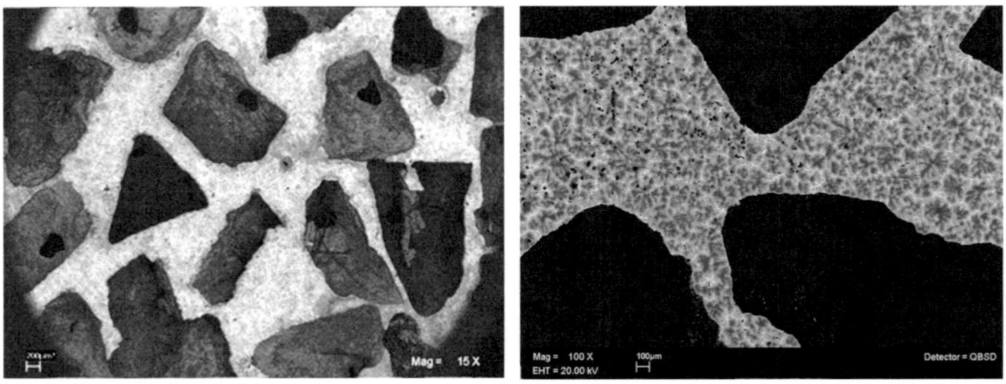

(a) (b)

Fig. 4. (a) NaCl grains in a zinalco matriz. (b) Microstructure of the zinalco matriz, in between the NaCl grains

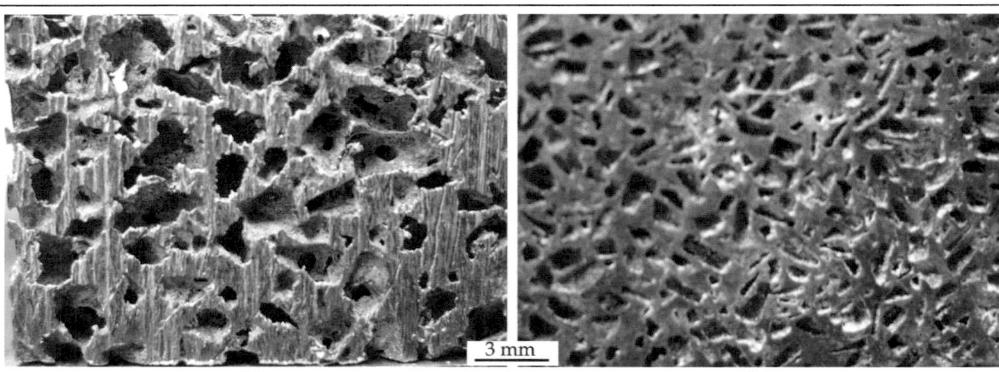

(a) (b)

Fig. 5. Images of the porous zinalco alloy produced varying initial salt grain size.

2.1 Mechanical properties

Compressive specimens with the dimensions of 15mmx15mmx30mm were prepared from a block of zinalco foam (Fig.5) with a diamond-saw cutting machine. The compressive tests were performed using a universal test Instron machine at a nominal strain rate of 5×10^{-3} s^{-1}. Figure 6 shows the compressive stress-strain curves of zinalco foams with different relative densities. It is found that the stress-strain curves exhibit two distinct regions: a linear elastic region at a very low strain and a plastic plateau region with slight stress fluctuation over a wide range of the strain. Other metallic foams show a third region called the densification region, which was not observed in our zinalco foams, before fracture. The first region is regarded as the elastic bending of cell struts and cell walls (Gibson & Ashby 1997). When the strain reaches about 2%, the failure takes place, and a rapid drop in the stress can be observed after the first peak stress is reached. The collapse stress (defined as the first peak stress) of zinalco foams rises with increasing relative density.

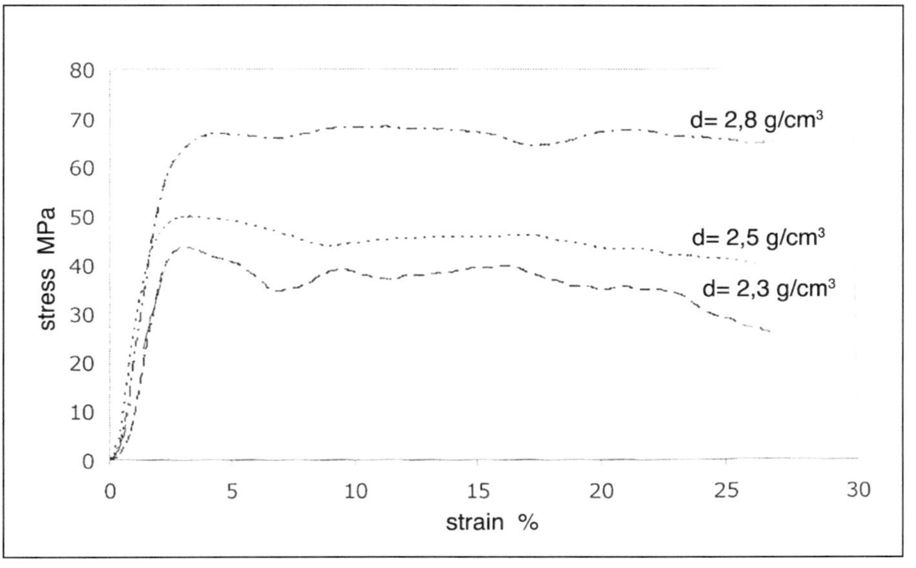

Fig. 6. Compression stress vs. strain for different densities (d) of the porous alloy with as cast microstructure.

The microstructure of the zinalco matrix can be changed by thermal treatment. Specimens annealed for 12 hours at 320°C and then water quenched, shows a fine grain microstructure (Fig.2b). This structure induces a higher ductility at room temperature. This thermal treatment has a remarkable effect on the deformation behavior of the zinalco foams. Fig.7. The stress-strain results indicate that the yield stress depends very much on the matrix structure. The foams with the as cast structure shows a higher collapse stress than foam with superplastic structure. This latter material shows a noticeable densification region after a extensive plateau region which is not observed when the matrix has the as-cast structure where a shorter plateau is observed, followed by a rather steep stress increase. A close observation of the deformed specimen with an as cast structure shows that the collapse of the cells takes place on a narrow region at 45° with the load axis (Fig. 8) the rest of the specimen remaining without appreciably deformation. This behavior is probably due to the low ductility (4-5%) of the as cast alloy. In the case of the material with superplastcic structure, a uniform deformation is observed.
Figure 9 shows the relationship between the relative plastic collapse stress and relative density for foams with an as-cast structure. The yield stress of the matrix was taken as 280 MPa and the density of the matrix as 5,4 g/cm³.

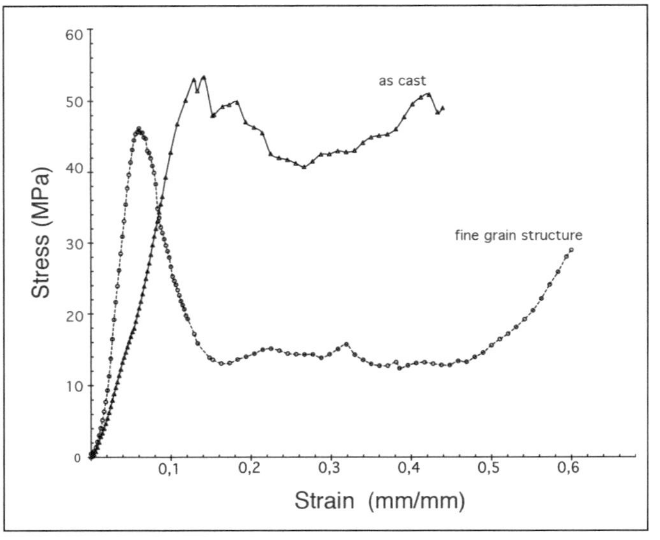

Fig. 7. Compression stress vs. strain behavior of zinalco foams with as cast and fine grain structure.

Fig. 8. Localized deformation in zinalco foam with as-cast structure.

Material	Porous size (mm)	Density (g/cm^3)	Compression stress (MPa)	Energy absorption at 20% strain. (MJ/m^3)
Foam A	3 ± 0,2	2,8	68 ± 5	3,1 (as cast)
Foam B	2,2 ± 0,2	2,5	48 ± 5	6,5 (as cast)
Foam C	1,5 ± 0,3	2,3	41± 3	25 (fine grains)

Table 3. Physical properties of zinalco foams studied in this work.

The plastic collapse stress increases with increasing relative density, in accordance with Gibson and Ashby model.
The energy absorption during the compressive deformation is defined as the energy necessary to deform a given specimen to a specific strain. The energy absorbed per unit volume (W) for a sample up to a strain (ε_0) can be evaluated by integrating the area under the stress-strain curve. For computing W, densification strain is taken to be 20%. The mechanical properties of the zinalco alloy foams along with the physical properties are summarized in Table 3. These values are much higher than those observed for low- density aluminum foams. Thus zinalco foams are competitive against the aluminum foams.

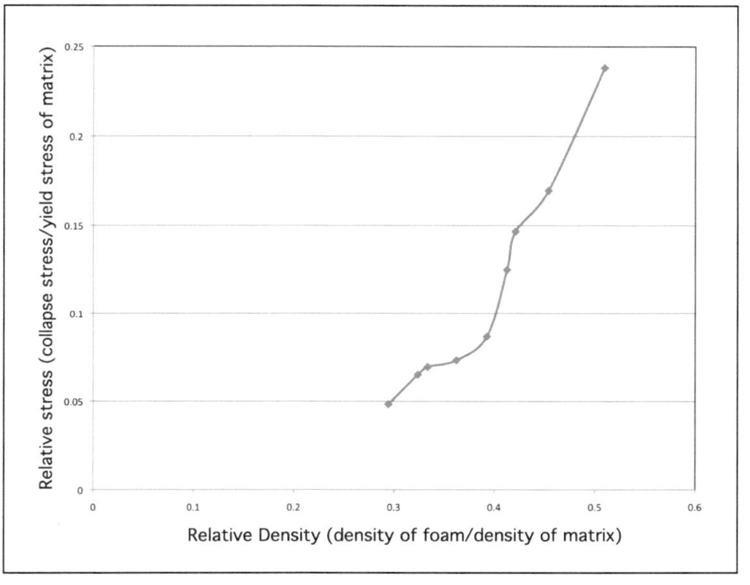

Fig. 9. Variation of the relative stress vs relative density.

3. Bimetallic materials

The realization that two metals can be joined together by cladding during rolling or extrusion has gained special importance for heavy metals. Copper sections clad with silver for electrical contacts are produced in this way in a wide variety of shapes. Recently, the plating of copper onto aluminum has been developed. The cladding is usually a layer on each side of the core alloy and usually comprising 2,5 to 5 % of the total thickness. The process requires clean and oxide-free contacting surfaces of the two metals forming the billet to be enlarged during deformation and new surfacesto be formed. These weld together under the action of the pressure and temperature. The flow stresses of the two metals must be similar (Kwon 2003). The aim of this work is directed to develop an article and a process that satisfy the need for an inexpensive, tarnish resistant, wear resistant, sufficiently hard, composite metal laminate material which is silver in color, does not require secondary annealing and has a composition so that the byproduct of its production process has economic value. The article comprises two commercially pure aluminum-cladding layers metallurgically bonded to a zinc alloy core. The combination of the zinc alloy core and the commercially pure aluminum cladding layers is ideal for the manufacturing of architectural profiles, keys or coin planchets. The zinc alloy and commercially pure aluminum are inexpensive, silver in color, and sufficiently hard. Zinc and zinc alloys are heavier than the commercially pure aluminum, so that the ratio of each may be adjusted to achieve an acceptable weight for a given application. Further, the tarnish resistance of the commercially pure aluminum cladding layers compensates for the zinc alloy's poor tarnish resistance. The zinc alloys used in the present work are Zn-20mass%Al-2mass%Cu (zinalco) and Zn-22mass%Al-1mass%Ag (zinag). These alloys are high resistant Zn alloys (yield strength ≈ 300 MPa.)with a moderate density (5.4 g/cm^3). Zinalco and zinag alloys, however, are not tarnish resistant. When they are exposed to the atmosphere, the surface of these materials becomes dull gray in color in a relatively short period of time. Because a metallic surface with a dull gray finish is not desirable or marketable, the zinc alloys core is combined with a cladding of aluminum that is tarnish resistant. The core material used in the present work was zinalco and zinag alloy prepared from aluminum and copper of commercial purity. The aluminum cladding layers (upper and lower) are composed of commercially pure aluminum, which is at least about 98,5% aluminum.

Commercially pure aluminum is silvery-white. Further more this aluminum is relatively inexpensive and is of sufficient hardness to ensure a long life and to allow that the products manufactured with it also have excellent resistance to corrosion and tarnishing, providing a long lasting lustrous finish. Besides, commercially pure aluminum is relatively lightweight. Zinag and zinalco alloys billet of 25mm wide, 100mm length and with a thickness of 5mm, were used as core material and aluminum sheet as the sheath material. The initial thickness of sheath material was 0.6mm. The aluminum sheet was fixed at both sides of the billet and then rolled. The range of rolling temperature was varied from 200 to 300 °C ; at this range of temperature zinc alloys shows a strength similar to that of the aluminum strength at this temperature. A thickness reduction of 50% was applied in one pass. After rolling the bonding strength test for the specimens produced under different rolling temperatures was performed. Another simple way to fix the aluminum to the surface of the alloy, was by using thermal spraying and then rolling. This last technique may be an easier way for commercially production of the bimetallic. The cladding thicknesses were from approximately 2% to 3 % of the final sheet thickness (0.4 mm). Tensile test specimens were cut from a composite material composed of Al + Zinag + Al. The samples were tested at 230°C in tension in an Instron universal testing machine at a strain rate of 10^{-2} s^{-1}. At this strain rate and temperature the zinag alloy hows a maximum deformation of 800%, similar to the zinalco alloy. Figure 10 is a side view of the resultant bimetallic material made of a zinc alloy (zinag) core with two aluminum cladding layers on its opposite sides. The zinag core shows two phases, an aluminum solid solution (α) which is dark, and a zinc solid solution (η) which is bright. Some grains of the η phase are elongated in the direction of the rolling. Zinc diffusion in aluminum is expected to take place during the deformation at 230°C. Figure 11 shows a representative plot of stress (MPa) against strain of the bimetallic material, when tested in tension at 230 °C. A maximum strain obtained in the bimetallic (110%) shows that aluminum non-superplastic can deform superplastically together with superplastic zinalco or zinag. This process is called "membrane forming" (Ward 1988). For the zinalco alloy thicknesses of over 0,5 mm have being formed successfully by this technique during the elongation of the aluminum layer.

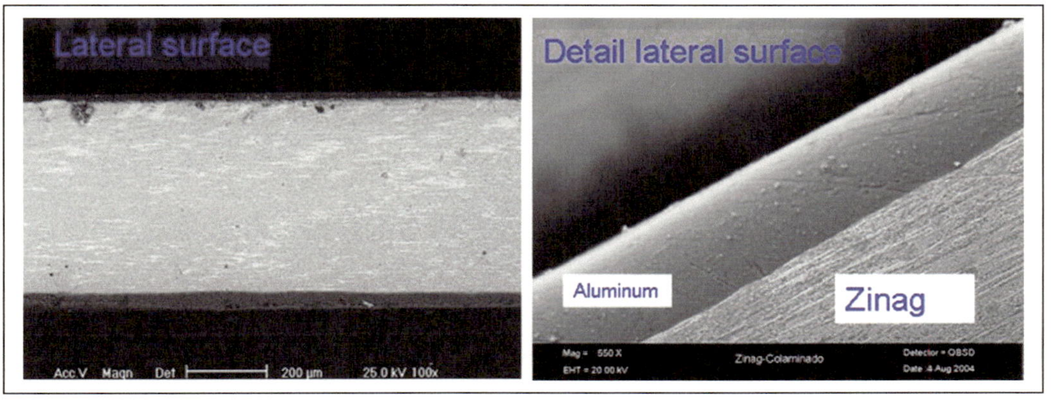

Fig. 10. Lateral view of the bonded Al-zinag-Al alloy interlayer.

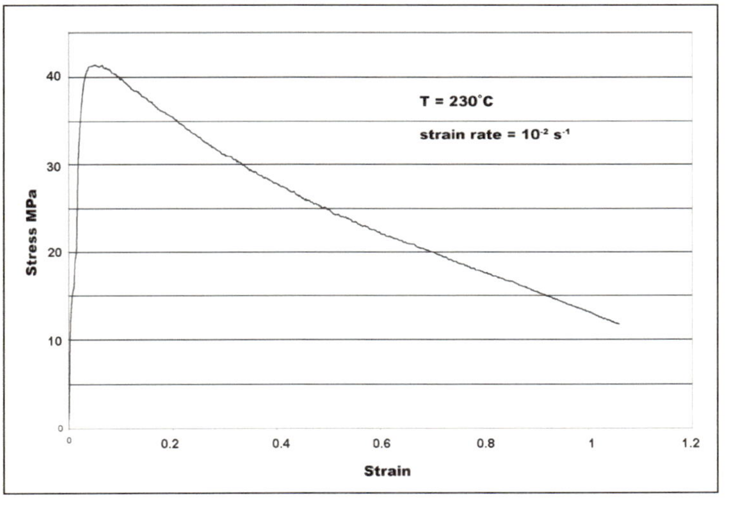

Fig. 11. Stress vs conventional strain tensile diagram of the bimetallic Al-zinag-Al

A detailed examination, of the microstructure of the pure aluminum layer after tensile testing to a strain of 100% at 230°C shows (Fig. 12) elongated chopped grains parallel to the vertical tension axis, imaged by atomic force microscope (AFM). The surface exhibited by the fractured elongated grains is roughly at 45° along the vertical tensile axis. This is known as the plane of maximum shear stresses of the clad material.

These observations can be viewed in figure 13b, in which a scan size of 20 µm was used. As a result of these fractures, the grain size of the aluminum layer is refined during the superplastic forming. In fig 13, a microestructural transversal section zone, typically deformation to both materials is obtained by conventional techniques of vacuum forming for the bimetallic. In the external region of the dome, where tension forces are active a distinctive structure is formed by fine equiaxed grains and black regions, which could be holes. Inspection showed that these round grains are not completely joined together but separated by empty regions. The mechanism of formation of these grains is possible by a fracture of the initial aluminum grains as was observed in the aluminum layer of the specimens deformed in tension. Figure 12a shows a region were it is possible to observe how a row of grains are fractured on a plane, which is possible at maximum shear stress.

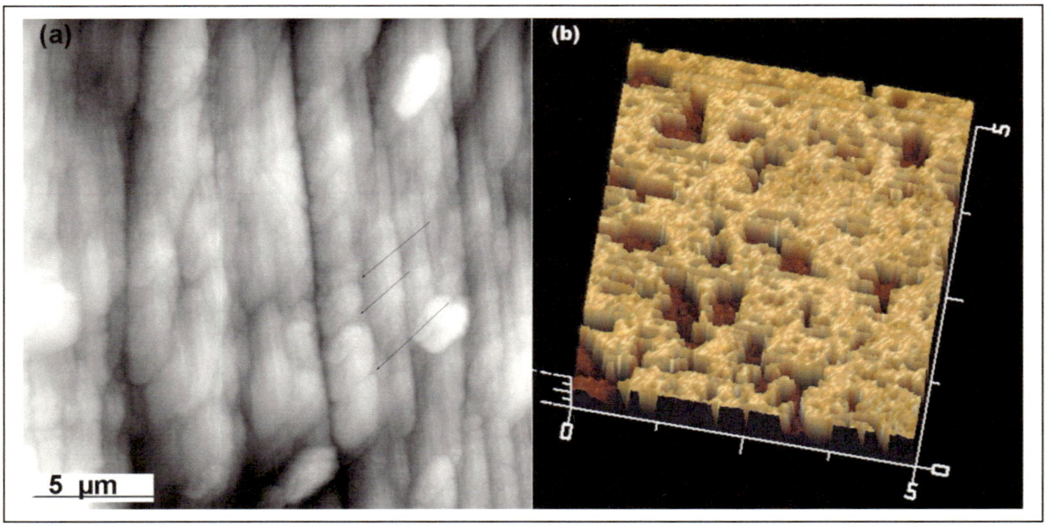

Fig. 12. Topographic view (AFM) of the aluminum layer after tensile deformation.

This plane runs up to the interface zinag-aluminum and then disappears in the superplastic material. Further details about the grains of the deformed discontinuous aluminum layer area are given in figure 12b. At higher magnifications the grains are separated by regions (holes) where new grains are emerging from the lower layers to fill the empty spaces produced when the fractured aluminum grains start to separate. At the boundary between superplastic zinag and aluminum exist a wide band of aluminum-zinc solid solutions, where we do not observe round grains of aluminum. At the initial steps of the deformation of the bimetallic material, the zinag core starts to deform by grain boundary sliding and the aluminum sheath deforms by conventional dislocation mechanisms, which originate work hardening on these grains and elongation in the tension direction. Further deformation fractures of the aluminum grains (shown in Fig.12a) maintain a fine grain structure in this layer. As the deformation proceeds, the superplastic zinag continues deforming by conventional GBS and the aluminum sheath, now composed of a structure of fine grains, starts to deform by GBS originating in the separation of the grains. New grains are emerging (Fig. 12b) from the lower layers to restore the surface. According to the results mentioned above, we consider that the superplastic deformation of the bimetallic material proceeds by a combination of GBS in the superplatic zinag core and an initial conventional deformation in the aluminum layer sheath which elongates the grains in the tension direction. This is followed by a particular superplastic mechanism, produced by the separation of fine grains formed by fractures during the plastic deformation of the elongated aluminum grains. The flow stress of aluminum was 30 to 15 MPa in pure aluminium and 35 to 25 *MPa* in zinalco or zinag alloy, between 230 °C and 350 °C. Therefore, the flow stress ratio is close to 1 at this temperature range. The required stress to tear out the aluminum sheath from the zinalco´s core, is 50*MPa* when rolling at 240 °C and decreases to 25*MPa* when rolling is performed at 350 °C (almost the same values for zinag). This suggests that the reducción in strength is due to the phase transformation that takes place in the Zn-alloy core material (around 275°C) during the cooling of the bimetallic sheet after rolling, affects the strength of the bonding. The yield strength of this alloy is around 320 MPa. at room temperature; hence the bimetallic Al-zinalco-Al give us a high strength-low density (5,2 g/cm^3) material, with a great variety of applications in the automotive industry, building and construction or the manufacturing of low cost keys.

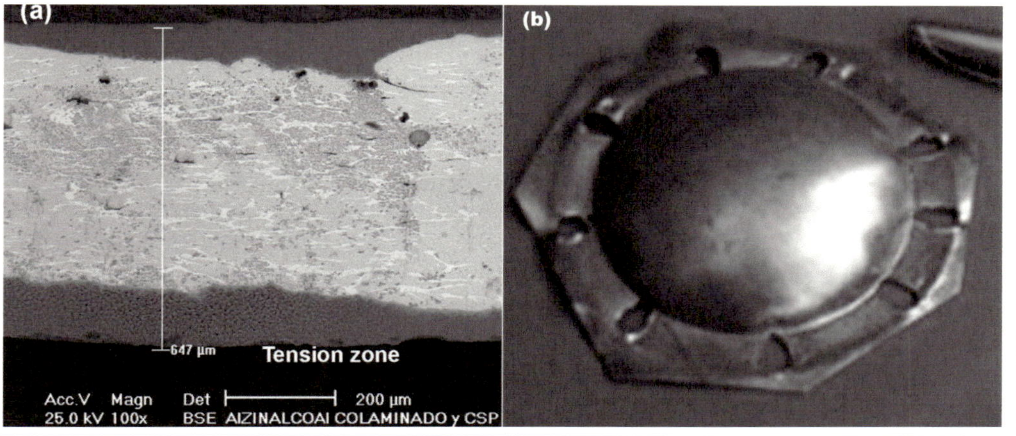

Fig. 13. Transverse section (a) of a dome vacuum formed (b)

Fig. 14. Medals elaborated with aluminum clad zinalco.

The bimetallic zinalco-aluminium is a moderately heavy and soft planchet with exellent aplications in the production of coins and tokens. Figure 14 is a top plan view of some products produced with this bimetallic. Scrap may be reused or sold. This economic benefit is obtained by heating the scrap above its melting point, which produces a zinc-aluminum alloy desired in the die-casting industry. Zinc-aluminum die-cast products are extensively used in the automotive, architectural, aeronautical, and other industries. In this study, the best rolling conditions to obtain an optimum bimetallic planchet were that the rolling temperature was around 240 °C ± 20 for the alloy Zn-20Al-2Cu. Under this rolling conditions the bonding strength was around 50*MPa* for aluminum- zinalco bimetallic planchet.

4. Metal-matrix composites

Composites possess significantly higher strength and stiffness than unreinforced materials zinalco alloy shows very attractive room-temperature properties, these properties decay very rapidly with temperature. Essentially the degradation of tensile strength and creep resistance occurs at temperatures around 100 °C. With this limitation, the extent of application of the alloy is restricted to those for ambient service temperature environments. An attractive possibility of improving room temperature and relatively high temperature properties (tensile strength, Young´s modulus, wear resistance) is by the reinforcement of particles to form a composite material. Powder metallurgy is a suitable process to produce materials with highly useful characteristics from metal powder alloy without passing trough the melt condition. Powder metallurgy provides important technical support for special shape process, principally for five important groups of materials: (1) particulate composite materials, (2) porous materials, (3) refractory materials, (4) frictionless materials and (5) high resistance material with improved properties. Metal powder used in the production of sintered parts can be characterized by three categories of properties: (a) metallurgical properties; (b) geometrical properties and (c) mechanical properties. All these powder properties are the result of the process by which the powder was produced. Parameters like microstructure and chemical composition of metallic powder have a great influence upon the final strength properties of the sintered parts. Impurities may have an adverse effect upon compressibility and upon the life of compacting parts. Particle size distribution, particle shape and particle porosity, determine the powder's specific surface which is the driving force of the sintering process and is directly related to compressibility. The compressibility of the powder is a fundamental factor in deciding the compacting pressure required to achieve a desired compact density. Some advantages in the use of powder metallurgy techniques are: (a) they allow the problems of wet between liquid metal and reinforced ceramic to be overcome, (b) better homogenization of composite materials, (c) high amount of reinforced particles will be attained, (d) a good control of particle size will be obtained, (e) parts with very simple form or with intricate shapes can be easily processed and sintered. Some disadvantages of this technique are: (a) they can be used only for material composites reinforced with particles, (b) the size of parts that can be formed are limited by technique and economic aspects. The forming of particulate material composites using powder metallurgy begins with the mixing of metal and ceramic particles. Powder mixes are compacted in a rigid die. In this operation, high pressure (about 650 MPa) is exerted upon the powder in the die cavity using one or more vertically moving compacting punches. Under the influence of such high compacting pressures, the powder particles are being squeezed together so closely that their surface irregularities interlock and a certain amount of cold welding takes place between their surfaces. The compaction process does not require high temperature, it can be performed at room temperature, therefore, there is low interaction between metal and reinforce particles, minimizing undesirable interfacial reactions to get better mechanical properties. This investigation examines the properties of Zinalco matrix combined with alumina, graphite, hidroxiapatite and intermetallic particles, fabricated by conventional cold powder pressing and hot solid state sintering techniques. In general, the goal of grinding could be: (a) to diminish the size of particles, (b) to mix the powder of raw material or (c) to study microstructural and morphology changes.

4.1 Characterization of reinforcement particles

Filings of Zinalco alloy were comminuted in a vibratory milling and in a high-energy planetary mill. The resultant particles are shown in figure 15.

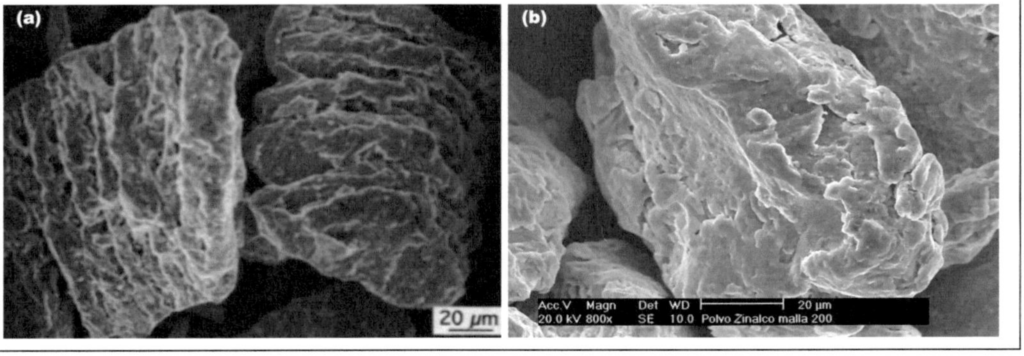

Fig. 15. (a) Zinalco powder obtained with a vibratory mill has flake shape and a groove surface, (b) rounded particles of zinalco powder obtained in a high energy planetary mill.

Figure 15a shows powder obtained by this process the particles have a flake shape with a grooved surface. Figure 15b, shows rounded shape powder obtained when a cast-alloy comminuted in a high energy planetary mill. Particle surface shows intense plastic deformation and grooves are less deep than particles obtained in the vibratory mill. Fig 16a shows the microstructure of the cold compacted powder, there can be observed α (black) and η (brigth) fine grains. X ray diffraction pattern of this material still shows the presence of the metastable ε phase. Powder obtained in vibratory mill, requires long periods of grinding, about 100 hours, for total ε phase transformation into the τ' phase, which is the stable phase at room temperature. Cold compacted zinalco powder was sintered at 200°C in air; the average size of equiaxed grains before sintering was 800 nm. After long periods of sintering, about 20 hours or more, the microstructure transformed into interconnected grains and an important increase in size was observed; see figure 16b. Both zinalco grinding processes, were performed without any protective atmosphere, unlike other powder metals such as aluminum and magnesium that can be explosive and require inert atmosphere. The x-ray diffraction pattern of both zinalco alloys compacted powder does not show reflections related to the presence of zinc oxide or aluminum oxide; this indicate that material do not present an appreciable oxidation. Ceramic particles as reinforce of metal matrix composites give the opportunity of combine the relative easily capacity of metals and alloys for deformation with the hardness, high strength and low density of ceramic particles. In the first stage of this work it was studied three different kinds of reinforce particles were studied. The first one was γ-alumina powder of commercial distribution with average particle size of 0.05 µm. average diameter.

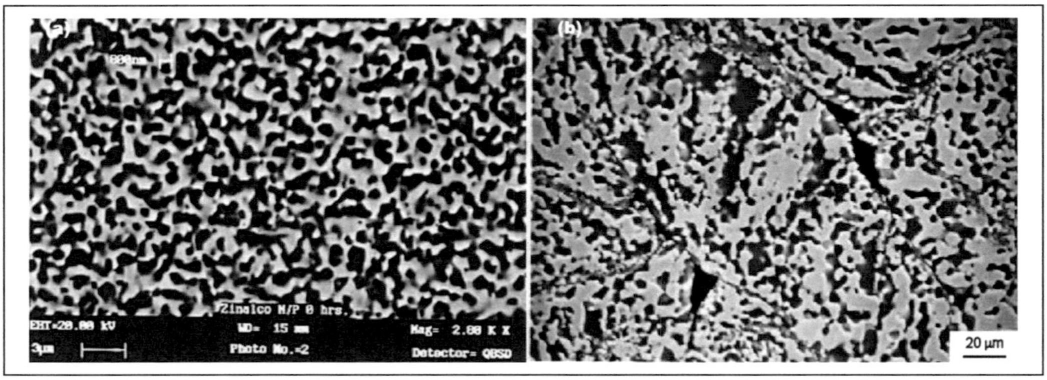

Fig. 16. (a) Microstructure of Zinalco flakes cold compacted . (b)Microstructure of zinalco compacts sintering at 473K during 80 hour in air atmosphere.

This material has been widely used to reinforce composites based on aluminum alloys, principally to improve its temperature resistance and to increase its tensile strength. In Zinalco-alumina composites we attempt to diminish the density of this material and improve its compression strength. Alumina particles gathered to form clusters with 10 µm eliminate. Graphite form plates or thin sheets eliminate these plates has a wide size distribution from 0,05 µm up to about 25 µm. Density of this material was 2.26 g/cm^3. The graphite presumably imparts improved tribological properties to the composites through the formation of a graphite-rich film on the tribo-surface which provides solid lubrication (Rohatgi 1992). Hidroxiapatite powder [$Ca_{10}(PO_4)_6(OH)_2$] forms clusters with a wide size distribution as shown y figure 17a. This is a crystalline material with applications as biomaterial. In a second stage of this work, it was used to rinforce reinforced the intermetallic 57mass%Cu-26mass%Al-17mass%Zn (phase t′). Ingots of as cast intermetallic were comminuted in a bar mill to form particles of about 1 mm diameter and then refined in a n horizontal ball mill until a size distribution between 30 and 45 µm was obtained, as show in figure 17b. The formation of a compound using powder metallurgy starts with a compaction process and then green compacts are subjected to a densification process. Regular compaction process are: (a) to apply pressure into a rigid die, (b) to apply isostatic pressure into a flexible mould, (c) rolling compaction of powder and (d) powder extrusion. For compaction it is necesary to apply higher pressures, about 650 MPa. In this work zinalco-ceramic composites were prepared by uniaxial compaction into a rigid die at room temperature while zinalco-τ′ composites were formed by both hot extrusion and hot forging process of the powder material. Physical and mechanical properties of composites material depends on homogenous mixture of powder raw materials. In all cases materials were manualy mixed and then blended in a mechanical mill.

Fig. 17. (a) Clusters of hidroxiapatite. (b) Particles of intermetallic t′ used as reinforce.

For the uniaxial compaction of powder it is important to controll die filling, frictional conditions, and velocity range of load application. After the compaction process it is necessary to determine the temperature and time of sintering. Forging and extrusion process were performed at 250 °C, at which temperature we found less porosity and higher hardness. Using powder metallurgy techniques zinalco-alumina, zinalco-graphite and zinalco-hidroxiapatite composites were prepared. A maximum of 20 mass% of alumina, 20 mass%of graphite and 15 mass% of hidroxiapatite were incorporated in the composites. Composites with 5 mass% alumina, 7 mass% graphite and 5 mass% hidroxiapatite show good mechanical characteristics. In all composites were observed a homogeneous distribution of reinforced particles. With other methods of preparation such as reocastig only a maximum of 1,2 mass% of alumina and 3 mass% of graphite could be incorporated (Muñoz-Lasso, 1992). In these composites it was observed that reinforced particles were not homogeneously distributed into the metal matrix and formed clusters. Zinalco-τ′ compounds were prepared by hot forging; a maximum of 50 mass%τ′ was incorporated. Figure 18a shows the microstructure of zinalco-10 mass%τ′ hot-forged material. The microstructure of the extruded material is formed by equiaxed grains and a preferential orientation in direction of extrusion is observed see figure 18b. An advantage of adding ceramic particles into a metallic matrix is to diminish the density of the resulting material. Table 4 shows the results of the density changes that occur after sintering the pure and reinforced alloy. The theoretical values were calculated using mixture rules. The actual density for the composites prepared under conditions of this work changes from 5,0 g/cm^3, for unreinforced material, to 3,83 g/cm^3, for composite with 27 vol. % alumina. This large changes observed in density, is due to porosity as well as the large amounts of alumina. No appreciable volume changes were observed during the sintering of these composites, density remained almost constant for all compositions and sintering time. At the sintering temperature used in this paper (200 °C) there was not expected any bonding between ceramic particles. Zinalco composite reinforced with 10 wt. % τ′ has an average density of 5,17 g/cm^3 in both hot-forging and hot-extruded materials.

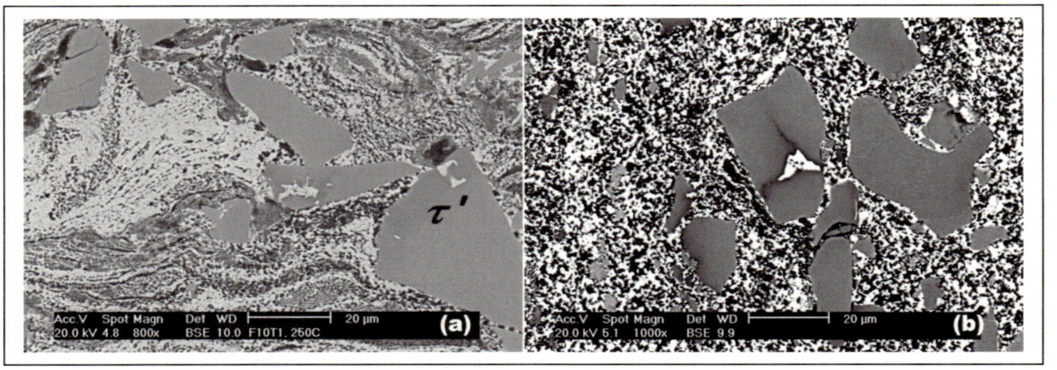

Fig. 18. (a) Zinalco 10 mass%τ' forged at 523 K in air, the alloy retains as-cast microstructure. Zinalco 10 mass%τ', extruded at 523 K in air, zinalco base alloy show equiaxed grains of α and β phases.

4.2 Mechanical properties

The base alloy yield stress in compression was measured in both green and sintered material. The compressive properties of ceramic composites tested in function of the sintering time are shown in figure 19. Results indicate that addition of 5 wt.% alumina particles into the zinalco matrix cause an improvement of the yield strength after sintering, reaching an increment of 13 %in yield stress after 20 hours of sintering. Yield strength remains almost constant for longer sintering times in all tests. Composites with 10 and 20 wt. % alumina and with 7 wt. % graphite fracture without yielding. Hardness (Fig. 20) diminishes when alumina content is 20 wt. % because a great amount of unsintered ceramic particles separates more easily when the load is applied. Hardness increases with sintering time for the unreinforced base alloy, for zinalco-5 wt.% hidroxiapatite and zinalco-5 wt.% alumina composites, hardness reach a maximum after 20 h of sintering time, then the hardness remains almost constant.

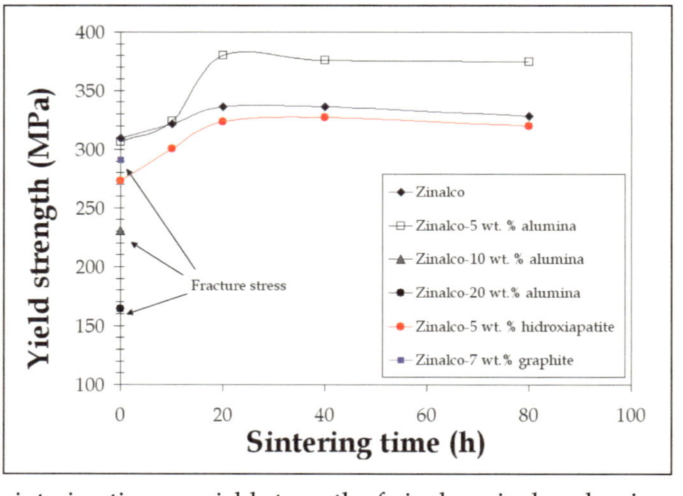

Fig. 19. Effect of sintering time on yield strength of zinalco, zinalco-alumina composites, zinalco-5 mass% hidroxiapatite and zinalco-7 mass% graphite. Tests were performed at strain rate 10^{-3} s^{-1}.

Material	Density (g/cm³)		Porosity (%)
	Theoretical	Actual	
Zinalco	5,40	5.00	7
Zinalco-5 mass% alumina	4,92	4,51	8
Zinalco-10 mass% alumina	5,15	4,44	14
Zinalco-20 mass% alumina	4,91	3,83	22
Zinalco-7 mass% graphite	4,63	4,43	4
Zinalco-5 mas% hidroxiapatite	4,85	4,57	6
Zinalco-10 mass% τ' hot forging	5,56	5,18	7
Zinalco-10 mass% τ' hot extruded	5,56	516	7

Table 4. Theoretical and actual density for zinalco-ceramic composites sintered 10 h at 473 K and zinalco 10 mass%τ' obtained by both hot forging and hot-extrusion at 523 K.

The hardness of the zinalco-7wt. % graphite composite remains almost constant with sintering time. Studies of the effect of strain rate over the mechanical properties were undertaken for zinalco-alumina composites, (Fig.21). Reinforced material with 5 wt. % alumina increases its ductility up to 10^{-3} s^{-1} and then starts to fail in a brittle manner. Green unreinforced zinalco behaves in a way similar to high strain rate superplastic metals (Nieh et al., 1984) reaching 113% of true strain at 10^{-1} s^{-1} and it may be possible to reach higher values at a higher strain rate, not measured in this work, (Fig.21b). This behavior changes completely with long sintering periods and maximum deformation achieved is lower in unreinforced and reinforced zinalco tested at any strain rate. Figure 23 shows the microstructure of green composite with 5 wt.% alumina deformed 113% in compression. It is possible to observe a plastic deformation of metal particles, which are elongated perpendicularly with respect to the compression axis. Metal particles slip one over the other because the alumina interface has been broken into small rounded pieces. The sintering process produces welding between metal particles and makes the boundary slip more difficult, reducing the ductility of this material. Composites with 10 and 30 mass% t' hot extruded and unreinforced hot extruded zinalco base alloy were tested in compression using two values of strain rate of 10^{-2} and 10^{-1} s^{-1}. Table 5 summarized yield stress values and hardness Rockwell B values for these materials

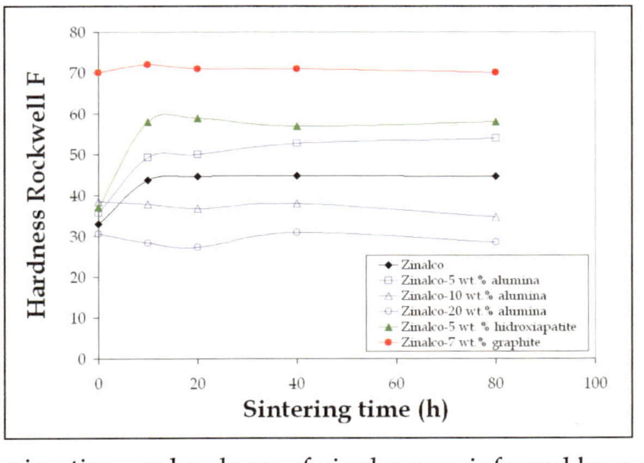

Fig. 20. Effect of sintering time on hardness of zinalco unreinforced base alloy and zinalco-ceramic composites.

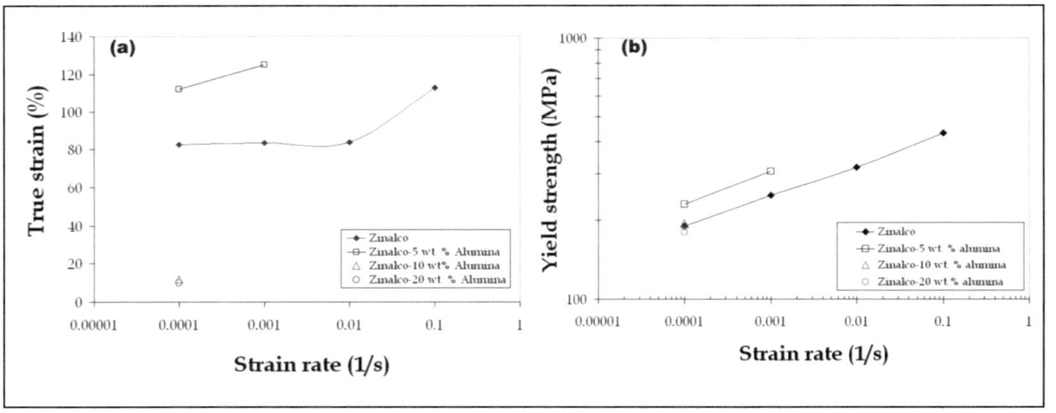

Fig. 21. Effect of strain rate on (a) yield strength and (b) true strain for zinalco unreinforced base alloy and zinalco-alumina green composites.

Hot extruded Material	Yield stress (MPa) $\dot{\varepsilon} = 10^{-2} s^{-1}$	Yield stress (MPa) $\dot{\varepsilon} = 10^{-1} s^{-1}$	HRB
Zinalco	238,3	252,8	32,17
Zinalco- 10 wt. % t'	238,8	253,8	38,92
Zinalco- 30 wt. % t'	291,0	189,7	60,92

Table 5. Yield stress values for hot extruded composite materials and unreinforced base alloy

4.3 Conclusions

The incorporation of ceramic and intermetallic particles in zinalco alloy modifies density and mechanical properties with respect to the unreinforced alloy. The low sintering temperature (200 °C) used in this work is not enough to produce sintering of ceramic particles, therefore under these experimental conditions, large amounts of ceramic reinforce produce brittle behavior and eliminate grain growth inside metal particles. During mixing and compaction operations, fine ceramic particles infiltrate into groves on the zinalco filling surface, making a smoother interface, resulting in a better accommodation of the distorted particles during deformation at low strain rates (< $10^{-3}s^{-1}$) in 5 mass% alumina unsintered composite and increasing total strain with respect to the unreinforced green zinalco alloy, see figure 22b. In green base alloy a linear dependence of yield stress with the strain rate was observed see figure 22(a). These values increase as function of strain rate. This behavior changes with sintering process in reinforced and unreinforced materials. In the sintered ones yield stress was observed only when the strain rate was below 10^{-3} s^{-1}; for higher strain rates (>10^{-3} s^{-1}) fracture occurs without yielding in both sintered materials, possibly due to the diffusion bonding between metal particles, which obstructs the free accommodation of these particles one over the other. Yield strength in the unreinforced material achieves a maximum after 20 hours of sintering and then the stress remains constant with longer sintering time. In reinforced material with 5 mass% alumina yield stress increases up to 20 hours. The initial increase of the yield stress in both reinforced and unreinforced materials is probably due to the diffusion bonding of the metal particles that obstructs the free accommodation of the deformed metal particles. A bigger yield stress is observed in the case when the alumina is present an increase in alumina weight fraction (10 and 20 mass.%), results in a poor contact between metal particles, reducing the possibility that diffusion bonding can occurs between them and hence the material behaves in a very brittle manner because particle/alumina decohesion occurs. Composites reinforced with τ' intermetallic particles were formed by hot-forging and hot extrusion process. Better mechanical properties were observed in the as-extruded composite because the microstructure consists of equiaxed grains produced by severe plastic deformation. It can be very interesting to study if superplasticity at high strain rates (>10^{-1}) can be achieved with powder metallurgy in Zn-Al alloys.

5. References

Davies G.J. & Zhen S. (1983). Metallic foams: their production, properties and applications. *Journal of Materials Science* Vol.18, No. 4 (July 1983) pp1899-1911, ISSN 1573-4803.

Daoud A. (2008). Synthesis and characterization of novel ZnAl22 syntactic foam composites via casting. *Materials Science and Engineering A* Vol.488, (2008), pp. 281-295

Gibson L.J. & M.F. Ashby M.F.,(1997). Cellular Solids: Structure and Properties (2nd ed) Cambridge University Press, Cambridge, UK (1997)

Kwon H.C., Jung T.K., Lim S.C. & Kim M.S..(2003) "Fabrication of copper clad aluminium wire" *Designing, Processing and Properties of Advanced Engineering Materials*, Korea 2003, Trans Tech Publications LTD pp 317-320.

Jiaan L., Sirong Y., Xianyong Z. Ming W., Song L., Yanru L. & Yaohui L. (2008) Effect of Al_2O_3 short fiber on the compressive properties of Zn-22Al foams. *Materials Letters* Vol. 62 (2008), pp 3636-3638

Kitazono K & Takiguchi Y. (2007). Strain rate sensitivity and energy absorption of Zn-22Al foams *Scripta Mater.* Vol.55 (2006),pp. 501-504

Muñoz-Lasso A. (1992). Estudio de las propiedades mecánicas de un compuesto con matriz metálica (Zinalco). *Tesis de Maestría en Ciencias.* Facultad de Ciencias. UNAM. México 1992.

Nayak A. K. (1973). Thermal and quantitativethermal analysis of Al-Zn alloys and determination of the equilibrium diagram of the binary system. Journal of the Institute of Metals. Vol. 101, (1973) pp 309-314.

Rohatgi P.K., Ray S. & Lin Y.(1992) Tribological properties of metal matrix graphite particle composites. *International Materials Review.* Vol.37, No. 3(1992), pp. 129-135

Sirong Yu, Jiaan L., Yanru L.& Yaohui L. (2007). Compressive behavior and damping propertie of ZA22/SiC_p composite foams. *Materials Science and Engineering A* Vol. 457 (2007), pp325-328.

Ward D.M. (1988) Forming non-superpastic materials with superplastic membranes. *Proceedings of an International Conference on Superplasticity and Superplastic Forming,* pp. 595-599. ISBN 0-87339-089-X, Blain ,Washington, August 1-4,1988.

Wei J, N., Cheng H.F., Gong Ch, L. Zhou Z. C., Li Z.B.,Han F.S. (2002). Grain boundary peak in a foamed Zn-Al Eutectoid alloy. *Chin.Phys.Lett.* Vol.19, No. 3, (2002),pp.381-384.

Yu S., Liu J.,Wei M., Luo Y., Zhu X., Liu., Y. (2009). Compressive property and energy absorption characteristics of open-cell ZA22 foams. *Materials and Design*.Vol. 30(2009),pp. 87-90.

8

Mechanical Behavior of Filled Thermoplastic Polymers

S. Bazhenov
Institute of Synthetic Polymeric Materials
Russian Academy of Sciences, Moscow
Russia

1. Introduction

Filled composites consist of a continuous matrix phase and of filler particles [1, 2]. There were three stages in their development. The first one started in late 1950s and was related to the development of solid rocket fuel which contained about 80% of rigid powder particles bonded by an elastic rubber matrix. The second stage began in 1970s with the development of composites based on thermoplastic polymers filled with rigid nonorganic particles. Fracture of these composites is essentially different from that of the rubber matrix composites. The third stage started in late 1980s with the creation of nanocomposites. Nanocomposites are not discussed here, but some mechanisms typical of composites filled by microparticles can be applied to them.

Rigid nonorganic particles are introduced into thermoplastic polymers mainly in order to increase the Young modulus of the material. Sometimes it is done to improve thermal resistance, electric conductivity, magnetic properties, to lower friction wear, flammability and so on [1, 2].

The main disadvantage of filled thermoplastic composites is their brittleness leading to a dramatic decrease in fracture strain in comparison with unfilled polymers [2-5]. An important goal is development of filled polymer composites remaining ductile at high filler contents. The goal of the review is to describe approaches to suppress brittle fracture of thermoplastics filled with particles.

2. Deformation mechanisms

If a polymer yields with necking, an increase in filler content leads to fracture during neck propagation, to quasibrittle fracture during formation of a neck and, finally, to true brittle fracture as illustrated by Fig. 1. At brittle and quasibrittle fracture modes fracture strain is roughly 100-fold lower than that of the unfilled polymer. In addition, filler may suppress necking and initiate yielding in craze-like zones if the bonding with a polymer is weak [6]. Thus, in filled polymers at least six modes of deformation behavior are distinguished: 1) brittle, 2) quasibrittle fracture during neck formation, 3) fracture during neck propagation, 4) stable neck propagation, 5) microuniform yielding, and 6) yielding in crazes [7].

The load versus elongation curve has specific features at each fracture mode (Fig. 2). A typical stress-strain curve of a necking polymer is characterized by the presence of a maximum. After reaching the yield point, the stress drops and remains constant while the neck propagates through the specimen. This is followed by a region of strain-hardening where the stress gradually increases. Quasibrittle composites reach a stress maximum, however, just past the yield point, as the stress decreases toward the draw stress, composite fractures. In brittle materials the stress-strain relationship is linear and the fracture strain is less then 2-3%. A curve with nonlinear monotonously increasing stress is typical for both uniform yielding and yielding in crazes.

Composites based on polymers with uniform yielding and those deforming by neck propagation behave differently. In polymers deforming without necking (rubber, polytetrafluorethylene or super high molecular weight polyethylene) fracture strain gradually decreases with an increase in filler content but the material remains ductile up to very high filler contents [7,8]. This behavior is different from that of composites based on matrices deforming by neck propagation [3]. Therefore behavior of composites based on matrices deforming uniformly and with necking will be discussed separately.

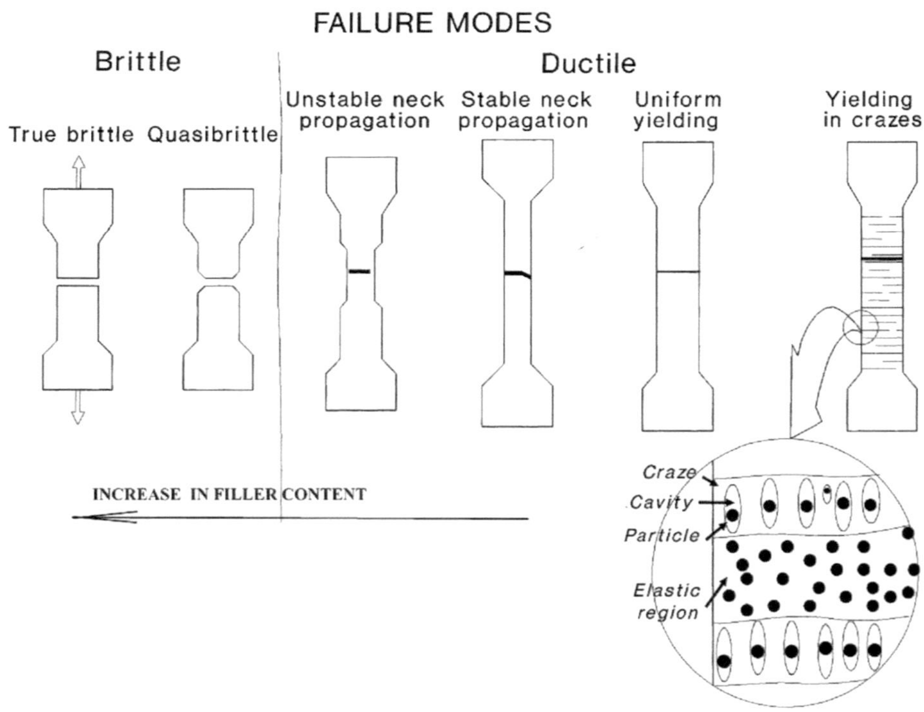

Fig. 1. Fracture modes of filled polymers.

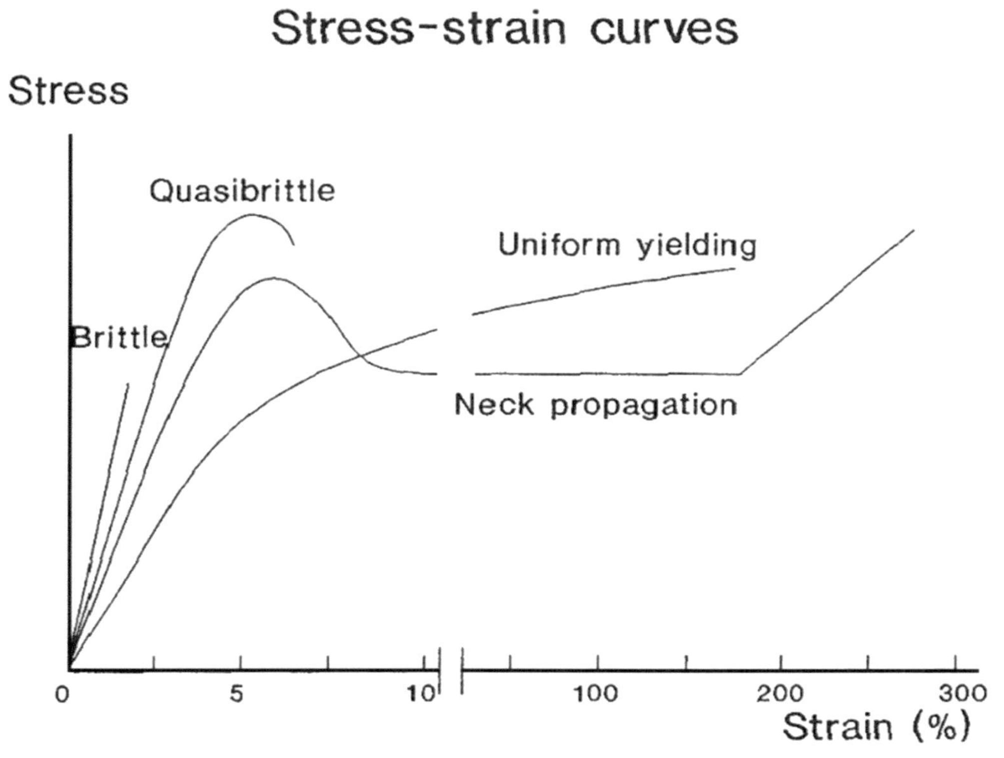

Fig. 2. Typical engineering stress versus strain curves at different fracture modes [5].

3. Deformation at fracture

3.1 Uniform matrix yielding

In polytetraphluorethylene (PTFE) and super high molecular weight polyethylene (SHMWPE) in tension the neck does not appear [7,8]. Both polymers have very high molecular weight and considerable strain strengthening which prevents necking. Due to very high molecular weight their viscosity is high even at elevated temperatures. As a result, they cannot be filled with particles by a traditional liquid technologies. However, SHMWPE was filled by in situ polymerization. A catalyst was put on aluminum particles and the matrix was polymerized [9-11].

Fig. 3 shows typical engineering stress σ - strain ε curves for unfilled SHMWPE and that filled with aluminum particles [8]. The stress-strain curve of unfilled SHMWPE is typical for ductile polymers. The stress gradually increases with an increase in strain. The unfilled polymer reached an engineering strain of approximately 270% before it fractured. The stress-strain curves of the SHMWPE filled with 57 vol.% of particles are similar to that of the unfilled polymer. The fracture strain decreases with an increase in the filler content. However, the fracture strain even at 57 vol.% of Al particles is approximately 100% and the composite remains ductile.

Qualitatively different behavior is observed at the filler content of 62 vol.% (curve 7) when in the material appear pores. These composites reach a yield point on the stress-strain curve, however, just in the yield point, the specimens fractured at the strain of approximately 2%. The large decrease in fracture strain between 57 and 62 vol.% is typical for the transition from ductile to brittle fracture.

Fig. 4 shows the relative fracture strain $\varepsilon_c/\varepsilon_0$ of filled SHMWPE (1) and PTFE (2) plotted against the volume fraction of particles V_f. Both composites remain plastic up to very high filler contents. Embrittlement of SHMWPE is observed at filler content of 62 vol.%. The decrease of fracture strain at lower filler content is moderate, and both composites remain ductile even at very high filler contents. Thus, embrittlement is not typical for polymers deforming without necking [8].

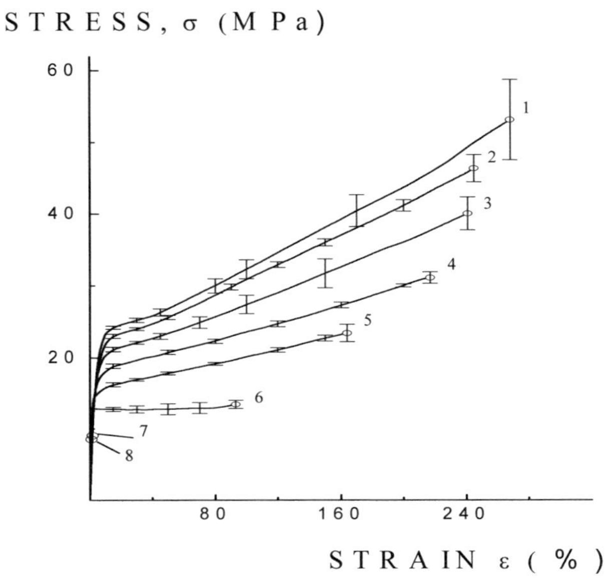

Fig. 3. Typical engineering stress σ versus strain ε curves for unfilled SHMWPE (curve 1) and SHMWPE filled with V = 4 (curve 2), 10 (3), 19 (4), 32 (5), 57 (6) 62 (7) and 63 vol.% (8) of spherical aluminum particles. The diameter of Al particles is ≈10 μm.

3.2 Matrices deforming by neck propagation

Fig. 5 shows typical engineering stress σ - strain ε curves for unfilled PETG and PETG filled with $CaCO_3$ [5] The character of the curves changes with an increase in the filler content. The stress - strain curve of unfilled PETG is typical of ductile polymers. After reaching the yield point, the stress dropped to the draw stress (σ_d) and remained constant while the neck propagated through the entire gauge length of the specimen. This was followed by a region of strain-hardening where the stress gradually increased as the necked material extended uniformly. The unfilled polymer reached an engineering strain of approximately 260% before it fractured. The stress-strain curves of the filled materials with the amount of $CaCO_3$, 2.4 and 7.2 vol.%, were similar to that of the unfilled polymer. However, the fracture strain decreased as the filler content increased and the length of the strain-hardening region also decreased until with 7.2 vol.% there was no strain-hardening. Different behavior was observed with higher filler contents, 14 and 24 vol.%. These compositions also reached a yield maximum in the stress-strain curve, however, just past the yield point, as the engineering stress was decreasing toward the draw stress σ_d, the specimens fractured. As a result, the fracture strain was in the range of 3-8% which is typical for brittle polymers. The drop in fracture strain between 7.2 and 14 vol.% was caused by the transition was from ductile to quasibrittle fracture.

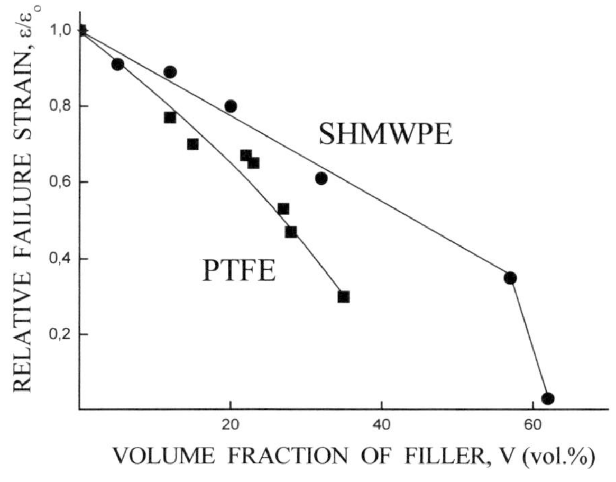

Fig. 4. Relative fracture strain $\varepsilon_c/\varepsilon_0$ of filled SHMWPE (1) and PTFE (2) plotted against the volume fraction of particles V. ε_0 - fracture strain of unfilled polymer. SHMWPE and PTFE are filled with Al and Cu particles respectively. The diameter of particles is ≈ 10 μm.

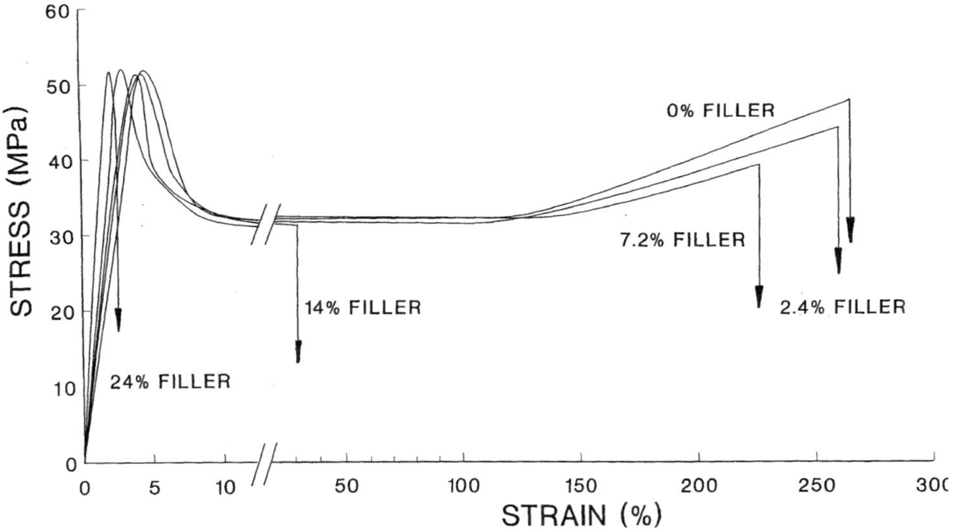

Fig. 5. Typical engineering stress σ - strain ε curves for unfilled PETG and PETG filled with $CaCO_3$ particles.

Photographs of two fractured specimens, one with 2.4 vol.% filler and the other with 14 vol.%, are shown in Fig. 6. The specimen with 2.4 vol.% filler fractured after the neck propagated through the entire gauge section. In contrast, the specimen with 14 vol.% filler showed only localized thinning at the fracture zone while the remainder of the gauge section was not plastically deformed. Necking initiated in both specimens, while the subsequent stability of the neck defined the transition from ductile to quasibrittle behavior. With the lower filler content, the neck was stable and propagated through the gauge length. At the higher filler contents, the neck formed but it was not strong enough to support the drawing, and fracture occurred in the neck. The ductile-to-quasibrittle transition was from propagation of a stable neck through the entire gauge length of the specimen to fracture in the neck without propagation.

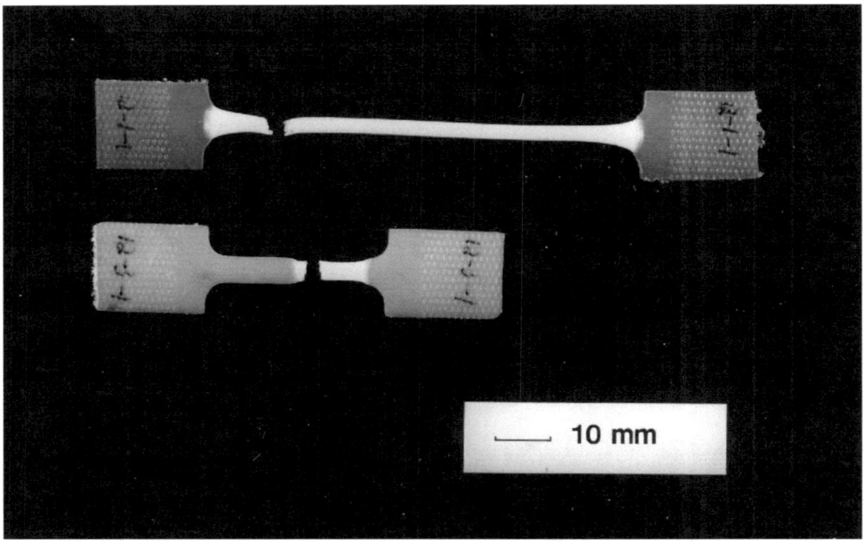

Fig. 6. Photographs of two fractured PETG specimens, one with 2.4 vol.% filler and the other with 14 vol.% of $CaCO_3$ particles.

Fig. 7 demonstrates the effect of particles on relative fracture strain $\varepsilon_c/\varepsilon_0$ of PVC, PET, HDPE and PP. All these polymers yield by neck propagation. The fracture strain depends on the polymer matrix. Most typical dependence is a step-like, and a moderate reduction in fracture strain at low filler contents, similar to that in Fig. 4, is followed by a sharp drop of fracture strain in a comparatively narrow interval of filler contents. The magnitude of strain decrease is approximately 100-fold. The drop in fracture strain is caused by the transition from a ductile to quasibrittle fracture [7].

Fig. 8 illustrates the mechanism of quasibrittle fracture of filled composite. The yielding of the matrix after necking is heterogeneous and localized in the narrow transition zone between the neck and the undeformed part of the sample. The local deformation in the forming neck can reach hundreds of percent. However, if the necking region is not strong enough the neck is not able to propagate along the sample. The length of the forming neck is short, and almost the entire sample is elastic everywhere except the neck zone. If the sample fractures in the forming neck, the average fracture strain is close to the strain of the elastic regions.

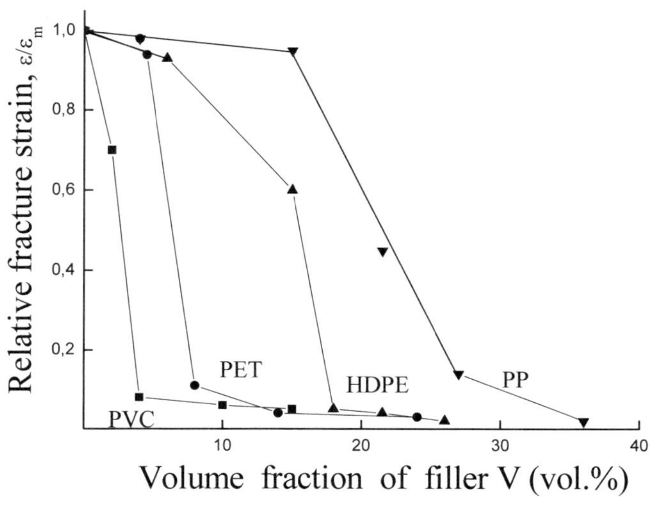

Fig. 7. Relative fracture strain of filled polymers, $\varepsilon_c/\varepsilon_m$ versus the volume fraction of particles V. ε_m - fracture strain of unfilled polymer. (▼) - PP/Al(OH)$_3$, 10 µm; (▲)-HDPE/Al(OH)$_3$, 10 ▲µm; (•) - PET/CaCO$_3$, 8 µm.

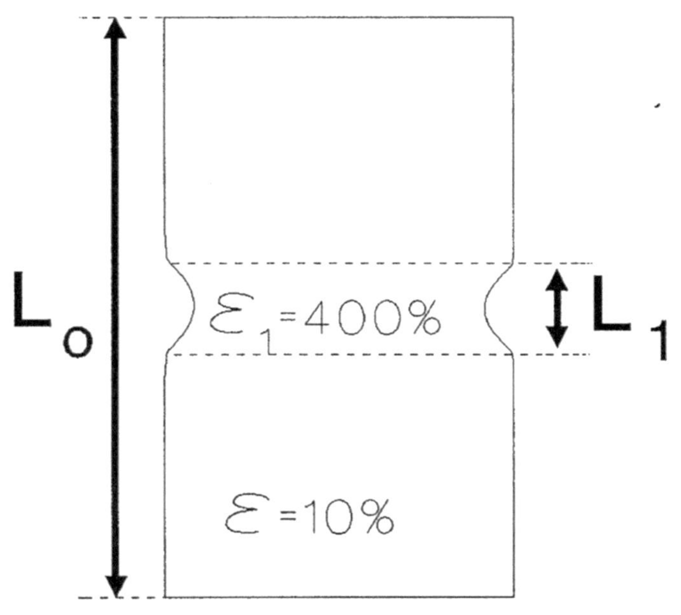

Fig. 8. Schematic of localized yielding of polymer in the narrow transition zone between the neck and the undeformed part of the sample. The local deformation in the forming neck in PP reaches 400% and the rest of sample is elastic. The length of the yield zone L_1 is short, and the composite is elastic on macroscopic scale.

Increase in filler content leads to a decrease in fracture strain in the forming neck as shown in Fig. 4. At low filler contents V this strain is higher than the strain in necking region. As a result, neck propagates through the sample and filled polymer is ductile. However, the composite fracture strain at some critical filler content becomes lower than the strain in the necking region, and the neck does not propagate through the sample. Transition to brittle fracture is accompanied by an approximately 100-fold decrease in the macroscopic fracture strain compared to that of the unfilled matrix. Thus, the material is macroscopically brittle despite local yielding in the forming neck. Hence, the composite on macrolevel is brittle, while on the microlevel in the fracture region it yields.

The neck is not able to propagate along the sample if the engineering draw stress σ_d is higher than the fracture stress. In the transitional point the composite strength is equal to its draw stress. Hence, the criterion of the transition from neck propagation to brittle fracture is described by equation [5]:

$$\sigma_d = \sigma_c \tag{1}$$

where σ_d is the draw stress and σ_c is fracture stress of filled polymer. On this reason below the effect of the particle content V on the draw stress σ_d and the strength σ_c is described.

4. Tensile strength

Fig. 9 shows the effect of particles on the relative strength of various polymers filled with inorganic particles [4]. Relative strength is comparatively insensitive to polymer type and its adhesion to particles. It depends only on the volume fraction of the particles V. This is explained by debonding of the particles in necking region both at week and strong adhesion (Fig. 10).

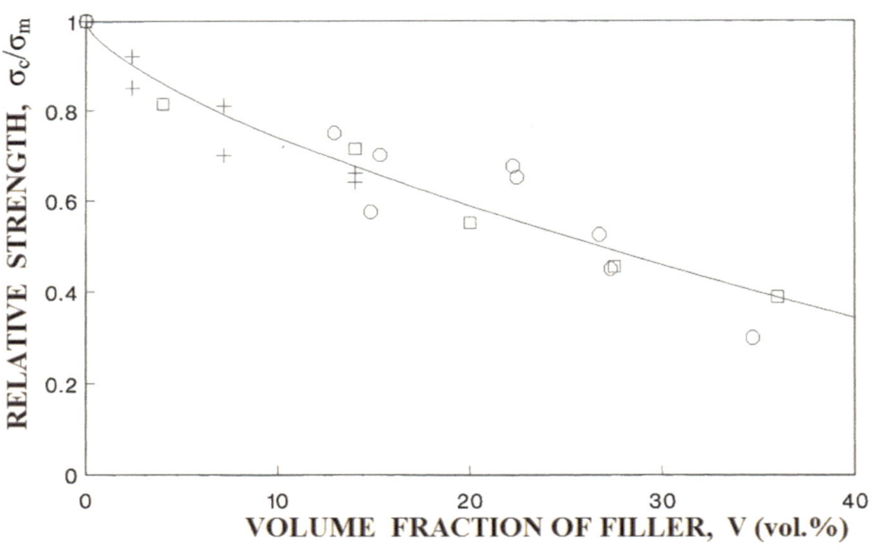

Fig. 9. Relative strength of filled polymer, σ_c/σ_m, versus the volume fraction of particles V. σ_m - strength of unfilled polymer. (o) - PTFE; (+) - PET/$CaCO_3$, diameter of particles 8 μm; (□) - HDPE/$Al(OH)_3$, 1 μm; (x) - HDPE/$Al(OH)_3$, 25 μm [7].

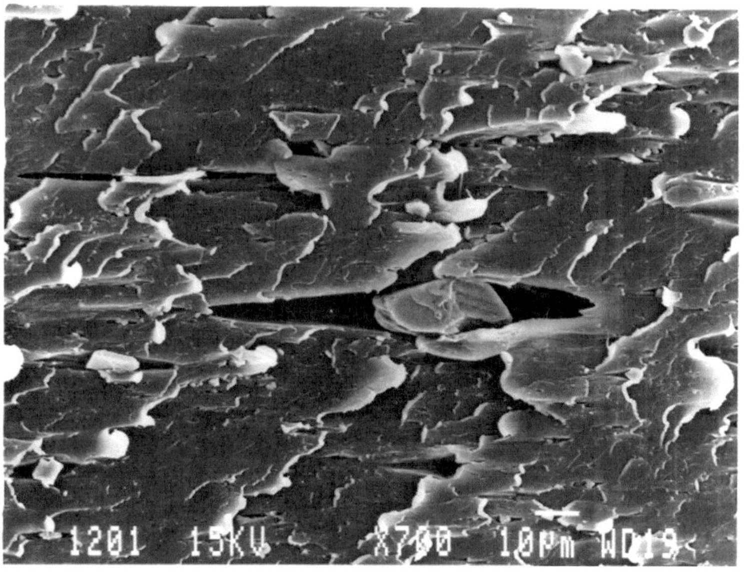

Fig. 10. Debonding of the well-bonded $CaCO_3$ particles from PETG matrix in neck. The sample was tensioned in the horizontal direction [5].

A number of attempts have been made to describe the effect of rigid filler particles on fracture stress and strain analytically. Smith introduced a model in which the composite was modeled by a cubic array of spherical particles [13] (Fig. 11a). He assumed that all particles debond from the matrix prior to fracture so that the crack grows through a plane weakened by the appearing pores (plane AA in Fig. 11). Neglecting stress concentration and assuming

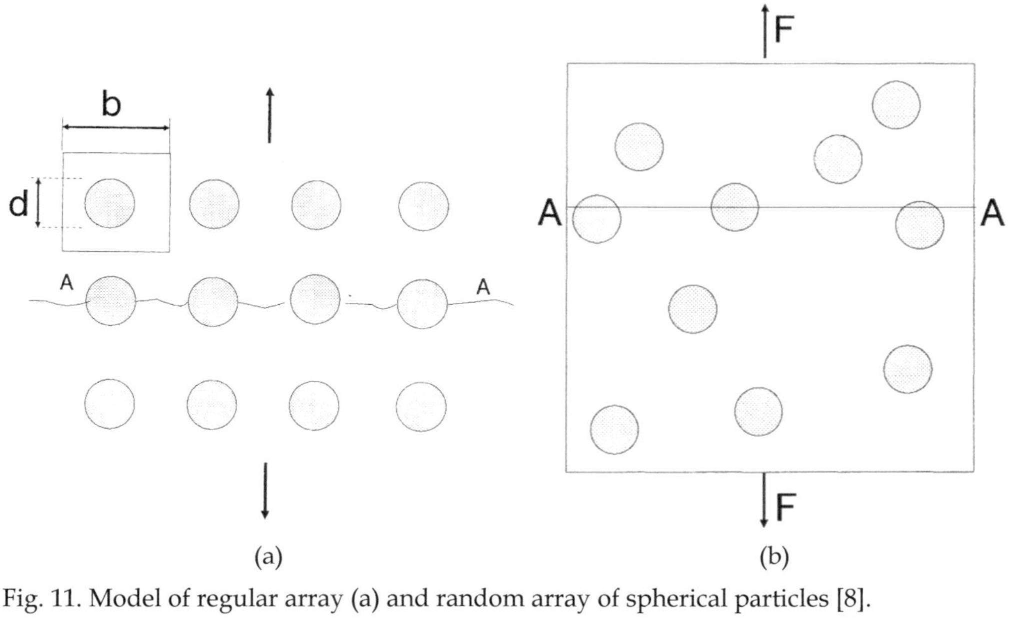

Fig. 11. Model of regular array (a) and random array of spherical particles [8].

that the strength of the composite is proportional to the cross-section of the matrix in the crack plane AA, the strength is given by equation:

$$\sigma_c = \sigma_m\left(1-\beta V^{2/3}\right), \tag{2}$$

where β = 1.21. For cubic particles Nielsen derived similar equation with β = 1 [14]. The coefficient β is sometimes called the shape factor. According to equation (2) the strength decreases with an increase in filler content. For example, the strength decrease at V = 40 vol.% is approximately threefold.

For the random particle distribution model (Fig. 11b) the following equation was derived [8]:

$$\sigma_c = \sigma_o \frac{1-V^{1/3}}{\sqrt{1+V^{1/3}+V^{2/3}}} \exp\left\{\sqrt{3}\left[\operatorname{arctg}\left(\frac{1+2V^{1/3}}{\sqrt{3}}\right)-\frac{\pi}{6}\right]\right\} \tag{3}$$

Expanding equation (3) in series by degrees of $V^{1/3}$, at low filler contents ($V^{1/3}$ <<1) it is simplified:

$$\sigma_c = \sigma_m\left(1-\frac{3}{2}V^{2/3}\right) \tag{4}$$

Fig. 12 shows the strength σ_c of filled SHMWPE plotted against aluminum particles V [8]. Equation (3) describes the strength of the composite (points) in the entire range of filling contents. Curve 1, corresponding to equation (2), describes experimental data only at V < 30 vol.%. Hence, the coefficient β = 1,21 is a fitting parameter adequately describing the experiment at filler contents V < 30 vol.%.

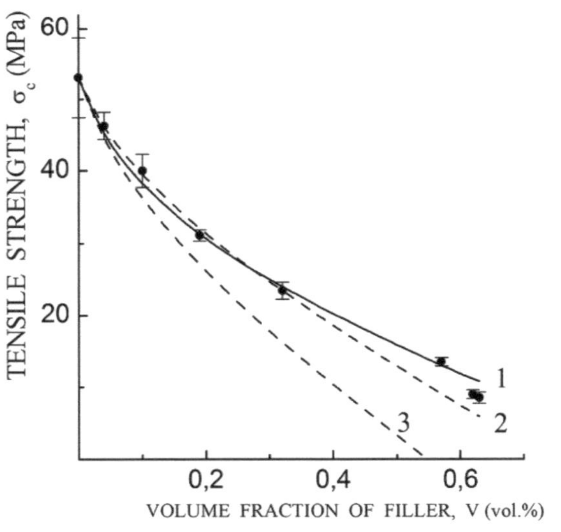

Fig. 12. Comparison of tensile strength σ_c of SHMWPE filled with Al particles. Theoretical values of strength in curves 1, 2 and 3 were calculated with equations (3), (2) and (4) respectively.

5. Draw stress

Fig. 13 shows the relative engineering draw stress σ_c of different filled polymers plotted against the filler content V [7]. In contrast to strength, the draw stress decreases linearly with an increase in the filler content V. The effects of filler on the strength and the draw stress of composite are functionally different. The draw stress depends on adhesion between the polymer and the particles. It is within the dashed fork region depending on adhesion between the polymer and the particles. The upper border of the fork, corresponding to well-bonded particles, is a constant equal to the draw stress of unfilled polymer. The lower border of the fork, corresponding to completely debonded in the yield point particles, is a straight line with the slope approximately equal to minus one. If particles are debonded partially, the draw stress is given by:

$$\sigma_d / \sigma_{dm} = 1 - \alpha V \tag{5}$$

where σ_{dm} is draw stress of unfilled matrix and α is fraction of debonded particles. If adhesion is good and particles are not debonded, $\alpha = 0$ and equation (5) gives $\sigma_d = \sigma_{dm}$. If all particles are debonded, $\alpha = 1$ [7].

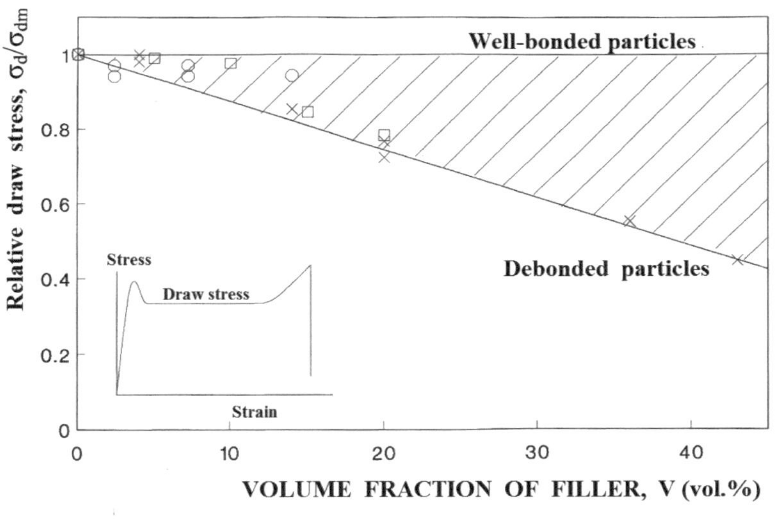

Fig. 13. Relative engineering draw stress of filled polymer, σ_d/σ_{dm}, versus the volume fraction of particles V. σ_{dm} – draw stress of unfilled polymer. (□) - PET/$CaCO_3$, 8 μm; (x) - PP/$Al(OH)_3$, 10 μm; (◊) - HDPE/$Al(OH)_3$, 1 and 25 μm [7].

6. The yield stress

Fig. 14 shows the effect of particles on the relative yield stress σ_y of filled polymers. The yield stress is within the dashed fork region depending on adhesion between the polymer and the particles. The upper border of the fork is a constant equal to the yield stress of the unfilled polymer σ_{ym}. The yield stress of filled polymer is equal to that of unfilled polymer if adhesion is strong and in the yield point particles are well-bonded with the polymer matrix. Reduction of adhesion leads to a decrease in the yield stress. The yield stress is described by the lower border of the fork if all particles debond before the yield point. If debonding is partial, the yield stress is inside the fork region, and is given by the equation [15]:

$$\sigma_y / \sigma_{ym} = 1 - 1{,}21 \alpha V^{2/3} \tag{6}$$

where σ_{ym} is the yield stress of unfilled matrix and α is fraction of debonded particles. If the adhesion between particles and the matrix is high, the particles in the yield point are not debonded, $\alpha = 0$ and $\sigma_y = \sigma_{ym}$ [3, 16].

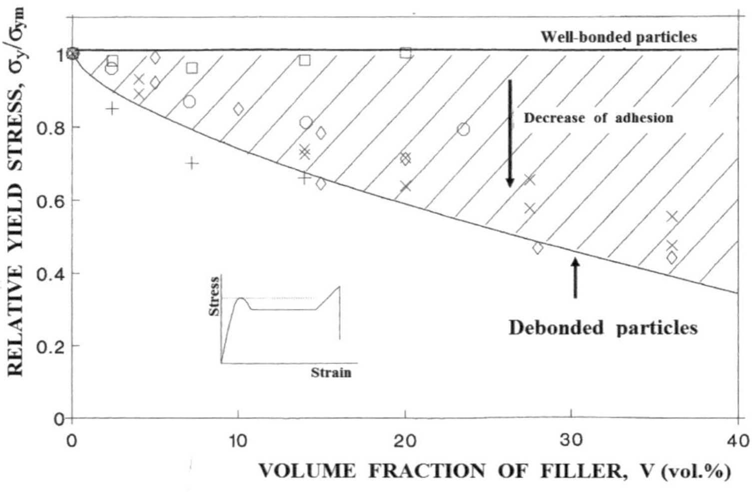

Fig. 14. Relative yield stress of filled polymer, σ_y/σ_{ym}, versus the volume fraction of particles V. σ_{dm} – draw stress of unfilled polymer. (□) - PET/$CaCO_3$, 8 μm [5]; (○) – PET filled with coated $CaCO_3$ particles; (x) - PP/$Al(OH)_3$, 1 and 25 μm; (◊) - HDPE/$Al(OH)_3$, 10 μm [7].

6.1 Transition to brittle fracture

6.1.1 Good adhesion

Fig. 15 shows schematically the composite strength and the draw stress plotted against the filler content V. The composite strength and the draw stress for well-bonded particles were calculated with equations (2) and (5). If the unfilled polymer yields with necking, at low filler contents the strength is higher than the draw stress. As a result, the neck propagates through the specimen, and filler does not change the deformation mode of the polymer. In contrast, at high filler contents the strength is lower than the draw stress, and the neck is not able to propagate through the sample. As a result, the composite fails during neck formation and a ductile-to-brittle transition is observed. The criterion of the transition from neck propagation to brittle fracture is calculated from equations (1), (2) and (5) at $\alpha = 0$ [5]:

$$V^* = \left(\frac{\sigma_m - \sigma_{dm}}{1.21\sigma_{dm}} \right)^{2/3} \tag{7}$$

where σ_m and σ_{dm} are the strength and the draw stress of the unfilled polymer.

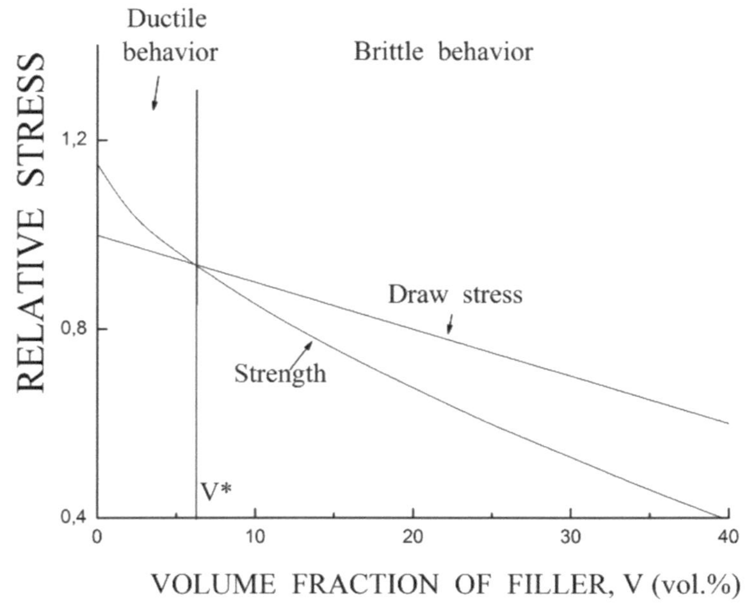

Fig. 15. Schematical drawing of the strength and draw stress of filled polymer plotted against the filler content, V. V* - the critical filler content of the transition from neck propagation to quasibrittle fracture.

Fig. 16 shows the embrittlement filler content V* plotted against the ratio of the strength to the draw stress σ_m/σ_{dm} of unfilled matrix. The critical filler content V* increases with an increase in the ratio σ_m/σ_{dm}. For example, the critical volume fraction V* for polyvinylchloride (PVC) is less than 0.05, and is not improved by toughening with rubber particles. If a polymer does not strain-harden, like PVC, even a small amount of filler leads to loss of composite ductility. The critical volume fractions for PP and for copolymer of ethyleneterephtalate (PET) are higher, from 0.1 to 0.2. The critical filler content of HDPE is from 0.2 to 0.25 and depends on particle size [2]. In general, the ratio σ_m/σ_{dm} describes the ability of the polymer to strain-hardening and hence the critical filler content V* increases with the polymer ability to strain-harden. The critical filler content calculated from equation (7) is plotted as a solid line in Fig. 16. The theoretical curve corresponds to experimental data for different matrices [7].

The polymer strain-hardening rises with an increase in polymer molecular weight. Hence the polymers with high molecular weight become brittle at higher filler contents then that with lower molecular weight. Unfortunately high molecular weight polymers have too high viscosity and are hardly mixable with rigid particles. Thus, requirements of high mechanical properties of composite contradict technological polymer viscosity requirements.

6.2 Weak adhesion

It is generally considered that a composite may have high mechanical properties only at good adhesion of the strengthening phase to a matrix. The example is fiber-reinforced plastics. However, filled composites are exclusion and week particle adhesion may suppress brittle fracture. At weak adhesion filler particles debond from the matrix before the yield point, and the yield and draw stresses are described by lower boundaries of the forks in Figs. 13 and 14.

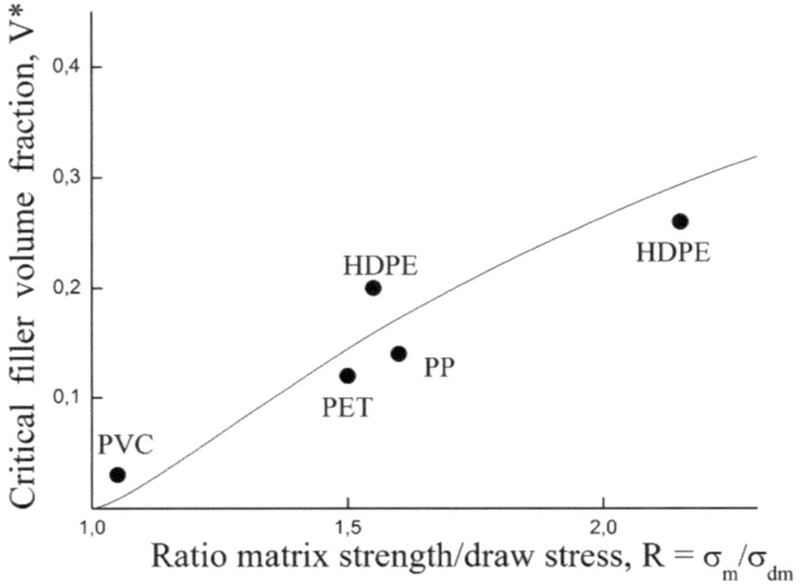

Fig. 16. The embrittlement filler content V* for well-bonded particles plotted against the ratio of the strength to the draw stress of unfilled matrix, $R = \sigma_m/\sigma_{dm}$. The points correspond to PVC, PET, PP and HDPE filled by particles with two different diameters, 8 and 25 μm.

6.2.1 $\sigma_m < \sigma_y$. Transition to brittle fracture

At week adhesion the draw stress corresponds to the lower border of the fork in Fig. 13, and the coefficient α in equation (5) is equal to 1 [7]. The critical filler content of the transition from neck propagation to brittle fracture is determined from criterion (1) and equations (2) and (5) at α = 1 [17]:

$$V^* = \frac{\beta^3 R^3}{27}[1 - 2\cos(\frac{\alpha}{3} + \frac{\pi}{3})]^3, \quad (8)$$

where β=1.21; $\alpha = \arccos[1 - \frac{27(R-1)}{2\beta^3 R^3}]$; $R = \sigma_m/\sigma_{dm}$; σ_m – strength and σ_{dm} – draw stress of the unfilled matrix.

Fig. 17 shows the embrittlement filler content V* plotted against the ratio of the strength to the draw stress of the matrix, σ_m/σ_{dm}. The critical filler content is determined by the matrix ability to strain-harden, similarly to the case of week adhesion. The critical filler content at good adhesion is higher than for well-bonded particles.

6.2.2 $\sigma_m > \sigma_y$. Transition to uniform yielding

Fig. 18 shows schematically the composite yield stress, strength and the draw stress plotted against the filler content, V_f. The strength, yield stress and the draw stress were calculated

with equations (2), (5) and (6). If the matrix strength is higher than its yield stress, the composite strength is higher than its yield stress at any filler content. At low filler contents the draw stress is lower than the yield stress and the strength of filled polymer. Hence, the neck propagates through the specimen, and filler does not change the deformation mode of the polymer. In contrast, at V>V** the yield stress is lower than the draw stress, and the yielding of the composite is uniform. Thus, fillers initiate transition from neck propagation to uniform yielding instead of brittle fracture [17]. This is a ductile to ductile transition allowing avoid the embrittlement of the composite. This behavior was observed in PP filled with $Al(OH)_3$ particles and glass spheres coated with release agent [18].
The criterion of the transition from neck propagation to uniform yielding is calculated from equation:

$$\sigma_d = \sigma_y \tag{9}$$

where σ_d and σ_y are the draw stress and the yield stress of filled polymer. Solution of equations (5), (6) and (9) gives the critical filler content V** of the transition from neck propagation to uniform yielding [17]:

$$V^{**} = \frac{\beta^3 R_1{}^3}{27}\left(1 - 2\cos\left(\frac{\alpha}{3} + \frac{\pi}{3}\right)\right)^3, \tag{10}$$

where β=1.21; $\alpha = \arccos\left(1 - \frac{27(R_1 - 1)}{2\beta^3 R_1{}^3}\right)$; $R_1 = \frac{\sigma_{ym}}{\sigma_{dm}}$, β=1,21; σ_{ym} – the yield stress and σ_{dm} – the draw stress of the matrix.
The critical filler concentration of the ductile to ductile transition is determined by the ratio of the yield stress to the draw stress of the unfilled matrix R_1. The ratio R_1 defines the height of the yield tooth. If the yield tooth is not high, R_1 value is close to 1. In this case the neck transition zone is not sharp. Therefore if the neck is not obvious transition occurs at a relatively low particle concentration. An increase in the yield tooth altitude results in an increase in R_1. Therefore to prevent embrittlement of filled composites the matrix should meet the following conditions: 1) the matrix strength should exceed the yield stress ($\sigma_m > \sigma_{ym}$) and 2) the yield point is not high.
Thus, fracture mode depends on matrix properties, particle content and their adhesion to the polymer matrix. Low amounts of filler do not change the fracture mode of a polymer. An increase in filler content leads to a transition in fracture mode. Usually this is the transition from ductile necking to brittle fracture. However, if the adhesion is weak and the strength is higher than the yield stress, the transition may be ductile-to-ductile from neck propagation to uniform yielding [17, 18].
Thus, the behavior of filled polymers depends on the matrix parameters which are not important for unfilled polymers. For example, the ratio of the strength to the draw stress and the ratio of the yield stress to the draw stress and even the yield point height.

6.3 Large particles

Fig. 19 shows a fracture strain of filled HDPE plotted against particle diameter D [19]. This dependence has maximum at the particle diameter of 10 μm. The decrease in fracture strain with an increase in the particle diameter is explained by initiation of cracks by the large particles.

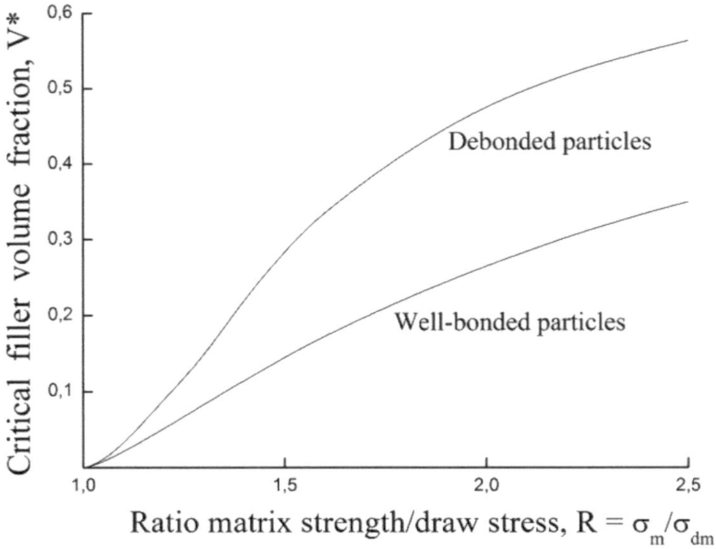

Fig. 17. The theoretical embrittlement filler content V* for well-bonded and debonded particles plotted against the ratio of the strength to the draw stress of unfilled matrix, R = σ_m/σ_{dm}. The curves were calculated with equations (7) and (8).

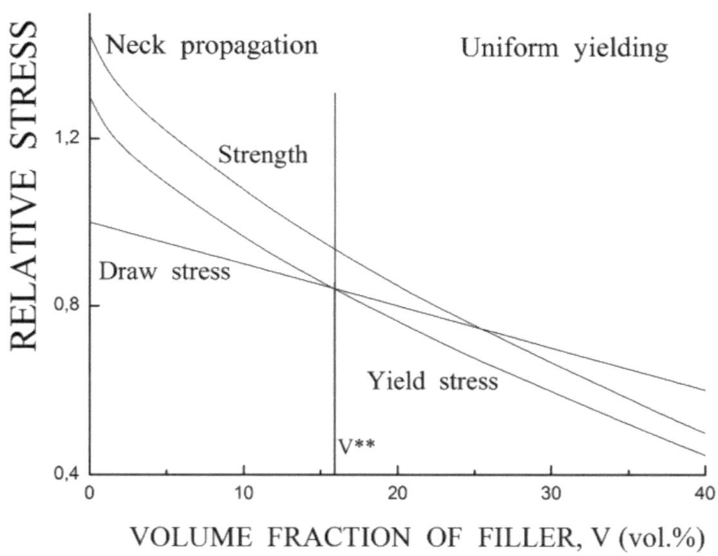

Fig. 18. Schematical drawing of the composite strength, draw stress and yield stress of filled polymer plotted against the filler content, V. V** - the critical filler content of the transition from neck propagation to uniform yielding [17].

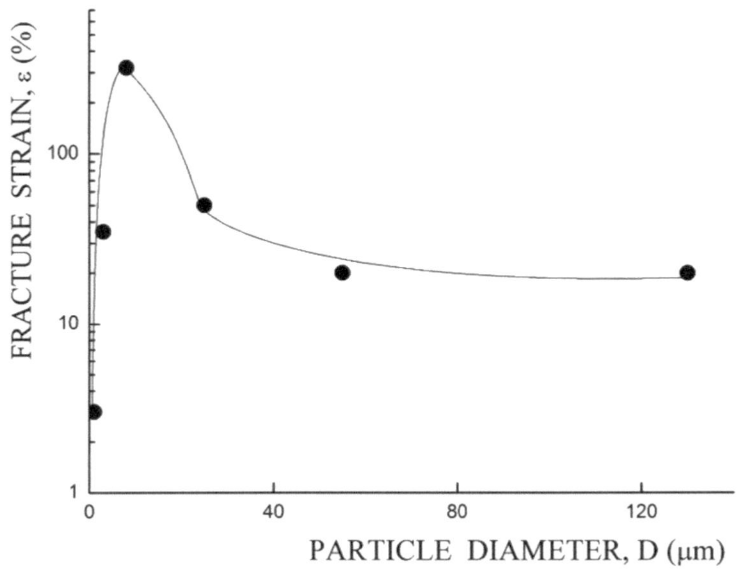

Fig. 19. Fracture strain of HDPE filled by rigid inorganic particles plotted against particle diameter, D [19].

Fig. 20 shows SEM images of fractured sample of HDPE filled by grinded rubber particles. Two types of pores in the composite are observed: large diamond pores (Fig. 20a) and small oval-shaped pores (Fig. 20b) [19]. Diamond pores appear only near large particles and oval pores near the small ones. Diamond pores are growing microcracks leading to material fracture. Pores of both types initiate as a result of particles debonding from the matrix.

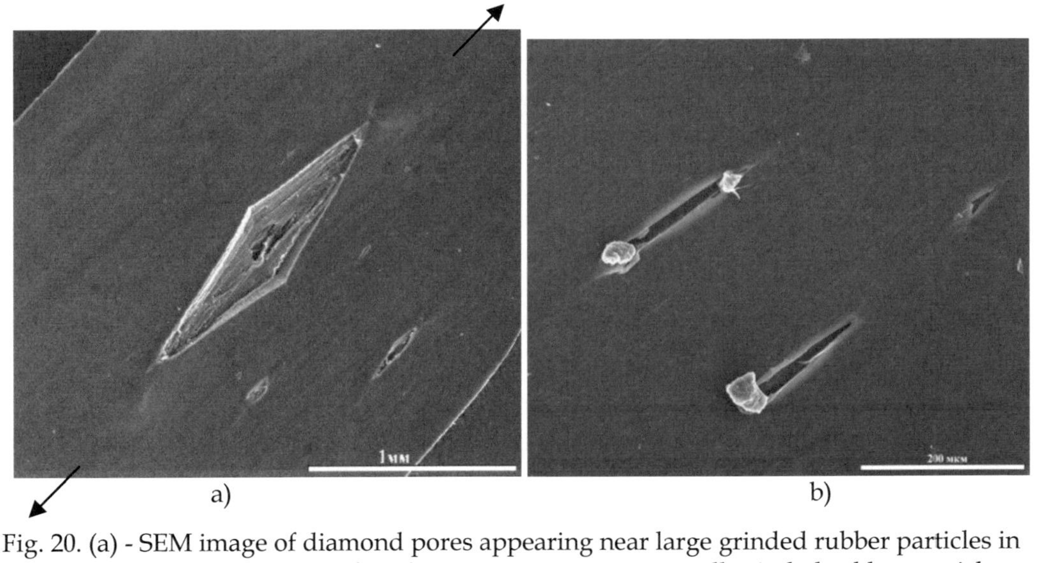

Fig. 20. (a) - SEM image of diamond pores appearing near large grinded rubber particles in filled HDPE. (b) - SEM image of oval pores appearing near small grinded rubber particles. Arrows show the elongation direction.

Initiation of pores is schematically illustrated in Fig. 21. Oval pores originate near small particles (a); at tension they elongate but remain oval until fracture. Diamond pores form near large particles (c). An intermediate case is presented by an oval pore transforming in a diamond one in tension. Consequently, use of too large particles is undesirable. Large particles initiate appearance of diamond pores and premature fracture of a composite.

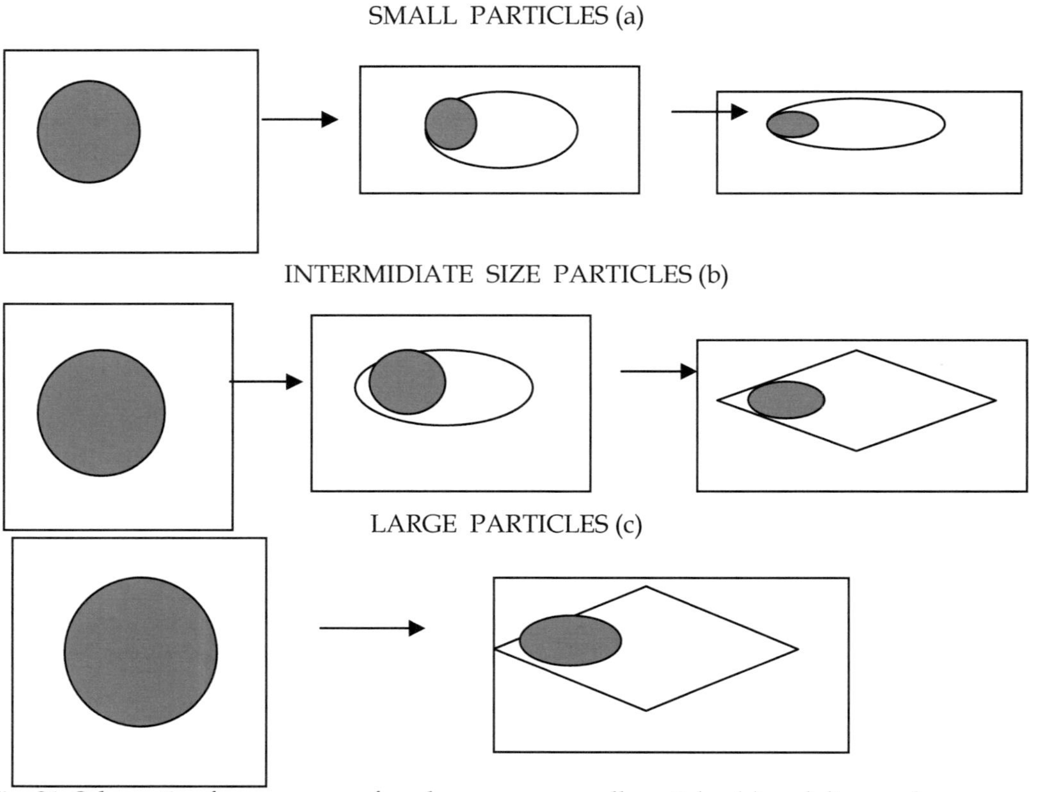

Fig. 21. Schematic of appearance of oval pores near small particles (a) and diamond pores near intermediate size (b) and large particles (c).

Behavior of a pore resembles a crack tip in a notched material. At small elongations the crack is blunted and its tip is round in shape (Fig. 22). Tension causes a gradual opening of the crack. At a certain point the geometry of its tip changes from rounded to V-shaped and the crack starts to grow. At further tension the size of the wedge grows through the sample. The wedge resembles a half of diamond pore.

The crack growth starts when its opening reaches a critical value of the crack opening δ_c [19]. Similarly an oval pore transforms into a diamond-shaped one when its opening reaches a critical value δ_c. Let us consider a pore initiated at elongation of a spherical particle with diameter D. Its length is $L = \lambda D$ (Fig. 22) and elongation is equal to the difference between the current length and initial size:

$$\delta = (\lambda - 1)\,D, \tag{11}$$

where λ is degree of the polymer matrix stretching.

Transformation of oval pore into a diamond one is described by the equation $\delta = \delta_c$, and the diamond pores appear only if the diameter of particles exceeds the critical value D*:

$$D^* = \delta_c/(\lambda - 1) \tag{12}$$

Equation (12) determines the elongation degree leading to initiation of diamond pores:

$$\lambda = \frac{\delta_c}{D} + 1 \tag{13}$$

Equation (13) determines transformation of oval pore into a diamond pore and the onset of fracture. The critical elongation is inversely proportional to the particle size. If the critical elongation λ is lower than the matrix elongation in the neck λ_d fracture is observed during the neck formation, and the composite is quasi-brittle. Therefore the critical particle size is

$$D^* = \delta_c/(\lambda_d - 1). \tag{14}$$

Fig. 23 shows that equation (14) quite accurately describes the critical diameter of particles initiating diamond pores. Large particles with the diameter exceeding D* should not be used in filled composites. Such particles initiate appearance of diamond pores and fracture of composite. The fracture toughness G_{Ic} is related with the critical crack opening δ_c by equation $G_{Ic} = \sigma_y \delta_c$, and (14) may be written:

$$D^* = \frac{G_{Ic}}{\sigma_y(\lambda_d - 1)} \tag{15}$$

where G_{Ic} is the fracture toughness and σ_y is the yield stress of the unfilled matrix. As shown in the previous sections, embrittlement is not typical for polymers deforming without necking. The effective method of suppressing necking is orientation of a polymer and, particularly, its rolling.

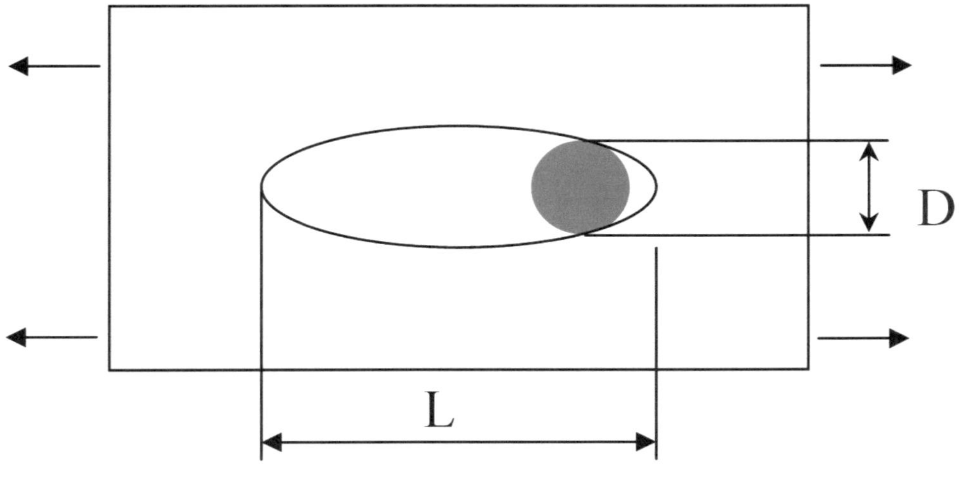

Fig. 22. Schematical drawing of oval pore appearing near a small particle (see Fig. 21a).

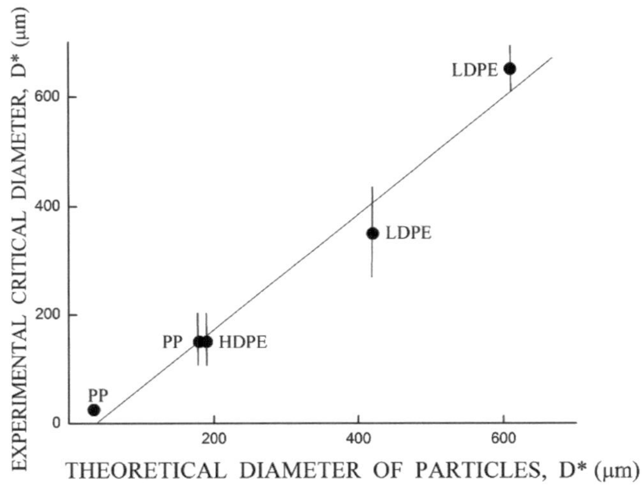

Fig. 23. Correlation between the experimental critical diameter of particles and theoretical values estimated with equation (15). Data are presented for HDPE, two types of LDPE and two types of PP [19].

7. Rolling

Fig. 24 shows typical engineering stress σ - strain ε curves for unoriented and rolled PET [20]. The increase in elongation degree after rolling leads to an increase in the ratio of the strength to draw stress. In addition, it suppresses necking and at the rolling degree $\lambda = 1.9$ the neck does not appear in the unfilled polymer. According to the above data, in this case brittle fracture of filled composite should not be observed and rolling suppresses embrittlement of filled polymers.

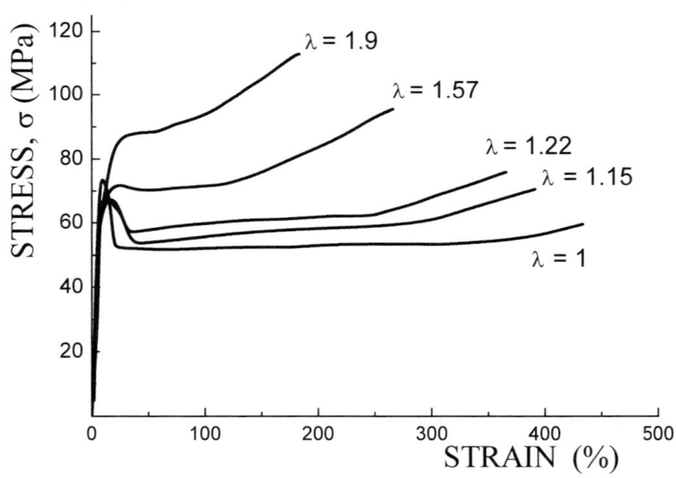

Fig. 24. Typical engineering stress σ versus strain ε curves of unoriented (λ=1) and rolled PET at rolling ratios λ=1.15; 1.22; 1.57 and 1.9. λ is equal to ratio of the sample length after rolling to its initial length.

Fig. 25 shows typical engineering stress σ - strain ε curves for unoriented and rolled PP filled with grinded rubber particles. An increase in rolling degree suppresses necking. The unoriented composite under tensile load is brittle (Fig. 25a) due to crazing of PP near large rubber particles. In contrast, rolled composites are ductile (Fig. 25b). At the rolling degree λ = 2.1 PP yields uniformely without necking. In addition, after rolling the crazes do not appear and filled composite is ductile. Hence, rolling effectively suppresses brittle fracture of filled polymers. Particularly, PP behaviour is ductile and crazes do not appear even at temperature -40°C [21, 22].

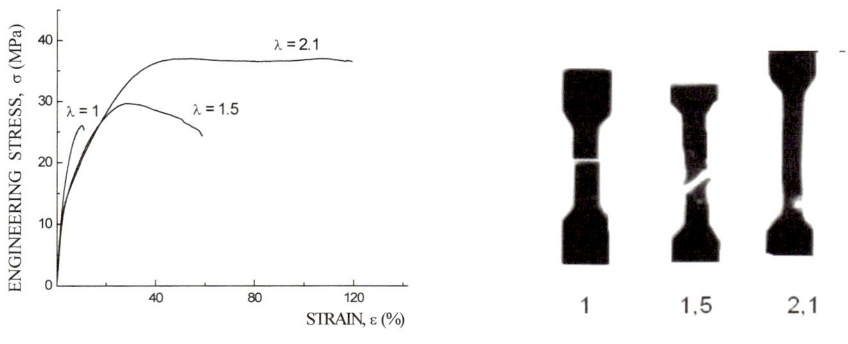

(a) (b)

Fig. 25. Typical engineering stress σ versus strain ε curves of PP filled with 8 vol.% of rubber particles at rolling ratios λ = 1 (unoriented), λ = 1.5 and 2.1. (b) - fractured samples.

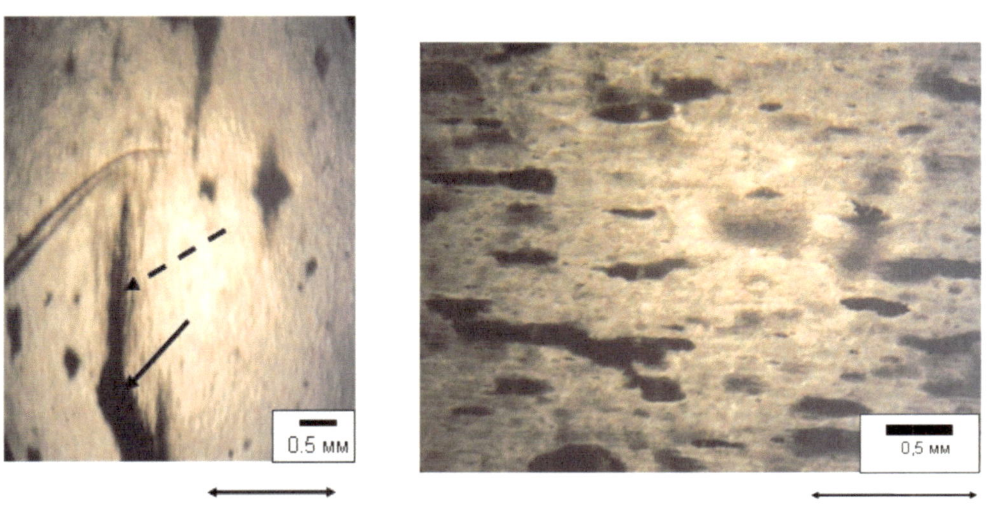

(a) (b)

Fig. 25. (a) - Crazes appearing near large rubber particles in unrolled PP under tension. (b) - Uniform deformation of PP filled with rubber particles after rolling to λ = 1.9. Arrows show the rolling elongation direction.

Fig. 26 shows engineering yield stress, draw stress and strength of PP plotted against the rolling ratio λ. At low rolling ratio λ the yield stress slightly decreases due to strain softening of the polymer. In addition, the difference between the yield stress and the draw stress decreases. Hence, rolling suppresses necking of the polymer. As a result, rolling allows avoid brittle fracture of filled polymer.
Fig. 27 illustrates the effect of rolling on the critical crack tip opening δ_c of HDPE. The critical crack tip opening δ_c sharply increases after rolling. Hence, rolling improves sensitivity of polymers to defects and, particularly, to filler particles. This also prevents brittle fracture of rolled composites.

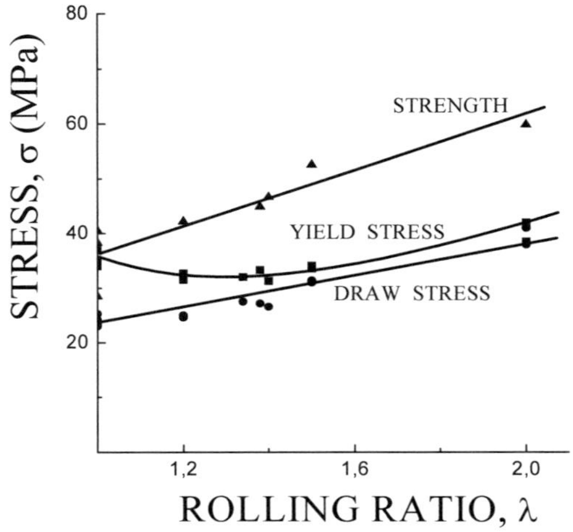

Fig. 26. Engineering yield stress, draw stress and strength of unfilled PET plotted against the rolling ratio λ [20].

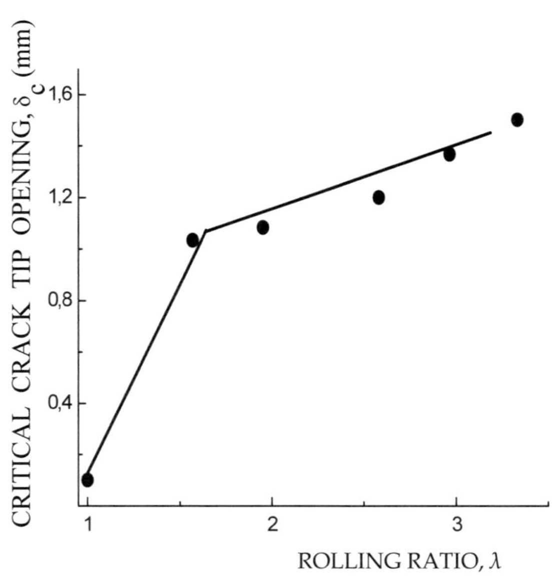

Fig. 27. Critical crack tip opening δ_c of HDPE plotted against the rolling ratio λ.

8. Conclusions

The main disadvantage of filled thermoplastic polymers is their brittle fracture. However, if unfilled polymer yields uniformly without necking, (PTFE, SHMWPE) composite remains ductile up to very high filler contents. Neck does not appear in polymers with very high molecular weight. Unfortunately, viscosity of these polymers is too high even at elevated temperatures, and they cannot be filled with particles by traditional liquid technologies.

The alternative method of necking suppression is orientation of polymer. Rolling is effective method of orientation of filled polymers. It allows avoid fracture during orientation, suppresses necking, improves ability to strain-hardening and leads to an increase in critical crack tip opening of polymer. As a result, rolled composites are ductile.

It is well known that orientation improves tensile strength and Young modulus of a polymer. In addition, it suppresses brittle fracture of unfilled and especially filled composites. Particularly, it suppresses crazing of polymers and composites.

The effect of particle size on fracture of particle filled composite is considered. Too large particles initiate appearance of diamond cracks and fracture of filled polymer. The critical particle size is determined by the critical crack tip opening or fracture toughness of the polymer matrix.

9. References

[1] Mathews F.L., Rawlings R.D. Composite materials: Engineering and science, Woodhead Publishing Limited, Cambridge, England (1999).

[2] Ashby M.F., Jones D.R.H., Engineering Materials, Elsevier Ltd. (2006).

[3] Tochin V.A., Shchupak E.N., Tumanov V.V., Kulachinskaya O.B., and Gai M.I., Mekh. Kompoz. Mater., 1984, p. 635.

[4] Berlin A.A., Topolkaraev V.A., Bazhenov S.L. Physical Aspects of Fracture and Deformation, Leningrad: Fizikotekh. Inst., 1987.

[5] Bazhenov S., Li J.X., Hiltner A., Baer E. J. Appl. Polym. Sci., 1994, 52, 243.

[6] Dubnikova I.L., Topolkaraev, V.A., Paramzina, T.V., and Diachkovskii, F.S., Visokomol. Soedin. 1, A32 (1990), 841.

[7] Bazhenov S.L. Fillers: their effect on the failure modes of plastics. // Plastics Additives. London, Chapmen and Hall. (1998). P.252.

[8] Bazhenov S.L., Grinev V.G., Kudinova O.I., Novokshonova L.A. Visokomol. Soedin. A52, 2010, p. 833.

[9] Novokshonova L.A., Meshkova I.N. Visokomol. Soedin. A36. 1994. P. 629.

[10] Grinev V.G., Kudinova O.I., Novokshonova L.A., Chmutin I.A., Shevchenko V.G. Visokomol. Soedin. A46. 2004. p. 1037.

[11] Novokshonova L.A., Meshkova I.N., Ushakova T.M., Grinev V.G., Ladigina T.A., Gultseva N.M., Kudinova O.I., De Boer S. J. Appl. Polym. Sci. 2003. V. 87. P. 577.

[12] Bazhenov S.L., Berlin A.A., Kulkov A.A., Ashmian V.G. Polymer Composite Materials. Intellect, 2010, p.347.

[13] Smith T.L. Trans. Soc. Reology. (1959), V.3, P.113.

[14] Nielsen L.E. J.Appl.Polym.Sci. 1966. V.10. P.97.

[15] Nicolais L., Narkis M. Polym. Eng.Sci. (1971), V.11, №3, P.194.

[16] Topolkaraev V.A., Tovmasian Y.M., Dubnikova I.L., Petrosian A.I., Meshkova I.N., Berlin A.A., Gomza Y.P., Shilov V.V. Composite Mechanics (1987), p.616.

[17] Bazhenov S., Polymer Engineering and Science, (1995), V.35, P.813.

[18] Dubnikova I.L., Topolkaraev V.A., Paramzina T.V., Gorokhova E.V., Diachkovsky F.S. Visokomol. Soedin. A32. 1990. p. 841.
[19] Bazhenov S.L., Serenko O.A., Berlin A.A., Dubnikova I.L., Berlin A.A. Doklady. 2003. V.393. p.336.
[20] Tunkin I.V., Bazhenov S.L., Efimov A.V., Kechekian A.S., Timan S.A. Visokomol. Soedin. A53, 2011 (to be published).
[21] Lin, Y. J.; Dias, P.; Chen, H. Y.; Hiltner, A.; Baer, E. Polymer, 49 (2008), 2578.
[22] Yang Y., Ponting M., Thompson G., Hiltner A. Baer E., J. Appl. Polymer Science (2011).

9

Porous Ceramic Matrix Al_2O_3/Al Composites as Supports and Precursors for Catalysts and Permeable Materials

S.F.Tikhov[1], N.A.Pakhomov[1], E.I.Nemykina[1], A.N.Salanov[1], V.A.Sadykov[1], V.E.Romanenkov[2] and Ya.Ya. Piatsiushyk[2]
[1]*Boreskov Institute of Catalysis RAS, Novosibirsk*
[2]*Powder metallurgy institute NAS Belarus*
[1]*Russia*
[2]*Belarus*

1. Introduction

Al/Al_2O_3 composites are widely regarded as promising construction materials (Yun et al., 2002; Watari et al., 2000; Saiz & Tomsia, 1998). Methods used for production of such materials traditionally include high-temperature processing. As a result, such materials have a low porosity and can not be used in adsorption and catalytic processes.

Meanwhile, permeable systems based on porous powder materials (PPM) - metals or oxides, are widely used as supports for catalysts and membranes (Sabirova et al., 2008; Ismagilov et al., 1997; Lee, 2003; Wang et al., 2004; Vityaz et al., 1987). When combined with catalysts (catalytically active components), permeable materials manufactured as membranes can substantially change selectivity to products of catalytic reactions (Rohde et al., 2005). Unfortunately, the amount of the active component that can be incorporated into traditional PPM is limited by poorly developed microporous and mesoporous structure of such materials. Porosity also declines during high-temperature processing required for production of mechanically durable products. For example, specific surface area of ceramic membranes based on alumina does not exceed 9 m^2/g with average diameter of macropores about 0.12 μm (Ismagilov et al., 1997). The specific surface of PPM and metal powders is significantly lower. Attempts of increasing the mesoporous component in PPM prepared from metal powders by filling macropores with suspensions resulted in substantial decrease of the average size of transport macropores and PPM permeability (Wang et al., 2004).

It was shown earlier that during hydrothermal oxidation (HTO) aluminum-containing powders are cemented into a robust macroporous monolith with relatively high specific surface area due to partial aluminum oxidation by water (Tikhov et al., 2005). During calcination the formed aluminum hydroxides are decomposed yielding nano(meso)porous component in Al_2O_3/Al ceramometals (cermets) localized in their oxide ceramic part. Particles of remaining aluminum are evenly distributed in the oxide matrix. A typical relief of the fracture face of Al_2O_3/Al cermet prepared by oxidation of aluminum powder by water at 100°C followed by calcination in air at 550°C is shown in Fig. 1. One can see that aluminum particles are covered by a porous oxide film.

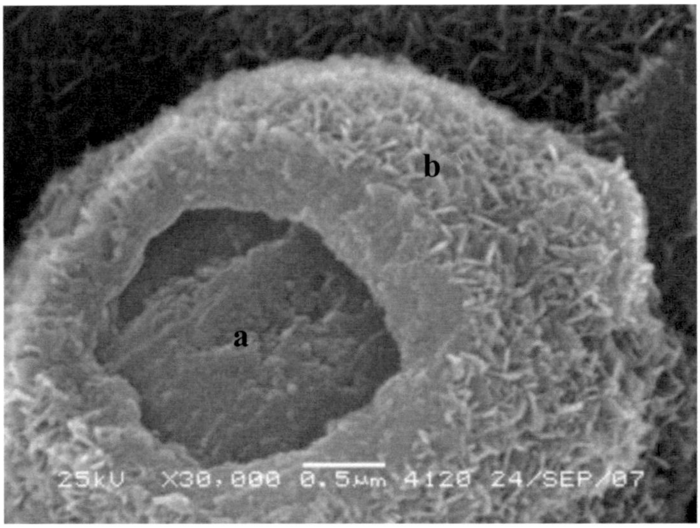

Fig. 1. SEM micrograph of Al_2O_3/Al cermet prepared from ASD-4 powder: a – aluminium core, b - porous alumina matrix.

Depending on the degree of aluminium oxidation during HTO and calcination temperature, the nano- and macroporous structure of the materials can be varied in a broad range (Tikhov et al., 2004; Rat'ko et al., 2004). The macroporous structure and permeability of the porous composites can be varied by changing the size and shape of the aluminum precursor particles. In addition, the mesoporous structure of aluminum oxide in the cermets can be tuned by mixing aluminum powder with aluminum oxides or hydroxides (Tikhov et al., 2004). In this paper we shall report the results of studying the properties of permeable cermets prepared from aluminum powders PAP-2, ASD-1 (Tikhov et al., 2004), ASD-4 and a mixture of aluminum powder PA-4 with the product of gibbsite thermal activation (Zolotovskii et al., 1997). The effects of the synthesis parameters on structural-mechanical and catalytic properties of cermets in isobutane dehydrogenation will be discussed.

2. Phase composition of Al_2O_3/Al composites

Typical XRD patterns of Al_2O_3/Al cermets are presented in Fig. 2. They include intense narrow peaks corresponding to 111, 220, and 200 reflections of aluminum phase. In addition, the XRD patterns contain weak broad peaks corresponding to 311, 400, 511, 440 reflections of aluminum oxide spinel phase [Tsybulya & Kryukova, 2008]. The phase composition of the aluminum oxide can not be characterized in more detail due to its poor crystallinity.

A comparison of the intensity of 440 peak of Al_2O_3 with that of 220 peak of Al shows (Fig. 2) that the content of aluminum oxide in the Al_2O_3/Al composites grows in the sequence: ASD-1<ASD-4<PAP-2. Quantitative estimates of the oxide content based on the earlier reported calibration curve (Tikhov et al., 2004) are shown in Table 1.

The differences of the oxide content in cermets are mostly due to different reactivity of the used aluminum powders at the initial hydrothermal oxidation stage (Tikhov et al., 2007). Aluminum is practically not subjected to any oxidation during calcination in air at 550°C. Hence, variation of the aluminum reactivity with respect to water allows to tune the content of the oxide phase in the Al_2O_3/Al cermets.

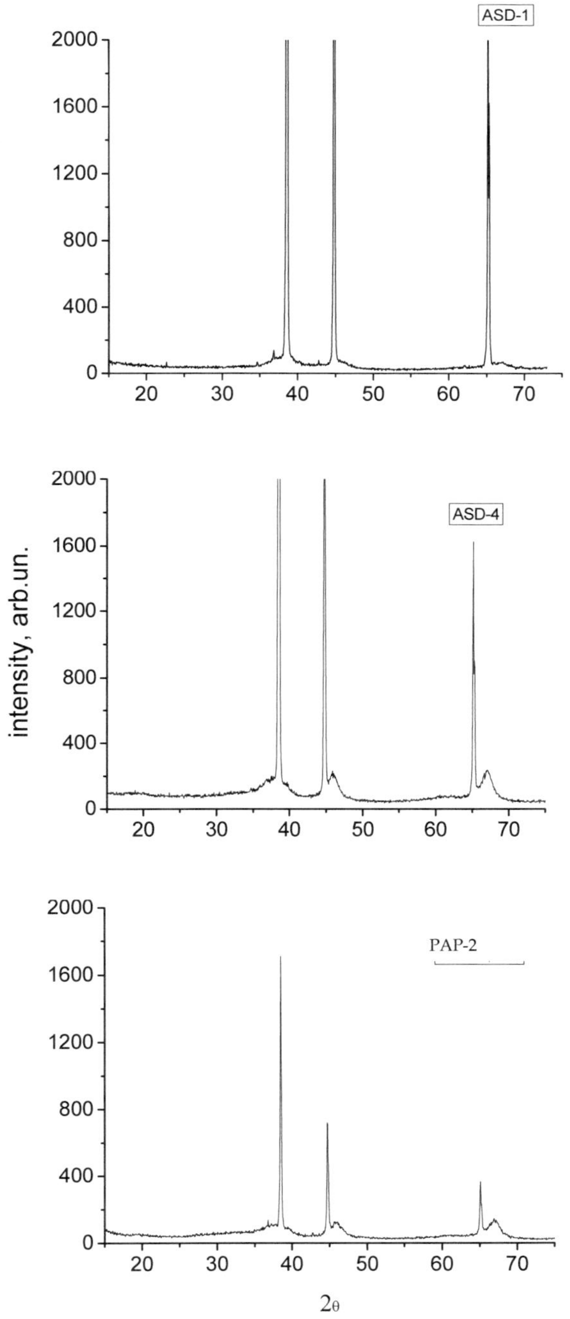

Fig. 2. XRD patterns of Al_2O_3/Al cermets prepared from different aluminum powders.

3. Macropore structure and permeability of cermets obtained from different aluminum powders

Qualitative analysis of the cermet macrostructure (Fig.3a,b,c) shows that it substantially depends on the type of the aluminum powder. The loosest packing of the monoliths is typical for cermets produced from PAP-2 powder. Averaged characteristics of the obtained materials estimated by the Darcy and "bubble point" methods (Vityaz et al., 1987; Khasin, 2005) using supports in the form of porous disks made of PTK-grade titanium foam (average pore size 120 μm, diameter 30 mm) containing the studied cermet deposited on their external surface by HTO followed by calcination are also very different (Table 1). The average pore diameter determined by this method increases from 1 to 22 μm in the series of cermets ASD-4<ASD-1<PAP-2. The maximum pore size varies in a similar order. The permeability coefficients of obtained materials also differ by more than a factor of 50 correlating with the average pore size (Table 1).

Fig. 3. SEM micrographs of porous Al_2O_3/Al composites prepared from different aluminum powders: (a) - ASD-1, (b) - ASD-4, (c) - PAP-2. (LEO 1455VP ,"Tescan"; JEOL JSM-6460V).

Comparison of the macropore structure parameters (Table 1) with a qualitative analysis of the macrotexture (Fig. 3) shows that a decrease of the average size of cermet particles aggregates decreases the average size of macropores and permeability. The average size and shape of particles in cermets are determined by the particles of the aluminum source powder used for synthesis. According to the data obtained by the Koulter method, the average particle size changes as follows: ~ 33 μm (ASD-1), ~ 12 μm (ASD-4), ~ 5 and ~ 35 μm (PAP-2) (Table 1). This trend qualitatively matches the results obtained by SEM (Table 1, Fig. 3). Larger average particle sizes obtained by the Coulter method for ASD-1 and ASd-4 samples in comparison with the SEM data are due to partial aggregation of the aluminum powder during its preparation or storage. Substantial variation of the particle size obtained for PAP-2 is, most likely, due to the flat shape of the particles that are also partially aggregated.

The differences in the shape and average sizes of porous cermet particles are quantitatively expressed in the loading density of aluminum powder in a die before HTO. For example, the loading density is about 1.6-1.8 g/cm³ for ASD-4, about 1.3-1.4 g/cm³ for ASD-1 and about 0.3-0.4 g/cm³ for PAP-2. In turn, the loading density is largely determined by the size and shape of aluminum particles. The flat shape of partially aggregated PAP-2 particles is known to result in substantially lower filling density compared to round particles (Tikhov,

2004 [8]). Round particles typical for ASD-1 and ASD-4 aluminum source (Tikhov et al., 2004) provide for a denser packing. Apparently, different particle packing is also preserved in the obtained cermet monoliths (Fig. 3). As the result, the average size of macropores in PAP-2 monolith is substantially larger. The composite obtained from ASD-4 has the smallest pore size. This is determined by the dimensions of cavities between the aluminum particles that are much larger for ASD-1. The degree of aluminum conversion to hydroxide determining the fraction of oxide in Al_2O_3/Al cermets has comparatively minor effect on the parameters of the macropore structure (Table 1).

Source aluminum powder	Average particle size of Al^0, μm	Permeability coefficient, $K\times10^{-13}$, m^2	Maximum macropore size, μm	Average macropore size, μm	Crushing strength, (σ), MPa	Al_2O_3, wt.%	Cermet porosity (ε),%
ASD-1	33*(25)**	~3.5	12.5	6.5	11	~24	~36
ASD-4	12(5)	0.4	4.5	~1	12	~44	~57
PAP-2	~5.4, 35	20	~63.5	22	23	~94	~42

* Coulter method.
()** According to SEM.

Table 1. Macropore structure, permeability and mechanical properties of Al_2O_3/Al composites prepared from different aluminum powders.

Thus, depending on the particle shape, the character of the macropore structure and permeability of the composite materials prepared using hydrothermal aluminum oxidation can be substantially varied.

A comparison with known permeable systems (Khasin, 2005) shows that smaller pores about 5-8 μm are formed when copper powder is strongly pressed together with combustibles. This pore size corresponds to permeability coefficient about 10^{-14}-10^{-13} m^2. The permeability coefficient of macroporous ceramic supports prepared by extrusion of pastes containing aluminum hydroxides and α-Al_2O_3 powder followed by calcination at 1200°C was estimated to be about ~10^{-14} m^2 (Ismagilov et al., 1997). Furthermore, the mechanical strength of ceramics (~0.8 MPa) proved to be substantially lower than that of PAP-2 monolith. This result opens great prospects for production of permeable materials with complex geometrical shapes from aluminum powder.

4. Mechanical properties of cermets

Dependence of crushing strength (σ) on porosity (ε) for PPM is expressed by Balshin's empirical relation (Leonov et. al., 1998; Balshin, 1972):

$$\sigma = \sigma_0(1-\varepsilon)^m , \quad (1).$$

Here m is the ratio of the total mass of the material to its mass subjected to the mechanical loading; σ_0 is the crushing strength of a non-porous body of the same composition. Equation (1) reflects a well-known trend that the strength of solids decreases when their porosity grows. This relation is valid for most porous cermets prepared using the HTO stage (Tikhov et al., 2000a; Tikhov et al., 2004a; Tikhov et al., 2004b). However, a comparison of the porosity and strength values reported in Table 1 suggests that the Balshin's relation is not valid for this PPM series. In particular, the cermet prepared from ASD-1 is approximately as strong as that made from ASD-4 despite substantially different porosity.

Most likely, this effect is related to a substantial difference in the character of contacts between particles forming PPM. According to (Rebinder et al., 1965), the crushing strength of a porous material (σ) is proportional to the number of contacts between particles (N), the surface area of the contact (S_i) and mechanical strength of the contact unit surface area (σ_i):

$$\sigma = N \cdot S_i \cdot \sigma_i \qquad (2).$$

Apparently, the Balshins's relation is valid only when one or several parameters of Eq. (2) vary only slightly. An abrupt change of parameters N, S, σ related to changes of the size or shape of particles forming a porous composite, or their chemical composition may lead to a substantial deviation from Eq. (1). In our system such deviation may be caused either by the change of the particle diameter from 20 to 5 μm or by change of the particle shape from spherical to flat. Moreover, it is very likely that the surface area of a singe contact in monolith PAP-2 is substantially larger due to the higher PAP-2 conversion at the HTO stage because these are the HTT products that cement the places of contact between the monolith particles (Rat'ko et al., 2004).

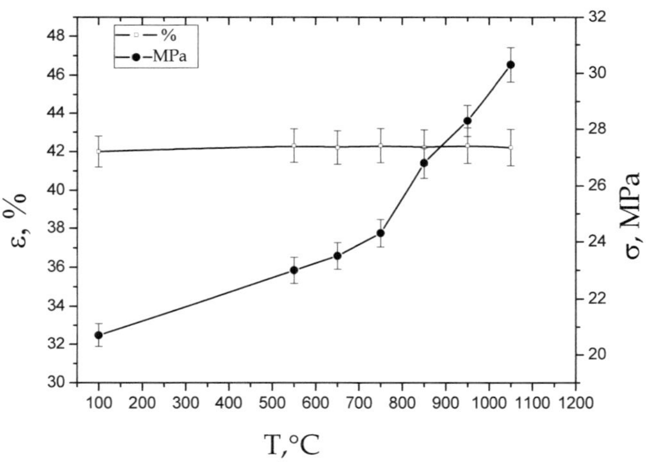

Fig. 4. Dependence of porosity (ε) and crushing strength (σ) on the calcination temperature for Al_2O_3/Al composite prepared from PAP-2.

Another example of a deviation from the Balshin's relation is a weak dependence of the crushing strength on porosity when the temperature of preliminary calcination of the ceramic granules increases (Fig. 4). In this case high-temperature sintering may also lead to the growth of contact surface area due to the surface diffusion or strengthening of a single contact due to its better crystallization.

Overall, one should take into account that porous ceramic supports, which are chemically the closest analogs of our materials, usually have substantially lower mechanical strength at comparable parameters of the macropore structure (Ismagilov et al., 1991). So, ceramometallic PPM with the aluminum oxide matrix may have great prospects in various applications.

5. Specificity of the nanopore structure of cermets

One of the most remarkable properties of composites prepared by cementing aluminum-containing powders under hydrothermal conditions is the presence of developed nano (micro, meso) porous structure formed by primary aluminum oxide nanoparticles and their aggregates. These particles are formed during thermal decomposition of aluminum hydroxides obtained from aluminum metal particles at the HTO stage. Qualitatively, differences in the size and shape of nanoparticles can be seen even in SEM images despite substantial limitations in resolution (Fig. 5). The largest aggregates of particles are typical for PAP-2. Meanwhile, the smallest particle size (< 0.1 μm in diameter) is observed for ASD-4. The aggregates of the oxide formed from ASD-1 have an intermediate size.

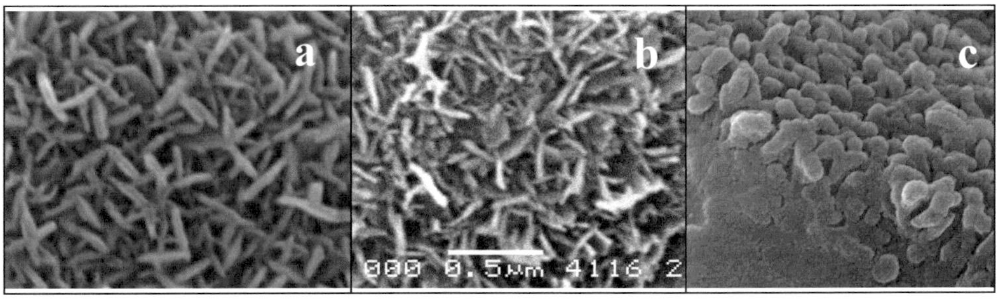

Fig. 5. Specificity of the microstructure of alumina aggregates on the surface of aluminum particles inside Al_2O_3/Al cermets prepared from different powders according to SEM data: (a) - ASD-1, (b) - ASD-4, (c) - PAP-2.

Averaged quantitative information on nanopores can be obtained from the analysis of adsorption-desorption isotherms. The isotherm presented in Fig. 6a and the pore size distribution (Fig. 6b) are typical for all PPM synthesized by HTO of aluminum powders and calcined at 500-700°C. According to the IUPAC classification (Gregg & Sing, 1982), the isotherms are close to type II, whereas the capillary condensation hysteresis loops are similar to type H3 (Fenelonov, 2002). The latter is typical to slit-like pores formed by flat parallel particles. The data reported in Table 2 show that the surface area and volumes of PPM nanopores may substantially vary depending on the type of the Al source powder and HTO conditions. At similar calcination temperatures the total surface area and volume of PPM nanopores usually increase when the fraction of highly porous aluminum oxide increases. Therefore, the highest values of these parameters were obtained for PPM prepared from PAP-2 that has ~90% aluminum conversion.
The characteristics of the alumina nanopore structure in the composites can be estimated more precisely with the account of the total specific surface area of the composite (S) and the fraction of oxide in it by calculating the specific surface area of the oxide (S_{Al2O3}) according to Eq. (3).

$$S_{Al2O3}=S_{sp}(1+\alpha X_0)/\alpha(1+X_0) \quad (3).$$

Here α is aluminum conversion to aluminum oxide, and $X_0 = 0.89$ is the relative weight change of the solid when all aluminum metal is oxidized (Tikhov et al., 2004a). The values of the aluminum oxide specific surface areas reported in parenthesis (Table 2) show that the specific surface areas of the oxide differ substantially for PPM prepared from ASD-1 and ASD-4 despite the fact that their total specific surface areas are similar. This result is largely related to different ratios of the aluminum hydrothermal oxidation rate to the ageing rate of the HTO products under hydrothermal conditions (Tikhov et al, 2000b). The particle dimensions should increase due to recrystallization of nanoparticles when the relative ageing rate increases, whereas the specific surface area should decrease. For ASD-1 and ASD-4 powders the ageing rates determined by the external HTT conditions were approximately equal. Meanwhile, the rates of Al consumption at the second diffusion-controlled stage of hydrothermal oxidation were substantially different. For ASD-4 this rate was higher by two orders of magnitude (Tikhov et al., 2007). Therefore, the smallest hydroxide (oxide) particles were obtained for this PPM. For PAP-2 the lowest oxidation rate was observed in the diffusion region. This led to a significant growth of the primary particle dimensions and decrease of the alumina specific surface area in comparison with the other samples.

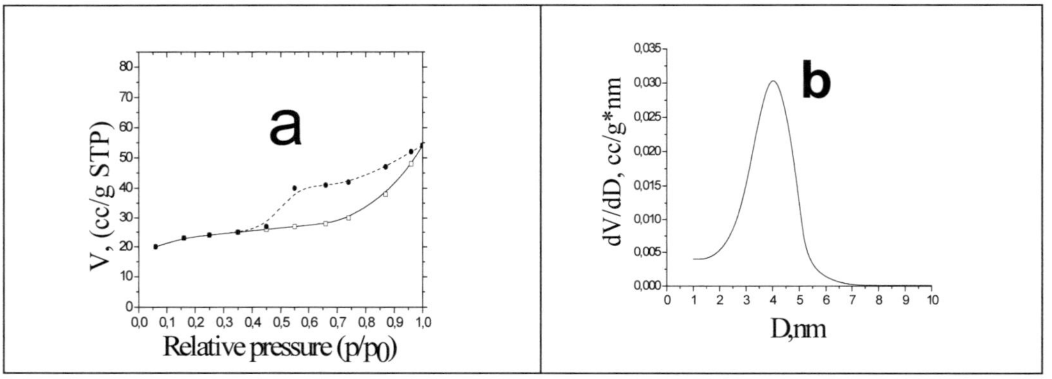

Fig. 6. Adsorption-desorption isotherm (a) and nanopore size distribution (b) for Al_2O_3/Al composite prepared from ASD-4.

Precursor aluminum powder	Specific surface area of Al_2O_3/Al (Al_2O_3), m^2/g	Nanopore volume (V), cc/g	Nanopore diameter, nm	Reference
ASD-1	52 (221)	0.05	3.8	(Tikhov et al., 2004a)
ASD-4	54 (318)	0.05	4.7	-
PAP-2	121 (~121)	0.16	4.7	-

Table 2. Parameters of the nanopore structure of Al_2O_3/Al composites.

The parameters of the nanopore structure are also substantially affected by the calcination temperature of cermets. The results obtained during investigation of the adsorption characteristics of PPM prepared from PAP-2 and calcined at increasing temperatures are reported in Table 3 and Fig. 7. The adsorption isotherms visibly change from type II to type III according to the IUPAC classification. The hysteresis loop practically disappears at high calcination temperatures (Fig. 7g). This fact is largely related to the changed pore size distribution (Fig. 7 b,d,f,h). Initially the sample had nanopores with a narrow pore size distribution around 4 nm. Additional mesopores with diameter ~ 5 nm appear starting from calcination temperature 850°C. At higher calcination temperatures the relative fraction of the nanopores with the size ~ 4 nm decreases whereas that of the larger pores increases. The average size of these pores grows to 8-10 nm.

The behavior of the specific surface area and volume of nanopores with increasing calcination temperature is somewhat unusual. The initial growth of the specific surface area after the heat treatment at 550°C is due to the thermal dehydration of aluminum hydroxide and formation of Al_2O_3. The pore size does not change during this process. However, the number of nanopores grows as it is indicated by a two-fold increase of the pore volume. Probably, this effect is due to a deviation of the aluminum hydroxide dehydration from an ideal pseudomorphic transformation (Tikhov et al., 2004a; Tikhov et al., 2004b).

Calcination temperature, °C	Specific surface area, m^2/g	Nanopore volume, cc/g	Nanopore diameter, nm
120	86	0.07	4.7
550	121	0.16	4.7
650	116	0.16	4.9
750	107	0.16	5.0
850	98	0.21	5.9
950	75	-	-
1050	24	0.12	9.5

Table 3. Influence of the calcination temperature on the nanopore characteristics of Al_2O_3/Al composite prepared from PAP-2.

Subsequently, continuous smooth increase of the specific surface area accompanied by the growth of the average pore size is observed with increasing calcination temperature. Until the calcination temperature of 750°C the volume of nanopores is approximately constant with a minor increase of the pore size. For the sample calcined at 850°C the pore volume substantially grows, decreasing at higher temperatures (Table 3). After calcination at 1050°C the pore volume decreases by almost a factor of two, whereas their diameter increases by the same value. Thus, the specific surface area of samples calcined at 550°C or higher temperatures continuously decreases with calcination temperature. Meanwhile, the changes in the volume of nanopores are not monotonous. Such effects are not typical for aluminum oxide powders. Usually, both the pore volume and specific surface area decrease with the calcination temperature (Khalil, 1998; Zhu et al., 2002). However, for granulated aluminum oxides with a large amount of macropores in some cases a small growth of the total pore volume with the temperature of calcination was observed, while their micropore volume and specific surface area decreased (Boreskov et al., 1951). Such effect is, most likely, due to the specific features of sintering of primary aluminum oxide nanoparticles in granulated ceramics. The mechanism of the nanoparticle sintering may be attributed to "internal" sintering due to surface diffusion (Geguzin, 1984), when the growth of particles and decrease of the free surface energy is not accompanied by the pore structure "densification", so that the total pore volume remains approximately constant. This process is possible due to the growth of nanoparticles and formation of separate particles (dense or low-porous) with sizes substantially exceeding those of nanoparticles present in the initial sample. Adjacent particles merge with such particles.

Thus, variation of the aluminum reactivity and calcination temperature allows one to modify the parameters of aluminum oxide nanostructure in PPM.

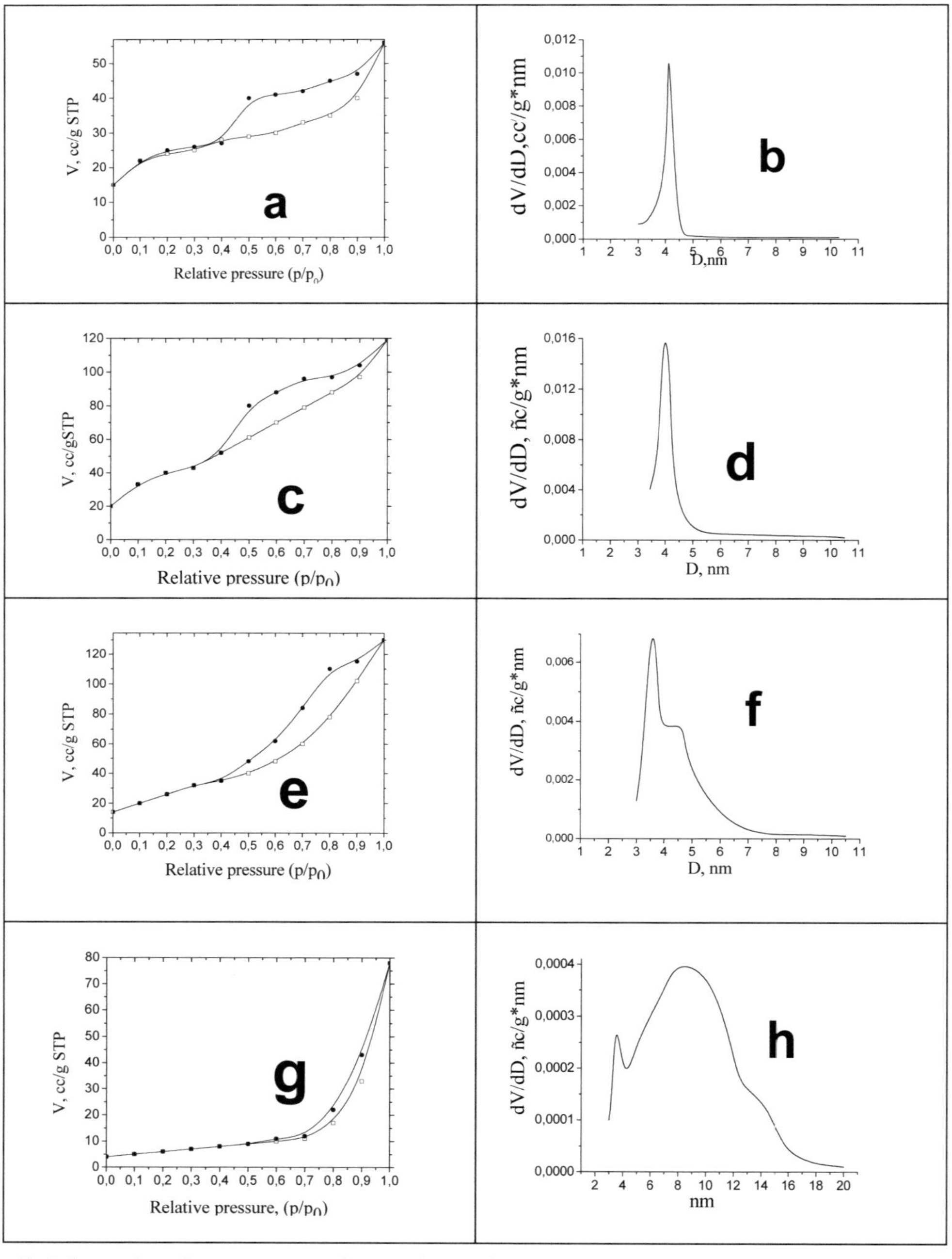

Fig. 7. Adsorption-desorption isotherms (a,c,e,g) and nanopore size distribution (b,d,f,h) of Al_2O_3/Al composite prepared from PAP-2 after different temperatures of calcination: (a,b) - 120°C; (c,d) - 650°C; (e,f) - 850°C; (g,h) - 1050°C.

6. Nanoporous structure of composites prepared from blends of aluminum with TCA and catalytic properties of the $CrO_x/Al_2O_3/Al$ catalysts in the dehydrogenation of isobutane

Despite the fact that variation of synthesis conditions and type of aluminum powders makes possible to change the properties of nanoporous oxide component in a broad range, their application is substantially restricted by economic and technological problems. In particular, the problems of cost and application safety are the most important for application of cermets as catalyst supports. In this respect, such powders as ASD-1 and similar to it PA-4 are more promising that PAP-2 or ASD-4. The main drawback of the PA-4 powder is a low activity in HTO, which leads to a low surface area and moisture capacity of cermets prepared from it. The last factors are exceptionally important for synthesis of catalysts by impregnation. To improve these parameters, we synthesized supports by blending aluminum powder with the product of gibbsite thermochemical activation (TCA). TCA is amorphous aluminum oxide (Al_2O_3x1.3 H_2O) synthesized by gibbsite dehydration in a pulse mode and characterized by a high chemical activity and ability to be rehydrated to form pseudoboemite in the presence of water vapor (Zolotovskii et al., 1997). During the cermet synthesis TCA is a precursor for additional nanoporous aluminum oxide incorporated in the cermet macropores.
The XRD patterns of the Al_2O_3/Al cermet prepared from a PA-4 + TCA blend and alumina prepared from TCA are shown in Fig. 8. The aluminum oxide concentration in this cermet is higher compared with those where all alumina was obtained by aluminum oxidation (compare Figures 2 and 8). The structure of aluminum oxide both in the composite and in pure aluminum oxide are close to that of γ-Al_2O_3 (Tsybulya & Kryukova, 2008).

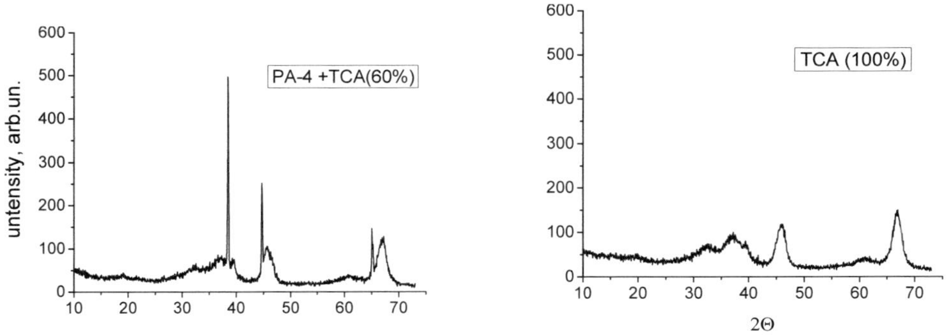

Fig. 8. XRD patterns of Al_2O_3/Al cermet prepared from blend of PA-4 + TCA and alumina from TCA.

Catalysts from Al_2O_3/Al cermets were prepared by incipient wetness impregnation of their calcined granules (3x3 mm cross-section and 3-5 mm length, Fig. 9) with solutions of active component precursors - chromic anhydride and promoters (Pakhomov et al., 2008). In this Chapter we shall report textural properties of the composite supports and catalytic properties of chromium-alumina catalysts in dehydrogenation of isobutane.
Comparison of Figures 1 and 10 proves that the surface texture of cermets prepared with TCA differs substantially from that of cermets without its addition. Primarily, this is a result of introducing a large amount of aluminum hydroxide with aggregates of irregular shape (Fig. 10a). Another reason for the differences in the surface shape is the high

calcination temperature (700°C) favoring aluminum melting from the nanoporous ceramic spheres formed during HTO and calcination (Fig. 10b).

Fig. 9. General view of granulated Al_2O_3/Al composites prepared from powdered TCA and PA-4 used as supports for dehydrogenation catalysts.

Fig. 10. SEM micrographs of the fracture surface of Al_2O_3/Al composite granules prepared from powdered TCA and PA-4.

However, this process did not lead to destruction of the cermet support granules or substantial mesoporosity degradation (Table 4). According to the data presented in Table 4, the increase of TCA concentration in the initial blend leads to a continuous increase of the specific surface area, volume and average diameter of nanopores.The total pore volume, which, in addition to nanopores, includes macropores and ultramacropores formed by voids between aluminum particles, aggregates of aluminum oxide formed during TCA decomposition and hollow spheres, also increases. Meahwhile, the fraction of nanopores in the total pore volume grows from 44 to 67%. The aluminum precursors alone (see ASD-1 sample with similar properties in Table 2) are not capable of providing for such developed nanopore structure in cermets.

Sample, TCA content in initial blend, wt.%	Specific surface area, m^2/g	Nanopore volume, cm^3/g	Total pore volume, cm^3/g	Average nanopore diameter, nm
60	97	0.25	0.35	7.3
50	87	0.20	0.34	6.8
40	78	0.16	0.33	5.7
30	64	0.12	0.27	5.9

Table 4. Nanopore textural characteristics of granulated composites prepared from powdered TCA and PA-4 (calcination temperature 700°C).

At the same time, the catalytic properties of chromium-alumina catalysts did not change continuously with variation of the TCA concentration in the precursor blend. Maximum isobutene conversion was observed for the sample containing 50% TCA (Fig. 11). This sample is also characterized by the highest selectivity to isobutene. Apparently, this composition has optimum pore structure of the support. In particular it is shown in Fig. 12 that the nanopore structure of the support includes both nanopores with dimensions 3-4 nm (narrow peak) and larger pores with diameters 6-8 nm. The improvement of all nanopore structure parameters is exclusively due to the larger nanopores as the fraction of the smaller ones decreases (Fig. 12). There is certain ratio between the two types of pores in the optimum 50% TCA sample that decreases for the 60% TCA sample (Table 4). It appears that such nanopore structure provides for the optimum structure or size of the catalyst active component.

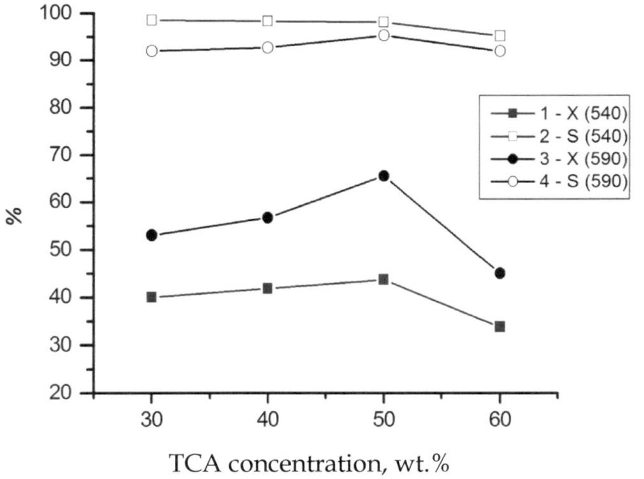

Fig. 11. Dependence of isobutane conversion X (1,3) and selectivity to isobutene S (2,4) at 540°C (1,2) and 590°C (3,4) on TCA concentration (wt.%) in the initial blends used for preparation of granulated composites $CrO_x/Al_2O_3/Al$.

The developed macropore structure of the support granules with 3-5 mm size excludes problems of internal diffusion limitations typical for granulated catalysts. Overall, this provides for high isobutene conversion close to the equilibrium one at this temperature

(Pakhomov, 2006) and unusually high selectivity (>94%) to the dehydrogenation product (Fig. 11).

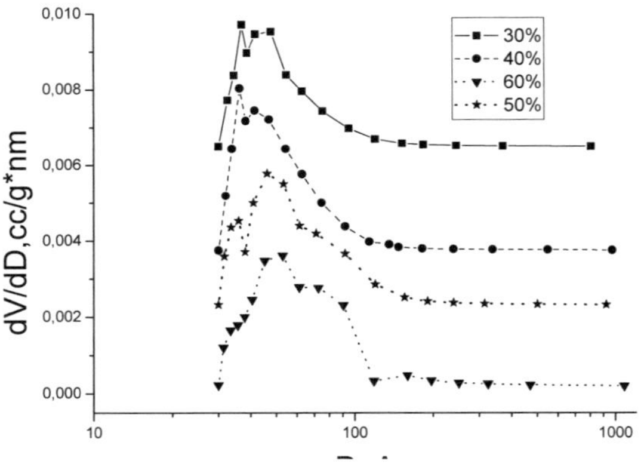

Fig. 12. Nanopore size distribution of Al_2O_3/Al composites prepared from blends with different TCA concentrations.

7. Conclusion

Hydrothermal treatment of blends with aluminum metal powders provides an efficient technique for preparation of mechanically strong monolith composite materials with developed nanopore structure along with a relatively high fraction of macropores. The developed macropore structure provides a high permeability and decreases diffusion limitations inside the porous composite. Changing the reactivity of Al powder particles allows to tune the oxide/aluminum ratio, while their shape affects the monolith permeability. These materials can be used as filters or membrane supports.
Incorporation of precursors of the nanoporous materials in a macroporous system leads to a substantial increase of the nanopore volume. Application of granulated Al_2O_3/Al composites prepared from powdered blends of aluminum and thermally activated gibbsite as catalysts supports allowed us to prepare catalysts with high activity and selectivity in dehydrogenation of isobutane.

8. Acknowledgment

This work was supported by Integrated project of SB RAS and NAS Belorus No 115_3 and T10C-023.

9. References

Balshin, M., (1972), Scientific bases of the powder and fiber metallurgy, Metallurgy Publ. House, Moscow, USSR [in Russian]

Boreskov, G., Dzisko, V., Borisova, M., Krasnopolskaya, V., (1951), „Influence of thermal treatment on the structure and catalytic activity of alumina", Journal of Physical Chemistry, Vol. 26., No. 4., pp.492-499 ISSN 1089-5639 [in Russian]

Fenelonov, V., (2002), Introduction into physical chemistry of formation of supramolecular structure of the adsorbents and catalysts. Publishing house of SB RAS, ISBN-5-7692-0647-0, Novosibirsk, Russia [in Russian]

Gregg, S., Sing, K., (1982), Adsorption, surface area and porosity,: Academic Press, ISBN-10 0123009561, London, United Kingdom

Geguzin, Ya., (1984), Physics of sintering, Nauka, Moscow, USSR [in Russian].

Ismagilov, Z., Shkrabina, R., Koryabkina, N., Kirchanov, A., Verinda, H., Pex, P., (1997), "Porous alumina as a support for catalysts and membranes. Preparation and study", Reaction Kinetics and Catalysis Letters, Vol. 60, No. 1, pp. 225-231, ISSN 0133-1736

Ismagilov, Z., Shepeleva, M., Shkrabina, R., Fenelonov, V., (1991), "Interrelation between structural and mechanical characteristics of spherical alumina granules and the initial hydroxide properties", Applied Catalysis, Vol. 69., No.1., pp. 65-74. ISSN 0926-860X

Khasin, A., (2005), Memrane reactors for the Fischer-Tropsch synthesis, In: "Sustainable Strategies for the Upgrading of Natural Gas: Fundamentals, Challenges, and Opportunities", E.G.Derouane et al. (Eds), 249-271, Springer, ISBN 101-4020-3309-5, Dordrecht, Netherlands

Khalil, K., (1998), „Synthesis of short fibrous boehmite suitable for thermally stabilized transition aluminas formation", Journal of Catalysis, Vol. 178., No.1, pp.198–206, ISSN 0021-9517

Lee, D., Lee, Y., Nam, S., Ohm, S., Lee, K., (2003), "Study on the variation of morphology and separation behavior of the stainless steel supported membranes at high temperature", Journal of Membrane Science, Vol. 220, No1, pp. 137-153. ISSN 0376-7388

Leonov, A., Smorigo, O., Romashko, A., Deshko, M., Ketov, A., Novikov, L., Tankovitch, V., (1998), "Comparative study of the properties of monolith supports of the honeycomb and foam structure for the use in the processes of the catalytic gas purification", Kinetics and Catalysis, 39., 691-700. ISSN 0023-1584

Pakhomov, N., (2006), Modern state and prospects of the dehydrogenation processes. In: Industrial catalysis in lections, Boreskov Indstitute of Catalysis, ISBN 978-5-89530-018-3, Vol. 6., pp. 53-98, Novosibirsk, Russia [in Russian]

Pakhomov, N., Molchanov, V., Zolotovskii, B., Nadtochii, B., Isupova, L., Tikhov, S., Kashkin, V., Kharina, V., Tanashev, Yu., Parakhin. O., (2008), „Development of catalysts for dehydrogenation of C3-C4 paraffins using gibbsite thermal activation product", Catalysis in Industry, Special issue, pp. 13-19, ISSN 1816-0387 [in Russian]

Rat'ko, A., Romanenkov, V., Bolotnikova, E., Krupen'kina, Zh., (2004), "Hydrothermal Synthesis of Porous Al2O3/Al Metal Ceramics: II. Mechanism of Formation of a Porous Al(OH)3/Al Composite", Kinetics and Catalysis, Vol. 45., No. 1., pp. 141-148, ISSN 0023-1584

Rebinder, P., Margolis, L., Schukin, E., (1964), "About mechanical strength of the porous dispersed bodies", Doklady of Physycal Chemistry, 154., 695-700. ISSN 086669-5652 [in Russian]

Rohde, M., Unruh, D., Schaub, G., (2005), "Membrane application in Fischer–Tropsch synthesis reactors—Overview of concepts", Catalysis Today, Vol. 106., No.1, pp. 143-148, ISSN 0920-5861

Sabirova, Z., Danilova, M., Kuzin, N., Kirillov, V., Zaikovskii, V., Kriger, T., Mescheryakov, V., Rudina, N., Brizitskii, O., Hrobostov, L., (2008), "Nickel catalysts on the base of porous nickel for the reaction of methane steam reforming", Kinetics and Catalysis., Vol. 49., No. 2., pp. 449-456, ISSN 0023-1584

Saiz, E., Tomsia, A., (1998), "Kinetics of Metal-Ceramic Composite Formation by Reactive Penetration of Silicates with Molten Aluminum", Journal of American Ceramic Society, Vol. 81., No. 10., pp. 2381-2393, ISSN 0002-7820

Tikhov, S., Fenelonov, V., Sadykov, V., Potapova, Yu., Salanov, A., (2000a), "Porous metalloceramics Al2O3/Al prepared by oxidation of powdered aluminum in hydrothermal conditions. I. Content and macrocharacterisics of composites", Kinetics and Catalysis, Vol.41., No. 3, pp. 907-912, ISSN 0023-1584

Tikhov, S., Zaikovskii, V., Fenelonov, V., Potapova, Yu., Kolomiichuk, V., Sadykov, V., (2000b), "Porous metalloceramics Al2O3/Al prepared by oxidation of powdered aluminum in hydrothermal conditions. II. Content and microtexture of composites", Kinetics and Catalysis, Vol. 41., No.3, pp. 916-922, ISSN 0023-1584

Tikhov, S., Romanenkov, V., Sadykov, V., Parmon, V., Rat'ko, A., (2004a), Porous composites based on oxide-aluminum cermets (synthesis and properties), Publ. House of SB RAS "Geo" Branch, ISBN-5-7692-0705-1, Novosibirsk, Russia

Tikhov, S., Potapova, Yu., Fenelonov, V., Sadykov, V., Salanov, A., Tsybulya, S., Melgunova, L., (2004), "Porous metalloceramics Al2O3/Al prepared by oxidation of powdered aluminum in hydrothermal conditions. IY. Influence of oxide additives on the content and textural properties of composites MOx/Al2O3./Al, Kinetics and Catalysis, Vol. 45., No.2, pp. 642-653, ISSN 0023-1584

Tikhov, S., Potapova, Yu., Sadykov, V., Fenelonov, V., Yudaev, I., Lapina, O., Salanov, A., Zaikovskii, V., Litvak, G., (2005), "Synthesis of alumina through hydrothermal oxidation of aluminum powder conjugated with surfactant-directed oriented growth", Materials Research Innovations, Vol.9., pp. 431-446, ISSN 1432-8917

Tikhov, S., Sadykov, V., Ratko, A., Kuznetsova, T., Romanenkov, V., Eremenko, S., (2007), "Kinetics of aluminum powder oxidation by water at 100°C", Reaction Kinetics and Catalysis Letters, Vol. 92., No.1., pp. 83-88, ISSN 0133-1736

Tsybulya, S.V., Kryukova G.N. (2008). Nanocrystalline transition alumina: Nanostructure and features of x-ray powder diffraction patterns of low-temperature Al2O3 polymorphs, Physical review, Vol.77B, No 024112, pp. 1-13, ISBN 1098-0121

Vityaz, P., Kapcevich, V., Sheleg, V., (1987), "Porous powdered materials and their products", Vysshaya Shkola, Minsk, USSR [in Russian]

Wang, D., Tong, J., Xu, H., Matsumura, Y., (2004), "Preparation of palladium membrane over porous stainless steel tube modified with zirconium oxide", Catalysis Today, Vol. 93-95., pp. 689-693, ISSN 0920-5861

Watari, T., Torikai, T., Tai, W., Matsuda, O., (2000), "Fabrication and mechanical properties of α-Al2O3/β-Al2O3/Al/Si composites by liquid displacement reaction", Journal of Material Science, Vol. 35., No. 2., pp. 515-520, ISSN 0261-8028

Yun, Y., Hong, S., Choi, S., (2002), "Metal penetration processing and mechanical properties of Al/Al2O3 composite system", Journal of Materials Science Letters, Vol. 21., No. 16., pp. 1297-1299, ISSN 0261-8028

Zhu, H., Riches, J., Barry, J. (2002), „γ-Alumina nanofibres prepared from aluminum hydrate with poly(ethylene oxide) surfactant", Chemistry of Materials, Vol. 14., No.8, pp. 2086–2093, ISSN 0897-4756

Zolotovskii, B., Buyanov, R., Bukhtiyarova, G., Taraban, E., Murin, V., Grunvald, V., Demin, V., Sayfullin, R., (1997), "Development of the technology and production of spherical alumina for catalysts supports and adsorbents", Journal of Applied Chemistry, Vol. 70., No.1, pp. 299-306, ISSN 0044-4618 [in Russian]

10

On Thermal Conductivity of an In-Situ Metal Matrix Composite - Cast Iron

J.K.Chen and S.F.Chen
National Taipei University of Technology/ Inst. Materials Science and Engineering
Taiwan, R.O.C.

1. Introduction

Cast iron is a typical in-situ metal matrix composite consisting of graphite, ferrite, and pearlite microstructures. It is well known that a great range of strength and ductility can be attained by controlling the shapes of graphite and alloy design in cast irons. With increasing applications of cast irons in high temperature environment, e.g. disc brakes, engine exhaust manifolds and cylinders, etc., thermal conduction properties of such materials are particularly important.

Modelling of thermal conductivity for composite, however, is not an easy materials property to deal with especially for a heterogeneuos material such as cast iron. The cast irons represent a combination of graphite with different shapes, ferrite, and pearlite. Furthermore, alloying elements can affect their thermal conductvity in two aspects: the inherent thermal conductivity of various phases and the stability of different phases due to compositional changes (Helsing and Grimvall 1991).

The current article treats cast iron as a metal matrix composite via application of effective-medium theories. The effects of three factors upon thermal conductivity of cast irons are discussed including matrix phases, shapes of graphite, and alloying elements. These effects are concerned with the anisotropic properties and morphology of second phase particles in composites. For example, pearlite stands for lamellar composite structure, whereas nodular and flake graphite represents different shapes of second phase particles. It is thus the objective of this article to discuss the effects of different phases and their morphology on thermal conductivity of cast irons through both measurement and theoretical predictions.

The study starts out with the review of theorems describing thermal conductivity of composites. Effective Medium Theory is of particular interest for applications in cast irons for its versatility in modeling different second phase particle morphology and lamellar structures. Both room temperature and high temperature thermal conductivity of cast irons and basic matrix structures are measured to compare with theoretically predicted values. Furthermore, nodular, compact, and flake graphite are formed by alloying and casting process controls.

Thermal conductivities of cast irons range widely from 20 to 70 W/m/°C. It is found that the calculated thermal conductivity matches well with the measured values. The grey irons with flake graphite greatly increase thermal conductivity due to its anisotropy. Nodular irons then have lower thermal conductivity. Ferrite structure has nearly twice the thermal conductivity of lamellar pearlitic matrix. Alloying elements then have two-fold effects on thermal conductivity of cast irons, one on the intrinsic thermal conductivity of different microstructures and the other on amount of constituent phases. It is observed that effects of alloying elements to thermal conductivity fall mainly on the amount of different phases formed rather than on phase compositions. The current work demonstrates an integrated path for thermal conductivity modelling in complex composite systems.

2. Theory

Concepts of effective medium theory or effective medium approximations were originated from the work by Maxwell-Garnett (1904) and Bruggeman (1935). Many extended formulas have since been developed with these bases (Choy, 1999). Such approximation was developed to describe the physical properties of heterogeneous bodies through combination of homogeneous isotropic phases. The combined composites thus give rise to an effective property corresponding to a homogeneous medium. As the development of effective medium theory was based upon the effects of polarization under influence of electromagnetic fields, the most common physical properties considered using effective medium theory are conductivity and dielectric constants among all.

In Cheng and Vachon's (1969) study, thermal conductivity of two and three phase solid heterogeneous mixtures were considered based upon Tsao's (1961) model. Tsao's model slices the composite bodies into parallel plates in perpendicular to heat transfer direction. Fractions of discontinuous phases in each plate attribute to the solutions of an effective quantity. The effective quantity such derived thus is inappropriate for describing the anisotropic phase such as graphite flakes in cast irons.

The methodology developed by Helsing and Helte (1991) is of interest in current article which treats a continuous medium as an isotropic aggregate of anisotropic grains. The concept was based upon Schulgasser's (1977) study on the conductivity of polycrystalline materials whereas each equiaxial crystals are aligned in varied orientation (Fig.1). Such model is especially advantageous in constructing ferritic and pearlitic matrix structures in cast irons. The anisotropic nature of graphite and pearlitic grains can thus be simulated.

2.1 Thermal conductivity of matrix phase

2.1.1 Thermal conductivity of ferritic matrix

For a ferritic matrix of varied compositions, it is necessary to consider both its phonon and electronic thermal conductivity. The electronic thermal conductivity can be derived using Lorenze function, L_e, relating the electronic thermal conductivity, K_e, and electrical resistivity, ρ_m, of pure iron by Wiedemann–Franz law:

$$K_{e,\text{pure iron}} = \left(\frac{\rho_m}{L_e T}\right)^{-1} \tag{1}$$

For ferritic alloys, the effects of alloying alloying elements must also be added by considering the electrical resistivity of ith element (Williams, 1981):

$$K_{e,\text{ferrite}} = \left(\frac{\rho_m}{L_e T} + \frac{1}{L_o T}\sum_i \rho_i^{'} C_i\right)^{-1} \tag{2}$$

where L_o is the Sommerfeld-Lorenz number of $2.44\text{x}10^{-8}$ $W\Omega°C^{-2}$, $\rho_i^{'}$ is relative electrical resistivity of *i*-th alloying element in comparison with pure iron, C_i is the concentration of *i*-

th element in molar fraction, and T is temperature. The L_e term represents the changes of Lorenz function of pure iron with temperature and is assumed to be independent of alloying elements (Table 1.) Therefore, Eqn.(2) comprises of temperature-dependent part of electrical thermal conductivity for pure iron and the second term considering variations caused by alloying elements.

Fig. 1. Schematics of polycrystals with anisotropic grains (from Schulgasser 1977)

	27°C	100°C	200°C	300°C
p_m μΩ cm	11.16	14.76	21.76	27.36
L_e x 10^8 WΩ°C^{-2}	2.03	2.09	2.22	2.31

Table 1. Lorenz functions and resistivity of pure iron at different temperature (Williams, 1981)

The contribution of lattice vibration to thermal conductivity in ferrite, $K_{p,ferrite}$, is then expressed by

$$K_{p,ferrite} = \left(\frac{1}{K_{p,pure\ iron}} + \sum_i A_i C_i\right)^{-1} \tag{3}$$

where $K_{p,pure\ iron}$ is the phonon contribution of thermal conductivity in pure iron, and the second term in Eqn.(3) represents the phonon scattering (A_i) due to impurity atoms, *i*.

	Si	Al	Cr	Mo
$\rho_i^{'}$ μΩ cm/at.%	7	6.4	4.6	4.8
10^3xA_i m°C/W/at.%	8	7	0.1	13

Table 2. Coefficients $\rho_i^{'}$ and A_i (Helsing and Grimvall, 1991) for different alloying elements

Therefore, the thermal conductivity of ferrite, $K_{ferrite}$, is simply the combination of electrical and phonon contributions,

$$K_{ferrite}=K_{e,ferrite}+K_{p,ferrite} \quad (4)$$

2.1.2 Thermal conductivity of pearlitic matrix

Pearlite structure is represented by ferrite and cementite in parallel lamellae. Apparently, such structure is anisotropic, while the aggregation of anisotropic grain should give rise to an effective thermal conductivity. Thermal conductivity of pearlite parallel to the lamellae can be written as

$$K_{pearlite,//}=f_{Fe_3C}\ K_{Fe_3C}+f_{ferrite}\ K_{ferrite} \quad (5)$$

and that of pearlite perpendicular to the lamellae is

$$K_{pearlite,\perp}=(f_{Fe_3C}\ /K_{Fe_3C}+f_{ferrite}\ /K_{ferrite})^{-1} \quad (6)$$

where f_{Fe3C} (=0.122) and $f_{ferrite}$ (=0.878) are the volume fractions of Fe_3C and ferrite, respectively. K_{Fe3C} is the thermal conductivity of cementite or 8 W/m/°C (Helsing and Grimvall, 1991). The effective isotropic thermal conductivity of pearlite, $K^{eff}_{pearlite}$, estimated by Effective medium approximation is thus given by (Helsing and Grimvall, 1991)

$$K^{eff}_{pearlite}=\frac{1}{4}[K_{pearlite,//}+\sqrt{K^2_{pearlite,//}+8K_{pearlite,//}K_{pearlite,\perp}}] \quad (7)$$

The $K_{pearlite,//}$ is calculated to be in the order of 58W/m/°C, whereas $K_{pearlite,\perp}$ is only around 35W/m/°C. The effective thermal conductivity of pearlite is thus in the order of 45~50W/m/°C.

2.1.3 Experimental and theoretical thermal conductivity of matrix

Thermal conductivities of bulk matrix materials are measured using interstitial free steel (C<50ppm) and AISI 1080 (C=0.80wt%) steels to simulate ferrite and pearlite structure. Fig. 2 shows the microstructures of these samples.

Thermal conductivities of ferrite and pearlite materials are measured by hot disk method (Gustafsson, 1991) using two specimens with size of 50mm dia × 20mm t. The predicted values by combining Eqns. (1)~(7) are shown to bear good agreement with the measured values as shown in Fig. 3.

It is observed that there is a great difference in thermal conductivity of ferritic and pearlitic matrices. This does not necessarily all arise from the contribution of cementite which account for about 6.9W/m/°C of difference only. The main reduction of thermal conductivity is due to the stacking of ferrite and cementite. The conductivity in orientation perpendicular to the pearlitic lamellae, $K_{pearlite,\perp}$ is only 60% of conductivity in orientation parallel to the pearlitic lamellae indicating that the lamellar structures stand for a regular and great barrier for both electron and phonon conduction.

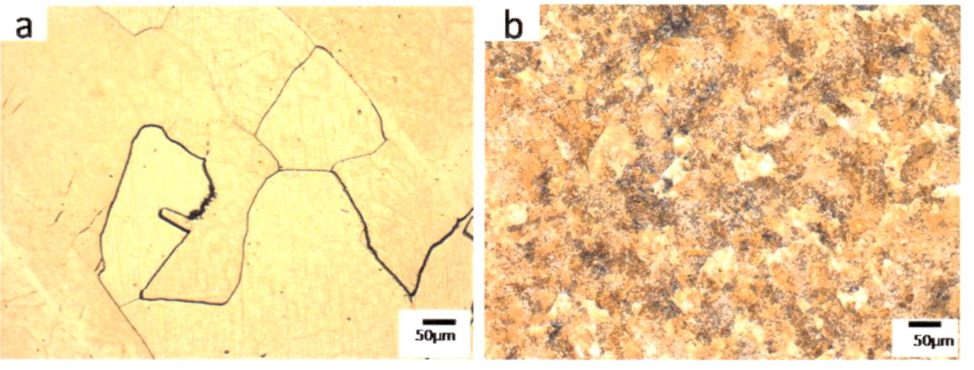

Fig. 2. Microstructures of (a) ferrite and (b) pearlite structures

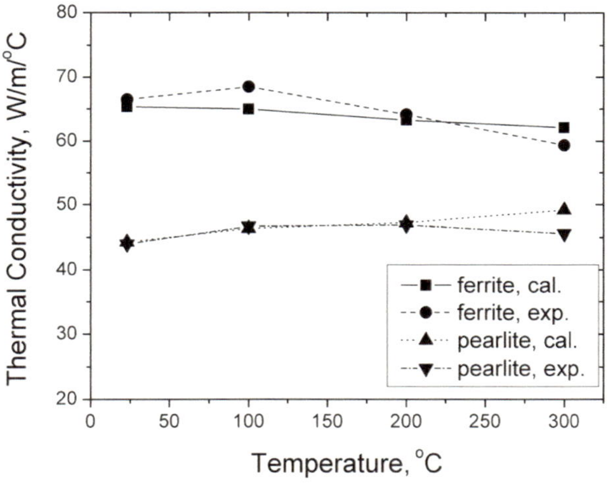

Fig. 3. Calculated (cal.) and measured (exp.) thermal conductivity of ferrite and pearlite structures.

2.2 Effects of graphite types on thermal conductivity

2.2.1 Thermal conductivities of cast irons with different graphite morphology

Different types of graphite affect the mechanical properties of cast irons greatly. There are two extreme shapes of graphite, namely spherical and flake graphite. In the current study, besides the ductile irons (FCD) with spherical graphite and the conventional grey irons (FC) with flake graphite, cast irons with compact graphite (CGI) that fall in between the two extreme cases is also considered.

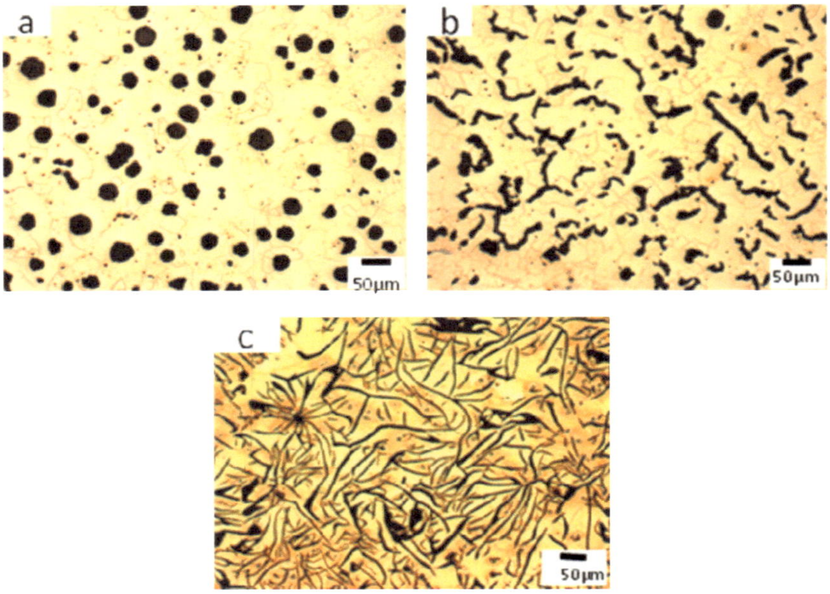

Fig. 4. OM microstructures of (a) nodular iron (FCD), (b) compact graphite iron (CGI), and (c) grey iron (FC) specimens

The FC specimen is a JIS FC250 grade cast iron with carbon equivalence of 3.83. The FCD sample is prepared using a JIS FCD450 grade iron (CE%=4.51) spheroidized using 1.1wt% of spheroidizer for 40s prior casting. And the CGI sample is then prepared by the same FCD450 grade iron spheroidized using 0.5wt% spheroidizer for 60s to deteriorate the nodular graphite formation.

From the microstructures shown in Fig. 4, it is apparent that the grey irons (Fig. 4c) with flake graphite give rise to highly anisotropic properties due to its characteristic planar shapes. The conductivity in orientations parallel to the flake surface is far higher than that in perpendicular direction. The effective thermal conductivity of grey iron is obtained by solving the following equation:

$$K_{\text{cast iron}}^{\text{eff}} = K_{\text{matrix}} + 3 f_{\text{other}} K_{\text{cast iron}}^{\text{eff}} \left(\frac{K_{\text{other}} - K_{\text{matrix}}}{2K_{\text{cast iron}}^{\text{eff}} + K_{\text{o}}} \right) + \frac{1}{3} f_{\text{graphite}} K_{\text{cast iron}}^{\text{eff}} \left[\frac{K_{\text{graphite},\perp} - K_{\text{matrix}}}{K_{\text{cast iron}}^{\text{eff}} + p_{\perp}(K_{\text{graphite},\perp} - K_{\text{cast iron}}^{\text{eff}})} + 2 \frac{K_{\text{graphite},//} - K_{\text{matrix}}}{K_{\text{cast iron}}^{\text{eff}} + p_{//}(K_{\text{graphite},//} - K_{\text{cast iron}}^{\text{eff}})} \right] \quad (8)$$

Here the shape factors in orientation parallel and perpendicular to graphite flakes, $p_{//}$ and $p_{\perp}$, are written as

$$p_{//} = \varepsilon(2 - 2\varepsilon^2)^{-1}[(1-\varepsilon^2)^{-\frac{1}{2}} \cos^{-1}(\varepsilon) - \varepsilon] \quad (9)$$

$$p_{\perp} = 1 - 2p_{//} \quad (10)$$

, respectively. The ε parameter represents the thickness-length ratio of graphite and is assumed to be 0.05 for FC specimens. The thermal conductivities of graphite in direction parallel and perpendicular to flake surface, $K_{graphite,//}$ and $K_{graphite,\perp}$, are taken to be 500 and 10 W/m/°C, respectively (Helsing and Grimvall, 1991).

For ductile cast irons with spherical graphite shape, thermal conductivity can be solved by the following:

$$K_{ductile\ iron}^{eff}=K_{matrix}+3f_{other}K_{cast\ iron}^{eff}(\frac{K_{other}-K_{matrix}}{2K_{cast\ iron}^{eff}+K_{o}})+f_{graphite} \\ (\frac{3}{2+\eta K_{graphite,\perp}/K_{ductile\ iron}^{eff}})(\frac{2K_{graphite,//}+\eta K_{graphite,\perp}}{2+\eta}-K_{matrix}) \tag{11}$$

where f_{other} and K_{other} are the volume fraction and thermal conductivity of other microconstituents such as austenite or carbides, and

$$\eta=\frac{1}{2}\left[(1+\frac{8K_{graphite,//}}{K_{graphite,\perp}})^{\frac{1}{2}}-1\right] \tag{12}$$

As for CGI specimens that have graphite morphology in between nodular and flake shape, an equation by combining Eqns. (8) and (11) linearly can be derived. Fractions of graphite with nodular shape and flake shape, $f_{nodular\ graphite}$ and $f_{flake\ graphite}$, are taken into account by the following equation:

$$K_{ductile\ iron}^{eff}=K_{matrix}+3f_{other}K_{cast\ iron}^{eff}(\frac{K_{other}-K_{matrix}}{2K_{cast\ iron}^{eff}+K_{o}})+\frac{1}{3}f_{flake\ graphite}K_{cast\ iron}^{eff} \\ \left[\frac{K_{graphite,\perp}-K_{matrix}}{K_{cast\ iron}^{eff}+p_{\perp}(K_{graphite,\perp}-K_{cast\ iron}^{eff})}+2\frac{K_{graphite,//}-K_{matrix}}{K_{cast\ iron}^{eff}+p_{//}(K_{graphite,//}-K_{cast\ iron}^{eff})}\right]+ \\ f_{nodular\ graphite}(\frac{3}{2+\eta K_{graphite,//}/K_{ductile\ iron}^{eff}})(\frac{2K_{graphite,\perp}+\eta K_{graphite,//}}{2+\eta}-K_{matrix}) \tag{13}$$

2.2.2 Calculated and measured thermal conductivity for cast irons with different graphite morphology

Fig. 5. demonstrates the calculated and measured thermal conductivity for different cast irons in Fig. 4. The compact graphite and grey iron samples show good agreement between the calculated and experimental values at temperatures below 100°C but diverge at temperatures over 200°C. Further observations note that the measured thermal conductivities are consistently lower with increasing temperatures. This is due to that the temperature dependence of graphite conductivity is not considered. Therefore, an estimated -3000ppm/°C of temperature coefficient of thermal conductivity for graphite is employed in Fig. 6. The predicted thermal conductivities apparently improve consistency with the measured values when temperature dependence of graphite thermal conductivity is considered.

It is also observed that the grey irons have the highest conductivity among all in comparison with nodular irons and compact graphite irons. The difference can be as large as 2~3 times.

This suggests that the anisotropic properties of grey irons improve their effective conductivity. It is attributed to the aligned graphite flake in random orientations to increase overall thermal conduction activities. Even though the compact graphite irons consist of graphite similar in shape with grey irons, their effective thermal conductivities are closer to those of nodular irons.

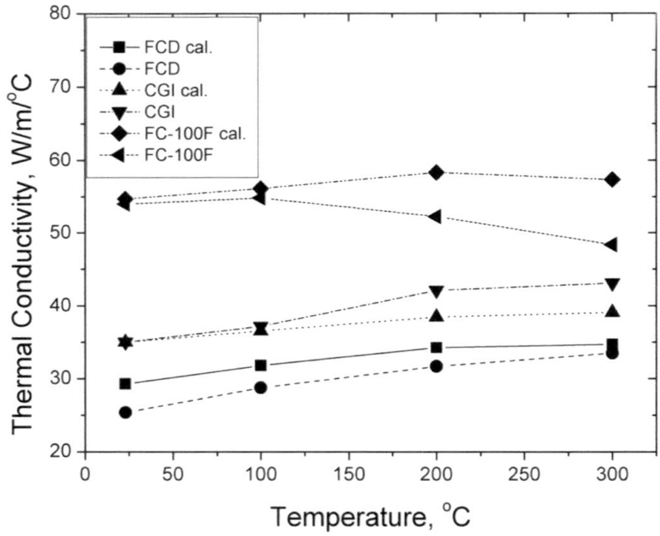

Fig. 5. Calculated (cal.) and measured thermal conductivity of grey irons (FC), compact graphite irons (CGI), and nodular irons (FCD).

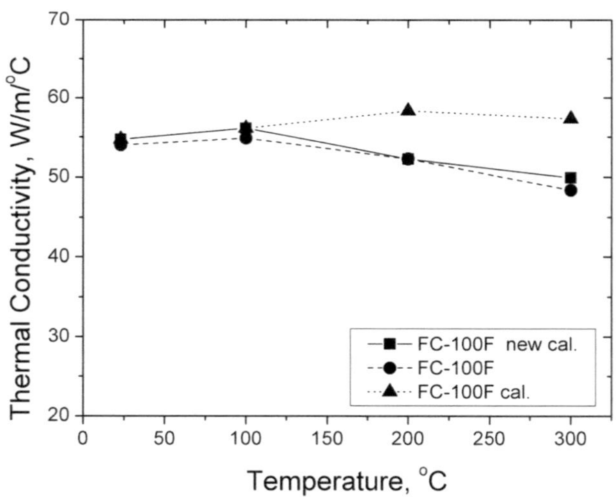

Fig. 6. Thermal conductivities of grey irons considering temperature dependence in thermal conductivity of graphite (new cal.) in comparison with non-temperature dependence predicted values (cal.) and measured values.

2.3 Effects of alloying elements on thermal conductivity

The alloying elements have two effects on thermal conductivity of cast irons. Different alloying elements can affect constituents of microstructures. Alloying elements such as Si and Cu are graphite stabilizer, while Cu also promotes pearlite formation (Figs.7a and 7b). On the other hand, Mo and Cr are carbide stabilizers. Therefore, different alloying elements induce different amount of graphite and matrix types (Fig. 7.)

Fig.6. demonstrates a series of FC samples with 1 additional wt.% of Si, Cu, Cr, Mo, and Al. For example, in Fig. 7b, the grey iron with 1wt.% of Cu addition indeed promote the matrix to form pearlite while graphite flakes appear to be thinner and smaller in comparison with the iron with 1wt.% Si addition in Fig. 7a.

For Cr and Mo added grey irons, the matrix apparently gives rise to many white areas which correspond to alloy carbides and would need to be taken into account in consideration of thermal conductivity. On the other hand, the 1wt.% aluminium added grey iron bear great amount of ferrite adjacent to the graphite. Alloying elements control the amount of different microconstituents formed and affect the thermal conductivities of alloyed grey irons.

Meanwhile, the alloying elements can also affect the compositions of different microstructures, such as ferrite and carbides. It is observed that the morphology of graphite varies with elements added. These will affect the values estimated by Eqn.(2) and thus demonstrate the dependence of thermal conductivity on alloying elements.

In considering the thermal conductivity of cast irons with alloying elements, quantitative metallography and SEM-EDS analyses are utilized to calculate the amount of different phases and to confirm the compositions of each microconstituent. Compositions of each constituent phase can thus be entered into the equations to estimate the thermal conductivity of each phase.

Fig. 8 demonstrates the dependence of thermal conductivity with alloying elements. In Fig.6, the base thermal conductivity of FC grey iron is ~55W/m/°C at room temperature. This value is compared with those measured or predicted in alloyed irons in Fig. 8. It is observed that all alloying elements pose a reduction effect on thermal conductivity. This is attributed to the scattering of both electrons and phonons by the impurity atoms in grey irons.

Among all alloying elements added, aluminum and copper has the least effects on reduction of thermal conductivity. In Table 2, it is noted that aluminum does not necessarily give a higher electrical resistivity and phonon scattering effects in comparison with other alloying elements. The lesser reduction effects on thermal conductivity are mostly likely from the microstructure factor. In particular, graphite stabilizer, such as copper and aluminum, can generate higher number of thinner graphite which compensates some negative effects of alloying elements upon thermal conductivity.

The carbide formers, chromium and molybdenum, have two-fold effects on the thermal conductivity. They not only affect the overall thermal conductivity by carbide formation but also reduce the graphite formation. Chromium has a relatively smaller effect on phonon scattering (Table 2) and shows a little higher conductivity value than molybdenum alloyed irons. Molybdenum gives the highest phonon scattering effects among the alloying elements and gives rise to the lowest thermal conductivity in Fig. 8.

In summary, reduction of thermal conductivity in cast irons is mostly attributed to the microstructures affected by the alloying elements. Amount of flake graphite, carbide, and ferrite can all contribute to the combined thermal conductivity. The composition in each phase, such as ferrite, affects less pronounced on the effective conductivity.

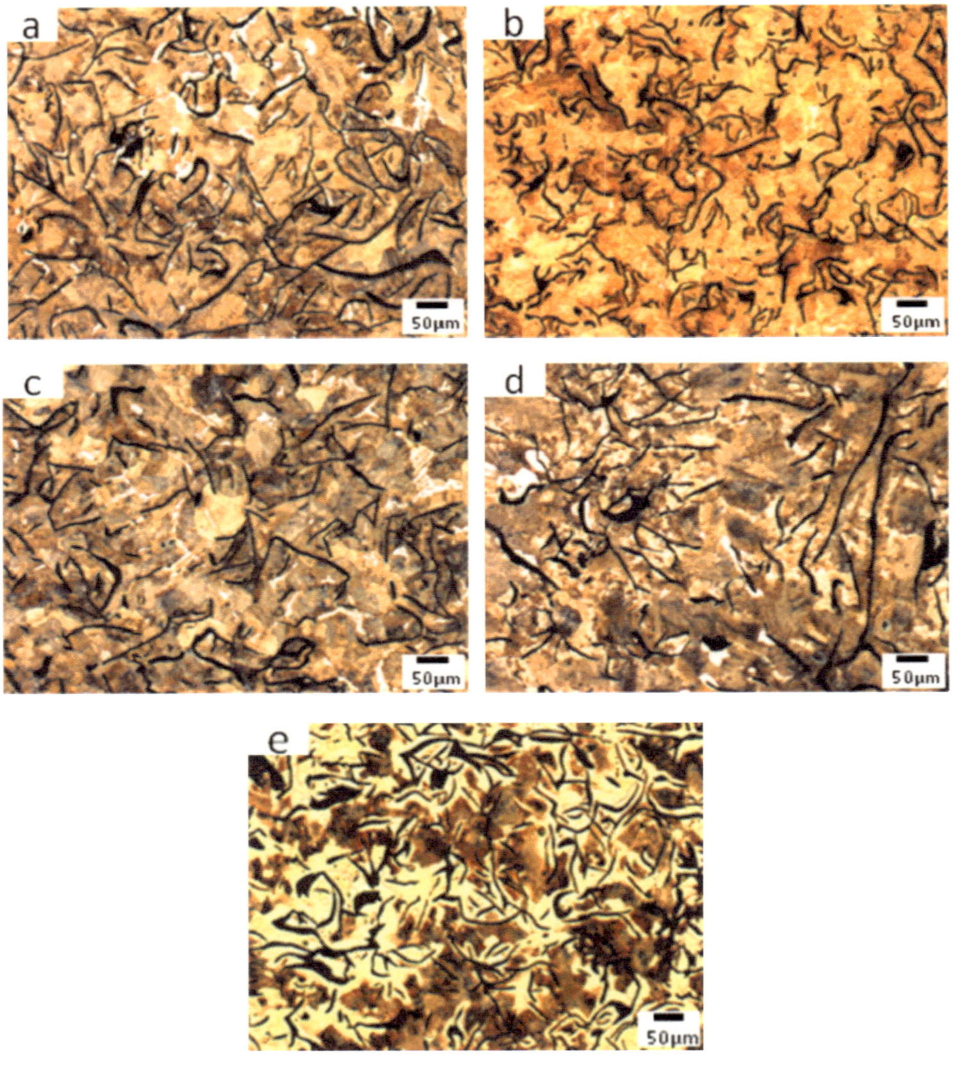

Fig. 7. Optical microstructures of (a) FC-1wt.%Si, (b) FC-1wt.%Cu, (c) FC-1wt.%Cr, (d) FC-1wt.%Mo, and (e) FC-1 wt.%Al specimens.

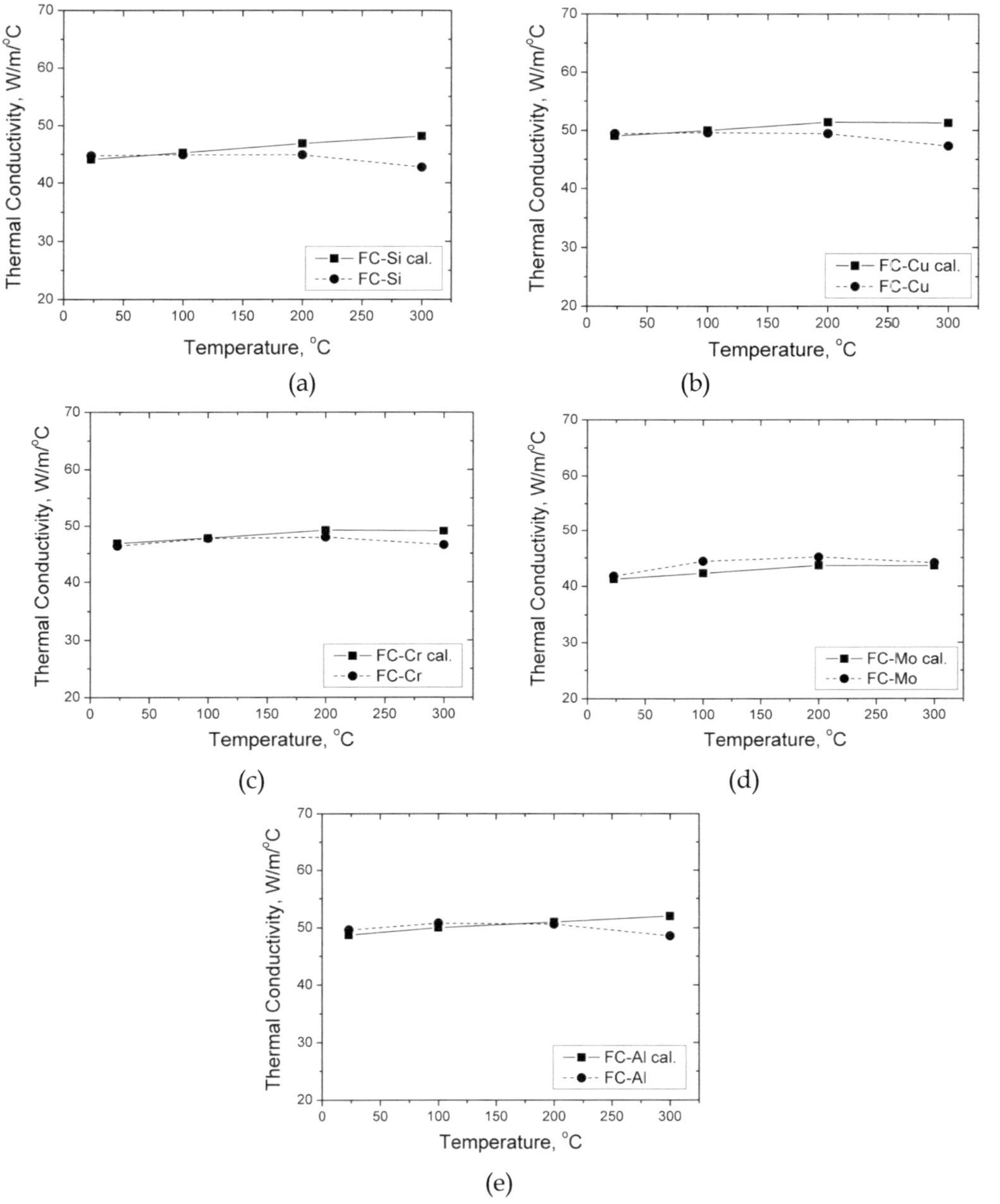

Fig. 8. Measured thermal conductivities of (a)FC-1 wt.%Si, (b)FC-1 wt.%Cu, (c)FC-1 wt.%Cr, (d)FC-1 wt.%Mo, and (e)FC-1 wt.%Al specimens and their predictions (cal.)

3. Conclusion

Current study reports the effects of microstructures, alloying elements, and graphite morphology on thermal conductivity of cast irons through both experimental measurements and theoretical estimations. Good agreement can be found in between the measured and estimated values using effective medium approximations. Reasons for great differences of thermal conductivity among cast irons of different types can thus be discussed. The following conclusions are drawn from these discussions.

1. Ferrite structure has as large as an order of magnitude of thermal conductivity higher than that of cementite. When the two structures are stacked in layers, the anisotropic properties cause the orientations in parallel and in perpendicular to the lamellae structure to have a 60% difference. This greatly affects the effective thermal conductivity of pearlite indicating that the interfaces between phases of great differences play an important role.
2. Graphite phase is the microstructure that bears the highest conductivity among all microconstituents in cast irons. It also has an even greater anisotropic nature in comparison with pearlite structure. This causes a great effect on the effective thermal conductivity of different graphite morphology. More specifically, as high as 55 W/m/°C is achieved in cast irons with flake graphite, whereas only 25W/m/°C of thermal conductivity is obtained by nodular irons. The anisotropic effect is obvious. Furthermore, cast irons with compact graphite or the graphite with shapes in between flake and spheroids are shown to have thermal conductivity (~35 W/m/°C) more similar to the nodular irons rather than grey irons, even though the compact graphite shape appears closer to the flake graphite.
3. Effects of alloying elements on cast irons are also discussed by fixing the graphite as flake morphology. It is well known that the alloying elements pose both electron and phonon scattering effects in thermal conductivity. Meanwhile, there appears also great difference in the phase constituents due to alloying elements. Specifically, copper and aluminium additions promote graphite to form higher numbers of thinner graphite and reduce alloying reduction effects on thermal conductivity of cast irons. On the other hand, chromium and molybdenum additions cause the formation of carbides which not only increase thermal resistivity but also reduced graphite formation and therefore reduce thermal conductivity most severely among all alloying elements considered. It is thus concluded that the alloying elements affect thermal conductivity more by the amount of constituent phases rather than by the compositions of each phase.

4. Acknowledgment

The completion of this research is partly supported by the grant of National Science Council of Taiwan through #NSC98-2221-E-027-031 project. The materials supplied by China Metal Products and assistance of thermal conductivity measurements by EMO center at National Taipei University of Technology are acknowledged.

5. Appendix – list of symbols

A_i: phonon scattering due to impurity atoms, *i*.
C_i: concentration of *i*-th element in molar fraction

ε: thickness-length ratio of graphite
f_{Fe3C} and $f_{ferrite}$: volume fractions of Fe_3C and ferrite in pearlite, respectively
f_{other} volume fraction of other microconstituents such as austenite or carbides
$f_{nodular\ graphite}$ and $f_{flake\ graphite}$: fractions of graphite with nodular shape and flake shape, respectively.
$K_{e,pure\ iron}$: electron contribution in thermal conductivity of pure iron
$K_{e,\ ferrite}$: electron contribution in thermal conductivity of ferrite
$K_{p,ferrite}$: contribution of lattice vibration to thermal conductivity of ferrite
$K_{ferrite}$: thermal conductivity of ferrite
K_{Fe3C}: thermal conductivity of cementite or 8 W/m/°C
$K^{eff}_{pearlite}$: effective isotropic thermal conductivity of pearlite
$K_{pearlite,//}$ and $K_{pearlite,\perp}$: thermal conductivity of pearlite in orientation parallel and perpendicular to the lamellae, respectively
$K_{graphite,//}$ and $K_{graphite,\perp}$: thermal conductivities of graphite in orientations parallel and perpendicular to the graphite surface, respectively
ρ_m: electrical resistivity of pure iron
K_{other}: thermal conductivity of other microconstituents such as austenite or carbides
L_e: Lorenze function relating $K_{e,pure\ iron}$ and ρ_m.
L_o: the Sommerfeld-Lorenz number of $2.44\text{x}10^{-8}$ $W\Omega°C^{-2}$
$\rho_i^{'}$: relative electrical resistivity of *i*-th alloying element in comparison with pure iron
$p_{//}$ and $p_{\perp}$: shape factors in orientations parallel and perpendicular to graphite flakes, respectively

6. References

Bruggeman, D.A.G. (1935). Dielectric constant and conductivity of mixtures of isotropic materials. *Ann. Phys. (Leipzig.)*, Vol.24, pp.636-679.

Choy, T.C. (1999). Effective Medium Theory, Principles and Applications, Oxford Science Publications, ISBN 0198518927, New York, USA.

Cheng, S.C. & Vachon, R.I. (1969). The prediction of the thermal conductivity of two and three phase solid heterogeneous mixtures, *Int. J. Heat Mass Transfer.*, Vol.12, pp.249-264.

Gustafsson, S.E. (1991). Transient plane source techniques for thermal conductivity and thermal diffusivity measurements of solid materials, *Rev. Sci. Instrum.*, Vol.62, No.3, pp. 797-804.

Helsing, J. & Grimvall, G. (1991). Thermal Conductivity of Cast Iron : Models and Analysis of Experiments. *Journal of Applied Physics*, Vol.70, No.3, pp.1198-1206.

Helsing, J. & Helte. A. (1991) Effective conductivity of aggregates of anisotropic grains, *J. Appl. Phys.*, V.69, No.6, pp.3583-3588.

Maxwell-Garnett, J.C. (1904). Colours in metal glasses and in metallic films. *Philos. Trans. R. Soc. A*, Vol.203, pp. 385–420.

Schulgasser, K. (1977) Bounds on the conductivity of statistically isotropic polycrystals, *J. Phys. C: Solid State Phys.*, Vol.10, pp.407-417.

Tsao, G.T.N. (1961).Thermal conductivity of two phase materials, *Ind. Engrg. Chem.*, Vol.53, No.5, pp.395-397.

Williams, R.K. ; Yarbrough, D.W. ; Masey, J.W. ; Holder, T.K. & Graves, R.S. (1981). Experimental determination of the phonon and electron components of the thermal conductivity of bcc iron, *J. Appl. Phys.*, Vol.52, No.8, pp.5167-5175.

11

Synthesis and Characterization of Ordered Mesoporous Silica Pillared Clay with HPW Heteropoly Acid Encapsulated into the Framework and Its Catalytic Performance for Deep Oxidative Desulfurization of Fuels

Baoshan Li[1], Zhenxing Liu[1], Jianjun Liu[1], Zhiyuan Zhou[2], Xiaohui Gao[2], Xinmei Pang[2] and Huiting Sheng[1]

[1]*State Key Laboratory of Chemical Resource Engineering*
Beijing University of Chemical Technology
[2]*Petrochina Petrochemical Research Institute*
P. R. China

1. Introduction

Because of the increasing environmental concern and legal requirements, it has been an urgent subject to find new processes for ultra-deep desulfurization of fuel oils. The conventional hydrodesulfurization (HDS) technique has some inherent problems in treating sulfur-containing aromatic hydrocarbon compounds, such as dibenzothiophene (DBT) and its derivatives. Thus, a large number of non-HDS processes, such as adsorption, extraction, oxidation, precipitation and bio-processes have been explored [1-7]. Among these methods, catalytic oxidative desulfurization (ODS) combined with extraction is regarded as the most promising and economical process [8-11]. The ODS process has been studied in various systems, including oxidant H_2O_2 with inorganic acids [12], organic acids [13-15], TS-1 [16], and heteropoly acids (HPA) catalysts [17-19]. HPA catalysts, especially those having the Keggin structure, have been determined to be very effective for the oxidation of sulfur-containing compounds in a liquid-liquid two-phase system. However, the separation and recovery of the catalysts are very difficult. For these reasons, the stable and water-tolerant solid acid catalysts have been developed. Li et al. [20], Lü et al. [21], and Gao et al. [22] reported $[(C_{18}H_{37})_2N(CH_3)_2]_3[PW_{12}O_{40}]$, $[C_{18}H_{37}N(CH_3)_3]_4[H_2NaPW_{10}O_{36}]$, and $Q[W(O)(O_2)_2(C_5H_4NCO_2)]$ ($Q=(C_4H_9)_4N$, $[(CH_3)_3N(C_{18}H_{37})]$ and $[(C_{18}H_{37})_2N(CH_3)_2]$) catalysts, respectively. Over these catalysts, DBT and its derivatives can be selectively oxidized into their corresponding sulfones using H_2O_2 as the oxidant in diesel oil. The surfactant-type catalysts provided a new process for deep ODS. However, the double long carbon chains of quaternary ammonium surfactant used in their system are very expensive.

In our previous works, monovacant lacunary Keggin-type polyoxometalate was introduced into mesoporous silica structure and showed high oxidation activity for ODS [23]. Furthermore, we recently prepared a series of ordered mesoporous clay materials known as silica pillared clay (SPC) materials and metal ion doped SPC materials [24-29]. These inorganic porous materials with a controlled pore structure have stimulated a great deal of interest for their potential application as selective catalysts, adsorbents, separating agents, and porous matrixes for encapsulation of specific functional molecules [30-33].

In this study, in order to combine the porous advantages of the SPC and the appropriate acidity of $H_3PW_{12}O_{40}$ (HPW) heteropoly acid, HPW was encapsulated into the ordered mesoporous SPC material by sol-gel method (denoted as HPW-SPC-SG) involving introduction of the HPW into clay interlayer template in an acidic suspension. The catalytic performance of the HPW-SPC-SG materials was studied through the ODS of the model oil. The results indicate that the HPW-SPC-SG materials exhibited high catalytic performance in ODS.

2. Experimental

2.1 Materials

All solvents and reactants are commercially available and were used without further purification. The natural montmorillonite clay (MMT) was obtained from Inner Mongolia and used as the starting clay without any further purification or ions exchanging. The raw clay had a basal (001) spacing of 15.4 Å and an anhydrous structural (layer) formula of $[Si_{7.86}Al_{0.14}][Al_{2.84}Fe_{0.30}Mg_{0.86}]O_{20}(OH)_4$, with a cation exchange capacity (CEC) of 91 meq/100 g [24].

2.2 Preparation

2.0 g of MMT was first suspended in 120 mL of deionized water in a round bottom flask, to which 4.0 g of cetyltrimethylammonium bromide (CTAB) dissolved in 10.0 mL of pure ethanol was added dropwise and stirred for 1 h and a gel mixture was gained. Then, the pH of the gel was adjusted by HCl solution to 2.0. Subsequently, the calculated amount of $H_3PW_{12}O_{40}.6H_2O$ was dissolved in 10.0 mL of water and added dropwise into the prepared gel under vigorous stirring. After the mixture was stirred for 4 h, 10.0 mL of tetraethyl orthosilicate (TEOS) was added, followed by stirring for 12 h at room temperature. And then, the mixture was put into an autoclave and heated for 24 h in a furnace at 110 ^oC. The autoclave was cooled, and the product was separated by filtration, throughly washed with deioined water, and dried in an oven at 110 ^oC. Then the dried samples were calcined at 500 ^oC for 6 h in a furnace. The samples were designated as *x*%HPW-SPC-SG (where *x*% represents the weight percentage of HPW in the samples), and analyzed by X-ray fluorescence analysis (XRF). The results show that the actual content of HPW are 6.65, 13.24, and 24.93 wt.% for the samples with nominal HPW content of 5, 15 and 25 wt.%, respectively.

The SPC sample was prepared following the same procedure as the HPW-SPC-SG samples without the addition of HPW. And the impregnated samples were also prepared by an incipient wetness impregnation method using SPC as the support [34-36]. The loading amount of HPW was varied in the range of 5 to 25 wt.% by changing the concentration of the HPW in water. Typically, an aqueous HPW solution volume of 5 ml was used per gram of freshly calcined SPC. After impregnation, the wet samples were dried at 110 oC in an oven. The final HPW/SPC samples prepared by impregnation were denoted as *x*%HPW-SPC-IM, where *x*% also represents weight percentage of HPW in the samples.

2.3 Catalytic performance in oxidative desulphurization

The model compound DBT was dissolved into *n*-octane to make a stock solution of model oil with sulfur content of 500 ppm. The reaction was performed in a three-neck glass flask with a water-bathed jacket. The mixture of the model oil (25 mL) and the catalyst (0.15 g) was heated under vigorous stirring. When the required reaction temperature was reached, 0.12 mL of 30% aqueous solution of H_2O_2 ($n_{H_2O_2}/n_S$=3) was added into the reactor under stirring for a certain time. Then the oxidized model oil was extracted three times by acetonitrile, and the volume ratio of the total solvent to model oil was 1:1. The amount of sulfur in the oil was determined by a Model WK-2D microcoulometric integrated analyzer (sulfur detection range from 0.2-5000 ppm, Jiangsu Jiang Fen Electroanalytical Instrument Co.).

2.4 Characterization

X-ray diffraction was performed on a Rigaku D/Max 2500 VBZ+/PC diffractometer using Cu-Kα radiation at low-angle range (2θ value 0.5^o-10°) and at wide-angle range (2θ value 3°-70°). XRF was performed on a Philips Magix-601 X-ray fluorescence spectrometer. Thermogravimetric and differential thermal analysis (TGDTA) was carried out on a HCT-1 thermal analyzer (Hengjiu Kexue Co.) using a heating rate of 10 ^{o}C min^{-1}. The FT-IR spectra were obtained in KBr pellets using a Bruker VECTOR 22 spectrometer in the range of 400-4000 cm^{-1}, and all spectra were collected at room temperature with a resolution of 4 cm-1. Nitrogen adsorption isotherms were obtained using a Micromeritics UNCALCINEDAP 2000 instrument. The samples were degassed at 115 ^{o}C for 8 h before the measurement. The specific surface area (S_{BET}) was estimated by the BET equation, and the pore size distribution and the mesopore analysis were obtained from the desorption branch of the isotherm using the Barrett-Joyner-Halenda (BJH) method. The scanning electron microscopy (SEM) micrographs were obtained on a Hitachi S-4700 microscope operated at 30 kV.

3. Results and discussion

3.1 The formation mechanism of the HPW-SPC-SG samples

Scheme 1 illustrates the proposed formation mechanism of the HPW-SPC-SG materials. Firstly, MMT was suspended in aqueous solutions and ion-exchanged with surfactant CTAB. During this process, the surfactant formed micelle in the interlayer regions. Then, the pH of the mixture was adjusted by HCl solution to 2.0 and an appropriate amount of HPW was introduced into the solution. $PW_{12}O_{40}^{3-}$ can substitute Br^- and Cl^- in the shell surrounding the micelle. Furthermore, the addition of HCl solution can keep the solution in a strong acidic environment to retain the Keggin-type HPW heteropoly acid intact. When TEOS was finally added into the gel mixture, it would intercalate into the clay interlayer regions by solvation and rapidly hydrolyze in acidic conditions to form the protonated $H_5SiO_4^+$ monomers. The silicate cations would interact with the anionic shell surrounding the CTAB micelle to trap both X^- (Cl^-, Br^-) and heteropoly anions, leading to the formation of a silica layer around the surfactant template. Thus, the heteropoly anions would be present at the interface between the silica and the CTAB micelle. During the long time stirring and hydrothermal treatment of the gel, the silica layer was polymerized and formed the Si-O-Si bonds between the interlayer regions. Washing and drying steps did not remove the heteropoly anions since they were trapped inside the clay interlayer. In the calcination step, the silica layer further condensed and yielded the completely cross-linked framework to strengthen the siloxanepillars and the mesopore structure, while the CTAB template decomposed and was eliminated from the pore system, but the HPW molecules remained fixed into the SPC frameworks [37].

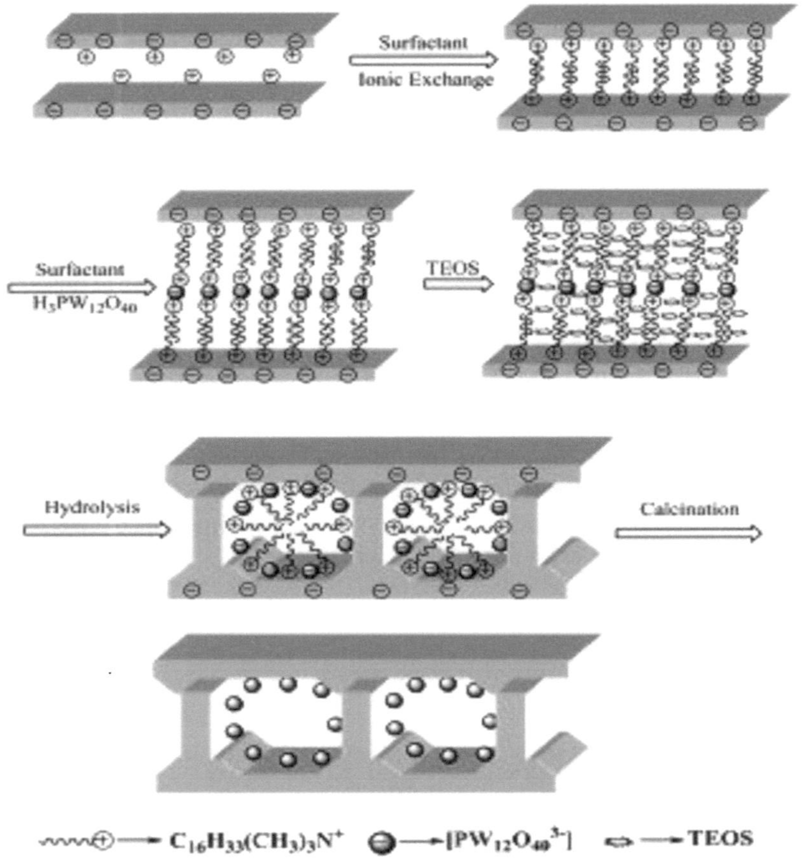

Scheme 1. Formation mechanism of the HPW-SPC-SG samples

3.2 The characterization of the samples

The small angle XRD patterns of the calcined MMT, SPC and HPW-SPC-SG samples are shown in Figure 1A. All the HPW-SPC-SG samples showed a broad characteristic (001) diffraction peak at about 2.0, which indicates that introducing HPW during the intercalation of CTAB did not destroy the SPC mesoporous lamellar structure. This result supports the proposed formation mechanism of the HPW-SPC-SG samples. The basal spacing and the gallery height of HPW-SPC-SG samples are summarized in Table 1. The HPW-SPC-SG materials exhibited refraction corresponding to a basal spacing of 4.38-4.47 nm. Since the thickness of the clay layer sheet is about 0.96 nm [38], the corresponding gallery heights are around 3.42-3.51 nm. Moreover, with theincrease of the HPW content, the refraction peak shifted to the low angle scope and the gallery height was found to increase gradually. This implies that the HPW has been encapsulated into the silica pillared frameworks of the SPC.

However, with the introduction of more HPW, the XRD peaks became less resolved, as the 25%HPW-SPC-SG sample gave no obvious diffraction peak. This is because that the HPW is encapsulated into the SPC frameworks, while the excessive amount of HPW clog the pore passages [39].

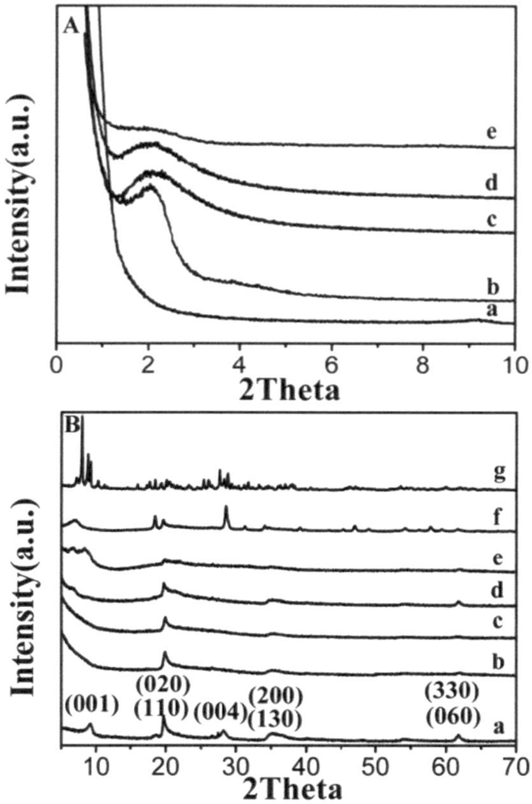

Fig. 1. A. Low-angle XRD patterns of samples: (a) calcined MMT, (b) SPC, (c) 5%HPW-SPC-SG, (d) 15%HPW-SPC-SG, and (e)25%HPW-SPC-SG.
B. Wide-angle XRD patterns of samples: (a) calcined MMT, (b) SPC, (c) 5%HPW-SPC-SG, (d) 15%HPW-SPC-SG, (e) 25%HPW-SPC-SG, (f) 15%HPW-SPC-IM, and (g) pure HPW.

The wide angle XRD patterns of the calcined MMT, SPC, HPW-SPC-SG, 15%HPW-SPC-IM and HPW samples are presented in Figure 1B. The SPC and the HPW-SPC-SG samples all showed the MMT characteristic peaks, which were assigned to (110), (020), (004), (130), (200), (330) and (060) diffractions [29, 40]. This ulteriorly approves that the laminated structure of MMT and the crystalline structure of the clay sheet have not been destroyed during the intercalation of HPW. However, with the increase of the HPW content, the HPW-SPC-SG samples gradually showed some characteristic peaks of HPW crystal, which was obvious for the 25%HPW-SPC-SG sample. The phenomenon indicates that the excess amount of HPW was introduced into the interlayer, but the HPW was not highly dispersed in the mesoporous SPC [41, 42]. On the other hand, the characteristic peaks of HPW in the pattern of 15%HPW-SPC-IM sample can be evidently seen. The results indicate that HPW is much more highly dispersed in the HPW-SPC-SG samples than that in the HPW-SPC-IM samples.

The TG-DTA curves of the SPC, HPW and HPW-SPC-SG samples before calcination in the region of 30-800 °C are shown in Figure 2. For all the samples, the weight loss below 100 °C was attributed to the loss of the physisorbed H_2O. There was a weight loss peak in the range of 210-500 °C for the SPC sample (Figure 2a), which was due to the decomposition of CTAB in the sample. The corresponding DTA curve indicated that CTAB template elimination was associated with a sharp exothermic peak centered at 323 °C [43]. Another endothermic peak in the DTA diagram centered at 706 °C was corresponding to the collapse of the mesoporous gallery structure [44].

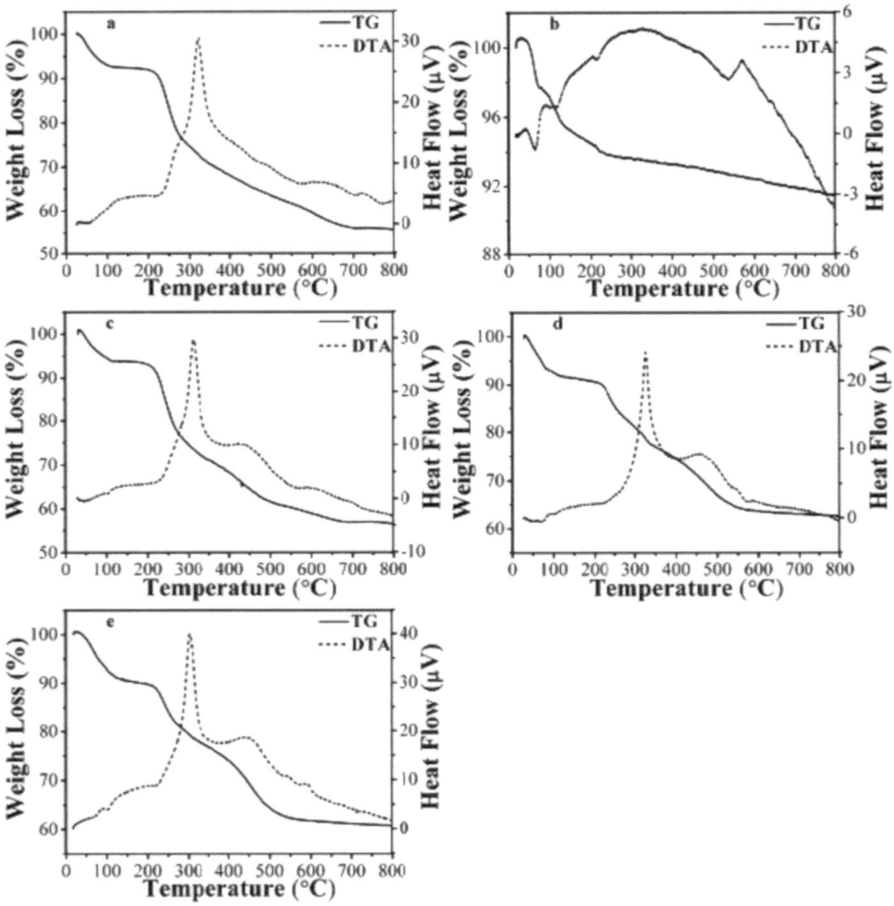

Fig. 2. TG-DTA curves of samples (a) SPC, (b) pure HPW, (c) 5%HPW-SPC-SG, (d) 15%HPW-SPC-SG, and (e) 25%HPW-SPC-SG

For the pure HPW (Figure 2b), there was an endothermic peak in the temperature range of 100-320 °C(weight loss 3.57%) centered at 215 °C in the DTA diagram, accounting for the loss of 6 H_2O molecules per Keggin unit of the HPW. The endothermic peak also appeared in the DTA diagrams of the HPW-SPC-SG samples (Figure 2c, d, e). The exothermic peak of HPW beginning at 538 °C in the DTA curve is assigned to the decomposition of heteropoly acid to form mixed oxides [45, 46]. The appreciable exothermic peak was also observed between 538-592 °C in the DTA diagrams of the HPW-SPC-SG samples (Figure 2c, d, e), indicating that the HPW-SPC-SG materials did not lower the thermal stability of the HPW heteropoly acid.

For the HPW-SPC-SG samples, besides the exothermic peaks at 323 °C, there was another exothermic peak centered at 407 °C in the DTA diagram (Figure 2c, d, e), which was also attributed to the decomposition of CTAB [47]. The decomposition of the organic cation extending to a higher temperature was due to the interactions of the CTAB and the HPW. Furthermore, the endothermic peak corresponding to the collapse of the mesoporous gallery structure also appeared in the DTA diagram of the HPW-SPC-SG samples (Figure 2c, d, e). This implies that the incorporation of the HPW did not destroy the interlayer gallery structures of the HPW-SPC-SG samples. The TG-DTA curves also show that the HPW-SPC-SG samples can be obtained after treating the gels at 500 °C for 6h.

The FT-IR spectra of the pure HPW, SPC and HPW-SPC-SG samples in the region of 1400-600 cm^{-1} are shown in Figure 3. The spectrum of the HPW showed a broadened band at 802 cm^{-1}, which was assigned to *v*as (W-Oc-W) in edge shared octahedral [48]. The spectrum of SPC showed a band at 795 cm^{-1}, which was attributed to the symmetric stretching frequency of Si-O-Si [49]. In FT-IR spectrum of HPW-SPC-SG, the peak was shifted to 816 cm^{-1}, which was due to the chemical interactions between HPW and SPC. As reported by Hu et al [50], the band at 816 cm^{-1} was attributed to the formation of Si-O-W bonds. Therefore, we can conclude that the HPW was encapsulated into the mesoporous silica framework and formed new Si-O-W bonds within the silica wall of the SPC material, which is consistent with the proposed formation mechanism of the HPW-SPC-SG samples. The characteristic bands of Keggin-type HPW were also observed at 1080, 980 and 890 cm^{-1}, which are usually assigned to *v*as(P-O), *v*as(W=O), and *v*as(W-Ob-W) in corner shared octahedral [48], respectively. Moreover, in the FT-IR spectra of HPW-SPC-SG samples, especially of the 25%HPW-SPC-SG sample, these bands were evidently strengthened compared to the bands of SPC, which further indicates the encapsulation of HPW in the SPC frameworks [51].

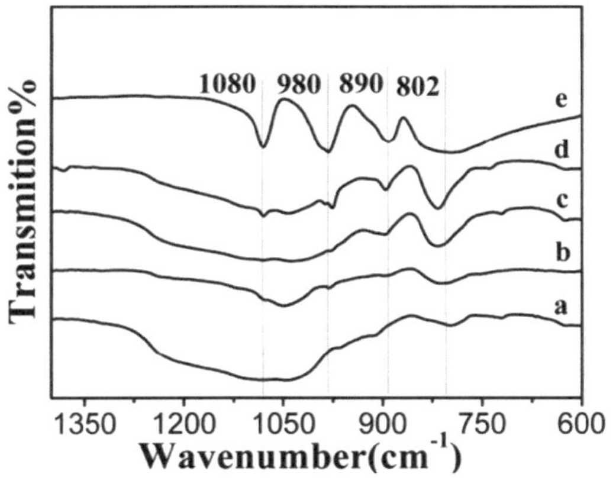

Fig. 3. FT-IR spectra of the samples: (a) SPC, (b) 5%HPW-SPC-SG, (c) 15%HPW-SPC-SG, (d) 25%HPW-SPC-SG, and (e) pure HPW.

The nitrogen adsorption/desorption isotherms and the mesoporous size distribution of the HPW-SPC-SG samples are shown in Figure 4. All samples presented type IV isotherm patterns with H3-type hysteresis loop starting at about 0.4 partial pressure, which are the characteristics of the mesoporous materials with the cylindrical pores formed in gallery regions [52]. For the 25%HPW-SPC-SG sample, the hysteresis loop of the isotherm was obviously smaller than that of the samples with relatively low HPW contents. Furthermore, the desorption branch of the sample extended to a relatively lower pressure, suggesting a partial loss of structural organization and the formation of some narrower slit-shaped pores [37]. Nevertheless, the main part of the hysterese was still between *P/P0* values of 0.50-0.95. The data shown in Table 1 demonstrate that the S_{BET} of the HPW-SPC-SG samples are about 324-444 $m^2\ g^{-1}$, which are lower than that of the SPC (538 $m^2\ g^{-1}$), but much higher than that of the HPW-SPC-IM samples (148-170 $m^2\ g^{-1}$) and the MMT (80 $m^2\ g^{-1}$). The results in Table 1 also indicate that the total pore volume (V_T) of the HPW-SPC-SG samples are slight lower than that of the SPC sample, but much higher than that of the HPW-SPC-IM samples and the MMT. The significant loss of S_{BET} and V_T of the HPW-SPC-IM samples can be related to the agglomeration of HPW molecules on the external surface of the materials resulting in pore blockage [36, 46]. The S_{BET} and V_T almost linear decreased with the increase of the HPW content both in the HPW-SPC-SG and the HPW-SPC-IM samples, which is logical since the heteropoly acid contributes to the sample weight.

The mesoporous pore size distribution is depicted in the inset of Figure 4. The results also indicate that all the HPW-SPC-SG samples have about the same pore diameters around 4.46-4.52 nm, which are slightly smaller than that of the SPC sample (5.06 nm). In the sol-gel synthesis method, the size of the majority of mesopores is formed by the CTAB micelle and the interclayer space, which is affected scarcely by the presence of HPW. This is in agreement with the proposed formation mechanism (Scheme 1). On the contrary, the pore diameters of the HPM-SPC-IM samples are reduced to about 4.33-3.82 nm, which indicates that HPW inside the mesopores occupied the space and decreases the pore width [37].

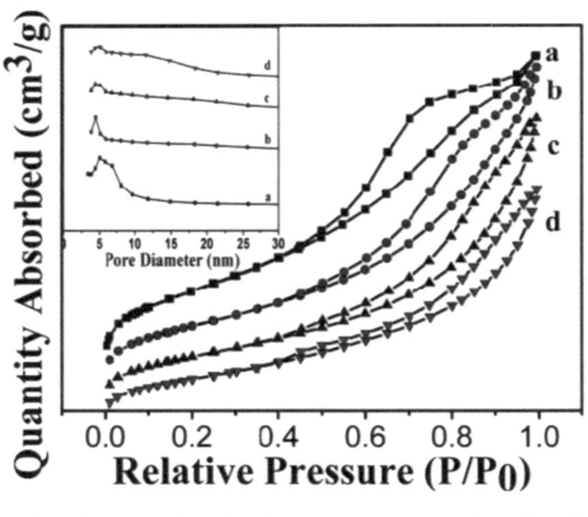

Fig. 4. Nitrogen adsorption/desorption isotherms and pore size distribution curves of samples: (a) SPC, (b) 5%HPW-SPC-SG, (c)15%HPW-SPC-SG, and (d) 25%HPW-SPC-SG.

Catalysts	Pore size[a] (nm)	S_{BET} ($m^2 \cdot g^{-1}$)	Basal spacing (nm)	Gallery height[b] (nm)	V_T ($cm^3 \cdot g^{-1}$)
MMT	4.30	64	1.54	0.58	0.14
SPC (0%HPW)	5.06	538	4.35	3.39	0.79
5%HPW-SPC-SG	4.46	444	4.38	3.42	0.73
15%HPW-SPC-SG	4.46	377	4.43	3.47	0.64
25%HPW-SPC-SG	4.52	324	4.47	3.51	0.51
5%HPW-SPC-IM	4.33	170	4.35	3.39	0.48
15%HPW-SPC-IM	3.86	152	4.35	3.39	0.44
25%HPW-SPC-IM	3.82	148	4.35	3.39	0.43

[a] The values were determined from N_2 desorption isotherm.
[b] The gallery height is defined as basal spacing minus the 0.96 nm thickness of MMT layer.

Table 1. Textural properties of MMT, SPC, HPW-SPC-SG and HPW-SPC-IM samples

The SEM images of the SPC, HPW-SPC-SG and 15%HPW-SPC-IM samples are shown in Figure 5. The SPC sample exhibited a slightly swelled nature mortmorillite plates morphology [27], while all the HPW-SPC-SG samples also exhibited plates structure, which was similar to the morphologies of the SPC sample. Though there were more fragmental particles with the increase of the HPW content in the HPW-SPC-SG samples, the main laminated structures of the HPW-SPC-SG samples were still retained, which indicates that the interlayer gallery structures were unaffected by the incorporation of the HPW. The 15 wt.% HPW-SPC-IM sample also retained a mortmorillite lamellar structure, though the HPW has an impact on the surface of the lamellar structure.

3.3 Catalytic performance of oxidizing modle oil

The effect of the reaction time on the ODS of the modle oil with sulfur content of 500 ppm over the HPW-SPC-SG catalysts is shown in Figure 6. It is obvious that the sulfur removal over 5%HPW-SPC-SG, 15%HPW-SPC-SG and 25%HPW-SPC-SG catalysts can reach up to 97.6%, 97.8% and 95.8% within 80 min, respectively. The results indicate that the HPW-SPC-SG samples are highly efficient in catalytic ODS of the DBT. But the 25%HPW-SPC-SG sample exhibited comparatively lower activity than the 15%HPW-SPC-SG sample, this is because the presence of high HPW contents leads to the decrease of the *S*BET of the 25%HPW-SPC-SG sample (Table 1). With the increase of the reaction time, the sulfur removal of the samples was remarkably raised in 80 min, and just slightly increased after 80 min. For the 15%HPW-SPC-SG sample, the sulfur content can be reduced to 7.2 ppm within 100 min.

The sulfur removal of the model oil over the 15%HPW-SPC-SG catalyst with reaction time at different temperatures is shown in Figure 7. The sulfur removal appeared to be unsatisfying at a temperature of 40 °C, which may be because H_2O_2 and the catalyst cannot work efficiently under low reaction temperature. Raising the reaction temperature from 40 to 60 °C led to a remarkable increase in the sulfur removal. However, further increase of the reaction temperature to 70 °C caused the decrease of the desulfurization efficiency. This is because the high temperature leads to the partial decomposition of H_2O_2, which brings down the oxidation effeciency. But the sulfur removal at 70 °C was still higher than that at 50 °C, which indicates that a proper increase of temperature favores the ODS of DBT.

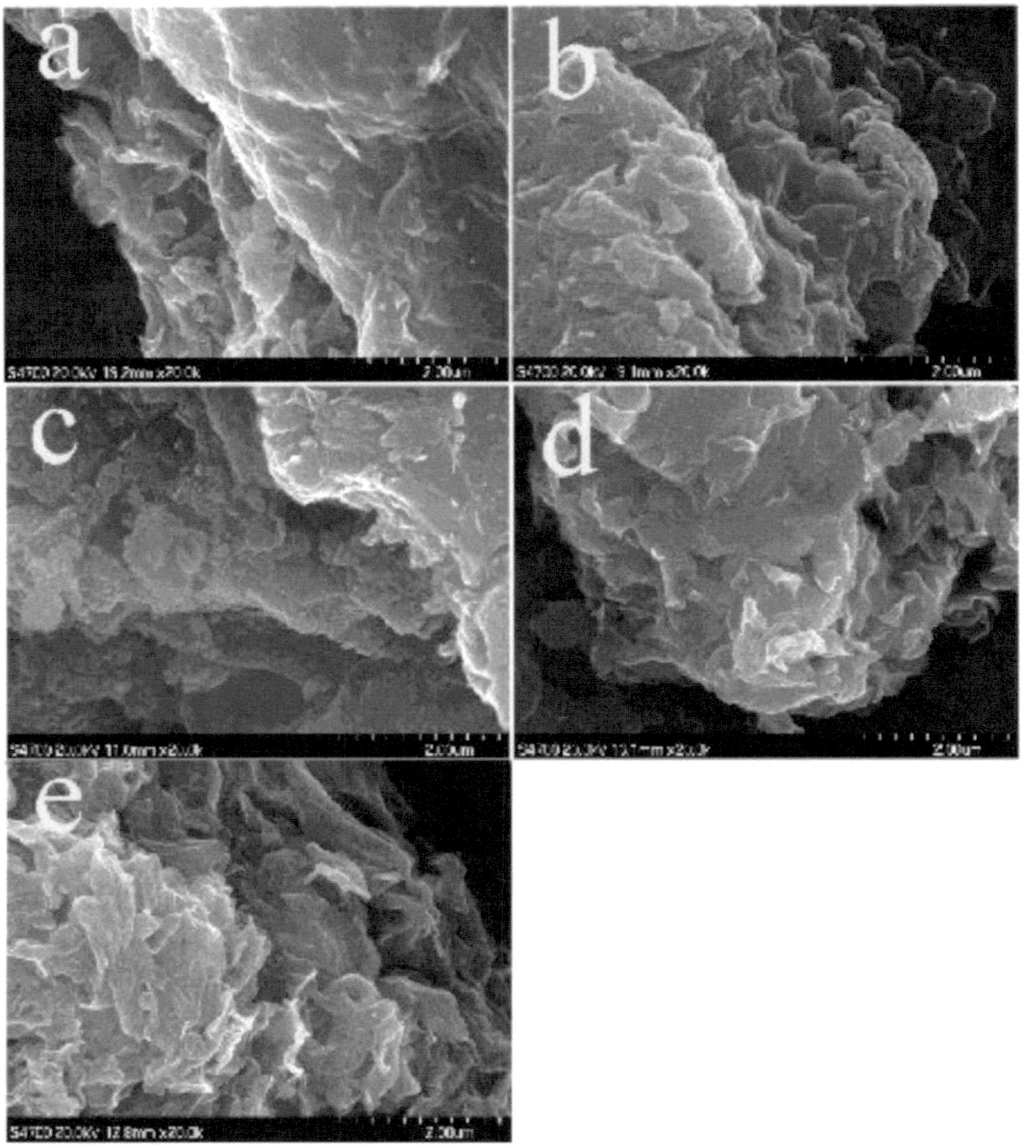

Fig. 5. SEM images of samples: (a) SPC, (b) 5%HPW-SPC-SG, (c) 15%HPW-SPC-SG, (d) 25%HPW-SPC-SG, and (e) 15%HPW-SPCIM.

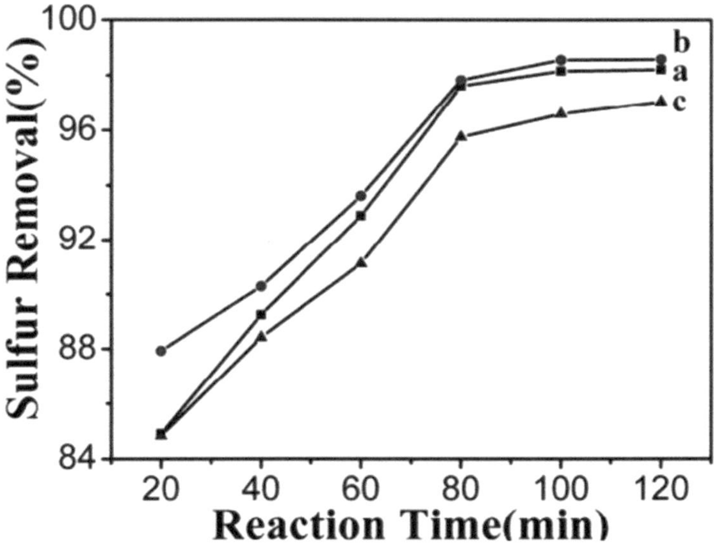

Fig. 6. Effect of reaction time on ODS of samples: (a) 5%HPW-SPC-SG, (b) 15%HPW-SPC-SG, and (c) 25%HPW-SPC-SG. Reaction temperature is 60 oC.

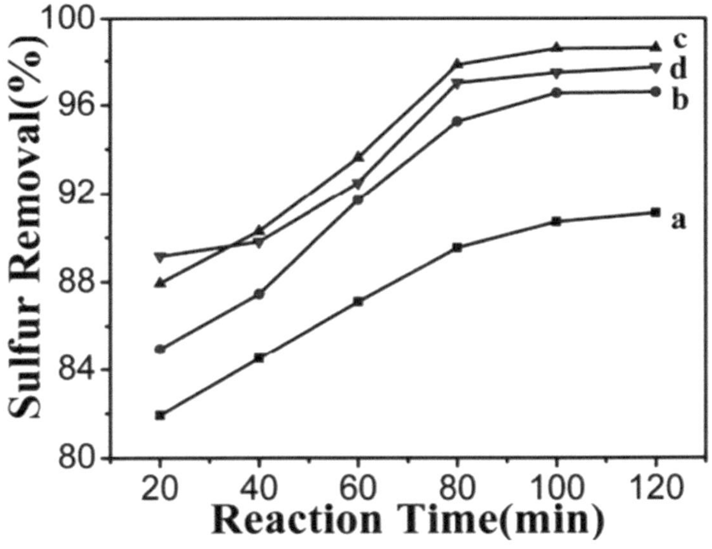

Fig. 7. Effect of reaction time and temperature on ODS: (a) 40 °C, (b) 50 °C, (c) 60 °C, and (d) 70 °C.

Table 2 shows the results of the sulfur removal of the model oil over different catalysts. The results indicated that direct extraction without oxidation resulted in lower sulfur removal. And the catalytic activity of the HPW-SPC-SG samples was better than that of the HPW-SPC-IM samples. This is because the *S*BET and pore size of the HPW-SPC-SG samples are much bigger than that of the HPW-SPC-IM samples. On the other hand, HPW has been encapulated into the silica pillared frameworks of SPC in the HPW-SPC-SG samples, which leads to the better stability and continuous catalytic ability of HPW.

Catalysts	S remaining (E)[a] (ppm)	S remaining (ppm)	S removal (%)
-	177.3	174.1	65.2
SPC (0% HPW)	154.3	148.8	70.2
5%HPW-SPC-SG	161.8	9.0	98.2
15%HPW-SPC-SG	164.2	7.1	98.6
25%HPW-SPC-SG	169.2	14.8	97.0
5%HPW-SPC-IM	168.3	146.7	70.7
15%HPW-SPC-IM	170.4	131.7	73.7
25%HPW-SPC-IM	173.0	116.9	76.6

a (E) means the extraction results of the samples without oxidation. Reaction temperature is 60 oC and reaction time is 120 min.

Table 2. Catalytic performance of SPC, HPW-SPC-SG, and HPW-SPC-IM catalysts

4. Conclusions

The HPW-SPC-SG materials were successfully prepared by sol-gel method. The catalysts showed a homogeneous dispersion of the HPW molecules even at 25 wt.% loading. HPW was encapsulated into the gallery silica framework structure in the HPW-SPC-SG samples. The HPW-SPC-SG materials showed better catalytic performance for ODS of the oil than that of the HPW-SPC-IM samples. Under the optimized conditions, the sulfur content in the model oil can be reduced from 500 to 7.1 ppm. The prepared HPW-SPC-SG materials are promising and efficient catalysts for ODS of fuel oils.

5. Acknowledgements

The work was financially supported by the National Basic Research Program of China (973 Program, Grant No. 2011CBA00506).

6. References

[1] R. T. Yang, A. J. Hernández-Maldonado, F. H. Yang, Science 301 (2003) 79.
[2] H. Tao, T. Nakazato, S. Sato, Fuel 88 (2009) 1961.
[3] Y. Shiraishi, K. Tachibana, T. Hirai, I. Komasawa, Ind. Eng. Chem. Res. 41 (2002) 4362.
[4] S. Murata, K. Murata, K. Kidena, M. Nomura, Energy Fuels 18 (2004) 116.
[5] I. V. Babich, J. A. Moulijn, Fuel 82 (2003) 607.
[6] E. Ito, J. A. Rob van Veen, Catal. Today 116 (2006) 446.
[7] C. Song, X. Ma, Appl. Catal. B: Environ. 41 (2003) 207.
[8] A. K. Sharipov, V. R. Nigmatullin, Pet. Chem. 45 (2005) 371.
[9] A. Sharipov, V. Nigmatullin, I. Nigmatullin, R. Zakirov, Fuels Oils 42 (2006) 451.
[10] A. Anisimov, A. Tarakanova, Russ. J. Gen. Chem. 79 (2009) 1264.

[11] J. M. Campos-Martin, M. C. Capel-Sanchez, P. Perez-Presas, J. L. G. Fierro, J. Chem. Technol. Biotechnol. 85 (2010) 879.
[12] P. S. Tam, J. R. Kittrell, J. W. Eldridge, Ind. Eng. Chem. Res. 29 (1990) 321.
[13] G. X. Yu, S. X. Lu, H. Chen, Z. Zhu, Carbon 43 (2005) 2285.
[14] F. Al-Shahrani, T. C. Xiao, S. A. Llewellyn, S. Barri, Z. Jiang, H. H. Shi, G. Martinie, M. L. H. Green, Appl. Catal. B: Environ. 73 (2007) 311.
[15] P. D. Filippis, M. Scarsella, Energy Fuels 17 (2003) 1452.
[16] L. Y. Kong, G. Li, X. S. Wang, B. Wu, Energy Fuels 20 (2006) 896.
[17] M. Te, C. Fairbridge, Z. Ring, Appl. Catal. A: Gen. 219 (2001) 267.
[18] H. M. Li, L. N. He, J. D. Lu, W. S. Zhu, X. Jiang, Y. Wang, Y. S. Yan, Energy Fuels 23 (2009) 1354.
[19] M. Arias, D. Laurenti, C. Geantet, M. Vrinat, I. Hideyuki, Y. Yoshimura, Catal. Today 130 (2008) 190.
[20] C. Li, Z. Jiang, J. Gao, Y. Yang, S. Wang, F. Tian, F. Sun, X. Sun, P. Ying, C. Han, Chem. Eur. J. 10 (2004) 2277.
[21] H. Y. Lü, J. Gao, Z. X. Jiang, F. Jing, Y. X. Yang, G. Wang, C. Li, J. Catal. 239 (2006) 369.
[22] J. B. Gao, S. G. Wang, Z. X. Jiang, H. Y. Lu, Y. X. Yang, F. Jing, C. Li, J. Mol. Catal. A: Chem. 258 (2006)261.
[23] Z. E. A. Abdalla, B.S. Li, A. Tufail, Colloids Surf. A: Physicochem. Eng. Aspects 341 (2009) 86.
[24] H. H. Mao, B. S. Li, X. Li, Z. X. Liu, W. Ma, Appl. Surf. Sci. 255 (2009) 4787.
[25] B. S. Li, H. H. Mao, X. Li, W. Ma, Z. X. Liu, J. Collid. Interface. Sci. 336 (2009) 244.
[26] H. H. Mao, B. S. Li, X. Li, Z. X. Liu, W. Ma, Mater. Res. Bull. 44 (2009) 1569.
[27] H. H. Mao, B. S. Li, X. Li, Z. X. Liu, W. Ma, Catal. Commun. 10 (2009) 975.
[28] H. H. Mao, B. S. Li, X. Li, L. W. Yue, Z. X. Liu, W. Ma, Ind. Eng. Chem. Res. 49 (2010) 583.
[29] H. H. Mao, B. S. Li, X. Li, L. W. Yue, Micropor. Mesopor. Mater. 130 (2010) 314.
[30] H. Ishida, S. Campbell, J. Blackwell, Chem. Mater. 12 (2000) 1260.
[31] R. A. Horch, T. D. Golden, N. A. D'Souza, L. Riester, Chem. Mater. 14 (2002) 3531.
[32] H. Y. Zhu, J. C. Zhao, J. W. Liu, X. Z. Yang, Y. N. Shen, Chem. Mater. 18 (2006) 3993.
[33] J. C. Juan, J. C. Zhang, M. A. Yarmo, J. Mol. Catal. A: Chem. 267 (2007) 265.
[34] A. Miyaji, T. Echizen, K. Nagata, Y. Yoshinaga, T. Okuhara, J. Mol. Catal. A: Chem. 201 (2003) 145.
[35] T. Sugii, R. Ohnishi, J. Zhang, A. Miyaji, Y. Kamiya, T. Okuhara, Catal. Today 116 (2006) 179.
[36] Q. H. Xia, K. Hidajat, S. Kawi, J. Catal. 209 (2002) 433.
[37] B. C. Gagea, Y. Lorgouiloux, Y. Altintas, P. A. Jacobs, J. A. Martens, J. Catal. 265 (2009) 99.
[38] M. Pluta, A. Galeski, M. Alexandre, M.-A. Paul, P. Dubois, J. Appl. Polym. Sci. 86 (2002) 1497.
[39] J. Toufaily, M. Soulard, J. L. Guth, J. Patarin, L. Delmote, T. Hamieh, M. Kodeih, D. Naoufal, H. Hamad, Colloids Surf. A: Physicochem. Eng. Aspects 316 (2008) 285.
[40] S. Bracco, P. Valsesia, L. Ferretti, P. Sozzani, M. Mauri, A. Comotti, Micropor. Mesopor. Mater. 107 (2008)102.
[41] A. Ghanbari-Siahkali, A. Philippou, J. Dwyer, M. W. Anderson, Appl. Catal. A: Gen. 192 (2000) 57.

[42] Á. Kukovecz, Zs. Balogi, Z. Kónya, M. Tobab, P. Lentzc, S. I. Niwa, F. Mizukami, Á. Molnárd, J. B. Nagyc, I. Kiricsi, Appl. Catal. A: Gen. 228 (2002) 83.
[43] F. Kleitz, W. Schmidt, F. Schüth. Micropor. Mesopor. Mater. 65 (2003) 1.
[44] M. L. Occelli, Ind. Eng. Chem. Prod. Res. Dev. 22 (1983) 553.
[45] I. V. Kozhevnikov, J. Mol. Catal. A: Chem. 262 (2007) 86.
[46] S. Ajaikumar, A. Pandurangan, J. Mol. Catal. A: Chem. 286 (2008) 21.
[47] J. A. Gamelas, F. A. S. Couto, M. C. N. TrovaÃo, A. M. V. Cavaleiro, J. A. S. Cavaleiro, J. D. Pedrosa de Jesus, Thermochim. Acta 326 (1999) 165.
[48] C. Rocchiccioli-Deltcheff, M. Foumier, R. Frank, Inorg. Chem. 22 (1983) 207.
[49] H. Y. Zhu, Z. Ding, J. C. Barry, J. Phys. Chem. B 106 (2002) 11420.
[50] L. H. Hu, S. F. Ji, Z. Jiang, H. L. Song, P. Y. Wu, Q. Q. Liu, J. Phys. Chem. C 111 (2007) 15173.
[51] A. Bordoloi, F. Lefebvre, S. B. Halligudi, J. Catal. 247 (2007) 166.
[52] F. Rojas, I. Kornhauser, C. Felipe, J. M. Esparza, S. Cordero, A. Dominguez, J. L. Riccardo, Phys. Chem. Chem. Phys. 4 (2002) 2346.

12

Biocomposites Based on New Monomer Systems Reinforced with Micro / Nanoparticles and Glass Fibers

Cristina Prejmerean[1], Tinca Buruiana[2], Teresa Nunes[3], Marioara Moldovan[1] and Loredana Colceriu[4]

[1]*Babes-Bolyai University, "Raluca Ripan" Institute of Research in Chemistry Cluj-Napoca*

[2]*"Petru Poni" Institute of Macromolecular Chemistry Iasi*

[3]*Centro de Química Estrutural, Complexo I, TU Lisbon*

[4]*"Iuliu Hatieganu" University of Medicine and Pharmacy Cluj-Napoca*

[1,2,4]*Romania*

[3]*Portugal*

1. Introduction

The resin composites were introduced for the first time in dentistry in the mid-1960s, for the restoration of anterior teeth, because of the necessity to eliminate the shortcomings of the restorative materials existing at that time. Because of their properties, aesthetic aspect, adhesion to hard dental tissues, high values of the physico-mechanical properties, they play an important role in dental practice (Bowen & Marjenhoff, 1992).

Today, resin composite materials are frequently used for direct restoration of anterior teeth, as pit and fissure sealants for the prophylaxis of caries with children, and in prosthetic dentistry for the preparation of composite veneers, inlays and onlays by the photo-barro-thermal polymerization. The composite veneers, inlays and onlays are fixed in the oral cavity using a luting agent, which is also a dual cured resin composite. The development of total polymeric restorations and the reinforcement of the resins with fibers for the obtaining of the prosthesis frameworks represent one of the actual achievements in the world.

The resin composites comprise three components: *a) the resin matrix* which consists of the monomer system, the initiator system for free radical polymerization, antioxidants and UV stabilizers, *b) the inorganic filler* consisting of particulates such as glass, quartz, and/or fused silica, or glass fibers, *c) the coupling agent* usually an organo-silane that chemically bonds the reinforcing filler to the resin matrix. When hardening is desired the polymerization of the monomers is initiated by a chemical or photochemical system. In the cured material, the inorganic filler particles are dispersed and chemically bonded in a crosslinked three-dimensional network formed by the polymerization of the di(poly)functional monomers. The properties, and hence the performances of the materials are dependent upon the nature and features of the components of the material. In spite of the progress in dental composite technology, loss of anatomic form in time, marginal leakage due to polymerization shrinkage, contraction stresses generally concentrated at the adhesive interface leading to marginal debondings, as well as chromatic instability and water sorption, are often cited as being the main problems of resin composites (Peutzfeldt, 1997).

The resin composites routinely employed in dental practice have a resin matrix based on dimethacrylate monomers, which have been used for over 40 years. The main component in the most of the present commercial dental resins is still Bis-GMA (2,2-bis[4-(2-hydroxy-3-methacryloxypropoxy)phenyl]propane, which was synthesized by Bowen in 1962 (Bowen, 1962). The reasons for its dominance are the positive properties, low volatility, fast hardening by radical polymerization, and relative low polymerization shrinkage. However, the main deficiencies of Bis-GMA are its high viscosity (1.0-1.2 kPa.s at 23^{0}C), water sensitivity and low degree of conversion (Mozner & Salz, 2001). A diluent, usually triethyleneglycol dimethacrylate TEGDMA, have to be added to reduce the viscosity of the resin. By adding TEGDMA to Bis-GMA, the water uptake and polymerization shrinkage and its associated stress of the cured material increase. The high viscosity and relatively high water sorption of Bis-GMA are due to the hydroxyl groups of its molecule. By converting the hydroxyl groups to other functional groups to avoid inter-molecular H-bonding thus minimizing viscosity, were obtained Bis-GMA substitutes, but they yield polymers with lower flexural strengths and modules (Holter et al., 1997). Most attention has been given to urethane derivatives of Bis-GMA , which exhibit lower viscosities, higher degree of vinyl conversion, and lower polymerization shrinkage at equivalent degrees of vinyl conversion than Bis-GMA (Moszner et al., 2008; Khatria et al., 2003; Pereira, et al., 2002).

Joining the research activities in the field, our studies have as the major objective to develop new resin composites with reduced polymerization shrinkage, high mechanical properties, decreased water sorption, chromatic stability and improved adhesion to tooth structure. The aim of the present work was to elaborate some particulate and fiber reinforced resin composites with new resin matrices and to investigate the influence of the resin composition upon the properties of the corresponding polymers and resin composite materials.

2. Materials and methods

2.1 The elaboration and characterization of a series of new experimental resins

The new experimental resins are formulated using mixtures of Bis-GMA type superior oligomers with high molecular weight (Prejmerean et al. 2000) and Bis-GMA analogues obtained by the addition of some blocking structures (e.g. methacryloyloxyethyl isocyanate) to the Bis-GMA glycidyl hydroxyl groups (Buruiana et al., 2008).

The formula for Bis-GMA0, Bis-GMA1, Bis-GMA2 and Bis-GMA3 are presented below:

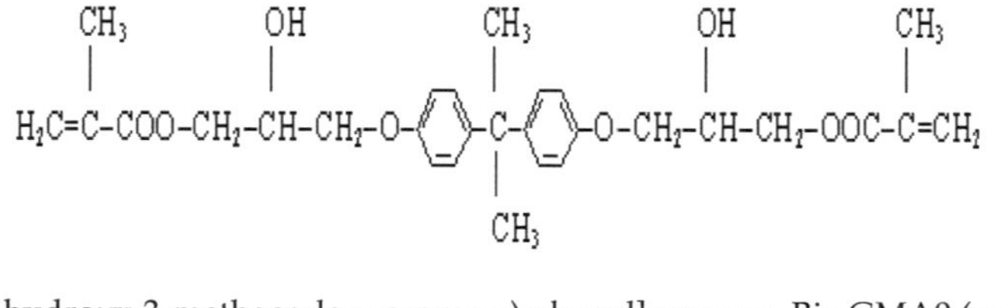

2,2-bis[4-(2-hydroxy-3-methacryloxypropoxy) phenyl]propane Bis-GMA0 (commercial)

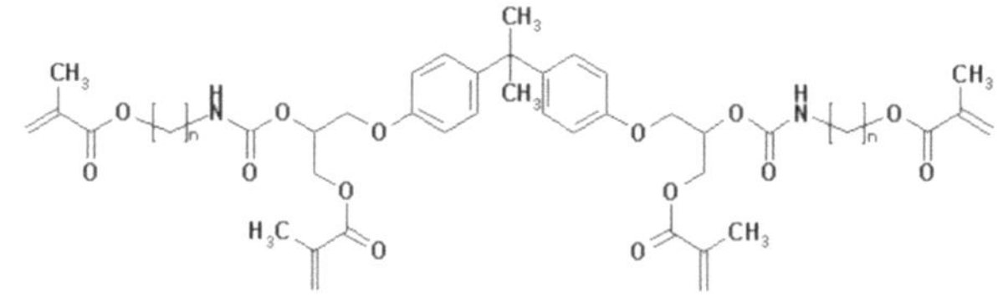

Bis-GMA1 (n=2)

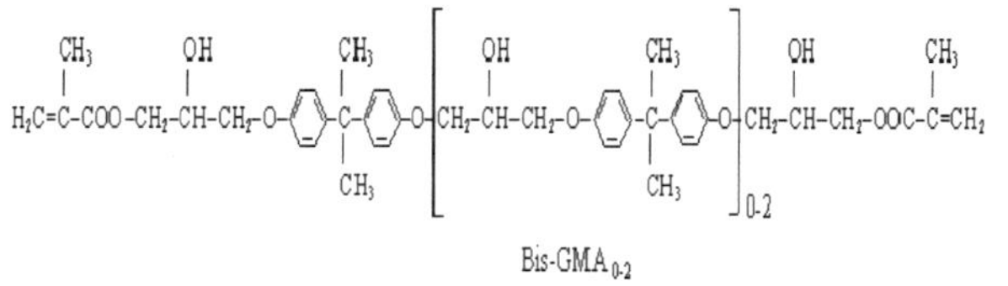

Bis-GMA2

(mixture of oligomers : 81% monomer, 17% dimer and 1% trimer)

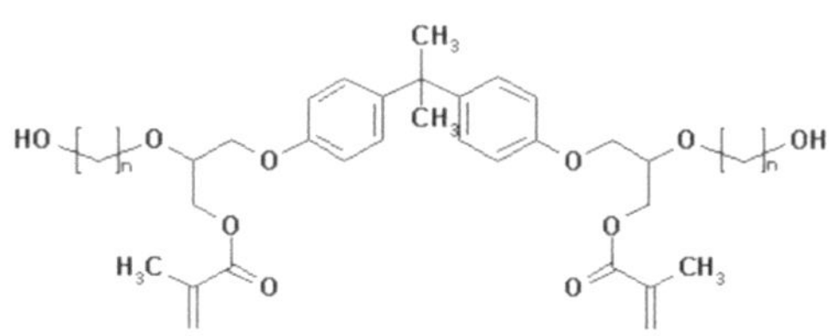

Bis-GMA3 (n=12)

Fig. 1. The experimental monomers used in the study

2.2 The preparation of the copolymers by chemical initiation of polymerization

In order to obtain the copolymers by chemical initiation of polymerization, two monomer mixtures were prepared for each copolymer (two resins), which had the same monomer amount, as presented in table 1: resins A and B. In resins A was dissolved 2% by weight of the polymerization accelerator (N,N-dihydroxyethyl-p-toluidine, DHEPT) and in resins B, was added 2% by weight of the polymerization initiator, benzoyl peroxide (POB). Butylated hydroxytoluene (BHT) was used as an inhibitor in an amount of 0.1% by weight. The copolymers were obtain by mixing equal amounts of A resins with corresponding B resins. The polymerization took place under a polyethylene band in order to prevent the contact with the atmospheric oxygen. The obtained copolymers were ground to fine powders.

2.3 The preparation of the copolymers by photochemical initiation of polymerization

Using the photochemical polymerization system of monomer mixtures presented in table 1, some copolymers were prepared. In the composition of the monomer mixtures, besides the methacrylic oligomers and monomers, a photosensitizer, camphorquinone (CQ) in an amount of 0.5% (by weight), and an accelerator dimethylaminoethyl-methacrylate (DMAEM), in an amount of 1% (by weight), were added. BHT was added in a quantity of 650 ppm related to the monomer mixture. The mixtures were hardened by exposing to a visible radiation in the wavelength range of 400-500 nm, for 30 seconds. The radiation was generated by an Optilux stomatological lamp (14V and 35 W), produced by Demetron Research Corporation USA. The copolymers were ground to fine powders.

2.4 Determination of the residual double bonds

The conversion of monomer mixtures was evaluated by determining the residual double bonds using the Infrared Spectroscopy Method. The quantity of unreacted methacrylate groups is determined in percent of the methacrylate groups originally present in the unpolymerized material. The decrease in the intensity of the methacrylate C=C absorbance (A_{meth}) at 1635 -1640 cm^{-1} was monitored. The phenyl absorbance (A_{arom}) at 1605-1610 cm^{-1} was used as an internal standard (Ruyter & Gyorosi, 1978).

The FTIR spectra of the liquid monomers mixtures prepared as liquid thin films between KBr plates and of the solid corresponding copolymers prepared as KBr pellets were recorded on a JASCO 610 FTIR Spectrophotometer.

2.5 Determination of volumetric polymerization contraction

Polymerization shrinkage was determined using the analysis of the network by stray-field MRI.

One-dimensional ^{1}H STRAFI images, projections along an axis, were acquired from the liquid and from the photopolymerized monomers using a Bruker MSL 300P spectrometer, under the static magnetic field gradient of 37.5 T/m generated near the edges of the 89 mm superconducting coil. A dedicated Bruker STRAFI probe-head was tuned to 123.4 MHz, which gives ^{1}H resonances at 2.9 T; this field strength was obtained just outside the bore of the magnet. Since only thin slices are observed at a time, 1D imaging of bulk specimens was conducted by moving the sample linearly in the direction of the field gradient to ensure that the volume under analysis experiences the same field conditions. Each liquid sample was introduced in a cylindrical glass vial (5 mm height and 6 mm inner diameter) filled up to 4 mm height. In order to obtain VPC, magnetization profiles were acquired prior irradiation (Optilux 401, Demetron Research Corp., Danbury, CT, USA) and after the irradiation period

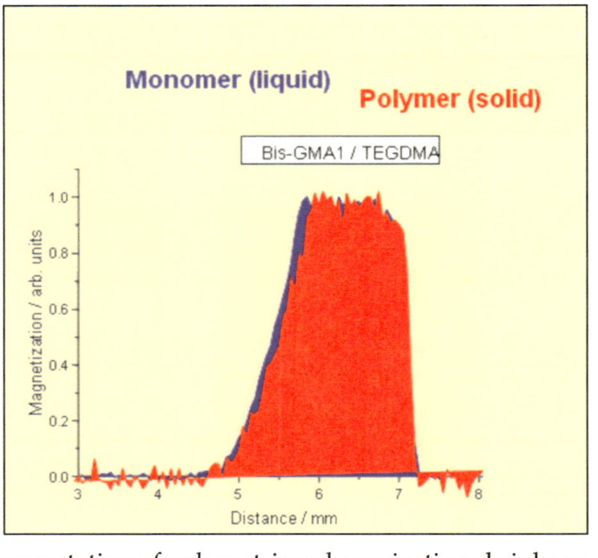

Fig. 2. Schematic representation of volumetric polymerization shrinkage

of 10 s (500 mW/cm^2). The photopolymerization was carried out at room temperature (about 22 °C). The magnetization was recorded as multiple 8 spin-echo trains. Each echo train was generated by the RF pulse sequence $90°_x$ -τ-[$90°_y$-τ-echo-τ]$_8$, where $90°_{x,y}$ represents a 90° pulse of relative phase x or y and τ is a short time delay; the RF pulse duration (t_p) and echo time (TE) were 10 μs and 35 μs, respectively (Nunes et al., 2002). Each data slice in the STRAFI profiles is the result of the summation of the 8 echoes.
The final volumetric polymerization shrinkage (VPC, in %) was determined for all monomers, according to a method previously described (Pereira et al., 2002; Nunes et al., 2005), as:

$$VPC = [\Sigma\,(MM) - \Sigma\,(MP) \times 100] / \Sigma\,(MM) \qquad (1)$$

where MM and MP represent the magnetisation of each slice of the comonomer mixture and of the system cured over 10 s at about 500 mW/cm^2, respectively; the highest intensity of MP was used as a reference for the intensity normalization of the magnetisation of MM and MP slices.

2.6 Determination of leachable components from experimental dental copolymers

The method of evaluating the quantities of residual monomer and of residual light stabilizer respectively, proposed in the present study is based on the accelerated extraction of the eluted components from the copolymers in chloroform, at the boiling temperature of chloroform, for ten hours. The residual monomer and residual light stabilizer amounts have been determined from the UV spectra of the chloroformic extracts as percent of the monomer (stabilizer) originally present in the unpolymerised materials. The experimental copolymers were transformed into fine powders and 0.5 g of powder was extracted into 100 ml chloroform.

2.7 Obtaining of the experimental light-curing composites

The experimental light-curing composites were prepared as monopastes from the resin matrices and the hybrid filler. The monomer mixtures in the light-curing composites consist of Bis-GMA-type monomer and TEGDMA. The inorganic phase consisted of 90% silanized barium glass (45% SiO_2, 10% Al_2O_3, 17% B_2O_3, 20% BaO, 8% NaF-CaF_2) and 10% silanized colloidal silica. The powder/liquid ratio was 4/1. The light-curing resin composites were cured by exposing to a visible radiation in the wavelength range of 400-500 nm for 40 seconds.

2.8 Determination of mechanical properties

2.8.1 Determination of compressive strength

The *compressive strength (CS)* was determined using specimens having cylindrical shape - with 8 mm height and 4 mm diameter. The specimens were illuminated in their respective molds by the light source (OPTILUX dental lamp, Demetron Research Corporation, USA) in steps of 2mm height, for 40 seconds each step. The measurements of the compressive strength were made using a Lloyd LR5K Plus mechanical testing apparatus at a 0.5 mm/min pressing speed. The compression strength CS (in MPa) was calculated applying formula (2) where F is the maximum applied load (N) and d is the specimen diameter (mm).

$$CS = F / 0.785d^2 \qquad (2)$$

2.8.2 Determination of diametral tensile strength

The *tensile strength* was measured using the diametral compression test (the indirect test of determining the tensile strength). The general technique of preparing and testing the composites specimens was similar to the one described when determining the compression strength, with the only difference that the cylindrical specimens had 4mm thickness (T) and 6mm in diameter (D). The specimens were subjected to compression along the cylinder generator. The F force that acts on the cylinder, caught between the pans of the apparatus, determines the appearance of the tensile efforts on the vertical diameter plane. The tensile strength DTS (in MPa) was calculated applying the formula (3):

$$DTS=2 \times F/\pi \times D \times T \quad (3)$$

2.8.3 Determination of flexural strength

The *flexural strength* was determined using specimens having rectangular form (length 25.0 mm, height 2.0 mm and width 2.0 mm), according to ISO 4049/2000. The general technique of preparing the composite specimens was similar with the one used at the determination of compressive strength. After the measurement of height *a* and width *b*, the specimens were subjected to three-point loading with l=20 mm between the supports. The crosshead speed of the testing machine was 1.0 mm/min. The flexural strengths FS were calculated as:

$$FS= 2 \times F \times l/2 \times b \times a^2 \quad (4)$$

2.9 Determination of color changes of the stabilized composites

In order to evaluate the color changes of the composites, the samples in form of disks (30 mm diameter x 2 mm thick) have been exposed to UV radiation (Narva lamp- HgE/2 (λ=250-450 nm)) . The light source used a mercury vapor lamp at high pressure. The samples were stored in water at 37°C. After water storage, the specimens were placed on a Unicam 4 UV-VIS spectrometer. The coordinates (**Δa^*, Δb^* ΔL^***) of reflection spectrum were recorded and the color differences **ΔE^*** in cielab system were calculated by formula:

$$\Delta E^*= (\Delta L^{*2}+\Delta a^{*2}+\Delta b^{*2})^{1/2} \quad (5)$$

ΔL^* means the difference of luminosity between initial and final samples; Δa^* means the difference between the "red- green" coloured coordinates of the initial and final sample; Δb^* means the difference between the " yellow-green" colored coordinates of the initial and final sample. Spectra were recorded after 1, 7 and 30 days exposure to the ultraviolet light source. The coordinates (**Δa^*, Δb^* ΔL^***) were determined also by analyzing the digital images of the initial and final samples using a specific software.

2.10 Determination of water sorption

Disk specimens of composites 15±1mm diameter and 1±0.1mm thickness were prepared in a mold by exposing each side to a visible radiation in the range of 400-500 nm for 40 sec using a dental lamp.

Water sorption values were determined according to ISO 4049/2000 at 7 days of storage in water. The values for water sorption (Wsp), for each of the specimens were calculated using the formula:

$$Wsp= M1-M2 / V \quad (6)$$

M1 - the mass of the specimen after immersion in water [μg],
M2 - the mass of the reconditioned specimen [μg],
V- volume of the specimen [mm^3]

2.11 Investigation of adhesion of the resin composites to dental tissues

For this purpose were used 24 human premolar teeth in which two class V cavities (3 mm x 4 mm x 1.5 mm) were prepared on facial and oral surfaces with a cervical margin in cementum (dentin). Resin composites with or without previous use of an adhesive were applied to the cavities. The teeth were sectioned on an Isomet Low Speed Saw-Buechler LTD machine. SEM (FEI Comp) photomicrographs of the interfacial region between tooth and resin composites were performed.

3. Results and discussion

3.1 Determination of residual double bonds (RDB) in copolymer matrices

A crosslinked three-dimensional network is formed by the polymerization of the monomer mixtures existing in the dental diacrylic resins. During the polymerization of dimethacrylate monomers, the gel effect occurs, leading to the closing of radicals and unreacted monomers in the crosslinked network. The commercial diacrylic resins, investigated in the literature have between 25 and 50% unreacted methacrylate groups, where a tenth part of them represents the residual monomer (Asmussen, 1982; Ferracane, 1994). Presence of residual monomer or unreacted double bonds in the harden matrix has a plasticizing effect on the polymer. More than that, residual double bonds can make the polymeric matrix more susceptible to degradative reactions (Grassie, 1966).

Composition of the experimental monomer mixtures and the quantity of residual double bonds of the corresponding dental copolymers are presented in table 1. The quantity of residual double bonds remained unreacted after the chemical and respectively photochemical polymerization, as a percent of the total quantity of double bonds existing before the polymerization (RDB%) of the experimental BisGMA-type base monomers/TEGDMA mixtures were shown. It can be noticed an increasing of RDB with the increasing of the content of Bis-GMA type monomer, hence to the decreasing of TEGDMA monomer, both for the chemically and photochemically polymerized mixtures. The value of RDB ranged between 34.12% and 47.12% in the case of chemically polymerized mixtures, and between 30.40% and 46.39% in the case of photochemical polymerization, for Bis-GMA-type monomer/TEGDMA ratios between 50/50 and 75/25.

The increasing of RDB percent with the increasing of the Bis-GMA-type monomer/TEGDMA ratio can be explained by the effect of physical-chemical characteristics of the mixtures upon the evolution of polymerization reaction. The mixtures with a high Bis-GMA-type monomer/TEGDMA ratio, characterized by a high viscosity because of the large rigid molecules with reduced mobility of Bis-GMA-type monomers, reach the gel state in a short time after the beginning of reaction. After their polymerization, a large number of double bonds remain unreacted, as pendant double bonds and as residual monomer respectively.

The more advanced conversion (lower percent of RDB) obtained in the case of the photochemical polymerization than in the case of chemical polymerization of monomer mixtures, for the same Bis-GMA-type monomer/TEGDMA ratios can be explained by the relatively fast release of polymerization initiating radicals in the light cured materials.

Comparing the monomer mixtures that differ by the nature of the Bis-GMA type monomer and have the same ratio between the Bis-GMA-type monomer and TEGDMA, one can observe that the lowest percent of residual double bonds are registered in the case of polymerization the Bis-GMA0-based mixture, followed in order by Bis-GMA1-based mixture, Bis-GMA3- based resin and Bis-GMA2-based resin. Comparing the monomer mixtures containing Bis-GMA0 and Bis-GMA1, it can be noticed that Bis-GMA1-based mixtures lead to a slight increase of the quantity of RDB. This behaviour can be explained by the fact that Bis-GMA1 molecule contain four methacrylic bonds comparing to two methacrylic bonds contained by Bis-GMA0, so after polymerization remains a greater number of RDB in the case of Bis-GMA1-based mixtures. Replacing of the Bis-GMA0 monomer with Bis-GMA2 monomer produces mixtures with increased viscosities, the existence of larger molecules with less mobility belonging to the dimer and trimer from Bis-GMA2 leads finally to achieve the gel state at lower times, and to a larger number of unreacted double bonds. Bis-GMA3 shows conversions of double bonds between the values obtained for Bis-GMA1 and Bis-GMA2.

No.	Bis-GMA %	Bis-GMA1 %	Bis-GMA2 %	Bis-GMA3 %	TEGDMA %	RDB % Chemical initiation	RDB % Photo-chemical initiation
R1.	75	-	-	-	25	37.25	36.89
R2.	60	-	-	-	40	36.89	32.18
R3.	50	-	-	-	50	34.75	30.40
R7.	-	75		-	25	40.37	37.33
R8.	-	60		-	40	38.43	35.2
R9.	-	50		-	50	34.12	32.11
R4.	-		75	-	25	47.12	46.39
R5.	-		60	-	40	41.30	38.21
R6.	-		50	-	50	38.53	35.53
R10	-	-	-	75	25	43.76	42.11
R11.	-	-	-	60	40	40.11	39.90
R12	-	-	-	50	50	37.12	36.17

Table 1. Composition of the monomer mixtures and the quantity of residual double bonds (RDB) of the corresponding copolymers

3.2 Determination of polymerization shrinkage

Resins and composite resins shrink during polymerization, mainly because the monomeric units of the polymer are then located within covalent bond distance of each other, that is, closer to one another than they were in the original monomer state. In fact, two van der Waals distances and one double bond are converted into two single bonds when a radical reacts with a methacrylate group; overall a total of about 3.74Å is shortened to about 3.08 Å. Therefore, polymerization reaction induces necessarily a volumetric shrinkage (VPC) of the monomer mixture. Clearly, a high degree of conversion determines a strong VPC and the best compromise must be found in order to obtain dental materials with optimum performance. Particularly for the monomer-based materials, unreacted monomers remain in the final product, which is mostly the case of surface layers in contact with atmospheric molecular oxygen, a well-known-radical scavenger. ^{1}H STRAFI enables measuring the volumetric contraction *in non-invasive and non-destructive modes,* in the presence of oxygen from the atmosphere; in the case of dental adhesives VPC is due to solvent evaporation and photopolymerization shrinkage (Nunes et al., 2005).

The polymerization shrinkage of the experimental new resins is presented. The resins contain 75% Bis-GMA type monomer (Bis-GMA0, Bis-GMA1, Bis-GMA2 and Bis-GMA3) and 25% TEGDMA. Figure 3 shows profiles of monomer mixtures (prior irradiation) and of monomer mixtures photopolymerized over 10 s at 500 mW/cm^2. It must be noticed that in this figure the profiles are rotated of 90 degrees and, consequently, the sample surfaces are shown on the left-side of the plots. The first general observation is the narrowing of the profiles after exposure to light. This effect is clearly associated with contractions during polymerization and hence overall volume reduction. Table 2 presents VPC data obtained for the indicated comonomer mixtures; these values show that the resin based on Bis-GMA1 monomer has a polymerization shrinkage that represents 2/3 of the value presented by Bis-GMA0-based resin. The evolution of curing was locally monitored by the changes observed to the ^1H magnetization signal intensity, which depend on hydrogen concentration but are strongly weighted essentially by proton mobility in the kHz frequency range. Therefore, intensity variations can be directly related to volumetric contractions (Nunes et al., 2001).

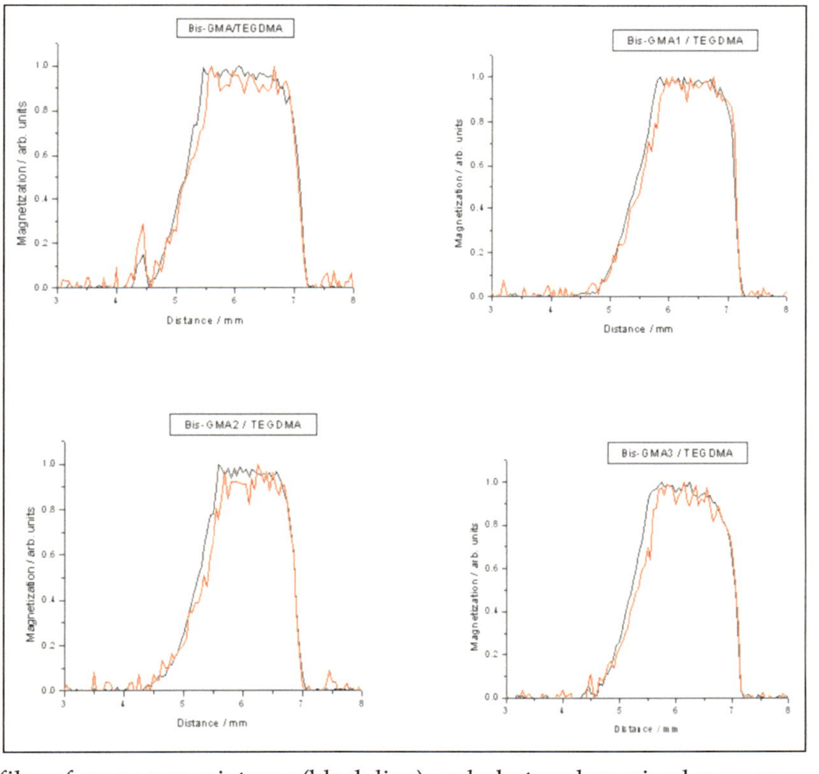

Fig. 3. Profiles of monomer mixtures (black line) and photopolymerized monomer mixtures over 10 s at 500 mW/cm^2 (red line); normalization was based on the assignment of 1 arb. unit to the highest magnetisation intensity obtained from both profiles.

Monomers with shorter molecular length (with identical functionality, hence with a high concentration of double bonds) are expected to react faster; in addition, monomers of higher flexibility (lower T_g) and lower viscosity are expected to reach a higher degree of conversion and VPC, even if the polymerization is carried out at ambient temperature. It was verified that TEGDMA and some Bis-GMA analogues followed the trend on VPC, measured using

STRAFI (S.G. Pereira, 2005). Overall, results were well correlated with an increase in the extent of polymerization due to the higher flexibility of the less viscous comonomer starting system and the hydrophobic character of the Bis-GMA analogues.

Monomer mixture (% by weight)	VPC (%)*
75% Bis-GMA0 / 25% TEGDMA	5.9
75% Bis-GMA1 / 25% TEGDMA	3.8
75% Bis-GMA2 / 25% TEGDMA	7.7
75% Bis-GMA3 / 25% TEGDMA	6.2

* an error of about 0.1 % should be assigned to these values.

Table 2. Volumetric polymerization shrinkage (VPC, %) of the indicated monomer mixtures obtained after an irradiation period of 10 s at about 500 mW/cm^2.

Bis-GMA and TEGDMA, generally used to prepare resins to be used as control, were employed in order to assess the role of the filler loading in the degree of double-bond conversion, polymerization shrinkage and molecular dynamics (Nunes et al., 2008). This study showed how treated filler loading influences the photopolymerization of the dimethacrylate comonomer mixture, regarding, in particular, shrinkage and inhibition under atmospheric oxygen, present in oral environment. Bis-GMA/TEGDMA (75/25 wt.%) resins were loaded (50 wt.%) with hybrid filler (Ba aluminosilicate glass and pyrogenic silica, treated with γ-methacryloxy(propyl)trimethoxysilane) and light cured over a total of 30 s (45 mW/cm^2). VPC, obtained using STRAFI (3.00 and 0.35 % ± 0.11% for the unfilled resin and for the filler loaded resin, respectively), showed a reduction on polymerization contraction and reaction inhibition at the composite resin surface with filler loading.

3.3 Determination of leachable components from experimental dental copolymers

The leaching of components (free molecules) from dental copolymers and composites has a potential impact on both the structural stability and the biocompatibility of the material. The latter is of much grater concern. Components may be eluted into salivary fluids and brought into contact with mucosal tissue. In addition, components may be extracted into dentine where they may diffuse toward the pulp (Ferracane, 1994).

3.3.1 Determination of residual monomer

The finding that a percent of the initial monomer remains unreacted while curing diacrylic dental resins, this being found as residual monomer trapped in the polymer matrix, has led some researchers to study the phenomenon of elution of these unbound molecules from the reinforced composites in different environments.

Ferracane studied the release of unreacted components from a dental resin composite (Silux, 3M, St. Paul, MN) and from a dimethacrylate unfilled resin with identical composition of organic matrix with Silux composite (50% Bis-GMA and 50% DMTEG), immersed in water immediately after exposure to visible radiation, reaching to results of 1.5-2% for the two materials into a period of 7 days. Rathburn et al immersed composites in organic solvents, ethanol, chloroform and toluene, immediately after curing, and extract 4-6%, 7.5% and 11% respectively, from their components in a period of 7 days.

It is desirable to remain the lower amount of residual monomer after polymerization of dimethacrylate monomers. By the method proposed, the total amount of unpolymerized monomer trapped in the crosslinked polymer network was determined. UV spectra of the residual monomer extracted in chloroform from copolymers based on Bis-GMA2 and TEGDMA are shown in figure 4:

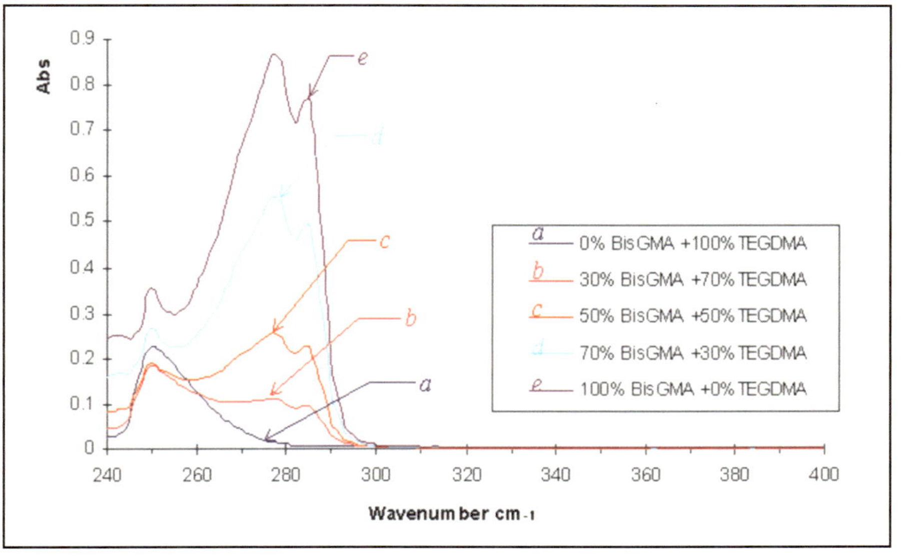

Fig. 4. UV spectra of the residual monomer extracted in chloroform from copolymers based on Bis-GMA2 and TEGDMA

Using the UV spectra of the residual monomer in chloroform and the calibrations curves of the two components, the quantities of the residual Bis-GMA2 monomer and of the TEGDMA monomer have been calculated as a percentage of the initial amount of the corresponding monomer present in the unpolymerized resin. The results are shown in table 3.

Bis-GMA2/TEGDMA	Extracted Bis-GMA2/Initial Bis-GMA2	Extracted TEGDMA/Initial TEGDMA
0/100	-	7.3
30/70	1.1	7.0
40/60	1.7	7.0
50/50	1.5	6.8
60/40	1.8	6.85
70/30	2.3	6.75
100/0	2.5	-

Table 3. The percent of the extracted monomers from the Bis-GMA2/TEGDMA copolymers

From the results presented in table 3 it can be seen a slight tendency of increasing the quantity of the residual Bis-GMA2 monomer and the decreasing of the quantity of the residual diluting monomer with the increasing of the Bis-GMA2/TEGDMA ratio. It can be also noticed that the residual TEGDMA percentage is about 3 up to 7 times higher than that of residual Bis-GMA2 monomer. This is due to the small and flexible molecules of TEGDMA which are more easily extracted from the polymeric matrix comparing to the rigid and large molecules of Bis-GMA2 monomer.

3.3.2 Determination of residual light stabilizer

Starting from the idea that low molecular weight UV stabilizers that are not chemically bonded to the polymer network of the cured diacrylic dental resins can migrate in time from the network, being extracted into saliva, and therefore their role being compromised, in the present paragraph a way of evaluating the effectiveness of the UV stabilizers from the point of view of their permanence in the polymer matrix is presented. In order to evaluate the UV stabilizers permanence, the quantity (percent) of extractable stabilizer from the polymer matrix was determined.

The investigated UV stabilizers were conventional and unconventional UV stabilizers (polymer-bound stabilizers) belonging to the hydroxybenzophenones, hydroxyphenylbenzotriazoles and sterically hindered amines (HALS) classes (Table 4). In order to calculate the amount of extracted UV stabilizers, there were determined the extinction coefficients of the stabilizers at every characteristic wavelength. The ultraviolet extinction coefficients of investigated stabilizers are shown in table 5. After the extraction of stabilized copolymer powders in chloroform, the UV spectra of the chloroformic extracts were recorded.

Chemical structure of the light-stabilizers	**Trade name/** Chemical name
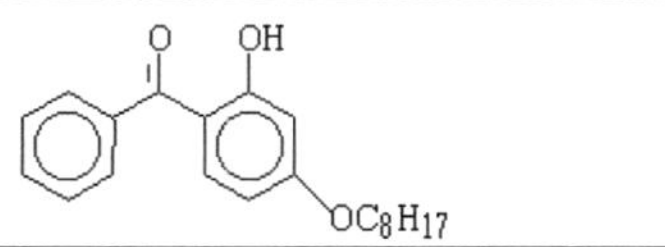	**Chimassorb 81** 2-hydroxy-4-n-octyloxybenzo-phenone
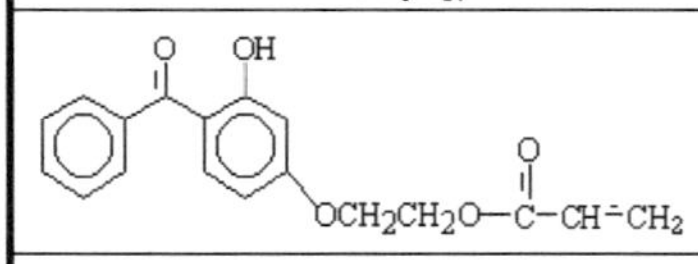	**Cyasorb UV 2098** 2-hydroxy-4-acryloyloxyethoxy-benzophenone
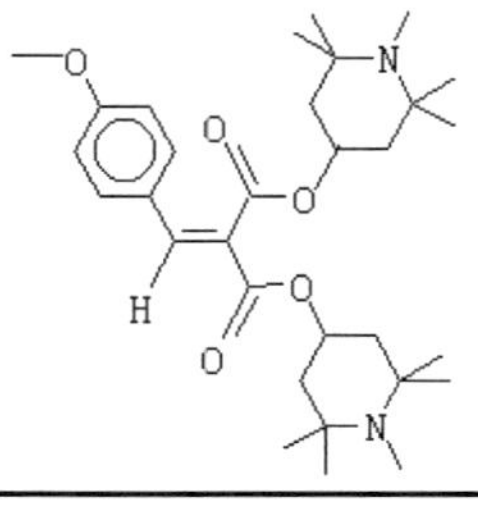	**VP Sanduvor PR-31** Propanedioic acid, [(4-methoxyphenyl) methylene]-, bis(1,2,2,6,6-pentamethyl-4-piperidinyl)ester
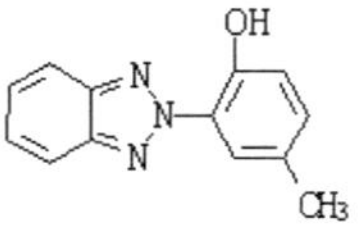	**Tinuvin P** 2-(2'-hydroxy-5'-methylphenyl)-2H-benzotriazole
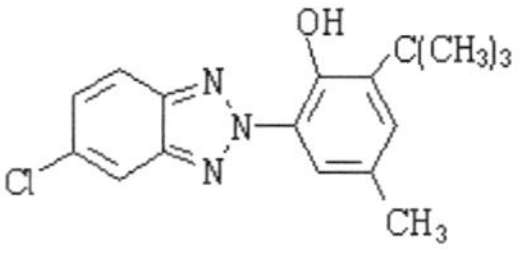	**Tinuvin 326** 2-(3'-tertbutyl-2'-hydroxy-5'-methylphenyl)-2H-5-chlorobenzotriazole
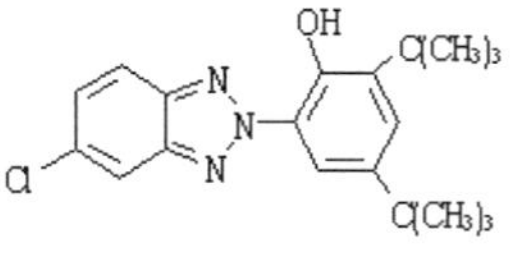	**Tinuvin 327** 2-(2'-hydroxy-3',5'-di-tertbutyl-phenyl)-2H-5-chlorobenzo-triazole

Table 4. The investigated light-stabilizers, conventional (Chimassorb 81, Tinuvin P, 326, 327) and polymerizable ones (Cyasorb UV 2098 and VP Sanduvor PR-31)

The spectra of chloroformic extracts corresponding to TEGDMA polymers and Bis-GMA2 polymers stabilized with Cyasorb UV 2098, VP Sanduvor PR 31 and Chimassorb 81, obtained by photochemical polymerization and chemical polymerization are presented in figure 5, 6, 7, 8. In order to determinate the UV stabilizer amount extracted from the polymer matrix it was used a characteristical wavelength of the stabilizer over 300 nm, where the residual monomer and other additives don't absorb.

The percent of extracted stabilizers resulted from the chemically and photo-chemically initiated polymers obtained by the polymerization of TEGDMA and Bis-GMA2 and by copolymerization of Bis-GMA2/TEGDMA are presented in tables 6 and 7.

No.	UV Stabilizer	λ_1	λ_2	λ_3	ε_1	ε_2	ε_3
A.	Chimassorb 81	242 nm	286 nm	325 nm	16233	24466	16266
B.	CyasorbUV 2098	244 nm	285 nm	325 nm	9660	15233	9600
C.	VP Sanduvor PR 31	312 nm			29960		
D	Tinuvin P	250 nm	301 nm	341 nm	8026	16984	20123
E	Tinuvin 326	253 nm	313 nm	354 nm	11020	27880	31040
F	Tinuvin 327	254 nm	315 nm	354 nm	2599	11013	12334

Table 5. Ultraviolet extiction coefficients of investigated UV stabilizers

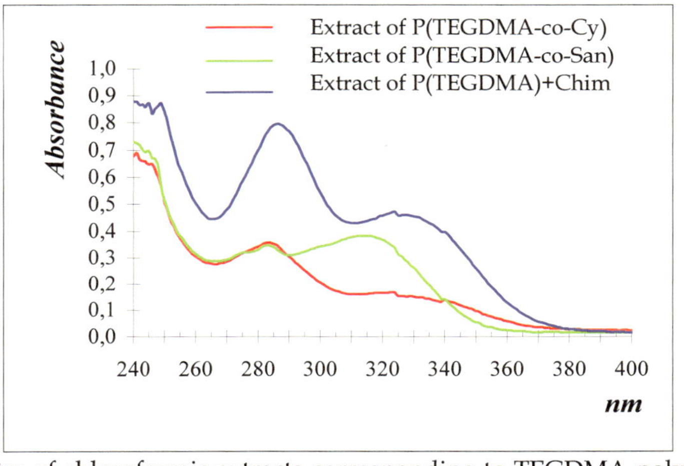

Fig. 5. The spectra of chloroformic extracts corresponding to TEGDMA polymers stabilized with Cyasorb UV 2098, VP Sanduvor PR 31 and Chimassorb 81, obtained by photochemical polymerization

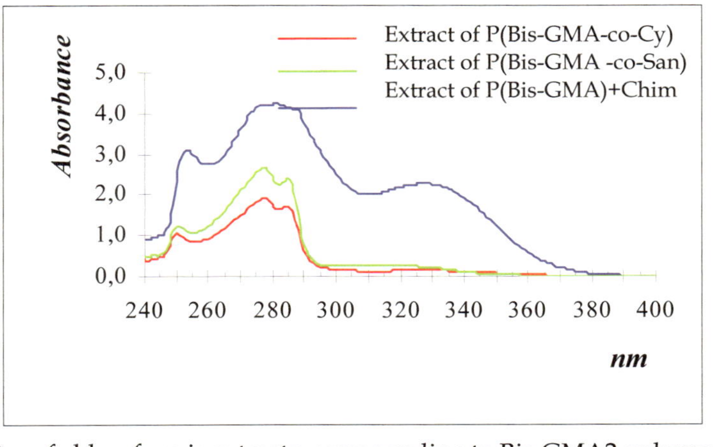

Fig. 6. The spectra of chloroformic extracts correponding to Bis-GMA2 polymers stabilized with Cyasorb UV 2098, VP Sanduvor PR 31 and Chimassorb 81, obtained by photochemical polymerization

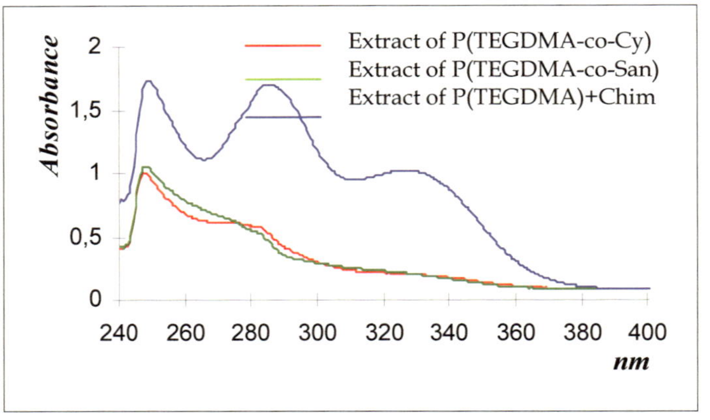

Fig. 7. The spectra of chloroformic extracts corresponding to TEGDMA polymers stabilized with Cyasorb UV 2098, VP Sanduvor PR 31 and Chimassorb 81, obtained by chemical polymerization

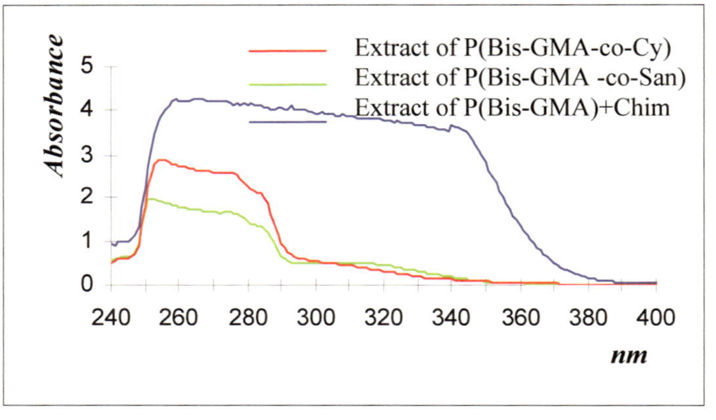

Fig. 8. The spectra of chloroformic extracts corresponding to Bis-GMA 2 polymers stabilized with Cyasorb UV 2098, VP Sanduvor PR 31 and Chimassorb 81, obtained by chemical polymerization

Chemically initiated polymerization	P(TEGDMA)	P(Bis-GMA2)	P(Bis-GMA2-co-TEGDMA)
Cyasorb UV 2098	6.5%	6.5%	7.15%
VP Sanduvor PR-31	3.56%	6.16%	5.85%
Chimassorb 81	20.22%	59.03%	48.75%
Tinuvin P	28.25%	40.3%	38.22%
Tinuvin 327	27.0%	35.7%	34.8%
Tinuvin 326	27.55%	39.8%	38.8%

Table 6. The percent of extracted stabilizer p% resulted from the chemically initiated polymers

Photohemically initiated polymerization	P(TEGDMA)	P(Bis-GMA2)	P(Bis-GMA2-co-TEGDMA)
Cyasorb UV 2098	5.43%	4.86%	5.01%
VP Sanduvor PR-31	3.65%	4.46%	3.65%
Chimassorb 81	30.2%	44.00%	37.88%
Tinuvin P	22.4%	38.44%	28.77%
Tinuvin 327	24.0%	39.02%	33.12%
Tinuvin 326	25.4%	38.5%	31.5%

Table 7. The percent of extracted stabilizer p% resulted from the photo-chemically initiated polymers

The results of the extraction tests have underlined the net superiority of the polymerizable stabilizers Cyasorb UV 2098 and VP Sanduvor PR-31 compared to the conventional stabilizers, the percent of polymerizable stabilizer extracted from the polymer matrix being 4 to 10 times smaller than the one of the conventional stabilizer.

3.3.3 Determination of mechanical properties

High conversion is not a goal in itself, however. If the monomer is very flexible and not sufficiently bulky, the degree of conversion will be high, but the mechanical properties will be poor. High mechanical properties are particularly important in the process of mastication, where the greatest loads are experienced by the tooth structures. Compressive strength (CS), diametral tensile strength (DTS) and flexural strength (FS) are important parameters for the comparison of dental restorative composites.
The composition of the resins of the experimental composites used for mechanical tests is presented in table 8:

No.	Bis-GMA0 %	Bis-GMA 1 %	Bis-GMA 2 %	Bis-GMA 3	TEGDMA %
R1.	65				35
R2.		65			35
R3.			65		35
R4.				65	35

Table 8. Composition of the resins in the experimental composites

Compressive strengths, diametral tensile strengths and flexural strengths of the new experimental dental composites as function of composition is shown in figure 9. The results presented in figure 9 show that the mechanical properties are influenced by the type of Bis-GMA monomer presented in resin composite.
The compressive strengths obtained for experimental composites were relatively high, and range between 240 and 279 MPa. Bis-GMA2-based composite present the highest values for CS, followed by Bis-GMA0-based composite and Bis-GMA1-based composite. Bis-GMA3 composite showed the lowest value for CS.
The diametral tensile strengths are about in the same ranges for all four composites (37.2 MPa for Bis-GMA3-based composite, 42.0 MPa in the case of Bis-GMA2-based composite and 45.0 MPa for Bis-GMA0-based composite and 45.3 MPa for Bis-GMA1 composite).

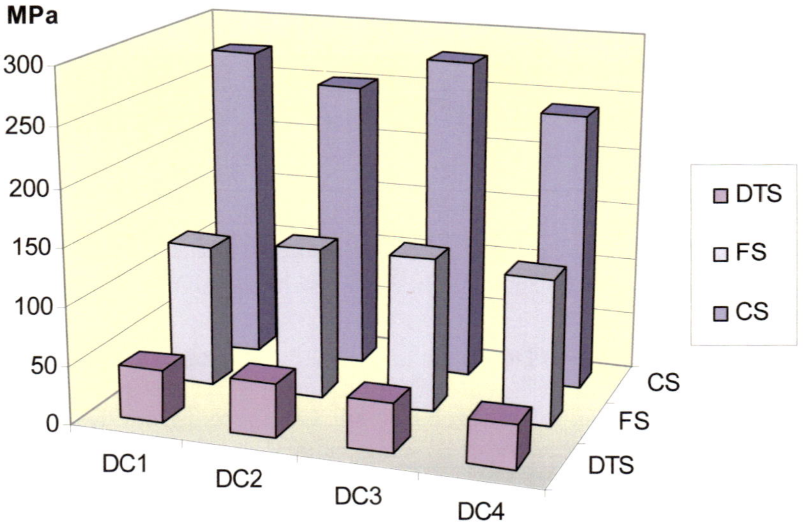

Fig. 9. The strength properties of experimental resin composites

The flexural strengths are high, being over 120 MPa for all experimental composites. The highest values were obtained for Bis-GMA2-containing composite (132.0 MPa), followed by Bis-GMA1-based composite (130.8 MPa). Among the synthesized Bis-GMA type oligomers, Bis-GMA2 leads to the obtaining of composites with highest CS and FS. The highest value for DTS was obtained in the case of Bis-GMA1-based composite (45.3 MPa).

Recently, relevant mechanical properties of Bis-GMA/TEGDMA and some Bis-GMA analogues/TEGDMA composite resins were evaluated as a function of filler loading and polymeric matrix composition (Pereira SG, 2007).

3.3.4 Determination of flexural strengths of fiber reinforced composites (FRCs)

The metal alloys used to make substructures that reinforced crowns and fixed prostheses are strong and rigid, but they are not esthetic and present corrosion phenomena, leading to an allergic reaction of some patients. FRCs materials could be used in any situation in which the main purpose is esthetics, because they present excellent esthetic properties. The mechanical properties increase very much because of the reinforcing fibers, getting ahead of the values obtained for the metal alloys, which are used today in dentistry (Freilich et al., 2000).

In the present work, a series of fiber reinforced composites (FRCs) using Bis-GMA2-based resin matrix and continuous unidirectional bundles, respectively woven E-glass fibers was obtained (table 9). The resins contained 65% Bis-GMA2 and 35% TEGDMA. The continuous unidirectional bundle contained about 800 E-glass fibers with 15±1μm diameter. Woven glass fibers contained about 200 fibers having 2 mm width and 9±1μm diameter. The fibers were silanated with A-1100 and A-174 silanes, and silanated and treated with PMMA, respectively. After initial photopolymerization, the specimens were postcured by barro-thermal treatment at 135^{0}C temperature and 60 psi pressure, for 20 minutes, using a 'belleGlass" oven. The flexural strengths for the light cured and post-cured experimental FRCs were determined (fig. 10 and fig. 11)

Code	Resin	E-glass fibers		Reinfor - cement % wt
		Type	Treatment	
Control	65% Bis-GMA2+ 35%TEGDMA	-	-	-
10FlRC-S	65% Bis-GMA2+ 35%TEGDMA	10 continuous unidirectional bundles	Silane	43.50
20FlRC-S	65% Bis-GMA2+ 35%TEGDMA	20 continuous unidirectional bundles	Silane	71.62
10FlRC-SP	65% Bis-GMA2+ 35%TEGDMA	10 continuous unidirectional bundles	Silane+ PMMA	55.77
15FlRC-SP	65% Bis-GMA2+ 35%TEGDMA	15 continuous unidirectional bundles	Silane +PMMA	78.98
2FwRC-S	65% Bis-GMA2+ 35%TEGDMA	2 woven E-glass fibers	Silane	29.67
2FwRC-SP	65% Bis-GMA2+ 35%TEGDMA	2 woven E-glass fibers	Silane +PMMA	42.54
3FwRC-SP	65% Bis-GMA2+ 35%TEGDMA	3 woven E-glass fibers	Silane+PMM A	47.28

Table 9. Composition of the experimental fiber-reinforced composites

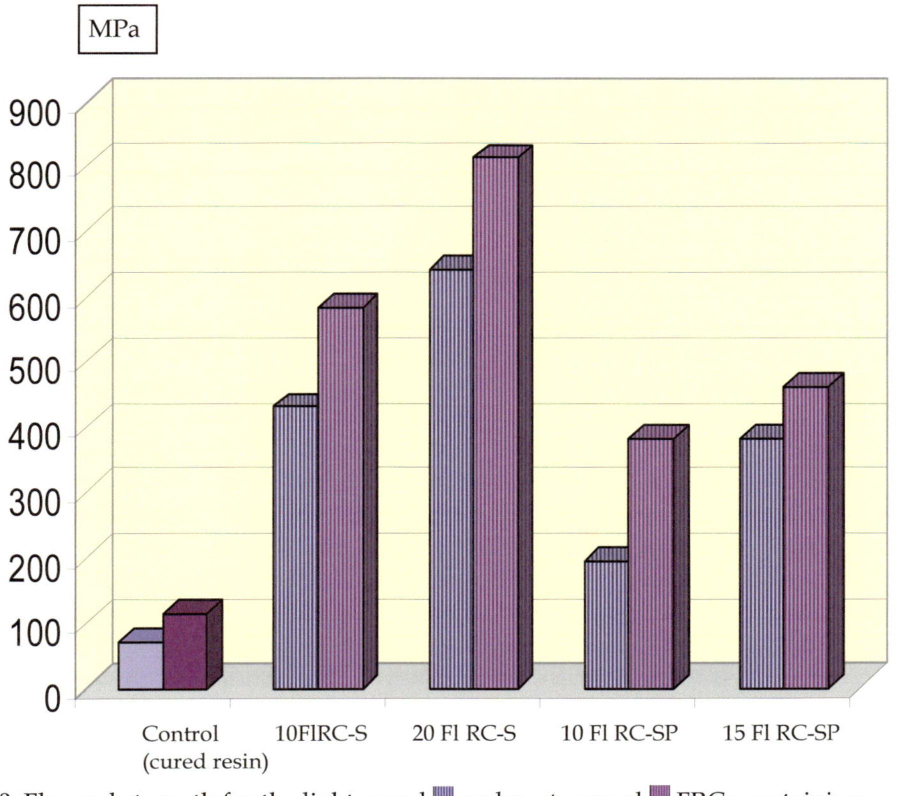

Fig. 10. Flexural strength for the light-cured and post- cured FRCs containing unidirectional glass fibers

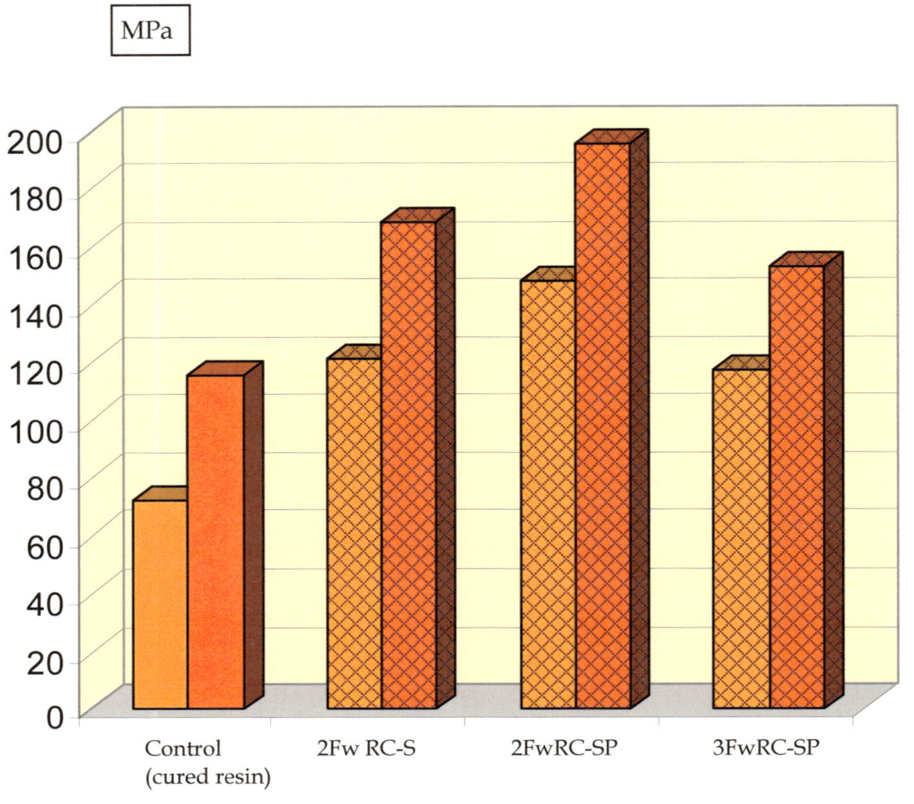

Fig. 11. Flexural strength for the light-cured and post- cured FRCs containing woven glass fibers

The amount and the type of fibers have a major influence upon the flexural strength values. The FRCs based on linear bundles of fibers recorded higher values for flexural strength than the FRCs based on woven E-glass fibers at the same reinforcement. All the post-cured FRCs presented higher flexural strength values comparing with the corresponding light-cured FRCs. The highest values for flexural strength were obtained in the case of post-cured FRC containing 20 silanated continuous unidirectional bundles of E-glass fibers (812 MPa).

The fracture appearances of the experimental FRCs as the consequence of the flexural stresses have been investigated by SEM. As one can see from the figure below, in the case of FRCs with woven fibers, the resin can not penetrate well enough between the fibers in order to ensure a strong connection of the fibers with the polymer matrix in the cured composites. This behavior was observed in the SEM images of fracture composites when the phenomenon of delamination of fibers from the polymeric matrix after the mechanical stress appeared. Within the limits of our experimental test, fibers with PMMA prepolymer coating were beneficial only for manipulative reasons. Mechanical studies and microscopic images revealed that the best mechanical properties were obtained in the case of silanized glass fibers without a PMMA preimpregnation

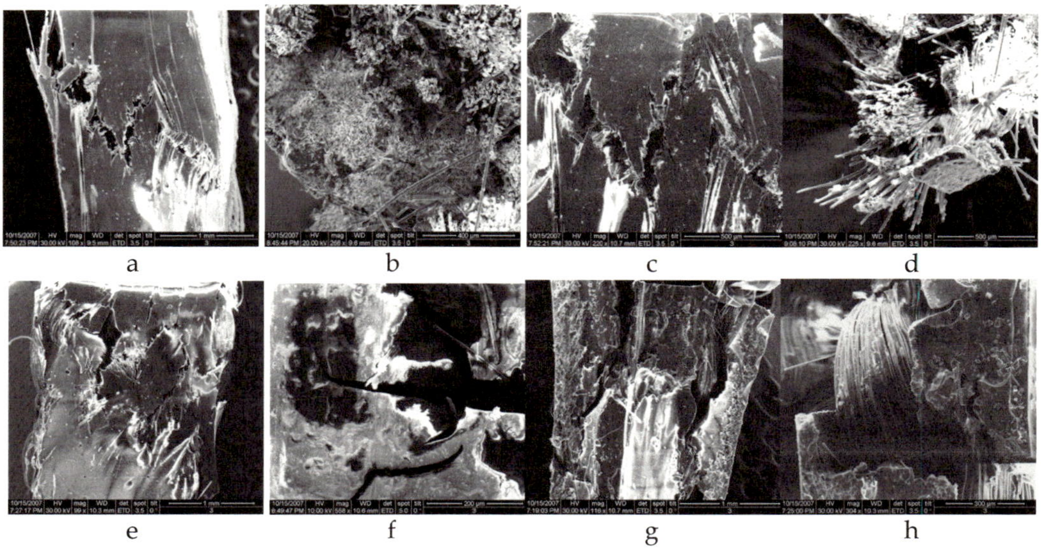

Fig. 12. SEM photomicrographs. Fracture appearance of 3FwRC-SP (a, b), 15FlRC-SP(c, d), 2FwRC-S (e, f), 3FwRC-SP(g, h)

3.3.5 Determination of color changes

Δa*, **Δb*** **ΔL***coordinates and **ΔE*** color changes of the stabilized samples of light-cured composites after exposure 1, 7 and 30 days to UV light are presented in figure 13a, b, c, d.

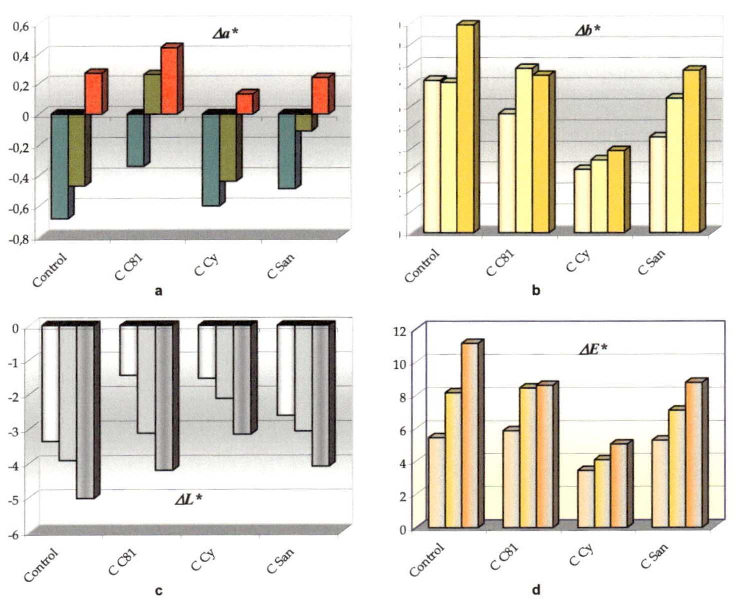

Fig. 13. **Δa*** (a), **Δb*** (b), **ΔL*** (c) coordinates and **ΔE*** (d) color changes of the stabilized samples of light-cured composites after exposure 1day (the first bar), 7 days (the second bar) and 30 days (the third bar) to UV light

The light curing resins were prepared using 65% Bis-GMA2 monomer and 35% TEGDMA. The resins have been stabilized using conventional and polymerizable stabilizers (Chimassorb 81, CyasorbUV 2098 and VP Sanduvor PR 31). As a control was used the resin composite that don't contain a stabilizer in the composition.

An increasing in b coordinate (yellowing) and a decreasing in L coordinate (darkening) of the experimental composite samples after the UV illumination was recorded. The color changes were more intense for the samples that don't contain a stabilizer than for the stabilized composites. Quantitatively, the color changes ranged from 2.3 to 11.6 CIELAB units. The samples stabilized with Chimassorb 81 and VP Sanduvor PR 31 presents relatively the same color change after 30 days of illumination.

The effectiveness of Cyasorb UV 2098 in time can be explained by the high compatibility of this stabilizer with the monomer mixture and by its capacity of polymerization with the monomers.

3.3.6 Determination of water sorption

The resin matrix used in dental materials absorbs saliva from the environment. Saliva is mostly water (more than 99 %) and less than 1% solids (macromolecules like proteins, sugars, calcium, sodium and chloride, amino acids, free glucose etc). Water can penetrate in the gaps of the polymer network, reaching around functional groups which are able to deliver hydrogen bonds. Sorption of saliva may produce physical and chemical processes and can lead to a change of structure and properties of polymer network. The sorption of saliva is accompanied by dimensional changes, deleterious effects on the structure and function of dental polymers and composites. This behavior affects the restoration life, therefore water sorption of these materials should be as low as possible.

Water sorption values for the experimental resin composites with the composition presented in the 3.3.3. Paragraph, after 7 days water storage are presented in table 9:

From the table 9 one can be noticed that the sorption values obtained for all resin composites are in the limits of ISO 4049 International standard. The lowest value was obtained for the Bis-GMA2-based composite. This behavior could be due to the lower quantity of hydroxyl groups from the Bis-GMA2 superior oligomers contained in the same amount of monomer mixture comparing to Bis-GMA0. Bis-GMA1-based composite and Bis-GMA3-based composite proved to absorb more water than Bis-GMA0 and Bis-GMA2.

Resin Composite	Water sorption $\mu g/mm^3$
DC1	9.45
DC2	11.88
DC3	7.36
DC4	19.25

Table 9. Water sorption of the experimental resin composites

3.3.7 Investigation of the adhesion of the resin composites to dental tissues

The polymerization shrinkage produces contraction stresses generally concentrated at the adhesive interface (Davidson & Feilzer, 1997; Davidson et al., 1984) leading to generating cuspal deflection (Causton et al., 1985; Cara et al., 2010) and cracks in the surrounding dentin and enamel (Jorgensen et al., 1975). Polymerization shrinkage stress may compromise the dental tissue/material interface leading to bacterial microleakage (Kemp-Scholte & Davidson, 1990) and ultimately to marginal discoloration, marginal debondings, secondary caries, pulpal inflammation, or necrosis (Letzel, 1989).

The basic mechanism of bonding the resin composites (RCs) to enamel and dentin is essentially an exchange process involving replacement of minerals removed from the hard dental tissue by resin monomers, which upon setting become micro-mechanically interlocked in the acid-induced porosities. Beside the micro-mechanical adhesion, the physical and chemical interactions with the tooth structure are also important. Because the RC pastes are very viscous, they are not able to adhere to the enamel and dentin surfaces, so it is necessary to use an adhesive system of low viscosity before applying the RCs.

Generally, an adhesive system involves three steps:

1. *Conditioning step* in which the smear layer is removed, the dentin and enamel are demineralized and the dentin tubules are opened;
2. *Priming step* in which the hydrophylic surface layer is transformed into a hydrophobic surface layer. The primer can be represented by M-R-X molecules, where M-methacrylic group, X-functional group that can create a chemical bond to the dentin surface, R-spacer. Theoretically, the primer should be applied in a monomolecular layer, should be chemically bond to dental tissues before the bonding is applied and should cover the surfaces with polymerizable groups.
3. *Application and curing the low viscosity adhesive resin* (bonding);

The present paragraph presents the adhesion of the experimental resin composites to dental tissues by investigating the tooth / RCs interfaces using the scanning electron microscopy method.

For the adhesion test, there were prepared three experimental adhesive systems containing the same etchant and the same primer but different bondings. The primer was based on the acrylic acid/itaconic acid copolymer grafted with methacrylic groups . The bondings were prepared for each composite material. Each bonding contain the base monomer that was present in the corresponding composite, Bis-GMA0, Bis-GMA1 or Bis-GMA2 respectively. As control was used the three step adhesive OptiBond FL from Sybron-Kerr Orange, CA, USA.

The main differences between the experimental adhesives and the control adhesive were the composition and the curing procedure of the primers and bondings. The bonding of OptiBond FL is a filled resin, unlike the experimental bondings that are unfilled resins. In the case of control adhesive, the primer was only dried and the bonding was light cured for 30 s. In the case of the experimental adhesives, the primer was light cured for 30 s in order to obtain an elastic layer fixed on the tooth surfaces that can reduce the stress at the interface, and the bondings were light cured for 60 seconds.

The composites were prepared using 75% Bis-GMA-type monomer and 25% TEGDMA mixture. Herculite resin composite was used as a control.

The composition and the procedure of applying the three steps adhesive systems were presented in table 11.

SEM photomicrographs of dentin surface treated with three step Bis-GMA1 based adhesive system and Bis-GMA1 based composite are presented in fig. 14 (1a-1c thick layer and 1d-1f thin layer).

Material /Manufacturer	Composition	Procedure
Control adhesive system OptiBond FL Sybron-Kerr Orange, CA, USA	*Etchant*- 37%phosphoric acid *Primer*:2-hydroxyethylmethacrylate (HEMA) Glycerophosphate-dimethacrylate (GPDM), Phtalic acid monomethacrylate (MMEP), ethanol, water, initiators *Bonding* : Bis-phenol A diglycidylmethacrylate (Bis-GMA) 2-hydroxyethylmethacrylate (HEMA) Glycerol-dimethacrylate (GDMA), initiators Barium-aluminium borosilicate glass, , Disodium hexa-fluoro-silicate Fumed silica	*Etchant*: apply 15s. rinse 15s. and dry *Primer*: apply one coat, dry 5 s. *Bonding*: apply one coat, light cure for *30 s*
3 Experimental adhesive systems	*Etchant*- 37% phosphoric acid *Primer*: Copolymer of acrylic acid/itaconic acid grafted with methacryloiloxiethylcarbamoyl-aminoethanol HEMA, water, ethanol, initiators *Bondings*: Bis-GMA0/Bis-GMA1/ Bis-GMA2, TEGDMA, HEMA initiators	*Etchant*: apply 15s. Rinse 15s. and dry *Primer*: apply one coat 15 sec., dry 5 s. , cure 30 s *Bonding*: apply one coat, light cure for 60 s

Table 11. The composition and the procedure of applying the three steps adhesive systems

SEM photomicrographs of dentin surface treated with three step Bis-GMA2-based adhesive system and Bis-GMA2- based composite are presented in fig. 15. SEM photomicrographs of dentin surface treated with three step Bis-GMA0-based adhesive system and Bis-GMA0-based composite are presented in fig. 16. SEM photomicrographs of dentin surface treated with Optibond FL adhesive system and Herculite composite are presented in fig. 17.

From figure 14 one can observe that the adhesive layer follows the shape of dentin surface having about the same width. Signs of strong interactions between the resin and dentin can be observed. From figure 15 it can be seen that the adhesive layers don't have the same width along the dentin surface. Some dislocations appeared. Figure 16 shows that the adhesive layer is not uniform. No interactions can be seen. From figure 17 one can observe that the adhesive layer is relatively uniform and the filler particles presented in the bonding can be seen.

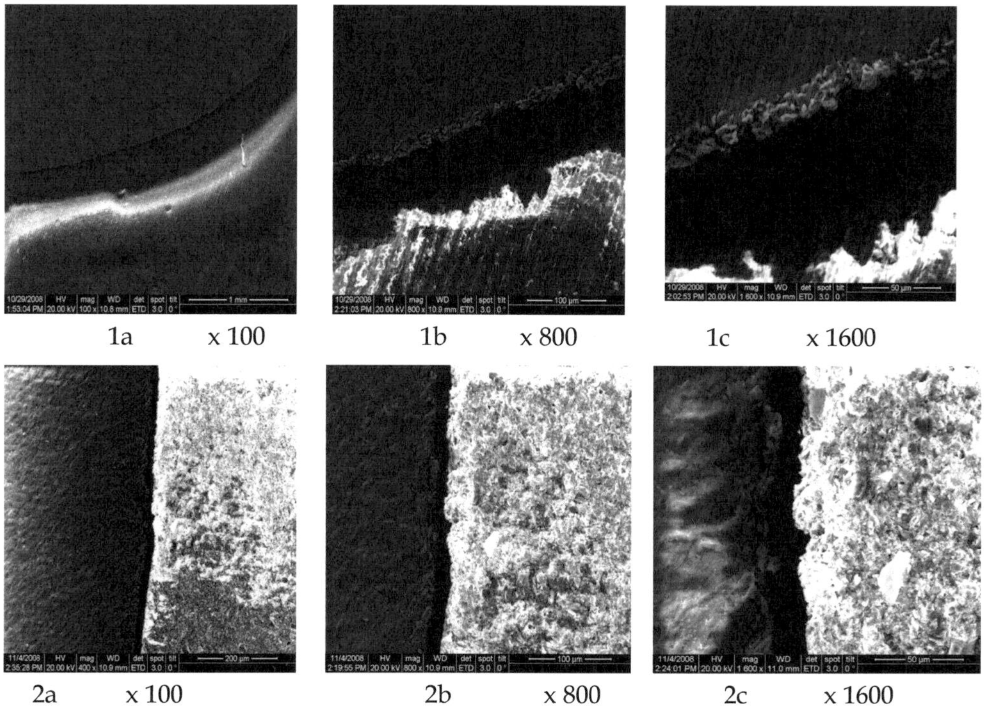

Fig. 14. SEM photomicrographs of dentin surface treated with three step Bis-GMA1 based adhesive system (1a-1c thick layer and 1d-1f thin layer) and Bis-GMA1 based composite

Fig. 15. SEM photomicrographs of dentin surface treated with Bis-GMA2 based adhesive and Bis-GMA2 based composite

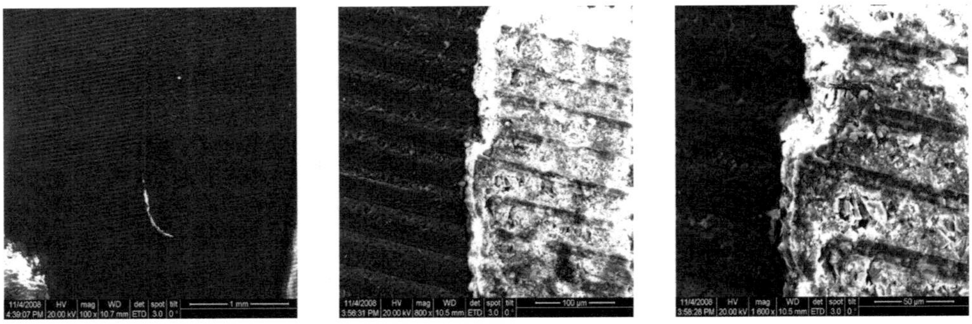

3a x 100 3b x 800 3c x 1600

Fig. 16. SEM photomicrographs of dentin surface treated with Bis-GMA0 based adhesive and Bis-GMA0 based composite

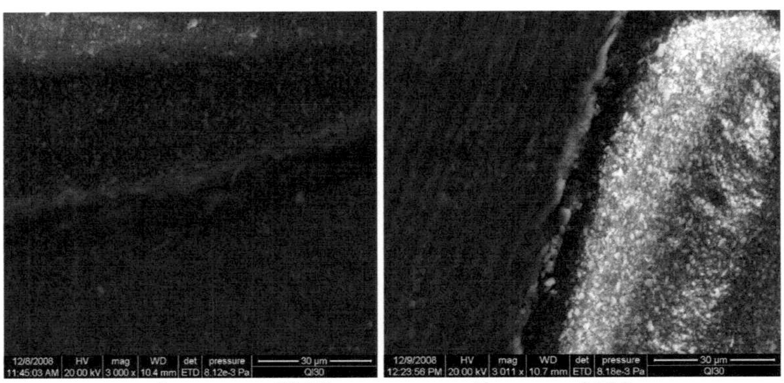

4a x 3000 4b x3000

Fig. 17. SEM photomicrographs of dentin surface treated with Optibond FL adhesive system and Herculite composite

4. Conclusions

Within the limits of our experiments, we came to the following conclusions.
Among the experimental Bis-GMA type oligomers, Bis-GMA2 leads to the obtaining of composites with highest compressive and flexural strengths and the lowest value for water sorption. The values obtained for polymerization shrinkage show that the resin based on Bis-GMA1 monomer has a polymerization shrinkage that represents 2/3 of the value presented by commercial Bis-GMA-based resin. This behavior can be correlated with the good adhesion to tooth tissues that was obtained in the case of using the adhesive system and the light curing composite based on Bis-GMA1 monomer. The polymerizable light-stabilizer Cyasorb UV 2098 was more effective in time than the conventional low molecular weight stabilizer Chimasorb 81. Further studies are needed to prepare resins with different Bis-GMA1/Bis-GMA2/TEGDMA ratio and different filler loading in order to obtain dental materials with optimum performance.

5. Acknowledgments

This work was supported by Romanian National Project PNCDI2 No. 72168/2008 (FLORODENT), Parteneriate Program .TGN dedicates her contribution to the memory of Prof. S. Kalachandra, an unforgettable friend and colleague.

6. References

Asmussen, E. (1982). Factors affecting the quantity of remaining double bonds in restorative resin polymers. *Scand J Dent Res*,Vol. 90, No. 6, pp. 490-496, ISSN 0029-845X

Bowen R. L., (1962). U. S. Patent, 3.066.112.

Bowen R. L., Marjenhoff W. A. (1992). Dental Composites/Glass Ionomers: The Materials. *Adv Dent Res*, Vol. 6, pp. 44-49, ISSN 2229-4112

Buruiana, T., Buruiana E. C., Melinte V., Pohoata V., Prejmerean C., Moldovan M. (2008). Newurethane dimethacrylates fortesting in dental applications. *J. Optoelectron. Adv. Mater.*, Vol. 10, No. 4, pp. 969-974, ISSN 1454-4164

Cara R.R., Nicola C., Prejmerean C., Sava S., Baciut G., Baciut M., Bran S., Bondor C., Prodan D., Moldovan M., Buruiana T., Patrascu I. (2010). Influence of Bis-GMA Derivative Monomer-Based Particulate Composite Resins on the Cuspal Deformation and Microleakage of Restored Teeth. Particulate Science and Technology, Vol. 28, pp. 191-206, Publisher Taylor &Francis, ISSN 0272-6351

Causton, B. E., Miller B., Sefton J. (1985). The deformation of cusps by bonded posterior composite restorations: An in vitro study. *Br. Dent. J.*, Vol. 159, pp. 397-400, ISSN 0007-0610

Davidson, C. L., Feilzer A. J. (1997). Polymerisation shrinkage and polymerisation shrinkage stress in polymer-based restoratives. *J Dent.*, Vol. 25, pp. 435-440, ISSN 0300-5712

Davidson, C. L., De Gee A. J., Feilzer A. J. (1984). The competition between the composite-dentin bond strength and the polymerization contraction stress. *J Dent Res*, Vol. 63, No. 12, pp. 1396-1399, ISSN 0022-0345

Ferracane JL. (1994.) Elution of leachable components from composites. *J. Oral Rehabil* 1994; 21: 441-452, ISSN 0305-182X

Freilich, M.A., Meiers, J. C., Duncan, J. P., Goldberg, A. J. (2000). Fiber-reinforced composites in clinical dentistry, Quintessence Pub. Co., ISBN 0867153733

Grassie N. (1966). Degradation, In : *Encyclopedia of polymer science and technology*, Mark HF, Gaylord NG, Bikales NM, vol. 4. pp. 647-716, John Wiley & Sons Inc., ISBN 10 0470569700, New York

Holter, D., Frey H., Mulhaupt R., (1997). Branched bismethacrylates based on Bis- GMA-A systematic route to low shrinkage composites. *Polym. Preprints*, Vol. 38, No.2, pp. 84-85, ISSN 0032-3934

Jorgensen, K. D., Asmussen, E., Shimokobe, H.. (1975). Enamel damages caused by contracting restorative resins. *Scand J Dent Res.* Vol. 83, pp. 120-122, ISSN 0029-845X

Kemp-Scholte, C.M., Davidson C.L. (1990). Marginal integrity related to bond strength and strain capacity of composite resin restorative systems. *J.Prosthet. Dent.*, Vol. 64, pp. 658-664, ISSN 0022-3913.

Khatria, C.A., Stansbury J. W., Schultheisz C. R., Antonucci J. M. (2003). Synthesis, characterization and evaluation of urethane derivatives of Bis-GMA. *Dent. Mater.* , Vol. 19, No.17, pp. 584-588, ISSN: 0109-5641

Letzel, H. (1989). Survival rates and reasons for failure of posterior composite restorations in multicentre clinical trial. *J. Dent.* Vol. 17, pp. 10-17, ISSN: 0300-5712

Mozner N., Salz U. (2001). New developments of polymeric dental composites. *Prog. Polym. Sci.*, 26, pp. 535-576, ISSN 0079-6700.

Moszner, N., U. K. Fischer, J. Angermann, and V. Rheinberger. (2008). A partially aromatic urethane dimethacrylate as a newsubstitute for Bis-GMA in restorative composites. *Dent. Mater.* 24: 694-699, ISSN: 0109-5641

Nunes, T.G., Pires, R., Perdigão, J., Amorim, A, Polido, M. , (2001). *Polymer,* Vol. 42, pp. 8051-8054, ISSN 0032-3861

Nunes, T.G., Guillot, G., Pereira, S.G., Pires, R. (2002). *Journal of Physics D-Applied Physics;* Vol. 35, No.11, pp.1251-1257, ISSN0022-3727.

Nunes T.G., Ceballos L., Osorio R., Toledano M. (2005). Spatially-Resolved Photopolymerization Kinetics and Oxygen Inhibition in Dental Adhesives, *Biomaterials*, Vol. 26, No. 14, pp. 1809, ISSN 0142-9612

Nunes T. G., Pereira S.G. , Kalachandra S. (2008). Effect of Treated Filler Loading on Dental Resin Photopolymerization Inhibition and Contraction, *J. Materials Science: Materials in Medicine*, Vol. 19, pp. 1881-1889, ISSN: 0957-4530

Pereira S.G., Reis N., Nunes T.G. (2005). Spatially resolved studies on the photopolymerization of dimethacrylate monomers, *Polymer*, Vol. 46, pp. 8034-8044, ISSN 0032-3861

Pereira, S. G., Nunes T. G., Kalachandra S. (2002). Low viscosity dimethacrylate comonomer composition [Bis-GMA and CH3 Bis-GMA] for novel dental composites, *Biomaterials* Vol. 23, pp. 3799-3806, ISSN 0142-9612

Pereira S.G., Osorio R, Cabrerizo-Vilchez, M.A., Toledano, M., Nunes T.G., Kalachandra, S., (2007). *Dent Mater*, Vol. 23, No.10, pp. 1189-1198, ISSN: 0109-5641

Peutzfeld, A. (1997). Resin composites in dentistry: The monomer systems, *Eur. J. Oral Sci.*, Vol. 105, pp. 97-116, ISSN 0909-8836

Prejmerean, C., Vezsenyi, M., Moldovan, M., Grecu, R., Musat, O., Brie, M., (2000). The synthesis of Bis-GMA oligomers and their effects upon the properties of some experimental dental composites, *Rev Roum Chim,* Vol. 45, No 6, pp. 567-577, ISSN 0035-3930

Ruyter, I.E., Gyorosi P.P., (1976). An infrared spectroscopic study of sealants, *Scand J Dent Res,* Vol. 84 pp.396-400, ISSN0029-845X

13

Wollastonite-Based Chemically Bonded Phosphate Ceramic Composites

Henry A. Colorado[1,2], Clem Hiel[3,4], Thomas Hahn[1] and Jenn Ming Yang[1]
[1]*University of California Los Angeles, Materials Science and Engineering*
[2]*Universidad de Antioquia, Mechanical Engineering*
[3]*Composites Support and Solutions*
[4]*MEMC-University of Brussels (VUB)*
[1,3]*United States*
[2]*Colombia*
[4]*Belgium*

1. Introduction

Chemically Bonded Ceramics (CBCs) have been extensively used many applications. These include: radiation shielding systems, nuclear waste solidification and encapsulation; high temperature structural applications; composites and biomedical applications. The objective of this chapter is to present the science, manufacturing and mechanical properties of the Wollastonite ($CaSiO_3$) based Chemically Bonded Phosphate Ceramics (Wo-CBPCs) composites, Fig. 1. In general, the CBPCs belong to the broader field of Chemically Bonded Ceramics (Della Roy, 1987; Jeong & Wagh, 2002; Wagh, 2004) where the name CBPC refers to ceramic materials that reach their final mechanical properties by chemical reactions at low temperatures (typically less than 300°C) instead of the high temperature processing (by thermal diffusion or melting) as is normally done in traditional ceramics and cements. The Wo-CBPCs are composite materials themselves, they belong to Chemical Bonded Ceramics (CBCs) consolidated by an acid base reaction (Wilson & Nicholson, 1993). Wo-CBPCs are multiphase materials with silica, Wollastonite and brushite grains in a matrix of amorphous calcium phosphates. Reinfocing materials such as graphite nanoplateles, glass fibers and carbon fibers have been successfully incorporated into - CBPCs, realizing high performance composites.

1.1 Wo-CBPCs and traditional cements

The manufacturing of CBPCs and Portland cement concrete has some similarities. In Portland cement concrete water, clinker and aggregates are used to consolidate the concrete, in CBPCs an acid formulation and a metallic oxide are mixed to fabricate the ceramic. In particular, for Wo-CBPC, a phosphoric acid formulation and Wollastonite powder are the main raw materials. However, there are several important advantages of using CBPCs as opposed to cements. Among them are better mechanical properties (compressive and flexural strength), lower density, ultra fast (controllable) setting time, and an environmentally benign process. Although cements are usually made at room temperature, clinker is processed at very high temperature (about 1450ºC). It is well known that between 900 and 1100 kg of CO_2 is emitted for every 1000 kg of Portland cement produced in the U.S. This quantity depends on the fuel type, raw ingredients used and the energy efficiency of the cement plant (EPA, 2005). Juxtaposed the Wo-CBPCs process is all performed at room temperature but other CBPCs, like traditional cements, may include high temperature processing of the raw materials (Wagh, 2004).

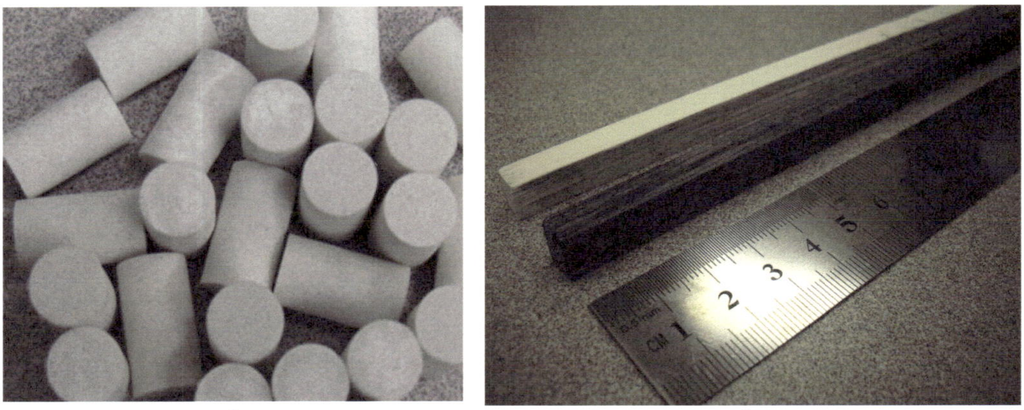

a b

Fig. 1. Wollastonite-based CBPCs, a) compression samples, b) pultruded bars of CBPCs reinforced with glass fibers (white) and with carbon fibers (black).

Fig. 2 illustrates the advantage of CBPCs compared Portland cement concrete. In ceramics and cements, the density has an important role in the mechanical response of the material. For Wo-CBPC it has been found that (Colorado et al., 2010a; 2010c) the porosity can be controlled by changing the mixing parameters. Changing the porosity is a way to change the density. It has been observed in CBPCs that density typically varies to values from 1.9 to 2.4 g/cm^3. This produces a variation in the compressive strength typically from 60 to 130 MPa (Colorado et al., 2010b). Similarly, Portland cement concrete has a density that typically ranges from 3 to 3.5 g/cm^3 which produces a variation in the compressive strength typically from 30 to 40 MPa. When these values are plotted as compressive strength over the density, we can see that Wo-CBPCs, and in general other CBPCs (it is possible to obtain pretty similar values), have superior advantages for applications of high strength and low density, in addition to the fabrication process. Then, CBPCs fill a gap between cements and ceramics. In particular, Wo-CBPCs have been developed for structural applications including fire environments (Colorado et al., 2010a). Sections below will present this material in detail as well as some of these applications.

The bonding in CBPCs is a mixture of ionic, covalent, and van der Waals bonding, with the ionic and covalent dominating; unlike in traditional cement hydration products, in which van der Waals and hydrogen bonding dominate. CBPCs can reach a compressive strength of 100 MPa in minutes, which can be controlled with additives and processing as will be presented later, whereas Portland cement based concrete reaches a compressive strength of about 20Mpa after 28 days. Furthermore, CBPCs do not require aggregates, which is important for applications where light weight is required. CBPCs density can range between 1.8 and 2.5 g/cm^3, which is particularly interesting when compared with Portland cement, for which in excess of 3.3 g/cm³. Even lower density values have been reported for other CBPCs.

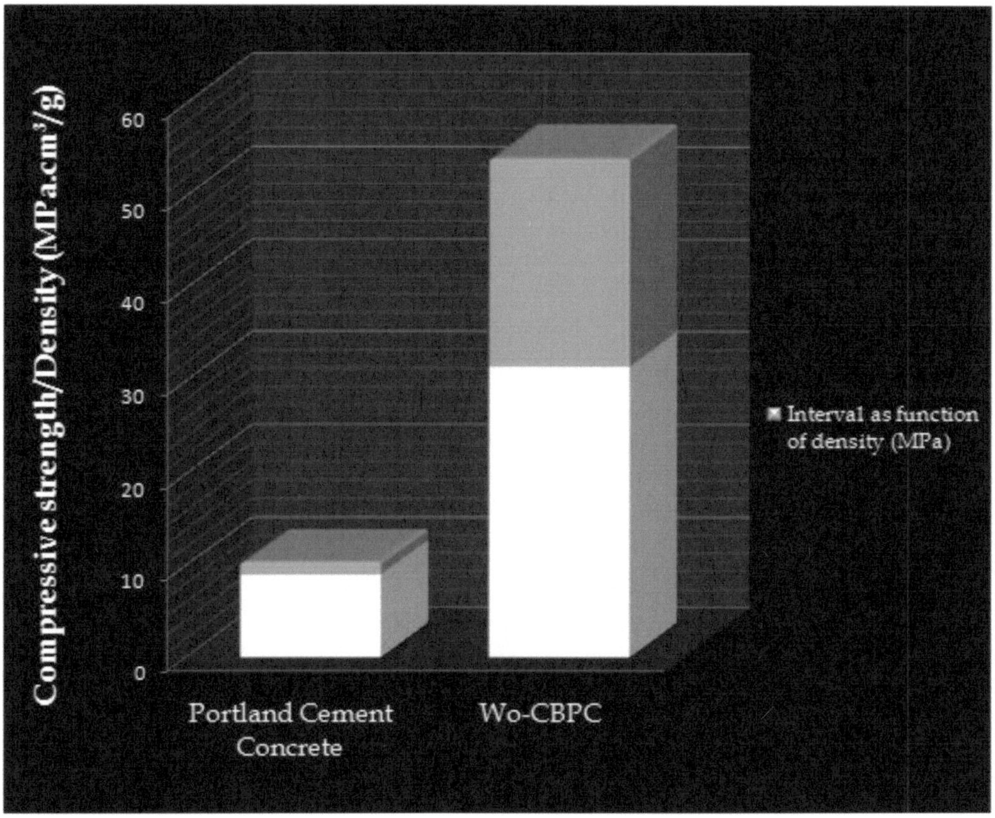

Fig. 2. Comparison of the ratio compressive strength over the density for Portland cement concrete and CBPCs.

Also, pH is an important factor for manufacturing of composite materials. We found that ratios of phosphoric acid to Wollastonite between 1.0 and 1.2 leads into a neutral product, with a pH of 7.0. This is particularly interesting when they are compared with Portland cement for which pH is in excess of 12, which does not allow the use of E-glass fibers as a reinforcement. The neutral pH is very convenient when either particles, fibers or any other reinforcements are introduced in the CBPCs. This opens up many applications related to infrastructure repair such as roads, bridges and pipes.

On the other hand, the pH can be controlled out of this range by putting more or less acidic solution, open applications in which a strong reaction is required. CBPCs can be used to encapsulate nuclear waste. There are several solutions for treatment of nuclear wastes, however, CBPCs are unique due that can set very fast minimizing the risk of contamination and irradiation, process can be conducted at room temperature (better when compare with a glass encapsulation in which manufacturing involves high temperature, which can leads in evaporation of hazardous materials), and pH can be controlled easily. By using different acids, oxides and adding different additives, different efficient materials for structural and shielding applications can be fabricated.

The chemical processing in CBC's allows them to be inexpensive in high volume production. Therefore CBCs are environmentally benign and fill the gap between cements and ceramics. The fabrication of conventional cements and ceramics involve energy consuming high temperature processes, which adversely affect the environment.
The sections below will present the manufacturing and science of the CBPCs with different reinforcements.

1.2 Structure formation of Wo-CBPCs

The CBPCs form by acid-base reactions between an acid phosphate and Wollastonite. This reaction is exothermic. When the phosphoric acid formulation and the Wollastonite powder mixture are stirred, the sparsely alkaline oxides dissolve and an acid base reaction is initiated. The result is a slurry that hardens in a ceramic product. The setting is the result of gelation by salt formation and the Ca^{+2} cations are extracted from the calcium silicate. Then, we have two main process involved, the release of cations from the Wollastonite and the interaction with the acidic solution. Details of this proposed mechanism of formation have been extended presented (Wagh & Jeong 2003a, 2003b and 2003c; Wagh, 2004).
When powders of metal oxides (for Wo-CBPC is $CaSiO_3$) are stirred in solvent such as an acid-phosphate solution (For Wo-CBPC is H_3PO_4), they dissolve slowly in the solvent and release cations (For Wo-CBPC are Calcium cations) in the acidic solution. These cations react with the phosphate anions within the solvent and form a precipitate of salt molecules.
Under the right conditions mainly controlled by the pH (in Wollastonite and phosphoric acid, the Wo-CBPC can be consolidated over a wide range of pH), these molecules form an ordered structure that grows into crystals that conform the CBPC. Thus, CBPC formation is a result of the following three steps (Wagh, 2004):

a. The acid phosphates dissolve in water, release phosphate anions, and form an acid-phosphate solution of low pH.
b. The oxides dissolve gradually in the low pH solution and release cations.
c. The phosphate anions react with the newly released cations and form a coordinated network and consolidate into a CBPC.

Several equations for the products when mixing Wollastonite with phosphoric acid have been proposed (Mosselmans et al., 2007) for molar ratios r of the phosphoric acid to Wollastonite powder between 0.39 and 1.66. For r between 0.39 and 1, only one type of calcium phosphate (brushite) is formed. The formation reaction of brushite ($CaHPO_4 \cdot 2H_2O$) is represented in Eq. (1).

$$CaSiO_3 + H_3PO_4 + (1+x)H_2O = SiO_2.xH_2O + CaHPO_4.\ 2H_2O \quad (1)$$

For r between 1 and 1.66, three different calcium phosphates (brushite, monetite and calcium dihydrogenphosphate monohydrate) can be formed, depending on the molar ratio r. The reactions leading to monetite ($CaHPO_4$) and calcium dihydrogenphosphate monohydrate ($Ca(H_2PO_4)_2.H2O$) are depicted in Eqs (2) and (3), respectively.

$$CaSiO_3 + H_3PO_4 = SiO_2.yH_2O + CaHPO_4 + (1-y)H_2O \quad (2)$$

$$CaSiO_3 + 2H_3PO_4 + zH_2O = SiO_2.zH_2O + Ca(H_2PO_4)_2.H_2O \quad (3)$$

It has been found that a ratio of an special phosphoric acid formulation to Wollastonite powder between 1.0 and 1.2 leads into a neutral product, with a pH of 7.0 (Colorado et al., 2010b).

2. Case of study: Wollastonite-based CBPCs

2.1 Raw materials

The Wo-based CBPCs are fabricated with Wollastonite (Calcium silicate, $CaSiO_3$) powders and a phosphoric acid formulation. Wollastonite is a mineral with a triclinic structure. It is usually white and may contain small amounts of iron, magnesium, and manganese substituting for calcium. There are many industrial applications for Wollastonite. These include reinforcements for structural ceramics products like bricks and pipes; filler for paintings with improved resistant to environment; filler for plastics. Typical Wollastonite powders used in structural applications are showed in Table 1 (from NYCO). These have mean size distribution ranging from 3.5 to 15 µm.

Powder reference	Mean particle size (µm)	Surface Area (m^2/g) (BET)	Moisture (%)
M200	15	1.1	0.05
M325	10	1.3	0.20
M400	8	1.6	0.20
M1250	3.5	2.9	0.25

Table 1. Properties of typical Wollastonite powders as received

On the other hand, Phosphoric acid (H3PO4) also known as orthophosphoric acid, is an inorganic acid used in many applications. These include chemical etching of silicon nitride (Si_3N_4) in microfabrication, hydroponics to lower the pH of nutrient solutions, rolyte in copper electropolishing, high-performance liquid chromatography, ingredient that gives the bite taste in Coca-Cola and Pepsi sodas and additive to stabilize acidic aqueous solutions within a wanted and specified pH range.
For Wo-CBPCs, the phosphoric acid formulation was prepared in Composites Support and Solutions-CS&S. Thus, an attenuated phosphoric acid with additives was used in the fabrication of all Wo-CBPCs.

2.2 Manufacturing

The fabrication of CBPCs is one of its main advantages when they are compared to traditional ceramics and cements since high temperature processing can be completely annulled in the entire manufacturing.
Mixing process: For mixing the raw materials, a good mixing is needed and several techniques have shown efficiency. The traditional Mechanical Blade Mixing is enough for most of the structural applications. This is the more inexpensive method and it is available everywhere. Also, it is ideal for applications where you need a material with some porosity such as in fire walls or thermal insulation. Another useful method is Resonant Acoustic Mixing (RAM). This method is particularly useful for CBPCs when a dried mixing is a better option. A good mixing of different Wollastonite powder size, particles or short fibers as reinforcement and solid additives have been demonstrated with this method. The RAM is also effective in removing bubbles and pores from the liquid in CBPC manufacturing, which is particularly important for structural applications. Finally, Planetary Centrifugal Mixing is also effective in removing bubbles and pores.
Moulds: Different materials for moulds have been investigated. Polymer based materials like Teflon in combination with inorganic mould release is the best solution. Metals in

general bond very well to the CBPC which prevent them to be used as mould in most of the cases.

Temperature: in order to extend and control the pot life of the resin, since the reaction is exothermic, cooling the raw materials can help to extend the life for hours.

Aging: Exposing the Wollastonite powder to air also contributes to increase the pot life of the resin. Moisture reduces the reactivity in the powder surface. However, it has been found for that when they are exposed for more than a year, the compression strength of the Wo-CBPC is reduced.

a

b

Fig. 2. a) Typical Wollastonite, b) blade mixing used for the manufacturing of Wo-CBPCs.

2.3 Wo-CBPCs and its composite

The Wo-CBPCs are composite materials itself as confirmed with X Ray Diffraction (XRD) in Fig. 3. Several crystalline (Wollastonite and Brushite) and amorphous phases (silica and amorphous calcium phosphates) are shown. From one highly crystalline mineral (Wollastonie) is obtained a composite material with amorphous and crystalline phases.

In Fig. 4, phases were identified in the SEM-EDAX by X-ray mapping for the cross section of CBPCs. The composition distribution of calcium, silicon and phosphorous is shown in Fig. 4b, c and d respectively. Fig. 4b shows evidence of some Wollastonite ($CaSiO_3$) grains which transformed into silica (by giving up calcium) and some which did not transform. Fig. 4c shows silica glass as well as Wollastonite grains. Fig. 4d shows the calcium phosphate matrix. Numbers 1, 2 and 3 correspond to silica, Wollastonite and calcium phosphates respectively.

Fig. 5a shows a general SEM image of the CBPC with microcracks. Smooth zones correspond to silica and Wollastonite (partially or un-reacted) grains.

In some of these silica and Wollastonite grains, interfacial cracks appear. However, most of the cracks grow and propagate in the calcium phosphate matrix until they are eventually stopped by the grains. The silica and remaining Wollastonite particles provide nuclei for crystallization and can act as a second phase that reinforces the matrix. As the drying process occurs, shrinkage and micro crack growth increase.

As a result, interfacial cracks appear in grain interfaces between both silica and Wollastonite grains with the calcium phosphate matrix. A representation of the phenomena is presented in Fig. 5b. The drying process is natural because of the excess water present in the aqueous phosphoric acid formulation.

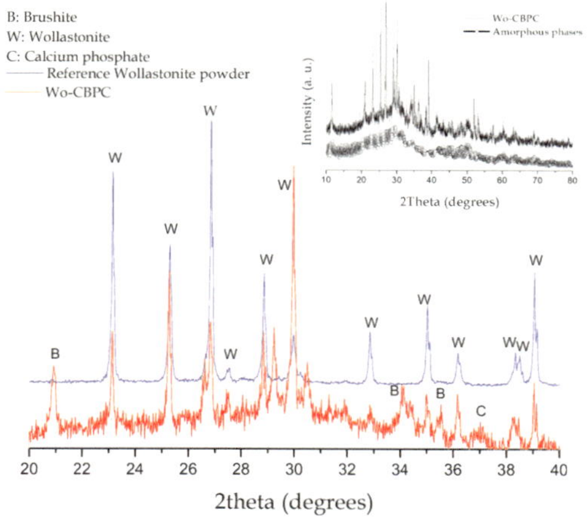

Fig. 3. XRD spectra for the Wo-CBPC, its Wollastonite raw material shown for comparison proposes and a separated plot showing crystalline and amorphous phases in Wo-CBPC.

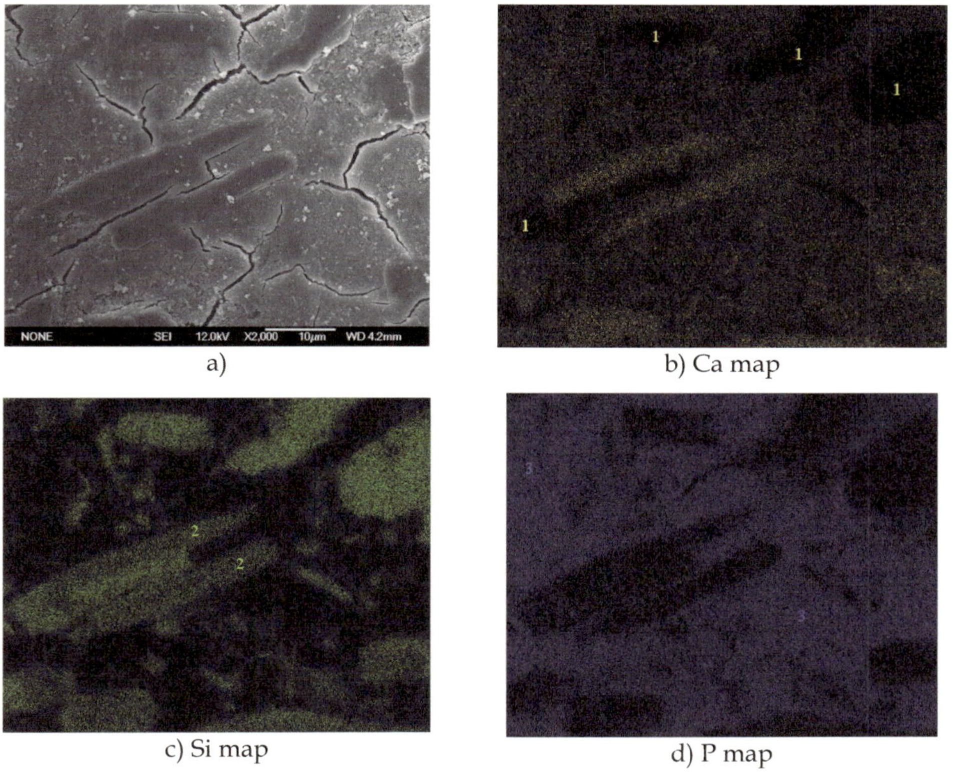

Fig. 4. X-ray maps of pultruded Wo-CBPC, a) topographical image; and b) Ca, c) Si and d) P concentration images. Numbers 1, 2 and 3 correspond to silica, Wollastonite and calcium phosphates respectively.

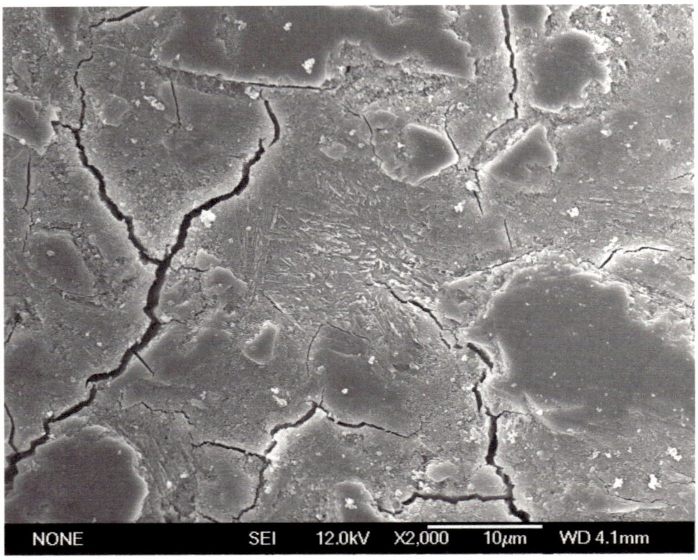

a

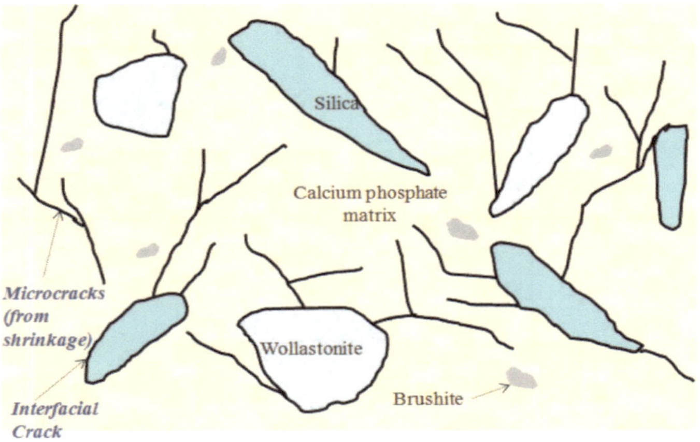

b

Fig. 5. Cross section view a) Wo-CBPC SEM image fabricated with M200 Wollastonite powder, b) representation of the Wo-CBPC before and after the shrinkage process starts.

2.4 Influence of aging and particle size distribution of wollastonite on the mechanical properties of CBPC

The compressive strength for the Wo-CBPCs made with new and two years old M200 Wollastonite powders are presented in Fig. 6a. The compressive strength for two year old Wo-CBPC made with M200 is 45% less than for the baseline material made with two week old M200. This effect has been associated with the moisture effect (Colorado et al., 2010b) on the Wollastonite surface, which decreases its reactivity when it is mixed with the aqueous phosphoric acid formulation.

The compressive strength for the CBPCs made with two years old M200, M400 and M1250 Wollastonite powders are presented in Fig. 6b. As expected, the highest values (not the mean) are observed for the small powder sizes (see Table 1). However, the error bars are slightly longer for the small powders sizes, which can be caused by increased porosity due to the high reactivity of the smallest powders. This is an effect of the increase in the surface area. As we explained before, the manufacturing problems associated with the fast reaction and viscosity can be reduced by cooling the raw materials at temperatures below 10°C. However, Fig. 6 indicates that the M400 particle size distribution creates a maximum compressive strength.

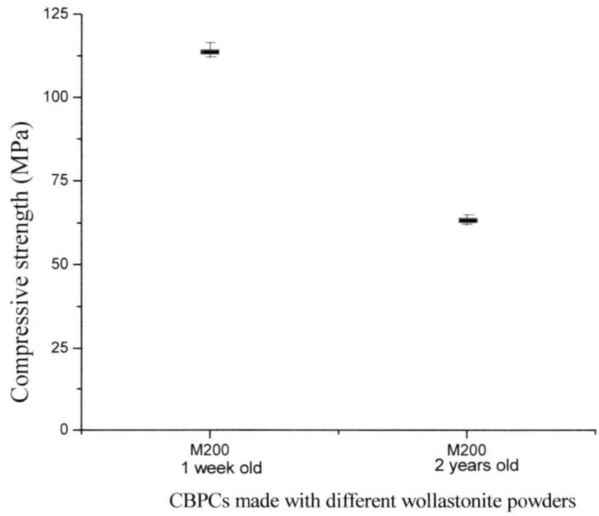

a

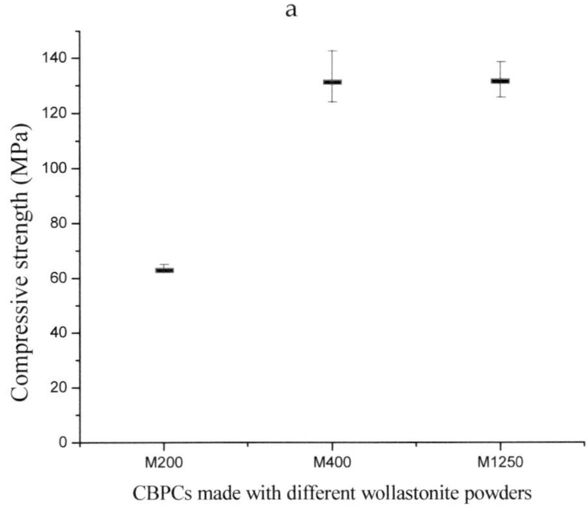

b

Fig. 6. Compressive strength for Wo-CBPCs made with a) M200 2 weeks and 2 years old Wollastonite powder; b) M200, M400 and M1250 Wollastonite powders all 2 years old.

3. Wo-CBPC composites for structural applications

3.1 Wo-CBPCs reinforced with graphite nanoplatelets (GNPs)

GNPs have been shown as effective to reinforce Wo-CBPCs matrix composites (Colorado et al, 2011). GNPs are shown in Fig. 7 a and b respectively. These GNPs have typically a diameter of 1μm and a thickness of 300nm. Two procedures were followed according to the GNP treatment, illustrated in Fig. 8.

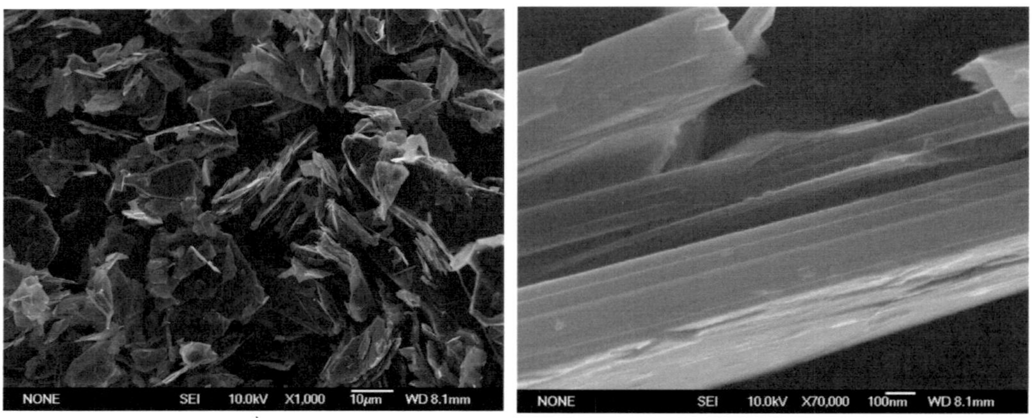

a) b)

Fig. 7. GNPs, a) group view at low magnification, b) side view at high magnification.

In the first process, GNPs were oxidized in open air at 450°C for 2 hours in a Barnstead Thermolyne 1400 furnace, using the method shown in Fig. 8a. A homogeneous mixture of 100g of wollastonite powder and 1.0 wt% of GNPs was obtained in the RAM at 60Hz after 5 min of mixing. Then 101g of these powders' mixture and 120g of the phosphoric acid formulation was kept in a closed container at 5° C to prevent water absorption. Then, the same procedure was followed in the Thinky mixer and the corresponding samples preparation.

In the second process, GNPs were functionalized in nitric acid at 115°C for 4 hours, using the method shown in Fig. 8b. Then filtered and 1.0 wt % of GNPs over the total mixture was mixed with phosphoric acid without washing the remaining nitric acid. It was found in this research that small concentrations of either sulfuric or nitric acid do not change the bending strength in CBPC significantly. Then, the same procedure was followed cooling the liquid and powder and, the mixing in the Thinky mixer and the corresponding samples preparation.

Fig. 9 shows GNPs (treated with nitric acid) reinforced Wo-CBPC matrix composite. The cracks generated in the shrinkage process (see black dashed arrows) were stopped by the GNPs grouped in the wide marked areas. A silica grain (primitive wollastonite powder) is free of cracks. However, cracks are around the entire interface with the matrix and some of them are growing far away, in the opposite direction to the GNPs' zone.

A representation of the Wo-CBPC reinforced with GNPs is shown in Fig. 10. The GNPs contribute to stop crack growing generated in the shrinkage process. Interfacial cracks in the silica and Wollastonite grains are also represented.

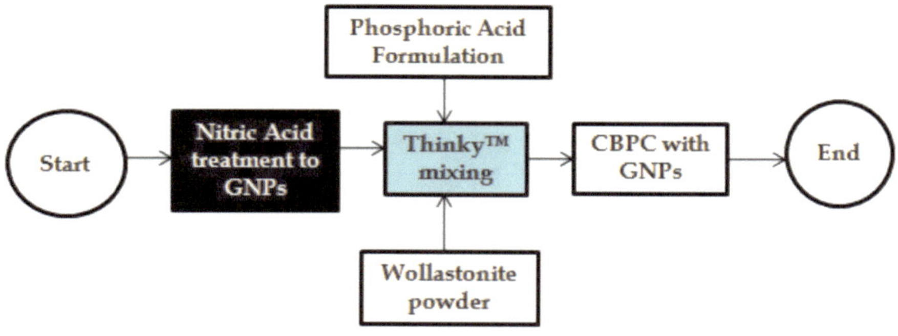

a

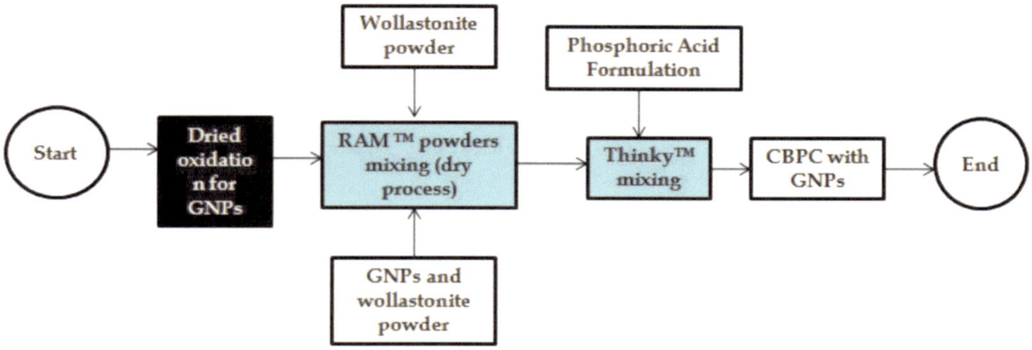

b

Fig. 8. Simplified flow charts of the processes for functionalize the GNPs, a) procedure using nitric acid, b) procedure using a dried oxidation on air.

The effect of GNPs in the bending strength of Wo-CBPC was investigated by the three point bending test. Results are presented in Fig. 11. A total of 60 samples were tested, 10 samples for the Wo-CBPCs (C-Ref) as reference , 20 samples for Wo-CBPCs with GNPs treated as dried oxidation (C-GNPs in DO), 20 samples for Wo-CBPCs with GNPs functionalized with nitric acid (C-GNPs in NA), and 10 samples of Wo-CBPCs with GNPs as-received (C-GNPs). The span length was 40mm, the samples dimensions were 90x8.0x6.0 mm^3 and the crosshead speed was 2.5 mm/min. Some samples were tested after drying at room temperature in open air for 5 days, and others after the drying process described above (50C-105-200C).

For all type of samples the bending strength decreased after the drying process. For GNPs oxidized in open air, maximum bending strength and mean for non dried samples were 26.7 and 23.3 MPa respectively. This result is almost three times the bending result from the CBPC. For GNPs functionalized with nitric acid, maximum bending strength and mean for non dried samples were 20.4 and 19.3 MPa respectively. The lower bending strength of CBPC reinforced with GNPs functionalized with nitric acid with respect to the GNPs oxidized in open air is believed related to the presence to a bigger amount of amorphous phases due to the 9.0 wt % of nitric acid added (over the total mixture), which is a fifth of the acid content in the mixture.

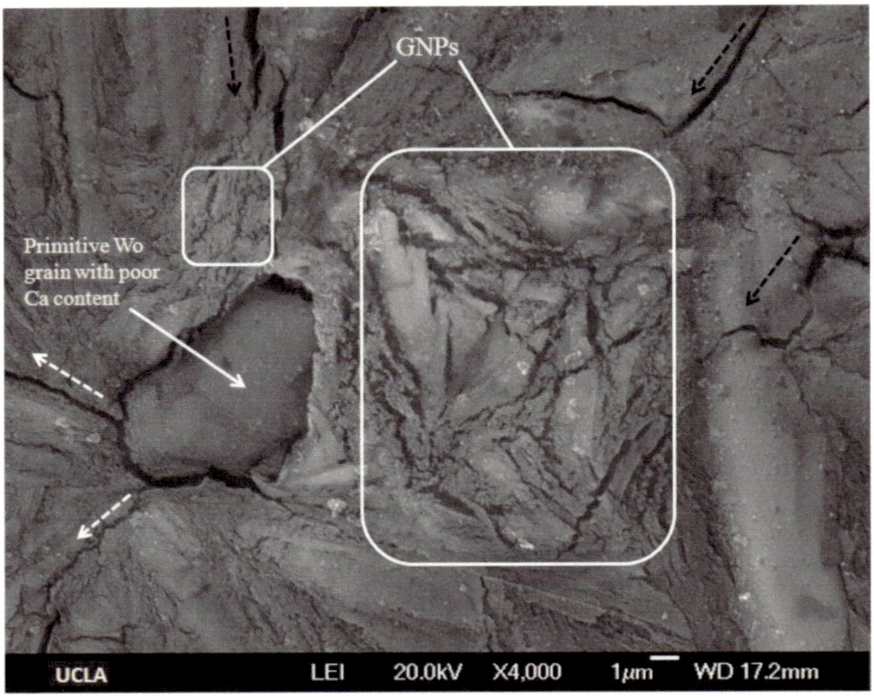

Fig. 9. SEM image of Wo-CBPC reinforced with GNPs showing crack growing stopped due to the GNPs.

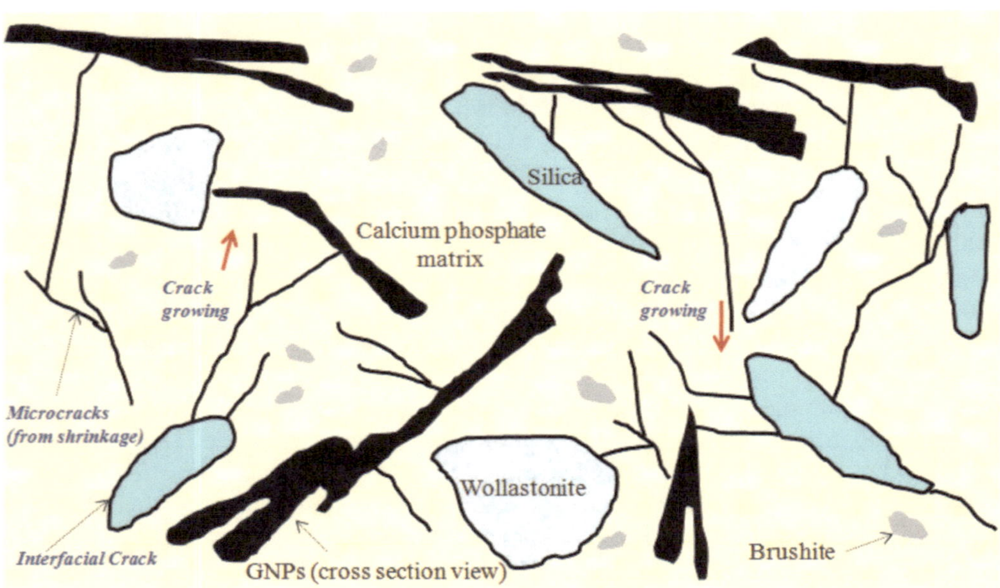

Fig. 10. Cross section view representation of the cracks (generated during the shrinkage process) growing and being stopped by GNPs.

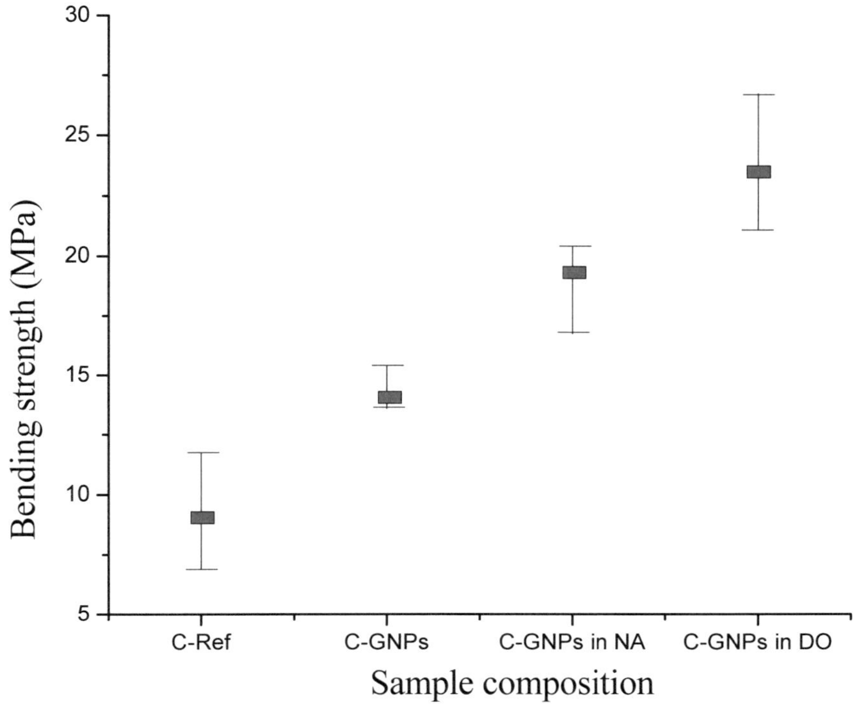

Fig. 11. Three point bending tests for Wo-CBPC and Wo-CBPC with GNPs

3.2 CBPCs reinforced with continuous fibers fabricated by pultrusion

For all samples, 120g of a phosphoric acid formulation and 100g of treated natural wollastonite powder (M325 from CS&S, see *Table 1*) were mixed to obtain a 1.2 ratio of liquid/powder. Resonant Acoustic Mixing (RAM) technique was used to mix the acidic solution with the powder.

The pultrusion process with both E-glass and carbon fiber reinforcement was developed at Composites Support and Solutions Inc (Colorado et al., 2010c). Composite Wo-CBPCs reinforced samples were fabricated with 15 volume % of fibers. The die-temperature was 110°C (230°F) and the pull speed was 30.5 cm/min (1 ft/min). Wo-CBPC, pultruded Wo-CBPC with glass fibers and pultruded Wo-CBPC with carbon fiber samples were tested in three point bending tests. or each type of sample, three specimens were tested through the three point bending test. These tests were performed on an Instron 4411 at a crosshead speed of 2.5 mm/min. A Teflon mold was used for the fabrication of the CBPC samples that were not pultruded, which was then covered with plastic foil to preserve humidity and minimize the shrinkage effects. Samples were released after 48 hours and then dried at room temperature in open air for 5 days. Pultruded samples were made in a metallic die as shown in Fig. 12a. All samples were kept in a furnace at 50°C for 48 hours, to stabilize the weight loss at similar values than those obtained for long times at 20°C. Fig. 12a shows the pultruded glass and carbon fiber composites with Wo-CBPC matrix fabricated with the die shown in Fig. 12a.

Fig. 13 shows cross section view images of the products shown in Fig. 12b.

a) b)

Fig. 12. a) Pultrusion die during manufacturing of Wo-CBPCs reinforced with glass fibers, b) pultruded bars of Wo-CBPCs reinforced with glass fibers (white) and with carbon fibers (black)

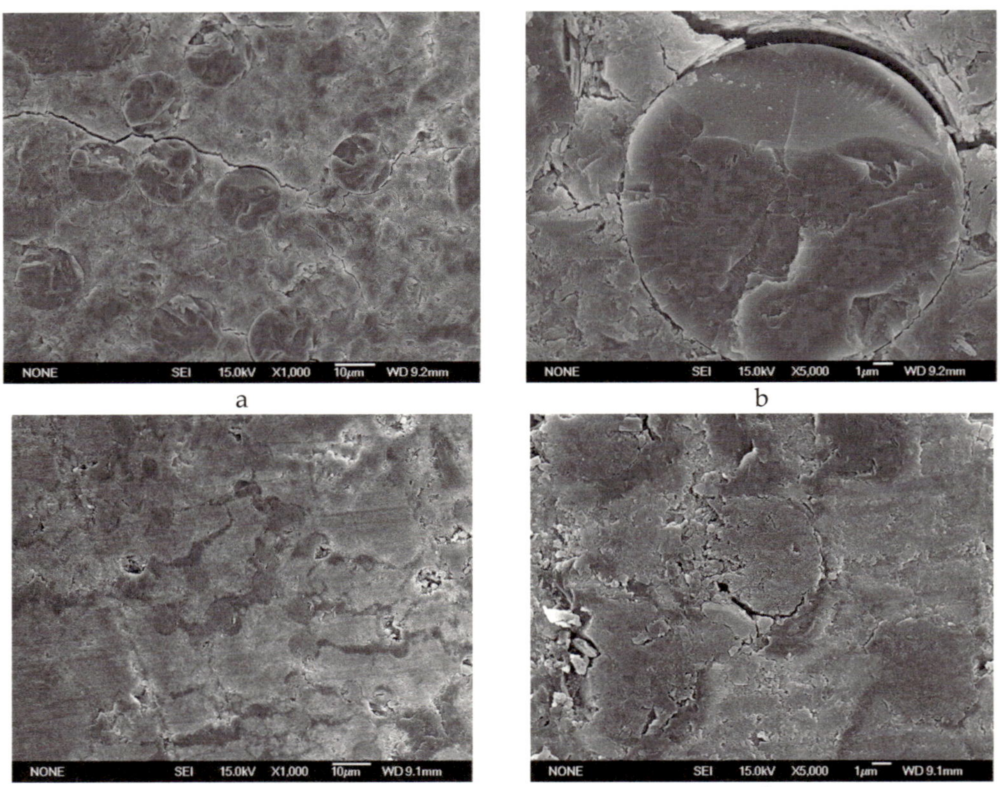

a b

c d

Fig. 13. SEM Cross section view images for Wo-CBPCs reinforced with a) and b) glass fibers; c) and d) carbon fibers; e) and f) basalt fibers; g) and h) SiC fibers.

Fig. 13a and b show the CBPC with glass fibers. Fig. 13a shows fibers homogeneously distributed and Fig. 13b, a magnification of Fig. 13a, shows CBPC fully impregnating the fibers. However, glass fibers appear almost debonded. Also, some cracks growing through the matrix and fiber-matrix interfaces.

On the other hand, Fig. 13c and d show the CBPC with carbon fibers. Fig. 13c is a cross section view image of the CBPC reinforced showing the carbon fibers homogeneusly distributed and a good matrix impregnation to fibers. Unlike the glass fibers reinforcement, debonding between the fiber and the matrix was not observed.. It is better shown by Fig. 13d, a magnification of Fig. 13c, which shows four fibers. Through whole sample, no cracks appear in matrix and matrix-fiber interfaces as well.

Fig. 14a and b show representation overviews of the pultruded glass fiber- and pultruded carbon fiber-reinforced chemically bonded phosphate ceramics respectively. The main difference between these two systems is basically the absence of interfacial cracks in the composite with carbon fibers.

Typical curves for the three point bending strength obtained for the Wo-CBPC pultruded fiber composites are presented in Fig. 16.

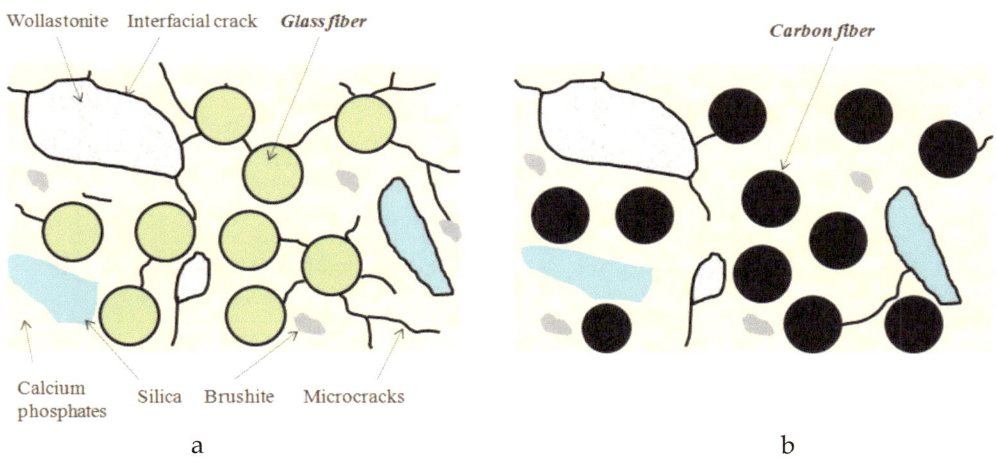

Fig. 14. Cross section view representation of the Wo-CBPCs a) reinforced with glass fibers, b)reinforced with carbon fibers

4. Composite applications for Wo-CBPCs

CBCs have been extensively used for multiple applications. They include: dental materials (Chow & Eanes, 2001), bone tissue engineering (Barinov & Komlev, 2008), shielding gamma and neutron radiation, nuclear waste solidification and encapsulation (Singh et al., 2000; Wagh, 2004, Chattopadhyay, 2003), electronic materials (Young & Dimitry, 1990), tooling for advanced composites (Miller & Wise, 1990), high temperature (Colorado et al., 2010a; Gulgun et al., 1994), composites with fillers and reinforcements (Wagh, 2004; Colorado et al., 2010a, 2010b, 2010c; Colorado et al., 2011). Above have been shown some of the Wo-CBPCs composites fabricated with GNPs and fiber reinforcements. Other structural applications involving high temperatures are the fire wall structures fabricated from Wo-CBPCs reinforced with different fibers. These are shown in Fig. 17. Electrical transformers are expensive equipments usually installed in groups with short space in between. If fire starts in one, it can be easily transferred to the other in a chain reaction. Wo-CBPCs have been successfully applied as a protection from fire in this case. In addition, because of the curing can be easily reduced to short periods and they are fabricated as cements, in situ fabrication and repair is also commercially feasible.

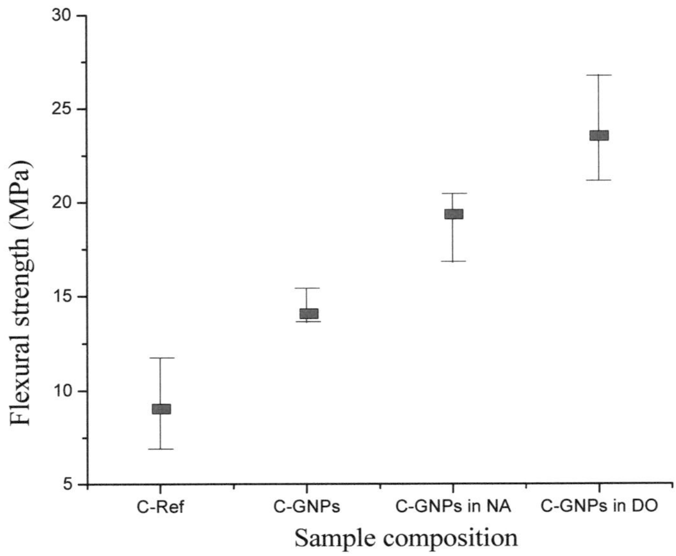

Fig. 15. Flexural strentgh results for the Wo-CBPCs pultruded fiber glass and graphite composites

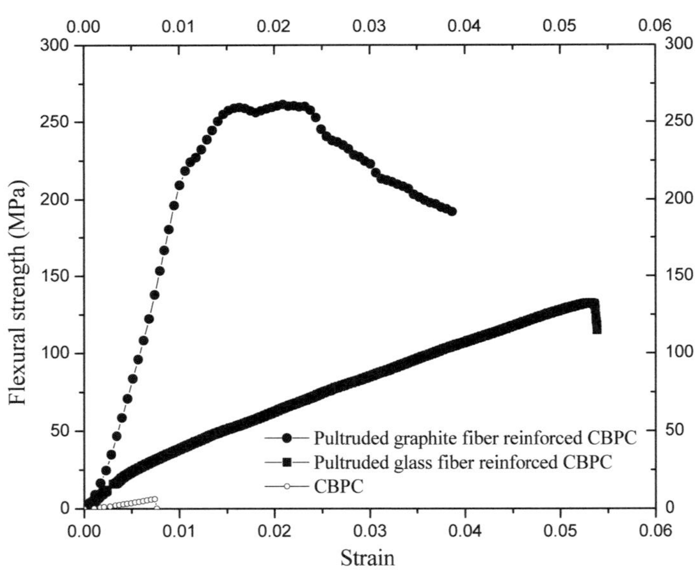

Fig. 16. Three point bending tests for pultruded Wo-CBPC with glass and carbon fibers

a b

Fig. 17. Composites fabricated from Wo-CBPCs, a) firewalls made of CBPCs composite panels reinforced with fibers for protection of electric transformers, b) detail of the panels installation.

Wo-CBPCs have also shown good adhesion properties with wood products and other ceramics and cements. This opens up the spectrum of applications for Wo-CBPC as structual materials like fire resistant houses or other wood-based constructions.

5. Summary

An overview of Wo-CBPCs composite materials have been presented in this chapter. Their formation, materials and main parameters controlling their manufacturing have been discussed. Also, it has been shown that Wo-CBPCs are composite materials itself with several crystalline (Wollastonite and Brushite) and amorphous phases (silica and amorphous calcium phosphates). Particle size distribution and Wollastonite age has a big impact in both setting time and compressive strength. Finally, progress in applications such Wo-CBPCs reinforced with graphite nanoplatelets and Wo-CBPCs with both carbon and glass fibers fabricated by a pultrusion process has been presented. These applications, in addition to all the advantages that CBPCs have with respect to traditional ceramics and cements, make possible Wo-CBPC as an ultimate material for structural applications.

6. Acknowledgment

The authors wish to thank to the NIST-ATP Program through a grant to Composites and Solutions Inc. (Program Monitor Dr. Felix H. Wu) and to Colciencias from Colombia for the grant to Henry A. Colorado.

7. References

Barinov, S. and Komlev, V. (2008). *Calcium Phosphate based bioceramics for bone tissue engineering*. Trans Tech Publications Ltd, Switzerland.

Chattopadhyay, S. (2003). *Evaluation of chemically bonded phosphate ceramics for mercury stabilization of a mixed synthetic waste*. National Risk Management Research Lab. Cincinnati, Ohio.

Chow, L. C. and Eanes, E. D. (2001). *Octacalcium phosphate. Monographs in oral science*, vol 18. Karger, Switzerland.

Colorado, H. A.; Hiel, C. and Hahn, H. T. (2010a). Chemically bonded phosphate ceramic composites under thermal shock and high temperature conditions. *Society for the Advancement of Material and Process Engineering*, May 17-20, 2010. Seattle, Washington USA.

Colorado, H. A.; Hiel, C. and Hahn, H. T. (2010b). Influence of Particle Size Distribution of Wollastonite on the Mechanical Properties of CBPC's (Chemically Bonded Phosphate Ceramics). *Materials Science & Technology, Conference and Exhibition*. October 17-21 in Houston, Texas USA. Ceramics Transactions.

Colorado, H. A.; Hiel, C. and Hahn, H. T. (2010c). Pultrusion of glass and carbon fibers reinforced Chemically Bonded Phosphate Ceramics. *Journal of Composites Materials*. In printing.

Colorado, H. A.; Hiel, C. and Hahn, H. T. (2011). Chemically bonded phosphate ceramics composites reinforced with graphite nanoplatelets. *Composites: Part A*, 42, 376–384.

Della Roy, M. (1987). New Strong Cement Materials: Chemically Bonded Ceramics, *Science*, Vol. 235: 651-58.

EPA. (2005). *Compilation of Air Pollutant Emission Factors,* Volume 1: Stationary Point and Area Sources, AP-42, Fifth Edition, U.S. Environmental Protection Agency, Office of Air Quality Planning and Standards, Research Triangle Park, North Carolina, January, 1995.

Gulgun, M. A.; Johnson, B. R. and Kriven, W. M. (1994). Chemically bonded ceramics as an alternative to high temperature composite processing. *Mat. Res. Soc. Symp. Proc.* Vol. 346. Materials Research Society.

Jeong, S. Y. and Wagh A. S. (2002). *Chemical bonding phosphate ceramics: cementing the gap between ceramics, cements, and polymers*, Argonne National Laboratory report.

Miller, L. and Wise, S. (1990). Chemical bonded ceramic tooling for advanced composites. *Materials and Manufacturing Processes*, Volume 5, Issue 2, 229-252.

Mosselmans, G.; Monique, B.; Willem, R.; Wastiels, J.; Leermakers, M.; Rahier, H.; Brughmans, S.; and Van Mele, B. (2007). *Journal of Thermal Analysis and Calorimetry* Vol. 88 3 723.

Singh, D.; Jeong, S. Y.; Dwyer, K. and Abesadze, T. (2000). Ceramicrete: a novel ceramic packaging system for spent-fuel transport and storage. Argonne National Laboratory. *Proceedings of Waste Management 2K Conference*, Tucson, AZ.

Wagh, A. S. (2004). *Chemical bonded phosphate ceramics*, Elsevier, Argonne National Laboratory, 283. ISBN: 0-08-044505-5, USA.

Wagh A. S. and Jeong, S. Y. (2003a). Chemically bonded phosphate ceramics: I, a dissolution model of formation, *J. Am. Ceram. Soc.*, 86 (11): 1838-44

Wagh A. S. and Jeong, S. Y. (2003b). Chemically bonded phosphate ceramics: II, warm-temperature process for alumina ceramics, *J. Am. Ceram. Soc.*, 86 (11): 1845-49.

Wagh A. S. and Jeong, S. Y. (2003c). Chemically bonded phosphate ceramics: III, reduction mechanism and its application to iron phosphate ceramics, *J. Am. Ceram. Soc.*, 86 (11): 1850-55.

Wilson, A. D. and Nicholson, J. W. (1993). *Acid based cements: their biomedical and industrial applications.* Cambridge, England, Cambridge University Press.

Young, J. F. and Dimitry, S. (1990). Electrical properties of chemical bonded ceramic insulators. *J. Am. Ceram. Soc.*, 73 [9] 2775-78.

14

Microstructure and High-Temperature Strength of 9CrODS Ferritic Steel

Shigeharu Ukai
Hokkaido University
Japan

1. Introduction

Oxide-dispersion-strengthened (ODS) ferritic steel is known to be candidates for advanced fission and fusion materials that require excellent radiation resistance and high-temperature capabilities (Odette et al., 2008). We focus on the development of 9CrODS ferritic steel with a composition of Fe-9Cr-0.13C-2W-0.2Ti-0.35Y_2O_3 (mass%) (Ohtsuka et al., 2004, 2005a, 2005b, 2006, 2007; Ukai et al., 1998, 2002a, 2002b, 2003, 2007, 2009, 2011; Yamamoto et al., 2010, 2011). This steel contains nano-size Y-Ti complex oxide particles. Its microstructures can be easily controlled by a reversible α/γ phase transformation with a remarkably high driving force of a few hundred MJ/m^3. The 9CrODS ferritic steel is recognized as a composite-like material composed of a residual ferrite and tempered martensite. The residual ferrite is a metastable phase, and high-temperature strength of 9CrODS steel is significantly ascribed to the presence of the metastable type of the residual ferrite. In this article, formation process of the composite-like structure and resultant superior high-temperature strength of the 9CrODS steel are reviewed with incorporating current result of manufacturing in terms of newly developed thermo-mechanical processing.

2. Formation of composite-like material

The composite-like structure of 9CrODS steel is controlled by the α/γ phase transformation of the steel, which yields the residual ferrite embedded in the tempered martensite. The nano-sized oxide particles are concomitantly distributed among both phases. This section describes the process governing a formation of composite-like structure.

2.1 Manufacturing process

Pure elemental powders of iron (99.5 mass%, 45 - 100 μm), carbon (99.7 mass%, 5 μm), chromium (99.9 mass%, under 250 μm), tungsten (99.9 mass%, 4.5 - 7.5 μm), and titanium (99.7 mass%, under 150 μm) were mechanically alloyed (MAed) together with Y_2O_3 powder (99.9 mass%, 20 nm) for 48 h in an argon gas atmosphere using a planetary-type ball mill (Fritsch P-6). Pots with a volume of $250 cm^3$ were rotated at a speed of 420 rpm; the pots contained the powders (35 g) and balls (total weight: 350 g); the weight ratio of the balls to the powders was set at 10:1. The standard chemical composition of MAed powders produced is Fe - 9Cr - 0.13C - 2W - 0.2Ti - 0.35Y_2O_3 (mass%). The excess oxygen (Ex.O) was measured to be 0.09 mass%, which was determined by subtracting the amount of oxygen contained in Y_2O_3 from the total oxygen content; this amount of oxygen is inevitable for the formation of a Y - Ti complex oxide.

The MAed powders were consolidated by spark plasma sintering (SPS) or hot-isostatic-pressing (HIP) at 1,150 °C for 1 h. The final heat treatment was conducted by normazing at 1,050 °C for 1 h and tempered at 800 °C for 1 h. Microstructure of the produced 9CrODS steel is shown in Fig.1. The structure is of the dual-phase, composed of a smooth surface phase and surrounding tempered martensite containing black dots type of chromium carbide (Ukai, 2011). This smooth surface phase belongs to a residual ferrite, of which formation process is mentioned in the next section.

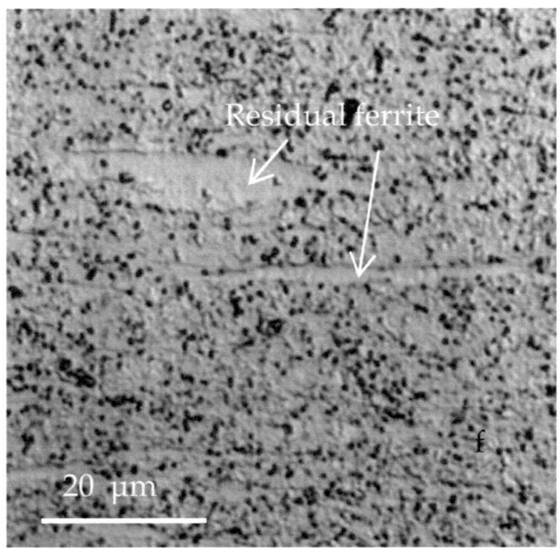

Fig. 1. Mirostructure of 9CrODS steel taken by laser-microscopy, showing a residual ferrite surrounded by the tempered martensite (Ukai, 2011).

2.2 Residual ferrite formation

The computed phase diagram of the Fe - 9Cr - 0.13C - 2W - 0.2Ti system without Y_2O_3 is shown in Fig. 2 (a) with respect to the carbon content (Yamamoto et al., 2011). For a nominal carbon content of 0.13 mass%, the single austenite γ-phase containing TiC carbide is stable at the normalizing temperature of 1050 °C. The equilibrium γ/γ+δ phase boundary at this temperature corresponds to a carbon content of 0.08 mass%, beyond which δ-ferrite is not stable. Fig. 3 shows SEM micrographs of the MAed powders for specimens with 0.1 mass%, 0.35 mass% and 0.7 mass% Y_2O_3 and without Y_2O_3 that were heat-treated by normalizing at 1050 °C and quenching in water (Yamamoto et al., 2010). The specimens without and with 0.1 mass% Y_2O_3 have the single martensite structure, which were obtained by inducing the transformation from the γ-phase by water quenching. This result is consistent with the computed phase diagram shown in Fig. 2(a). Nevertheless, the specimens with 0.35 mass% and 0.7 mass% Y_2O_3 exhibit a dual phase comprising martensite and smooth surface phases.

Digital image analyses showed that the area fraction of the smooth surface phase was approximately 0.2 for both specimens with 0.35 mass% and 0.7 mass% Y_2O_3. No obvious difference was observed between X-ray diffraction peaks of the martensite-phase and smooth surface phases at room temperature.However, high-temperature X-ray diffraction measurement at 950 °C showed a considerable difference, as shown in Fig. 4; the specimen without Y_2O_3 had diffraction peaks corresponding only to the austenite γ-phase, whereas specimens with 0.35 mass% and 0.7 mass% Y_2O_3 had diffraction peaks corresponding to the

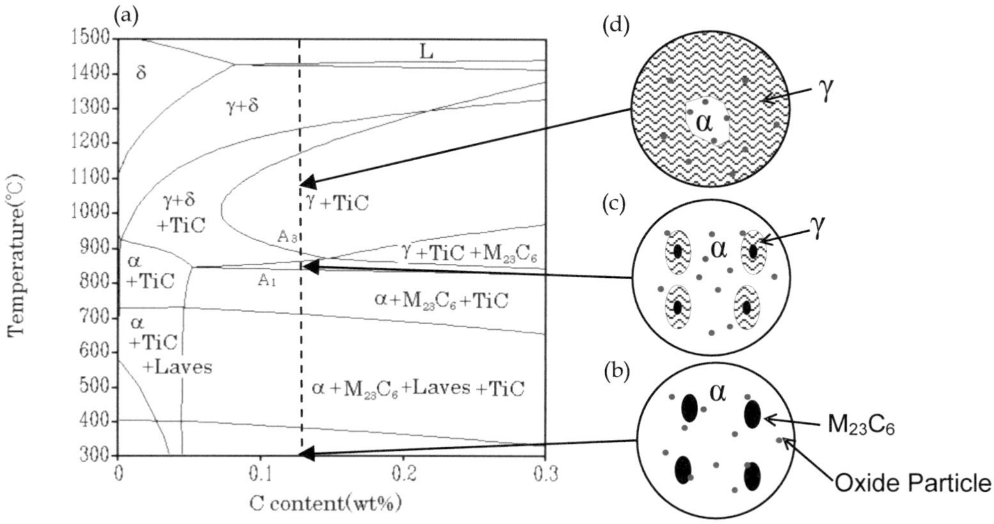

Fig. 2. (a) Computed phase diagram with respect to carbon content for Fe–9Cr–xC–0.2Ti–2W without oxide particle and schematic image at, (b) room temperature, (c) A_{C1} point, and (d) 1050 °C for system with oxide particle (Yamamoto et al., 2011).

austenite γ-phase and the ferrite phase (Yamamoto et al., 2010). The austenite γ-phase transformed to the martensite-phase, but the ferrite phase remained unchanged by water quenching. These results prove that the smooth surface phase is composed of ferrite. Considering that the ferrite phase is formed only in the specimens containing 0.35 mass% and 0.7 mass% Y_2O_3 and that four types of ODS steels shown in Fig. 3 have an identical chemical composition except for Y_2O_3 content, it seems that the presence of a Y_2O_3 particle favors the retention of the ferrite phase at 950 °C and 1050 °C.

Dilatometric measurement was also carried out from the room temperature to 1100 °C with a heating rate and a cooling rate of 5 °C/min. Fig. 5 shows the results of the dilatometric measurement when Fe - 9Cr - 0.13C - 2W - 0.2Ti was heated without and with 0.35 mass% Y_2O_3 (Yamamoto et al., 2010). In the case of the specimen without Y_2O_3, the linear thermal expansion coefficients started decreasing from A_{C1} point of 850 °C to A_{C3} point of 880 °C because of the reverse transformation of the martensite-phase to the γ-phase, which was in reasonably good agreement with the computed phase diagram shown in Fig. 2(a). The addition of 0.35 mass% Y_2O_3 induced an increase of up to the A_{C3} point of 930 °C. By comparing both the curves, we found that the specimen with 0.35 mass% Y_2O_3 exhibited a smaller amount of reduction in the linear thermal expansion during the reverse transformation of the martensite-phase to the γ-phase; this observation indicates that the entire martensite-phase was not transformed to the γ-phase. The untransformed ferrite phase corresponds to that observed by the X-ray diffraction peaks in the specimen with 0.35 mass% Y_2O_3 at 950 °C. This untransformed ferrite phase was designated as the residual ferrite. More detail consideration is discussed from a viewpoint of difference in the thermal expansion coefficient in reference (Yamamoto et al., 2010)

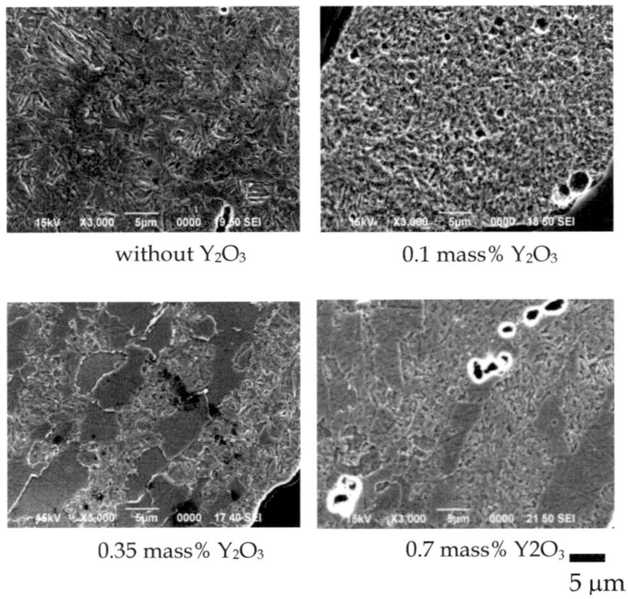

Fig. 3. SEM photographs of the MAed powders for 0 mass%, 0.1 mass%, 0.35 mass% and 0.7 mass% Y_2O_3 specimens after normalizing at 1050 °C and subsequently quenched to water (Yamamoto et al., 2010).

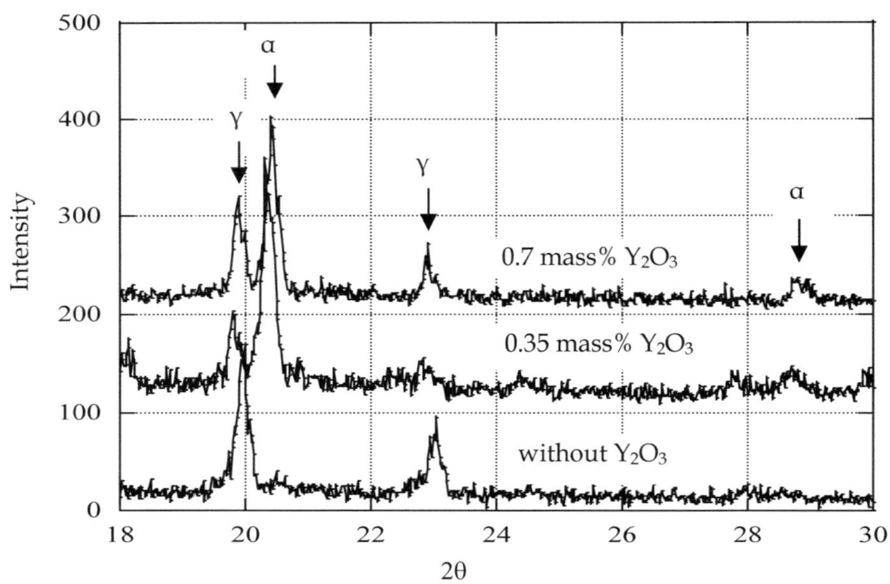

Fig. 4. Results of X-ray diffraction measurement at 950 °C for 0 mass%, 0.35 mass% ,and 0.7 mass% Y_2O_3 specimens (Yamamoto et al., 2010).

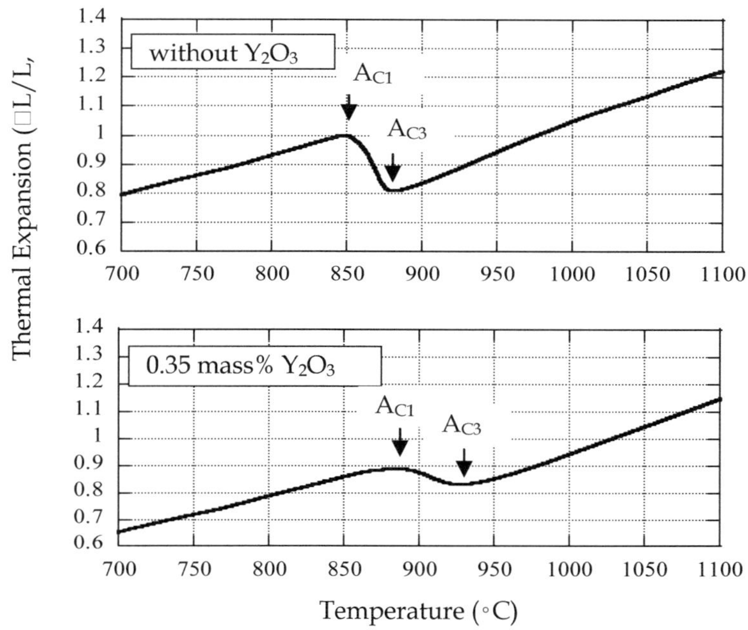

Fig. 5. Results of thermal expansion measurement between 700 °C and 1100 °C at temperature rising of 0.33 °C/s for 0 mass% and 0.35 mass% Y_2O_3 specimens (Yamamoto et al., 2010).

2.3 Thermodynamic analyses

Alinger et al. carried out a small angle neutron scattering (SANS) experiment for a U14YWT specimen (Fe - 14Cr - 0.4Ti - 3W - $0.25Y_2O_3$). The neutron scattering cross-section and scattering vector were measured for the MAed powder after hot-isostatic-pressing (HIP) or thermal annealing at 700 °C, 850 °C, 1000 °C, and 1150 °C. Their results indicate that the MAed powder annealed at 700 °C shows the smallest radius and highest density for the Y - Ti complex oxide particles (Alinger et al., 2004). Assuming that the Y_2O_3 particles were decomposed during mechanical alloying, the subsequent annealing resulted in the formation and precipitation of Y - Ti complex oxide particles at elevated temperatures of 700 °C or higher. Since the reverse transformation of the ferrite α-phase to the austenite γ-phase took place at the elevated temperature of over 850 °C, which is higher than the precipitation temperature of the Y - Ti complex oxide particles, it is possible to consider that the partial retention of the residual ferrite can be attributed to the presence of the Y - Ti complex oxide particles in the 9CrODS ferritic steels; these particles should block the motion of the α/γ interface, thereby partly suppressing the reverse transformation of the α-phase to the γ-phase.

The chemical driving force (ΔG) for the reverse transformation of the α-phase to the γ-phase in the Fe - 0.13C - 2W - 0.2Ti system without Y_2O_3 can be evaluated in terms of Gibbs free energy vs. carbon content curves at each temperature; these curves were derived by using the Thermo-Calc code and TCFE6 database. The pinning force (*F*) against the motion of α/γ interface can be expressed as equation (1), which was obtained from the modified Zener equation (Nishizawa et al., 1997).

$$F = \frac{3}{8}\frac{\sigma \cdot f_p^{\,2/3}}{r} \quad (1)$$

where σ (J/m^2) is the interfacial energy between α and γ-phase and its value was selected to be 0.56 J/m^2 (Martion & Doherty, 1976). r (m) is the radius of the oxide particles in the α-phase. The chemical form of the oxide particles was determined to be $Y_2Ti_2O_7$ by TEM, and their diameter was estimated to be about 3 nm in the specimen of the same composition. Thus, r was set as 1.5 nm. f_p is the volume fraction of dispersed oxide particles, and it was derived on the basis of the experimental evidence that oxide particles consist of $Y_2Ti_2O_7$. The velocity of the α/γ interface motion (v) is proportional to the difference between F and ΔG, as shown in equation (2):

$$v = M \times (\Delta G - F) \quad (2)$$

M is the mobility of the interface. ΔG and *F* are competitive, and ΔG> F indicates a positive velocity for the interface motion, i.e., the reverse transformation of the α-phase to the γ-phase. On the other hand, ΔG< F indicated that the α/γ interface could be pinned by oxide particles, and thus the α-phase was retained.

When the chemical driving force (ΔG) for the reverse transformation of the α-phase to the γ-phase is derived, we should consider a change of carbon content in the matrix due to carbide dissolution, because carbon content greatly affects the Gibbs energy of γ-phase. At A_{C1} point, $M_{23}C_6$ carbide dissolves into matrix and carbon concentration at carbide-matrix interface becomes relatively high. For example, carbon content of 0.2 mass % leads to ΔG higher than pinning force F by oxide particle at 1050 °C as shown in Fig. 6 (Yamamoto et al.,

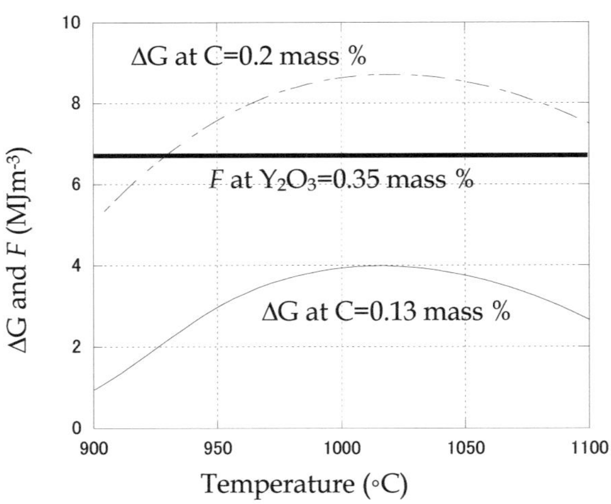

Fig. 6. Comparison of the driving force (ΔG) for α to γ reverse transformation and the pinning force (*F*) by oxide particle: C denotes carbon content (Yamamoto et al., 2011).

2011). This result suggests that α to γ reverse transformation proceeds from the carbide-matrix interface, where $M_{23}C_6$ carbide dissolves into matrix. This situation is schematically illustrated in Fig. 2(c). When temperature rises up beyond A_{C3} point, $M_{23}C_6$ carbide completely dissolves into matrix and thereby average carbon concentration approaches to the initial content of 0.13 mass %, where pinning force F is higher than driving force ΔG, as shown in Fig. 6. This result of calculation implies that α-ferrite can partially remain, as schematically shown in Fig. 2(d). Besides, on the basis of the above calculation, the formation and volume fraction of the residual ferrite is able to be controlled by appropriate amount of carbon in the matrix. We have successfully adjusted the volume fraction of the residual ferrite by controlling a direct addition of carbon or oxygen that affects the matrix carbon content through either TiO_2 or TiC formation (Ukai et al., 2007).

2.4 Oxide particle dispersion

The size distribution of oxide particles within the residual ferrite and the tempered martensite matrix was separately evaluated. In this study, both ODS fully ferritic steel and ODS fully martensitic steel were produced to simulate the residual ferrite and martensite phases, respectively, where the microstructure of the fully ferrite is completely the same as residual ferrite, because both phases are not transformed to γ-phase at the temperature above A_{C3} point. Fig. 7 shows the dispersed oxide particles in full martensite and full ferrite ODS steels, observed by TEM (Ukai, 2011). It is obvious that a few nm size oxide particles are finely dispersed in the fully ferrite ODS steel, whereas their size is coarser in the

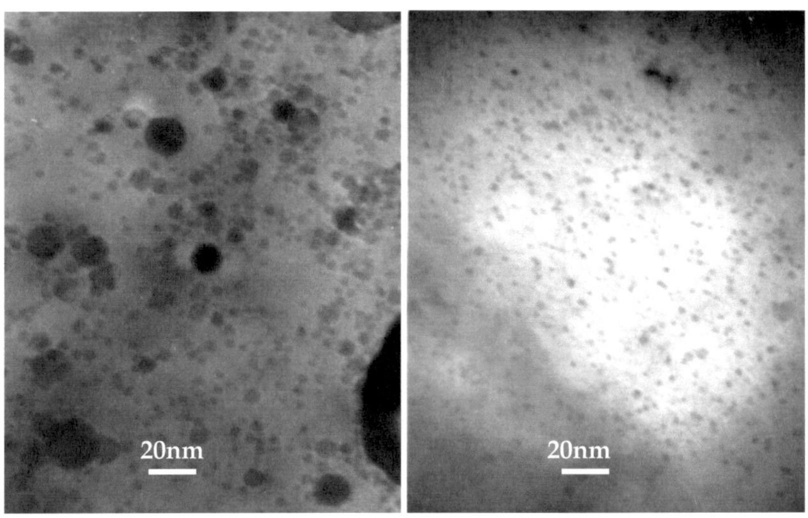

Fig. 7. TEM micrographs showing oxide particles in (a) ODS fully martensitic steel (b) ODS fully ferritic steel, residual ferrite (Ukai, 2011).

martensite ODS steel. The image analysis of the TEM micrographs is shown in Fig. 8; the mean size of oxide particle in full ferrite and full martensite ODS steels is 3 nm and 7 nm, respectively (yamamoto et al., 2011). From Fig. 8, the oxide particles dispersed in the ODS fully martensitic steel, i.e. martensite matrix, seems to shift to a coarser size compared to the dispersion in ODS fully ferritic steel, i.e. residual ferrite. One of the main reasons for this behavior could be associated with α/γ reverse transformation. Yazawa et al. reported that the coherent to incoherent transition at precipitation interfaces sometimes induces coarsening of precipitates (Yazawa et al., 2004). Similar behavior is expected in the ODS martensite matrix. Our consideration is schematically shown in Fig.9. The oxide particles

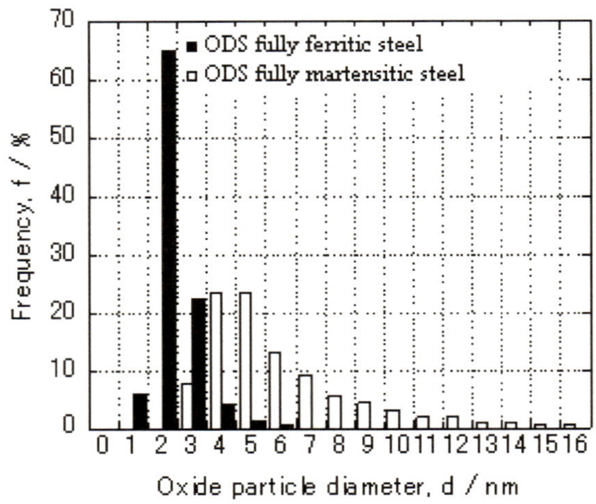

Fig. 8. Oxide particle size distributions in ODS fully ferritic steel and ODS fully martensitic steel (Yamamoto et al., 2011)

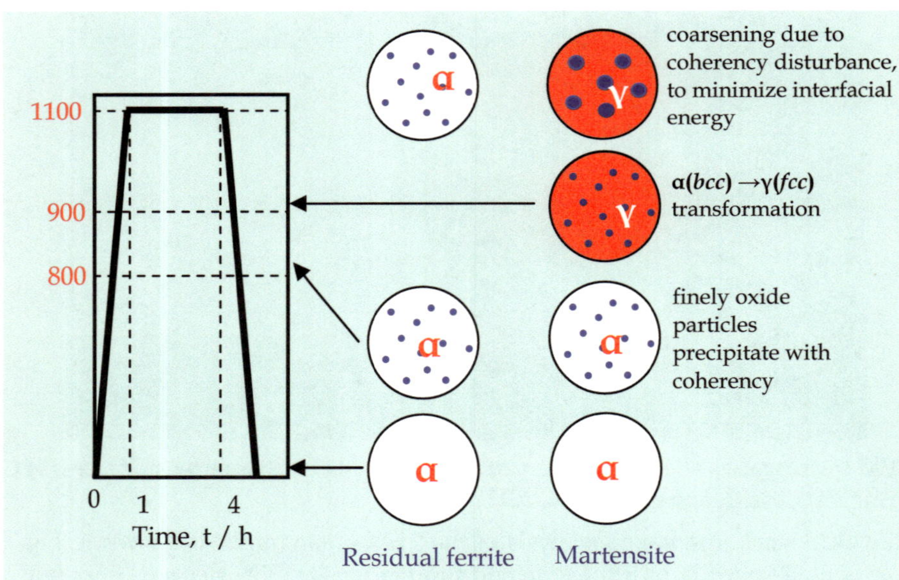

Fig. 9. Schematic representation of a process of the oxide particle coarsening in the martensite phase, compared to the residual ferrite with fine oxide particles by keeping coherency.

are formed by a process of precipitation during hot-pressing at 1100 °C after their decomposition was induced by mechanical alloying. This precipitation leads to partially coherency between oxide particles and the α-ferrite matrix, since precipitation takes place at temperatures below the A_{C1} point (Alinger et al., 2004). Subsequent change of the matrix phase from α-bcc crystal structure to γ-fcc by the reverse transformation could disturb the interfacial coherency; the oxide particles coarsen to minimize the interfacial energy during staying at γ-phase. Therefore, martensite phase transformed from γ-phase by cooling contains larger size of oxide particles, whist the residual ferrite contains finer oxide particles due to keeping coherency.

3. Creep deformation characterization

The creep deformation of the 9CrODS steel is unique and so different from that of the existed heat-resistant steels. The packet and block boundaries also play an important role for deformation. This section characterizes a deformation mechanism of the 9CrODS steel from a viewpoint of composite-like behavior.

3.1 Nano-hardness of residual ferrite

With the aim of evaluating the mechanical property of the residual ferrite, nanoindentation measurements were conducted using a Berkovich tip-type indentator at a loading of 3 mN. The trace of a Berkovich tip of sub-micron size can be placed within the interiors of the residual ferrite regions, while the conventional micro-Vickers diamond tips using 100 mN loads cover 7 x 7 micron-meters. Considering that the plastic zone is only two to three times larger than the contact area of the Berkovich tip, the mechanical property of the residual ferrite can be adequately evaluated by nanoindentaion measurements. Fig.10 shows the hardness change in the individual phases as a parameter of the tempering conditions (Ukai et al., 2009). The tempering parameters are shown in equation (3) in in terms of a Larson-Miller parameter (LMP; C = 20);

$$LMP = T\,(20 + \log t) \tag{3}$$

where T is temperature in kelvin and t is time in hours. The time at different temperature can be derived so as to satisfy the same LMP value.. The hardness decrease is significantly restricted in the residual ferrite with increasing tempering conditions compared with that of the martensite phase. The overall hardness measured by the micro-Vickers tester is also shown by the broken line; this covers both the residual ferrite and martensite, thereby representing the average hardness of both phases. The hardness H_v is correlated with the yield stress σ_y using the following relationship (Tabor, 1951):

$$\sigma_y = 1/3\, H_v\,. \tag{4}$$

By tempering at 800 °C for 58 h, which is equivalent to tempering at 700 °C for 10,000 h based on the LMP, the hardness can be converted to the yield stress at room temperature for the individual phases: 1360 MPa for the residual ferrite and 930 MPa for the martensite. One of the main reasons for increasing hardness at the residual ferrite is come from the dispersion strengthening of oxide particles that are finely distributed in the residual ferrite as mentioned in the previous section. It is worth noting, therefore, that 9CrODS steel behaves like composite material comprising of hard and soft elements.

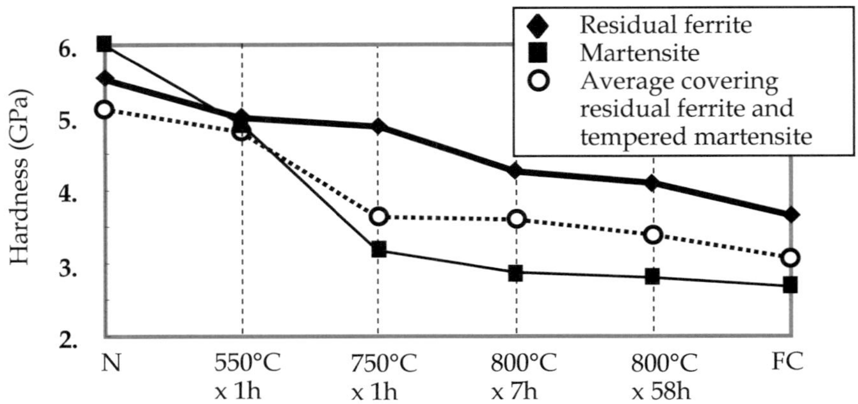

Fig. 10. Hardness change as a function of tempering conditions for the individual phases of residual ferrite and martensite measured by nanoindentator as well as average covering both measured by micro-Vickers; N: normalizing, FC: furnace cooling (Ukai et al., 2009).

3.2 Creep strain curve

Fig. 11 shows the typical creep curve of 9CrODS steel after normalizing and tempering (NT) for stress loadings of 155 MPa and 230 MPa at 700 °C (Ukai et al., 2009). The optical microstructure of this specimen is shown in the left upper of Fig. 12 (Ukai et al., 2009). This steel is composed of the residual ferrite and the tempered martensite. The secondary creep is obviously decelerated with strain accumulation, and the specimen ruptured without an accelerated creep region at 155 MPa. A stepwise increase in the primary creep strain can be observed with stress increase up to 230 MPa. These behaviors are completely different from those of conventional heat-resistant steels. In conventional steels, the rupture occurs after the accelerated creep deformation, where coarsening of carbides, movement of the lath and block boundaries and a local recovery of dislocations become dominant (Abe, 2004, 2008).

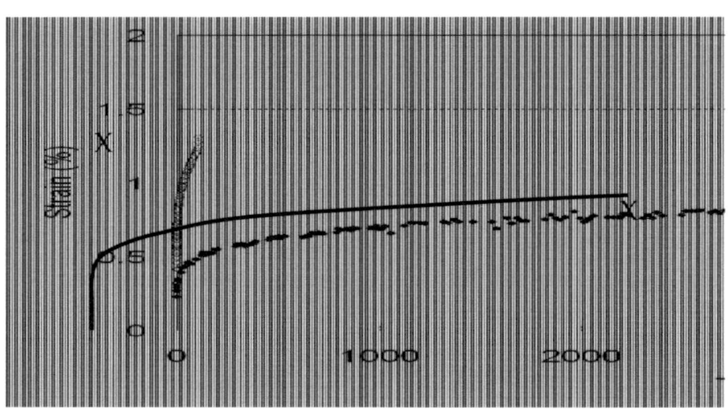

Fig. 11. Creep strain curves of 9CrODS steel after NT in the stress loading of 155 MPa and 230 MPa at 700 °C (Ukai et al.,2009).

Such dynamic recovery and structure coarsening should be suppressed by the dispersed nano-size oxide particles in 9CrODS steel. Their dominant fracture mode could be deformation near block and packet boundaries. The stress induced by this deformation can't be easily relaxed by the dislocation creep, since nano-size oxide particles can block the dislocation movement. This is the reason why the creep strain rate gradually decreases with increasing strain, and work hardening dominantly occurs during creep deformation.

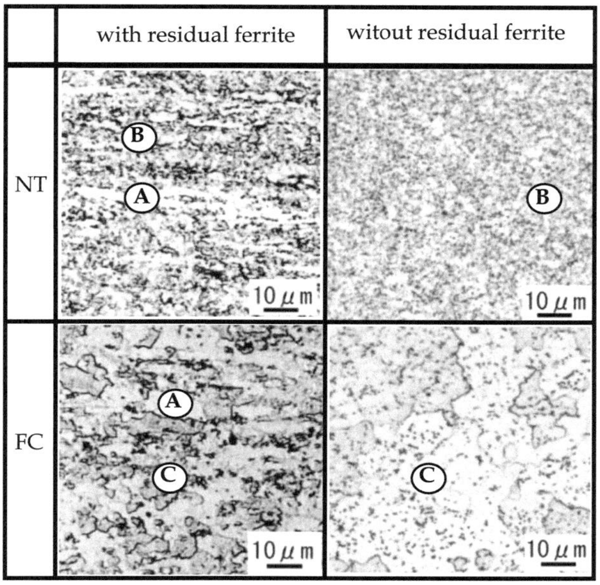

Fig. 12. Optical micrographs of 9CrODS steels with and without residual ferrite in the heat treatment conditions of normalizing and tempering (NT) and furnace cooling (FC). A: residual ferrite, B: tempered martensite, C: transformed ferrite (Ukai et al.,2009).

3.3 Creep rupture property

The optical microstructures of 9CrODS steels with and without the residual ferrite are shown in Fig. 12 for NT and furnace cooling (FC) conditions. The 9CrODS steel containing the residual ferrite has a standard chemical composition, whilst higher content of excess oxygen was added for the free residual ferrite one, where the titanium combines with oxygen rather than carbon, thus increase of matrix carbon content leads to complete α to γ transformation. The detail is mention in reference (Ukai & Ohtsuka, 2007). Based on the continuous cooling transformation (CCT) diagram, NT treatment induces the tempered martensite due to a rapid cooling rate of 7000 °C /h, while FC treatment causes a diffusional transformation to the α-phase by slow cooling at a rate of 30 °C /h.

Fig.13 shows their creep rupture strength (Ukai et al.,2009). The following two characteristic features were observed in creep rupture properties of 9CrODS steels. The first is that the creep rupture strength of NT with residual ferrite (TM+RF) increases stepwise from that of NT without residual ferrite (TM). This is attributed to the effects of the harder residual ferrite strengthening the 9Cr-ODS steels, therefore. The second feature is that the furnace cooling (FC) heat treatment for with or without residual ferrite induces stepwise increases in the creep rupture strength in comparison with the NT heat treatment. This result means that the α-phase has superior strength to that of martensite. Martensitic phase transformation produces the packet and block boundaries, where grain boundary deformation and fracture can be preferably originated. The strength improvement in the α-phase induced by FC could be, therefore, due to the suppression of the deformation at packet and block boundaries.

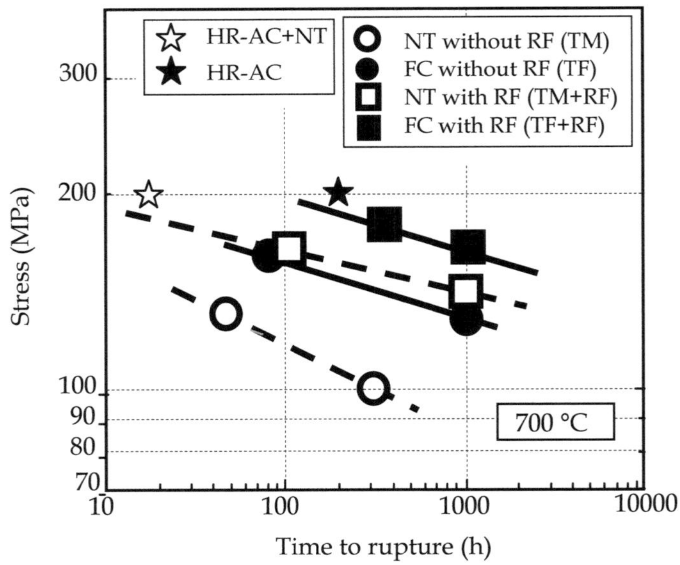

HR: hot-rolling
NT: normalizing and tempering
FC: furnace cooling
TM: tempered martensite
TF: transformed ferrite
TM+RF: tempered martensite and residual ferrite
TF+RF: transformed ferrite and residual ferrite

Fig. 13. Creep rupture data of 9CrODS steels with and without residual ferrite after NT and FC; creep rupture data of HR-AC and HR-AC-NT are also superimposed at 700 °C (Ukai et al.,2009).

3.4 Creep strain rate vs. stress relation

The threshold stress for deformation at 700 °C is estimated on the basis of equation (5) (Scattergood & Bacon, 1975).

$$\sigma = \frac{AMGb}{2\pi\lambda}\left[\ln\frac{D}{r_0} + B\right], \quad (5)$$

where, G is the shear modulus (50.6 GPa at 700 °C); M, the Taylor factor (3.0) (Stoller & Zinkle, 2000); b, the magnitude of the Burgers vector (2.48×10^{-10} m). The parameter r_o is the inner cut-off radius of a dislocation core, which was assumed to be the magnitude of the Burgers vector. In addition, λ is the average face-to-face distance between particles on a slip plane; it is given in terms of the average particle radius r_s and the average center-to-center distance l_s between the particles :

$$\lambda = 1.25\, l_s - 2r_s \tag{6}$$

The values of l_s and r_s are determined using the particle radius r:

$$l_s = \sqrt{2\pi \overline{r^3}} \,/\, \sqrt{3f\bar{r}} \tag{7}$$

$$r_s = \pi / 4 \cdot (\overline{r^2} / \bar{r}) \tag{8}$$

where f is the volume fraction of the dispersed particles, which is estimated from the oxide particle density (4.84 × 10^3 Kg/m^3). The D corresponds to harmonic mean of $2r_s$ and λ. At the elevated temperature of 700 °C, the interface between the particles and matrix could tend to slide. Under such condition, an attractive interaction between a dislocation and the particles can be applied, as proposed by Srolovitz (Srolovitz et al., 1983). The A and B values are expressed by following formula.

$$A = \left[1 - \frac{\nu}{1-\nu}\sin^2\phi\right]\cos\phi, \quad B = 0.7 \qquad \text{for edge dislocation (9)}$$

$$A = \left[\frac{1+\nu\sin^2\phi}{1-\nu}\right]\cos\phi, \quad B = 0.6 \qquad \text{for screw dislocation (10)}$$

where ϕ is the critical angle at which the dislocation detaches from the particles: φ = 46° for a screw dislocation and ϕ = 19° for an edge dislocation. ν is the Poisson's ratio (0.334). Substituting these dispersion parameters into equation (5), the threshold stress σ for the average of edge and screw dislocations is derived to be 265 MPa. Fig. 14 shows the strain rate vs. stress relation (Ukai et al., 2009). The deformation occurs at a stress far below the threshold stress of 265 MPa. The deformation is, therefore, induced at weaker regions such as grain boundaries rather than being intra-grain themselves, which is strengthened by the dispersed oxide particles. The strain rate vs. stress relation is almost parallel, and the slope corresponding to a stress exponent n is estimated to be ranging from 10 to 14. It is well known that the typical value of the stress exponent is significantly higher (around 40) for deformation that is dominated by particle-dislocation interaction (Arzt, 1991). Therefore, the accelerated deformation at a lower stress level could have arisen from the local deformation associated with packet and block boundaries. The cause of this stress increase at which creep strain initiates by furnace cooling can be explained by a disappearance of packet and block boundaries, as described above. The residual ferrite also enlarges the critical stress for initiating creep strain. Concerning this phenomenon, the creep constitutive equation of 9CrODS steel should be formulated by analogy with a fiber composite material (Sakasegawa et al., 2008). From the stress distribution between the hardened residual ferrite and the softened tempered martensite, the stress loading on the tempered martensite is derived by the following equation:

$$\sigma_m / \sigma_t = 1/\{1 + (D - 1)\, f_r\}, \tag{11}$$

where σ_t and σ_m are the total stress loading and stress onto the individual tempered martensite regions, respectively. The parameter f_r is the volume fraction of the residual ferrite grains, and D is the stress distribution coefficient expressed as a ratio of the shear modulus of residual ferrite μ_r to that of the tempered martensite μ_m:

$$D = \mu_r / \mu_m. \tag{12}$$

Assuming that the shear modulus ratio is approximately expressed as the hardness ratio of both phases, D becomes equal to 1.46 from Fig.10. Thus, based on Equation (12), σ_m/σ_t is equal to 0.88 using f_r = 0.3, as determined by microstructure observation. This estimation implies that the stress loading onto the softened tempered martensite decreases to 0.88, which leads to increase in the stress corresponding to the initiation of creep strain.

As a summary of the creep deformation of 9CrODS steel, Fig. 15 schematically illustrates a configuration of the hardened residual ferrite and softened tempered martensite. Under stress loading, softened tempered martensite is preferentially deformed. Dislocations piled up at the packet and block boundaries could induce the localized deformation near these boundaries, since the intra-grain deformation is hardly taken place due to finely dispersed nano-sized oxide particles. The stress accumulated at such boundaries could lead to fracture. The hardened residual ferrite provides not only sharing the stress loading onto the softened tempered martensite, but also geometrically restricting the deformation of the softened tempered martensite. It can be said that deformation and fracture of the 9CrODS steel behaves like composite material.

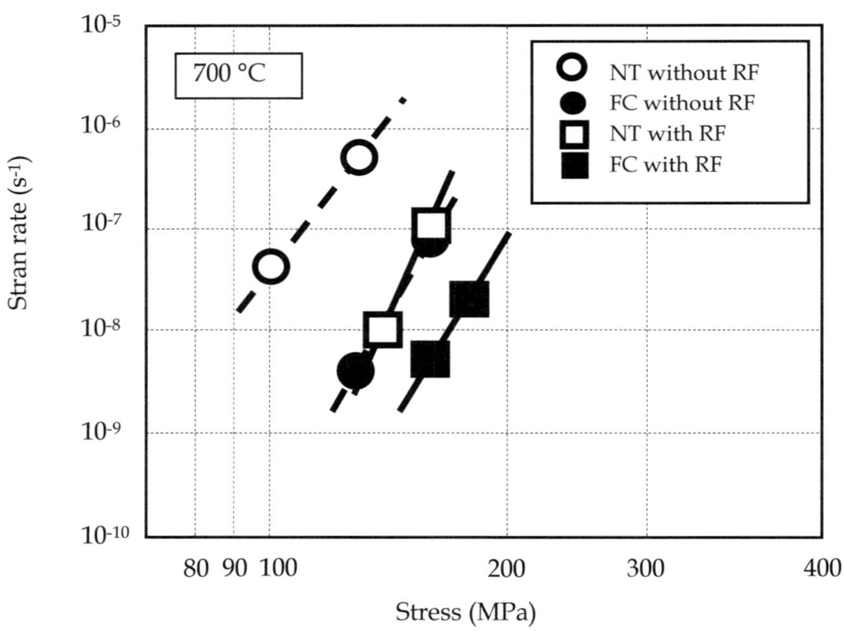

Fig. 14. Strain rate vs. stress relationship for 9CrODS steels with and without residual ferrite after NT and FC heat treatment (Ukai et al., 2009).

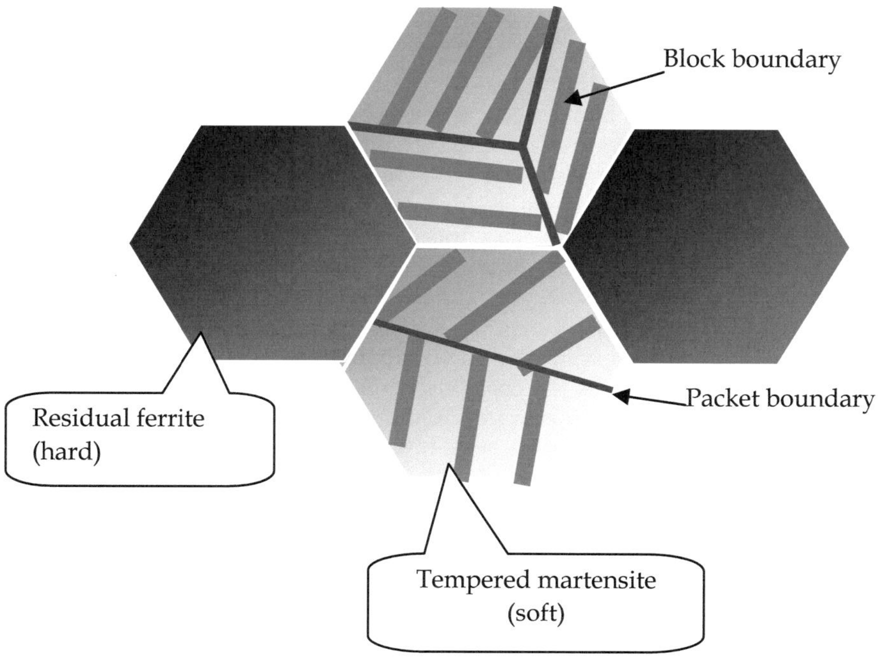

Fig. 15. Illustration of the hardened residual ferrite and softened tempered martensite.

4. Thermo-mechanical processing

Modification of a microstructure is required for the tempered martensite in 9CrODS steel, to suppress the localized deformation at packet and block boundaries. Hot-rolling was recently processed for this purpose. The results of currently developed thermo-mechanical processing and resultant high-temperature strength are assessed in this section.

4.1 Microstructure

The 9CrODS steel powders in the standard composition and high carbon content (0.41 mass%) were mechanically alloyed by the planetarium type boll mill, where high carbon specimen was prepared to produce the free residual ferrite 9CrODS steel. MAed powders were consolidated by spark plasma sintering (SPS) at 1100 °C and 44 MPa for 1 h; the hot-rolling (HRing) was conducted in the total reduction rate of 84 % at the austenitic γ phase region. The detail condition is as follow;

1. keeping temperature at 1,100 °C for 1 min in the furnace and taking out from the furnace.
2. hot-rolling at three times in the total reduction ratio of 52 %; temperature decreases to 930 °C
3. back to furnace to keep temperature at 1,100°C for 10 min and taking out from the furnace
4. again hot-rolling at three times in the total reduction ratio of 67 %; temperature decreases to 840°C at the end of hot-rolling, and air-cooled (AC) to room temperature in the cooling rate of 4000 °C/h.

As a parameter, normalizing (N) by water quench and tempering (T) heat-treatments were subsequently carried out at 1050 °C × 1 h and 800 °C × 1 h, respectively, to evaluate structure change of the HRed specimens.

Fig. 16 shows SEM micrographs of high carbon specimen (0.41mass%): (a) HR-AC and (b) subsequent NT heat treatment. Both structures are significantly different and the coarsened grains are formed just after HR-AC, whilst subsequent NT heat treatment induces finely homogeneous and the typical tempered martensite that is usual structure of 9CrODS steel without residual ferrite similar to the right upper optical photo of Fig. 12. The larger grains in HR-AC belong to ferrite that was transformed from the γ-phase during the air cooling after HR. A surrounding area is martensite structure. The microstructure of the standard 9CrODS steel is shown in Fig. 17. The larger grains are also observed in HR-AC specimen, and subsequent NT one has fine grains that involve the residual ferrite and tempered martensite similar to the left upper photo of Fig. 12. The coarsened grains in HR-AC specimen shown in Fig. 17(a) could be composed of the residual ferrite and transformed ferrite, however, it is difficult to separate them by SEM photograph, since both are the same ferrite phases.

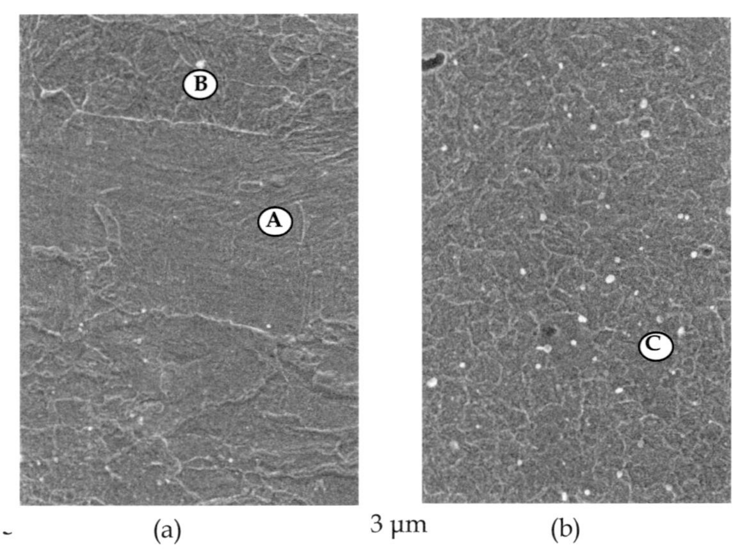

Fig. 16. SEM micrographs of 0.41 mass% carbon 9CrODS steel: (a) hot-rolling and air-cooling (HR-AC), (b) HR-AC and normalizing and tempering (NT); A: transformed ferrite, B: martensite, C: tempered martensite

Concerning a formation of the coarsened ferrite grains, it is obvious that these grains do not correspond to the coarsened residual ferrite, since coarsened grains are seen even in high carbon specimen without the residual ferrite as shown in Fig. 16(a). It is considered that the finer γ-grains made by hot rolling at 1000 °C accelerate the ferrite formation rather than martensite transformation during air cooling, from a viewpoint of a continuous cooling transformation (CCT) diagram of 9CrODS steels (Ukai & Ohtsuka, 2007). Furthermore, accumulation of severe strain energy in γ-grains induced by hot-rolling could be responsible for coarsening of the transformed ferrite grains.

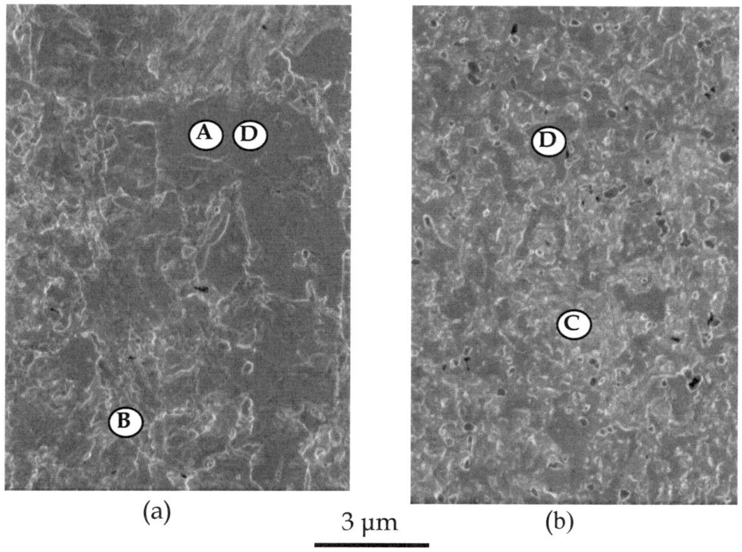

Fig. 17. SEM micrographs of standard 9CrODS steel: (a) hot-rolling and air-cooling (HR-AC), (b) HR-AC and normalizing and tempering (NT); A: transformed ferrite, B: martensite, C: tempered martensite, D: residual ferrite

4.2 High-temperature mechanical properties

The results of the tensile test at 700 °C are shown in Fig. 18. This test was conducted for the standard 9CrODS steel, varying with just HR-AC and subsequent NT heat treatment. The NT specimen exhibits the same level of strength and ductility as the conventional 9Cr-ODS steels [6]. This result can be understood by considering that the structure induced by HR completely disappears by normalizing heat treatment and its structure is composed of the residual ferrite and tempered marutensite as shown in Fig. 17(b). For HR-AC condition, the tensile strength and rupture elongation are recorded more than 500 MPa and 20 %, respectively. It is to be noticed that these tensile properties are adequately improved by HR-AC treatment far beyond those of the conventional 9CrODS steels.The excellent tensile performance of HR-AC specimen is ascribed to an existing of the transformed ferrite grains in the coarser size, which is shown in Fig. 17(a). The block and packet boundaries in martensite structure sometimes give appropriate sites for fracture, and thus lead to premature rupture (Abe, 2004). The coarsened ferritic grains could suppress the grain boundary deformation, and thus improve the tensile strength and ductility.

The creep rupture data of the standard 9CrODS steel for HR-AC and subsequent NT specimens were superimposed by star symbols in Fig. 13, where data are shown in the stress-rupture time correlation at 700 °C. The creep rupture strength of HR-AC specimen (solid star symbol) obtained in this test at 700 °C and 200 MPa approximately lies on the extrapolated line of FC specimen, but it is slightly higher. This improved strength over FC could be owing to the coarsening of the ferrite grains made by hot-rolling. Subsequent NT specimen also shows slightly higher strength than NT condition of the previous specimen containing the residual ferrite. Based on those finding, it is to be noticed that the ferrite grains in coarser size adequately improve the creep strength. This is due to a suppression of the block and packet boundary deformation, as mentioned in the tensile test.

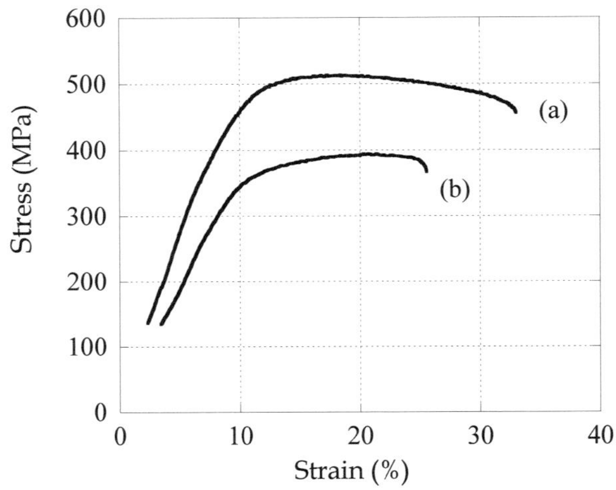

Fig. 18. The results of tensile tests of standard 9CrODS steel: (a) hot-rolling and air-cooling (HR-AC), (b) HR-AC and normalizing and tempering (NT).

5. Conclusion

The 9CrODS steel is an attractive composite-like material consisted of the residual ferrite and tempered martensite. Formation of both phases can be controlled by the process of α-γ phase transformation. The residual ferrite is a metastable phase and involves extremely finer nano-size oxide particles that are responsible for harder hardness; on the contrary, the tempered martensite contains coarser oxide particles and thus being softer hardness. Mechanical response of 9CrODS steel is completely different from that of the existed heat-resistant steels. Under stress loading, softened tempered martensite is preferentially deformed, in particular at the packet and block boundaries. The stress accumulation could give rise to a localized fracture along the packet and block boundaries in the tempered martensite. The microstructure of the tempered martensite was modified in terms of thermo-mechanical processing, and high-temperature strength was significantly improved through changing microstructure from the martensite to the transformed ferrite. The hardened residual ferrite plays a key role that provides a geometrical restriction for the deformation of softened tempered martensite. The best microstructure of 9CrODS steel is considered to be composed of hardened residual ferrite and softened transformed ferrite with free block and packet boundaries.

6. Acknowledgment

The author wishes to acknowledge many colleagues: in particular Mr. Takeji Kaito and Dr. Satoshi Ohtsuka of Japan Atomic Energy Agency (JAEA) and Dr. Takanari Okuda of Kobelco Research Institute Co. and Dr. Tsukasa Azuma of Japan Steel Works (JSW). The author would like to also thank many graduate students of my laboratory.

7. References

Abe, F. (2004). Bainitic and Martensitic Creep-Resistant Steels. *Curr. Opin. Sol. State Mater. Sci.* 8, 305 - 311

Abe, F. (2008). *Creep-Resistance Steels*, Edited by F. Abe, T-Ulf Kern and R. Viswanathan, The Institute of Materials, Minerals & Mining

Alinger,M. J., Oddett, G.R. & Hoelzer, D.T. (2004). The Development and Stability of Y–Ti–O anoclusters in Mechanically Alloyed Fe–Cr Based Ferritic Alloys. *J. Nucl. Mater.* 329 - 333, pp.382 - 386

Arzt, E. (1991). Creep of Dispersion Strengthened Materials: A Critical Assessment. *Res. Mech.* 31, pp.399 - 453

Martion, J.W. & Doherty, R.D. (1976). *Stability of Microstructure in Metallic Systems*, Cambridge University Press, Cambridge, pp. 173.

Nishizawa, T., Ohnuma, I. & Ishida, K. (1997). Examination of the Zener Relationship between Grain Size and Particle Dispersion. *Mater. Trans.* 38, 11, pp.950 - 956.

Odette, G.R., Alinger, M.J. & Wirth, B.D. (2008). Recent Developments in Irradiation-Resistant Steels. *Annu. Rev. Mater. Res.* 38, pp.471 - 503

Ohtsuka, S., Ukai, S., Fujiwara, M., Kaito, T. & Narita, T. (2004). Improvement of 9Cr-ODS Martensitic Steel Properties by Controlling Excess Oxygen and Titanium Contents. *J. Nucl. Mater.* 329 - 333, pp.372 - 376

Ohtuska, S., Ukai, S., Fujiwara, M., Kaito, T. & Narita, T. (2005a). Improvement of Mechanical Properties of 9CrODS Martensitic Steel by Controlling Titanium Concentration. *Mater. Trans.* 46, 3, pp.487 - 492

Ohtsuka, S., Ukai, S., Fujiwara, M., Kaito, T. & Narita, T. (2005b). Nano-structure Control in ODS Martensitic Steels by Means of Selecting Titanium and Oxygen Contents. *J. Phys. Chem. Solids.* 66, pp.571 - 575

Ohtsuka, S., Ukai, S. & Fujiwara, M. (2006). Nano-Mesoscopic Structural Control in 9CrODS Ferritic/Martensitic Steels. *J. Nucl. Mater.* 351, pp.241 - 246

Ohtsuka, S., Ukai, S., Fujiwara, M., Sakasegawa, H., Kaito, T. & Narita, T. (2007). Nano-Mesoscopic Structural Characterization of 9Cr-ODS Martensitic Steel for Improving Creep Strength. *J. Nucl. Mater.* 367 - 370, pp.160 - 165

Sakasegawa, H., Ukai a, S., Tamura, M., Ohtsuka, S., Tanigawa, H., Ogiwara, H., Kohyama, A. & Fujiwara, M. (2008). Creep Constitutive Equation of Dual Phase 9Cr-ODS Steel. *J. Nucl. Mater.* 373, pp.82 - 89

Scattergood, R.O. & Bacon, D.J. (1975). The Orowan Mechanism in Anisotropic Crystals. *Phil. Mag.* A 31, pp.179 - 198

Srolovitz, D.J., Petkovic-Luton, R.A & Luton, M.J. (1983). Edge Dislocation-Circular Inclusion Interactions at Elevated Temperatures. Acta Metall. 31, pp.2151 - 2159

Stoller, R.E & Zinkle, S.J. (2000). On the Relationship between Uniaxial Yield Strength and Resolved Shear Stress in Polycrystalline Materials. *J. Nucl. Mater.* 283 - 287, pp.349 - 352

Tabor, D. (1951). *The Hardness of Metals*, Oxford University Press, Oxford

Ukai, S., Nishida, T., Okuda, T. & Yoshitake, T. (1998). R&D of Oxide Dispersion Strengthened Ferritic/Martensitic Steels for FBR. *J. Nucl. Mater.* 258–263, pp.1745–1749

Ukai, S. & Fujiwara, M. (2002a). Perspective of ODS Alloys Application in Nucrlear Environment. *J. Nucl. Mater*. 307-311, pp.749-757

Ukai, S., Mizuta, S., Fujiwara, M., Okuda, T. & Kobayashi, T. (2002b). Consolidation Process Study of 9Cr-ODS Martensitic Steel Tubes. *J. Nucl. Mater*. 307-311, pp.758-76

Ukai, S., Kaito, T., Otsuka, S., Narita, T., Fujiwara, M. & Kobayashi, T. (2003). Production and Properties of Nano-Scale 9Cr-ODS Martensitic Steel Claddings. *ISIJ Int*. 43, 12, pp.2038-2045

Ukai, S. & Ohtsuka, S. (2007). Nano-mesoscopic Structure Control in 9Cr-ODS Ferritic Steels. *Energy Mater*. 2, 1, pp.26 - 35

Ukai, S., Ohtsuka, S., Kaito, T., Sakasegawa, H., Chikata, N. & Hayashi, S. (2009). High-Temperature Strength Characterization of Advanced 9Cr-ODS Ferritic Steels. *Materials Science and Engineering*. A 510-511, pp.115-120

Ukai, S. (2011). Oxide Dispersion Strengthened Steels. *Comprehensive Nuclear Materials*. Elsevier, in press

Yamamoto, M., Ukai, S., Hayashi, S., Kaito, T. & Ohtsuka, S. (2010). Formation of Residual Ferrite in 9Cr-ODS Ferritic Steels. *Materials Science and Engineering* A 527, pp.4418 - 4423

Yamamoto, M., Ukai, S., Hayashi, S., Kaito, T. & Ohtsuka, S. (2011). Reverse Phase Transformation from α to γ in 9Cr-ODS Ferritic Steels. J. Nucl. Mater., in press

Yazawa, Y., Furuhara, T. & Maki, T. (2004). Effect of Matrix Recrystallization on Morphology, Crystallography and Coarsening Behavior of Vanadium Carbide in Austenite. *Acta Metall*. 52, pp.3727 - 3736

15

Spider Silk Composites and Applications

Yang Hsia, Eric Gnesa, Felicia Jeffery, Simon Tang and Craig Vierra
University of the Pacific, Department of Biology
USA

1. Introduction

In the past few years a number of reviews, books and journal articles featuring silks have been published. Silks are produced by over 41,000 species of spiders (class Arachnida) and by many insects (terrestrial and aquatic), particularly in the order Lepidoptera (Foelix, 1996); they are defined as externally spun fibrous material generated from protein secretions. The cocoon silk from the domesticated silkmoth, *Bombxy mori*, represents one of the best characterized silks. Cocoon silks are natural composite materials that contain two core silk proteins, along with an outer adhesive protein dubbed sericin. Sericin can be removed by heat and alkaline treatment, leading to about 300 to 1,200 meters of usable fiber from each cocoon. Silk from spiders have also generated considerable scientific potential in recent years. Over the past two decades, scientists have been attempting to unravel the molecular details of spider silks because these composite materials offer a broader range of diverse mechanical properties. Spider silk fibers have remarkable mechanical properties and certain threads are superior to Nylon, Kevlar and high-tensile steel. As advances in the processes for their synthesis, processing, and extrusion continue to emerge it presents scientists with exciting opportunities for developing new classes of biomaterials. Spider silk proteins are synthesized from specialized abdominal glands that function as biofactories to produce large quantities of silk fibroins; these fibroins are spun into silks with different properties, compositions and morphologies. Spider silks have been reported to be used in the South Pacific for gill and dip nets, fishing lures, and in weaving ceremonial dresses. They have also been used clinically as sutures for centuries due to their biocompatibility, slow degradability and high tensile strength. Silks also have been shown to have piezoelectric properties and stability over a range of temperatures (Ando et al., 1980). Although cocoon silk from silkmoths has been harvested by the Chinese for over 5,000 years, the domestication of spiders for large-scale silk collection is impractical due to their cannibalistic nature. Thus, because the development of spider silk farms is improbable, the interest has shifted to the development of recombinant DNA methodologies to clone spider silk genes for expression in transgenic organisms. The cob weaver black widow spider, *Latrodectus hesperus*, along with the orb weaver, *Nephila clavipes*, are rapidly becoming model organisms for retrieving the genetic blueprints from spider silk gene family members for expression studies. The long-term goal of the silk community is to express and purify recombinant silk proteins for a wide-range of applications, including medicine, engineering and defense. Here we cover the following topics: the diversity of spider silks, the composition and mechanical properties of silks, and the natural silk extrusion pathway. By drawing upon these biochemical data and processes, we describe the development of spider silk fiber composites using truncated recombinant spider silk proteins blended with regenerated dragline silks. The application of mixing regenerated spidroins with recombinantly expressed spider silk proteins to produce composite fibers has not been explored by the silk community and this book chapter investigates the aspect of spinning artificial spider silk fibers for biomimetics.

2. Diversity of spider silks

As scientists unravel the secrets of spider silk, it has become evident that a broad range of diverse fiber types has evolved to serve different ecological purposes. These threads have a host of different functions: locomotion, wrapping prey, protection of developing eggs and web construction. The structural diversity of spider silk fibers can be attributed to over 400 million years of natural selection. Anatomical studies of typical orb or cob-weavers reveal the presence of seven distinct silk-producing glands: the major and minor ampullate glands (locomotion and web frame), tubuliform (egg case silk), flagelliform (spiral capture silk), aciniform (prey wrapping silk), aggregate (sticky adhesives on spiral capture or gumfooted lines), and the pyriform glands (attachment disc and joining fibers) (Fig. 1). Molecular and biochemical analyses demonstrate that the different silk-producing glands are highly specialized, expressing specific members of the spider silk gene superfamily (Table 1). Through the evolution of different silk-producing glands, spiders can express gland-specific silk proteins with different sequences, generating a broad range of different fiber types with diverse functional attributes (Guerette et al., 1996).

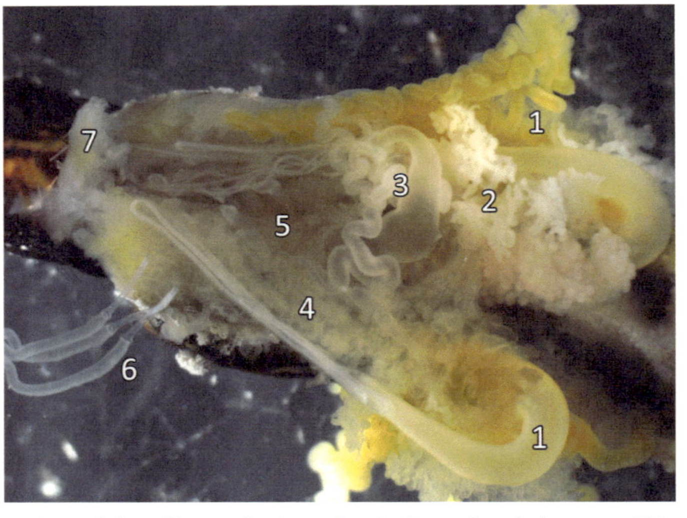

Fig. 1. Microdissection of the silk-producing glands from the abdomen of *N. clavipes* indicates several specialized structures that manufacture silk fibroins. Numbers indicate each gland: 1 (top and bottom – major ampullate), 2 (flagelliform), 3 (minor ampullate), 4 (aggregate), 5 (aciniform), 6 (tubuliform) and 7 (pyriform).

2.1 Major ampullate silks

The major ampullate gland has been the principal focus for researchers investigating the natural silk production pathway. This bulb-shaped gland is found in pairs in spiders and is responsible for producing dragline silk (Fig. 1). Dragline silk is referred to as the 'safety line' of spiders and it is often spun in pairs and is accompanied with two minor ampullate silk threads (Fig. 2A). During locomotion, dragline silk fibers are immobilized to surfaces by structures known as attachment discs. Attachment discs are comprised of small diameter fibers spun from the pyriform gland that are extruded in a sticky substance that dries rapidly, sealing dragline silk to wood, concrete, glass, plants and other materials (Fig. 2B). The mechanical properties of major ampullate silk are dependent upon the amino acid sequences of the spun fibroins and the reeling speed during extrusion. When major ampullate silks are spun rapidly, it produces a stiffer fiber that efficiently supports the weight of spiders. In contrast, spinning major ampullate silk at a slower rate during the web frame construction leads to a fiber with higher extensibility, a necessity for dissipating the energy of prey impact on the web. Transmission electron microscopic analysis reveals dragline silk contains four different layers: core, skin, glycoprotein layer and a lipid coat (Sponner et al., 2007). These studies also demonstrate that dragline silk threads lack adhesive coatings on their surface.

Spider Silks and their Natural Functions in Cob weavers

Silk Gland	Location of Silk	Fiber Proteins
Major ampullate	Web frame, safety line (dragline)	MaSp1; MaSp2
Minor ampullate	Web reinforcement , safety line (dragline)	MiSp1-like
Flagelliform	Spiral capture silk*	Flag*
Aciniform	Wrapping silk, small diameter egg case fiber	AcSp1-like
Tubuliform	Large diameter egg case fiber	TuSp1; ECP-1; ECP-2
Aggregate	Glue coating for gumfooted lines	Unknown
Pyriform	Attachment disk and joining fiber	PySp1

Table 1. Summary of silk-producing glands, fiber types and silk proteins found in the threads of orb- or cob-weavers. (*Only identified in orb-weaver's spiral capture silk)

In orb-weaving spiders the functions for dragline silk include web frame construction and locomotion. Black widow spiders also use dragline silk to build three-dimensional webs for prey capture. We have shown by MS/MS analyses that solubilized proteins from scaffolding, gumfooted and dragline silk fibers, followed by tryptic digestion, produce overlapping peptide fingerprints that match protein sequences within the major ampullate fibroins (unpublished data). These studies demonstrate scaffolding, gumfooted lines and dragline silk consists of major ampullate silks. Because scaffolding silk fibers represent the bulk of a black widow spider's three-dimensional web, it provides an excellent opportunity to collect dragline silk from black widow spiders housed in cages. The collection of large amounts of scaffolding silk can be done rapidly from spiders stored within the laboratory. Milligram quantities can be obtained from a single cage that is 1′ x 1′ x 1.5′. In order to produce regenerated major ampullate silk, we have taken advantage of the fact that scaffolding fibers are a rich source of full-length protein sequences for the major ampullate silk fibroins.

2.2 Egg sac silks

Of the seven different silk types spun from spiders, egg case silks are the easiest to collect. When egg sacs are spun from female spiders, they must endure a variety of different environmental factors. Some of these factors include predators and invasion by parasites or temperature and humidity fluctuations. Black widow spider egg sacs contain two thread types: aciniform and tubuliform silks (Fig. 2C-D). Aciniform fibers have diameter sizes that are approximately 500 nm, which is one order of magnitude smaller than the average 5 μm diameter sizes for tubuliform silks. These fibers are also coated with small droplets, but the chemical contents of the liquid material have not been determined (Fig. 2C-D). In addition to the structural proteins in the fibers, MS/MS analysis has identified the presence of two spider coating peptides, which have been named SCP-1 and SCP-2. Intriguingly, these peptides can be efficiently removed from egg sacs soaked in water, suggesting these peptides are water soluble, polar compounds (Hu et al., 2007). The biological significance of the coating peptides is presently unknown, but one hypothesis is these molecules function as antimicrobial agents that help prolong the longevity of the fibers. Quantitative real-time PCR analysis has demonstrated that SCP mRNA levels are elevated in the flagelliform gland (Blasingame et al., 2009).

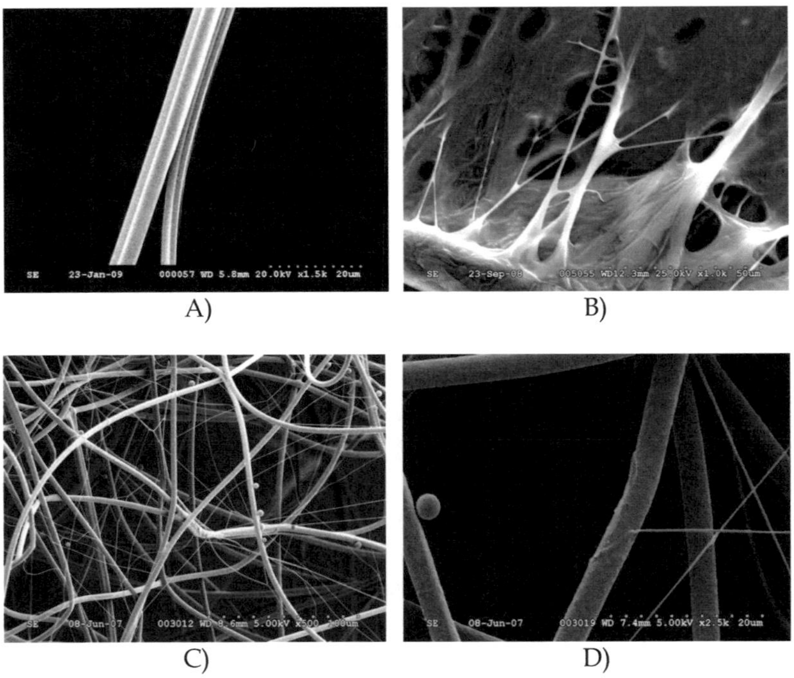

Fig. 2. Scanning electron microscopy of dragline silk, attachment disc fibers and egg sacs from the black widow spider, *L. hesperus.* A) Dragline silk; B) Attachment discs; C) Egg sacs at 500X magnification; D) Egg sacs at 2500X magnification.

In addition to aciniform silks being constituents of egg cases, these fibers are also used for prey capture (Hayashi et al., 2004; Vasanthavada et al., 2007). Because detailed molecular data are available for dragline and tubuliform silks and these fiber types are available in large quantities, we are attempting to study the mechanical behavior of synthetic composite spider silk threads spun from regenerated dragline and egg sac silk protein mixtures. Additionally, we are supplementing regenerated dragline silk fibroins with different recombinantly expressed spider silk proteins. This approach has many advantages, including the ability to have native-sized full-length fibroins in artificially spun fibers with truncated recombinant proteins.

3. Properties and compositions of spider silk

Spider silks represent natural composite materials that are biopolymers. Biochemical analyses reveal that the different silk types contain at least 2-3 distinct structural proteins, commonly referred to as fibroins. Table 1 summarizes the major proteins stored within the luminal contents of several silk-producing glands within *L. hesperus*. With respect to molecular characterization, dragline silk has received the most attention from the scientific community.

3.1 Molecular components of dragline silk

The major and minor ampullate fibroins contain the highest levels of glycine and alanine relative to other spider silk family members; these levels approach >50% of the total amino acid content (Casem et al., 1999; Lombardi & Kaplan, 1990). Biochemical experiments show dragline silk is a composite material largely composed of two structural proteins or spidroins (contraction of the words spider and fibroin) called MaSp1 and MaSp2 (Xu and Lewis, 1990; Hinman and Lewis, 1991). Structural studies demonstrate the major ampullate spidroins form the core of the fiber that is wrapped inside a glycoprotein coat. Although the identities of the constituents of the glycoprotein layer remain unknown, experimental evidence supports this layer is added in the ampulla prior to extrusion (Casem et al., 2002; Sponner et al., 2005). The molecular sequences coding for the dragline silk fibroins were the first to be identified from *N. clavipes* (Xu & Lewis, 1990). Recently, the complete genetic blueprints for MaSp1 and MaSp2 were determined from *L. hesperus* (Ayoub et al., 2007). The predicted sequences for these fibroins encode large molecular weight proteins that are approximately 3,500 amino acids in length. These spidroins are highly modular, each containing internal repetitive block repeats that are flanked by N- and C-terminal non-repetitive ends comprised of approximately 100 amino acids. The internal block repeats are rich in glycine and alanine; these regions form polyalanine or polyalanine-glycine stretches that are interrupted by glycine-rich regions. The polyalanine segments form beta-sheet crystal domains and are responsible for the high tensile strength while the glycine-rich regions adopt 3_1-helix type structures and beta-turns that link the crystalline domains (Simmons et al., 1996). These interconnecting glycine-rich regions constitute the semi-amorphous regions and have been implicated in the extensibility of the fibers. Extensibility of dragline silk fibers also has been attributed to glycine-proline-glycine-X-X (GPGXX) repeats within the MaSp2 protein sequence and the formation of beta-spirals. MaSp2 proteins have been shown to be tightly packed in certain core regions of fibers from *N. clavipes*, whereas MaSp1 appears to be uniformly distributed along the radial axis (Sponner et al., 2005). These data demonstrate that MaSp1 and MaSp2 are not evenly distributed down the long axis of natural fibers. The biochemical mechanisms that modulate their differential localization are not well understood, but could be explained by differences in expression levels and/or their protein sequences that control partitioning during extrusion. Interestingly, in natural fibers, MaSp1 and MaSp2 ratios have been shown to vary between different species (Tso et al., 2005); these differences have been linked to the diet and environment of spiders (Craig et al., 2000). It has been suggested that the synthesis of MaSp2 is more energetically expensive because of the higher cost associated with proline biosynthesis. Therefore, it would appear that spiders synthesize cheap silk when resources are limiting, perhaps producing fibers that contain predominantly MaSp1 molecules. By testing the mechanical properties of composite spider silk fibers spun from regenerated dragline silk fibroins combined with different ratios of recombinant MaSp1 and MaSp2, it will help address the structural roles of the MaSp1 and MaSp2 fibroins. It will also reveal how this spider silk composite behaves from a mechanical perspective. To a large extent, this has not been fully explored because obtaining large amounts of purified recombinant silk proteins for the spinning process is a difficult task.

3.2 Molecular components of tubuliform silks

Tubuliform fibroins have been shown to contain considerably lower levels of glycine relative to dragline silk. Similar to pyriform silks, tubuliform silks incorporate amino acids with more polar side chain groups (Blasingame et al., 2009). Although dragline silk fibers have received much attention in the scientific community, the molecular and structural details for egg case silk fibers are rapidly emerging. Egg sacs are produced by female spiders during reproductive periods. For black widow spiders, egg cases are composite materials containing two different silk types: tubuliform and aciniform (Vasanthavada et al., 2007). The presence of two different fiber types implies tubuliform and aciniform fibers provide a unique role that is central to the protection of spiderlings during development. Biochemical studies have shown that aciniform silks contain the fibroin AcSp1-like (Vasanthavada et al., 2007) and the tubuliform silks contain at least three different proteins: TuSp1, ECP-1 and ECP-2 (Hu et al., 2005; 2006b). Scanning electron microscopy, along with amino acid composition analyses performed on the egg sacs, support that tubuliform silks, based upon volume, constitute the bulk of egg sac fibers. Additionally, mass spectrometry studies (MS/MS) demonstrate that TuSp1 is one of the major constituents of tubuliform silks and the non-repetitive N-terminus is retained in the spun fibroin. Manual inspection of the amino acid sequence of TuSp1 reveals high levels of serine and alanine, but lower levels of glycine. This fibroin lacks the traditional A_n, $(GA)_n$, $(GGX)_n$ and $(GPGXX)_n$ repeats, but it contains new amino acid motifs that include S_n, $(SA)_n$, $(SQ)_n$, and GX (X represents A, D, F, I, N, Q, V and Y). Internal block repeats are ~184 amino acids in length and are extremely homogenous in nature. Short polyalanine stretches are highly iterated in the TuSp1 sequence and ^{13}C NMR spectroscopy demonstrates that the majority of alanine is in a beta-sheet structure in post-spun egg case silk (Hu et al., 2005).

The ECPs (ECP-1 and ECP-2) are also components of tubuliform silks (Hu et al., 2006b). Based on Q-PCR studies, TuSp1 mRNA levels are ~20-fold higher relative to the ECPs. Both ECP-1 and ECP-2 contain large amounts of alanine, glycine and serine. These residues account for more than 50% of all of the amino acids in their translated sequences. Translation of the ECP-1 and ECP-2 cDNAs predict proteins with 932 and 825 amino acids, respectively. The N-terminal 182 residues of ECP-1 and ECP-2 are 70% identical, with 16 cysteine residues showing identical spacing patterns and 100% conservation. Whether these cysteines participate in disulfide bond formation with each other or the N-termini of TuSp1, which has two cysteine residues within its non-repetitive conserved terminus, is unknown. The ECPs lack traditional internal block repeats commonly found in spidroin family members, however, ECP-2 contains scattered, internal blocks of polyalanine and a C-terminus rich in GA repeats. To date, the orb-weaver ECP orthologues have not been reported. The precise role the ECPs serve in tubuliform silks is unclear, but their rich alanine and glycine content, similar to dragline silk fibroins, suggests structural importance.

4. Mechanical properties of spider silk

The mechanical properties of spider silks have been shown to be highly renowned relative to some of the best man-made materials, displaying a combination of high strength, extensibility and toughness. With respect to energy absorption prior to breaking, spider silks are unmatched in the world of synthetic and natural fibers (Table 2). Different spider silk types produce distinct stress-strain curves, demonstrating that spiders spin a broad range of fibers with diverse mechanical properties (Gosline et al., 1986; Hu et al., 2006a). Similar to other protein polymers, there are numerous factors that influence the strength and the load-deformation response of silk fibers, which includes temperature, humidity, and extrusion rate. The majority of mechanical studies have focused on dragline silk. Dragline silk fibers have been subject to tensile testing from a wide range of different spider species. Natural dragline silk fibers show both extremely high tensile strength and toughness (often measured as "Energy to Break"). In man-made polymers, such as polyethylene, increasing strength would often compromise fracture toughness. Yet in the dragline silk, both strength and toughness can be maintained at high levels. The variations in the mechanical behavior in dragline silk fibers across different species suggest that these silks have been fine-tuned by each spider for their nuanced needs. Other spider silk fiber types have also been studied using tensile testing, including aciniform silks (Hayashi et al., 2004), tubuliform silks (Hu et al., 2006a), flagelliform silks (Lewis, 2006) and minor ampullate silks (Liivak et al., 1997).

Fiber	Elongation (%)	Strength (GPa)	Energy to Break (MJ/m^3)
Araneus dragline	27	1.1	160
L. hesperus dragline	34	1	nd
B. mori cocoon silk	18	0.6	150
Flagelliform	>200	1	150
Minor ampullate	30	0.346	nd
Tubuliform	71.7	0.629	nd
Aciniform	80	0.7	290
Nylon fiber	18	0.95	80
Kevlar 49 fiber	2.7	3.6	50
High-tensile steel	0.8	1.5	6

*Data was collected from several sources: Gosline et al. 1999; Hayashi et al., 2004; Livak et al., 1997; Lawrence et al., 2004; Hu et al. 2006a. nd = no data

Table 2. Mechanical properties of spider silks in comparison to other fibers.

4.1 Dragline silk

Dragline silk is five times stronger by weight than steel and three times tougher than the man-made fiber Kevlar, a synthetic fiber used in body armor (Table 2). Correlations between the reeling speed and the tensile strength of dragline silk fibers have been observed experimentally, supporting that spiders can tailor their thread properties by adjusting the draw rate from their spinnerets. By controlling the reeling speeds during extrusion, spiders can tune beta-sheet crystallite size and orientation of the protein polypeptide chain network. These parameters are important structural factors that modulate the tensile strength of silks (Du et al., 2006). Studies investigating the effects of reeling speeds demonstrate 3 nm nanocrystals are optimal for producing fibers that have exceptionally high tensile strength.

Consistent with this observation, molecular modeling stress-strain simulations predict that 3 nm crystals, which correlate to the natural size found in dragline silk, have higher ultimate strengths and toughness relative to fibers with larger nanocrystal sizes (Nova et al., 2010). This implies that spiders can manipulate the material properties of their fibers by changing the reeling speed to tune the size of the beta-sheet nanocrystals.

Spider dragline silk fibers have been shown to be thermally stable to about 230^{0}C on thermal gravimetric analysis. In addition, these fibers have been shown to be able retain their morphology, structure and mechanical properties when autoclaved (Hedhammar et al., 2010). Unlike tubuliform silks, natural dragline silk fibers have been shown to experience supercontraction when exposed to water, resulting in shrinking by up to 50% in length. For certain silk types, supercontraction may provide an important mechanism to tailor silk properties for synthetic composite spider silks.

4.2 Minor, aciniform and flagelliform silk

Similar to major ampullate silks, minor ampullate silks have been shown to display a high Young's modulus (Liivak et al., 1997). In addition, mechanical studies have demonstrated that aciniform silks represent the toughest spider silk type (Hayashi et al., 2004), a likely evolutionary adaptation to help prevent the escape of immobilized prey (Table 2). Previous biochemical analyses performed by our laboratory have demonstrated that aciniform silks are also constituents of egg case silk in black widow spiders. The presence of aciniform silks in egg sacs is intriguing, but it may suggest these fibers function to prevent predators from gaining access to the enclosed eggs. Flagelliform silks, which have only been studied from orb-weavers, are highly extensible fibers that constitute spiral capture silk (Lewis, 2006). Spiral capture silk fibers are strategically positioned in orb webs to absorb and dissipate large amounts of mechanical energy introduced during the collision of aerial insects (Lin et al., 1995). These fibers are extremely compliant, which results in energy dissipation through thread displacement and stretching (Table 2). Whether a functional equivalent of spiral capture silk fibers is present in black widow spiders is unclear. Black widow spiders cast three-dimensional webs, instead of two-dimensional webs that are characteristic of orb weavers. Therefore, it would appear that black widow spiders have replaced FLAG silk, which comprises the spiral capture silk fibers, with scaffolding silk or dragline silk. This observation supports that spiders have evolved specialized silk fibers and different web types to optimize prey capture.

4.3 Tubuliform silks

Tubuliform silks have been shown to be highly extensible, but display lower tensile breaking strength relative to dragline silk fibers (Hu et al. 2006a; Zhao et al., 2006). However, because tubuliform silks have increased extensibility relative to dragline silks, tubuliform fibers display similar toughness (Table 2). To our knowledge, no experiments have been reported to examine the mechanical properties of regenerated tubuliform silk.

5. Natural extrusion of silk

Many laboratories have focused on studying the extrusion process of dragline silk from the major ampullate gland, primarily because this tissue is the easiest to identify and remove. This structure is also the largest relative to the other silk-producing glands. Morphological and histological studies demonstrate the major ampullate gland consists of various components: it has a tail, ampulla, funnel and extrusion duct (Fig. 3).

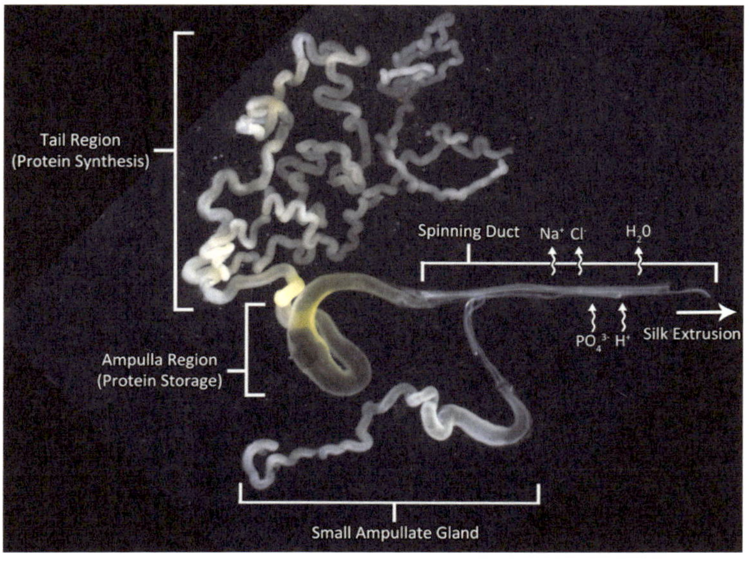

Fig. 3. The major ampullate gland of *N. clavipes* with tail, ampulla and the extrusion duct (spinning duct). Shown connected is the small ampullate gland. Biochemical events known to occur during the extrusion process are also labeled.

Large quantities of major ampullate spidroins are manufactured from columnar epithelial cells residing in the tail, transported to the ampulla through vesicles, followed by extrusion through a duct that tapers down, finally exiting from a spigot positioned on an abdominal spinneret. In our microdissections of spiders, we also observe the presence of a small ampullate gland (SAG) that is connected to the major ampullate gland (Fig. 3) (Ortiz et al., 2000). The SAG is distinct from the minor ampullate gland (not shown) and it has received little attention in scientific literature. Intriguingly, the SAG has its own tail, ampulla, funnel and spinning duct that is fused to the extrusion duct of the major ampullate gland (Fig. 3). The specific biological function of the SAG or the molecular contents it synthesizes, stores and spins is unclear.

5.1 Biochemical processes during silk extrusion

To be able to produce synthetic spider silk composite fibers, scientists must first understand the natural spinning process. Biophysical studies of the liquid contents stored within the major ampulla, also referred to as the spinning dope, reveal these proteins are unfolded and have a disordered secondary structure (Knight et al., 2000; Lefevre et al., 2008). The spinning dope has been shown to represent a highly concentrated aqueous mixture of major ampullate spidroins. During extrusion, these spidroins undergo changes that alter their conformation and orientation. This process, which involves converting native liquid silk into insoluble fibers, is modulated by chemical and mechanical events (Fig. 4).

These events include changes in pH, salt concentration and elongational flow (Chen et al., 2002; Dicko et al. 2004; Knight & Vollrath, 2001). This course of action gives rise to a hierarchically organized semicrystalline material. The crystalline region consists of short polyalanine segments that form beta-sheet nanocrystals, whereas the amorphous phase is formed from glycine-rich segments that likely adopt 3_1-helices and a few alpha helices.

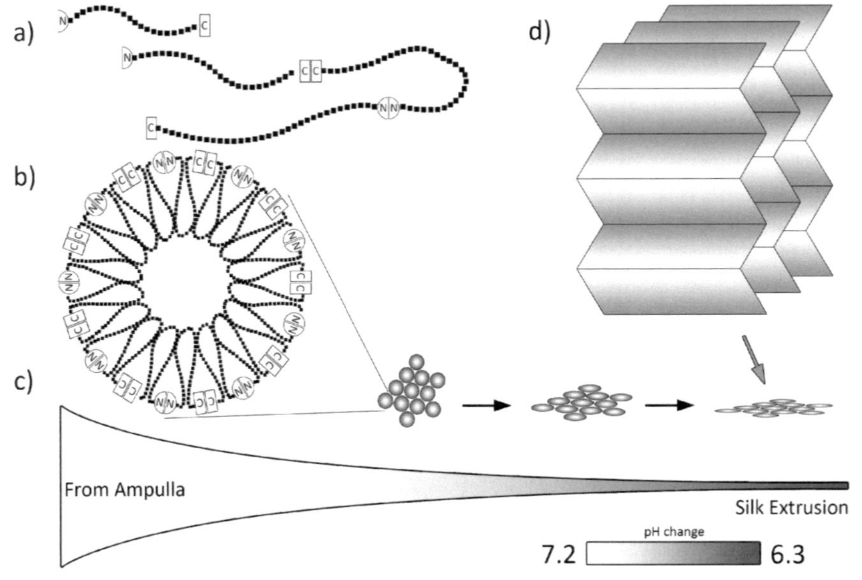

Fig. 4. Biochemical events that occur during the silk extrusion pathway (a-d). a) monomers assembling into multimers via their N- and C-termini; b) micelle and globule-like structure formation; c) fibroin concentration and hydrogen ion levels increase (left to right) as the proteins transition from a liquid to solid state, traveling from the ampulla through the spinning duct; d) Beta-sheet structure of crystallite regions.

Histological data reveal that H^+ pump activity increases steadily along the length of the spinning duct (Vollrath et al., 1998). Studies using a pH-sensitive microelectrode probe support this observation, demonstrating the pH decreases from 7.2 in the major ampullate gland to 6.3 in the extrusion duct (Dicko et al., 2004). Other chemical changes have been implicated in the transition from liquid to solid during fiber production. These include a decrease in Na^+ ions, coupled with increased K^+ and phosphate ion concentrations during extrusion (Huemmerich et al., 2004) (Fig. 3). The pH decrease in the extrusion duct has been associated with changes in the ionization state of side chain residues positioned within the conserved N- and C-termini of spidroins, promoting dimerization and fiber polymerization. Although the N- and C-termini both contain conserved non-repetitive regions, their amino acid sequences are dissimilar. However, despite the differences in amino acid sequences, there are some similarities in their conformational flexibility and secondary structures. Recent structural studies of the MaSp1 N-terminal domain (NT) from the nursery web spider, *Euprosthenops australis*, demonstrate this region mediates protein assembly (Askarieh et al., 2010; Hagn et al., 2010). Using recombinant NT mini-spidroins, it has been shown that charge-dependent self-assembly occurs at pH values around 6.3 and promotes dimeric formation, whereas at pH values around 7, the NT domain delays protein aggregation and the spidroins remain in a monomeric state. Although pH changes have been implicated in the assembly process, it is unclear whether other non-fibroin proteins also participate in enhancing fibroin solubility or folding. Peroxidases have been detected in the luminal contents of the major ampulla, but their biological function and contribution to fiber assembly has not been established (Pouchkina et al., 2003). Future studies of the glandular contents using mass spectrometry, combined with transcriptome analyses of the different silk-producing glands, should help lead to the identification of new molecular components that modulate the assembly process. These findings will undoubtedly have a substantial impact on manufacturing strategies that integrate silk proteins in composite materials.

The mechanisms of fiber extrusion in the other silk-producing glands have not been thoroughly investigated at the biochemical level. Analyses of the morphological features of these silk-producing glands reveal profound structural differences, aside from the architecture of the minor ampullate gland, which closely resembles the major ampullate gland (Jeffery et al., 2011). These data imply that the mechanisms for fiber extrusion are distinct in the different glands that produce silk. Raman spectroscopy studies performed on the glandular contents of the tubuliform gland support this supposition, which demonstrate proteins are stored in a globular conformation with a high degree of alpha helical nature; this differs significantly from the conformations of fibroins stored within the major ampullate gland (Lefevre et al., 2011). However, after spinning, the tubuliform silks are structurally similar to major ampullate fibers, containing a high degree of beta-sheet structure. NMR studies have been performed to analyze the secondary and tertiary structures of *N. antipodiana* TuSp1. Recombinant proteins containing either the N- or C-termini attached to one repetitive domain spontaneously assemble into micelle-like structures that can be converted into fibers at ≥ 37°C (Lin et al., 2009). The N- and C-termini, which are less hydrophilic relative to the internal block repeats, assemble and form the outer region of the micelles, whereas the internal block repeats become buried on the inside of the micelles; these internal block repeats exist in an unfolded state (Lin et al., 2009). During extrusion the concentration of silk fibroin in the gland gradually increases to form micelles, which further form globule-like structures and gels. These structures then experience shear force, resulting in the conversion of random coil protein structures into beta-sheet structures. Solid state NMR conducted on egg case silk fibers is consistent with this transition and supports the presence of beta-sheet structures in spun fibers.

6. Synthetic composite spider fibers from recombinant proteins

Although it is feasible to harvest milligrams of natural dragline silk by forcibly "milking" spiders or retrieving large amounts of tubuliform silks through the collection of egg sacs, these are expensive and time-consuming endeavors. For example, a 3.4 m textile weaved from natural dragline silk took more than one million *N. madagascarenis* spiders and the efforts of over 70 individuals working 4 years; the total cost exceeded over $500,000. Moreover, attempts to farm spiders have proven unsuccessful given their intrinsic cannibalistic nature. Because of these difficulties, scientists have turned to using recombinant DNA technologies to express spider silk proteins in transgenic systems. Expression of silk proteins in heterologous systems has much promise; however, several challenges still need to be resolved before artificial silk fibers can be spun from recombinant proteins on an industrial scale. These include the development expression systems that manufacture large quantities of full-length recombinant silk protein in a fast, cost effective manner, and the second involves the development of spinning apparatuses that produce silk fibers with mechanical properties approaching natural threads, preferably without the accumulation of toxic waste products during the manufacturing process. Since harvesting large quantities of silk from domesticated spiders for industrial applications is impractical, recombinant spider silk production in either prokaryotic and/or eukaryotic organisms will need to be developed to ensure large amounts of soluble protein are available for artificial fiber production.

6.1 Silk fibroin expression

Both prokaryotic and eukaryotic expression systems have their advantages and disadvantages. For example, many eukaryotic proteins require posttranslational modifications for functional activity, such as glycosylation and phosphorylation. Although the significance of posttranslational modifications (PTMs) to fibroins remains unclear, amino acid residues have been shown to be subject to phosphorylation and glycosylation within fibroin sequences (Michal et al., 1996). Because PTMs are largely restricted to eukaryotic cells, expression systems equipped with the cellular machinery for PTMs offer benefits to potentially achieving fibroin states that resemble *in vivo* conditions. There have been several successful reports of partial or truncated fibroins expressed in prokaryotes, in particular *Escherichia coli*. However, the expression of full-length fibroin cDNAs have not been reported in either prokaryotic or eukaryotic expression systems. Both expression and replication of spider silk cDNAs has been hampered by their long, repetitive and guanine-cytosine-rich nature. These intrinsic properties have presented technical challenges with sequencing spider silk genes, the ability to maintain intact transcription units for expression studies, and the translation of long mRNAs dominated by alanine and glycine codons. The inability to express full-length fibroins has been attributed, in part, to the depletion of the tRNA pools, low solubility of the fibroins during expression and proteolysis of the recombinant silk proteins. Several different cloning strategies have been applied to overcome these barriers, including codon optimization of silk genes, specialized growth media and conditions, as well as use of a variety of different host organisms for recombinant silk production (Fahnestock and Irwin, 1997; Schmidt et al., 2007). Various plants have been explored as model systems for spider silk expression, including *Arabidopsis* (Yang et al., 2005), tobacco and potato (Scheller et al., 2001). Mammalian cell culture systems have shown some promising success for expressing large molecular weight fibroins using the immortalized cell lines from bovine mammary epithelial alveolar cells and baby hamster kidney cells (Lazaris et al., 2002). Transgenic mice and goats that express spidroins also have been generated, but these animals have resulted in the expression of low levels of fibroins and have not proved to be cost effective for large-scale production of silk proteins (Xu et al., 2007). Efforts to use the methylotrophic yeast *Pichia pastoris*, a robust eukaryotic expression system that allows for recombinant proteins to be secreted into liquid growth media, have also been explored but with limited success (Fahnestock & Bedzyk, 1997). Perhaps the most success for producing recombinant spider silk proteins that approach native sizes has come from the development of a metabolically engineered *E. coli* strain (Xia et al., 2010). However, despite this most promising case, the levels of expressed proteins that can be readily purified for the spinning process still remain too low for industrial scale production.

6.2 Artificial fiber production

Despite the broad range of expression systems used to synthesize recombinant spider silk proteins, only a handful of research groups have spun artificial silk fibers using the recombinant spidroins (Geurts et al., 2010; Lazaris et al., 2002; Teule et al., 2009). Even fewer research labs have subsequently used these synthetic fibers to compare their mechanical properties to the natural fibers (Xia et al., 2010). For most labs that have participated in the production of recombinant silk fibroins, strategies have included using several block repeats and/or the non-repetitive C-terminal domain for expression studies. The majority of these studies have used either the major ampullate spidroin 1 or 2 (MaSp1 or MaSp2). *Araneus diadematus* factor 3 and 4 (ADF-3 and ADF-4), two variants of MaSp2, have been used in a number of studies. Recombinant MaSp1 spidroins have been produced using natural cDNA sequences or multimerized synthetic gene pieces assembled from oligonucleotides (Fahnestock & Bedzyk, 1997). In both cases overexpression of the truncated recombinant proteins has often resulted in products that are water insoluble or precipitate out of solution during protein purification. Several different strategies have been attempted to increase the fibroin solubility during expression. One approach has been to express fusion proteins that contain N- or C-terminal solubility-enhancing fusion partners such as thioredoxin (Stark et al., 2007). Other strategies have included the incorporation of enzymatic phosphorylation/dephosphorylation switches (Szela et al., 2000; Winkler et al., 2000) or the integration of oxidation/reduction residues into the fibroin amino acid sequence to control the assembly process (Winkler et al., 1999). Although these methods have led to improved solubility, premature aggregate formation has been difficult to control during expression and protein purification. Currently, solubilization of aggregates is often achieved using hexafluoroisopropanol (HFIP), urea, guanidine hydrochloride, LiBr or formic acid. Treatment of the aggregates with these solvents has become necessary to obtain sufficient amounts of solubilized recombinant silk proteins for the artificial spinning process, which requires 15-50% of spinning dope. However, because these solvents are hazardous and removal of them from the spun threads is challenging, alternative approaches that integrate spinning fibers from a water-based solvent will need to be developed and applied to ensure the successful use of fibers for biomimetics.

6.2.1 Composite silk fibers spun from regenerated major ampullate silk spidroins blended with recombinant proteins

Although synthetic silk fibers have been spun from recombinant spider silk proteins, silk filaments have also been reported from dissolving natural dragline silk and re-spinning the solubilized fibroin extract, a process that leads to a material known as regenerated silk. Regenerated dragline spider silk has been described in the scientific literature just a few times. These experiments have demonstrated reconstituted dragline spider silk self-assembles into filaments from *N. edulis* (Shao et al., 2003). Additionally, these fibers were shown to contain similar amino acid compositions relative to natural silks, suggesting the molecular assembly process could be mimicked *in vitro*. For these studies, the investigators focused on gleaning the molecular details about the spinning process.

In our attempt to enhance the properties of regenerated dragline spider silk, we have supplemented this silk with recombinantly expressed truncated dragline silk fibroins, MaSp1 and MaSp2. These truncated forms, referred to as $1RCT_{MaSp1}$ and $1RCT_{MaSp2}$, consisted of one block repeat fused to the non-repetitive, conserved C-terminus. To facilitate recombinant protein solubility during induction in *E. coli*, we added an N-terminal thioredoxin tag. Recombinant proteins were purified by using a 6x-his tag placed on the C-terminus. Following purification, different ratios of $1RCT_{MaSp1}$ and $1RCT_{MaSp2}$ were combined with regenerated scaffolding silk fibroins and spun into synthetic fibers. To obtain regenerated fibroins, scaffolding fibers were collected from cages and dissolved overnight in hexafluroisopropanol (HFIP) at a final concentration 15% (w/v). HFIP is a commonly used spinning solvent. Reconstituted scaffolding fibers and blended filaments were spun using a previously published protocol (Teule et al., 2009). Mixing different ratios of the recombinant proteins led to reconstituted composite fibers that morphologically resembled fibers spun from regenerated spider silk alone (Fig. 5A-D). These fibers had a fibrillar structure and contained diameter sizes ranging from 33-49 μm. The morphology of the regenerated as well as the blended fibers was also similar to native dragline silk fibers (Compare Fig. 2A with 5A-D). We also examined the fiber interior core after breaking the threads. Analyses of the fractured regenerated fibers showed a solid, smooth interior (Fig. 5E) whereas the blended silk fibers were more fibrous-like (Fig. 5F). The more fibrous-like interior may be the result of fibroin assembly from truncated proteins.

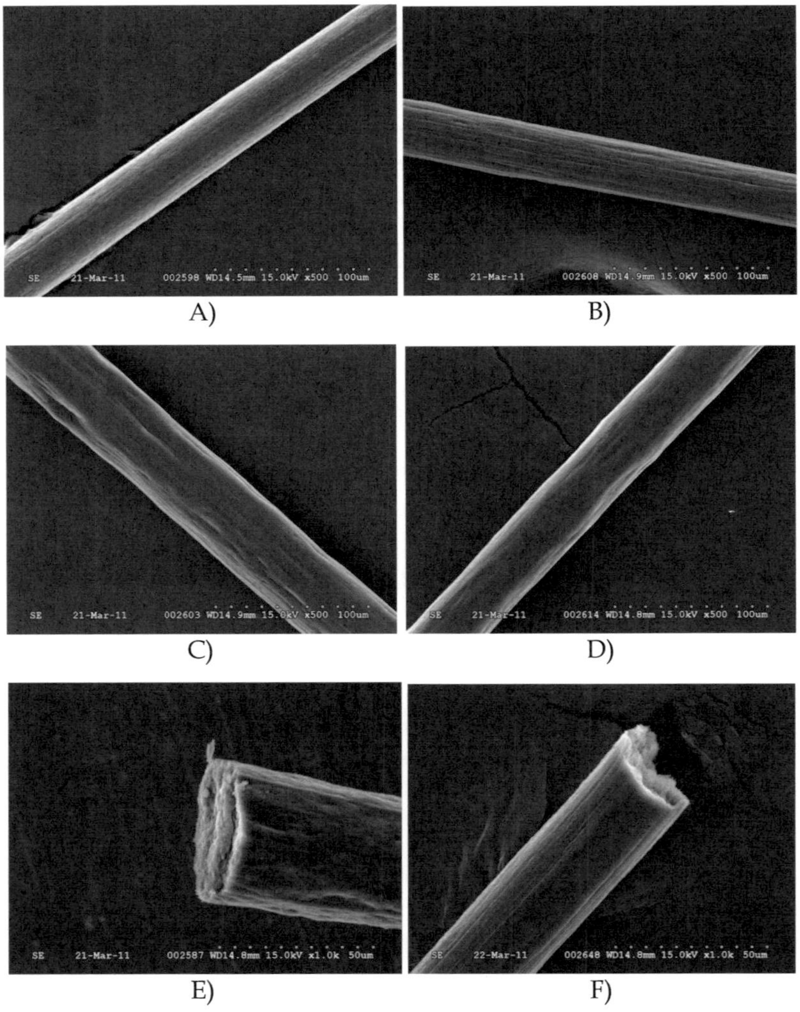

Fig. 5. SEM images of the surface of regenerated dragline silk fibers and blended fibers supplemented with recombinant MaSp1, MaSp2 or both spidroins. A) Regenerated dragline silk; B) Regenerated $+1RCT_{MaSp1}$ (1:1); C) Regenerated + $1RCT_{MaSp2}$ (1:1); D) Regenerated $+1RCT_{MaSp1}$ $+1RCT_{MaSp2}$ (2:1:1). Fibroin mixture ratios are indicated in parentheses. All images were collected at a magnification of 500-1000X.

In addition to morphological differences, the fibers of varying formulations also exhibit markedly different mechanical behavior. When compared with the native dragline, the regenerated dragline silk has reduced modulus, extensibility, and toughness (Fig. 6). Supplementing the regenerated dragline silk with either recombinant $1RCT_{MaSp1}$ or $1RCT_{MaSp2}$ increases the elastic modulus, tensile strength, and toughness over the regenerated dragline silk alone. However, supplementing with both recombinant $1RCT_{MaSp1}$ and $1RCT_{MaSp2}$ appears to diminish the mechanical enhancements relative to either fibroin alone, suggesting that specific combinations of fibroins may provide optimal mechanical behavior. The molecular basis is currently being investigated. Taken together, our synthetic composite dragline silk fibers display higher tensile strength and extensibility relative to previous reports involving artificial fibers spun from truncated recombinant MaSp1 and MaSp2 lacking their C-termini (Brooks et al. 2008). This supports that the C-terminus contributes to enhancing the mechanical properties of the blended fibers.

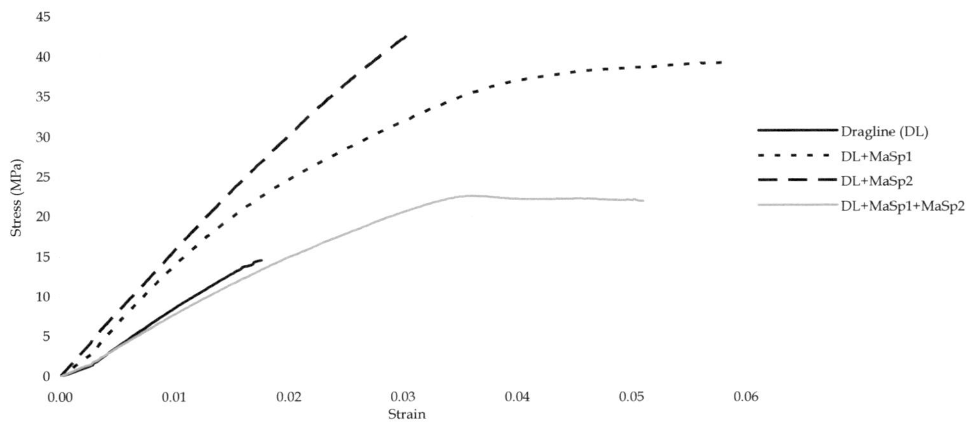

Fig. 6. Stress-strain curve performed using regenerated dragline and blended fibers containing MaSp1 and/or MaSp2. Plotted data are from a single fiber, but are representative of at least 4 different fibers.

6.2.2 Composite silk fibers spun from regenerated aciniform and tubuliform silk proteins

Regenerated spider silk fibers spun from reconstituted egg sacs have not been reported in the scientific literature. Here we demonstrate that solubilization of black widow egg sacs with HFIP, followed by wet-spinning of the fibers into an isopropanol coagulation bath, can lead to a composite filament containing the fibroins AcSp1-like, TuSp1 and the ECP silk proteins. These artificial spider silk composites were spun from 20% (w/v) solutions of dissolved egg sacs. These are the first reported composite silk fibers that include native sized aciniform and tubuliform silk proteins blended together into a single filament. SEM studies demonstrated that the reconstituted egg case fibers contained diameter sizes approximately 90 μm (Fig. 7A). Analysis of the surface morphology of the fibers revealed more irregularities relative to the natural tubuliform silks (Compare Fig. 2D with 7A). The interior of fractured regenerated egg case fibers showed a solid core, suggesting that fibroin assembly was not impaired during the spinning process (Fig. 7B). Stress-strain curves performed on individual threads demonstrated these composite fibers were 2-fold lower and an order of magnitude below the tensile strength of natural tubuliform silk (Fig. 7C). The % strain and tensile strength was 30% and 83 MPa, respectively. In the future, we plan on testing the biocompatibility of these fibers to determine their suitability for applications, such as tissue engineering.

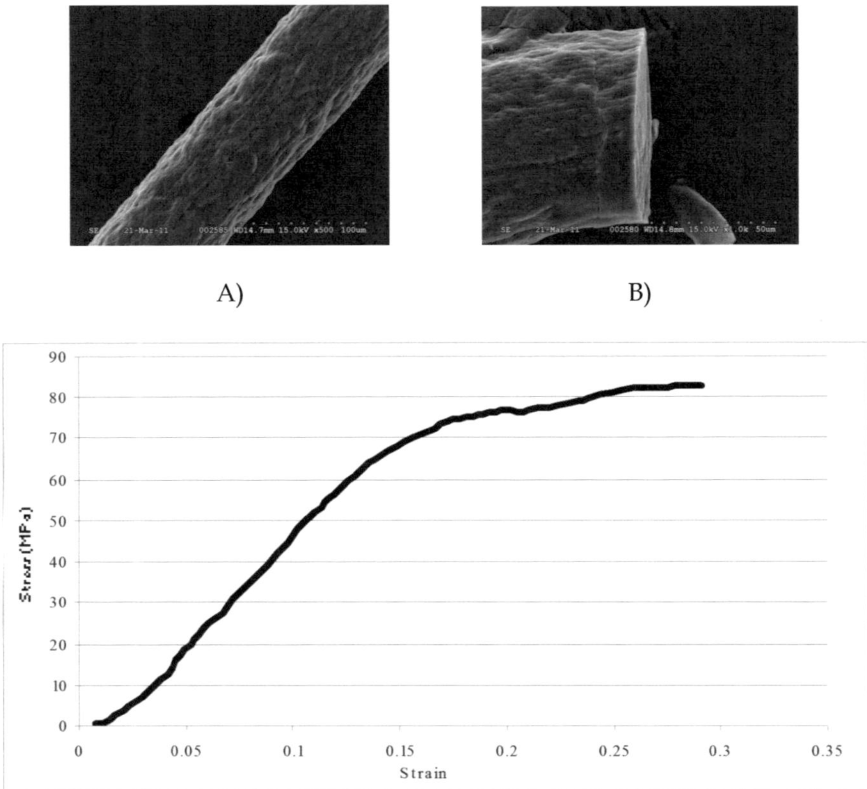

C)

Fig. 7. Synthetic composite spider silk threads spun from regenerated egg case silk proteins. A-B) SEM analysis of regenerated and fractured egg case silk fibers, respectively; C) Stress-strain curve performed on regenerated egg case silk fibers.

7. Other applications

Methodologies to spin artificial silk fibers on a large-scale for industrial, medical, military and commercial use are progressing rapidly. In addition to spinning silk threads for a host of applications other uses for spider silk proteins are being pursued, including drug delivery and scaffolding materials. Both dragline silk fibers and egg sacs have been studied for their biocompatibility in a few human cell culture systems (Allmeling et al., 2006; Gellynck et al., 2008). Primary chondrocytes have been shown to attach and survive for several weeks on natural dragline silk and egg sacs (Gellynck et al., 2008). Natural dragline silk fibers have also been used to culture human primary Schwann cells and demonstrate these cells can adhere and elongate (Allmeling et al., 2006). *In vivo* biocompatibility of native dragline silks has also been demonstrated by subcutaneous implantation in pigs (Vollrath et al., 2002), mice and rats (Gellynck et al., 2008). These studies have revealed that spider silk is well tolerated and generates little, if any, inflammatory response in the host animal.

Thus far, few *in vitro* studies have investigated the biocompatibility of recombinant spider silk fibroins. Because spidroin family members have different amino acid sequences within their block repeats, each member will require testing to determine whether certain family members are more suitable relative to the others. Advances in molecular biological techniques have facilitated the production of spider silk fibroins in sufficient quantities for the fabrication of various scaffolds. Three-dimensional scaffolds fabricated from recombinant dragline silk proteins have been shown to allow the attachment and migration of mouse fibroblast cells into the deep layers of the scaffolds within 7 days, implying that dragline silk fibers have excellent biocompatibility, stability and pore interconnectivity (Agapov et al., 2009). Additionally, recombinant spider silk proteins engineered with RGD cell-binding domains have been used successfully as biomaterial matrixes to enhance the differentiation of human bone marrow derived mesenchymal stem cells to bone-like tissue (Bini et al., 2006). Other uses for spider silk proteins include the production of recombinant spider silk particles as drug delivery vesicles (Lammel et al., 2011). Therefore, the use of spider silk fibroins for a host of different applications is currently being pursued in the scientific community.

8. Conclusion

This chapter highlights how regenerated dragline and tubuliform silk fibroins can be used to generate a spinning dope suitable for synthetic fiber production. It also explores the use of blending truncated recombinantly expressed dragline silk fibroins MaSp1 and MaSp2 with reconstituted dragline silk fibroins to generate composite spider silk fibers. With the identification of different spider silk cDNAs, coupled with the advancements in heterologous protein expression systems and the spinning process, the timing to explore the production of new engineering materials that integrate different mixtures of recombinant silk proteins and reconstituted silk proteins provides an exciting time for silk biology research. We have taken a novel approach, combining regenerated silk fibroins with recombinant proteins, to study the material properties of synthetic composite spider silk fibers. Our studies demonstrate that infusion of recombinant proteins into solubilized regenerated silk fibroin mixtures is a valid method to study fibroin function in spun composite silk fibers. It is also a method to generate a vast number of diverse synthetic composite fibers with a broad range of mechanical functions.

9. Acknowledgment

We thank Anne Moore for the donation of her tensometer for the mechanical studies. In addition, we thank Kaitlen Colafrancesco and Coby La Mattina for their critical comments on the book chapter. We also are indebted to Dr. Mark Brunell at the University of the Pacific, Department of Biology, for his assistance with the scanning electron microscope. This work was supported by a NSF RUI Grant MCB-0950372 entitled *Molecular Characterization of Black Widow Spider Silks*.

10. References

Agapov, I., Pustovalova, O. L., Moisenovich, M. M., Bogush, V. G., Sokolova, O. S., Sevastyanov, V. I., et al. (2009). Three-dimensional scaffold made from recombinant spider silk protein for tissue engineering. *Dokl Biochem Biophys, 426*, 127-130.

Allmeling, C., Jokuszies, A., Reimers, K., Kall, S., & Vogt, P. M. (2006). Use of spider silk fibres as an innovative material in a biocompatible artificial nerve conduit. *J Cell Mol Med, 10*(3), 770-777.

Ando, Y., Okano, R., Nishida, K., Miyata, S., & Fukade, E. (1980). Piezoelectric and related properties of hydrated silk fibroin. *Rep Prog Polymer Physics Japan, 23*, 775-778.

Askarieh, G., Hedhammar, M., Nordling, K., Saenz, A., Casals, C., Rising, A., Johansson, J., & Knight, S. (2010). Self-assembly of spider silk proteins is controlled by a pH-sensitive relay. *Nature, 465*(7295), 236-238.

Ayoub, N. A., Garb, J. E., Tinghitella, R. M., Collin, M. A., & Hayashi, C. Y. (2007). Blueprint for a high-performance biomaterial: full-length spider dragline silk genes. *PLoS ONE, 2*, e514.

Bini, E., Foo, C. W., Huang, J., Karageorgiou, V., Kitchel, B., & Kaplan, D. L. (2006). RGD-functionalized bioengineered spider dragline silk biomaterial. *Biomacromolecules, 7*(11), 3139-3145.

Blasingame, E., Tuton-Blasingame, T., Larkin, L., Falick, A. M., Zhao, L., Fong, J., Vaidyanathan, V., Visperas, A., Geurts, P., Xu, X., La Mattina, C., & Vierra, C. (2009). Pyriform spidroin 1, a novel member of the silk gene family that anchors dragline silk fibers in attachment discs of the black widow spider, *Latrodectus hesperus*. *J Biol Chem, 284*(42), 29097-29108.

Brooks, A. E., Nelson, S. R., Jones, J. A., Koenig, C., Hinman, M., Stricker, S., & Lewis, R. (2008). Distinct contributions of model MaSp1 and MaSp2 like peptides to the mechanical properties of synthetic major ampullate silk fibers as revealed in silico. *Nanotechnol Sci Appl, 1*, 9-16.

Casem, M. L., Turner, D., & Houchin, K. (1999). Protein and amino acid composition of silks from the cob weaver, *Latrodectus hesperus* (black widow). *Int. J. Biol. Macromol., 24*(2-3), 103-108.

Casem, M. L., Tran, L. P., & Moore, A. M. (2002). Ultrastructure of the major ampullate gland of the black widow spider, *Latrodectus hesperus*. *Tissue Cell, 34*(6), 427-436.

Chen, X., Knight, D. P., & Vollrath, F. (2002). Rheological characterization of *Nephila* spidroin solution. *Biomacromolecules, 3*(4), 644-648.

Craig, C. L., Riekel, C., Herberstein, M. E., Weber, R. S., Kaplan, D., & Pierce, N. E. (2000). Evidence for diet effects on the composition of silk proteins produced by spiders. *Mol. Biol. Evol., 17*(12), 1904-1913.

Dicko, C., Vollrath, F., & Kenney, J. M. (2004). Spider silk protein refolding is controlled by changing pH. *Biomacromolecules, 5*(3), 704-710.

Du, N., Liu, X. Y., Narayanan, J., Li, L., Lim, M. L., & Li, D. (2006). Design of superior spider silk: from nanostructure to mechanical properties. *Biophys J, 91*(12), 4528-4535.

Fahnestock, S. R., & Bedzyk, L. A. (1997). Production of synthetic spider dragline silk protein in *Pichia pastoris*. *Appl. Microbiol. Biotechnol., 47*(1), 33-39.

Foelix, R. (1996). *Biology of spiders*. New York: Oxford University Press.

Gellynck, K., Verdonk, P. C., Van Nimmen, E., Almqvist, K. F., Gheysens, T., Schoukens, G., Van Langenhove, L., Kiekens, P., Mertens, J., & Verbruggen, G. (2008). Silkworm and spider silk scaffolds for chondrocyte support. *J Mater Sci Mater Med, 19*(11), 3399-3409.

Gosline, J. M., DeMont, M. E., Denny, M. W. (1986). The structure and properties of spider silk. *Endeavour, 10*, 31-43.

Gosline, J.M., Guerette, P.A., Ortlepp, C.S., & Savage, K.N. (1999). The mechanical design of spider silks: from fibroin sequence to mechanical function. *J. Exp. Biol.*, 202(Pt 23), 3295-3303.

Geurts, P., Zhao, L., Hsia, Y., Gnesa, E., Tang, S., Jeffery, F., La Mattina, C., Franz, A., Larkin, L., & Vierra, C. Synthetic spider silk fibers spun from Pyriform Spidroin 2, a glue silk protein discovered in orb-weaving spider attachment discs. *Biomacromolecules, 11*(12), 3495-3503.

Guerette, P. A., Ginzinger, D. G., Weber, B. H., & Gosline, J. M. (1996). Silk properties determined by gland-specific expression of a spider fibroin gene family. *Science, 272*(5258), 112-115.

Hagn, F., Eisoldt, L., Hardy, J. G., Vendrely, C., Coles, M., Scheibel, T., & Kessler, H. (2010). A conserved spider silk domain acts as a molecular switch that controls fibre assembly. *Nature, 465*(7295), 239-242.

Hayashi, C. Y., Blackledge, T. A., & Lewis, R. V. (2004). Molecular and mechanical characterization of aciniform silk: uniformity of iterated sequence modules in a novel member of the spider silk fibroin gene family. *Mol. Biol. Evol., 21*(10), 1950-1959.

Hedhammar, M., Bramfeldt, H., Baris, T., Widhe, M., Askarieh, G., Nordling, K., Aulock, S., & Johansson, J. (2010). Sterilized recombinant spider silk fibers of low pyrogenicity. *Biomacromolecules, 11*(4), 953-959.

Hinman, M. B., & Lewis, R. V. (1992). Isolation of a clone encoding a second dragline silk fibroin. *Nephila clavipes* dragline silk is a two-protein fiber. *J. Biol. Chem., 267*(27), 19320-19324.

Hu, X., Lawrence, B., Kohler, K., Falick, A. M., Moore, A. M., McMullen, E., Jones, P., & Vierra, C. (2005). Araneoid egg case silk: a fibroin with novel ensemble repeat units from the black widow spider, *Latrodectus hesperus*. *Biochemistry, 44*(30), 10020-10027.

Hu, X., Vasanthavada, K., Kohler, K., McNary, S., Moore, A. M., & Vierra, C. A. (2006a). Molecular mechanisms of spider silk. *Cell Mol Life Sci, 63*(17), 1986-1999.

Hu, X., Kohler, K., Falick, A. M., Moore, A. M., Jones, P. R., & Vierra, C. (2006b). Spider egg case core fibers: trimeric complexes assembled from TuSp1, ECP-1, and ECP-2. *Biochemistry, 45*(11), 3506-3516.

Hu, X., Yuan, J., Wang, X., Vasanthavada, K., Falick, A. M., Jones, P. R., La Mattina, C. & Vierra, C. (2007). Analysis of aqueous glue coating proteins on the silk fibers of the cob weaver, *Latrodectus hesperus*. *Biochemistry, 46*(11), 3294-3303.

Huemmerich, D., Helsen, C. W., Quedzuweit, S., Oschmann, J., Rudolph, R., & Scheibel, T. (2004). Primary structure elements of spider dragline silks and their contribution to protein solubility. *Biochemistry, 43*(42), 13604-13612.

Jeffery, F., La Mattina, C., Tuton-Blasingame, T., Hsia, Y., Gnesa, E., Zhao, L., Franz, A., & Vierra, C. (2011) Microdissection of black widow spider silk-producing glands. *J Vis Exp*(47).

Knight, D. P., Knight, M. M., & Vollrath, F. (2000). Beta transition and stress-induced phase separation in the spinning of spider dragline silk. *Int. J. Biol. Macromol., 27*(3), 205-210.

Knight, D. P., & Vollrath, F. (2001). Changes in element composition along the spinning duct in a *Nephila* spider. *Naturwissenschaften, 88*(4), 179-182.

Lammel, A., Schwab, M., Hofer, M., Winter, G., & Scheibel, T. (2011) Recombinant spider silk particles as drug delivery vehicles. *Biomaterials, 32*(8), 2233-2240.

Lawrence, B.A., Vierra, C.A., & Moore, A.M. (2004). Molecular and mechanical properties of major ampullate silk of the black widow spider, *Latrodectus hesperus*. *Biomacromolecules,* 5(3), 689-695.

Lazaris, A., Arcidiacono, S., Huang, Y., Zhou, J. F., Duguay, F., Chretien, N., Welsch, E., Soares, J., & Karatzas, C. (2002). Spider silk fibers spun from soluble recombinant silk produced in mammalian cells. *Science, 295*(5554), 472-476.

Lefevre, T., Boudreault, S., Cloutier, C., & Pezolet, M. (2008). Conformational and orientational transformation of silk proteins in the major ampullate gland of *Nephila clavipes* spiders. *Biomacromolecules, 9*(9), 2399-2407.

Lefevre, T., Boudreault, S., Cloutier, C., & Pezolet, M. (2011). Diversity of molecular transformations involved in the formation of spider silks. *J Mol Biol, 405*(1), 238-253.

Lewis, R. V. (2006). Spider silk: ancient ideas for new biomaterials. *Chem Rev, 106*(9), 3762-3774.

Lin, L. H., Edmonds, D. T., & Vollrath, F. (1995). Structural engineering of an orb-spider's web. *Nature, 373*, 146-148.

Lin, Z., Huang, W., Zhang, J., Fan, J. S., & Yang, D. (2009). Solution structure of eggcase silk protein and its implications for silk fiber formation. *Proc Natl Acad Sci U S A, 106*(22), 8906-8911.

Liivak, O., Flores, A., Lewis, L., & Jelinski, L. W. (1997). Conformation of the polyalanine repeats in minor ampullate gland silk of the spider *Nephila clavipes*. *Macromolecules, 30*, 7127-7130.

Lombardi, S., & Kaplan, D. L. (1990). The Amino Acid Composition of Major Ampullate Gland Silk (Dragline) of *Nephila clavipes* (Araneae, Tetragnathidae). *J. Arachnol., 18*, 297-306.

Michal, C. A., Simmons, A. H., Chew, B. G., Zax, D. B., & Jelinski, L. W. (1996). Presence of phosphorus in *Nephila clavipes* dragline silk. *Biophys J, 70*(1), 489-493.

Nova, A., Keten, S., Pugno, N. M., Redaelli, A., & Buehler, M. J. (2010). Molecular and nanostructural mechanisms of deformation, strength and toughness of spider silk fibrils. *Nano Lett, 10*(7), 2626-2634.

Ortiz, R., Cespedes, W., Nieves, L., Robles, I. V., Plazaola, A., File, S., et al. (2000). Small ampullate glands of *Nephila clavipes*. *J Exp Zool, 286*(2), 114-119.

Pouchkina, N. N., Stanchev, B. S., & McQueen-Mason, S. J. (2003). From EST sequence to spider silk spinning: identification and molecular characterisation of *Nephila senegalensis* major ampullate gland peroxidase NsPox. *Insect Biochem Mol Biol, 33*(2), 229-238.

Scheller, J., Guhrs, K. H., Grosse, F., & Conrad, U. (2001). Production of spider silk proteins in tobacco and potato. *Nat. Biotechnol., 19*(6), 573-577.

Schmidt, M., Romer, L., Strehle, M., & Scheibel, T. (2007). Conquering isoleucine auxotrophy of *Escherichia coli* BLR(DE3) to recombinantly produce spider silk proteins in minimal media. *Biotechnol Lett, 29*(11), 1741-1744.

Shao, Z., Vollrath, F., Yang, Y., & Thogersen, H. (2003). Structure and Behavior of Regenerated Spider Silk. *Macromolecules, 36*, 1157-1161.

Simmons, A. H., Michal, C. A., & Jelinski, L. W. (1996). Molecular orientation and two-component nature of the crystalline fraction of spider dragline silk. *Science, 271*(5245), 84-87.

Sponner, A., Unger, E., Grosse, F., & Weisshart, K. (2005). Differential polymerization of the two main protein components of dragline silk during fibre spinning. *Nat. Mater., 4*(10), 772-775.

Sponner, A., Vater, W., Monajembashi, S., Unger, E., Grosse, F., & Weisshart, K. (2007). Composition and hierarchical organisation of a spider silk. *PLoS ONE, 2*(10), e998.

Stark, M., Grip, S., Rising, A., Hedhammar, M., Engstrom, W., Hjalm, G., et al. (2007). Macroscopic fibers self-assembled from recombinant miniature spider silk proteins. *Biomacromolecules, 8*(5), 1695-1701.

Szela, S., Avtges, P., Valluzzi, R., Winkler, S., Wilson, D., Kirschner, D., & Kaplan, D. (2000). Reduction-oxidation control of beta-sheet assembly in genetically engineered silk. *Biomacromolecules, 1*(4), 534-542.

Teule, F., Cooper, A. R., Furin, W. A., Bittencourt, D., Rech, E. L., Brooks, A., & Lewis, R. (2009). A protocol for the production of recombinant spider silk-like proteins for artificial fiber spinning. *Nat Protoc, 4*(3), 341-355.

Tso, I. M., Wu, H. C., & Hwang, I. R. (2005). Giant wood spider *Nephila pilipes* alters silk protein in response to prey variation. *J. Exp. Biol., 208*(Pt 6), 1053-1061.

Vasanthavada, K., Hu, X., Falick, A. M., La Mattina, C., Moore, A. M., Jones, P. R., Yee, R., Reza, R., Tuton, T., & Vierra, C. (2007). Aciniform spidroin, a constituent of egg case sacs and wrapping silk fibers from the black widow spider *Latrodectus hesperus*. *J Biol Chem, 282*(48), 35088-35097.

Vollrath, F., Wen Hu, X., & Knight, D. P. (1998). Silk production in a spider involves acid bath treatment. *Proc. R. Soc. B, 263*, 817-820.

Vollrath, F., Barth, P., Basedow, A., Engstrom, W., & List, H. (2002). Local tolerance to spider silks and protein polymers in vivo. *In Vivo, 16*(4), 229-234.

Winkler, S., Szela, S., Avtges, P., Valluzzi, R., Kirschner, D. A., & Kaplan, D. (1999). Designing recombinant spider silk proteins to control assembly. *Int J Biol Macromol, 24*(2-3), 265-270.

Winkler, S., Wilson, D., & Kaplan, D. L. (2000). Controlling beta-sheet assembly in genetically engineered silk by enzymatic phosphorylation/dephosphorylation. *Biochemistry, 39*(41), 12739-12746.

Xia, X. X., Qian, Z. G., Ki, C. S., Park, Y. H., Kaplan, D. L., & Lee, S. Y. (2010). Native-sized recombinant spider silk protein produced in metabolically engineered *Escherichia coli* results in a strong fiber. *Proc Natl Acad Sci U S A, 107*(32), 14059-14063.

Xu, M., & Lewis, R. V. (1990). Structure of a protein superfiber: Spider Dragline Silk. *Proc. Natl. Acad. Sci., 87*, 7120-7124.

Xu, H. T., Fan, B. L., Yu, S. Y., Huang, Y. H., Zhao, Z. H., Lian, Z. X., Dai, Y.P., Wang, L.L., Liu, Z.L., Fei, J., & Li, N. (2007). Construct synthetic gene encoding artificial spider dragline silk protein and its expression in milk of transgenic mice. *Anim Biotechnol, 18*(1), 1-12.

Yang, J., Barr, L. A., Fahnestock, S. R., & Liu, Z. B. (2005). High yield recombinant silk-like protein production in transgenic plants through protein targeting. *Transgenic Res., 14*(3), 313-324.

Zhao, A., Zhao, T., Sima, Y., Zhang, Y., Nakagaki, K., Miao, Y., et al. (2005). Unique molecular architecture of egg case silk protein in a spider, *Nephila clavata*. *J. Biochem. (Tokyo), 138*(5), 593-604.

Zhao, A. C., Zhao, T. F., Nakagaki, K., Zhang, Y. S., Sima, Y. H., Miao, Y. G., Kajiura, Z., Nagata, Y., Takadera, M., & Nakagaki, M. (2006). Novel molecular and mechanical properties of egg case silk from wasp spider, *Argiope bruennichi*. *Biochemistry, 45*(10), 3348-3356.

16

Natural Silkworm Silk-Epoxy Resin Composite for High Performance Application

K.M. Kelvin Loh and C.K. Willy Tan
School of Applied Science, Republic Polytechnic
Singapore

1. Introduction

Synthetic (man-made) fibres are important reinforcement materials in modern composites for high performance applications primarily due to their high specific strength and high specific modulus. However, the environmental unfriendly production method of synthetic fibres is an area of concern. Amid the growing global awareness and push for environmentally friendly products, the need to seek viable alternatives materials for synthetic fibres has been rising. As such, natural animal-based fibres are looked into as potential candidates. Modification of natural fibres to rival the mechanical performance of synthetic fibres has been looked into by many research. In this chapter, we will also explore the use of a low cost method to enhance the mechanical properties of silkworm silk during the sericulture stage. In addition, the feasibility to fabricate natural silkworm silk-epoxy composite will be evaluated.

2. Synthetic fibres as reinforcements in composites for high performance applications

The development of composites accelerated in mid 20th century, where the occurrences of the World Wars and Cold War rapidly pushed for the development of synthetic fibres like fibreglass, carbon fibres, boron fibres and aramid fibres, as countries vied for aerospace dominance. Since then, high performance industries including aerospace, automobile, protective gear and sports equipments are the major users of synthetic fibres. Fibre-reinforced composites have also found their way into consumer items like bicycles and golf clubs. This has empowered cyclists to cycle faster and golfers to swing further with less effort than before. However, the manufacturing process of common synthetic fibres like aramid and carbon creates environmental problems. Synthesis of aramid fibres utilise petroleum-based precursors with hot, concentrated sulphuric acid as solvent, while the synthesis of carbon fibres also require petroleum-based precursors and a series of high temperatures processing steps. Moreover, due to rising costs of energy, transportation, and raw materials, production costs of synthetic fibres have increased.

3. Natural fibres as an alternative

To tackle this imminent problem, the development and use of "greener" natural alternatives is gaining momentum to replace synthetic fibres in composite fabrication. Currently, the

Germans used a total of 30,000 tonnes of natural fibre reinforced polymers in their automotive industry in 2005. (Karus et al., 2006, as cited in Mussig, 2010). The use of natural fibres in composites is nothing new to human. Evidence of natural fibre reinforced composites could be found in pottery containing hemp fibre which dated back to as early as 10 000BC. (Rowell, 2008, as cited in Mussig, 2010).

3.1 Different types of natural fibres

Natural fibres can be categorised into plant fibres and animal fibres. Plant fibres include flax, hemp, jute, abaca, sisal, coir and cotton. Animal fibres include silk and wool. Numerous silk spinning animals like caterpillars, silkworms and spiders exist in nature, but spider silk is the strongest of them all. Silk is spun by animals for various reasons, but most of which are for the strength that their silk provides. Silkworm spins silk for protection of the larvae during pupation, while spider spins silk for shelter, protection of young spiders and trapping of prey. The high tensile strength of spider silk rivals that of high grade steel and aramid fibres such as Kevlar. Spider silk is also lighter than steel and elongates more than aramid. Thus spider silk seems to be the perfect candidate to replace synthetic fibres. However, the carnivorous and cannibalistic nature of spider makes harvesting of its silk difficult to be implemented on a large scale. Since *Bombyx Mori* silkworm silk also possesses a high tensile strength after spider silk, silk harvesting from this highly domesticated silkworm is an easier alternative. The mechanical properties of some common natural and synthetic fibres are shown below. (Table 1).

Types of Fibre	Material	Density (g/cm3)	Tensile Strength (MPa)	Young's Modulus (GPa)	Elongation at Failure (%)
Natural Fibres	Spider Silk	1.3	1300-2000	30	19-30
	Enhanced *B.Mori* Silk	1.3-1.38	600-700	12.2	30-35
	B. Mori Silk	1.3-1.38	500	8.5-8.6	15
	Flax	1.45	500-900	50-70	1.5-4.0
	Hemp	1.48	350-800	30-60	1.6-4.0
Synthetic Fibres	Kevlar 49	1.44	3600-4100	130	2.8
	Carbon	1.4	4000	235	2
	E-glass	2.5	3100-3800	76-79	4.8
	Dyneema	0.97	2300-3500	550	2.7-4.5
	High Grade Steel	7.8	1000	200	30

Table 1. Mechanical Properties of Common Natural and Synthetic Fibres.

4. Silkworm silk and its enhancement during sericulture

Different species of silkworms exist in nature that can be reared for their silk. These include *Bombyx Mori*, *Antheraea Mylitta*, *Philosamia Synthia Ricini* and Antheraea Assamensis. The domesticated *Bombyx Mori* silkworm, which feeds only on mulberry leaves, is easier to rear as compared to the other wild species. Being a natural biodegradable material, *Bombyx Mori* silkworm silk fibers generate no toxic waste in production and are cheaper to produce than synthetic fibres. This makes them a very attractive alternative to synthetic fibres. Although it may not be comparable in strength, *Bombyx Mori* silkworm silk has the advantage of being more ductile than synthetic fibres. As such, it may still be used in applications which is less stringent on tensile strength but require higher elongation.

4.1 Characteristics of silkworm

The life cycle of *Bombyx Mori* silkworm is as below. (Fig. 1).

Fig. 1. Life Cycle of *Bombyx Mori* Silkworm

After hatching from the eggs, *Bombyx Mori* silkworm larvae will undergo a total of four rounds of moulting before they start to spin silk to form cocoons. On cocooning_for a couple of days, the pupae are then stifled by heat to prevent them from emerging out of their cocoons, so that the silk can be reeled into yarns, and woven into fabric.
The structure of Bombyx Mori silk consists of two main chains of fibroins protein made up of parallel bundles of nanofibrils. Fibroin is coated with the gummy sericin protein. The chemical composition of fibroin consists of the amino acids glycine, alanine and serine present in the form of beta sheets. High tensile strength of silkworm silk is attributed to two main reasons. Firstly, the large amount of small sized glycine molecule produces a tightly and efficiently packed structure. Secondly, the extensive network of strong hydrogen bonds within the fibroins chains renders the silk fibres high strength and elongation. The gummy sericin is responsible for conferring the properties of anti-oxidation, antibacterial, UV resistance and hydrophilicity (Zhang *et al.*, 2002, as cited in Hakimi, 2006).

4.2 Enhancement of silkworm silk

Silkworm silk is inferior in mechanical properties as compared to spider silk. Numerous methods have been attempted to enhance the mechanical properties of silkworm silk including force silking and most recently, genetic modification. Artificial spider silk produced by transgenic silkworms was claimed to achieve similar mechanical properties to that of the original spider silk. As for force silking of *Bombyx Mori* silkworm, an enhanced tensile strength of 650 MPa was reportedly obtained at an optimum silking speed of 240 cm min^{-1}. The enhanced tensile strength was attributed to a greater conformation of the beta sheet molecular structure and a smaller silk diameter obtained by the accelerated silking speed (Morikawa et al., 2008). However, these two methods have yet to achieve large-scale commercial feasibility due to high cost of genetic modification and difficulty in obtaining large length of silk from force silking.
Another method of potential commercial viability is to enhance the mechanical properties of silkworm silk with a low cost and easily implemented technique. This method subjects the *Bombyx Mori* silkworm to electric field stimulation when the adult silkworm starts to spin silk to form a cocoon. The resultant silk fibre reportedly gained a 40% in tensile strength (682 MPa), 100% in breaking strain (34.4%) and 200% in breaking energy (124kJ/kg) as compared to the ordinary silkworm silk. In addition, optical light stimulation was also reported to have a slight improvement on the mechanical properties of the silk fibre. Analysis of the enhanced silkworm silk fibre found that it possessed greater crystallite alignment than an untreated silkworm (Liu & Du, 2008).
As this method is an environmentally friendly solution to enhancing the mechanical properties of silkworm silk without the need of expensive and sophisticated machines, it presents a viable alternative material to replace synthetic fibers in composites where high breaking strain and breaking energy are more important. This will be especially appropriate for ballistic applications where the high breaking energy of fabrics is needed to absorb the high projectile energy.

5. Optimization of yarn and weaving parameters for enhanced silk

Likewise for synthetic fibres, silkworm silk can be processed into yarns and weaved into fabrics. The mechanical properties of *Bombyx Mori* silkworm silk, as a potential replacement for synthetic fibres in composite, can be possibly further enhanced by optimization of the yarn and weaving parameters for its specific application. Research on the optimization of silk for incorporation into composites may not be as extensive as that conducted for synthetic fibres like Kevlar and carbon, however the results can potentially be adapted and used on silk material too.

5.1 Optimization of yarn parameters

Tensile strength, one of the most important mechanical properties of yarn for high performance applications, is influenced by many parameters including yarn diameter, degree of twisting and yarn irregularities. Tensile properties of silk yarn can be determined by ASTM D2256 (Standard Test Method for Tensile Properties of Yarns by the Single-Strand Method).

Commercial reeled silk yarn is normally available in the sizes 20/22 Denier or 40/44 Denier which consist of 10 and 20 silk filaments respectively. Doubling to increase the number of filaments of silk is used to increase the yarn diameter and minimise irregularities inherent in the length of a single ply of silk yarn. Furthermore, a thin silk yarn is too weak to withstand the stress present in the fabric weaving process. An increase in yarn diameter usually results in higher average strength of the yarn, as irregularities along the length are averaged out.

Twisting of two or more single silk yarns, by introducing spiral turns to hold them together to form plied yarns by interfacial contact, is an important step in a yarn configuration Thinner yarns require more twist than thicker yarns. In general, the maximum tensile strength and strain of yarns tends to increase with the increase in twist factor. However in some cases, an optimum value will be reached after which there will be a drop in mechanical properties.

In a woven fabric, the warp yarns that make up the length of a fabric require more twist than the weft yarns that make up the width of a fabric. The higher degree of twist in the warp yarns allow them to withstand the tension imposed on them by the weaving machine. Some high twist yarns might tend to unravel when left free. In order to minimise untwisting of these high twist yarns, the yarns may be conditioned at high temperature and moisture to heat set the twist incorporated into the yarns. Excessive twisting of yarns might not be desirable for woven fabric, as shrinkage in yarn and fabric after weaving produces an unbalanced fabric with localised stress concentration.

The direction of twist is also important in producing a balanced fabric. Twist direction present in a yarn may be either of a S or Z spiral. This does not influence the strength of the yarn, but is crucial in providing a balanced yarn structure to prevent kinking of the yarn and fabric.

In particular for protective woven fabrics used in ballistic applications, ballistic resistance is dependent on friction, modulus and tensile strength of yarns. Yarns with high modulus and tensile strength are the main influences for better ballistic performance. Friction between yarns served to maintain the integrity of weaved fabrics (Keefe, 2009). Typically, ballistic fabrics are constructed from low-twist, low crimp yarns, often with equal thread densities in the warp and weft (Cork, 2005).

5.2 Optimization of weaving parameters

In addition to yarn parameters playing an important role to the overall composite mechanical properties, fabric weave structure is also as important. Weaving patterns represents the interlocking arrangement of warp yarns and weft yarns. In the optimization of fabric stability, crimp, drapeability and mechanical performance of the fabrics, weaving patterns must be looked into. Tensile properties of differently woven silk fabric can be determined by ASTM D5035 (Standard Test Method for Breaking Force and Elongation of Textile Fabrics (Strip Method).

Plain weaving pattern is typically used for weaving reinforcement fabrics due to its balanced structure. However, the low drapeability of plain weave will require basket, twill, satin weave when conformance to special geometry is required.

Stitching is another commonly used method to improve mechanical properties of weaved fabrics. When multiple layers of weaved fabrics are incorporated into composites, the through thickness properties are weak. Stitching of multiple layers of woven fabrics together can improve the interlaminar fracture toughness. Plain woven Kevlar fabrics-rubber composite were demonstrated to have higher ballistic performance than that of unstitched fabrics, due to diamond stitching of the Kevlar fabrics, (Ahmad, 2007).

6. Fabrication of natural silkworm silk fabric-epoxy resin composite

The feasibility to fabricate natural silkworm silk fabric-epoxy composite will be explored in this section. Wet layup method will be used to create silk-epoxy composite. Compaction of the composite will be done via the hot press method with vacuum bagging to increase the volume percentage of reinforcement in the composite and lower the void content. Mechanical testing of the composite will thereafter be performed to analyze and evaluate the experimental results.

In theory, 61% reinforcement by weight is considered to be the optimum ratio for standard composites. There are several predicted issues with the fabrication of silk-epoxy composites for actual commercial purposes, such as delamination, adhesion problems of epoxy with silk, selection of epoxy to used, fracture analysis, design of the composite, and design of the silk. In this section of our study, we will aim to explore some of these issues and improve them in future works.

6.1 Analysis of silk fabric

A typical un-dyed silkworm silk woven fabric was obtained from China. No alteration to the silkworm sericulture or silking process was done. A small sample of the silk fabric was analyzed under the scanning electron microscope (SEM). Analysis of the fabric at 100X magnification revealed that it was of a satin weave with 10 untwisted warp yarns interlocked with 10 twisted weft yarns. (Fig.2).

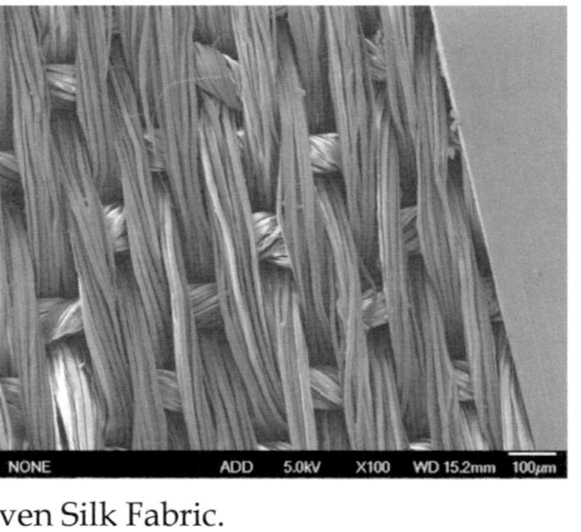

Fig. 2. SEM Analysis of Woven Silk Fabric.

Satin weave possesses the advantages of good drape, low porosity, low crimp, but brings along the negativity of poor stability, balance and symmetry.

6.2 Wet-layup of composite

Composite fabrication involved wet-layup of the silk fabrics with epoxy. The silk fabric serves as the continuous reinforcement phase while the epoxy is the matrix that binds with the reinforcement to form the composite together. Epoxy is chosen as the matrix due to its high structural strength.

The two component epoxy system consists of a resin and a hardener. The resin contains epoxide functional group, formed from the reaction between epichlorohydrin and bisphenol-A. The hardener contains polyamine monomers with amine functional groups. After mixing the resin and hardener according to the ratio of 10:6, the amine groups will react with the epoxide groups to form covalent bonds, resulting in a strongly cross linked polymer. Curing can be fine-tuned through temperature adjustments, mixing ratios and selection of resin and hardener. The selected viscosity of this epoxy system is one that allows adequate time for the operator to perform wet-layup of multiple pieces of silk fabric.

A square metal mould was first treated with a suitable mould-release agent for easy release of the silk epoxy composite panel from the metal mould after curing. Multiple layers of silk fabric were cut to suit the size of the mould. The hardener component was added to the resin component before use. (Fig.3).

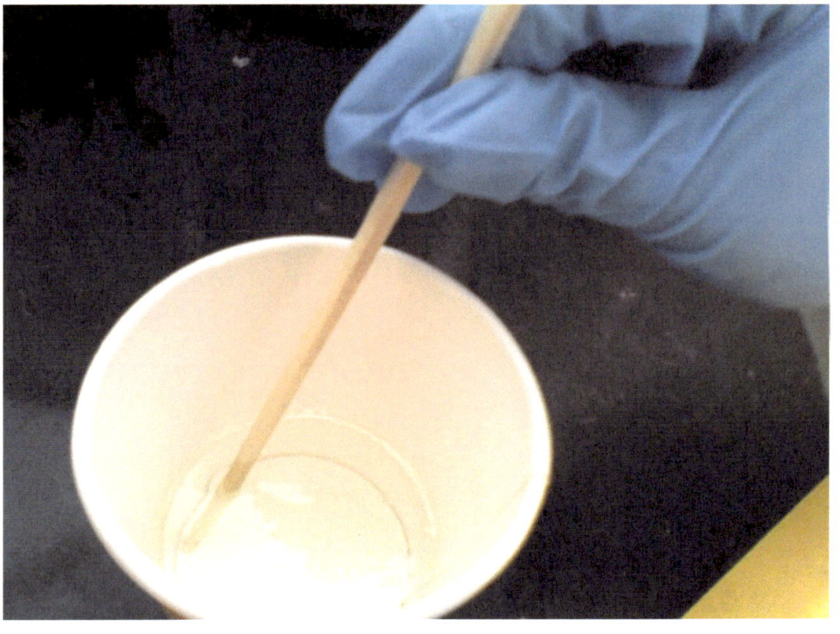

Fig. 3. Mixing of epoxy resin and hardener

Silk fabric and epoxy were alternatively applied onto the metal mould. A metal roller was used to remove air bubbles that were trapped in between the fabric layers and remove excess epoxy from the composite. (Fig. 4).

Fig. 4. Wet layup of silk fabric

After the stipulated numbers of layers of silk fabric were laid on the metal mould, another piece of square metal mould treated with mould-release agent was place on top of the composite, before vacuum bagging.

6.3 Vacuum bagging setup

The whole setup was transferred into a vacuum bag consisting of a high temperature vacuum bagging film, breather/bleeder fabric and sealant. The breather/bleeder fabric (glass fiber cloth) is placed at the vacuum port to better allow the air to be sucked out, by providing a small gap between the vacuum bag and composite. It also acts as a filter to prevent the leakage of epoxy to the vacuum pump by absorbing any excess resin that overflows during hot pressing. The sealant serves to seal off the edges of the vacuum bagging film, by maintaining a continuous airtight seal. (Fig. 5).

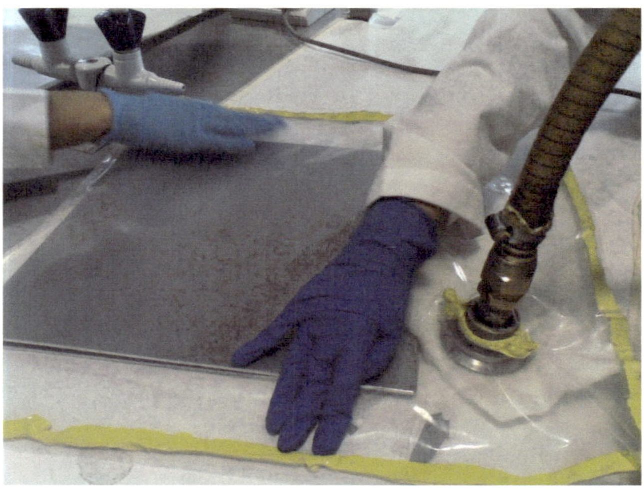

Fig. 5. Vacuum Bagging Setup

Next, the vacuum bag was connected to a vacuum pump via a plastic hose and vacuum port to remove air from within and decrease the occurrences of void and air bubbles inside the silk-epoxy composite. A vacuum pressure of at least 600-800MPa, as indicated by the vacuum gauge was obtained, before the vacuum bag was transferred to the hot press.

6.4 Hot press setup

In the absence of external applied stress on the silk-epoxy composite panel during its curing stage, the compactness of the final cured composite will only be dependent on the atmospheric pressure exerted on its surfaces and the rolling that was used to squeeze out excess epoxy during the wet-layup. In order to increase the volume percentage of silk fibre in the composite, an external applied stress need to be applied onto the composite panel at the curing stage to remove excess epoxy from the composite, while sufficient epoxy remains to bond the silk reinforcement in the composite. The use of hot press allows the application of external pressure and heat to enhance densification, compressing a laminate into a compact shape and squeezing out any excess material. The higher temperature also expedite the curing time of the epoxy system, which will otherwise take at least 24 hours to cure at room temperature.

Both the top and bottom platens of the hot press were set at 60 degrees Celsius. The vacuum bag holding the uncured silk-epoxy composite was placed between the two platens. The two platens were closed and a pressure of 1.38 MPa was applied in between the two platens onto the silk-epoxy composite. (Fig.6).

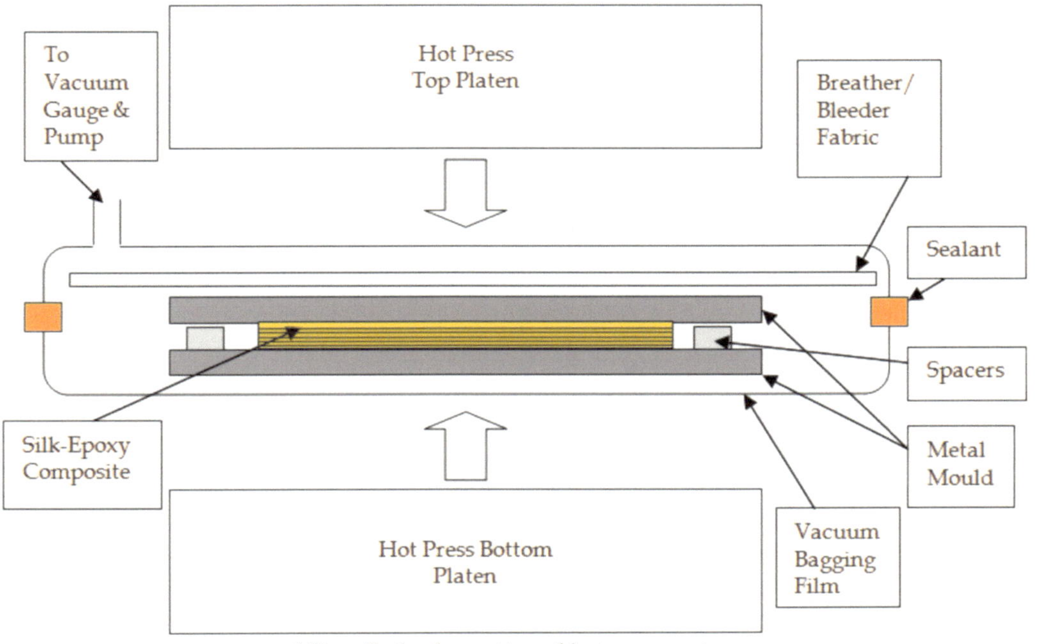

a) Detailed schematics of hot press setup

b) Actual hot press setup

Fig. 6. Hot Press Setup

The setup was left in this condition for 3 hours in accordance with the recommended elevated curing time of the epoxy system. At the end of the 3 hours, the cured composite panel was removed from the mould by gently forcing the pieces apart. The composite panel was left to cool at room temperature for at least 24 hours before any tests were performed on it. (Fig. 7).

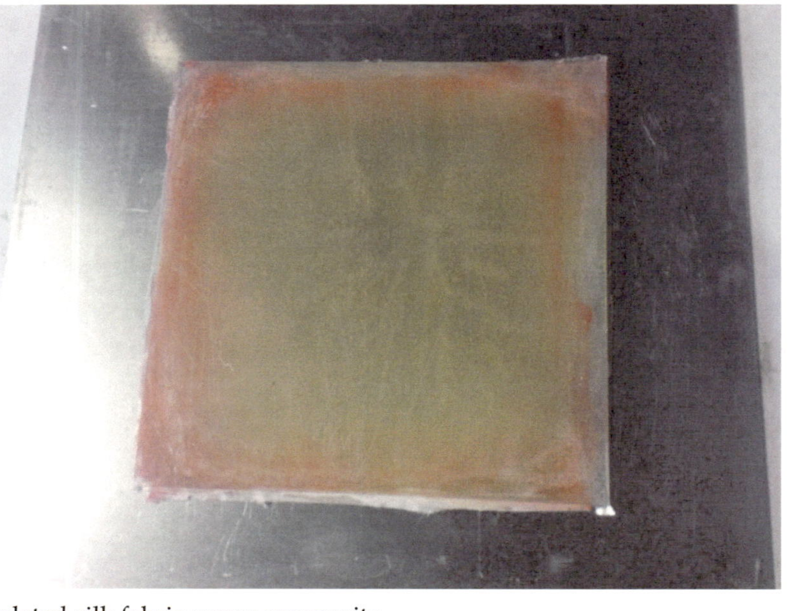

Fig. 7. Completed silk fabric epoxy composite

6.5 Surface treatment of silk

Delamination is most often the crippling cause of failure for fibre-reinforced composites. Good surface interaction of fibres with the matrix in a composite is dependent on the physical and chemical characteristics on the fibres' surface. Silk fabric can possibly be treated with suitable surface coupling agents to minimize interlaminar delamination and make it adhere better to the epoxy before being used to make the composites. The use of silane coupling agents to treat silk fibre was found to increase the interlaminar fracture toughness for woven silk-epoxy composites. (Zulkifli, et. al., 2009). In this section, surface treatment of silk fabric with silane coupling agent will be performed. Mechanical tests on silanated silk-epoxy composites will be conducted thereafter to gauge its effectiveness.

Silane coupling agent was originally developed for glass fibres. Silane coupling agents are usually used to bind organic and inorganic materials together. Resistance to moisture uptake by fibres in wet or dry conditions can also be enhanced by silane treatment. Silane molecule contains a central silicone atom, with organic functional group (R) like amino or epoxy vinyl, and a second functional group (X) like methoxy or ethoxy groups. A coupling effect occurs when the R functional group attaches to an organic substrate while the X functional group attaches to an inorganic substrate. Hydrolysis of the X functional group forms silanol which will react with the inorganic substrate to form a siloxane bond. The R functional group will react with the organic substrate to form a covalent bond. (Fig. 8).

Silquest A-1100 (Gamma-Aminopropyltriethoxysilane) with the chemical formula of $H_2NCH_2CH_2CH_2Si(OCH_2CH_3)_3$ was used in our study. A 95% ethanol / 5% water solution was adjusted to pH 4.5–5.5 with acetic acid. Silane was added with stirring to yield a 2% final concentration. Layers of silk fabric were then dipped into the solution, agitated gently, and removed after 5 minutes to allow for hydrolysis and silanol formation.. They are rinsed free of excess materials by dipping briefly in ethanol. Curing of the silane layer was performed for 20 minutes at 60°C or 24 hours at room temperature (60% relative humidity). The silanated silk fabrics were subsequently used to fabricate composite.

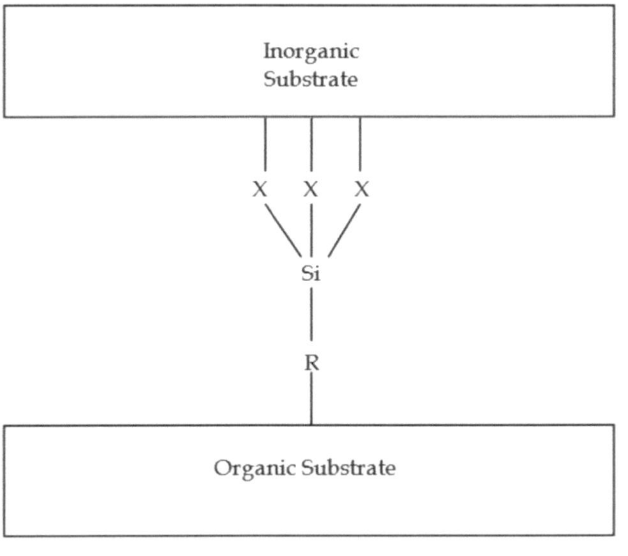

Fig. 8. Silane Bonding

7. Mechanical testing & evaluation of fabricated composite

The completed silk-epoxy composite panels were subjected to three point flexural bend test based on ASTM D790 to determine the mechanical properties like flexural strength and flexural modulus. Flexural tests tell us the behaviour of materials subjected to simple beam loading tests and indicates the composite ability to resist deformation under load.

7.1 Procedure for flexural test

The test specimens were first cut according to the required dimensions in ASTM D790 by a diamond cutter. For blank epoxy, the test specimens were prepared in a Teflon mold of dimensions 88mm x 22mm x 5.5mm. The hot pressed silk-epoxy composite were of differing thickness, depending on the number of layers of silk incorporated. Thus, each batch of silk composite have to be cut into similar rectangular dimensions with a minimum span to depth ratio of 16:1 or 32:1. The length of each test specimen was cut long enough such that 10% of test specimen was left hanging over at each support. All uneven surfaces of the cut specimens were ground flat with a sanding machine. (Fig. 9).

Five test specimens per composite were tested. Each batch of test specimens was from a single panel of its respective number of silk fabric layers incorporated in the composite. The dimensions of the test specimens are shown below. (Table 2).

The test specimens were conditioned at 23°C ± 2°C and 50% ± 5% relative humidity for not less than 40 hours prior to test. Test specimens were supported on two points with load applied from the top at mid-length. Constant strain rate of 0.01mm/mm/min is applied until maximum flexural stress was obtained or test specimen had fractured. (Fig.10).

Fig. 9. Machining test specimen to size according to ASTM D790

No. of Silk Fabric Layers in Composite	Length (mm)	Span (mm)	Width (mm)	Depth (mm)	Span : Depth ratio
0 (blank)	105.60	88	22	5.5	16
15	92.16	64	16	2	32
20	96.00	80	20	2.5	32
20 (Silanated)	96.00	80	20	2.5	32
25	99.84	83.2	20.8	2.6	32
30	115.20	96	24	3	32

Table 2. Dimensions of flexural test specimens

In this study, calculation of volume fibre percentage was as below. Voids were assumed to be insignificant as hot pressing to achieve high compactness was used in the composite fabrication.

$$\text{Volume \%silk fibre} = (\text{volume of silk} / \text{volume of composite}) \times 100\%$$

$$= (\text{mass}/\text{density})_{\text{silk}} / (\text{L} \times \text{B} \times \text{W})_{\text{composite}} \times 100\% \quad (1)$$

In accordance with ASTM D790, the flexural strength and and modulus were calculated as below:

$$\text{Flexural Strength: } \sigma_f = 3PL/2bd^2 \quad (2)$$

$$\text{Tangent Modulus of Elasticity: } E_B = L^3 m/4bd^3 \quad (3)$$

where:
D: maximum deflection of the center of the beam (mm),
d: depth (mm),
L: support span (mm),
P: load at a given point on the load-deflection curve (N),
b: width of beam tested (mm),
m: slope of the tangent to the initial straight-line portion of the load-deflection.

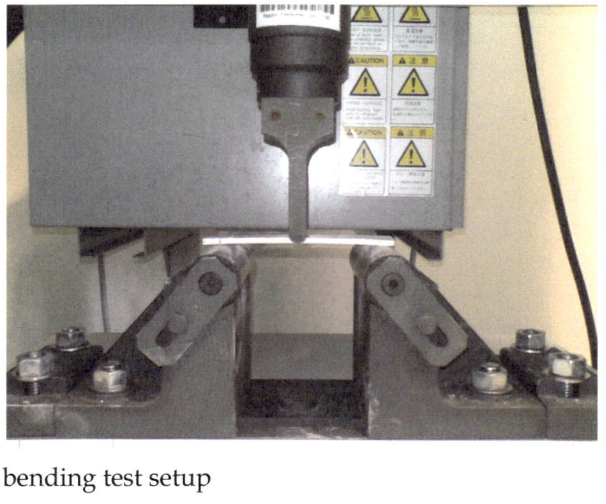

Fig. 10. Three point bending test setup

7.2 Results

The three point flexural bend test results for the composites of different volume percentage of silk fibre were calculated and compared against the flexural strength and modulus. (Table 3). (Fig. 11).

No. of Silk Fabric Layers in Composite	Volume % of Silk Fibre (%)	Average Flexural Modulus (MPa)	Average Flexural Strength (MPa)
0	0.00	1387.99	53.31
20	33.09	4083.71	123.75
20 (Silanated)	33.09	3162.11	106.08
25	40.41	3795.79	118.57
30	41.62	3459.63	91.53

Table 3. Flexural Modulus & Strength vs Vol % of Silk Fibre

It can be seen from the results that as the volume percentage of silk fibre increases, the average flexural strength and flexural modulus of the silk composite increase. However, both properties peaked at the point where about 33% of silk fibre was added into the composite. This indicated that the optimum percentage of reinforcement was almost reached. The reason that composite strength starts to decrease at high fiber volume fractions can be attributed to the shortage of resin to wet all fibers in the composite as more fibers are added.

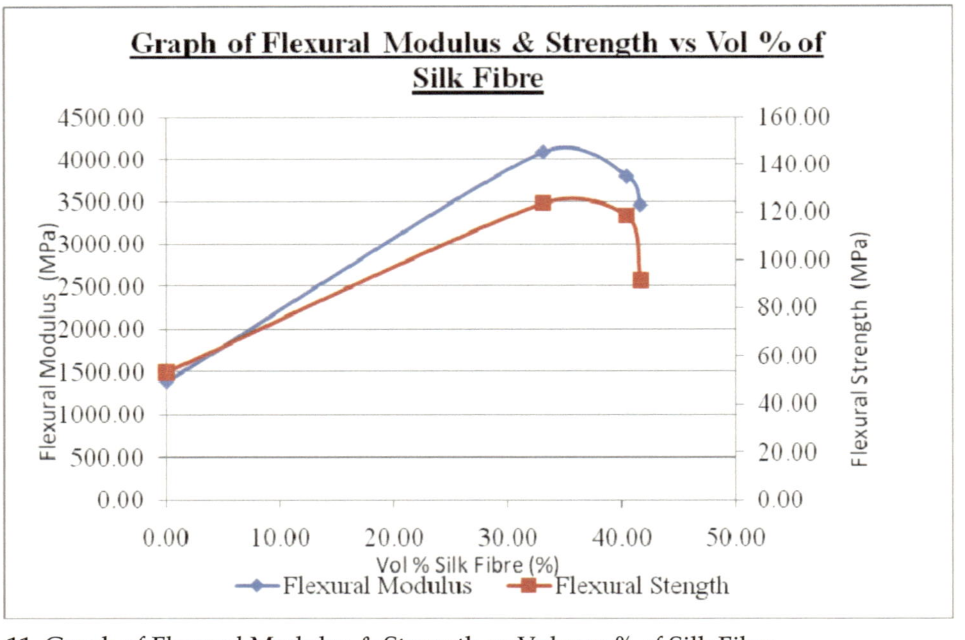

Fig. 11. Graph of Flexural Modulus & Strength vs Volume % of Silk Fibre

The silanated (20 layers) silk fabric composite exhibited lower flexural strength and modulus than the non-silanated (20 layers) silk fabric composite. This suggested that the silane treatment did not increase adhesion between silk and epoxy, but rather decrease it. Silk is organic and few sites to interact with the inorganic functional group of silane. As such, silane can only bond with epoxide functional groups in the epoxy resin, which reduces the available epoxide-bonding sites for silk amine groups to bond with epoxy groups. Thus, the flexural strength and modulus of silanated silk-epoxy composite were compromised, instead of improved.

8. Conclusion

Wet layup, vacuum bagging and hot press were successfully employed to fabricate the silk fabric-epoxy composite in this study. Based on the three point flexural bend test results, incorporating silk fabric into epoxy was found to increase the flexural strength of the composites up till a optimum volume percentage of 33% silk fibre reinforcement content. This favorable result indicated the potential to further explore and utilize natural silkworm silk-epoxy resin composite for actual high performance applications.

Future work to enhance the flexural strength of the silk–epoxy composite will include optimization of the yarn, weaving parameters and also utilize stitched multilayer fabrics to improve interlaminar strength. Enhancement of the silkworm silk fibre by electric field and optical light stimulation during the sericulture stage will also be tested. Additional mechanical tests like tensile, impact, fracture toughness will be performed on the silk composite in order to better compare with other synthetic fibres, such as carbon fibers and Kevlar. Due to the high elongation of silk fibres, the ductility of the silk-epoxy composite could not be covered in the flexural test conducted. This will subsequently be determined by tensile test. In addition, in order to further increase the volume percentage of silk fibres in the composite, higher pressing pressure will be explored.

9. References

Ahmad M.R., Ahmad W.Y.W., Salleh J., Samsuri A. (2007). Effect of fabric stitching on ballistic impact resistance of natural rubber coated fabric systems. Materials and Design 29, pp. 1353–1358.

Bourzac K. (October 2010). Transgenic Worms Make Tough Fibers, In: *Kraigslabs,* 01.03.2011, Available from: http: / / www.technologyreview.com / biomedicine / 26623.

Cork C.R., Foster P.W. (2005). The ballistic performance of narrow fabrics. International Journal of Impact Engineering 34, pp. 495–508.

Hakimi O, Knight D.P., Vollrath F., Vadgama P. (2007). Spider and mulberry silkworm silks as compatible biomaterials. Composites: Part B 38 pp. 324–337.

Keefe M., Rao M.P., Duan Y., Powers B.M., Bogetti T.A. (2009). Modeling the effects of yarn material properties and friction on the ballistic impact of a plain-weave fabric. Composite Structures 89 pp. 556-566.

Khan Md.M.R., Morikawa H., Gotoh Y., Miura M., Ming Z., Sato Y., Iwasa M. (2008). Structural characteristics and properties of *Bombyx mori* silk fiber obtained by different artificial forcibly silking speeds. International Journal of Biological Macromolecules 42, pp. 264–270.

Liu X.Y., Du N. (2008). Patent Number: WO 2008/033104 A1, Enhanced Silk Protein Material Having Improved Mechanical Performance and Method of Forming the Same.

Mussig J (2010). (Ed.). *Industrial Applications of Natural Fibres*, John Wiley & Sons, Ltd, pp. 386, ISBN 978-0-470-69508-1, United Kingdom.

Zulkifli R., Azhari C.H., Ghazali M.J., Ismail A.R., Sulong A.B. (2009). Interlaminar FractureToughness of Multi-Layer Woven Silk/EpoxyComposites Treated with Coupling Agent.European Journal of Scientific Research Vol.27No.3, pp. 454-462.

17

Noncrystalline L-Phenylalanine-Silica Hybrid Composite Materials for High Selective Reversed Phase Liquid Chromatography

M. Mizanur Rahman[1,2], Makoto Takafuji[2] and Hirotaka Ihara[2]
[1]*Department of Applied Chemistry and Chemical Engineering*
Faculty of Engineering and Technology, University of Dhaka
[2]*Department of Applied Chemistry and Biochemistry*
Faculty of Engineering, Kumamoto University
[1]*Bangladesh*
[2]*Japan*

1. Introduction

Research developments in analytical techniques explore novel silica hybrid materials with interesting and useful functionalities as hybrid inorganic-organic composite materials have attracted increasing attention because of their potential ability to combine the properties of both organic and inorganic materials. There has also been renewed attention in synthetic polypeptides because of their promising application as biodegradable and biomedical polymers, as well as their ability to form highly ordered hierarchical structures through non-covalent forces such as hydrogen bonding. Self-assembling of low molecular compounds have been attractive as a construction tool for nanotechnology because they are essentially generated through three-dimensional network formation from nano-sized one-dimensional fibrillar aggregates, constructed of highly-ordered assemblies like aqueous lipid bilayer membranes. The formation of assemblies from small molecules requires a hierarchically organized aggregation through noncovalent bonds to yield elongated assemblies that further self-organize into fibers that percolate the solution. It is also known that a number of chiral compounds form nano-sized one-dimensional unique aggregates with large curvatures such as nanotubes, nanohelices, and nanoribbons (Terech & Weiss, 1997; Esch & Feringa, 2000; Ihara et al. 2004).

Self-assembly is a "bottom-up approach" in which materials are assembled molecule by molecule to produce supra-molecular structures for appropriate functions (Abdallah & Weiss, 2000; van Esch & Feringa, 2000). For example, molecular self-assembly from solution states often leads to the formation of one-dimensional (1D) solid fibers. This process has attracted attention because 1D fibrous solid, ranging from the submicrometer to nanometer scale, are easily obtained by simple self-assembly through noncovalent interactions The underlying mechanisms of intermolecular binding interactions include hydrogen bonding, ionic interactions, and π–π stacking in a variety of solvents.[3] The aggregation or network formation of the fibers in the solvents leads to the occurrence of physical gels in which the molecules that drive physical gelation are called as gelators (Kobayashi et al, 2002) If a functional moiety is introduced into the backbone of a gelator, a new functional 1D object would be obtained. These aggregates are morphologically interesting and applicable as templates for the creation of peculiar nanostructures, such as inorganic nanomaterials (Suzuki et al., 2006) synthesized by condensation of alkoxysilane or chlorosilane on self-assembled fibrous/helical aggregates (Jung et al. 2000) Equally important that a template method has been also applied to prepare various metal (Hatano et al. 2004) nanomaterials.

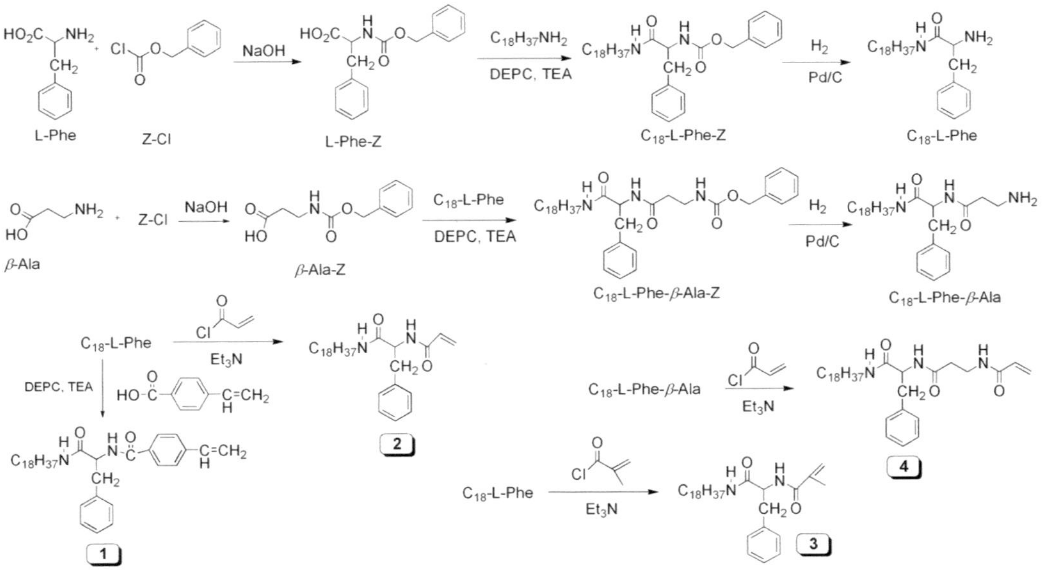

Scheme 1. Synthesis scheme for polymerizable octadecyl L-phenylalanine derivatives

Combining organic and inorganic materials on a nanometer/micrometer scale, to create functional hybrid materials is receiving a great deal of attention. Organic modification of inorganic materials such as silicas and/or silicates, not only permits a precise control of surface properties for specific applications, but also can affect their bulk properties. In fact, organic features can introduce flexibility into the framework, while the inorganic components can provide structural, mechanical or thermal stability (Stein et al., 2001; Anwander, 2001; De Vos et al., 2002; Tao et al., 2000). One of the most important classes of composites is the materials obtained by the incorporation of various kinds of polymers of controlled physico-chemical properties into porous inorganic material. The insertion of the polymers into the inorganic host cavities can be obtained carrying out the polymerization process inside the cavities themselves (Cardin, 2002). If the initiator is covalently bound to inner surface of the host the polymer can be synthesized directly inside the host pores (Spange et al., 2001). Applications of hybrid materials in analytical chemistry are not new, particularly in the area of separations science. For example, since the advent of silica as an appropriate chromatography material for separations, several hybrids have been successfully employed, such as the hydrocarbon-silica materials prepared from sol-gel synthesis process. In the fields of separation sciences, sol-gel hybrid materials are often used as alternative stationary phases for liquid chromatography, to improve column efficiency, stability and selectivity. Conventional strategies to introduce functional groups in silica stationary phases for chromatography applications include physisorption and covalent attachment of desired compounds via surface grafting methods to pre-formed silica

materials (Guo & Becker 1995, Crego et al. 1993) These works have led to successful application of organic functionalized silica phases in reversed phase liquid chromatography (RPLC) in particular, those of C8 and C18 phases (Guo & Becker 1995; Crego et al. 1993).
In recent years, there has been renewed interest in synthetic polypeptides, because of their potential application as biodegradable and biomedical polymers (Kolk et al. 2000) as well as their feasibility to create highly-ordered hierarchical structures through non-covalent forces such as hydrogen bondings (Li et al. 2003; Hamley et al. 2005). Incorporation of a high degree of amino acid functionality and chirality in polymer chains can enhance the potential to form secondary structures (α-helix and β-sheet) and higher ordered structures (Oishi et al. 2002). These synthetic polymers can be useful as chiral recognition stationary phases for HPLC (Lekchiri et al. 1987) metal ion absorbents, drug-delivery agents, and biocompatible materials (Nagai et al. 2004) Such characteristic self-assembled structures and the potential applications of polymers derived from amino acids have attracted researchers to develop new synthetic routes to prepare a variety of amino acid based polymers using various polymerization techniques. The application of these self-assembling polymeric systems is of great interest to scientists dealing with the preparation of new packing materials for HPLC. Extremely enhanced selectivity was obtained for shape-constrained solutes, e.g., PAHs, aromatic positional isomers, and nucleic acid constituents by using silica-selfassembled dialkyl glutamide derived composite materials. (Rahman et al. 2008) This finding encouraged us to develop polymeric peptide lipid type stationary phases. With this back ground herein we will discuss four octadecylated L-phenylalanine derived polymerizable monomers as shown in Scheme 1 of those monomer 1 and 4 can aggregate in organic solvents. The polymerization/telomerization of all four monomers are done by both conventional radical polymerization followed by grafting on to inorganic silica particles ("grafting to" method) and also by controlled radical polymerization method ("grafting from" method) as given in scheme 2 to prepare hybrid composite materials. One practical application of these composite materials for separation of toxic PAHs will also be discussed for high selective reversed phase liquid chromatography. The chromatographic performance for the octadecyl L-phenylalanine derived hybrid composites prepared by "grafting to" and "grafting from" method will also be compared.

2. Self-assembling properties of the octadecylated L-phenylalanine derivatives

The monomers were synthesized and characterized by previously reported methods. (Rahman et al. 2008, 2009). The morphological properties of these compounds were studied and the results demonstrated that 1 and 4 can self-assemble in different organic solvents and can form fibrillar aggregation. The FT-IR and ^{1}H NMR spectroscopies are powerful tools to study hydrogen bonding interactions. Extensive precedent indicates that secondary amide groups (N-H) engaged in the standard amide-amide hydrogen bonds (C=O··H-N) display stretch bands in the range 3370-3250 cm^{-1} while stretch bands in the range 3500-3400 cm^{-1} are attributed to "free" secondary amide groups that are not involved in hydrogen bonding. Detailed FT-IR studies of gel state of compound 1 and 4 also shows long alkyl chains in these compounds also played key role as van der Waals interaction among the alkyl chains also plays an important role in the self-assembly of these octadecyl L-phenylalanine derivatives(Rahman et al. 2008). The morphological studies by ^{1}H NMR and UV aggregation of compounds 1 and 4 and non aggregation of 2 and 3. ^{1}H-NMR studies of compound 1 in

$CDCl_3$ at different temperature clearly shows the aggregation of compound 4 at low temperature, however at 70°C the peaks for proton in alkyl chain are very much visible as like as the peaks in solution state. This indicates the aggregation of compound 1 at low temperature which does not exists at higher temperature as high as 60-70°C.

The FTIR, NMR and UV-visible measurements demonstarted that intermolecular hydrogen bonding among the amide moieties, van der Waals interaction between the alkyl chains and p-p interactions of the aromatic groups played the most important role in the self-assembly of these compounds. A possible mechanism for self assembling of octadecylated L-phenylalanine is given in Fig. 2

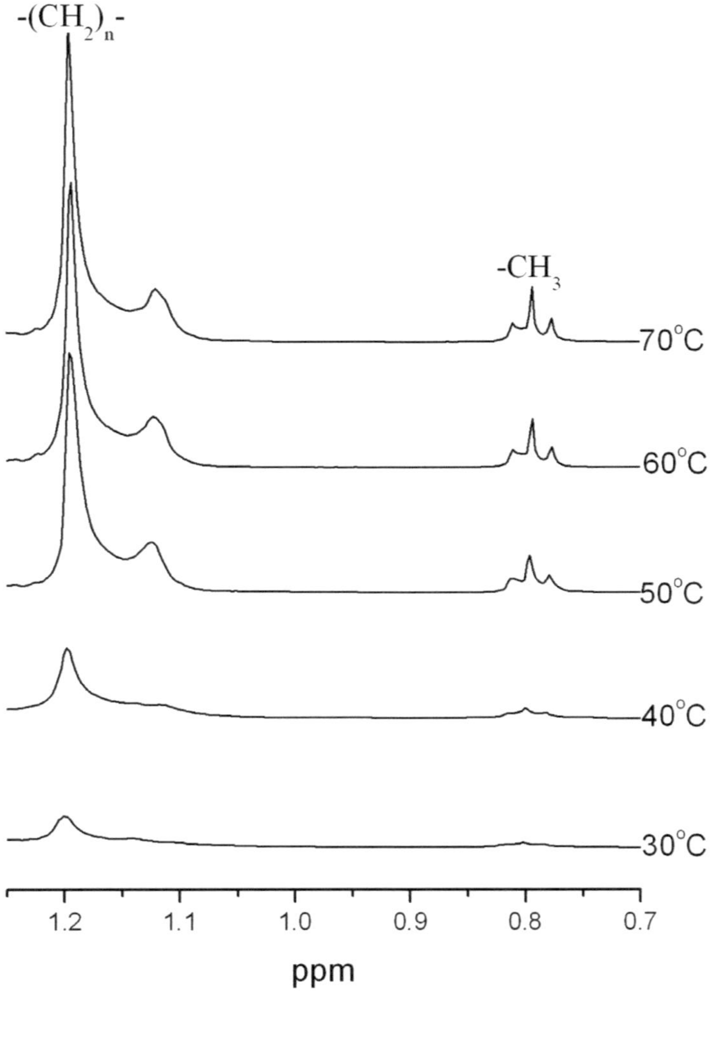

Fig. 1. Partial ^{1}H-NMR spectra of compound 1 in $CDCl_3$ at different temperature

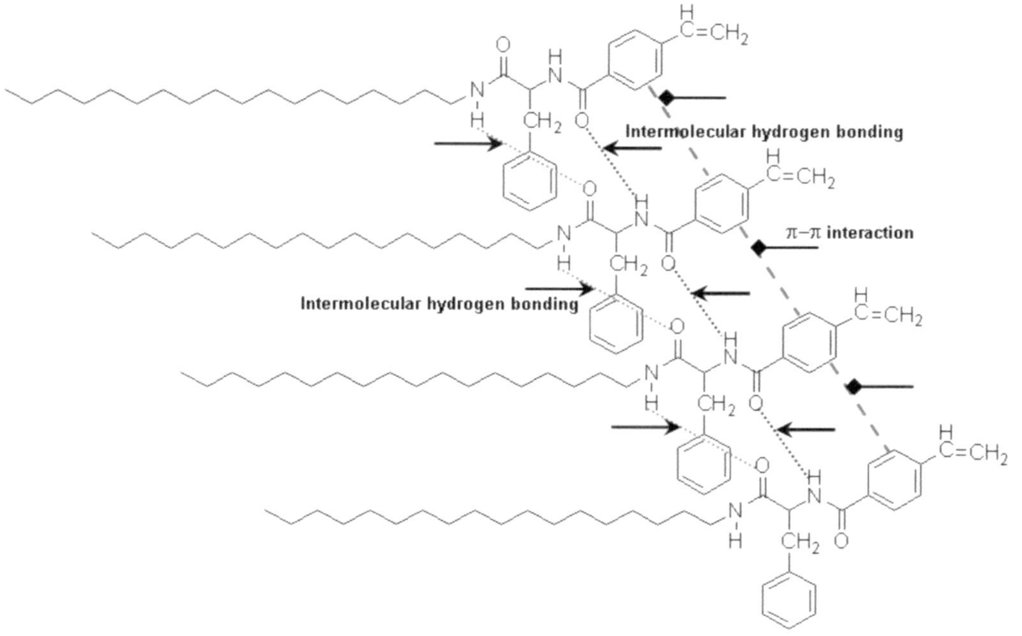

Fig. 2. Mechanism of gelation of compound 1 in different organic solvents

3. Preparation of silica-polymer hybrid composite materials

The surfaces of inorganic materials are functionalized with polymer chains either chemically (through covalent bonding) or physically (by physisorption). Physisorption involves absorption of block copolymers with sticky segments. The non-covalent adsorption makes the adsorption reversible especially during processing and is not a favored technique.
Covalent grafting techniques are preferred to maximize a stable interfacial compatibility between the two phases. Covalent grafting techniques involve either the "grafting to" or "grafting from" methods as shown in Scheme 2. "Grafting-to" involves the reaction of reactive, preformed macromolecules with compatible surface groups. Though experimentally simple, the "grafting to" method has a number of limitations; most notably low grafting densities because of steric crowding of reactive sites by previously attached polymers (Lyatskaya & Balazs 1998). The limitations of physisorption and "grafting to" can be circumvented by the "grafting from" technique. In this process the surface of the particle is modified with an initiator monolayer followed by polymerization under conventional conditions. Using the right system and techniques, one can control the functionality, density and thickness of the polymer brushes. The "grafting from" technique results in significantly higher grafting density because the steric barrier to incoming polymers imposed by the in situ grafted chains does not limit the access of smaller monomer molecules to the active initiation sites ([Pruker & Ruhe, 1993). This polymerization technique is also commonly referred to as surface initiated polymerization (SIP) (Edmondson et al., 2004). Conventional free radical polymerizations are difficult to control and are generally characterized by broad molecular weight distributions, poor control of molecular weight and chain end

functionality, and the inability to synthesize well-defined block copolymers. However, many of these drawbacks can be overcome by using controlled free radical polymerization. The use of surface initiated controlled/living radical polymerization has proven to be a versatile approach for incorporating different types of organic polymers with varied architecture on the silica surfaces. By using this technique, one can manipulate the structure of the resultant polymer shell through changes in grafting density, composition and molar mass. In controlled radical polymerization, the life-time of the growing radical can be controlled resulting in the synthesis of predefined molar mass, polymers with low polydispersity, controlled composition, and functionality. In general controlled/living polymerization can be achieved by stable free radical polymerization, e.g. nitroxide-mediated processes (NMP), metal catalyzed atom transfer radical polymerization (ATRP) and degenerative transfer, e.g. reversible addition-fragmentation chain transfer (RAFT). ATRP Matyjaszewski (Wang and Matyjaszewski 1999) and Sawamoto (Kato et al. 1995) simultaneously reported ATRP in the mid 1990's. The ATRP initiator (typically a α-halo ester or a α-benzyl moiety) is activated in the presence of metals such as Cu, Ru, Fe, and others. The compounds were polymerized by grafting to and grafting from method. Grafting to involves telomerization with silane cou0pling agent MPS followed by grafting onto silica particles to prepare composite materials. The controlled radical polymerizations were done for monomers 1 and 3 by atom transfer radical polymerization method. The compounds are also polymerized by grafting from via ATRP method. A typical "grafting to" method for compound 1 is 4.0 g (7.32 mmol) of compound 1 was dissolved in 50 ml anhydrous toluene by heating and 0.15 g (0.73 mmol) of 3-mercaptopropyltrimethoxysilane (MPS) and 40 mg AIBN were added into the solution and purged with N_2 gas. The mixture was degassed by three-freeze-pump thaw cycles. It was then placed in oil bath and heated at 60 °C for 24 hrs to complete the polymerization reaction. The reaction mixture was concentrated and dissolved in minimum amount of chloroform and re-precipitated from methanol. The white crystal appeared were filtered washed with methanol several times and dried under vacuum to obtain telomer. For grafting on to silica surface 4.0 g of telomer and 4.0 g silica were taken in a 100 mL three-necked flask and 40 mL dry toluene was added. The suspension was refluxed for 72 hours to complete the grafting process. After filtration the silica particles were washed with hot toluene, methanol, and chloroform repeatedly and finally washed with diethyl ether and dried under vacuum to get telomerized silica-polymer composite materials as shown in scheme 2. To synthesize polymer-grafted silica particles by ATRP method we have synthesized radical polymerization initiator (11-(2-bromo-2-methyl)propionyloxy)undecyltrichlorosilane by previously reported literature. ATRP initiator was grafted onto silica in dry toluene in the presence of triethylamine under nitrogene atmosphere. The reaction between the surface accessible OH group of silica and the anchoring group of the initiator (-$SiCl_3$) resulted in the formation of chemical bound between the silica surface and initiator. ATRP processes have been carried out using Sil-initiator as macroinitiator suspended in the mixture of monomer and toluene in the presence of CuBr and 1,1,4,7,7-pentamethyldiethylenetriamine (PMDETA) as catalyst precursors After the ATRP process the prepared composite materials were purified by repeated washing in different solvents to remove the non-grafted polymers from the surface. The grafted hybrid composite materials were characterized by different spectroscopic (DRIFT, NMR) and thermoanalytic methods (TGA).

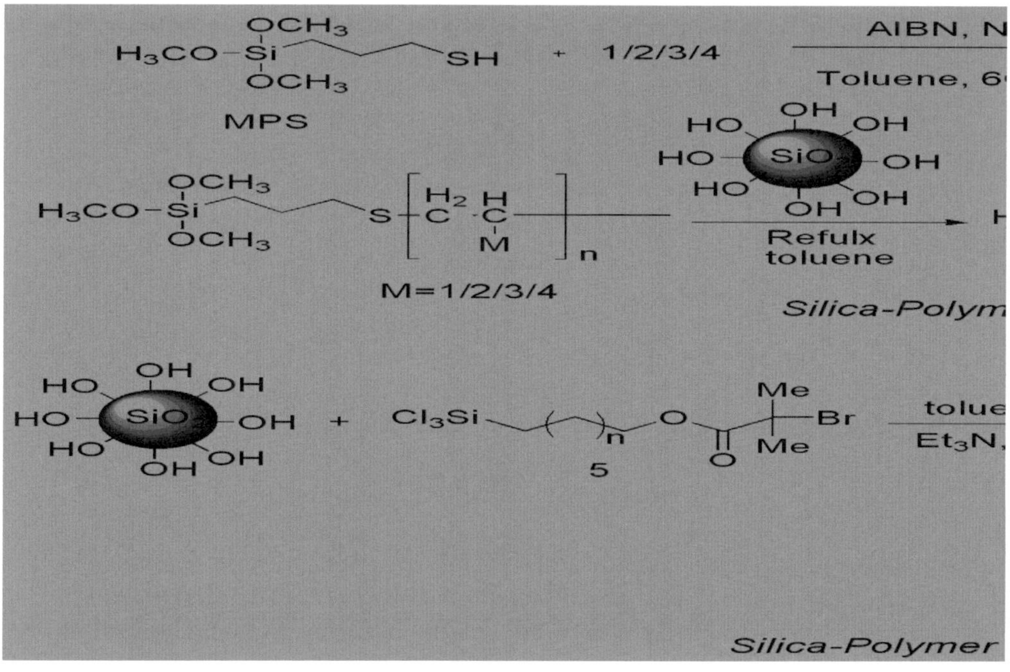

Scheme 2. Preparation of silica L-phenylalanine composite by different grafting processes

4. Characterization of hybrid-composite materials

Diffuse reflectance fourier transform infrared spectroscopy (DRIFT IR) is a suitable method to study the effectiveness of both immobilization of organic molecules onto silica surface and surface-initiated polymerization. DRIFT IR and FTIR spactra were recorded in the range of 3750-1370 cm^{-1} at room temperature using 4 cm^{-1} resolution and number of scans of 64. Comparison of DRIFT IR spectra of bare silica, ATRP initiator-grafted silica, Silica-grafting from and silica grafting to polymer furthermore FTIR spectrum of telomer are also presented in Figure 3.

The spectra of b and c are nearly identical providing an evidence that fairly high amount of telomere has been grafted onto silica particles. Successful immobilization of ATRP initiator is confirmed by the two strong bands at 2924 cm^{-1} and 2853 cm^{-1}, respectively attributed to the asymmetric and symmetric CH_2 stretching furthermore the band at 1717 cm^{-1} arising from carbonyl group of the ester moiety. Analysis of spectrum e shows a broad band at 1633 cm^{-1} with a shoulder at 1656 cm^{-1} derived from the overlapped amide carbonyl stretching vibrations and a band at 3290 cm^{-1} due to the N-H stretching vibration indicating the presence of polymer on silica surface.

These findings clearly proved that silica particles could be coated using two approaches, namely grafting to technique and grafting from technique (Surface initiated ATRP).

^{29}Si CP/MAS NMR spectroscopy is well suited for assessing the surface reaction of silanes with silica. Resonances for the silica species appears from -90 to -110 ppm. The ^{29}Si CP/MAS NMR spectra of the polymer grafted silica for sil-grafting from polymer and sil-grafting to polymer are shown in Fig. 4. The ^{29}Si CP/ MAS NMR spectrum for bare silica and ATRP initiator grafted silica (sil-initiator) are also included for comparison. The two polymer grafted silica showed that large extent of silanol groups remain unfunctionalized by both grafting methods.

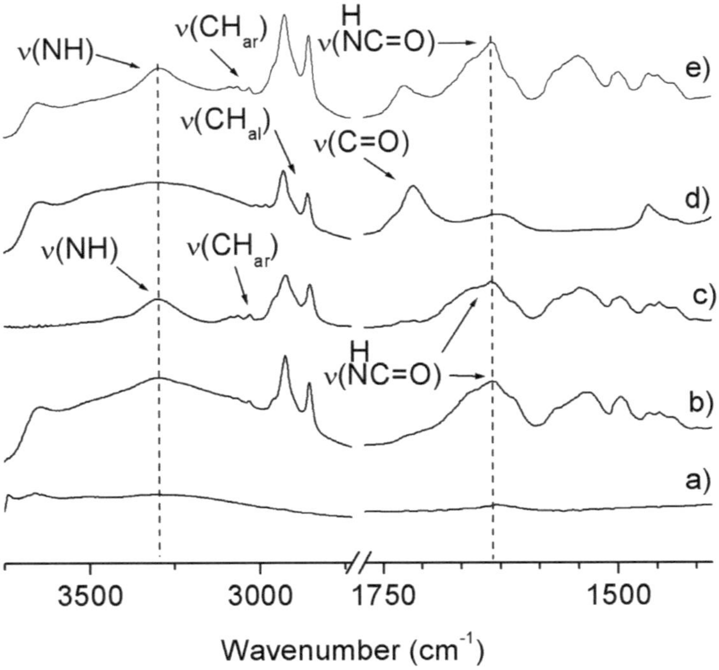

Fig. 3. DRIFT IR spectum of a) bare silica, b) Sil-grafting to polymer of 1, c) Telomer of 1, d) Sil-ATRP initiator, e) Sil-grafting from polymer of 1.

Figure 4 shows the differentiation between free silanol groups (Q^3) and geminal silanol groups (Q^2) besides the siloxane groups (Q^4) that are indicated by signals at -92, -102 and -111 ppm, respectively. In ATRP polymer-grafted silica the signal corresponding to residual geminal silanols (Q^2) is not seen while its intensity is very less in ATRP initiator grafted silica (sil-initiator), however this signal is quite intense for d) which was prepared by "grafting to" method. In the spectra after immobilization of ATRP initiator (b) and after polymerization (c) emphasize signals for T^n species that are related to the number of siloxane bonds. When the initiator was reacted with silica surface a large amount of cross-linked T^2 type silicon species (-57 ppm) was observed while polymer grafting increased cross-linked surface indicating by the appearance of T^3 signals (-65 ppm). The disappearance of signal for T^1 in the spectrum (b) and (c) is due to the reaction of the silanol with self-assembling monolayer formed by ATRP initiator. For d) the signal corresponds to T^2 and T^3 are completely absent however, T^1 signal is slightly appeared which indicates that relatively higher amount of silanol groups remain unfunctionalized in d) than c).

In liquid- or suspended-state NMR, only those molecules or parts of molecules are detectable that has very fast rotational motions. Motion must be in such a fast range that it can average out dipolar coupling and chemical shift anisotropy until related NMR peaks becomes narrow enough to be detected. The suspension-state ^1H NMR of sil-grafting from polymer was measured from 25 to 50 °C. Neither half-height width (line width) of methylene groups nor spin-spin relaxation time (*T*2) showed any significant change with temperature (20-50 °C) for both hybrid materials prepared by grafting from and grafting to method. We observed that intensity of the NMR peaks representing terminal methyl and methylene groups of octadecyl moieties increased slightly and detectable when a very high vertical scale was used for graphical presentation. These results indicate that the organic phase on the silica surface is in a solid state at room temperature.

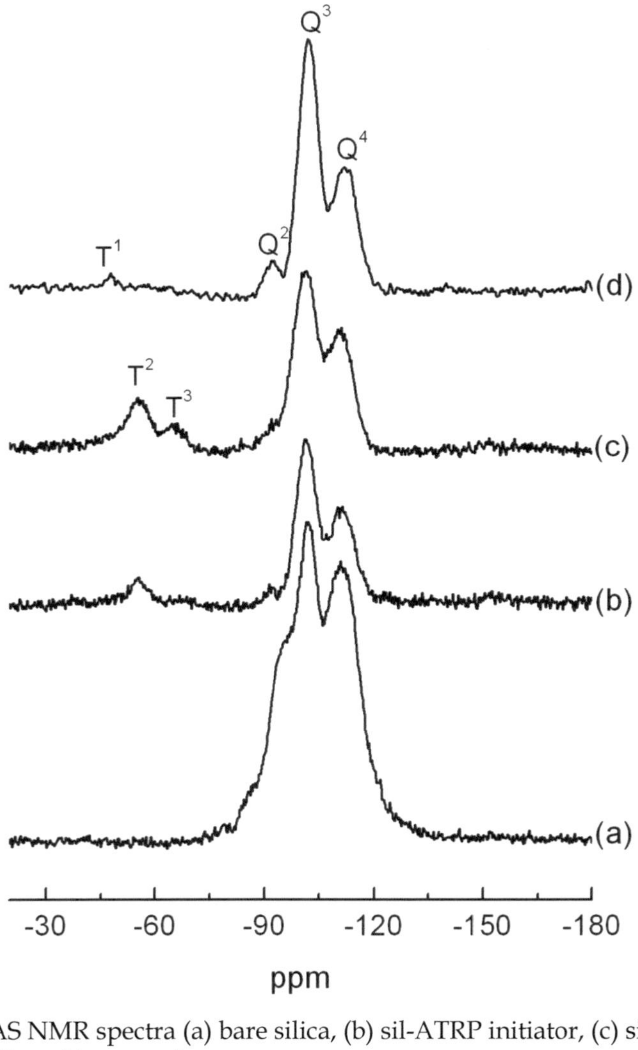

Fig. 4. ^{29}Si CP/MAS NMR spectra (a) bare silica, (b) sil-ATRP initiator, (c) sil-grafting from polymer and (d) sil-grafting to polymer

Under the condition of magic angle spinning and dipolar coupling of protons, the chemical shift of methylene groups in ^{13}C CP/MAS NMR spectroscopy depends largely on the conformation of octadecyl chains. For each central carbon atom in an octadecyl chain with the *trans* conformation, a chemical shift of around 33 ppm is expected. On the other hand, for conformations with rapid changes between *gauche* and *trans*, a chemical shift of about 30 ppm is expected.Solid-state ^{13}C CP/MAS NMR offers useful information about the chemical composition of the polymer grafted silica in addition, it gives evidence about the conformation and dynamics of immobilized alkyl chains (Grant & Cheney, 1967). Under the condition of magic angle spinning and dipolar coupling of protons, the chemical shift of methylene groups in ^{13}C CP/MAS NMR spectroscopy depends largely on the conformation of alkyl chains - $(CH_2)_n$-. In general, different conformations of the long alkyl moieties are detectable in two signals; for a rigid and ordered *trans* conformation, a chemical shift of around 33 ppm is expected, conversely, for mobile, amorphous and less ordered *gauche* conformations a chemical shift of about 30 ppm is anticipated. ^{13}C CP/MAS NMR spectra for polymer grafted silica were recorded at variable temperature 30 °C to gain insight of the conformational form of the alkyl chain (Fig. 5). The intense signal at 30.1 ppm is attributed due to *gauche* conformation however, the signal at 32.3 ppm due to *trans* conformations are very low. Generally, as the temperature is decreased more *trans* conformations are observed due to the lower mobility and higher rigidity of alkyl chain (Ansarian et al. 2005) The shows the alkyl chains of the grafted polymeric materis exists as gauche form which indicates the noncrystalline nature of the polymers also the immobilized alkyl chain stayed mobile, amorphous and less ordered form on silica surface.

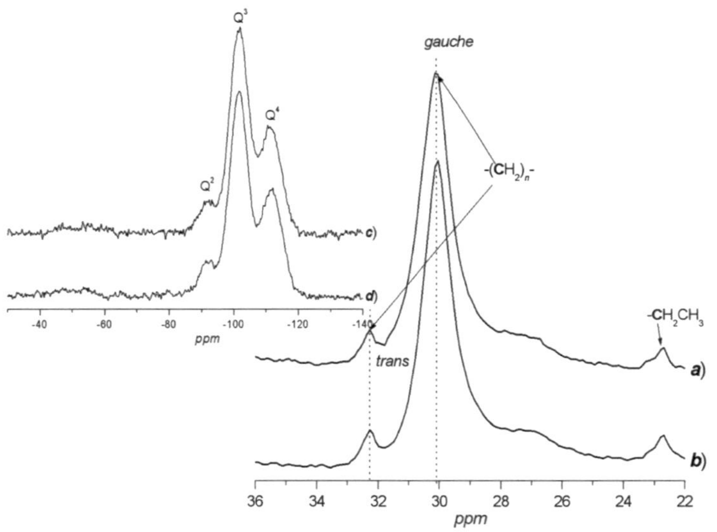

Fig. 5. Partial ^{13}C and ^{29}Si CP MAS NMR spectrum of hybrid composite prepared materials (a) and (c) are for Silica-polymer by "grafting from" method and (b) and (d) are for Silica - polymer by "grafting to" method.

The organic content of the grafted silica particles has been determined by thermogravimetric analysis. TGA runs were conducted at a constant heating rate of 10 °C /min in air using an empty crucible as reference. Heating process was carried out up to 800 °C that has been demonstrated to be sufficiently high to degrade all surface bonded organosilanes. and the resulted thermograms of a) bare silica, b) Sil-ATRP initiator, c) Silica-grafting to polymer d) silica-grafting from polymer and e) telomere in Fig. 6. All weight retention profiles was observed to reach a plateau at 650 °C confirming that there is no organic material remained on silica at 800 °C. Considering the TGA curve of bare silica particles as reference the weight of the immobilized initiator can be calculated as 9.6 % which was translated to grafting density an average 0.61 initiator per nm^2. Similarly, the thermogravimetric analysis revealed that 25.5 % ATRP polymer is attached onto silica surface if the weight retention of sil-ATRP initiator was considered as reference at 800 °C. Comparison between the weight retention of bare silica particles and sil-grafting to polymer showed 13.5 % of grafted telomere that is only 53 % of the immobilization obtained by surface initiated ATRP or than that of grafting from process.

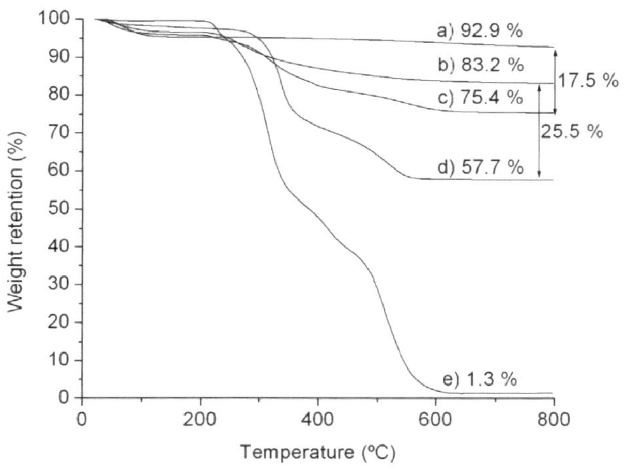

Fig. 4. Thermogravimetric analysis (TGA) data of hybrid composite materials prepared by both "grafting from" and "grafting to" method; a) bare silica, b) Sil-ATRP initiator, c) Silica-grafting to polymer d) silica-grafting from polymer and e) telomer

5. Application of hybrid composites for separation by HPLC

The hybrid composite materials were packed into stainless steel column in order to use stationary phase for high performance liquid chromatography. The chromatographic evaluation was also performed using the Tanaka test mixture containing hydrophobic probes which give information about shape and methylene selectivity, ion-exchange capacities from acidic and neutral media and hydrogen bonding capacities (Tanaka et al. 1989). The chromatogram obtained for monomer 1 by both grafting techniques is shown in Fig. 5 and Fig. 6 and is compared with conventional polymeric ODS column (ODS-p) stationary phase. This characterization protocol is a well-developed approach that is recommended to obtain information about functionality of the silylant reagent, and the methylene selectivity as well as to establish the repeatability and reproducibility of the separation behavior of commercially available stationary phase's ODS-p. The chromatogram showed the separation of two homologous alkylbenzenes as well as planar and nonplanar PAHs.

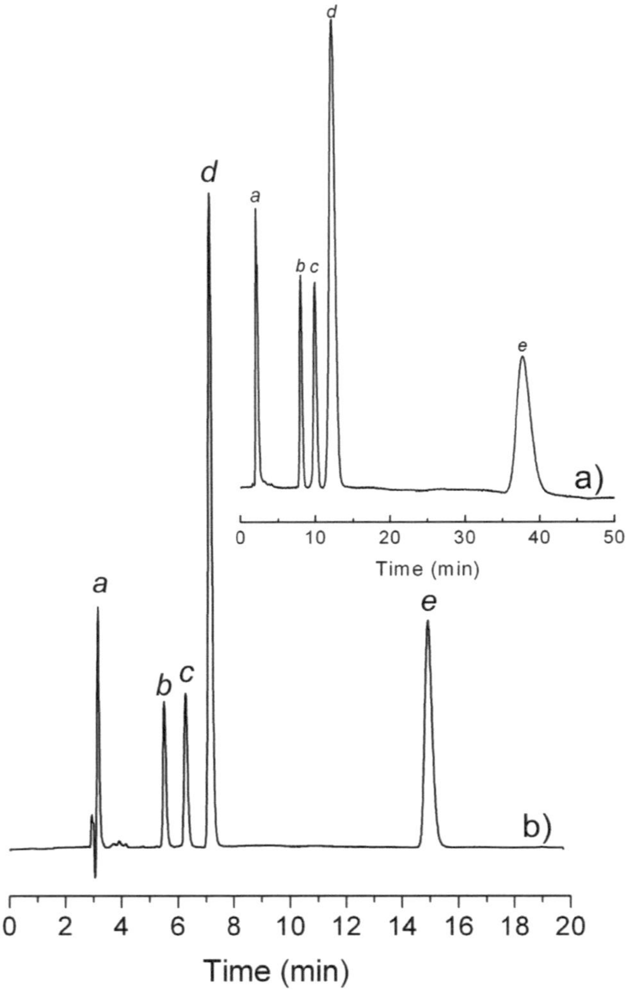

Fig. 5. Chromatogram for Tanaka test mixture for hybrid composite materials obtained from monomer 1 by a) "grafting from" and b) grafting to method. Elutes: a- uracil; b-butylbenzene; c-amylbenzene; d- *o*-terphenyl; e-triphenylene. Mobile phase: methanol: water = 90:10, column temperature 30 °C, flow rate: 1.0 $mLmin^{-1}$

6. Selectivity for PAHs

The shape selectivity of a given stationary phase can be measured by the separation factor (α) value for pairs of test solutes having the same carbon number but different shape, such as perylene/ 1, 1´-binaphthyl, triphenylene/ *o*-terphenyl, chrysene/ pyrene and *trans*-stilbene/ *cis*-stilbene. On a typical monomeric C_{18} phase (C_{18}-M) the separation factor $\alpha_{\text{triphenylene}/o\text{-terphenyl}}$ ranges from 1.0 to 1.7, while the value on a polymeric C_{18} (C_{18}-P) phase lies between 2.0 to 2.7. Jinno et al. [56,57] suggested that a separation factor (α) of triphenylene/ o-terphenyl ≥2.0 is an indication of significant solute planarity recognition.

Table 1 lists some of the α values measured on columns packed with silica-grafting to and silica-grafting from polymer as well as commercial monomer C_{18} or ODS (Octadecyl silane) phase. It is clear that both the hybrid composite prepared from L-phenylalanine shows selectivity factor α(triphenylene/o-terphenyl) of 3.30-3.20 possesses enhanced shape selectivity for PAHs. The selectivity factor α (perylene/1,1′-binaphthyl)) 16.5 and 13.5 on the sil-grafting to and sil-grafting from further exemplifies this extremely high planarity recognition. Interestingly, for the isomeric aromatic pair, chrysene and pyrene, the sil-grafting to phase yielded a selectivity factor of α(chrysene/ pyrene)= 1.55 while sil-grafting from yielded slightly lower value 1.53 demonstrated that on both phases shape recognition includes not only planarity but also the topology of solutes. However all the chromatographic data suggested that the chromatographic performance does not differ for the composite materials prepared by "grafting from" and "grafting to" method.

	separation factor (α)							
	silica-grafting to [b]	silica-grafting to [c]	silica-grafting from [b]	silica-grafting from [c]	C_{18}-P [b]	C_{18}-P [c]	C_{18}-M [b]	C_{18}-M [c]
triphenylene/ *o*-terphenyl	3.30	2.19	3.20	2.84	2.28	2.27	1.29	1.28
chrysene/ pyrene	1.55	1.33	1.52	1.43	1.40	1.40	1.22	1.22
perylene/ 1,1-binaphthyl	13.0	8.10	16.5	14.8	4.5	4.55	2.10	2.12
trans-stilbene/ *cis*-stilbene	1.58	1.28	1.52	1.41	1.22	1.22	1.06	1.06

mobile phase: [b] methanol/ water (90/10), [c] methanol / acetone / water (70/20/10), column temperature: 30°C and flow rate: 1.00 mLmin^{-1}

Table 1. Separation factors of PAH sample sets on different columns after adding acetone to the mobile phase

7. Conclusion

In conclusion we can say, by "grafting from" method significantly higher amounts of polymer can be grafted on to porous silica gels i.e. high grafting density than "grafting to" methods usually enhance the properties of composite materials. In contrast for chromatographic point of view there is no significant differences on selectivity of shape constrained solutes or toxic organic pollutants of different sizes and shapes.

8. References

Abdallah JP & Weiss RG (2000) *Adv Mater 12*, 1237-1247.
Ansarian HR, Derakhsan M, Rahman MM, Sakurai T, Takafuji M, Taniguchi I, Ihara H (2005) *Anal Chem Acta* 25, 179.
Anwander R (2001) Chem Mater 13, 4419
Cardin DJ (2002) *Adv Mater* 14,553
Crego AL, Diezmasa JC, Dabrio MD (1993) *Anal Chem* 65, 1615.

De Vos DE, Dams M, Sels BF, Jacobs PA (2002) *Chem Rev* 102, 3615
Edmondson S, Osborne, VL and Huck, WTS (2004) *Chem Soc Rev* 33, 14.
Grant DM, Cheney BV, *J Am Chem Soc* (1967) 89, 5315.
Guo Y, Colon LA, (1995) *J. Microcolumn Sep.* 7 485.
Hatano T, Bae AH, Takeuchi M, Fujita N, Kaneko K, Shinkai SM (2004) *Chem Eur J 10*, 5067-5075
Hamley IM, Ansari A, Castelletto V, Nuhn H, Rosler A, Klok HA, *Biomacromolecules* (2005) *6*, 1310-1315.
Ihara H, Takafuji M, Sakurai T (2004) *Encyclopedia of Nanoscience and Nanotechnology*; H. S. Nalwa, Ed.; American Scientific: California, CA. Vol. 9, 473-495.
Jung JH, Ono Y, Shinkai S (2000) *Langmuir 16*, 1643–1649
Kato M, Kamigaito M, Sawamoto M and Higashimura T (1995) *Macromolecules* 28, 1721
Kobayashi K, Hamasaki N, Suzuki M, Kimura M, Shirai H, Hanabusa K, (2002) *J Am Chem Soc 124*, 6550-6551
Klok HA, Langenwalter JF, Lecommandoux S (2000) *Macromolecules 33*, 7819-7826
Li BS, Cheuk KKL, Ling L, Chen J, Xiao X, Bai C, Tang BZ (2003) *Macromolecules 36*, 77-85
Lekchiri A, Morcellet J, Morcellet M (1987) *Macromolecules 20*, 49-53.
Lyatskaya Y & Balazs A C (1998) *Macromolecules* 31, 6676
Nagai A, Sato D, Ishikawa J, Ochiai O, Kudo H, Endo T (2004) *Macromolecules, 37*, 2332-2334.
Oishi T, Lee YK, Nakagawa A, Onimura K, Tsutsumi H, *J. Polym. Sci., Part A: Polym. Chem.*(2002) *40*, 1726-1741.
Pruker O & Ruhe J (1993) *Mater. Res. Soc. Symp. Proc.*, 304, 1675
Rahman MM, Czaun M, Takafuji M, Ihara H (2008) *Chem Eur J* 14, 1312.
Rahman MM, Takafuji M, Ihara H (2008) *Anal Bioanal Chem* 392, 1197.

Part 2

New Materials with Unique Properties

18

High Dispersion Power of Cardo-Typed Fluorene Moieties on Carbon Fillers

Shinichi Kawasaki[1], Fengzhe Jin[2] and Toshikazu Takata[3]
[1]*Energy Technology Laboratories, Osaka Gas Co., Ltd., Konohana, Osaka*
[2]*SANYU Chemical Co., Ltd., Itami, Hyogo*
[3]*Department of Organic and Polymeric Materials*
Tokyo Institute of Technology, Ookayama, Meguro, Tokyo
Japan

1. Introduction

Composite materials containing carbon fillers are well used in a variety of material fields including electronic and optical device fields. The carbon fillers are capable of endowing conductivity, heat-releasing ability etc. to the composites. Property indispensable for such matrix resins, as expected for most matrix resins, is to perform high dispersion of the fillers. Surface modification of fillers is one of the straightforward approaches to produce the excellent composite materials. However, if the matrix resin itself can possess good carbon filler dispersing power, it can simply achieve the fine dispersion of a variety of carbon fillers including non-surface modified fillers, making the composite fabrication easy.

Currently, much attention has recently focused on the filler-dispersing power of fluorene moieties. In the recent study on polymer materials based on fluorene based polymer (FBP), we have found that FBP shows not only a much filler-incorporating ability but also an excellent filler-dispersing ability for carbon fillers, suggesting that FBP is a highly suitable matrix resin for carbon fillers.

In this chapter, we discussed the interesting nature of FBP as matrix resin that shows extremely high carbon filler-incorporating and dispersing powers to yield high quality composite materials, along with the investigation on the interaction based on the structural characteristics of 9,9-diarene-substituted fluorene skeleton of FBP.

2. Carbon filler

2.1 Allotrope of carbon

Atomic carbon is a very short-lived species and, therefore, carbon is stabilized in various multi-atomic structures with different molecular configurations called allotropes. The three relatively well-known allotropes of carbon are amorphous carbon, graphite, and diamond. Once considered exotic, fullerenes are nowadays commonly synthesized and used in research; they include buckyballs, [Samal, S. & Ebbesen, T.W.] carbon nanotubes,[Dresselhaus, M.S. et al] carbon nanobuds [Nasibulin, A. G;] and nanofibers.[Nasibulin, A.& Vieira, R.] In this paper, we will focus on two carbon filler in the market: carbon black and carbon nanotube.

2.2 Carbon black

Carbon black [C.A.S. NO. 1333-86-4] is a material produced by the incomplete combustion of heavy petroleum products such as FCC tar, coal tar, ethylene cracking tar, and a small amount from vegetable oil. Carbon black is a form of amorphous carbon that has a high surface-area-to-volume ratio, although its surface-area-to-volume ratio is low compared to that of activated carbon. Its physical appearance is that of a black, finely divided pellet or powder. Its use in tires, rubber and plastic products, printing inks and coatings is related to properties of specific surface area, particle size and structure, conductivity and color.
In this paper we are focused on conductive carbon black, and evaluated dispersion power of fluorene moieties on conductive carbon black.

2.3 Carbon nanotube

Carbon nanotubes (CNTs) are allotropes of carbon with a cylindrical nanostructure. Nanotubes have been constructed with length-to-diameter ratio of up to 132,000,000:1,[Wang, X. et al] significantly larger than any other material. Nanotubes are categorized as single-walled nanotubes (SWNTs) and multi-walled nanotubes (MWNTs).

2.3.1 Mechanical properties of carbon nanotube

Carbon nanotubes are the strongest and stiffest materials yet discovered in terms of tensile strength and elastic modulus respectively. This strength results from the covalent sp2 bonds formed between the individual carbon atoms.[Yu, M.F.; Collins, P. G.; Jensen, K; Belluci, S.; Chae, H.G.; Meo, Michele; Sinnott, S.B.; Demczyk, B.G; ASTM; & Wagner, H. D.]

	Young's modulus (TPa)	Tensile strength (GPa)	Elongation at break (%)
SWNT E	~1 (from 1 to 5)	13–53	16
Armchair SWNT T	0.94	126.2	23.1
Zigzag SWNT T	0.94	94.5	15.6–17.5
Chiral SWNT T	0.92		
MWNT E	0.2~0.95	11~150	
Stainless steel E	0.186 ~0.214.	0.38. ~1.55	15–50
Kevlar-29&149 E	0.06~0.18	3.6~3.8	~2

Table 1. Mechanical properties of carbon nanotube

2.3.2 Electrical properties of carbon nanotube

Because of the symmetry and unique electronic structure of graphene, the structure of a nanotube strongly affects its electrical properties. For a given (n,m) nanotube, if $n = m$, the nanotube is metallic; if $n - m$ is amultiple of 3, then the nanotube is semiconducting with a very small band gap, otherwise the nanotube is a moderate semiconductor. Thus all armchair ($n = m$) nanotubes are metallic, and nanotubes (6,4), (9,1), etc. are semiconducting.[Lu, X; Chen, Z. et al]

However, this rule has exceptions, because curvature effects in small diameter carbon nanotubes can influence strongly electrical properties. Thus, a (5,0) SWCNT that should be semiconducting in fact is metallic according to the calculations. Likewise, *vice versa*-- zigzag and chiral SWCNTs with small diameters that should be metallic have finite gap (armchair nanotubes remain metallic).[Hong, S.H.; et al] In theory, metallic nanotubes can carry an electric current density of 4×10^9 A/cm^2 which is morethan 1,000 times greater than metals such as copper,[Takesue, I. et al] where for copper interconnects current densities are limited by electromigration. Multiwalled carbon nanotubes with interconnected inner shells show superconductivity with a relatively high transition temperature Tc = 12 K. In contrast, the Tc value is an order of magnitude lower for ropes of single-walled carbon nanotubes or for MWNTs with usual, non-interconnected shells. [Pop, E. et al]

2.3.3 Thermal properties of carbon nanotube

All nanotubes are expected to be very good thermal conductors along the tube, exhibiting a property known as "ballistic conduction", but good insulators laterally to the tube axis. Measurements show that a SWNT has a room-temperature thermal conductivity along its axis of about 3500 W • m−1 • K−1;[Sinha, S.L. et al] compare this to copper, a metal well-known for its good thermal conductivity, which transmits 385 W • m−1 • K−1. A SWNT has a room-temperature thermal conductivity across its axis (in the radial direction) of about 1.52 W • m−1 • K−1, [Thostenson, E. et al] which is about as thermally conductive as soil. The temperature stability of carbon nanotubes is estimated to be up to 2800 °C in vacuum and about 750 °C in air. [Ando, W.]

By challenging on fine dispersion of CTNs in polymer matrix, polymer composites with promising mechanical properties, electrical properties and thermo-conductive properties can be expected.

3. Fluorene moieties

3.1 Basic properties of fluorene moieties

Fluorene is a member of polycyclic aromatic hydrocarbon (PAH). Two benzene rings are fused to cyclopentane ring. It emits violet fluorescent color. It is not synthesized commercially but is obtained from middle oil fraction of coal tar. It is insoluble in water; soluble in ether and acetone; melting point 116-117°C. It plays important part in metallocene catalysts as a ligand. It is used in the formation of polyradicals for resins. It is used in manufacturing antimalaria drugs and other pharmaceuticals.[Stephan, O. et al]

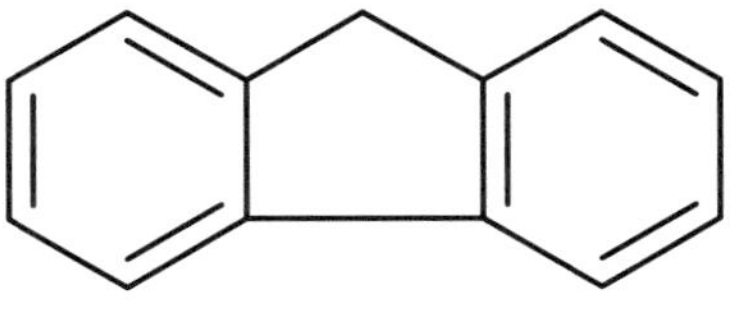

Fig. 3.1. Fluorene

Fluorene family compounds are base materials for dyes and optical brightening agents. They have useful functions such as light and temperature sensitivity, heat resistance, conductivity, emittability, corrosion resistance and detection of amino groups.

They are used in the applications of thermo and light sensitizer, liquid crystal chemistry, luminescence chemistry, spectrophotometric analysis, molecular chemistry, organometallic-complexes and biochemorphology industry.

3.2 Fluorene polymers

3.2.1 Derived from 2.7-disubstituted fluorene

9,9-disubstituted and polymerized in 2,7 disubstited fluorene-polymers had its promissing photo-sensitive applications. The application areas of polymers derived from 2,7 disubstited fluorene covers polarized photo- and electroluminescence materials [Grell, M.; Lim, E.; Morin, J.-F.; Jin, S. H.; Yang, R.; Mo, Y.; & Lee, P.I.], White-Light-Emitting Diodes by copolymerization with other funtional unit. [Salim, T.]; Solar cell application of Polyfluorene Copolymer by blending with Poly(3-hexylthiophene) Nanofiber was also reported. [Teramoto, T.]

3.2.2 Polymers derived from 9,9-bis(4-hydroxyphenylfluorene)

Because 9 position of fluorene with its high reactivity, it was easy to be oxidized by oxygen, and convert to fluorenone. With further reaction with the compounds with aryl group, can design and achive many kinds of 9,9-diaryl fluorene. [Teramoto, T.; Papava, G. S; Morgan, P. W.; Yamada, M.; Yamada, M.; Li, Y.J.; & Korshak, V.V]

These kinds of diaryl fluorene, because they have multiple aromatic rings in one molecule, fluorene unit and aryl group can had to be positioned in same planar due to high stereo hindrince of them. The structure based on 9 position of fluorene similar to wings and body of butterfly so called cardo structure. [Sakurai, K. et al]

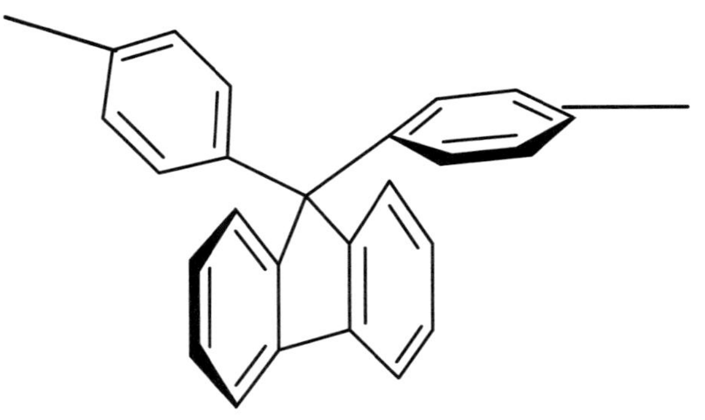

Fig. 3.2.2. *Cardo* structure of fluorene based compound.

3.2.2.1 Polymers in commercial base

Fluorene based polyester (FBP) possess high refractive indices (1.64 in average) as well as extremely low birefringence, so they are used in high fidelity lens materials such as digital cameras and cell phones with cameras, and can also be used as promising optical film materials due to its high flexibiliy and low retardation.

In genaral, organic polymers like polycarbonate and polyethyleneterephthalate have their high refractive index due high ratio of aromatic unit in their molecular chain. But, if the aromatic chains in same plane, anisotropy of unit with high dielectricity will cause optical anistropy in moleculars, finaly results in high birefringence of molecules. But in the case of

fluorene molecules with cardo-structure, even the molecule had many aromatic ring, due to the rings not in the same plane, anistropy of entile molecule was reduced, finaly the achivement the molecule with high rafractive index and low birefringence. [Yao, K.]
Fluorene based polyester (FBP) with cardo structure in it, regardless of it's polyester structure, they demonstrated as armouphous materials. FBP had its refractive index over 1.6 which was higher than polycarbonate, and lower birefringence than PMMA. Figure 3.2.2.1 shows injection molded polycarbonate, PMMA, and FBP under polarized plates. [Kawasaki, S. et al]

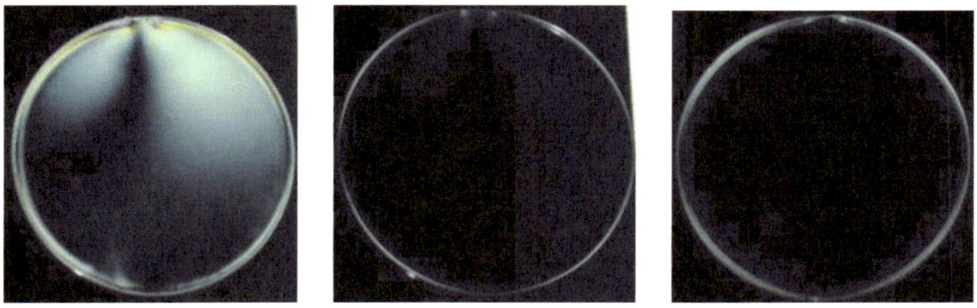

PC PMMA FBP

Fig. 3.2.2.1. Birefringence property of optical polymers.

Epoxy and Acrylates resin derived from 9,9-bis(4-hydroxyphenylfluorene) possess useful advantages such as high refractive indices and high heat resistance, and are used in hard coating materials and sealants and high refractive index coating films.[Liu, W. B.; Xiong, Y.Q.; & Dai, Z.]
FBP can also forming polymer alloys with commercial available polycarbonate and polyethylene naphthalene, and the final products with promissing transparency and low regardations compared to other alloys. [Kawasaki, S. et al]

3.2.2.2 Polymers in development

Except the polymers already available in the market, there are still many kinds of new applications on challenging by applying fluorene unit in the main chain.Polymers based on fluorene unit, like polyimide[Yang, C.P. et al], polyetheretherketone[Kawasaki, S. et al], polythioether[Hayashi, H. & Seesukphronrarak, S] were synthesized to increase solubiliy and process ability of high performance polymers; silicon contained polymers to increase heat resistance.[Seesukphronrarak, S et al] Spiro unit contained polymers to increase refractive index of fluroene polymers[Seto, R. et al] now under challenging too. Some nanocomposites based on organic-inorganic hybridyzation of fluorene with organic metallic compounds were also published recently.[Matsukawa, K.; & Suzuki, A.]

3.3 Potential power of fluorene moieties to disperse carbon filler

Since the first finding by Chen F. et. al that fluorene-based polymers are able to selectively wrap the single-walled carbon nanotubes (SWNTs) with certain chiral angles or diameters depending on their chemical structures, many report regarding selective dispersion of CNTS in fluorene based polymers were published in the world. [Chen, F.; & Ozawa, H.]
The polymers investigated in the research focused on 2,7-linked polymerization of fluorene diaryl substitutes. No report hade been published before the finding of fine dispersion of

9,9-bis (4-hydroxyphenyl) substitute derivatives on carbon fillers. In chapter 4, detailed finding of 9,9-bis (4-hydroxyl phenyl) substituted fluorene derivatives on carbon filler dispersion will be discussed. [Inada, T.; & Kawasaki, S.]

4. Dispersing power of carbon fillers in fluorene moieties

4.1 Dispersing power of fluorene moieties on carbon black in fluorene moieties

4.1.1 Introduction

Although the surface modification of carbon fillers is one of the established methods for producing excellent composite resins containing carbon fillers, a matrix resin that can disperse carbon fillers on its own will be a more efficient way to achieve high performance composite materials.

In this section, we compare the dispersion of carbon black (CB) in fluorene-based epoxy (FBE) resin with bisphenol-based epoxy (PBE) resin that does not contain a fluorene skeleton. The dispersity of carbon black in thermal plastic resin containing 9,9-cardo fluorene structure compared with the common thermal plastic resin polycarbonate are also discussed in this section.

4.1.2 Composites of CB/FBE

It was found that the L value of CB/FBE composite decreased with an increase in the content of CB. For the CB/FBE composites, the L value was lower than that of CB/PBE composites in three cases (containing 3, 5, and 7 wt% of CB in CB/FBE resin), especially for the CB/FBE composite containing high CB content in them (Figure 4.1.1).

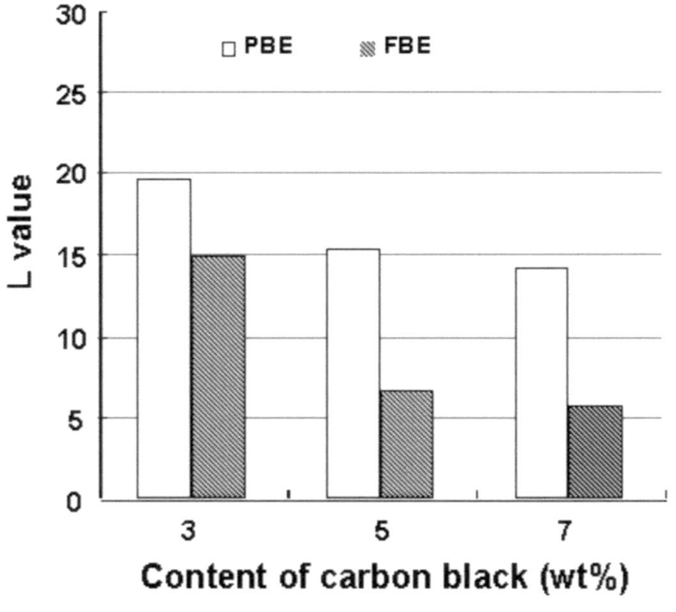

Fig. 4.1.1. Colorimetric analysis of coating film in which carbon black was dispersed in the epoxy resin.

The result of the reflectance between 400 nm and 700 nm was shown in Figure 4.1.2. The reflectance of CB/FBE composite was lower than PBE at either CB concentration. Generally, the optical density that was calculated from the degree of reflection was regarded as an index of the blackness. The optical density (OD) is defined in the equation,

$$OD = \log (1 / T)$$

where T is the transmittance.

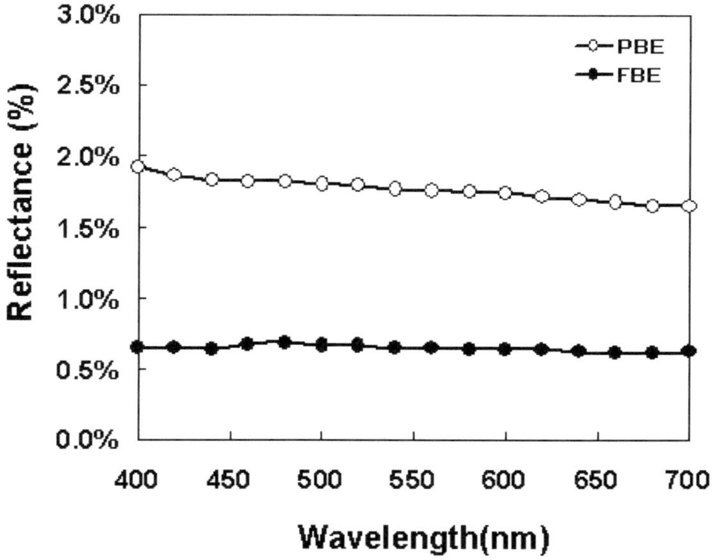

Fig. 4.1.2. Reflectance of CB/FBE composite sheets and CB/PBE sheets containing 7 wt% of CB in the composites.

In the case of CB/FBE composite containing 7 wt% content of CB, the optical density of the composite was 2.18, on the other hand, the optical density of CB/PBE composite was 1.75. This means that the degree of blackness of CB/FBE composite was higher than that of CB/PBE composite.

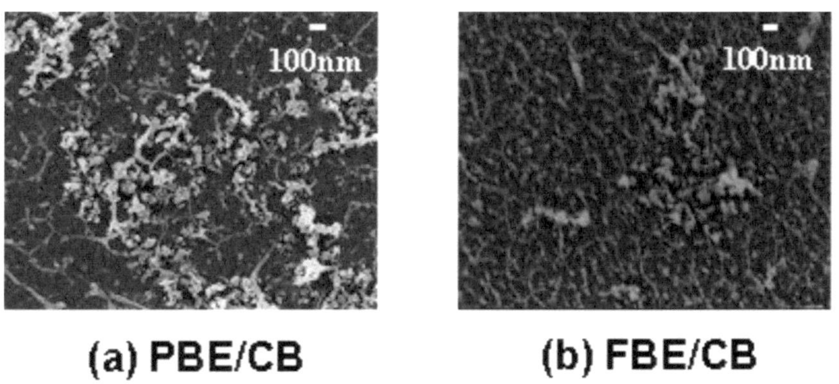

Fig. 4.1.3. FE-SEM images of the epoxy resin film (containing CB, 7 wt%).

The dispersion state of CB observed by FE-SEM is shown on Figure 4.1.3. In the case of CB/PBE composite, there are large aggregations of CB observed. On the other hand, in the case of CB/FBE composite, aggregate size was much smaller than in the CB/PBE composite.

4.1.2 Fine dispersion of CB in FBP

Even a 50% addition of carbon black (CB) was possible for the CB/FBP composite. High amount of CB in FBP did not cause a reduction in the efficiency of FBP. On the other hand, for the CB/PC composites only 10 wt% CB was available to be dispersed into the PC matrix. High dispersion amount of CB in FBP is obvious evidence for the high compatibility of FBP with CB.

4.1.2.1 Effect of CB addition on the mechanical properties and thermal properties of FBP

The effect of CB addition on CB/FBP composites is shown in Table 4.1.1. Storage modulus of CB/FBP composites was increased with the addition of CB in the composites. Enhancing ratio of CB on FBP was about two times above the PC. It means that CB interacted more strongly with FBP than with PC.

	FBP	PC
Without CB(A)	1.9×10^9 Pa	2.2×10^9 Pa
With 5wt% CB(B)	4.9×10^9 Pa	3.4×10^9 Pa
Effect of CB on CB/Polymer composites (B/A)	2.6	1.5

Table 4.1.1. Results of Dynamic Mechanical Analysis

The softening point of FBP was increased to about 22°C by the addition of CB; however, the effect of this addition was quite low on PC (about a 4°C increment). This was the other evidence supporting the high interaction between CB and FBP compared with the low interaction of CB and PC.

As shown in Figure 4.1.1, the degree of blackness for the CB/FBE composite was much higher than that of the CB/PBE composite. The aggregation of CB in the CB/FBE composite was much smaller than that in CB/PBE composite. According to the results shown above, it is obvious that the dispersion of CB in the CB/FBE composite was much finer than in the CB/PBE composite.

In addition, CB had a fine dispersion in FBP when compared with PC. With the same amount of CB addition, enhancement of mechanical properties and the increase in the Tg of FBP is much higher than that of PC (Table 4.1.2)

All the above results indicate that 9,9-substituted fluorene may have a strong interaction with carbon black. The strong interaction results in the fine dispersion of CB in FBP and the enhancement of mechanical properties as well as an increase in Tg.

	FBP	PC
Without CB(A)	148℃	166℃
With 5 wt% CB(B)	170℃	170℃
Effect of CB on CB/Polymer composites (B-A)	22℃	4℃

Table 4.1.2. Results of Glass Transition Temperature

4.1.3 Summarize of section 4.1

CB can be dispersed finely in CB/FBE composites. Compared with the aggregations of CB in PBE, there was no obvious aggregation in CB/FBE observed. The fine dispersion of CBs in FBE can be regarded as a result of strong π–π interaction between CB and FBE molecules. Also, because of the strong interaction, FBP molecules are firmly held by CB molecules by finely dispersing in them. This restriction of FBP molecules appears as a result of an increase in the Tg of FBP.

4.2 High dispersion ability of fluorene-based polyester as polymer matrix for carbon nanotubes

4.2.1 Introduction

The high electric conductivity of CNTs make them suited for use in electronic fields and makes it possible to prepare polymer composites using CNTs that have particular electric conductivity or electrostatic discharge properties. Although surface modification of carbon fillers is an established method for producing excellent composite resins, developing a matrix resin in which carbon fillers without such modification could be dispersed would provide a more efficient method of achieving high-performance composite materials.

The *cardo* structure of fluorene-based polymers, demonstrated in structure 1, makes it easy to achieve the fine dispersion of fillers such as carbon black, CNT and organic pigments. In a previous report, the fine dispersion ability of carbon black in FBP was proved.

This section discusses the fine dispersion of carbon fillers having a high aspect ratio in FBP. MWCNT was used as high aspect fillers for FBP. Commercially available polycarbonate (PC) was used for comparison.

4.2.2 Surface resistivity of MWCNT/polymer composites

A fine dispersion of MWCNT in matrix polymer will result in a low surface resistivity. In contrast, a covering of the matrix polymer on the MWCNT, will disturb the electrical pathways between MWCNTs — thus MWCNT covered with matrix polymer should have a higher surface resistivity than uncoated or 'naked' nanotubes.

Figure 4.2.1 shows the surface resistivity of MWCNT/polymer composites prepared by the solution casting method. Both of the composites of MWCNT with PC and FBP had a low percolation ratio, of between 0.5 ~ 1.0 wt%. The high dispersion power of PC and FBP on

MWCNT induced the low surface resistivity of the composites, and due to the covering of FBP on 'naked' MWCNTs, a higher surface resistivity of MWCNT/FBP-HX was achieved than for MWCNT/PC. The reason for this phenomenon was the covering of FBP-HX on MWCNT. It resulted in both the higher surface resistivity in MWCNT/FBP-HX composites and the higher percolation ratio of MWCNT/FBP-HX composites when compared with the values for MWCNT/PC composites.

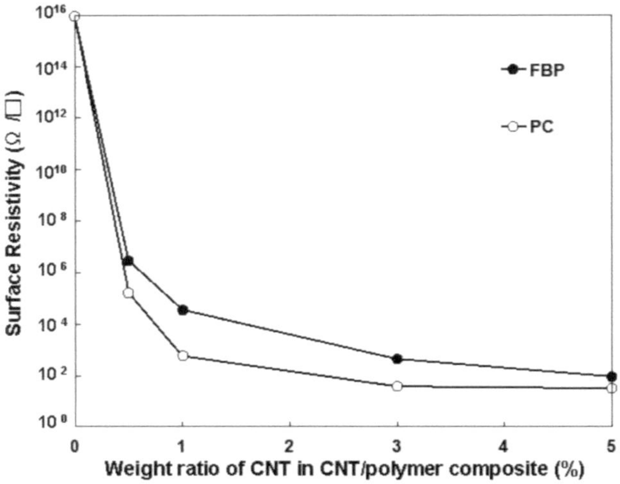

Fig. 4.2.1. Surface resistivity of MWCNT/FBP-HX composites prepared by solution casting, with different weight ratios of MWCNT (● MWCNT/FBP-HX composite, ○ MWCNT/PC composite).

4.2.3 Fine dispersion of MWCNT in FBP-HX observed by SEM

MWCNTs in composite materials as prepared by the solution method showed significant differences in the PC and the FBP-HX polymer matrices. The fluorene structure of FBP-HX had a strong π–π interaction with the graphite structure on the surface of the MWCNT. This improved the compatibility of MWCNT with the FBP-HX polymer through a process that is not yet clear. The MWCNTs were separated at the nanometer level in FBP-HXs. The fine dispersion of MWCNTs in FBP-HX and PC resulted in the low surface resistivity of both kinds of MWCNT/polymer composites, even those which contained low weight ratios (0.5 ~1.0 wt%) of MWCNT.

SEM observation of the cross-section of the MWCNT/FBP-HX composite indicated that the FBP-HX matrix polymer covered the surface of the MWCNTs (Figure 4.2.2b). However, this was not true for the MWCNT/PC composite, indicating that PC had a lower compatibility with the MWCNTs (Figure 4.2.2a, Figure 4.2.2c).

Figure 4.2.2a, 4.2.2b demonstrate the differences between the status of MWCNT in PC and FBP-HX. In figure 2b, almost none of the MWCNTs are separated from the matrix resin. In figure 4.2.2a, 'naked' MWCNT can be observed spilling out from the matrix resin.

The MWCNTs in FBP-HX, as observed by SEM, had diameters of 30 nm (Figure 4.2.2d). This was about 1.5 times that of the 'naked' MWCNTs (20 nm), providing more evidence for the fine dispersion of MWCNT in FBP-HX and the strong interaction between MWCNT and FBP-HX. It means that FBP-HX layer with 5 nm thick were covered on MWCNTs. The covered layers of FBP-HX became swollen in chloroform. The diameters of the MWCNT covered with FBP-HX increased to 60-70 nm after becoming swollen in $CHCl_3$ (Figure 4.2.2e).

All the above evidence shows that FBP-HX can cover the surface of MWCNT without further treatment. The fine dispersion of the MWCNT was the result of the strong interaction between the FBP-HX and the MWCNT.

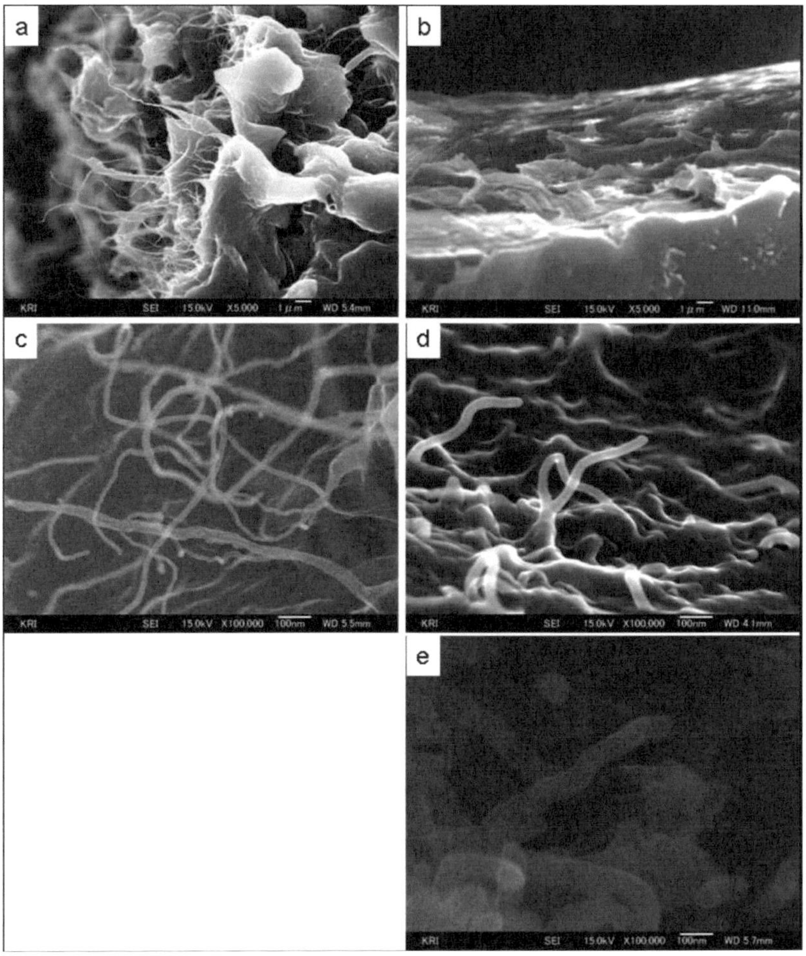

Fig. 4.2.2. Cross-sections of MWCNT/polymer composites observed by scanning electronic microscopy. a: MWCNT/PC composites with 3 wt% MWCNT, × 5000 magnification; b: MWCNT/FBP composites with 3 wt% MWCNT, × 5000 magnification; c: MWCNT/PC composites with 3 wt% MWCNT, × 100000 magnification; d: MWCNT/FBP composites with 3 wt% MWCNT, ×100000 magnification; e: swollen MWCNT/FBP composites with 3 wt% MWCNT, × 100000 magnification.

4.2.4 Variation of glass transition temperature (*Tg*)

The glass transition temperature, *Tg*, is the temperature at which an amorphous solid, such as a glass or polymer, becomes brittle on cooling, or soft on heating. In polymers, *Tg* is often expressed as the temperature at which the Gibbs free energy is such that the activation energy for the cooperative movement of 50 or so elements of the polymer is exceeded. From this definition, we can see that the introduction of relatively stiff materials (such as carbon nanotubes) will interfere with the flow process and hence increase *Tg*. [IUPAC]

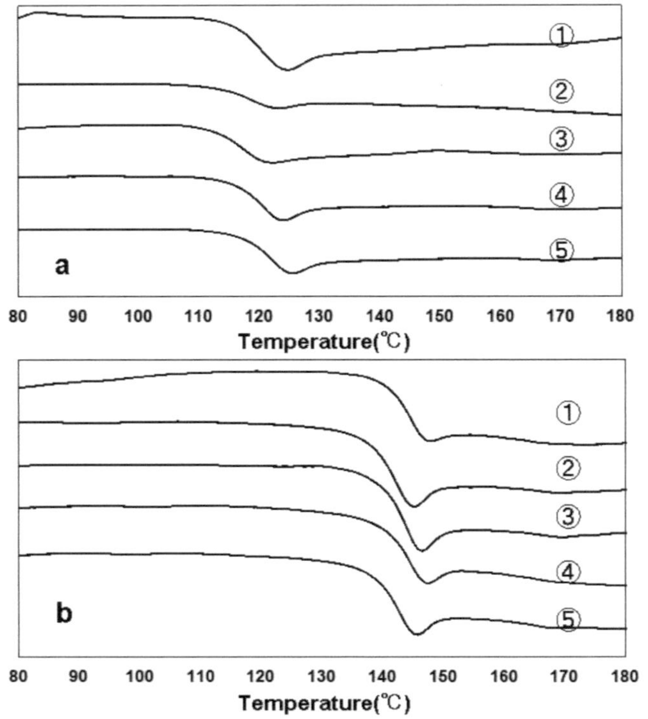

Fig. 4.2.3. Shift in the glass transition temperature of the matrix resin due to the addition of MWCNT. Figure 3a and 3b represent FBP-HX and PC composites with different amounts of MWCNT dispersed in them. ① Neat resin; ② composite with 0.5 wt% MWCNT in matrix resin; ③ composite with 1.0 wt% MWCNT in matrix resin; ④ composite with 3.0 wt% MWCNT in matrix resin; ⑤ composite with 5.0 wt% MWCNT in matrix resin.

It is shown in Figure 4.2.3 and Table 4.2.1 that the increase in *Tg* for FBP-HX in MWCNT/FBP-HX composites was due to the physical wrapping of the MWCNTs by the matrix resin. Although not the chemical modification of CNT, the immobilization of FBP molecules around CNT occurred. So, the matrix polymer (FBP-HX) was restricted by CNT. The reinforcements of the filler with high aspect ratio have been explained by the network forming method [Favier, V. et al]. According to Favier's theory, the incomplete nature of the network of the filler induces a reduction in the polymer's properties. This is reflected by the data we obtained at low additions (0.5~1.0 wt%) of CNT in polymer matrix. Above the ratio of network forming, the further addition of filler results in the greater reinforcing effect of the filler in the polymer matrix.

For the MWCNT/PC composites, the matrix polymer PC had a low interaction with the MWCNT. Therefore, it underwent only a minor change in its glass transition temperature.

Although the addition of MWCNT resulted in the improvement of the thermal dynamic properties for both the MWCNT/polymer composites, the driving force for the wrapping of MWCNT by PC and FBP is small. Therefore, the effect of MWCNT on the thermal dynamic properties was not that obvious, as demonstrated by the DSC data.

	Tg(°C)	$Tg_{Composite}/Tg_{Matrix}$
PC	143.6	1
PC-CNT 0.5wt%	140.9	0.98
PC-CNT 1.0wt%	142.5	0.99
PC-CNT 3.0wt%	143.2	0.997
PC-CNT 5.0wt%	143.6	1
FBP-HX	119.4	1
FBP-CNT 0.5wt%	118.2	0.99
FBP-CNT 1.0wt%	116.2	0.973
FBP-CNT 3.0wt%	119.2	0.998
FBP-CNT 5.0wt%	120.7	1.011

Table 1. Variation of glass transition temperature of MWCNT/polymer composites

4.2.5 Raman spectra of MWCNT/polymer composites

Raman spectra of MWCNT/polymer composites are shown in figure 4.2.4. For the MWCNT/PC composites (Figure 4.2.4a), a 1 wt% addition of MWCNT was enough to homogenize the mobility of the materials in the composites. A peak at around 1780 cm^{-1}, which is known to result from the vibration of the carbonate groups of PC, disappeared due to the addition of MWCNTs in PC. The Raman spectra of MWCNT/PC composites with MWCNT weight ratios higher than 1.0 wt%, were similar to the spectrum of MWCNT alone (1340 cm^{-1} for the defects of MWCNT and 1590 cm^{-1} for the G-band of MWCNT). This shows that the MWCNTs in PC were separated out from the resin and mainly covered the surface of the composites, without a covering of PC on them.

In the case of MWCNT/FBP-HX composites (Figure 4.2.4b), a specific peak arising from the stretching of the phenyl groups on fluorene moiety was observed around 1450~1500 cm^{-1}. The peak remained up till 3 wt% addition of MWCNT in FBP-HX. In the MWCNT/FBP-HX composites with a weight ratio higher than 5 wt%, this specific peak had disappeared

Because of the covering of FBP-HX on MWCNT, the mobility of the polymer and the MWCNT was almost the same; smaller amounts of 'naked' MWCNTs moved to the surface of MWCNT/FBP-HX composites. As per the increased diameter of MWCNT in FBP-HX compared to 'naked' MWCNT in PC, the Raman spectra provide more evidence for the fine dispersion of MWCNT in FBP-HX because of the strong interaction between the FBP-HX matrix and the MWCNT filler.

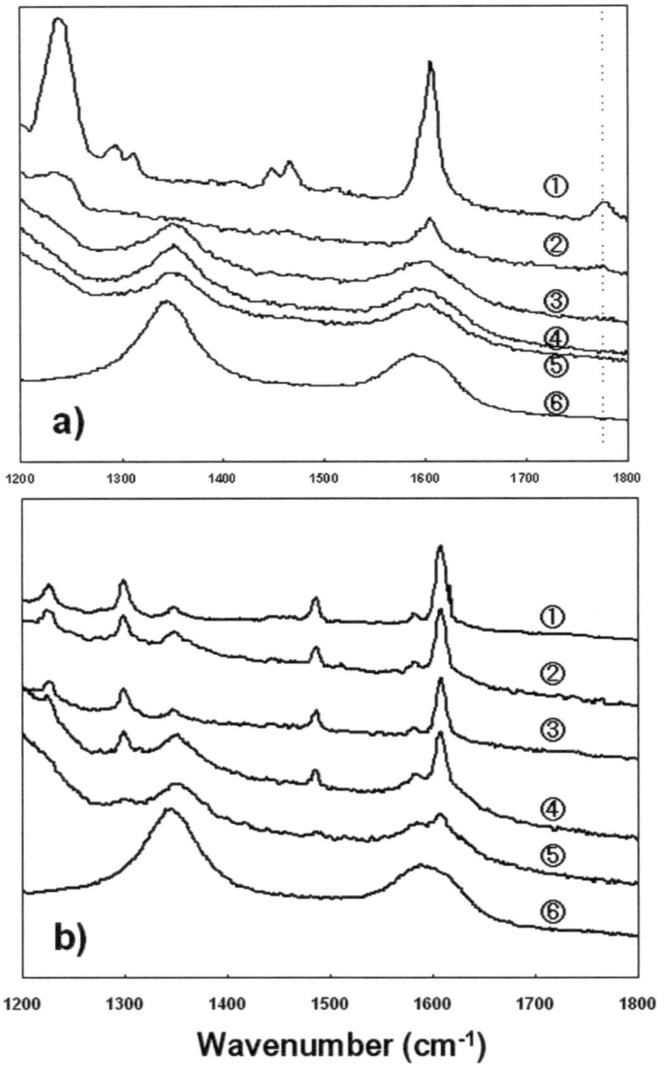

Fig. 4.2.4. Raman spectrum of MWCNT/polymer composites. Figure 4a and 4b represent PC and FBP-HX composites with different amounts of MWCNT in them. ① Neat resin; ② composite with 0.5 wt% MWCNT in matrix resin;③ composite with 1.0 wt% MWCNT in matrix resin; ④ composite with 3.0 wt% MWCNT in matrix resin; ⑤ composite with 5.0 wt% MWCNT in matrix resin; ⑥MWCNT.

4.2.6 Summarize of Section 4.2

The fine dispersion of MWCNTs in FBP-HX and PC resulted in low surface resistivity of both kinds of MWCNT/polymer composites, even those containing low weight ratios (0.5

~1.0 wt%) of MWCNT. FBP-HX covered MWCNTs were observed in MWCNT/FBP-HX composites; however, for the MWCNT/PC composite, the MWCNTs remained separate from the matrix polymer. Because of that, a higher surface resistivity and a higher percolation ratio were achieved for the MWCNT/FBP-HX composites, compared with the values for the MWCNT/PC composites. The *Tg* of the matrix polymer was lower at lower concentrations of MWCNTs in polymer. At higher concentration ratios, that is above 5 wt% of MWCNT incorporated with the polymer, an increase in the *Tg* was observed. The relative increase in the *Tg* for FBP-HX was higher than that for PC for the MWCNT/polymer composites. Raman spectroscopy of MWCNT/polymer composite sheets with 3 wt% addition of MWCNT indicated that the MWCNTs in the MWCNT/PC composites covered the surface of the sheets in their 'naked' form, but the MWCNTs in MWCNT/FBP-HX sheets were distributed homogeneously in the bulk and were covered with FBP-HX.

5. Discussion of dispersing power of fluorene moieties on carbon fillers

All the results in section 4 indicate that 9,9-substituted fluorene may have a strong interaction with carbon black. The strong interaction results in the fine dispersion of Carbon Plate in FBP and the enhancement of mechanical properties as well as an increase in Tg. To prove our assumption, we performed computational simulations on both Carbon Plate (Carbon plate with C96 structure was designated to represent carbon filler). /FBE and Carbon Plate models to determine their respective stabilizing energy (Carbon Plate /FBE, Figure 5d) and Carbon Plate /PBE, Figure 5e)). It was shown that the most stable structure of FBE-M was more stable than that of PBE-M at about 10 kJ/mol. This indicated that the compound containing a fluorene skeleton was able to interact with the carbon compound more strongly than that containing bisphenol. This was one of the reasons for the high dispersion of the carbon compound in the epoxy resin containing fluorene structure. On the other hand, 9,9-substituted phenyl groups on fluorene moieties also interacted with Carbon Plate through π–π interactions. This strong interaction was also related to the fine dispersion of Carbon Plate in FBE.

6. Conclusions

High dispersion power of *cardo*-typed fluorene moieties on carbon fillers was discussed in this chapter.

CB can be dispersed finely in CB/FBE composites. Compared with the aggregations of CB in PBE, there was no obvious aggregation in CB/FBE observed. The fine dispersion of CBs in FBE can be regarded as a result of strong π–π interaction between CB and FBE molecules. Also, because of the strong interaction, FBP molecules are firmly held by CB molecules by finely dispersing in them. This restriction of FBP molecules appears as a result of an increase in the Tg of FBP.

MWCNTs covered with FBP-HX were observed in MWCNT/FBP-HX composites; however, for the MWCNT/PC composite, the MWCNTs remained separated from the matrix polymer. Because of that, lower concentration of MWCNT is sufficient to induce electrical conductivity in MWCNT/PC composites than MWCNT/FBP-HX composites. The Tg of the matrix polymer decreased in lower concentration of MWCNTs in polymer. At higher concentration ratio above 5 wt% of MWCNT incorporated with polymer, increasing of Tg is observed. The increase in Tg for FBP-HX was higher than that for PC in MWCNT/polymer composites. Raman Spectrum indicates that MWCNT in MWCNT/PC composites were covered on the surface of composites sheets with naked state, compared to covered FBP-HX on MWCNT in MWCNT/FBP-HX composites.

Computer simulation of the graphene model combined with the FBE and PBE models also proved that the graphene/FBE model has a lower minimum energy than the graphene/PBE model. It meas that fluorene carbon model had strong interaction than other without fluorene. It also resulted fine dispersion of carbon fillers in fluorene moieties.

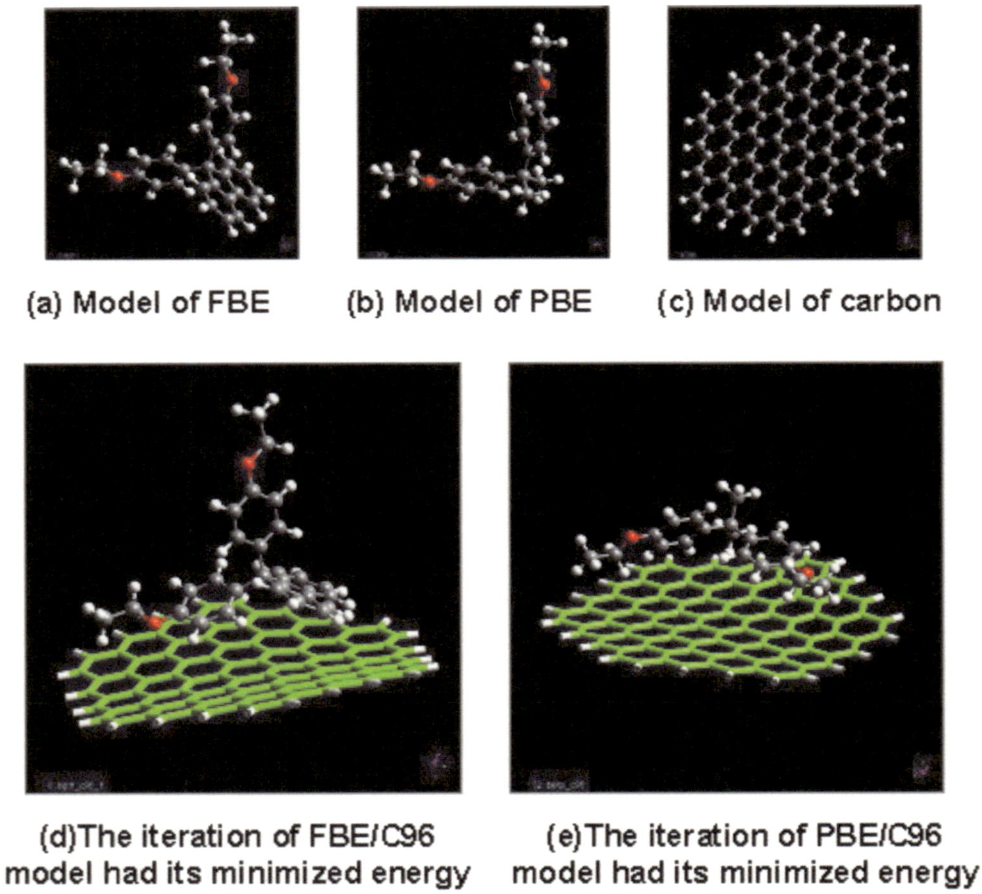

Fig. 5. Model structure and the iteration of FBE/Carbon plate and PBE/Carbon Plate models that have their minimum energy . Here, a, b, and c referred to the model of FBE, PBE, Carbon Plate (C_{96}), respectively; d and e referred to the iteration of FBE/ Carbon Plate and PBE/ Carbon Plate model that has its minimum energy.

7. References

Ando, W. (1987) Asakura Book Store, Encyclopedia of Organic Compounds.

ASTM (1965) Advances in the Technology of Stainless Steels and Related Alloys American Society for Testing Materials ISBN-EB: 978-0-8031-6001-9

Belluci, S. (19 January 2005). Carbon nanotubes: physics and applications. Physica Status Solidi (c) 2 (1): 34–47.

Chae, H.G.; Kumar, S. (2006). Rigid Rod Polymeric Fibers. Journal of Applied Polymer Science 100 (1): 791–802.

Chen, F.; Wang, B.; Chen, Y.; Li, L.J. (2007). Toward the extraction of single species of single-walled carbon nanotubes using fluorene-based polymers Nano Letters 7, 3013.

Chen, F.; Zhang, W.; Jia, M.; Wei, L.; Fang, X.F.; Kuo, J.L.; Chen, Y.; Park, M.B.C.; Li, L.J. (2009). Energy transfer from photo-excited fluorene-based polymers to single-walled carbon nanotubes J. Phys. Chem C 113, 14946.

Collins, P. G. (2000). "Nanotubes for Electronics" Scientific American: 67–69.

Dai, Z.; Li, Y.F.; Yang, S.G.; Zhao, N.; Zhang, X.L. and Xu, J. (2009). Kinetics and thermal properties of epoxy resins based on bisphenol fluorene structure. Eur. Poly. J., 45, 1941.

Demczyk, B.G; Wang, Y.M; Cumings, J.; Hetman, M.; Han, W.; Zettl, A.; Ritchie, R.O. (2002). Direct mechanical measurement of the tensile strength and elastic modulus of multiwalled carbon nanotubes. Materials Science and Engineering A 334 (1–2): 173–178.

Dresselhaus, M.S.; Dresselhaus, G.; Avouris, Ph., ed. (2001). Carbon nanotubes: synthesis, structures, properties and applications. Topics in Applied Physics80. Springer, ISBN 3540410864, Berlin

Ebbesen, T.W. ed. (1997). Carbon nanotubes—preparation and properties.: CRC Press. ISBN 0849396026, Boca Raton, Florida

Favier, V.; Chanzy, H.; Cavaille, J. Y. (1995). Macromolecules, 28(18), 6365.

Grell, M.; Bradley, D. D. C.; Ungar, G.; Hill, J.; Whitehead, K. S. (1999). Macromolecules, 32, 5810-5817.

Hayashi, H.; Kawasaki, S.; Kobori, K.; Koyama, Y.; Asai, S.; Takata, T. (2009). Synthesis and Properties of Polysiloxanes Possessing 9,9-Diarylfluorene Structure in the Main Chain. Polym. J., 41, 272.

Hong, S.H.; Myung, S. (2007). Nanotube Electronics: A flexible approach to mobility. Nature Nanotechnology 2 (4): 207–208.

Inada, T.; Masunaga, H.; Sakurai, K.; Kawasaki, S.; Yamada, M.; Kobori, K. (2005). Small-angle X-ray scattering from multi-walled carbon nanotubes (CNTs) dispersed in polymeric matrix. Chem. Lett., 34, 524.

IUPAC Recommendations 1994 PAC, 66, 577.

Jensen, K; Mickelson, W.; Kis, A.; Zettl, A. (2007). Buckling and kinking force measurements on individual multiwalled carbon nanotubes. Phys. Rev. B 76, 195436.

Jin, S. H.; Park, H.J.; Kim, J.Y.; Lee, K; Lee, S. P.; Moon, D. K; Lee, H. J; Gal, Y. S. (2002). Macromolecules, 35, 7532-7534.

Kawasaki, S.; Yamada, M. (2005). Densizairyo.Kogyochosakai. 7.

Kawasaki, S.; Yamada, M.; Kobori, K.; Jin, F. Z.; Takata, T. (2010). High dispersion ability of fluorene-based polyester as a matrix polymer for carbon nanotubes. J. Appl. Polym. Sci., 118, 2690.

Kawasaki, S.; Yamada, M.; Kobori, K.; Jin, F.Z.; Kondo, Y.; Hayashi, H.; Suzuki, Y.; Takata, T. (2007). Synthesis and Chemical, Physical, and Optical Properties of 9,9-Diarylfluorene-Based Poly(ether-ether-ketone). Macromolecules, 40, 5284.

Kawasaki, S.; Yamada, M.; Kobori, K.; Kakumoto, T.; Jin, F. Z.; Tarutani, A.; Takata, T. (2007) Extraordinary high carbon filler-incorporatingand dispersing ability of 9,9-diarylfluorene-based polymers as matrix resins. Polym. J., 39, 115

Kawasaki, S; Kato, M; Kobori, K; Sakamoto, H; Kondo, Y; Takata, T. (2009). New concept of reducing a birefringence of poly by a novel alloy with fluorene-based polyester. Polym.Eng.Sci., 49, 2374.

Kawasaki, S; Yamada, M; Kobori, K; Sakamoto, H; Kondo, Y; Jin, F. Z; Takata, T. (2009). Preparation of the novel alloy composed of fluorene-based polyester and poly(carbonate) and their properties for optical use. J. Appl. Polym. Sci., 111, 461.

Korshak, V.V.; Vinogradova, S.V.; Vygodskii, Y.S. (1974). J, Macromol. Sci-Rev., Macromol. Chem., C11, 45, 142.

Lee, P.I., Hsu, S.L.C.; Lin, P.Y. (2010) White-Light-Emitting Diodes from Single Polymer Systems Based on Polyfluorene Copolymers with Quinoxaline Derivatives Macromolecules, 43, 8051

Li, Y.J.; Yamada, M.; Wang, Y.F.; Chen, T.M.; Nakaya, T. (1996). Macromolecular Repord. Macromol, Rap, A33, 65-70.

Lim, E.; Jung, B. J.; Shim, H. K.. (2003). Macromolecules, 36, 4288-4293.

Liu, W. B; Wang, T.; Qiu, Q. H; Zhang, M. L. (2008). Syinthesis and characterization of an epoxy resin containing fluorene moieties and its cured polymer. Pigment & Resin Technology, 37, 389.

Lu, X.; Chen, Z. (2005). Curved Pi-Conjugation, Aromaticity, and the Related Chemistry of Small Fullerenes (<C60) and Single-Walled Carbon Nanotubes. Chemical Reviews 105 (10): 3643–3696.

Matsukawa, K.; Matsuura, Y.; Nakamura, A.; Nishioka, N.; Murase, H.; Kawasaki, S. (2007). Preparation of titania/Acrylate hybrid thin films and properties their refractive indices. J. Photopolym. Sci. Technol., 20, 307.

Meo, M.; Rossi, M. (2006). Prediction of Young's modulus of single wall carbon nanotubes by molecular-mechanics-based finite element modelling. Composites Science and Technology 66 (11–12): 1597–1605.

Mo, Y.; Jiang, X.; Cao, D. (2007). Org. Lett., 9, 4371-4373.

Morgan, P. W. (1970). Macromolecules, 3, 536.

Morin, J.-F.; Leclerc, M. (2001). Macromolecules, 34, 4680-4682.

Nasibulin, A. G; Pikhitsa, P.V.; Jiang, H; Brown, D.P.; Krasheninnikov, A.V.; Anisimov, A.S.; Queipo, P.; Moisala, A. et al. (2007). A novel hybrid carbon material. Nature Nanotechnology 2 (3): 156–161.

Nasibulin, A.; Anisimov, A.S.; Pikhitsa, P.V.; Jiang, H.; Brown, D.P.; Choi, M.S.; Kauppinen, E.I. (2007). Investigations of NanoBud formation. Chemical Physics Letters 446: 109–114.

Ozawa, H.; Fujigaya, T.; Niidome, Y.; Hotta, N.; Fujiki, M.; Nakashima, N. (2011). Rational concept to recognize/extract single-walled carbon nanotubes with a specific chirality. J Am Chem Soc. 133(8):2651-7

Papava, G. S.; Maisuradze, N. A.; Zarkua, Z. L.; Dokhturishvili, N. S.; Sarishvile, Z. M.; Razmadze, G. B.; Vinogradova, S. V.; Korshak, V. V. (1988). Acta Polymerica, 8, 445.

Pop, E.; Mann, D.; Wang, Q.; Goodson, K.; Dai, H.J. (2005). Thermal conductance of an individual single-wall carbon nanotube above room temperature. Nano Letters 6 (1): 96–100.

Sakurai, K.; Fuji, M. (2000). Polym J, 32, 676.

Salim, T.; Sun, S.Y.; Wong, L.H.; Xi, L.F.; Foo, Y.L.; Lam, Y.M. (2010) The Role of Poly(3-hexylthiophene) Nanofibers in an All-Polymer Blend with a Polyfluorene Copolymer for Solar Cell Applications J. Phys. Chem. C, 114, 9459.

Samal, S.; Sahoo, S.K. (1997) An overview of fullerene chemistry. Bulletin of Materials Science 20(2), 141-230

Seesukphronrarak, S.; Kawasaki, S.; Kobori, K.; Takata, T. (2007). Synthesis of fluorene-based high performance polymers. I. Poly (arylene thioether)s with excellent solubility and high refractive index. J. Polym. Sci. Part A, 45, 3073.

Seesukphronrarak, S.; Takata, T. (2007) Novel Fluorene-based Biphenolic Monomer: 9,9-Bis(4-hydroxyphenyl)-9-silafluorene. Chem. Lett., 36, 1138.

Seto, R.; Kojima, T.; Hosokawa, K.; Koyama, Y.; Konishi, G.; Takata, T. (2010). Synthesis and Property of 9,9′-Spirobifluorene-containing Aromatic Polyesters as Optical Polymers with High Refractive Index and Low Birefringence. Polymer, 51, 4744.

Seto, R.; Sato, T.; Kojima, T.; Hosokawa, K.; Koyama, Y.; Konishi, G.; Takata, T. (2010). 9,9′-Spirobifluorene-Containing Polycarbonates: Transparent Polymers with High Refractive Index and Low Birefringence. J. Polym. Sci. Polym Chem., 48, 3658.

Sinha, S.L.; Barjami, S.; Iannacchione, G.; Schwab, A.; Muench, G. (2005). Off-axis thermal properties of carbon nanotube films. Journal of Nanoparticle Research 7 (6): 651–657.

Sinnott, S.B.; Andrews, R. (2001). Carbon Nanotubes: Synthesis, Properties, and Applications. Critical Reviews in Solid State and Materials Sciences 26 (3): 145–249.

Stephan, O.; Vial. J. C. (1999). Synthetic Metals, 106, 115-119.

Suzuki, A.; Ando, S. (2010). Preparation and Characterization of Polyimide/ZnO Nano-hybrid Films Exhibiting High Refractive Indices. Journal of Photopolymer Science and Technology, 23, 521.

Takesue, I.; Haruyama, J.; Kobayashi, N.; Chiashi, S.; Maruyama, S.; Sugai, T.; Shinohara, H. (2006). "Superconductivity in Entirely End-Bonded Multiwalled Carbon Nanotubes" Physical Review Letters 96: 057001.

Teramoto, T. (1995). Industrial Materials, 43, 79.

Teramoto, T.; Kiyotaka, K.; Takayama, T.; Koyama, T. (1992). Aromatics. 44, 337.

Thostenson, E.; Li, C; Chou, T. (2005). Nanocomposites in context. Composites Science and Technology 65: 491–516.

Vieira, R. (2004). Synthesis and characterisation of carbon nanofibers with macroscopic shaping formed by catalytic decomposition of C2H6/H2 over nickel catalyst. Applied Catalysis A 274: 1–8.

Wagner, H. D. (2002). Reinforcement Encyclopedia of Polymer Science and Technology. John Wiley & Sons.

Xiong, Y.Q.; Liu, H.P.; Ou, E.C.; Zeng, X.L.; Zhou, W.; Xu, W.J. (2010) Crystal Structure, Curing Kinethics, and Thermal Properties of Bisphenol Fluorene Epoxy Resin. J. Appl. Polym. Sci., 118, 827.

Yamada, M.; Okimi, K.; Ogata, K. (1998). Kinki Chemical Sciety, 5, 14.

Yamada, M.; Sun, J.; Suda, Y.; Nakata, T. (1998). Chemistry Letters, 10, 1055.

Yang, C.P.; Lin, J.H. (2009). New poly (amide-imide) syntheses. XII. Preparation and properties of poly(amide-imide)s based on the diimide-diacid condensed from 9,9-

bis[4-(4-aminophenoxy)phenyl] fluorene and trimellitic anhydride. J. Polym. Sci. Part A: Polym. Chem., 32, 2653.

Yang, R; Tian, R; Hou, Q; Yang, W; Cao, Y. (2003). Macromolecules, 36, 7453-7460.

Yao, K. (1994). Reports Prepared for the Society of Polymer Science, Japan (annual session), 43.

Yao, K. (1995). Reports Prepared for the Society of Polymer Science, Japan (discussion session), 44.

Yu, M.F.; Lourie, O.; Dyer, M. J.; Moloni, K.; Kelly, T.F.; Ruoff, R. S. (2000). Strength and Breaking Mechanism of Multiwalled Carbon Nanotubes Under Tensile Load. Science 287 (5453): 637–640.

19

Optical Properties and Some Applications of Plasmonic Heterogeneous Materials

Sergey Moiseev
[1]*Ulyanovsk Branch of V.A. Kotelnikov Institute for Radio Engineering and Electronics of Russian Academy of Sciences*
[2]*Ulyanovsk State Technical University*
[3]*Ulyanovsk State University*
Russia

1. Introduction

Metamaterials with extraordinary optical properties can find a variety of applications, in particular for manufacturing superlenses, prisms, nonreflecting (absorptive) materials, as well as controlling the optical beam intensity and propagation direction, etc. (Cai & Shalaev, 2010; Hutter & Fendler, 2004). In recent years, a large number of structures have been proposed and theoretically and experimentally investigated, which possess negative, high or small refractive indices (Moiseev et al., 2007; Oraevskii & Protsenko, 2000; Shalaev et al., 2005; Shen et al., 2005; Sukhov, 2005; Yuan et al., 2007), selective absorption of or transparency to optical light (Kachan et al., 2006; Kravets et al., 2008; Protsenko et al., 2007).

The promising candidates for a design of structures with extraordinary optical properties are plasmonic materials. Collective electronic excitations called plasmons lead to strong dispersion and absorption of light in such artificial media. For practical applications, it is important that the optical properties of composite medium incorporating metallic nanoparticles can be controlled by careful selection of geometric parameters of nanoparticles (Moiseev, 2004, 2009; Oraevskii & Protsenko, 2000). By selecting distributions of the form of metallic nanoparticles, it is possible to achieve absorption of electromagnetic radiation in specified spectral regions of visible or near IR radiation (Protsenko et al., 2007). An additional effective way of the plasmon resonance control is by spatial arrangement of metal-dielectric nanocomposites, for example, forming the multilayer systems consisting of metallic nanoparticles separated by dielectric layers (Kachan et al., 2006). It was proposed that such composite materials with controlled absorption of light can be used as cut-of filters and high-absorbing coatings.

In this work, a more detailed investigation of optical properties of a matrix metal-dielectric medium with spheroidal silver inclusions is performed, and the possibility to realize plasmonic structures with beneficial effects in the visible region is considered. In so doing, it is assumed that the characteristic size of inclusions and their volume concentration satisfy the conditions of applicability of the effective medium Maxwell–Garnett model.

2. Main formulae

2.1 Size-dependent dielectric function of silver nanoparticles

From the optics of metal nanosized particles it is well known that their optical properties significantly depend not only on their chemical composition but also on their size, so the permittivity of metal nanoparticles differs from the bulk permittivity of the medium (Khlebtsov, 2008). Throughout this communication, silver nanoparticles are considered (they are chosen because of their lowest absorption in the optical range). In order to estimate variation in the dielectric permittivity ε of a metal nanoparticle in comparison with that of a bulk sample, let us use a classical model taking into account the limitation in the electron free path length due to its collision with the particle boundary (Kreibig & Vollmer, 2004). According to this model, the finite size of a metal particle leads to a change in the relaxation rate of conduction band electrons:

$$\gamma = \gamma_0 + v_F / r , \tag{1}$$

where γ_0 is the electron relaxation rate in the metal, v_F is the mean electron velocity at the Fermi surface, r is the characteristic particle radius along the electric field direction of external electromagnetic wave. Then, the size-dependent permittivity of the nanoparticle has the form (Yannopapas et al., 2002):

$$\varepsilon(\omega, r) = \varepsilon_b(\omega) + \frac{\omega_p^2}{\omega(\omega + i\gamma_0)} - \frac{\omega_p^2}{\omega(\omega + i\gamma)} , \tag{2}$$

where ω is the frequency of the incident field, ω_p is the plasma frequency, ε_b is the experimentally determined permittivity for a bulk sample. The tabular data for bulk silver are $v_F = 1.4 \cdot 10^6$ m/s, $\hbar\gamma_0 = 0.02$ eV, $\hbar\omega_p = 9.2$ eV (Yannopapas et al., 2002) and the spectral dependence values of the macroscopic dielectric permittivity $\varepsilon_b(\omega)$ are presented in Ref. (Ordal et al., 1983).

Calculations show that the correction taking into account the finite size of the particle leads mainly to a variation of the imaginary part of the permittivity ε (see Fig. 1), while the real

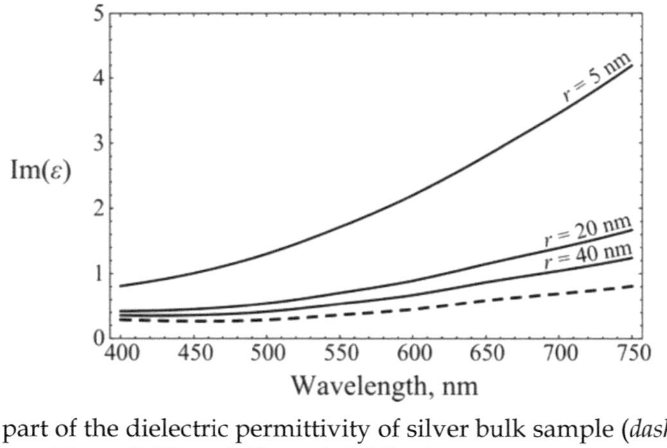

Fig. 1. Imaginary part of the dielectric permittivity of silver bulk sample (*dashed line*) and silver nanoparticles with r = 5, 20, and 40 nm (*solid lines*).

part of ε changes slightly for the particle with radius as large as 5 nm. For instance, for a wavelength of external radiation $\lambda = 400$ nm, the value of ε_b for a bulk silver is equal to $-3.72 + 0.29i$, while calculations via formulas (1) and (2) give the following results: $\varepsilon = -3.72 + 0.36i$ for a spherical particle with the radius $r = 40$ nm, $\varepsilon = -3.72 + 0.42i$ for $r = 20$ nm, and $\varepsilon = -3.68 + 0.81i$ for $r = 5$ nm. It should be noted that in the case of a nonspherical particle the size correction of the Drude model leads to the dependence of the permittivity ε on the direction of the external field.

2.2 Maxwell-Garnett approach

Let the light wavelength significantly exceed the size of particles suspended in a dielectric matrix. If the interference effects on the nanoparticles are ruled out (the spatial positions of the nanoparticles don't have to be periodic and can be random) and their volume fraction is as small as 1/3, it is considered that Maxwell-Garnett approach can be used successfully to analyse the optical properties of such composite medium (Golovan et al., 2007; Moiseev et al., 2007; Moiseev, 2009, 2010). The Maxwell - Garnett relation assumes a generalization for the case of shape anisotropy of the inclusion (particles), thus providing a possibility of analyzing optical characteristics of a composite medium with non-spherical inclusions based on analytical expressions.

For composite medium with uniformly oriented metallic spheroids in a dielectric matrix, the effective permittivity ε_{eff} according to Maxwell-Garnett approach is determined by the following expression (Golovan et al., 2007; Maxwell-Garnett, 1904, 1906):

$$\frac{\varepsilon_{eff} - \varepsilon_m}{L(\varepsilon_{eff} - \varepsilon_m) + \varepsilon_m} = \eta \frac{\varepsilon - \varepsilon_m}{L(\varepsilon - \varepsilon_m) + \varepsilon_m}, \tag{3}$$

where ε_m is the matrix permittivity, ε is the permittivity of inclusions (Eq. (2)), η is the volume concentration of the nanoparticles (filling factor), L is the geometrical factor (factor of depolarization) that accounts for the shape of a particle (Bohren & Huffman, 1998). In the long-wavelength limit, factor L is a real value, that depends on the ratio ξ of the length of polar semi-axis a and equatorial semi-axis b of spheroid, $\xi = a / b$. The value of factor L can be written as

$$L_{||} = \frac{1}{1-\xi^2}\left(1 - \xi \frac{\arcsin\sqrt{1-\xi^2}}{\sqrt{1-\xi^2}}\right) \tag{4}$$

for the field directed along the axis of revolution of spheroid and as

$$L_{\perp} = \left(1 - L_{||}\right) / 2 \tag{5}$$

for the field directed perpendicular to this axis. The case where $\xi < 1$ corresponds to disc-like nanoparticles, $\xi > 1$ - to needle-like nanoparticles, $\xi = 1$ - to spherical nanoparticles, with $L_{\perp} = L_{||} = 1/3$ in the latter case.

By a comparison with exact electrodynamical calculation, it was shown in (Moiseev et al., 2007; Moiseev, 2009, 2010) that for matrices with a moderate content of inclusions ($0.01 < \eta < 0.3$) results obtained within the Maxwell-Garnett model are in fairly good agreement with the results of exact electrodynamic calculation.

3. Spectral characteristics of Maxwell-Garnett composite with plasmon resonances

Eq. (3) is written for a composite medium with uniformly oriented spheroids, where the field vector is directed either along or perpendicular to the ellipsoid axis. For non-spherical particles, depolarization factors (4), (5) and dielectric permittivity (2) depend on the electric field direction. Therefore in the general case we deal with an anisotropic composite, which possesses properties of uniaxial crystal having an optical axis collinear with the polar axis *a* of spheroids. A light ray in this medium is ordinary if its propagation direction is collinear with the polar axis of spheroids; in this case, the light polarization vector lies in the equatorial plane of spheroids.

In the case of a spheroid and an applied electric field oriented along a spheroid axis, the static polarizability of nanoparticle is

$$\alpha = \frac{V}{4\pi}\frac{\varepsilon - \varepsilon_m}{L(\varepsilon - \varepsilon_m) + \varepsilon_m}, \tag{6}$$

where V is the particle volume. As follows from Eq. (6), the plasmon resonance frequency depends on the form of the particle. If a transparent dielectric is used as surrounding host medium, the resonance wavelength λ_{res} of metal spheroid can be estimated by using simple algebraic equation

$$\mathrm{Re}\left[\varepsilon(\lambda_{res})\right] + \frac{1-L}{L}\varepsilon_m(\lambda_{res}) = 0\,. \tag{7}$$

In the case of spheroidal ($\xi \neq 1$) silver nanoparticle, Eq. (7) has two roots corresponding to two polarizations of the electromagnetic field. One root tends to longer wavelengths due to the absorption of light polarized parallel to the long axis of the spheroid (the case of parallel polarized light). The other root shifts to shorter wavelengths due to the absorption of light polarized parallel to the short axis of the spheroid (the case of perpendicular polarized light). The more different shapes of nanoparticles from the spherical ones, the more is difference in the resonance wavelengths corresponding to two polarizations of the electromagnetic field. Dependence of the plasmon resonance wavelength of a silver spheroid on its aspect ratio is shown at Fig. 2.

Plasmon resonances of nanoparticles give rise to 'extraordinary' values of the effective permittivity $\varepsilon_{eff} = (n_{eff} + ik_{eff})^2$ of mixture. Shown in Fig. 3 are the results of calculation of spectral dependences of the effective refractive index

$$n_{eff} = \sqrt{\frac{\sqrt{(\mathrm{Re}[\varepsilon_{eff}])^2 + (\mathrm{Im}[\varepsilon_{eff}])^2} + \mathrm{Re}[\varepsilon_{eff}]}{2}} \tag{8}$$

and the effective extinction coefficient

$$k_{eff} = \sqrt{\frac{\sqrt{(\mathrm{Re}[\varepsilon_{eff}])^2 + (\mathrm{Im}[\varepsilon_{eff}])^2} - \mathrm{Re}[\varepsilon_{eff}]}{2}} \tag{9}$$

obtained from the Maxwell–Garnett relation (3). As a matrix, we chose a non-absorbing medium with the refractive index $n_m = \sqrt{\varepsilon_m} = 1.5$. One can see from Fig. 3 that due to plasmon resonance of silver nanoparticles the effective refractive index of composite medium differs considerably from the one of matrix. Moreover, in the vicinity of the plasmon resonance the effective refractive index assumes the values $n_{eff} < 1.25$ not observed in natural materials. From the practical point of view it is interesting that n_{eff} can achieve values equal or less than one as well as values much greater than one. Unfortunately, these 'extraordinary' values of n_{eff} are observed together with the effective extinction coefficient different from zero. An increase in the concentration of inclusions extends the range of attainable values of n_{eff}, but at the same time gives rise to higher values of k_{eff} in the resonance region.

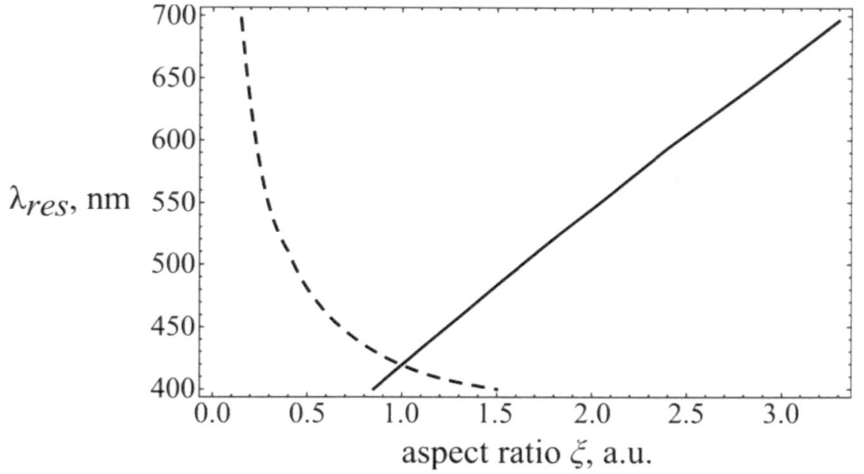

Fig. 2. Dependence of the plasmon resonance wavelength λ_{res} of a silver spheroid on the aspect ratio ξ obtained from Eq. (7) for parallel (*solid line*) and perpendicular (*dashed line*) polarized light. Dielectric function of the matrix ε_m= 2.25.

4. Applications of plasmonic heterogeneous materials

4.1 Heterogeneous composite with the unite refractive index

Let us have a closer look at the conditions where a unit effective refractive index ($n_{eff} = 1$) is observed in a composite medium with a comparatively low extinction coefficient ($k_{eff} << 1$). This medium is attractive from the point of view that a light ray falling on it from the air is nearly non-reflective (Sukhov, 2005). Indeed, using the Fresnel formulas we can readily obtain the following expression for reflectance: $R \approx (k_{eff} / 2)^2 << 1$. In this case, the reflectance depends only on the absorption coefficient of the medium and decreases very fast with k_{eff}. A comparatively thick film ($h \geq \lambda / k_{eff}$) manufactured from this material near the plasmon resonance frequency would possess properties of a weakly reflecting absorbing coating.

The conditions where an effective refractive index is observed could be derived analytically directly from the Maxwell - Garnett relation (3). For a fixed refractive index $n_{eff} = 1$, the real and imaginary parts of Eq. (3) form a system of two equations, wherein it makes sense to consider quantities k_{eff} and η as independent variables. We do not cite the solutions to this system of equations due to their awkward form but merely present the dependences obtained with them, which are shown in Fig. 4.

The dependences plotted in Figs. 3 and 4 imply that the conditions $n_{eff} = 1$, $k_{eff} << 1$ cannot virtually be satisfied at the same time, if the inclusions are spherical ($\xi = 1$). For the volume fractions of nanoparticles larger than 0.06, the condition $n_{eff} = 1$ is fulfilled near the external radiation wavelength 400 nm, but for a comparatively large extinction coefficient $k_{eff} > 0.6$. For a smaller fraction of silver, the unit refractive index cannot be obtained within the entire optical range.

The situation is different when inclusions are nonspherical ($\xi \neq 1$). In such composite media the influence of the dispersion subsystem on the effective optical characteristics is stronger, and the condition $n_{eff} = 1$ is fulfilled for a much smaller (by a few factors or even an order of magnitude) effective extinction coefficient. To illustrate this, let us consider the results of calculations presented in Fig. 3c: in the short-wave spectral region the effective refractive index acquires the values close to unity for all ξ, while for the spherical, flattened and elongated particles the absorption coefficient is different, $k_{eff} > 1$, $k_{eff} \sim 1$, and $k_{eff} < 0.1$, respectively. It should be noted that in those cases where the unit value of the effective refractive index is attained for several values of k_{eff} , we consider the smallest of these values.

The dependences presented in Fig. 4 also demonstrate that k_{eff} assumes small values predominantly in the short-wave spectral region and in the case of nonspherical inclusions only. Unlike the case where $\xi = 1$, an increase in the concentration of nonspherical inclusions, rather than giving rise to an increase in absorption of a composite medium, results in a decrease in the absorption of a medium with a unit refractive index, however, to a certain limit, since the function $k_{eff}(\eta)$ exhibits a local minimum falling within the visible region for certain $\xi \neq 1$. In the case corresponding to Fig. 4c, k_{eff} assumes minimum values at $\lambda = 434$ nm. For these parameters, the reflectance of the composite would be less than 10^{-3} .

4.2 Transparent Maxwell-Garnett composite with the unite refractive index

It follows from the Maxwell-Garnett formula (3) that because the permittivity of the metal particles is complex, condition of optical transparency

$$\mathrm{Im}(\varepsilon_{eff}) = 0 \tag{10}$$

cannot be satisfied exactly for the matrix with real ε_m. Let us try to compensate the effect of the imaginary part of the metal permittivity on the optical properties of the entire composite medium by choosing an amplifying medium as the matrix. For this purpose, we will simulate the optical parameters of the amplifying medium by adding an imaginary part to the permittivity (Moiseev et. al., 2007):

$$\varepsilon_m = n_m^2 - g^2 - 2in_m g , \tag{11}$$

where g>0 is the gain (extinction coefficient), which is equal to the gain coefficient multiplied by $\lambda / 2\pi$.

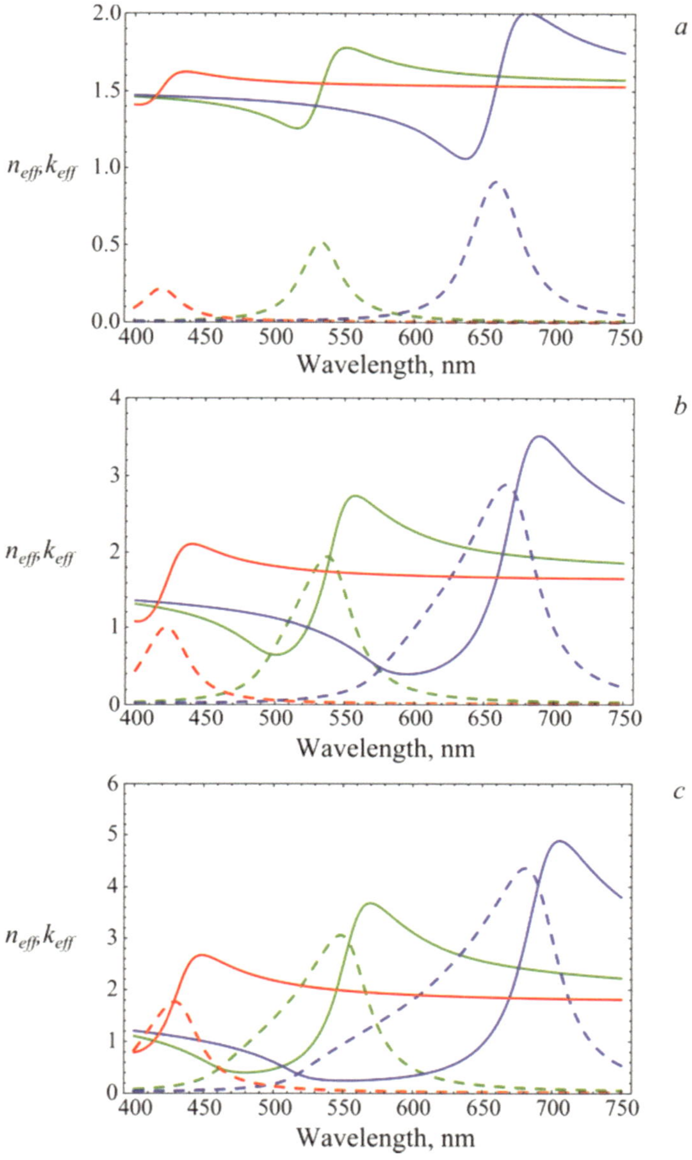

Fig. 3. Spectral dependences of n_{eff} (*solid lines*) and k_{eff} (*dashed lines*) of composite media with spherical inclusions ($\xi = 1$, *red lines*), inclusions shaped as flattened ($\xi = 1/3$, *green lines*) or elongated ($\xi = 3$, *blue lines*) spheroids for the volume fractions of inclusions 0.01 (a), 0.05 (b) and 0.1 (c). The radius of the spherical particle is equal to 7 nm, the lengths of the spheroid semi-axes are selected such that their volume is equal to that of a spherical particle. In the case of elongated spheroids, the light wave field is directed along the axis of rotation of particles, while in the case of flattened spheroids, the light vector lies in their equatorial plane.

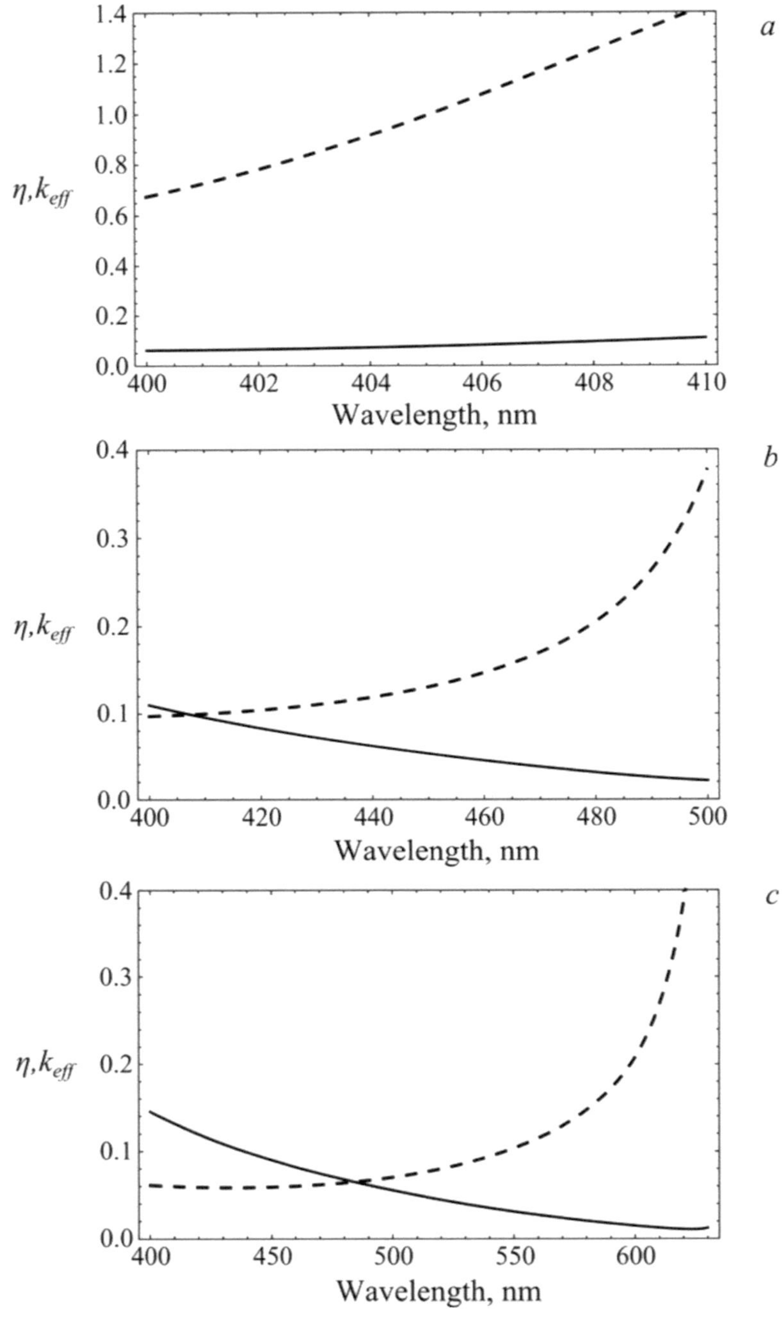

Fig. 4. Effective extinction coefficient k_{eff} (*dashed lines*) and the volume fraction of inclusions η (*solid lines*) of a composite medium with the unit effective refractive index. Silver inclusions are spherical particles (*a*) and flattened (*b*) and elongated (*c*) spheroids. Particle parameters are the same as given in Fig. 3.

Consider the dependence of the gain g, which is required for compensating the absorption of external electromagnetic radiation, on geometric and material parameters of composite medium. When the gain of the matrix is much smaller than its refractive index, from equation (3) one can obtain

$$g \approx \frac{\eta}{1-\eta} \frac{n_m^3}{2\left((1-L)n_m^2 + L\,\mathrm{Re}(\varepsilon)\right)^2 + 2(1-L)L\left(n_m^2 - \mathrm{Re}(\varepsilon)\right)^2 \eta} \mathrm{Im}(\varepsilon) . \qquad (12)$$

Unlike other metals, the permittivity of silver has comparatively small imaginary part. Thus, having taken silver inclusions we result in decreasing in the required gain. Decrease in the volume concentration of the nanoparticles contributes to decrease in the gain g, although this way is not worth considering because $n_{eff} \approx n_m$.

From equation (12) it follows that g depends on the factor of depolarization L. The variation of L leads to the shift of the plasmon resonance, and correspondingly, the wavelength λ, on which the unique index of refraction is observed. Thus, analysis of the transparency conditions is to be taken basing on predefined value of n_{eff}. Here we consider the case $n_{eff} = 1$ ($\varepsilon_{eff} = 1$), when the composite medium is invisible.

The conditions of invisibility of Maxwell-Garnett composite with inclusions of different shapes are presented in Fig. 5. Horizontal axis shows the region of wavelengths, on which the unit

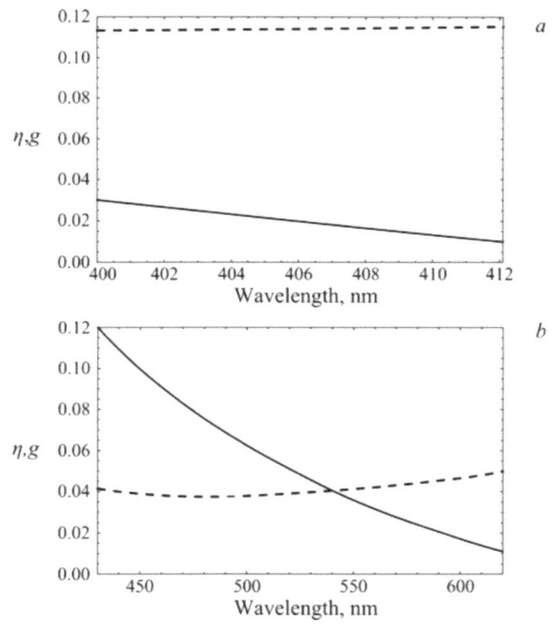

Fig. 5. Gain g of matrix (*dashed lines*) and the volume concentration η (*solid lines*) of spherical (*a*) and needle-like (*b*) nanoparticles are required to produce a transparent composite material with the unit index of refraction. The radius of the spherical particle is equal to 7 nm, the volume of the spheroid with $\xi = 3$ is equal to that of a spherical particle. The external field is oriented parallel to the spheroid revolution.

index of refraction can be obtained. The dependence obtained allows defining the filling factor, whereby the gain required for the transparency of composite medium is minimal. Contrasting parts (*a*) and (*b*) of Fig. 5, the following conclusions can be made. Firstly,composite media with nonspherical nanoparticles require less value of gain of the active component. It can be explained in the following way. In spite of volume of particle being equal the characteristic diameter of the needle-like particle in the direction of the field is bigger, and consequently, the imaginary part of the permittivity ε is less than that of spherical nanoparticles. Therefore in the case (*b*) the values of the gain necessary for compensation of absorption are smaller. Secondly, the condition $\varepsilon_{eff} = 1$ can be fulfilled over a wide range of wavelengths covering almost the entire visible spectrum, but exclusively for inclusions of nonspherical forms.

4.3 Heterogeneous medium for anti-reflection coating application

Optical properties of heterogeneous metal-dielectric composites can be efficiently tailored by nanoparticle sizes, shapes, and concentration. As it is shown in Section 4.1, for nonspherical inclusions the effective coefficient of composite extinction is several times smaller than for spherical ones. Let us determine the values of the structural parameters at which a composite slab with nonspherical inclusions can be used as an interference antireflection coating in the visible spectral range.

The use of the effective-medium model significantly simplifies the study of the dispersion characteristics of composite coating: disregarding the discrete-continuous structure of the composite, one can calculate the reflectance and transmittance using the Airy formulas (Born & Wolf, 1999). According to these formulas, for normal incidence of light the reflection from dielectric is completely suppressed by a layer of material with a complex refractive index $n_{eff} + ik_{eff}$ if its imaginary part $k_{eff} << 1$, the refractive index n_{eff} and thickness h satisfy the conditions

$$n_{eff} \approx \sqrt{n_s}\left(1 - \frac{\pi(n_s - 1)}{4n_s} k_{eff}\right), \tag{13}$$

$$h \approx \frac{\lambda}{4\sqrt{n_s}}\left(1 + \frac{4}{\pi(n_s - 1)} k_{eff}\right), \tag{14}$$

where n_s is the refractive index of substrate. The values of the composite parameters at which condition (13) is satisfied simultaneously with the inequality $k_{eff} << 1$ can be determined from relation (3), taking into account (2), (4), and (5).

The anisotropy of optical properties of ordered composite imposes certain limitations on the antireflection coating design. To make the composite coating reflectance and transmittance independent of the light vector orientation for normal incidence of light, the optical axis of the composite must be perpendicular to the interfaces, i.e., the equatorial plane of spheroids must be parallel to the plane of composite plate. With allowance for this circumstance, we will consider the optical properties of the composite only for an ordinary ray. For definiteness, we consider a transparent dielectric with a refractive index $n_m = \sqrt{\varepsilon_m} = 1.5$ as a matrix (light glass, polycarbonate, and other widespread optical materials have similar refractive indices).

Figure 6 shows the spectral dependencies of the effective refractive index and extinction coefficient of the composite, calculated from (3). The peaks in the curve $k_{eff}(\lambda)$ correspond to plasmon resonances of nanoparticles. The resonances for $\xi > 1$ and $\xi < 1$ are always at different sides from the plasmon resonance of spherical particles (the plasmon resonance in Fig. 6(b) blueshifts). Therefore, oblate silver nanospheroids (nanodisks) are most appropriate for our purposes: due to the redshift of the plasmon resonance, the range of small refractive index $n_{eff} < 1.3$ is in the visible spectral range at a relatively low extinction coefficient $k_{eff} << 1$.

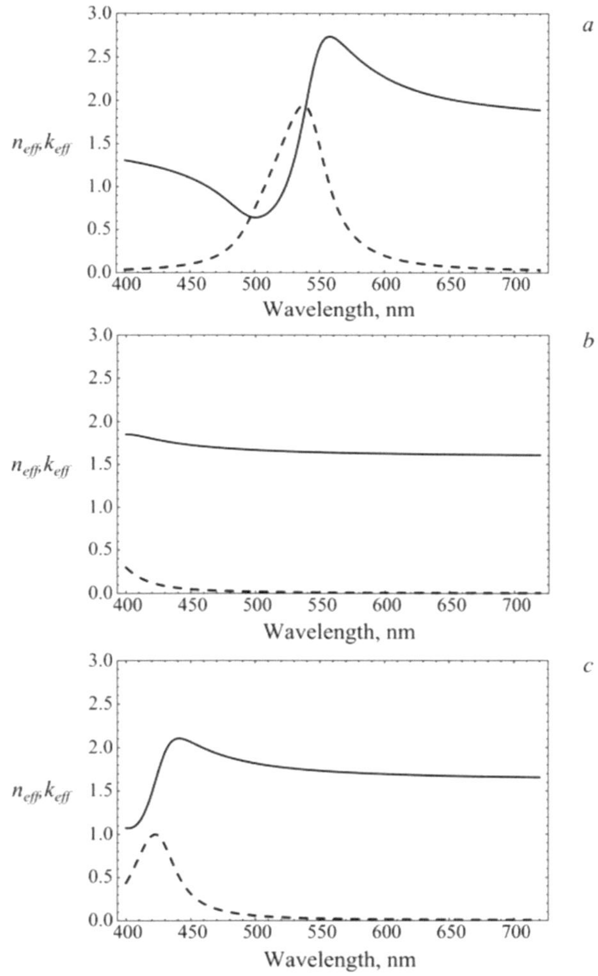

Fig. 6. Dispersion relations of the refractive index n_{eff} (*solid lines*) and extinction coefficient k_{eff} (*dashed lines*) for a composite with (*a*) spherical inclusions and (*b*, *c*) inclusions in form of (*b*) elongated and (*c*) oblate spheroids. The radius of spherical particles is 7 nm, and each spheroid has a volume equal to that of spherical particle. The volume fraction of silver nanoparticles is 0.05.

Thus, the antireflection coating design is determined by the features of composite optical properties. According to the results obtained, the composite coating for natural light must be formed by nanoparticles in the form of oblate spheroids, whose polar axis is oriented perpendicular to the surface of underlying medium.

The calculation shows that a composite with a moderate volume fraction of inclusions ($\eta \sim 0.01-0.1$) satisfies all necessary conditions for the visible spectral range. Figure 7 shows the reflectance and transmittance of the antireflection composite layer with the following parameters: $\xi = 0.1$ and $\eta = 0.05$. The layer thickness, calculated from formula (14), is $h = 93$ nm.

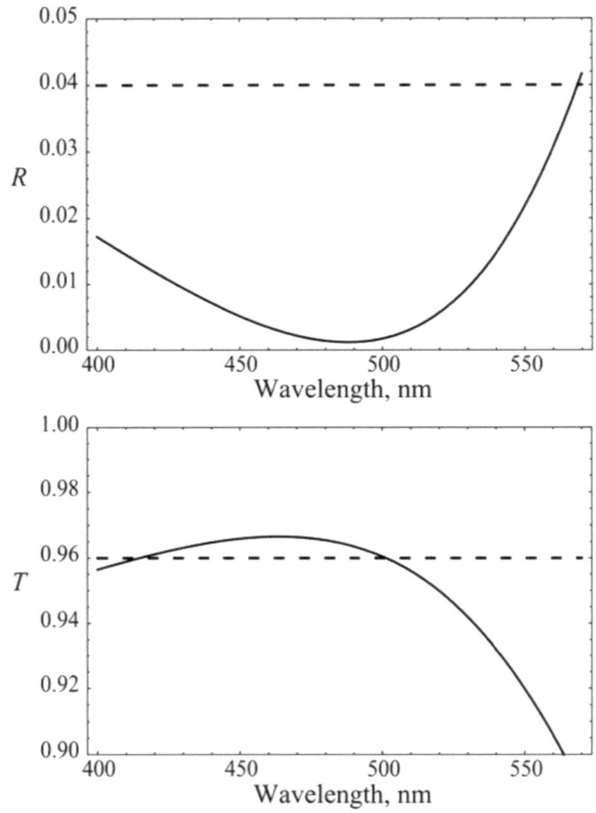

Fig. 7. Reflectance R and transmittance T of a composite slab with silver nanoparticles for normal incidence of light, calculated within the effective-medium model. The reflectance and transmittance of the clean dielectric surface are shown for comparison by dashed lines.

The dependencies in Fig. 7 show that coating of dielectric surface by a composite layer gives a positive effect. In a wide (>100 nm) spectral range the total intensity of reflected light decreases by a factor of more than two, and the minimum reflectance of the composite coating is lower than that of the dielectric by a factor of 20. Unfortunately, the refracted wave intensity increases only slightly in this case, and even decreases in comparison with the initial dielectric at $\lambda > 500$ nm. The latter circumstance can be explained as follows: some part of the light wave energy spent on excitation of free-electron oscillations in composite nanoparticles is transformed into heat. Thus, the light wave energy is partially absorbed by the dispersed subsystem of the composite coating, as a result of which the surface of transparent material cannot be made totally antireflective.

4.4 Thin-film composite polarizing splitter

As it is shown in Section 3, the absorption of the composite medium incorporating metallic nanoparticles can be controlled by careful selection of geometric parameters of nanoparticles. Here we investigate the possibility to realize a thin-film plasmonic polarizing splitter in the visible region. For definiteness, we consider a transparent dielectric with a refractive index $n_m = 1.5$ as a matrix. In our design, the high polarization contrast of composite film is obtained by using uniformly oriented silver nanoparticles of ellipsoidal shape.

The shape of nanoparticles can be selected so as to observe only one plasmon resonance in the visible region of light. According to Fig. 2, this condition complies with $\xi < 0.85$ or $\xi > 1.5$. In this paper, we consider the case of prolate particles of aspect ratio $\xi = 3$. In this case, the plasmon resonance in the visible region corresponds to the field component polarized parallel to the polar semi-axis of the spheroid.

Let us assume that a layer of the composite material with uniformly oriented prolate silver spheroids has been deposited on surface of transparent medium. Let light be incident normally on this layer from vacuum. To provide maximum optical anisotropy, nanoparticles have to be oriented parallel to the plane interface between media. Using the relation (3) and Airy equation (Born & Wolf, 1999), the reflectance and transmittance can be calculated for arbitrary values of filling factor η and layer thickness h.

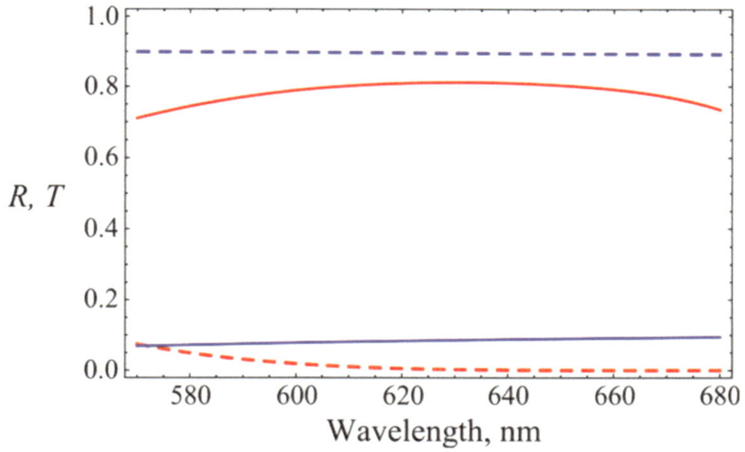

Fig. 8. Reflectance R (*solid lines*) and transmittance T (*dashed lines*) of parallel (*red lines*) and perpendicular (*blue lines*) polarized light of composite slab with uniformly oriented prolate silver spheroids on glass substrate.

Results of calculation show that for $600\,\text{nm} < \lambda < 650\,\text{nm}$, values of $\eta \sim 0.1$ and $h \approx 100 - 150$ nm are optimal from the point of view of the posed problem. Indeed, as one can see from Fig. 8, for η = 0.1 and h = 130 nm composite layer exhibits high polarization contrast. For radiation of the region from 570 nm to 680 nm, this composite slab reflects the parallel polarized light, and for the perpendicular polarized light this coating is transparent. From the practical point of view it is interesting that high reflectance corresponds with low transmittance and vice versa. According to the results shown in Fig. 8, the reflectance and transmittance of this composite slab possess the values either more than 0.7 or less than 0.1. So, such composite structure can be used as a polarizing beam splitter with high performance in transmission and reflection. By selecting the proper shape of the nanoparticles or choosing matrices with different refractive indexes it is possible to achieve the polarization contrast at specified spectral regions.

5. Conclusion

We explore the utility of effective medium representation to simplify the electromagnetic analysis of composite system, and demonstrate the use of this simplification in solving of the boundary problem under consideration. This approach allows us to easily control the parameters of a system and predictably change its optical properties, expressing the necessary conditions in an analytical form. The data obtained with the help of formula (3) show that a heterogeneous medium with plasmonic impurities, such as silver nanoparticles with a concentration of 10^{-2}–10^{-1} per unit volume, is new and interesting object of research with many perspectives for applications. Such plasmonic medium can be used as a transparent anti-reflection coating, weakly reflecting light-absorbing filter, or polarizing beam splitter with high performance in transmission and reflection. It is shown also that an extraordinary refractive index $n_{eff} = 1$ in a heterogeneous medium with metal nanoparticles at transparency can be reached with the compensation of absorption at metal nanoparticles by the amplification in the matrix, and in the case of nonspherical nanoparticles the values of the gain necessary for compensation of absorption are smaller. Provided that silver nanoparticles are located not on the surface but in the bulk of glass or some other sufficiently strong and chemically stable optical material, proposed nanocomposite devices would have high operating qualities, in particular, high stability to contaminations, mechanical effects, and corrosive media.

6. Acknowledgments

The author expresses gratitude to Dr. S.V. Sukhov for introducing him to the fantastic field of optics of nanocomposite media. This work was supported by the Ministry of Education of the Russian Federation through project contracts within the framework of the Federal Target Program 'Science, Academic and Teaching Staff of Innovative Russia for 2009-2013'.

7. References

Bohren, C. F., Huffman, D. R. (1998). Absorption and Scattering of Light by Small Particles, Wiley, ISBN 0471293407, New York, USA

Born, M., Wolf, E. (1999). Principles of Optics, Cambridge University Press, ISBN 9780521642224, Cambridge, United Kingdom

Cai, W., Shalaev, V. (2010). Optical Metamaterials: Fundamentals and Applications, Springer Science + Business Media, ISBN 978-1-4419-1150-6, New York, USA

Golovan, L.A., Timoshenko, V.Yu., Kashkarov, P.K. (2007). Optical properties of porous-system-based nanocomposites. Phys. Usp., Vol.50, No.6, pp.595-612, ISSN 1468-4780

Hutter, E., Fendler, J. H. (2004). Exploitation of Localized Surface Plasmon Resonance. Adv. Mater., Vol.16, pp. 1685-1706, ISSN 1521-4095

Kachan, S., Stenzel, O., Ponyavina, A. (2006). High-absorbing gradient multilayer coatings with silver nanoparticles. Appl. Phys. B, Vol.84, pp.281-287, ISSN 1432-0649

Khlebtsov, N.G. (2008). Optics and biophotonics of nanoparticles with a plasmon resonance. Quantum Electronics, Vol.38, No.6, pp. 504-529, ISSN 1468-4799

Kravets, V. G., Schedin, F., Grigorenko, A.N. (2008). Plasmonic blackbody: Almost complete absorption of light in nanostructured metallic coatings. Phys. Rev. B, Vol.78, pp.205405-10, ISSN 1550-235x

Kreibig, U., Vollmer, M. (1995). Optical Properties of Metal Clusters, Springer-Verlag, ISBN 3-540-57836-6, Berlin, Germany

Maxwell-Garnett, J. (1904). Colours in Metal Glasses and in Metallic Films. Philos. Trans. Roy. Soc. London,Vol.203, pp. 385-420

Maxwell-Garnett, J. (1906). Colours in Metal Glasses, in Metallic Films, and in Metallic Solutions. Philos. Trans. Roy. Soc. London, Vol.205, pp. 237-288

Moiseev, S.G. (2004). Optical Resonances in One-Dimensional Chains of Dipole Centers. *Jour. of Appl. Spectr.*, Vol.71, No.2, pp. 257-262 ISSN 1573-8647

Moiseev, S.G., Pashinina, E.A., Sukhov, S.V. (2007). On the problems of transparency of metal-dielectric composite media with dissipative and amplifying components, Quantum Electroics, Vol.37, No.5, pp. 446-452, ISSN 1468-4799

Moiseev, S.G. (2009). Optical Properties of a Maxwell-Garnett CompositeMedium with Nonspherical Silver Inclusions. Russ. Phys. J., Vol.52, No.11, pp.1121-1127, ISSN 1573-9228

Moiseev, S.G. (2010). Active Maxwell–Garnett composite with the unit refractive index. *Physica B: Condensed Matter*, Vol. 405, pp. 3042–3045, ISSN 0921-4526

Oraevskii, A.N., Protsenko, I.E. (2000). High refractive index and other optical properties of heterogeneous media, JETP Lett., Vol.72, No.9, pp. 445-448, ISSN 1090-6487

Ordal, M.A., Long, L.L., Bell, S.E., et al. (1983). Optical properties of the metals Al, Co, Cu, Au, Fe, Pb, Ni, Pd, Pt, Ag, Ti, and W in the infrared and far infrared. Appl. Opt., Vol.22, No. 7, pp. 1099–1119, ISSN 2155-3165

Protsenko, I.E., Zaimidoroga O.A., Samoilov V.N. (2007). Heterogeneous medium as a filter of electromagnetic radiation. J. Opt. A: Pure Appl. Opt., Vol.9, pp.363 - 368, ISSN 1741-3567

Shalaev, Vladimir M., Cai, Wenshan, Chettiar, Uday K., Yuan, Hsiao-Kuan, Sarychev, Andrey K., Drachev, Vladimir P., Kildishev, Alexander V. (2005). Negative index of refraction in optical Metamaterials. Optics Lett., Vol.30, No.24, pp. 3356-3358, ISSN 1539-4794

Shen, J., Catrysse, P., Fan, S. (2005). Mechanism for Designing Metallic Metamaterials with a High Index of Refraction. Phys. Rev. Lett., Vol.94. pp.197401-19740, ISSN 1079-7114

Sukhov, S.V. (2005). Nanocomposite material with the unit refractive index. Quantum Electronics, Vol.35, No.8, pp. 741-744, ISSN 1468-4799

Yannopapas, V., Modinos, A., Stefanou, N. (2002). Scattering and absorption of light by periodic and nearly periodic metallodielectric structures. Opt. Quantum Electron., Vol.34, pp. 227-234, ISSN 1572-817X

Yuan, H.-K., Chettiar, U. K., Cai W. et al. (2007). A negative permeability material at red light. Optics Express, Vol.15, No.3, pp. 1076 - 1083, ISSN 1094-4087

20

Polymer-Bioglass Composite Coatings: A Promising Alternative for Advanced Biomedical Implants

Floroian Laura, Popescu Andrei,
Serban Natalia and Mihailescu Ion N.
*Transilvania University of Brasov /
National Institute for Lasers, Plasma and Radiation Physics
Romania*

1. Introduction

Human bone can easily regenerate in normal conditions, so in case of fractures some support devices are used to take the load from the new forming bone. However, in case of diseases, ageing and large traumas, the bone has to be helped to regenerate. Usually, large bone defects are filled with natural material from a donor site from the same patient (''autografting"). Unfortunately the transplantable bone is limited and the procedure requires the extension of the operating site (Lahav et al., 2006).

The concept of an artificial device able to help human parts to regenerate or to replace entirely one of the body functions, stimulated numerous research teams around the globe to test materials and combinations of materials aiming towards the "holly grail" of biocompatibility: the biomimetism.

There are evidences that prove early tests on bone repair using metal parts, dating from antiquity (Bliquez, 1996). Up to present time, metals have been the elements of choice for manufacturing implants or prostheses in various shapes and dimensions. Tests and clinic reports have all agreed in time, that the best metal for such devices is Ti (Emsley, 2001). It is bio tolerated by the human body and it has a low density and mechanical properties that do not negatively affect the bone (Oshida, 2006).

Orthopedics as a medicine branch does not consider bio tolerated devices as satisfactory. The general aim of an orthopedic device is to stimulate the bone to regenerate, in other words it has to be bioactive. Ceramic materials made of calcium phosphates (CaP) in various formulas, pure or doped have been extensively studied in literature for their resemblance to the mineral part of the bone (León & Jansen, 2010). Because ceramics brittle when submitted to high tensions, devices made completely out of these materials are not convenient. The common consensus is to use for the bulk implant metals as main components, covered by a very thin and adherent ceramic layer, thus making the surface bioactive (Park, 2008).

This chapter is dedicated to bioactive glasses (BG), "intelligent materials" superior in bioactivity to CaP, made of various metal oxide combinations. They change their composition in vivo after prolonged contact with body fluids, transforming in an equivalent of the mineral bone (Hench & Wilson, 1993).

2. Bioglasses and bioglass composite

2.1 Bioactive glasses

2.1.1 General considerations

The Bioglass® was invented by Larry Hench, Professor at the University of Florida, at the beginning of the 1970s. The 45S5 Bioglass composition in wt% as proposed by Hench was consisting of 45% SiO_2, 24.5% Na_2O, 24.5% CaO and 6% P_2O_5 (Hench et al., 1972).
The SiO_2 content was chosen after numerous tests that evidenced a radically different behavior of the BG in relation to bone, function of this parameter. When they have less 40% SiO_2 in their composition, the glasses do not interact with bone and do not form a connection with it. Between 40 and 50% SiO_2 glasses are highly reactive to aqueous medium. In the vicinity of the implant, ions dissolution and precipitation will take place, resulting in generation of a matrix of SiO_2 that sticks to the bone, serving as precipitation ground for new compounds synthesized in vitro, extremely similar to the inorganic part of the bone. Over 60% SiO_2 content, glasses behave as bioinert materials, being covered by a fibrous membrane in vivo (Greenspan, 1976; Gross, 1980).
Hench described a sequence of five reactions that result in the formation of a hydroxyl-carbonate apatite layer when BG are used in vivo: 1) Na, K, Ca ion migration in the extracellular fluids, 2) hydrolysis of Si-O-Si bonds, resulting in Si-OH generation, 3) total transformation of the initial layer in a Si-OH gel depleted of Ca and alkaline ions 4) precipitation of an amorphous CaP layer on the gel, and 5) mineralization of the CaP layer and its transformation into crystalline carbonated and doped hydroxyapatite (HA) similar to the mineral phase of vertebrate bones (Hench, 1981).
Related to their bioactivity, silicate glasses exhibit several advantages in comparison to other bioactive ceramics, e.g. sintered HA. It has been demonstrated that dissolution products from bioactive glasses upregulate the expression of genes that control osteogenesis (Xynos et al., 2000), which explains the higher rate of bone formation in comparison to other inorganic ceramics such as HA (Wheeler et al., 2001).
The all new concept of bioglasses bioactivity and the enormous possible composition tuning, stimulated research efforts to find and test new formulations, to add new elements to their structure, to produce composite materials bioglass-polymer/metal/ceramic in the search for biomimetic behavior (Sepulveda, 2001; Oki, 2004; Lao, 2009).
Even if countless variations of bioactive glasses can be found in literature, the original 45S5 composition remains one of the most active and efficient, being used as a comparison by almost all the papers referring to BG and still being commercially sold and used in implantology and prosthetics.

2.1.2 Other practical uses

BG are not only active for the osseous tissue regeneration, but also for some soft tissues development. Nanoparticles of such materials were proven to increase the secretion of vascular endothelial growth factor (VEGF) in vitro and to enhance vascularization in vivo (Day, 2004; Gorustovich, 2010; Leu, 2008).
By mixing BG with Ag or Zn ions, different research groups obtained antibacterial materials (Vitale-Brovarone, 2008; Haimi, 2009).
In drug delivery, particles of BG can dissolve in the body fluids releasing metal ions or medicines (Azevedo et al., 2008). Magnetite release from bioactive glasses for hyperthermia treatment of cancer (Bretcanu et al., 2006) or Co ions release to stop cells hypoxia (Gomez-Vega et al., 2000) have been reported.

2.2 BG composites

2.2.1 State of the art

Scaffolds for cells proliferation have two main components: a porous structure usually made of polymer, and a bioactive material, covering the scaffold, that stimulates cell differentiation and growth. This is the main use of BG in conjunction with polymers and reports in literature have shown excellent results in terms of efficiency. Other uses of BG-polymer mix as bone filler highly malleable composite materials, or bioactive, anticorrosive thin films for prostheses coating have been reported in literature.

As filler material for bone defects and cavities treatment, BG lack cohesiveness in bulk. Chan et al (Chan et al., 2002) showed that the addition of medium molecular weight dextran improves the malleability of the ceramic. For bioactivity comparison, they filled standardized bony defects in the lateral femoral condyles of adult New Zealand rabbits with autogenous bone, Bioglass particulates, a mixture of Bioglass® and autogenous bone and Bioglass®-dextran composite.

Postoperative healing was observed after periods of 2 days, 1, 2, 3, 6 and 12 weeks. Histologically, Bioglass® mixed with dextran produced a response essentially similar to that of Bioglass® alone. Dextran produced no adverse tissue reaction and results indicated that it was removed from the site within the first week after implantation.

Verrier et al (Verrier et al., 2004) studied biodegradable PDLLA (poly(dl-lactic acid)) /Bioglass® composites for soft-tissue and hard-tissue scaffolds. In vitro tests using A549, a human epithelial-like lung carcinoma cell line and MG63, human osteosarcoma cells, proved a strong connection between the Bioglass® content in these composites and the cells response. The authors observed 100% MG63 cell adhesion (relative to plastic control) on pure PDLLA scaffolds (0% Bioglass®) and on PDLLA composite scaffolds containing 5 wt% Bioglass®, while a 20% increase in cell adhesion was obtained on composites with 40 wt% Bioglass®. The A549 adhered 100% only on substrates containing 40 wt% Bioglass®. However, in terms of cell proliferation, after 4 weeks, A549 cells developed two times faster on PDLLA foams containing 5 wt% Bioglass® when compared to the proliferation on foams with 40 wt% Bioglass®.

Misra et al (Misra et al., 2010a) tested as scaffold for cells growth, a composite made of poly(3-hydroxybutyrate) (P(3HB)) foams of highly interconnected porosity (85% porosity) and 45S5 Bioglass® particles, introduced in the scaffold microstructure, both in micrometer and nanometer sizes (~10% bioglass content). In vitro, on cultures of MG63 osteoblast like cells, a comparison between scaffolds of P(3HB) and P(3HB)/Bioglass® evidenced a higher proliferation (by 13%) in favor of the simple polymer scaffold. However, the P(3HB)/Bioglass® composite scaffolds are not to be disregarded due to an amazing antibacterial effect verified on S. Aureus. The S. Aureus drastic reduction in the presence of composite scaffolds in vitro was explained by the pH change around the implant. The optimum living medium for bacteria colonies was of pH 7, while around the scaffold, it went up to 10 in 48 h. The composite foams were further on implanted in rats and proved to be non-toxic after one week of implantation.

The same team added multiwall carbon nanotubes (MWCNTs) to the optimal P(3HB)/ bioglass composition (Misra et al., 2010b). They proved that in vitro MWCNTs (2 wt%) improved the MG63 cell attachment and proliferation. However, by increasing the nanotubes concentration to 4-7% had negative influence on the cells development. By electrical I-V analysis in SBF, the samples with 2% nanotubes were shown to facilitate the production of natural apatite (the signal decreased due to the isolating nature of the newly formed ceramic layer).

Yao et al. (Yao et al., 2005) synthesized poly (lactic-co-glycolic acid) (PLGA) and PLGA/BG (30% content) microparticles by an emulsification technique. When immersed in SBF for two weeks, the composite particles were covered in their entirety by a CaP rich layer. On porous scaffolds made of simple PLGA and composite PLGA/BG microparticles, rat mesenchimal stem cells were seeded in vitro. The alkaline phosphatase activity (indicative of new bone formation) was monitored and found to be ten times higher in case of the CaP covered scaffolds. The 3D porous PLGA-30% BG composites demonstrated significant potential as a bone replacement material, both supporting cell growth and promoting differentiation of marrow stromal cells to osteoblast cells.

Zhang et al. (Zhang et al., 2004) prepared by phase separation, porous (pores>100 μm) poly(l-lactide)(PLLA)/BG (average particle size: 1.5 μm) composites. A silane pretreatment of the glass resulted in better glass incorporation in the polymer matrix. The higher the glass content, the higher was the elastic modulus of the composites, but their tensile strength and break strain decreased. Glass pretreatment with silane prevented however the decrease in tensile strength with the increase of the glass content. Composites soaked in simulated body fluid (SBF) at body temperature formed bone-like apatite inside and on their surfaces. The lower the BG content, the faster was the apatite generation on the composite surface. After 1 week composites containing 9% bioglass were entirely covered by apatite, while no apatite was formed on composites with 29% bioglass. The silane pretreatment of glass particles delayed the in vitro apatite formation due to the superior incorporation of the glass in the polymer matrix and to the fewer exposed sites to SBF.

2.2.2 Synthesis of bioglass composite nanostructures

There are a few methods allowing in situ synthesis of BG and their mix with other compounds on site. Each technique has its own advantages and the type of application should recommend the use of a specific method, depending on the foreseen final results: high synthesis efficiency for mass production, high purity of final products at lower rates and/or unusual shape of BG nanostructures for new applications (Boccaccini et al., 2010).

SOL-GEL The method is largely used to obtain nanoparticles of numerous metallic and oxidic materials by precipitation following chemical reactions in liquid medium. BG in the system $CaO–P_2O_5–SiO_2$ in nanometric structures have been synthesized by sol-gel for antibacterial applications in combination with Ag, or for scaffolds coating on HA porous structures (Chen, 2009; Esfahani, 2008).

The major advantage of the sol-gel technique is its versatility. The morphology and size of BG nanoparticles could be tailored by varying the synthesis conditions and the feeding ratio of reagents (Couto, 2009; Hong, 2009).

However, sol-gel can be a time consuming process, due to the necessity of additional steps. Usually, the obtained nanostructures have to be exposed to heat-treatment in order to remove organic additives. It is also a method limited in terms of compositions that can be produced. Because of the numerous steps involved, batch-to-batch variations in terms of compositions are highly possible.

FLAME SPRAY SYNTHESIS The metalorganic deposition uses metal-organic precursor compounds to produce nanoparticles at temperatures above 1000 ^{0}C. The flame spray synthesis, a derivative of this technique, starts from a liquid compound that is pushed by a high flux of oxygen through a nozzle, forming a spray which is ignited. As the spray is burning, the organic constituents of the liquid precursor completely burn, decomposing in H_2O and CO_2, and metal constituents oxidize to form the nanoparticles. Consequently, the process allows the production of any kind of nanoparticulate mixed-oxides with high chemical homogeneity. By using flame spray synthesis, therefore, the preparation of nanoparticles of different BG compositions has become possible.

Mixtures of 2-ethylhexanoic acid salts of calcium and sodium, hexamethyldisiloxane, tributyl phosphate and fluorobenzene were used for BG nanoparticles synthesis (Brunner et al., 2006). The primary particles produced are usually spherically shaped with different degrees of agglomeration.

LASER SPINNING The team of Quintero et al used the laser spinning technique to obtain glass fibers with diameters in the nanometer to micrometer scale (Quintero et al., 2007a).

Large quantities of nanofibres can be produced with good control over chemical compositions and without the necessity of any chemical additives or post heat treatments. The process is very fast; nanofibres being produced in several microseconds. It involves the quick heating and melting of a small amount of the precursor material up to high temperatures using a high power laser. At the same time, a supersonic gas jet is injected into the melt volume to blow the molten material (Quintero, 2007a, 2007b, 2009a). The molten material is quickly spread and cooled by the supersonic gas jet (Quintero et al., 2009a). Long fibers with large length to diameter ratios can be produced by the elongation of the viscous molten material. The obtained BG is amorphous because of the high cooling speed.

Quintero et al. (Quintero et al., 2009b) developed 45S5 and 52S4.6 BG nanofibers and investigated the possibility of using these nanofibers as scaffolds or as reinforcement in polymeric matrix composites. Bioactivity test results were promising, the fibers forming very fast apatite coatings when immersed in SBF.

PULSED LASER TEHNOLOGIES Pulsed laser technologies emerged in late '80 of the last century as a major alternative for ablation and deposition of stoichiometric compounds in form of nanoparticles, nanofibres and nanostructured thin films. In pulsed laser depositions (PLD) the substance is expulsed from target under the action of pulse laser radiation and transferred to a nearby substrate. PLD presents numerous advantages in respect with other deposition technique among which we mention: accurate control of the stoichiometry of the deposited material; reduced film contamination due to the use of laser light; promotion of the growth of crystalline structures for desired applications; energy source independent of the deposition environment; relative simplicity of the growth facility offering great experimental versatility (multilayers, doping); control of the film thickness (with a precision of > 10^{-2} Å/pulse).

PLD was applied to the synthesis of BG nanoparticules and their deposition well- separated or assembled in these films (Floroian, 2008, 2010). These structures were converted in bioapatite after prolonged immersion in SBF.

However, PLD failed when extended to composite polymer - bioactive glass (**PBG**) transfer and deposition. The polymer was irreversibly photochemically decomposed and damaged under direct action of intense UV laser pulses (Pique, 2007). The only solution was conferred by another pulse laser deposition technology – the matrix assisted pulse laser evaporation (MAPLE).

We will review in the next sections recent results obtained in MAPLE deposition of polymer (polymethyl methacrylate, PMMA) in form of nanoparticle thin films.

3. MAPLE for thin films synthesis and deposition

3.1 MAPLE: definition and working procedure

The deposition of polymer, biopolymer and protein materials needs a special protection, which could be provided by MAPLE. Basically, the difference between PLD and MAPLE is target preparation and laser target interaction.

Specific to MAPLE is the use of a composite cryogenic target consisting of a mixture of the organic active material dissolved in a volatile solvent. The solvent must be carefully selected so that it absorbs majority the UV radiation. It has to present a high freezing point and the organic/polymeric material must to present a high solubility in solvent to be deposited (Cristescu et al., 2006). No chemical reaction is permitted between solute (organic polymer) and solvent. Two processes develop under the action of incident laser pulse: the frozen target is evaporated and the organic material is released. The organic molecules reach sufficient kinetic energy by collective collisions with the evaporating solvent molecules to be transferred in gas phase to the collecting substrate (Mihailescu et al., 2010). Finally, the volatile solvent is gradually evacuated by the vacuum system during the deposition of the solute on the substrate in form of thin film. A schematic representation of the MAPLE process is given in Fig. 1.

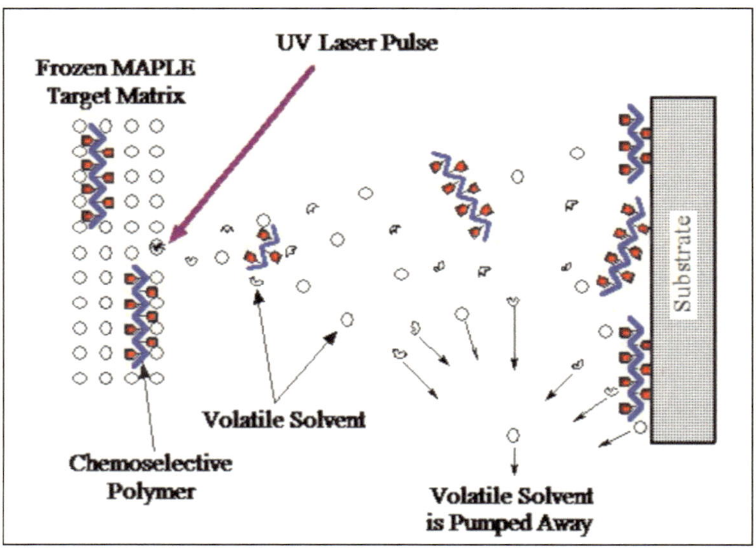

Fig. 1. Schematic of the MAPLE evaporation process (Mihailescu et al., 2010)

3.2 MAPLE experiment

To types of bioglasses were used for deposition: BG57 whit the composition 56.5% SiO_2, 11% Na_2O, 3% K_2O, 15% CaO, 8.5% MgO, 6% P_2O_5 and BG61 having the composition 61.1% SiO_2, 10.3% Na_2O, 2.8% K_2O, 12.6% CaO, 7.2% MgO, 6% P_2O_5 (Veljovic et al., 2009). Biomedical grade 4 Ti was used as substrate because it exhibits a good biocompatibility and is resistant to corrosion, properties which make it a good candidate for fabrication in orthopedic implants. The composite consisted of BG powder mixed with PMMA grains and the mixture was dissolved in chloroform (matrix material). The target was prepared from 0.5 g PMMA dissolved in 19.3 ml chloroform with addition of 0.08 g of either BG57 or BG61. The obtained suspension was frozen at liquid nitrogen temperature and kept during experiments at low temperature using a cooling device (cooler).

A KrF* excimer laser source (model COMPexPro from Coherent) (λ =248 nm, τ_{FWHM} =25 ns) was used for target evaporation. The depositions were conducted in a stainless steel chamber which was first evacuated down to residual pressure of 10^{-4} Pa (Fig. 2).

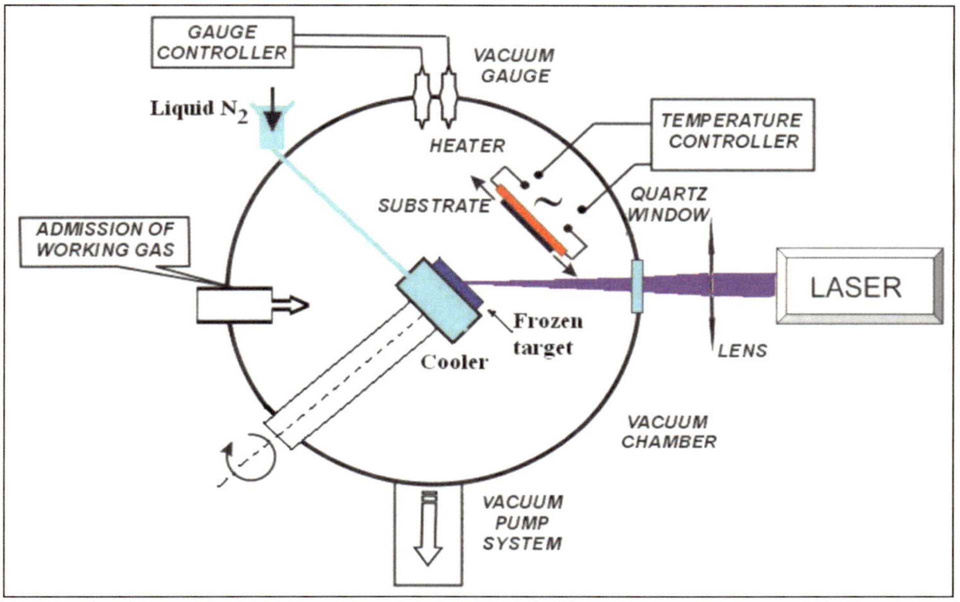

Fig. 2. MAPLE experimental device (Mihailescu et al., 2010)

The dynamic pressure inside chamber during deposition was maintained at 2.7 Pa. The laser beam was incident at 45° on target's surface. The expulsed material was collected onto the Ti substrates placed parallel to the target at a separation distance of 3 cm. Before the deposition the chemically etched Ti (5 mm diameter and 0.5 mm thick) substrates were carefully cleaned with deionized water in a TRANSONIC T 310 ultrasonic bath. The fluence was set for each deposition at 0.55 J/cm^2 and the irradiation spot was of 16 mm^2 area.

In order to improve films adherence, during the depositions the substrate was moderately heated at 35 °C. For the deposition of each film, 5000 subsequent laser pulses have been applied.

During the film growth, the target was continuously rotated at 0.3 Hz and translated along two orthogonal axes in order to avoid drilling and to ensure a uniform deposition.

3.3 Post deposition investigations

3.3.1 Morphological and structural characterization

The PBG deposited structures were characterized by Fourier transform infrared spectrometry (FTIR), energy dispersive X-ray spectroscopy (EDS), atomic force microscopy (AFM), scanning electron microscopy (SEM) and confocal scanning laser microscopy (CLSM).

FTIR analyses were performed using a Nicolet 380 apparatus. The spectra were taken in the absorbance mode. EDS cartography and SEM images were recorded with a Hitachi S2600N

microscope. AFM investigations were carried out in a semi-contact mode with AFM NTEGRA Vita equipment.
For nondestructive investigation of specimens by CLSM, a Leica TCS SP system was used, equipped with a He-Ne laser emitting at 633 nm wavelength and a set of PL Fluotar (40X magnification, 0.7 numerical aperture) objectives. The images were obtained in reflection mode. Data processing and displaying were made by Leica software.

3.3.2 Biocompatibility and bioactivity evaluation

To test the nanomaterials biocompatibility and bioactivity we evaluated the cells adhesion, the cellular viability, cellular proliferation and the capacity of the films to form a biological active layer of apatite (BHA) on surface.
For this purpose human osteoblast cells were harvested then maintained in a McCoy's culture medium supplemented with 10% inactivated bovine fetal serum, 50 U/ml penicillin, 50 mg/ml streptomycin and 1% L-glutamine. The cells were split every 2 days by a ratio of 1:3 and incubated at 37 °C in 5% CO_2. All studies were done 72 hours after cell cultivation on the studied films using FACS Calibur flow cytometer (BD Biosciences) and Leica TCS SP system in fluorescence mode.
The bioactivity of the films was assessed in vitro by soaking the composite films into simulated body fluid (SBF) followed by FTIR spectroscopic and CLSM analyses to determine the extent of BHA formation on surface.

3.3.3 Electrochemical measurement

The samples were immersed in SBF for 7, 14, 28, 35 or 42 days and the potentiodynamic measurements were performed with Palm Sens potentiostat (Palm Instrument) having three electrodes configuration with an Ag/AgCl reference electrode and a platinum wire as the counter electrode. As the working electrodes small discs of bare Ti or Ti substrates coated with nanocomposite films were used. In order to warrant reliable results, the potentiodynamic polarization measurements were repeated three times in SBF in each case.
The curves were recorded when scanning the working potential from (-1 to +2 V) with a scan rate of 0.002 V/s. The electrochemical impedance (EIS) measurements were carried out in an electrochemical cell with a PC-controlled Autolab frequency response analyzer using FRA 4.9 software (Autolab PGSTAT 100 Eco Chemie). A sinusoidal voltage perturbation of amplitude 0.01 V was applied, scanning from 10,000 Hz to 0.1 Hz with 10 points per frequency decade (auto-integration time 5 s).
The fitting to electrical equivalent circuits was performed with the FRA 4.9 software.

4. Polymer (PMMA) - bioglass thin films obtain by maple: physical, chemical and biological characterization

4.1 Chemical characterization: structure and composition

The chemical composition of PBG thin films was investigated by different techniques: FTIR, EDS and XPS. FTIR analyses revealed identical molecular bindings in powder and obtained films and the absence of any other peaks due to impurities (Fig. 3). These features make a strong case for MAPLE preservation of the chemical composition of the base material after transfer.
PBG structure depends on the wt ratio between P and BG. Thus for a ratio larger than 7.5 the polymer peaks prevail, while for a ratio inferior to 7.5 are higher BG peaks (Fig. 3).

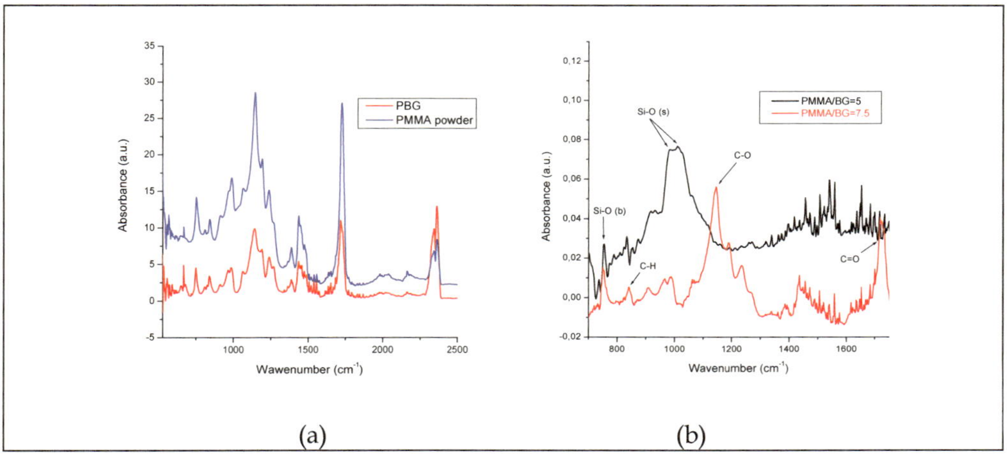

Fig. 3. Main vibration bands present in the FTIR spectra of PBG coatings obtained by MAPLE depending on PMMA/BG wt ratio

The EDS cartography of the PBG coating (Fig. 4) shows that all bioactive glass elements have been preserved in deposited nanostructures and were homogeneously spread over the whole investigated area.

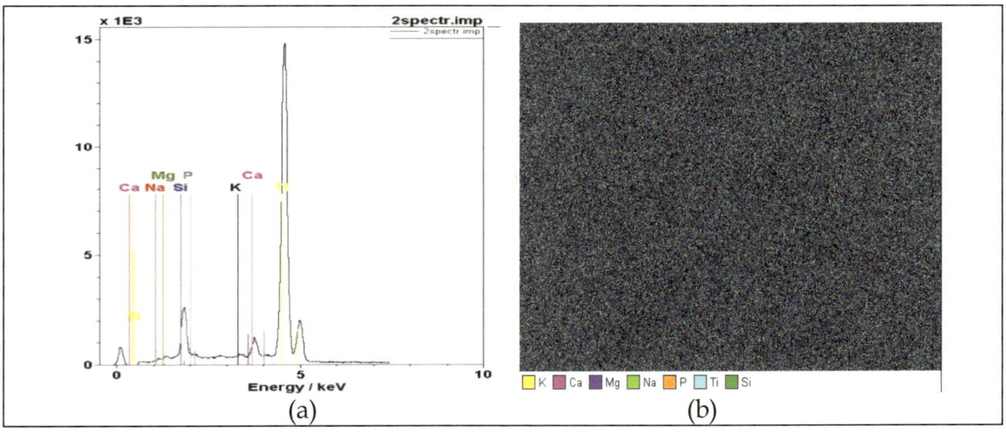

Fig. 4. EDS spectrum (a) and EDS cartography (b) of a PBG film deposited by MAPLE

FTIR and EDS analyses were performed on a large number of samples, and the similar results obtained point to the reproducibility of the method also confirming the stoichiometric transfer of material by MAPLE.

4.1.1 Morphological investigations

The morphology of PBG coatings was investigated by SEM, AFM and CLSM. We observed by SEM (Fig. 5 a, b) that PBG structures are rather uniform and compact.
They consist of droplets of variable size, characteristic to MAPLE or PLD depositions (Gyorgy, 2007; Tanaskovic, 2007, 2008).

The droplets are in the micrometric range, and are randomly spread across the surface. These features are considered favorable (Nelea et al., 2004) to the growth and proliferation of cells which can easily and deeper infiltrate their pseudopodia into surface imperfections and pores. These observations were supported by AFM evidence in 2D and 3D images respectively (Fig. 6 a, b).

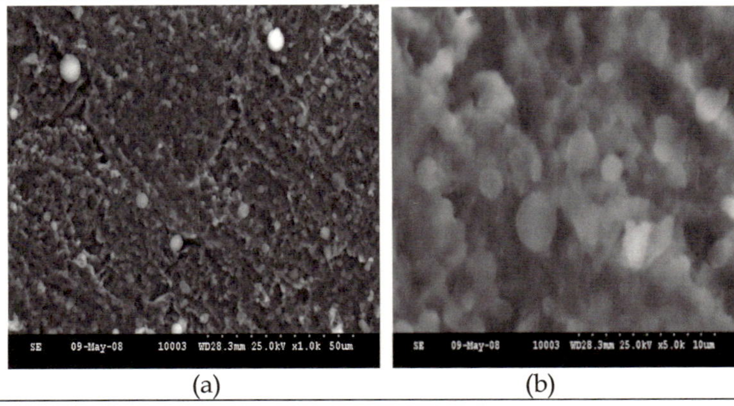

Fig. 5. SEM micrographs showing a typical surface morphology of a PBG coating: (a) general view; (b) detail from (a)

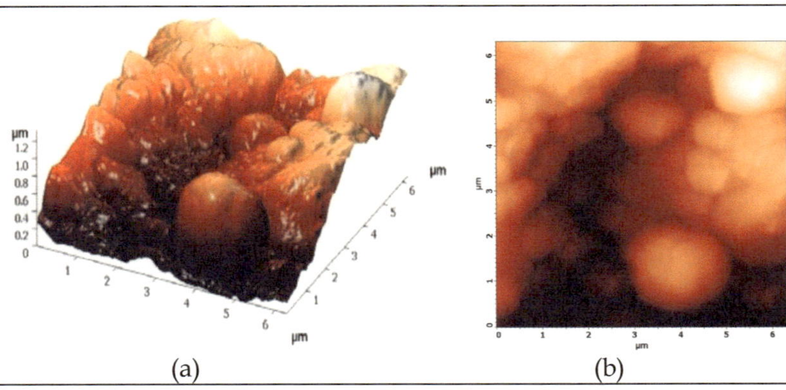

Fig. 6. 3D (a), 2D (b) typical AFM images recorded in case of PBG coating.

The confocal microscopy study also shows a uniform topography of PBG coatings (Fig. 7).
3D images of the deposited PBG thin films (Fig. 8 a, b) revealed the formation of a structure with a characteristic configuration, consisting of a great number of protuberances with few micrometers maximum height. Such features favor biocompatibility because it involves a significant increase of specific area of the deposited biofilms and the proliferation of viable cells (Nelea, 2004).
All these aspects are confirmed by the images below (see Fig. 9), which show the film surface profile compared with the profile of titanium substrate. Initial titanium surface profile shows that this area has a lot of scratches, thin grooves 1-2 µm wide, some reaching up to 13-14 µm deep and some small, with depth less than 1 µm, because titanium substrate

was polished and chemically etched to increase the active surface. This titanium area is ideal for implants: it has a suitable morphology for good cell adhesion and growth of bone and an appropriate roughness for increasing of bone-implant interaction activity.

After PBG thin film deposition all irregularities become lower, having 0.2-5 µm depth but the surface remains rough "friendly" for applications in the field of implantology.

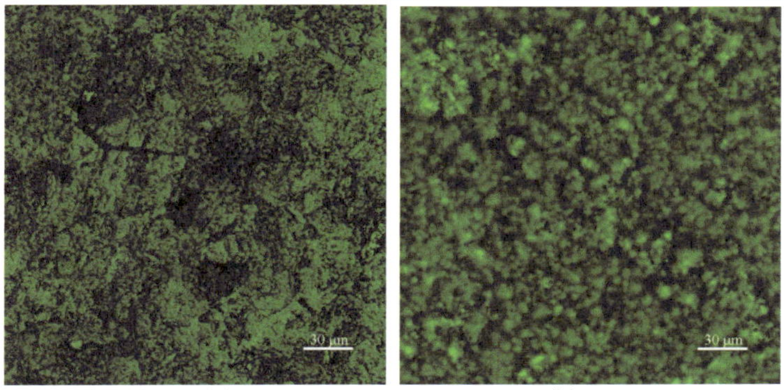

(a) (b)

Fig. 7. Topography (40X magnification, 0.7 aperture, zoom 1) for: a) titanium substrate, b) PBG film

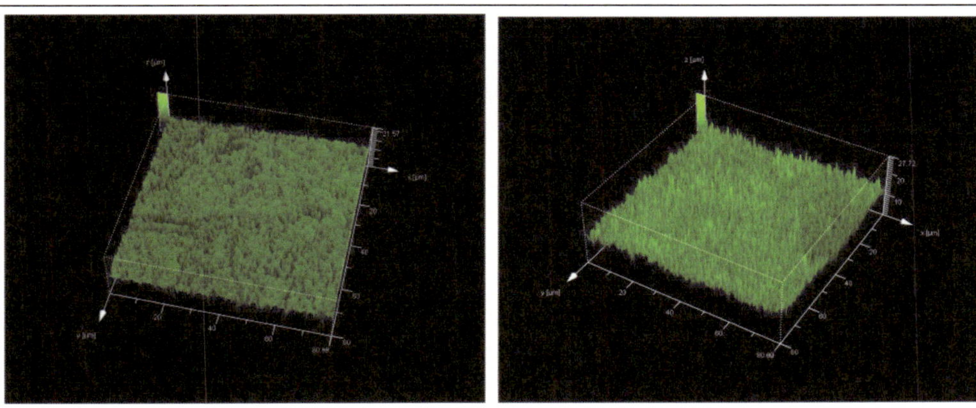

a) b)

Fig. 8. 3D images (40X magnification, 0.7 aperture, 3.12 zoom) for: a) titanium substrate, b) PBG film

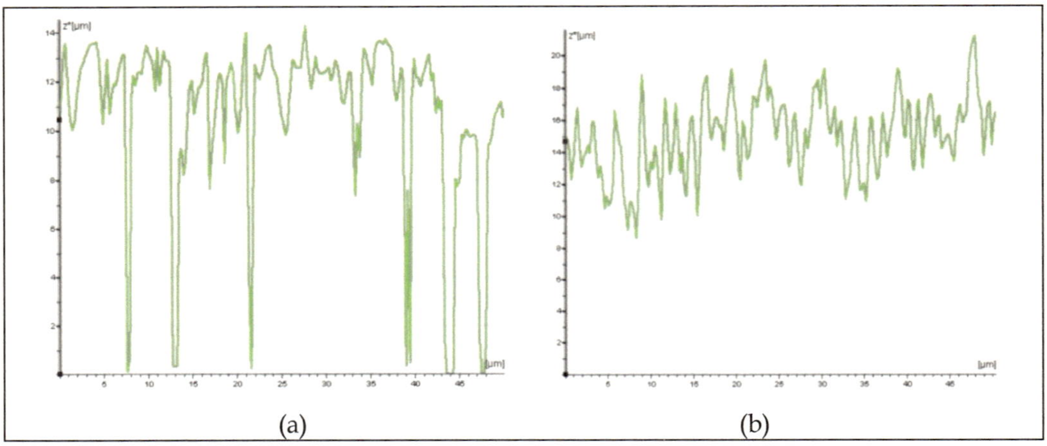

Fig. 9. Surface profiles along a random 50 µm length zone in case of (a) titanium substrate and (b) PBG films, made with a CLSM microscope, 40X magnification, 0.7 aperture, 3.12 zoom

4.1.2 Evaluation of film - substrate adherence

To estimate the adherence of PBG films to metallic substrate we applied scratch-test method to thin films of identical thickness of two types of BG (57 and 61) and two types of PBG nanostructures (with 57 and 61 bioactive glass). The critical loads for the PBG samples have similar values (see tab. 1), with a little higher value in case of PBG61, which shows that this film is more resistant to mechanical stress. The critical load values of PBG films are with an order of magnitude lower than for pure glass films because PMMA is much softer than glass. The critical load for more rough samples is bigger, pointing to a better adhesion to substrate. Despite the high roughness, the results are very repeatable as they show the low value of standard deviation.

Lc (N)	BG57	BG61	PBG57	PBG61
Data 1	0.55	0.82	0.036	0.048
Data 2	0.68	0.67	0.032	0.052
Data 3	0.63	0.61	0.028	0.056
Media	0.62	0.70	0.032	0.052
Standard dev.	0.11	0.12	0.004	0.004

Table 1. Critical load values of BG and PBG structures deposited on Ti substrate.

All peculiarities evidenced by the physical-chemical investigation of PBG coatings were similar for nanostructures contain BG57 or BG61 glass. This sustains the assertion that the two coatings are identical from the point of view of their physical-chemical parameters.

4.2 Biocompatibility and bioactivity assay

4.2.1 In vitro biotests

The osteoblast cells cultivated on PBG surface showed a good viability (Fig. 10), greater even than in the control experiment performed on a borosilicate glass: 85.92% viable cells on PBG surface versus 81.03% viable cells on control surface, the number of events recorded being statistically relevant.

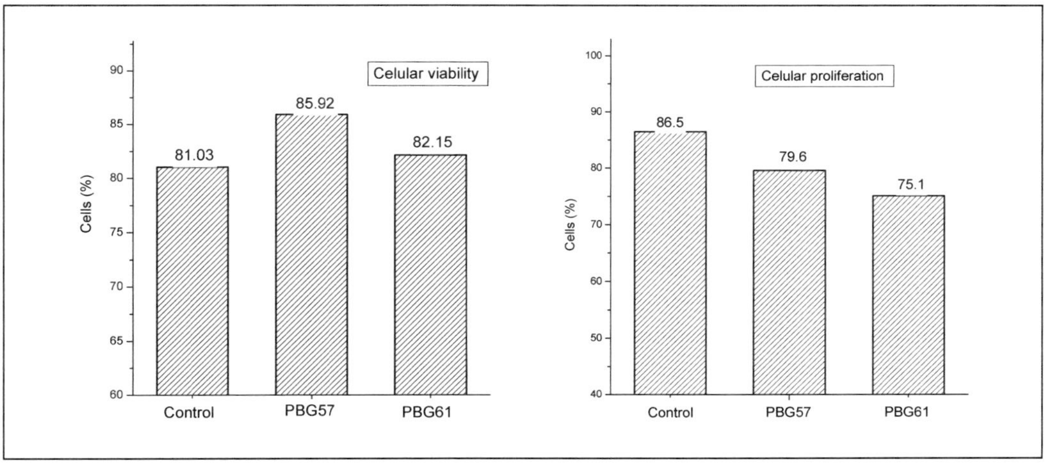

Fig. 10. Viability and proliferation at 72 hours after cell cultivation on control surface and on MAPLE deposited structures

The proliferation experiments were also performed by flow citometry. The primary osteoblasts were scored on day 0 with carboxifluorescein diacetat (CFDA), which has the ability to freely diffuse into cells. It becomes fluorescent when the acetate groups are cleaved by intracellular esterase, which gives rise to a fluorescent CFSE (carboxifluorescein succinimil ester) compound that emits in green at 517 nm. The succinimil ester groups react with the intracellular amines, forming fluorescent conjugates that are retained in the cells. Following division, the daughter cells inherit half of the fluorescence precursor cells, each generation being highlighted as a peak in FACS histogram. The osteoblast cells cultivated on the PBG surface proliferate extensively (see fig. 11), all these results showing a very good surface reaction from the cells.

The cellular adhesion is one of the essential initial events for the cell proliferation and the subsequent differentiation of the bone before the formation of the bone tissue. The analysis of the cellular distribution of the cytoskeleton proteins (vinculina) that mediate the interactions between the actin filaments and integrin (specific receptors of adhesion plaques) can provide useful information regarding the degree of contact between osteoblasts and the nanostructured films.

The samples were washed with phosphate buffer saline, fixed with 4% paraformaldehyde, and permeabilised with 0.2% Triton X-100 for 3 min. The actin filaments were labeled with Alexa Fluor 594 phalloidin (Invitrogen) and vinculina with anti-vinculina antibody (mouse monoclonal from Sigma) followed by Alexa Flour 488 goat anti-mouse antibody (Invitrogen). The primary osteoblasts grown on PBG structure were visualized with a fluorescence microscope (Fig.11). The staining of the two proteins of interest was made with fluorescent marks visualized on the green channel for vinculina and respective red channel for actin. The cells' nuclei were stained with DAPI and they were visualized on the blue channel of fluorescence. The overlap of images shows the spatial localization of vinculina and actin on the film surfaces, with some co-localization of vinculina with actin filaments ends. This suggests that vinculina was activated and was able to induce the generation of focal adhesion points.

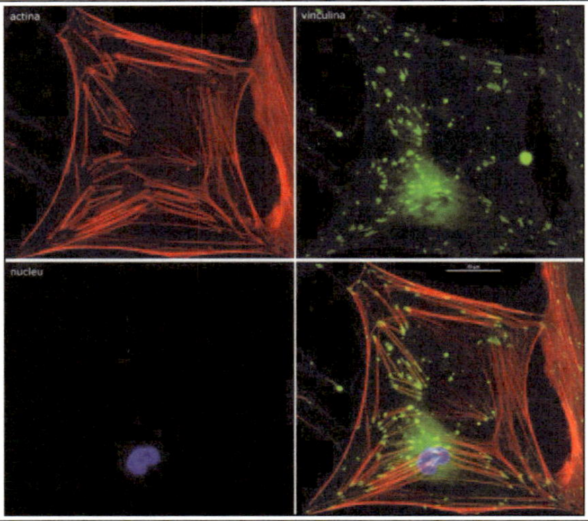

Fig. 11. Fluorescence image of PBG film surface: red staining for actin, green staining for vinculina and blue staining for nucleus, after 3 days of culture (400X) magnification

The osteoblasts are well spread out on the PBG films, with numerous contact points with the PBG surface indicating a good cell adhesion. The actin filaments are well highlighted in all cells and evidence the cytoskeleton organization. The cells present an irregular disposal due to the irregular surface of the film, with the pseudopods infiltrating into the depth of the material which increases cellular adhesion. The cells are well spread out and attached to the film on a large surface which is an evidence of the films biocompatibility. The bioactivity of the films was assessed in vitro by soaking the composite films into SBF followed by FTIR spectroscopic and CLSM analyses to determine the extent of BHA formation on surface. The SBF was prepared according to Kokubo prescription (Kokubo et al., 1990) and it has an ionic concentrate on equal to human plasma. Table 2 collects the ionic concentrations of SBF and blood plasma.

4.2.2 PBG films immersion in SBF

The titanium disks coated with the sintered films were tested by immersion into 25 ml SBF for different times then analyzed. We investigated the bioactivity for two types of PBG films obtained from targets which had a different PMMA/BG wt ratio: PMMA/BG=5 and PMMA/BG =7.5. For the first composition the target contained 0.6 g PMMA reinforced with 0.12 g BG57 bioactive glass particles dissolved in 19.3 g chloroform. For the second was used less glass, i.e. 0.08 g.

Composition (mM)	Na^+	K^+	Mg^{2+}	Ca^{2+}	Cl^-	HPO_4^{2-}	SO_4^{2-}	HCO_3^-
SBF	142	5	1.5	2.5	147.8	1	0.5	4.2
Blood plasma	142	5	1.5	2.5	103.0	1	0.5	27.0

Table 2. Comparison of ionic concentrations of SBF and blood plasma

In case of first composition, after one day of immersion into SBF solution, we can notice the growth of amplitude for all peaks (see Fig. 12 a– red plot) that indicates the forming of a rich superficial layer, where all elements have a greater concentration. This is consistent with other experimental observations (Hench, 1991; Saiz, 1998) and it is accompanied by the loss of soluble silica into the solution, accentuated by the depreciation of the surface's quality observed by microscopically surface investigation.

After one week of immersion into SBF, major transformations were clearly noticed on the surface of the coating (see Fig. 12 a- green plot). They indicate the decrease of silica concentration and the formation of the BHA layer on surface coating. The corresponding peaks for BHA are well observed at 630, 865, 961, 1031 and 1090 cm^{-1}. They correspond to asymmetric (peaks at 1031 and 1090 cm^{-1}) and respective symmetric (peak at 961 cm^{-1}) stretching of P–O bond in $(PO_4)^{3-}$, to $(HPO_4)^2$ (peak at 865 cm^{-1}) and also to the vibrational mode of OH (peak at 630 cm^{-1}). At the same time the peaks belonging to the glass disappeared.

After two weeks of immersion, only the peaks corresponding to BHA remain visible (see Fig. 12 a- blue plot). Their intensity increase in time suggesting the growth of BHA layer.

These certify two things: the dissolution of the bioactive glass and the formation on surface of a freshly growing BHA layer.

This behavior is similar to that reported for bulk glasses in the SiO_2-Na_2O-K_2O-CaO-MgO-P_2O_5 system, immersed into SBF (Gomez-Vega, 1999; Hill, 1986; Ogino, 1980; Saiz, 1998) and it is in accordance with the mechanism of apatite formation described by Hench for Bioglass® (Hench, 1991).

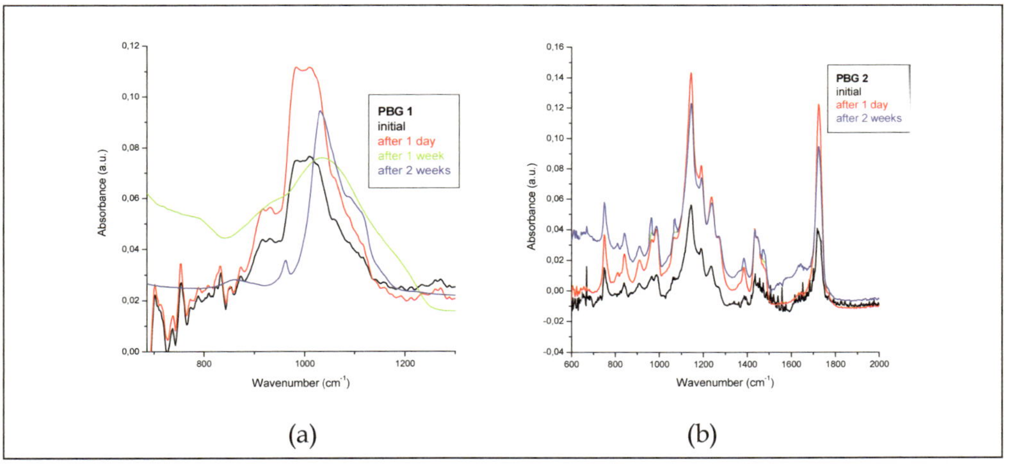

Fig. 12. FTIR spectra for different PBG compositions: PMMA/BG=5 (a) and PMMA/BG=7.5 (b) after their immersion into SBF at 37 °C

However the bioactivity of material is lower than that of pure bioactive glass BG films (Floroian, 2008, 2010) where the growing of BHA layer starts earlier, after few days of immersion. PMMA addition reduces the bone bonding ability of bioactive glasses. This is in good agreement with the studies of the effect of other materials doping in bioactive glasses published by Ohura and Anderson (Anderson, 1988; Ohura, 1982). PMMA does not dissolve into physiological fluids but remains on the metallic substrate. This way, the implant corrosion and the release of metallic ions into the body are prevented while the structure preserves the glass bioactivity. The more glass they contain, the more bioactive the layers are.

The second PBG composition showed a different behavior. No change is visible in his FTIR spectrum before two weeks of immersion (see Fig. 12 b). It presents a very low activity and the studies described below were made on films having the first composition (wt ratio PMMA/BG=5, noted PBG. 3D CLSM images for PBG thin films after 1, 7, 14 immersion days into SBF show changes in time of the surfaces morphology and of their roughness. The decrease of surface roughness within 7 days of immersion (see Table 3) suggests the attenuation of surface irregularities due to the dissolution of the glass. However, the roughness increases after 14 days of immersion that indicates the covering of surface with a thin BHA layer.

Immersion time	RMS (μm)
0 days	2,365
1 day	2,152
7 days	2,042
14 days	2,417

Table 3. Time evolution of surface roughness of PBG structures in SBF

Thus 3D images (see Fig. 13), topographical images (see Fig. 14) and surface profiles images (see Fig 15) sustain observations by FTIR .

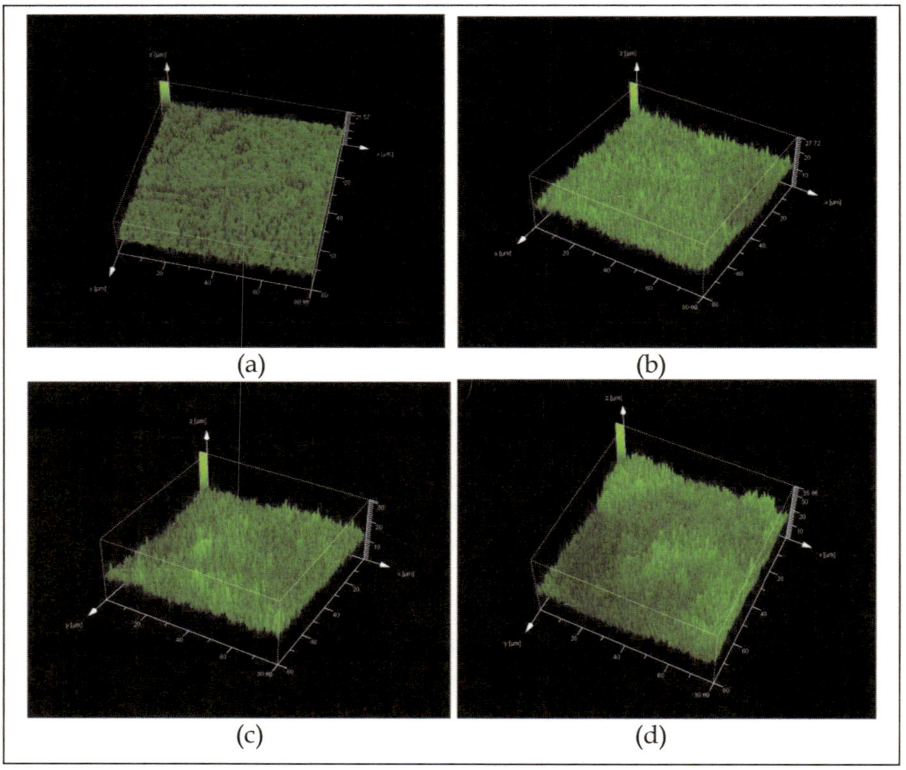

Fig. 13. 3D images (40X magnification, 0.7 aperture, 3.12 zoom) for: (a) titanium substrate, (b) PBG initial film,(c) PBG film after 1 week of immersion into SBF, (d) PBG film after 2 weeks of immersion into SBF

The CLSM analyses certify the glass dissolution and the appearance on the surface of a freshly BHA layer after two weeks of immersion into SBF. In figures 14 and 15 we can see particulates of variable size in the micrometric range that are randomly distributed over the whole investigated area. Surface root mean square roughness becomes RMS=2.417 μm.

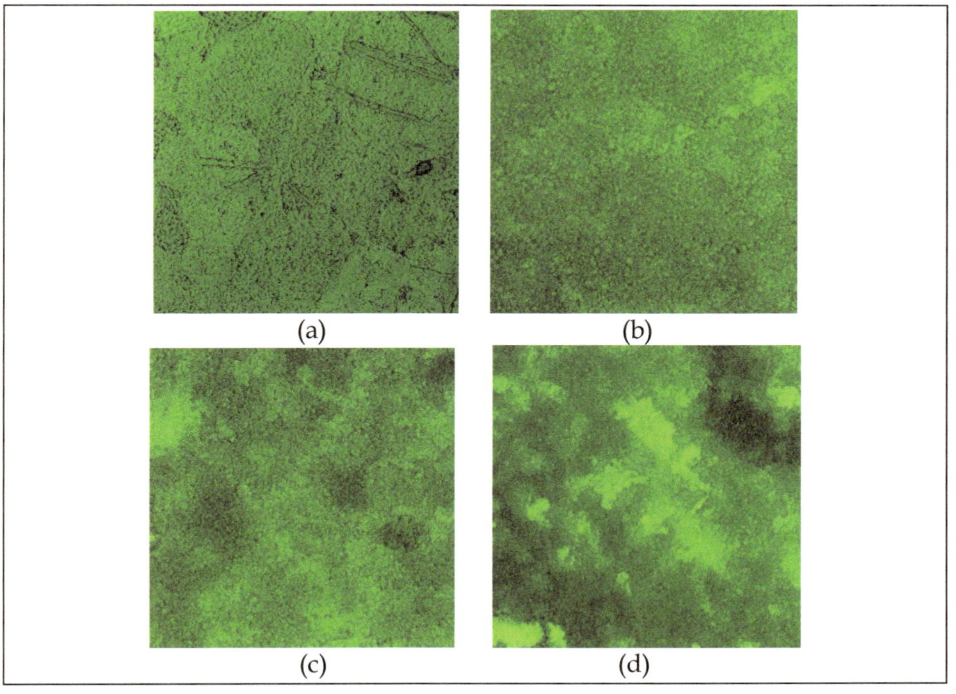

Fig. 14. Topography (40X magnification, 0.7 aperture, 1 zoom) for: titanium substrate (a) and PBG film: initially (b), after 1 week (c), and after 2 weeks of immersion into SBF (d)

4.2.3 Corrosion resistance assessment

The physiological fluids influence upon different coatings was investigated by electrochemical methods, largely spread nowadays because of their high sensibility (Karthega, 2008; Oliveira Brett, 2003).

The corrosion resistance parameters were comparatively studied by potentiodynamic polarization and electrochemical impedance spectroscopies (EIS). The electrochemical parameters of the involved processes were estimated and the electrical parameters of the circuits were verified by fitting the experimental data using equivalent electrical circuits.

Electrochemical polarization investigations were performed for bare uncovered Ti samples (Ti), samples covered with bioglass only (BG57 and BG61) or with the nanocomposite (PBG57 and PBG61). Some of samples PBG were immersed for 28 days in SBF and were covered by a BHA thin layer. The small discs of Ti, BG, PBG and BHA samples were used as working electrodes.

From the potentiodynamic polarization plots and the Tafel diagrams (Fig. 16), the corrosion parameters (the corrosion potential E_{corr}, the corrosion rate i_{corr}, the passive domain E_{pass} and the breakdown potential E_{bd}) and their average values (Table 4) have been obtained.

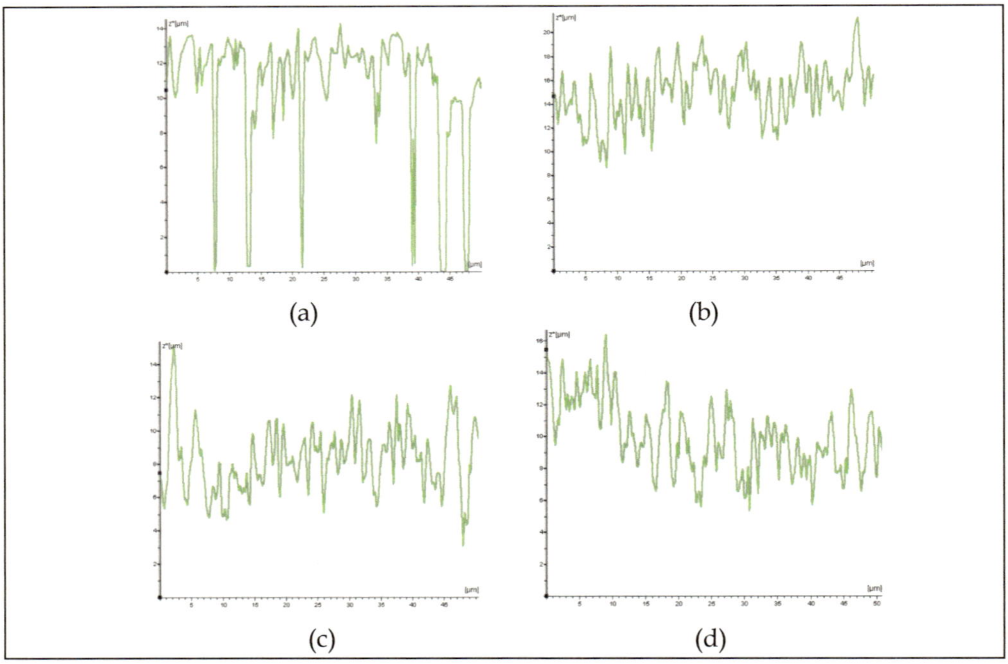

Fig. 15. Surface profiles along a random 50 µm length zone recorded with a CLSM microscope, 40X magnification, 0.7 aperture, 3.12 zoom) for: titanium substrate (a) and PBG film: initially (b), after 1 week (c), and after 2 weeks of immersion into SBF (d)

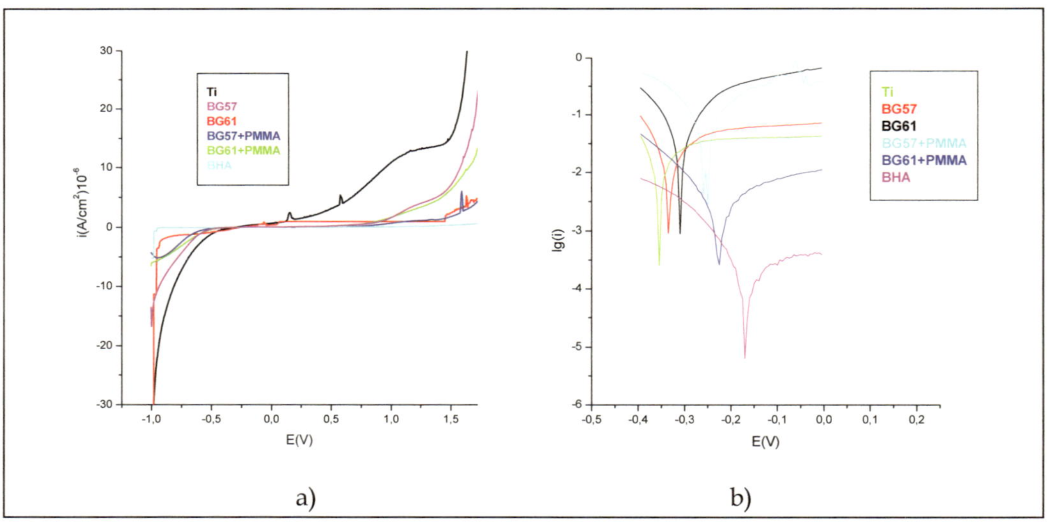

Fig. 16. Polarisation curves (a) and Tafel diagrams (b) into SBF for Ti, BGs, PBGs and BHA

The corrosion rate of uncoated Ti was i_{corr} =1.32 µA/cm². This small value justifies the preference given to Ti as compared to other metals for implants manufacturing. i_{corr} drops 2.3 times for both BG sample and 25 times for both PBG samples. This means that the bioactive glass and especially PBG layers protect very well the Ti implant against corrosion. Further arguments in support of this behavior come from the significant increase of E_{bd} and E_{pass}.

Film	E_{cor} (mV)	i_{cor} (μA/cm²)	E_{bd} (mV)	E_{pass} (mV)
Ti	-357	1,320	141	541
BG57	-335	0,560	535	1407
PBG57	-251	0,053	733	1246
BG61	-310	0,260	671	2303
PBG61	-224	0,022	1454	1512
BHA	-166	0,025	1080	1792

Table 4. Average values of corrosion parameters for uncoated Ti (Ti), samples coated with bioglass only (BG) and composite material (PBG). Corrosion parameters are also indicated for PBG sample after 28 days of immersion in SBF (BHA).

These PBG coatings are not only efficiently isolating the metal from corrosive fluids, but also introduce a very high electrical resistance, of about $10^{10}\,\Omega$. This interposes a resistance of the order of $10^{6}\,\Omega$ between the anodic and cathodic centers on the metal surface that become really isolated from each other, so that the corrosion current density gets neglectable (Perez, 2008).

The prolonged immersion in SBF is resulting in a further enhancement of anti-corrosive action of the layers. Indeed, i_{corr} dropped 2.1 times while E_{corr}, E_{bd}, E_{pass} increased ~ 1.5 times in case of samples BHA as compared to samples PBG. One can say that according to these electrochemical measurements the PBG layer stands for the best anti-corrosion protection of Ti implant surface, especially after prolonged immersion in SBF.

The electrochemical impedance data for the PBG samples after different times of immersion in SBF at room temperature were recorded at 0.1 V and they are shown in the complex plane in figure 17. These plots aim to visualize the differences in the adsorption behavior and charge separation of the interfacial region in different experimental conditions. The fitting of spectra was done using an equivalent electrical circuit, which comprised R_S, the cell (solution) resistance, in series with a parallel R_p-CPE configuration (see Fig. 18). The constant phase element (CPE) was necessary because of the non-homogeneous surface and was modeled as a non-ideal capacitor of capacitance C and roughness factor n (n = 1 corresponds to a perfectly smooth surface). R_p stands for the charge transfer resistance. The values of R_s, R_p, CPE and n obtained based upon the model fitting, are given in Table 5. The maximum phase angle was determined in each case using Bode plots and is presented in the same table.

For the initial coating, the phase angle is 75° suggesting the existence on surface of a highly stable film, with a behavior close to pure capacitive impedance. The Nyquist diagram (Fig. 17 a) is a beginning of a semicircle with a large radius, and both the solution resistance and charge transfer resistance have high values. This proves that the thin PBG layer is a good insulator. The roughness factor is 0.84 because the surface is not perfectly flat, being mechanical polished and chemical etched before deposition in order to increase the surface active area and the bioactivity. CPE value was 1.46 nF which corresponds to a 400 nm thick of PBG film.

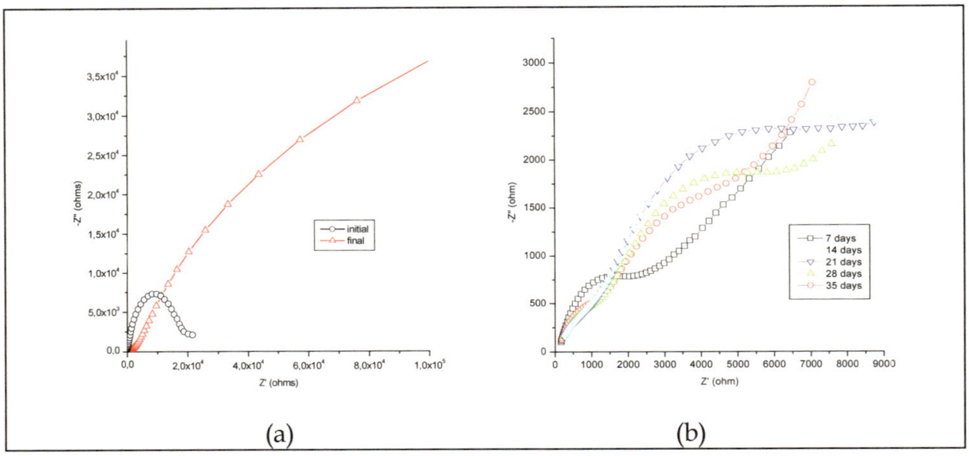

Fig. 17. Impedance spectra in complex plane for PBG61 sample: (a) initially, after 6 weeks immersion in SBF and b) after 1, 2, 3, 4, 5 weeks immersion in SBF at room temperature

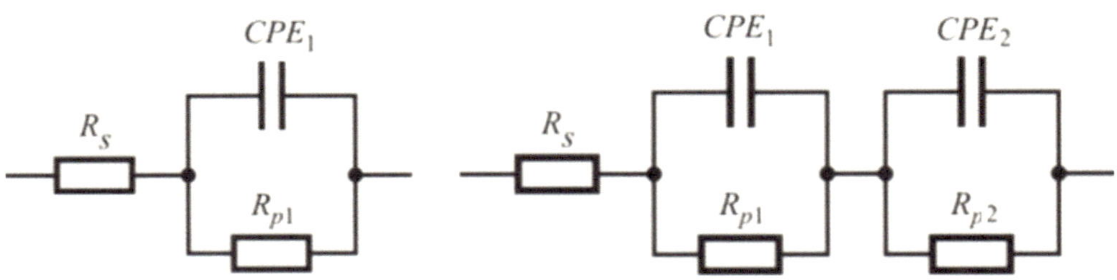

(a) (b)

Fig. 18. Equivalent electrical circuits for PBG61 sample: (a) initially and after 6 weeks immersion in SBF and b) after 1, 2, 3, 4, 5 weeks immersion in SBF at room temperature

PBG61	R_s (Ω)	R_{p1} (kΩ)	CPE_1 10^{-6}(F)	n_1	φ_1 (°)	R_{p2} (kΩ)	CPE_2 10^{-6}(F)	n_2	φ_2 (°)
Initial	1111	18.44	0.00146	0.84	-75.0	-	-	-	-
1 week	103.8	15.36	2.10	0.82	-43.0	-	-	-	-
2 weeks	75.4	12.41	38.58	0.56	-43.0	0.48	12.02	0.58	-26.0
3 weeks	49.1	11.13	43.19	0.52	-43.0	0.87	8.76	0.58	-28.0
4 weeks	47.9	10.00	45.93	0.55	-42.0	0.90	8.03	0.65	-31.2
5 weeks	30.27	9.07	55.62	0.51	-42.0	1.125	2.88	0.67	-31.5
6 weeks	1799	-	-	-	-	$44 \cdot 10^4$	0.0072	0.95	-72.0

Table 5. Electrochemical impedance spectroscopy data of PBG61 sample after various immersion times in SBF at room temperature.

After immersion in SBF both the charge transfer resistance and solution resistance became smaller, the maximum phase angle decreased and another peak appeared in the Bode plot at 26° as an effect of the adsorption process (see Table 5). The BG dissolution leads to the flat surface degradation and the roughness factor gets much smaller. The Nyquist diagram (Fig. 17 b) shows two well defined time constants. These facts suggest that the two processes of the BG dissolution in SBF and the adsorption of some electrolyte ions on surface take place simultaneously causing the formation of a new liquid - layer interface. These results are in good agreement with experimental evidence obtained by FTIR and CLSM (Floroian, 2008; Sima, 2009) and support the formation on the metal surface of a new BHA phase, rather similar to the human bone composition.

The fitting of impedance spectra over the whole frequency range as generated the ($R_s[R_{p1}CPE_1]$ $[R_{p2}CPE_2]$) equivalent circuits with two parallel R-CPE elements in series (Fig. 18). The parallel group $[R_{p1}CPE_1]$ corresponds to an inner layer composed of PBG barrier layer while the parallel group $[R_{p2}CPE_2]$ corresponds to the outer layer consisting of the new forming BHA phase.

Large changes were found after every week of immersion in SBF (see Fig. 17 and Table 5). After 6 weeks the maximum phase angle reaches 72° and both the adsorption and diffusion processes were stopped. The charge transfer resistance and solution resistance were very high. The CPE value was 7.25 nF which corresponds to an ≈ 100 nm thick BHA film. Nyquist plot evidenced an almost intact coating, proving that the superficial BHA layer is coating the entire surface, completely shielding it against next fluid action. The roughness factor was now higher (0.95), suggesting that the BHA layer is compact. It has very high impedance and excellent corrosion properties. The metal ions release by corrosion or wear processes that may induce aseptic loosening after long term implantation into human body are thus avoided.

PBG57 and BGs samples exhibited a similar behavior with these PBG61 samples, except that the diffusion and adsorption processes disappeared after 5 weeks and after 2-3 weeks of immersion in SBF at room temperature, respectively. This means that the compact BHA layer was faster formed on these surfaces compared with PBG61 samples, when the BHA film appears after 6 weeks of immersion only. This delayed formation of the BHA film could be the effect of the increased bioactivity of BG57 samples in respect with PBG57 ones. PMMA addition has reduced the bone ability to bond to BG, but the advantage of using the composite PBG is that PMMA those not dissolved in SBF remaining on implant (Ti) surface. The BG bioactivity is however preserved and even enhanced by the formation of the new BHA phase on top surface of the layer (Floroian et al, 2010). As a consequence, the resistance Rp of PBG samples get higher values in the end and acts therefore as a better shield against corrosion of the Ti implant (substrate).

4.2.4 Influence of the working potential upon the PBG samples

The effect of the electrical potential applied upon the recorded EIS of the Ti coated samples (Fig. 19), was studied. In case of PBG61 samples well defined semicircles were recorded for all applied potentials. The increase of the semicircle diameter with the applied potential can be accounted for by the calcium and phosphorus ions adsorption from solution on sample surface. The adsorption blocked the electrode surface and rendered the charge transfer rather prohibitive. Consequently, the charge transfer resistance increased. A similar behavior was observed for other biological compounds when adsorbed on glassy carbon electrodes (Oliveira Brett, 2002, 2003; Brett, 1994).

In case of PBG57 samples similar semicircles were observed at +0.1 V, but largely differing ones at 0 V and particularly at -0.1 V. A purely capacitive behavior with constant phase angle characteristics were noticed at -0.1 V for the PBG57 sample. The shift of the angle downwards from 90° with respect to the real axis is due to the rather high value of the roughness factor of 0.8 - 0.85 (angle of ≈ 80°), as well as to the low curvature. This evolution is congruent with the existence of an almost intact coating. The formed superficial BHA layer has covered the entire surface thus completely isolating the Ti substrate. These studies demonstrated that the application of an electrical potential has accelerated the chemical processes between electrodes, leading to the formation in a matter of minutes of a compact continuous BHA layer. Thus, the higher activity of PBG57 sample in SBF was confirmed by the formation of the bioactive layer at -0.1 V. Conversely, in case of PBG61 sample, the BHA layer formation has not initiated at -0.2 V yet.

Finally, we advance the hypothesis that the faster repair of a fracture or the osteointegration could be achieved by the proper application of an electrical potential on hybrid PBG coatings covering Ti prostheses.

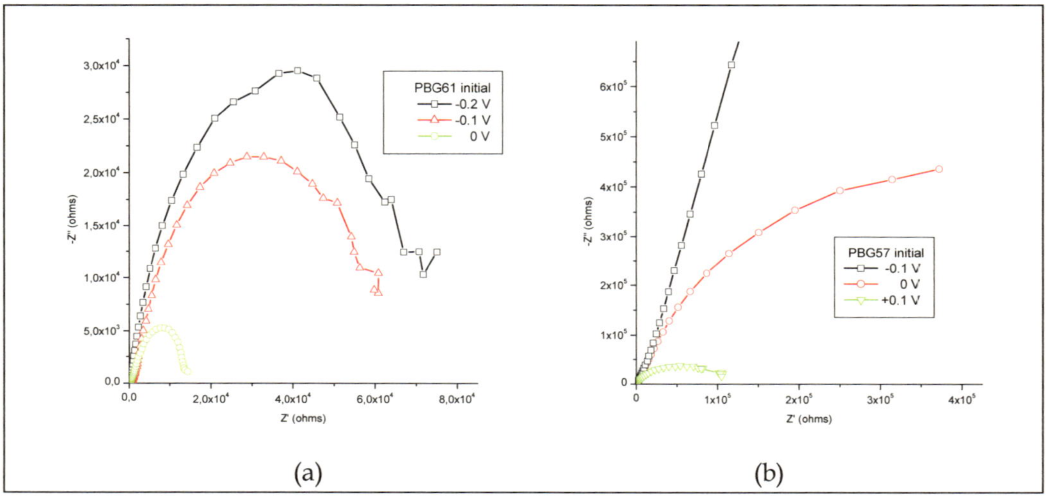

Fig. 19. EIS in the complex plane for PBG61 (a) and PBG57 (b) samples at different applied potentials in SBF at room temperature

5. Conclusions

An area of major interest in biomedical engineering is presently the development of new materials for medical implants. The bioactive glasses and glass ceramics have been extensively investigated in the past three decades for biomedical application due to their ability to form a tight bond to living bone.

Large research efforts have been directed to the new cations inclusion or mixture with biomaterials, thus providing novel properties to bioactive glasses. We reviewed in this chapter new results on the incorporation of the bioglass particles into a PMMA matrix in order to improve the mechanical and corrosion properties of bioglass coatings. The thin films of composite polymer-bioglass (PBG) were obtained by matrix assisted pulsed laser evaporation (MAPLE) - an advanced deposition technique which allows for the protected, safe transfer of delicate substances sensible to damage under direct action of laser and plasma.

The biocompatibility and bioactivity of the composite films was evaluated in vitro, in two modes. First, the films transformation when immersed in SBF was evaluated by FTIR, CLSM and electrochemical studies. Next, biological tests have been performed using human osteoblasts cultured on films. One important evidence was the formation of a bioapatite layer by ions exchange with the fluid concurently with the bioglass decomposition. The polymer is not dissolved in SBF but is deposited on the surface. The electrochemical impedance spectroscopy studies demonstrated that as an effect of immersion in biological fluids, the implant corrosion is slowed down and the metal ions release in to the body is prohibited.

Human osteoblastes cultivated on the composite coatings that were immersed for several days in SBF remained very adherent to the substrate, exhibiting a very good viability and proliferation confirming the films biocompatibility and the stable fixation to bone.

The presence of the nanostructured PBG coatings on the metallic implants surface made the implant bioactive and more resistant to corrosion. The use of the MAPLE deposited nanocomposite PBG layers is suggested to be a promising alternative for development of a new generation of implants and prostheses.

6. Acknowledgment

A.P., N.S., and I.N.M. acknowledge we thanks the financial support of this work by the UEFISCSU (Romania) under the project IDEI, 511/ 2008.

7. References

Andersson, O. H.; Karlsson, K.H.; Kangasniemi, H. & Yli- Urpo, A. (1988). Models for physical properties and bioactivity of phosphate opal glasses. Glastechnische Berichte, Vol. 61, No.10, pp. 300-305, ISSN 0017-1085

Azevedo, M., Jell, G., Hill, R. & Stevens, M.M., (2008), Novel hypoxia mimicking bioactive materials for tissue engineering, Tissue Engineering Part A, Vol.14, No. 5, (May 2008), pp. 889 - 891, ISSN 1937-3368

Bliquez, LJ., (1996), Prosthetics in classical antiquity: Greek, Etruscan and Roman prosthetics. In: Aufstieg und niedergang der Römischen welt II, Haase W., Temporini H. (Eds.), pp. 2640-2676, Walter de Gruyter, ISBN 3-11-015714-4, Berlin & New York

Boccaccini, A.R., Erol, M., Stark, W.J., Mohn, D., Hong, Z. & Mano, J.F., (2010), Polymer/bioactive glass nanocomposites for biomedical applications: A review, Composites Science and Technology, Vol. 70, No. 13, (November 2010), pp. 1764 - 1776, ISBN 0266-3538

Bretcanu, O., Spriano, S., Vitale C.B. & Verne E., (2006), Synthesis and characterization of coprecipitation-derived ferrimagnetic glass–ceramic, Journal of Materials Science, Vol. 41, No. 4, (February 2006), pp. 1029 - 1037, ISSN 0022-2461

Brett, C. M. A.; Oliveira Brett, A. M. & Serrano, S. H. P. (1994).On the adsorption and electrochemical oxidation of DNA at glassy carbon electrodes. J Electroanal Chem, Vol.366, pp. 225-238, ISSN: 0022-0728

Brunner, T.J., Grass, R.N. & Stark, W.J, (2006), Glass and bioglass noanopowders by flame synthesis, Chemical Communications, Vol. 13, (2006), pp. 1384 - 1386, ISSN 1359-7345

Chan, C., Thompson, I., Robinson, P., Wilson, J. & Hench, L., (2002), Evaluation of Bioglass/dextran composite as a bone graft substitute, International Journal of Oral & Maxillofacial Surgery, Vol. 31, No. 1, (February 2002); pp. 73–77, ISSN 0901-5027

Chen, X., Lei, B., Wang, Y. & Zhao, N., (2009), Morphological control and in vitro bioactivity of nanoscale bioactive glasses, Journal Non- Crystalline Solids, Vol. 355, No. 13, (May 2009), pp. 791 - 796, ISSN 0022-3093

Couto, D.S., Hong, Z. & Mano, J.F., (2009), Development of bioactive and biodegradable chitosan-based injectable systems containing bioactive glass nanoparticles, Acta Biomaterialia, Vol. 5, No. 1, (January 2009), pp. 115 - 123, ISSN 1742-7061

Cristescu, R.; Mihailescu, I. N.; Jelinek, M. & Chrisey, D. B. (2006). Functionalized Thin Films & Structures Obtained by Novel Laser Processing Issues, In: Functionalized Properties of Nanostructured Materials, R. Kassing, P. Petkov, W. Kulisch & C. Popov (Eds.), 211-226, NATO Science Series by Springer, Series II: Mathematics, Physics and Chemistry, ISBN 1-4020-4595-6

Day, R.M., Boccaccini, A.R., Shurey, S., Roether, J.A., Forbes, A. & Hench, L.L., (2004), Assessment of polyglycolic acid mesh and bioactive glass for soft-tissue engineering scaffolds, Biomaterials, Vol. 25, No. 27, (December 2004), pp. 5857 - 5866, ISSN 0142-9612

Emsley, J., (2001), Nature's Building Blocks: An A-Z Guide to the Elements, Oxford University Press, ISBN 0-19-850340-7, Oxford, England, UK

Esfahani, S.I.R., Tavangarian, F. & Emadi, R., (2008), Nanostructured bioactive glass coating on porous hydroxyapatite scaffold for strength enhancement, Materials Letters, Vol. 62, No.19, (July 2008), pp. 3428 - 3430, ISSN 0167-577X

Floroian, L.; Savu, B.; Stanciu, G.; Popescu, A. C.; Sima, F.; Mihailescu, I. N.; Mustata, R.; Sima, L. E.; Petrescu, S. M.; Tanaskovic, D & Janackovic, D. J. (2008). Nanostructured bioglass thin films syntherized by pulsed laser deposition: CLSM, Ftir investigations and in vitro biotests. Appl. Surf. Sci. Vol.255, pp. 3056-3062, ISSN: 0169-4332

Floroian, L.; Sima, F.; Florescu, M. , Badea, M.; Popescu, A.C.; Serban, N. & Mihailescu, I.N. (2010). Double layered nanostructured composite coatings with bioactive silicate glass and polymethylmetacrylate for biomimetic implant applications. J Electroanal Chem, Vol.648, pp. 111-118, ISSN: 0022-0728

Floroian, L.; Mihailescu, I. N.; Sima, F.; Stanciu, G. & Savu, B. (2010) Evaluation of biocompatibility and bioactivity for pmma - bioactive glass nanocomposite films obtained by MAPLE. UPB Sci. Bull. Series A, Vol.72, Iss.2, pp. 133-148, ISSN: 1223-7027

Gomez-Vega, J. M.; Saiz, E. & Tomsia, A. P. (1999). Glass-based coatings for titanium implant alloys. J Biomed Mater Res., Vol.46, pp. 549-559, ISSN 0021-9304

Gomez-Vega, J.M, Saiz, E., Tomsia, A.P., Marshall, G.W. & Marshall, S.J. (2000), Bioactive glass coatings with hydroxyapatite and Bioglass ® particles on Ti-based implants, Biomaterials, Vol. 21, No. 2, (January 2000), pp. 105 - 111, ISSN 0142-9612

Gorustovich, A.A., Roether, J.A. & Boccaccini, A.R, (2010), Effect of bioactive glasses on angiogenesis: in-vitro and in-vivo evidence, Tissue Engineering Part B: Reviews, Vol. 16, (April 2010), pp.199–207, ISSN 1937-3368

Greenspan, D.C. & Hench, L .L., (1976), Chemical and mechanical behavior of bioglass-coated alumina, Journal of Biomedical Materials Research, Vol. 10, No. 4 (July 1976), pp. 503 - 509, ISSN 1549-3296

Gross, U.M. & Strunz V., (1980), The anchoring of glass ceramics of different solubility in the femur of the rat, Journal of Biomedical Materials Research, Vol. 14, No. 5, (September 1980), pp. 607 - 618, ISSN 1549-3296

Gyorgy, E.; Grigorescu, S.; Socol, G.; Mihailescu, I. N.; Figueras, A.; Janackovic D.; Palcevskis, E.; Zdrentu L. E. & Petrescu S. (2007). Bioactive glass and hydroxyapatite thin films obtained by pulsed laser deposition. Appl. Surf. Sci., Vol.19, pp. 7981 - 7986, ISSN: 0169-4332

Haimi, S., Gorianc, G., Moimas, L., Lindroos, B. & Huthala, H., (2009), Characterization of zinc-releasing three-dimensional bioactive glass scaffolds and their effect on human adipose stem cell proliferation and osteogenic differentiation, Acta Biomaterialia, Vol. 5, No. 8, (October 2009), pp.(3122 - 3231, ISSN 1742-7061

Hench, L.L., Clark, A.E., Schaake, J R. & Schaake H.F., (1972), Effects of microstructure on the radiation stability of amorphous semiconductors, Journal Non- Crystalline Solids, Vol. 8–10, (June 1972), pp. 837 - 843, ISSN 0022-3093

Hench, L.L. (1981), Fundamental Aspects of Biocompatibility, D. F.Williams (Ed), CRC Press, ISBN 0849366119, United States

Hench, L. L. (1991). Bioceramics: from concept to clinic. J. Am. Ceram.Soc., Vol.74, pp. 1487-1510, ISSN 0002-7820

Hench, L.L. & Wilson J., (October 1993), An Introduction to Bioceramics (Advanced Series in Bioceramics-vol. 1), World Scientific Publishing Company, ISBN 981-02-1400-6, Singapore

Hill R. (1996). An alternative view of the degradation of Bioglass. J Mater Sci Lett, Vol.15, pp. 1122-1125, ISSN 0261-8028

Hong, Z., Merino, E.G., Reis, R.L. & Mano, J.F., (2009), Novel Rice-shaped Bioactive Ceramic Nanoparticles, Advanced Engineering Materials, Vol.11, No. 15, (May 2009), pp. B25 -B29, ISSN 1438-1656

Karthega, M.; Tamilselvi, S. & Rajendran, N. (2008). Effect of pH on the corrosion behavior of Ti6Al4V alloy for dental implant application in fluoride media. Trends Biomater.Artif.Organs, Vol.20, No.1, pp. 31-39, ISSN 0971-1198

Kokubo, T.; Kushitani, H.; Sakka, S.; Kitsugi, T. & Yamamuro, T. (1990). Solutions able to reproduce in vivo surface-structure changes in bioactive glass-ceramic A-W3. J. Biomed. Mater. Res., Vol.24, pp. 721-734, ISSN 1549-3296

Lahav, A., Burks, RT., Greis, PE., Chapman, A.W., Ford, G.M. & Fink, B.P., (2006), Clinical outcomes following osteochondral autologous transplantation, Journal of Knee Surgery, Vol. 19, No. 3, (July 2006), pp. 169-173, ISSN 15388506

Lao, J., Nedelec, J.M. & Jallot, E. J., (2009), New strontium-based bioactive glasses: physicochemical reactivity and delivering capability of biologically active dissolution products, Journal of Materials Chemistry, Vol. 19, No. 19, (2009), pp. 2940 - 2949, ISSN 0959-9428

León, B. & Jansen, J., (2010), Thin Calcium Phosphate Coatings for Medical Implants, Springer, ISBN 978-0-387-77718-4, New York

Leu, A & Leach, J.K., (2008), Proangiogenic potential of a collagen/bioactive glass substrate, Pharmaceutical Research, Vol. 25, (May 2008), pp. 1222 - 1229, ISSN 0724-8741

Mihailescu, I. N.; Ristoscu, C.; Bigi A. & Mayer, I. (2010). Advanced biomimetic implants based on nanostructured coatings synthesized by pulsed laser technologies, In Laser-Surface Interactions for New Materials Production Tailoring Structure and Properties, A. Miotello & P.M. Ossi (Eds.), 235 - 260, Springer Series in Materials Science, ISSN: 0933-033X

Misra, S.K., Ohashi, F., Valappil, S.P., Knowles, J.C., Roy, I., Silva, S. R.P., Salih, V. & Boccaccini, A.R., (2010), Characterization of carbon nanotube (MWCNT) containing P(3HB)/bioactive glass composites for tissue engineering applications, Acta Biomaterialia, Vol. 6, No. 5, (March 2010), pp. 735 - 742, ISSN 1742-7061

Misra, S.K., Ansari, T.I., Valappil, S.P., Mohn, D., Philip, S.E., Stark, W.J., Roy, I., Knowles, J.C., Salih, V. & Boccaccini, A.R., (2010), Poly(3-hydroxybutyrate) multifunctional composite scaffolds for tissue engineering applications, Biomaterials, Vol. 31, No. 10, (April 2010), pp. 2806 - 2815, ISSN 0142-9612

Nelea, V.; Morosanu, C.; Iliescu, M. & Mihailescu, I.N. (2004). Hydroxyapatite thin films grown by pulsed laser deposition and radio-frequency magnetron sputtering: comparative study. Appl. Surf. Sci., Vol. 228, pp. 346–356, ISSN: 0169-4332

Ohura, K.; Nakamura, T.; Yamamuro, T.; Ebisawa, Y.; Kokubo, T.; Kotoura, Y. & Oka, M. (1992). Bioactivity of CaO SiO2 glasses added with various ions. J Mater Sci, Vol.3, pp. 95-100

Ogino, M.; Ohuchi, F. & Hench, L. L. (1980). Compositional dependence of the formation of calcium phosphate films on Bioglass. J Biomed Mater Res., Vol.14, pp. 55-64, ISSN 0021-9304

Oki, A, Parveen, B., Hossain, S., Adeniji, S. & Donahue, H, (2004), Preparation and in vitro bioactivity of zinc containing sol-gel–derived bioglass materials, Journal of Biomedical Materials Research A, Vol. 69A, No. 2, (May 2004), pp. 216 - 221, ISSN 1549-3296

Oliveira Brett, A. M.; Silva, L. A.; Farace, G.; Vadgama, P. & Brett, C. M. A. (2003). Voltammetric and impedance studies of inosine-5-monophosphate and hypoxanthine. Bioelectrochemistry, Vol.59, pp. 49-57, ISSN: 1567-5394

Oliveira Brett, A. M.; Silva L. A. & Brett, C. M. A. (2002). Adsorbtion of guanine, guanosine and adenine at electrodes studied by differential pulse voltammetry and electrochemical impedance. Langmuir, Vol.18, pp. 2326-2342

Oshida, Y., (2007), Bioscience and Bioengineering of Titanium Materials, Elsevier Science, ISBN 978-0-08-045142-8, Oxford

Park, J., (2008), Bioceramics: Properties, Characterizations, and Applications, Springer, ISBN 978-0-387-09544-8, New York

Perez, N. (2008). Electrochemistry and corrosion science. Norwell. USA: Kluwer Academic Publishers.

Pique, A. (2007). Deposition of Polymers and Biomaterials Using the Matrix-Assisted Pulsed Laser Evaporation (MAPLE) Process, In: Pulsed Laser Deposition of Thin Films: Applications-Led Growth of Functional Materials, R. Eason (Ed.), John Wiley & Sons, Inc., ISBN-13: 978-0-471-44709-2

Quintero, F., Mann, A.B., Pou, J., Lusquinos, F. & Riveiro, A., (2007), Rapid production of ultralong amorphous ceramic nanofibers by laser spinning, Applied Physics Letters, Vol. 90, No. 15, (March 2007) pp. 153109, ISSN 0003-6951

Quintero, F., Pou, J., Lusquinos, F. & Riveiro., A., (2007) Experimental analysis of the production of micro- and nanofibres by Laser Spinning, Applied Surface Science, Vol. 254, No. 4, (December 2007), pp. 1042 - 1047, ISSN 0169-4332

Quintero, F., Dieste, O., Pou, J., Lusquinos, F. & Riveiro, A., (2009), On the conditions to produce micro- and nanofibres by laser spinning, Journal of Physics D: Applied Physics, Vol. 42, No. 6, (March 2009), pp. 065501, ISSN 0022-3727

Quintero, F., Pou, J., Comesana, R., Lusquinos, F., Riveiro, A., Mann, A.B., Hill, R.G., Wu, Z.Y. & Jones, J.R., (2009), Laser Spinning of Bioactive Glass Nanofibers, Advanced Functional Materials, Vol. 19, No. 19, (August 2009) pp. 3084 - 3090, ISSN 1616-3028

Saiz, E.; Tomsia, A.P. & Pazo, A. (1998). Bioactive coatings on Ti and Ti6Al4V alloys for medical applications, In: Tomsia AP, Glaeser AM, (Eds), Ceramic microstructures: control at the atomic level, Berkeley, Plenum Press, pp. 543–550, ISBN 0-306-45817-9

Saiz, E.; Tomsia, A.P. & Pazo, A. (1998). Silicate glass coatings on Ti-based implants. Acta Mater, Vol.46, pp. 2551-2558, ISSN 1359-6454

Sepulveda, P., Jones, J.R., & Hench, L.L., (2001), Characterization of melt-derived 45S5 and sol-gel–derived 58S bioactive glasses, Journal of Biomedical Materials Research, Vol. 58, No. 6, (June 2001), pp. 734 - 740, ISSN 1549-3296

Sima, F.; Ristoscu, C. Popescu, A.; Mihailescu, I.N.; Kononenko, T.; Simon, S.; Radu, T.; Ponta, O.; Mustata, R.; Sima, L. E. & Petrescu, S. M. (2009). Bioglass -polymer thin coatings obtained by MAPLE for a new generation of implants. Journal of Optoelectronics and Advanced Materials, Vol.11, pp. 1170-1179, ISSN 1454-4164

Tanaskovic, D.; Jokic, B.; Socol, G.; Popescu, A.; Mihailescu, I. N.; Petrovic, R. & Janackovic D. (2007). Synthesis of functionally graded bioactive glass - apatite multistructures on Ti substrates by pulsed laser deposition. Appl. Surf. Sci. Vol.254, pp. 1279-1282, ISSN: 0169-4332

Tanaskovic, D.; Veljković, Dj.; Petrović, R.; Janaćković, Dj.; Mitrić, M.; Cojanu, C.; Ristoscu, C. & Mihailescu, I.N. (2008). Double layer bioactive glass coatings obtained by PLD. Key Engin. Mater. Vol.361-363, pp. 277-280, ISSN: 1013-9826

Vallet-Regi, M., Salinas, A.J., Roman, J. & Gil, M., (1999), Effect of magnesium content on the in vitro bioactivity of CaO-MgO-SiO2-P2O5 sol-gel glasses, Journal of Materials Chemistry, Vol. 9, No. 2, (1999), pp. 515 - 518, ISSN 0959-9428

Veljovic, Dj.; Jokic, B.; Petrovic, R.; Palcevskis, E.; Dindune, A.; Mihailescu, I. N. & Janackovic D. (2009). Processing of dense nanostructured HAP ceramics by sintering and hot pressing. Ceramics International, Vol.35, pp. 1407–1413, ISSN 0272-8842

Verrier, S., Blaker, J.J., Maquet, V. &Hench, L.L., (2004), PDLLA/Bioglass® composites for soft-tissue and hard-tissue engineering: an in vitro cell biology assessment, Biomaterials, Vol. 25, No. 15, (July 2004), pp. 3013 - 3021, ISSN 0142-9612

Vitale-Brovarone, C., Miola, M., Balagna, C. & Verne, E., (2008), 3D-glass-ceramic scaffolds with antibacterial properties for bone grafting, Chemical Engineering Journal, Vol. 137, No. 1, (March 2008), pp. 129 - 136, ISSN 1385-8947

Wheeler, D.L., Montfort, M.J. & McLoughlin, S.W., (2001), Differential healing response of bone adjacent to porous implants coated with hydroxyapatite and 45S5 bioactive glass, J Biomed Mat Res, Vol. 55, No. 4, (June 2001), pp. 603 - 612, ISSN 1549-3296

Xynos, I.D., Hukkanen, M.V.J., Batten J.J., Buttery L.D., Hench L.L. & Polak J.M., (2000), Bioglass ® 45S5 Stimulates Osteoblast Turnover and Enhances Bone Formation In Vitro: Implications and Applications for Bone Tissue Engineering, Calcified Tissue International, Vol. 67, No. 4, (March 2000), pp. 321-329, ISSN 0171-967X

Yao, J., Radin, S., Leboy, P.S. & Ducheyne, P., (2005), The effect of bioactive glass content on synthesis and bioactivity of composite poly (lactic-co-glycolic acid)/bioactive glass substrate for tissue engineering, Biomaterials, Vol. 26, No. 14, (May 2005), pp.1935 - 1943, ISSN 0142-9612

Zhang, K., Wang, Y., Hillmyer, M.A., Francis, L.F., (2004), Processing and properties of porous poly(Image -lactide)/bioactive glass composites, Biomaterials, Vol. 25, No. 13, (June 2004), pp. 2489–2500, ISSN 0142-9612

21

Effects of Filler Content on Mechanical and Optical Properties of Dental Composite Resins

Seyed Mostafa Mousavinasab
Department of restorative dentistry. Dental school
Isfahn university of medical sciences, Isfahan
Iran

1. Introduction

Filler –matrix coupling determines, to a large extent, the mechanical strength and clinical longevity of dental composites. Incorporation of filler into resin matrix greatly influences and improves material properties provided that filler particles are bonded to polymer matrix or otherwise it may actually weaken the resin.
Benefits of filler content presence are increased hardness, strength, radiopacity and decrease in polymerization shrinkage, thermal expansion and contraction, water sorption, softening, staining and finally improved workability.
Formulation of the monomer and filler, shear rate and temperature greatly influence characteristics of resin composite.(1)
There are significant differences in number of filler particles, size and area occupied for the fine particle composite resins of different brands.(2)

2. Physical and mechanical properties

There is a correlation between physical and mechanical properties and filler content weight and size in composite resins. Following increasing filler particle size an increase in stress concentration and decrease in flexural strength is observed.(3)
In addition to correlation existed between elastic moduli and filler content fraction of composite resin, the shape and size seem to be fine-tuning factors for young's modulus.(4-7)
Flexural strength and modulus, hardness as well as fracture toughness are influenced by both filler morphology and filler loading.(8)
flexural strength and modulus of elasticity values are different among different universal hybrid composites whereas the microfine composite resins present lowest filler weight and mechanical properties compared to universal hybrid and the nanofilled resins present intermediary results .(9)
Mechanical properties of microfilled composite resins as well as polyacid- modified resin composites are dependent on the inorganic filler content and properties of matrix resin and whether matrix resin is hydriphilic or hydriphobic.(10-12)
Significant exponential relations are found between filler content of the resins and their flexural moduli and flexural strength .(13)Although some studies demonstrate that there is no correlation between fracture toughness and filler content by volume of flowable composites but hardness and fracture toughness tend to be linearly proportional to filler

content in the composite resin and resin composites demonstrate a range of fracture toughness values.(14-16)
There is a linear relationship between elastic module and filler loading but correlation of fracture strength and fatigue data to filler fraction could not be proved, therefore materials providing high initial strengths don't obviously reveal the best fatigue resistance.(17)
Condensable composite and hybrid composite resins with similar concentration of inorganic particles may show different flexural strength.(18)
Presence of spherically- shaped filler particles affect the microfracture mechanisms of dental resin composites and increase the bending strength and fracture toughness with a much higher rate for elastic modulus.(19-20)
Percent of submicron silica level in hybrid resin composites has direct effect on physical strength.(21)
The particle size of the filler appear to have a moderate influence on the properties of resin composite.(22)
Presence of nano-filler particles in resin -based restorative materials produce superior performance compared to micro-particles.(23)It also greatly influences grindability of composite resin adhesives.(24)
The flexural strength of metal –resin composite restorative materials containing 4-META treated particles is mainly affected by filler content and immersion time(25)and is increased by the particle size and content of Ag-Sn and Ag-In alloy particles as filler and 4-META as coupling agent.(26-30)
An experimentally prepared metal –resin composite using Ag-Cu particles as filler in which metal particles are involved in the polymerization initiation system has the potential to be used as a dental restorative material,(31)
The bending properties: such as maximum stress and bending modulus, increase with filler content.(32)
Uncontrolled aggregation of amorphous calcium phosphate particulate fillers and their uneven distribution within polymer matrices can have adverse effects on the properties of composite containing this kind of fillers.(33)
Adding apatite and titanium nanotubes to resin based cements will increase the fracture toughness, flexural and compressive strength, hardness and modulus without changing radiopacity and biocompatibility.(34-35)
Beside degree of cure other factors like filler content and monomer type affect the color stability, hardness, compressive strength, stiffness and flexural strength of composites.(36-37)
Coefficient of linear thermal expansion and water absorption of glass-fiber reinforced resins depends on the inorganic filler content or glass-fiber content.(38)
There is an underlying relationship among the composition, component stability and post polymerization properties of flowable composites.(39)
Discrepancy between filler and matrix, filler content, particle size and the ability of the polishing systems to abrade filler may contribute to polished surface characteristics of resin composite.(40)
Decrease in filler particle size to less than 1 micrometer and a lower filler loading permit the clinical development and maintenance of smooth surface with microfilled compared to conventional composite restorative resins.(41)
Fluoride release from filled resins containing CaF_2 particles as filler in the range of 9.09 mass% concentration is independent of PH solution and may help to reduce the occurrence of both secondary caries and restoration fracture.(42-44)

3. Wear resistance and polymerization shrinkage

Wear resistance is a major concern about composite resins. In general it is suggested that composite resins with smaller particles wear less and filler components characterize the wear patterns especially in the occlusal contact area. (45-47)

Wear resistance of composite resins is enhanced due to presence of higher filler volume and functional silane treated microfiller particles. (48)

The abrasion resistance of heat curing composites is also controlled by the filler size and filler content.(49)

The effect of filler content on toothbrush wear resistance may vary with different resin matrices.(50)

Increasing the filler content offers characteristics like bulk curing with less polymerization shrinkage, decreased wear and packability to the composite resin.(51)

Polymerization shrinkage:

Filler loading reduces polymerization contraction and reaction inhibition under atmospheric oxygen at the composite resin surface. (52)

Low shrinkage composites including Ormocers and cationic ring –opening systems despite their higher filler mass show no difference regarding mechanical properties compared to highly filled methacrylate-based materials.(53)

There is an inverse linear relationship between filler content and polymerization shrinkage (54)

Higher inorganic content is associated with lower polymerization stress values which is in direct relation with reduced shrinkage.(55)

Increased filler content in packable composite resins assures of improved wear-resistance, strength, longevity, postoperative sensitivity and esthetics.(56)

Compositional interactions between the filler and matrix influence viscosity and flow characteristics of resin composite.(57)

Higher filler loaded flowable composites are preferred to flowable composites with lower filler content when incremental technique is used in conjunction with conventional hybrid composites for 2-mm deep cavities to achieve better marginal integrity.(58)

Marginal adaptation is significantly related to the amount of inorganic component and also to the volume of the prepolymerized inorganic filler content.(59)

Both organic and inorganic composition of the composite resin influence its rate of cure .The rate of polymerization is increased with the level of HEMA and TEDDMA in the monomer composition.(60-61)

Substitution of Bis-GMA , UDMA and TTEGDMA with alternative monomers results in increased flexural strength. Increased hygroscopic expansion and reduced shrinkage are achieved using a very hydrophilic monomer.(62)

Optical properties:

One interesting aspect of filler content effect is on optical properties of the resin. When a ray of light interacts with composite surface, some of the light may be partly reflected and some partly refracted. The density of the filler determines how strongly the light is scattered within the material. The low filler proportion regions show lower scattering than denser filler regions. With the addition of filler to unfilled resin matrices (UDMA and TEGDMA based composite resins) a significantly higher transmitance value will be seen and there is a linear correlation between percentage of Bis-GMA in the resin matrix and the total and diffuse translucency.

Higher transmittance of Bis-GMA compared to UDMA and TEGDMA with the addition of filler positively influence translucency of composite resin. (63)
Improvement of surface treatment material of filler and composition of filler makes it harder to absorb the microwave energy which is dependent on the size of filler.(64)
There is a linear relation between optical scattering and filler concentration of resin composite and also efficiency of optical scattering is related to size and shape of the filler.(65)
Higher translucency is achieved by decontamination of the added fillers to the matrix of composite resins.(66)_
Filler type influences the color difference values and translucency of composite resin artificial teeth.(67) The pigments are usually metal oxides. Certain fluorescent agents are added to resin composite in order to give the materials a natural -looking, tooth like structure. Metal oxides such as titanium oxide are added to composite to produce opaque composite. The volume fraction of the filler and matrix in composite resins influence scattering and absorption as a result the color of composite resin.(68)
Adding fillers such as TIO2 particles to resin matrix will lead to increasing opallescence of resin composite, decrease in translucency parameter but no effect on fluorescency. Presence of TIO2 nanoparticles produce human enamel like appearance.(69)
The orientation of fillers affects the absorption and scattering coefficient differences in fiber reinforced composite resins.(70)
Some of particulate filled composite resins that can be fabricated into metal-free crowns are stable in both translucency and color during storage period in a media such as water.(71)
Darker shades of resin composite contain darker pigments that absorb more light. Refraction index of resin composite is an important factor in attaining a color matching between composite and dental tissues. To prevent light scattering at the resin-filler interface, the refractive indices of the different components (filler-matrix-coupling agent) shouldn't vary too much, otherwise resulting in a material that looks opaque with reduced transparency for light.
Opalescence parameter varies by the size and amount of filler and translucency parameter decreases as the amount of same filler size increases.(72)
Refractive index match between resin/filler linearly rises with the conversion of composite resin and increase in light transmission during conversion is greater for increasing filler levels.(73)
It has been shown by researchers; light scattering is related to filler particle size in the resin composite, that is maximized when the filler Particle size is one half the wavelength of activation light, resulting in a lower transmission coefficient and smaller depth of cure. Transmission Coefficient is influenced by the wavelength of light, refractive indices of the resin and fillers, and type and amount of filler particles.
The depth of cure of a composite resin is affected by the amount of light that reaches the photo initiator.
Light intensity decreases as it passes through the material. Fillers and pigments strongly influence the intensity of the incident light, limiting the depth of cure. Both intensity of the light source and attenuating power of the material influence the degree of conversion. Filler/resin refractive index has significant interaction on cure depth and color matching of composite resin.(74)

4. References

[1] Lee JH, Um CM, Lee IB. Rheological properties of resin composites according to variations in monomer and filler composition. Dent Mater. 2006 Jun;22(6):515-26.

[2] Jaarda MJ, Lang BR, Wang RF, Edwards CA. Measurement of composite resin filler particles by using scanning electron microscopy and digital imaging. J Prosthet Dent. 1993 Apr;69(4):416-24.
[3] Tanimoto Y, Kitagawa T, Aida M, Nishiyama N.Experimental and computational approach for evaluating the mechanical characteristics of dental composite resins with various filler sizes. Acta Biomater. 2006 Nov;2(6):633-9.
[4] Masouras K, Silikas N, Watts DC.. Correlation of filler content and elastic properties of resin-composites. Dent Mater. 2008 Jul;24(7):932-9.
[5] Masouras K, Akhtar R, Watts DC, Silikas N. Effect of filler size and shape on local nanoindentation modulus of resin-composites. J Mater Sci Mater Med. 2008 Dec;19(12):3561-6.
[6] Kawaguchi M, Fukushima T, Horibe T, Watanabe T.[Effect of filler system on the mechanical properties of light-cured composite resins. II. Mechanical properties of visible light-cured composite resins with binary filler system]. Shika Zairyo Kikai. 1989 Mar;8(2):180-4.
[7] Kawaguchi M, Fukushima T, Horibe T, Watanabe T. Effect of filler system on the mechanical properties of light-cured composite resins. I. Effect of various types of silica fillers on the mechanical properties of the composite resins]. Shika Zairyo Kikai. 1989 Mar;8(2):174-9.
[8] Kim KH, Ong JL, Okuno O.The effect of filler loading and morphology on the mechanical properties of contemporary composites.J Prosthet Dent. 2002 Jun;87(6):642-9.
[9] Rodrigues Junior SA, Zanchi CH, Carvalho RV, Demarco FF.Flexural strength and modulus of elasticity of different types of resin-based composites.Braz Oral Res. 2007 Jan-Mar;21(1):16-21.
[10] Hirasawa T, Hirano S, Hirabayashi S, Harashima I, Nasu I, Kurosawa T. [Mechanical properties of microfilled composite resins (author's transl)].Shika Rikogaku Zasshi. 1981 Apr;22(59):187-95.
[11] Raptis CN, Fan PL, Powers JM.Properties of microfilled and visible light-cured composite resins.J Am Dent Assoc. 1979 Oct;99(4):631-3.
[12] Schulze KA, Zaman AA, Söderholm KJ.Effect of filler fraction on strength, viscosity and porosity of experimental compomer materials. J Dent. 2003 Aug;31(6):373-82.
[13] Inoue M, Finger WJ, Mueller M.Effect of filler content of restorative resins on retentive strength to acid-conditioned enamel. Am J Dent. 1994 Jun;7(3):161-6.
[14] Yamaga T, Sato Y, Akagawa Y, Taira M, Wakasa K, Yamaki M..Hardness and fracture toughness of four commercial visible light-cured composite resin veneering materials. J Oral Rehabil. 1995 Dec;22(12):857-63.
[15] Bonilla ED, Mardirossian G, Caputo AA.Fracture toughness of posterior resin composites.Quintessence Int. 2001 Mar;32(3):206-10.
[16] Bonilla ED, Yashar M, Caputo AA.Fracture toughness of nine flowable resin composites.J Prosthet Dent. 2003 Mar;89(3):261-7.
[17] Lohbauer U, Frankenberger R, Krämer N, Petschelt A.Strength and fatigue performance versus filler fraction of different types of direct dental restoratives.J Biomed Mater Res B Appl Biomater. 2006 Jan;76(1):114-20.
[18] Adabo GL, dos Santos Cruz CA, Fonseca RG, Vaz LG.The volumetric fraction of inorganic particles and the flexural strength of composites for posterior teeth. J Dent. 2003 Jul;31(5):353-9.
[19] Kim KH, Park JH, Imai Y, Kishi T.Fracture behavior of dental composite resins.Biomed Mater Eng. 1991;1(1):45-57.

[20] Kim KH, Park JH, Imai Y, Kishi T. Microfracture mechanisms of dental resin composites containing spherically-shaped filler particles. J Dent Res. 1994 Feb;73(2):499-504.
[21] Suh BI, Ferber C, Baez R.Optimization of hybrid composite properties. J Esthet Dent. 1990 Mar-Apr;2(2):44-8.
[22] Li Y, Swartz ML, Phillips RW, Moore BK, Roberts TA. Effect of filler content and size on properties of composites. J Dent Res. 1985 Dec;64(12):1396-401.
[23] Rüttermann S, Wandrey C, Raab WH, Janda R.Novel nano-particles as fillers for an experimental resin-based restorative material. . Acta Biomater. 2008 Nov;4(6):1846-53.
[24] Iijima M, Muguruma T, Brantley WA, Yuasa T, Uechi J, Mizoguchi I. Effect of mechanical properties of fillers on the grindability of composite resin adhesives. Am J Orthod Dentofacial Orthop. 2010 Oct;138(4):420-6.
[25] Urapepon S, Ogura H. Metal-resin composite restorative material using powder-liquid system.Dent Mater J. 1999 Sep;18(3):278-94
[26] Kakuta K, Urapepon S, Miyagawa Y, Ogura H, Suchatlampong C, Rittapai A. Development of metal-resin composite restorative materia. Part 1. Experimental composite using silver-tin alloy as filler and 4-META as coupling agent. Dent Mater J. 1999 Mar;18(1):1-10.
[27] Urapepon S, Kakuta K, Miyagawa Y, Ogura H, Suchatlampong C, Rittapai A.Development of metal-resin composite restorative material. Part 2. Effects of acid and heat treatments of silver-tin filler particles on flexural properties of metal-resin composite.Dent Mater J. 1999 Jun;18(2):144-54.
[28] Urapepon S, Kakuta K, Ogura H, Suchatlampong C, Rittapai A. Development of metal-resin composite restorative material. Part 3. Flexural properties and condensability of metal-resin composite using Ag-Sn irregular particles.Dent Mater J. 2000 Jun;19(2):186-95.
[29] Kakuta K, Urapepon S, Miyagawa Y, Ogura H, Yamanaka M, Suchatlampong C, Rittapai A.Development of metal-resin composite restorative material. Part 4. Flexural strength and flexural modulus of metal-resin composite using Ag-In alloy particles as filler. Dent Mater J. 2002 Jun;21(2):181-90
[30] Urapepon S, Kakuta K, Ogura H.Development of metal-resin composite restorative material. Part 5: Evaluation of the bonding between Ag-Sn particle and 4-META coupling agent of the metal-resin composite.Dent Mater J. 2003 Jun;22(2):137-45.
[31] Soma H, Miyagawa Y, Ogura H.Setting and flexural properties of metal-resin composite using Ag-Cu particles as filler and chemical accelerator. Dent Mater J. 2003 Dec;22(4):543-55.
[32] Tanimoto Y, Nishiwaki T, Nemoto K, Ben G.Effect of filler content on bending properties of dental composites: numerical simulation with the use of the finite-element method. J Biomed Mater Res B Appl Biomater. 2004 Oct 15;71(1):188-95.
[33] Antonucci JM, Liu DW, Skrtic D.Amorphous Calcium Phosphate Based Composites: Effect of Surfactants and Poly(ethylene oxide) on Filler and Composite Properties. J Dispers Sci Technol. 2007;28(5):819-824.
[34] Okazaki M, Ohmae H. Mechanical and biological properties of apatite composite resins.Biomaterials. 1988 Jul;9(4):345-8.
[35] Khaled SM, Miron RJ, Hamilton DW, Charpentier PA, Rizkalla AS.Reinforcement of resin based cement with titania nanotubes.. Dent Mater. 2010 Feb;26(2):169-78.

[36] St Germain H, Swartz ML, Phillips RW, Moore BK, Roberts TA.Properties of microfilled composite resins as influenced by filler content. J Dent Res. 1985 Feb;64(2):155-60.
[37] Braga RR, Cesar PF, Gonzaga CC. Mechanical properties of resin cements with different activation modes. J Oral Rehabil . 2002 Mar;29(3):257-62.
[38] Fujii K, Arikawa H, Kanie T, Hamano T, Nishi Y, Nagaoka E.Dynamic viscoelastic properties of commercial glass-fibre reinforced resin used for crowns and bridges.J Oral Rehabil. 2002 Sep;29(9):827-34
[39] Tanoue N, Mikami A, Atsuta M, Matsumura H. Effects of monomer composition and original filler content on filler loading in the resulting centrifuged composites.Dent Mater J. 2007 Jul;26(4):501-5.
[40] Yap AU, Lye KW, Sau CW. Surface characteristics of tooth-colored restoratives polished utilizing different polishing systems.Oper Dent. 1997 Nov-Dec;22(6):260-5.
[41] Xu HH, Moreau JL, Sun L, Chow LC. Strength and fluoride release characteristics of a calcium fluoride based dental nanocomposite. Biomaterials. 2008 Nov;29(32):4261-7.
[42] Anusavice KJ, Zhang NZ, Shen C..Effect of CaF2 content on rate of fluoride release from filled resins.. J Dent Res. 2005 May;84(5):440-4
[43] Xu HH, Moreau JL, Sun L, Chow LC. Novel CaF(2) nanocomposite with high strength and fluoride ion release. J Dent Res. 2010 Jul;89(7):739-45.
[44] Okamoto A, Sekiya K, Fukushima M, Iwaku M.In vivo wear pattern of experimental light-cured hybrid composite resins. Dent Mater J. 1993 Dec;12(2):225-32.
[45] Lang BR, Jaarda M, Wang RF. Filler particle size and composite resin classification systems. J Oral Rehabil. 1992 Nov;19(6):569-84.
[46] Sekiya K, Okamoto A, Fukushima M, Iwaku M.In vivo wear pattern of experimental composite resins containing different filler components. Dent Mater J. 1994 Jun;13(1):36-46
[47] Lim BS, Ferracane JL, Condon JR, Adey JD.Effect of filler fraction and filler surface treatment on wear of microfilled composites. Dent Mater. 2002 Jan;18(1):1-11.
[48] Yuasa S. Influences of composition on brush wear of composite resins. Influences of particle size and content of filler. Shika Zairyo Kikai. 1990 Jul;9(4):659-78
[49] Frazier KB, Rueggeberg FA, Mettenburg DJ. J Esthet Comparison of wear-resistance of Class V restorative materials.Dent. 1998;10(6):309-14.
[50] Munoz-Viveros CA. An advance in condensable composites. Compend Contin Educ Dent Suppl. 1999;(23):S3-5.
[51] Nunes TG, Pereira SG, Kalachandra S.Effect of treated filler loading on the photopolymerization inhibition and shrinkage of a dimethacrylate matrix. J Mater Sci Mater Med. 2008 May;19(5):1881-9.
[52] Leprince J, Palin WM, Mullier T, Devaux J, Vreven J, Leloup G. Investigating filler morphology and mechanical properties of new low-shrinkage resin composite types.J Oral Rehabil. 2010 May
[53] Razak AA, Harrison A.The effect of filler content and processing variables on dimensional accuracy of experimental composite inlay material. J Prosthet Dent. 1997 Apr;77(4):353-8.
[54] Gonçalves F, Kawano Y, Braga RR.Contraction stress related to composite inorganic content.Dent Mater. 2010 Jul;26(7):704-9.
[55] Poss SD. Using a new condensable composite for posterior restorations.Compend Contin Educ Dent Suppl. 1999;(23):S14-8.

[56] Taylor DF, Kalachandra S, Sankarapandian M, McGrath JE.Relationship between filler and matrix resin characteristics and the properties of uncured composite pastes.Biomaterials. 1998 Jan-Feb;19(1-3):197-204.
[57] Ikeda I, Otsuki M, Sadr A, Nomura T, Kishikawa R, Tagami J. Effect of filler content of flowable composites on resin-cavity interface. Dent Mater J. 2009 Nov;28(6):679-85.
[58] Yukitani W, Hasegawa T, Itoh K, Hisamitsu H, Wakumoto S.Marginal adaptation of dental composites containing prepolymerized filler.Oper Dent. 1997 Nov-Dec;22(6):242-8.
[59] Ellakwa A, Cho N, Lee IB.The effect of resin matrix composition on the polymerization shrinkage and rheological properties of experimental dental composites. Dent Mater. 2007 Oct;23(10):1229-35.
[60] Asmussen E, Peutzfeldt A.Influence of composition on rate of polymerization contraction of light-curing resin composites. Acta Odontol Scand. 2002 Jun;60(3):146-50.
[61] Rüttermann S, Dluzhevskaya I, Grosssteinbeck C, Raab WH, Janda R.Impact of replacing Bis-GMA and TEGDMA by other commercially available monomers on the properties of resin-based composites. Dent Mater. 2010 Apr;26(4):353-9.
[62] Azzopardi N, Moharamzadeh K, Wood DJ, Martin N, van Noort R. Effect of resin matrix composition on the translucency of experimental dental composite resins. Dent Mater. 2009 Dec;25(12):1564-8.
[63] Urabe H, Nomura Y, Shirai K, Yoshioka M, Shintani H..Effect of filler content and size to properties of composite resins on microwave curing. J Mater Sci Mater Med. 1999 Jun;10(6):375-8.
[64] Campbell PM, Johnston WM, O'Brien WJ Light scattering and gloss of an experimental quartz-filled composite. J Dent Res. 1986 Jun;65(6):892-4
[65] Yoshida Y, Shirai K, Shintani H, Okazaki M, Suzuki K, Van Meerbeek B.. Effect of presilanization filler decontamination on aesthetics and degradation resistance of resin composites. Dent Mater J. 2002 Dec;21(4):383-95
[66] Imamura S, Takahashi H, Hayakawa I, Loyaga-Rendon PG, Minakuchi S.. Effect of filler type and polishing on the discoloration of composite resin artificial teeth. Dent Mater J. 2008 Nov;27(6):802-8.
[67] Lee YK. Influence of scattering/absorption characteristics on the color of resin composites. Dent Mater. 2007 Jan;23(1):124-31.
[68] Yu B, Ahn JS, Lim JI, Lee YK. Influence of TiO2 nanoparticles on the optical properties of resin composites. Dent Mater. 2009 Sep;25(9):1142-7
[69] Chirdon WM, O'Brien WJ, Robertson RE. Diffuse reflectance of short-fiber-reinforced composites aligned by an electric field. Dent Mater. 2006 Jan;22(1):57-62.
[70] Nakamura T, Saito O, Mizuno M, Tanaka H. Changes in translucency and color of particulate filler composite resins. Int J Prosthodont. 2002 Sep-Oct;15(5):494-9.
[71] Lee YK. Influence of filler on the difference between the transmitted and reflected colors of experimental resin composites. Dent Mater. 2008 Sep;24(9):1243-7.
[72] Howard B, Wilson ND, Newman SM, Pfeifer CS, Stansbury JW. Relationships between conversion, temperature and optical properties during composite photopolymerization. Acta Biomater.2010 Jun;6(6):2053-9.
[73] Shortall AC, Palin WM, Burtscher P. Refractive index mismatch and monomer reactivity influence composite curing depth. J Dent Res. 2008 Jan;87(1):84-8.

Part 3

Applications of New Materials

22

Tuneable Composites Containing Magnetic Microwires

L. Panina[1], M. Ipatov[2], V. Zhukova[2], J. Gonzalez[2] and A. Zhukov[2,3]
[1]*School of Computing, Communications and Electronics, University of Plymouth*
[2]*Dpto. Fisica de Materiales, Fac. Quimicas, San Sebastian*
[3]*IKERBASQUE, Basque Foundation for Science, Bilbao*
[1]*United Kingdom*
[2,3]*Spain*

1. Introduction

This chapter provides a comprehensive review on tuneable electromagnetic properties of magnetic wire composites. In these systems, the microwave permittivity and permeability (response to the electric and magnetic field in the wave, respectively) can be controlled by a weak magnetic field, mechanical stress, and temperature. The underlying physics involves the combination of collective frequency dispersive effects and giant magnetoimpedance (GMI) effect in amorphous microwires. In particular, the emphasis is placed on specific magnetic structures in amorphous magnetic wires, which makes it possible to achieve high sensitivity of the surface impedance to external stimuli; modelling the microwave spectra in magnetic wire composites with different microstructure, and experimental results on tuneable impedance and scattering spectra.

In order to face future developments in microwave technology with applications in such vital areas as wireless communication, antenna engineering, non-destructive testing of civil structures, multifunctional structural materials, and biomedical engineering, the investigations into innovative designs of electromagnetic materials continue to be an important issue. The problems difficult to overcome are related with unbalanced electric and magnetic properties of conventional matters, the lack of needed relationships between the refractive index and wave impedance and between the quality factor and tunability ratio. Here, we are examining diluted composites with magnetic metallic wires that can have both effective permittivity ε_{ef} and permeability μ_{ef} at microwave frequencies. A special feature of these composites is that both parameters can demonstrate a strong tunability with respect to varying magnetic structure in wires with such external stimuli as magnetic field, mechanical load and heat. Furthermore, incorporating arrays of magnetic wires in fibre-reinforced polymer composites has also a potential to engineer materials with required structural and electromagnetic functionalities.

Large values of permittivity can be engineered utilising ferroelectric or conducting elements. The latter could be preferable since very large values of ε_{ef} are obtained for small volume concentrations and various frequency and spacial dispersions of permittivity are realised. Another advantageous feature of metallic composites is that due to a low concentration of the metallic phase it could be combined with other subsystems to obtain, for example, magnetic and structural integrity properties. Metallic wire systems as artificial dielectrics were recognized as early as in 1960s (Brown, 1960) and were used in such applications as beam shaping systems and broadband absorption systems. Recently arrays of continuous wires gained much attention as systems with negative real part of the effective permittivity to constitute the materials with left-handed properties by combining the wire arrays and ring resonators (Pendry et al, 1998; Smith et al, 2000). In composites with short-cut wires the length of which is comparable with the wavelength, the effective permittivity may have a resonance dispersion in the GHz range due to induced dipole moments of wires resonating at half wavelength condition (Lagarikov & Sarychev, 1996). This differs greatly from natural dielectrics, where the charge oscillation resonances become important only at optical frequencies.

The magnetic properties of composites materials could be originated by incorporating a ferromagnetic phase. A constraint here is commonly related to response-bandwidth product known as Snoek's relationship (Snoek, 1948). To avoid this limitation, it was proposed to design magnetic activity from conducting elements, which is based on magnetic moments of electric currents. However, the ferromagnetic-like behaviour in systems of current-loops can be realised in rather narrow frequency band. Yet, this approach allows the magnetic activity to exist at very high frequencies up to infrared and visible spectral bands (Panina et al, 2000). At frequencies of 1-10 GHz, ferromagnetic components in the form of thin-films often provide an optimal response due to their high magnetisation saturation, reduced demagnetisation effects and weak skin effect. Similar performance could be achieved utilising magnetic wires with special circumferential magnetic anisotropy. Then, this magnetic subsystem could be easily integrated with the conducting wire arrays generating the electrical properties. The difference between these two subsystems is that the magnetic one should be much dense. Here we consider the use of magnetic wires for both electric and magnetic subsystems. It is demonstrated, that the effective permittivity of magnetic wire arrays shows strongly tunable behaviour. Such composites could be of interest for reconfigurable microwave devices as well as for sensory materials (Reynet et al, 2002; Makhnovskiy et al, 2003; Panina et al, 2005; Makhnovskiy et al, 2006).

Adjustability of electromagnetic properties is important for many applications, especially in communication, defence and non-destructive testing. This will be highly needed in realization of reconfigurable local network environment, beam steering antennas, and microwave methods of remote sensing and control. Several methods were proposed based on biased ferroelectric, ferrite or magnetic composite substrates (Adenot et al, 2000; Yashchyshyn & Modelski, 2000) and reconfigurable resonant elements implementing active devices (Sievenpiper et al, 2002) or a system of micro actuators such as the divergent dipoles (Barlevy, 1999). These technologies each have its advantages and limitations such as high power consumption, low operational speed, limited frequency band and high cost. Here we introduce a relatively new technology of magnetic wire arrays to manipulate the collective electric response from composite systems. Therefore, the overall response will be decided by interplay between structural design and magnetic behaviour of individual wires.

In thin conducting wires the currents that are responsible for effective permittivity are constrained with the associated resonances determined by the geometrical parameters. The current resonances are damped due to the wire impedance which may increase greatly when the wire magnetisation is changed. This is known as giant magnetoimpedance (GMI) effect (Panina & Mohri, 1994; Beach, & Berkowicz, 1994). In soft magnetic amorphous wires subjected to an external magnetic field GMI is in the range of 100% even at frequencies of few GHz. Increase in magnetic losses results in increase in the relaxation parameter which determines the frequency dispersion of the effective permittivity. In the case of plasmonic wire arrays, this will result in considerable decrease in the absolute value of the permittivity and will enhance the wave propagation. Similarly, in cut-wire composites, the increase in relaxation broadens the permittivity dispersion which may even show transformation from resonance to relaxation behaviour. Therefore, in composites containing ferromagnetic wires exhibiting GMI effect at GHz frequencies the effective permittivity will depend on the wire magnetic properties via the corresponding dependence of its impedance. Applying a magnetic field H_{ex} larger than the magnetic anisotropy field H_K in wires which is just in the range of few Oersted (fraction of mTesla), rotates the magnetisation towards the axis, and strongly increases its impedance (for frequencies from MHz towards 10 GHz). Then the permittivity behaviour is damped. This will result in dramatic changes in the wave propagation through this material. Thus, in composites with $Co_{68}Fe_4Cr_3B_{14}Si_{11}$ glass coated amorphous wires a change in transmission spectra of nearly 15 dB by applying a field of just 5.8 Oe was reported (Makhnovsky et al, 2006). Furthermore, this concept was demonstrated by investigating magnetic field dependent permittivity of Co-based magnetic microwires and fibre-reinforced 913 E-glass prepreg (Peng et al, 2009).

Another benefit of using magnetic microwires is that it will be possible to engineer low density materials with relatively high values of the effective magnetic permeability originated from natural magnetic properties of the wires with a circumferential magnetic anisotropy. The magnetic field in the incident wave along the wire will generate substantial magnetic activity as it will be in the orthogonal position with respect to a static magnetisation. The demagnetising effects will not deteriorate the axial permeability as the ac magnetisation could lay in the tangential position to the wire surface. Integrating electric and magnetic wire arrays, composites with relatively large values of both ε_{ef} and μ_{ef} could be realised. For example, it will be possible to achieve a negative index of refraction. As comparing to other types of left handed materials, an enhanced performance in terms of tunability, simple internal structure, reduced losses and low cost is anticipated.

This chapter is organised as following. In Section 2, the analysis of the effective permittivity and permeability in magnetic-wire media - random or periodic arrays of ferromagnetic wires is given. Emphasis is placed on such effects as permittivity dispersion dependent on the wire magnetic structure, and tunable negative refraction. Section 3 deals with the science and technology of thin magnetic wires with a diameter 10-50 microns (microwires). Emphasis is placed on tailoring their magnetic structure to achieve high dynamic permeability tensor and high sensitivity of GMI at microwave frequencies (few GHz). Here as constituent elements of wire metamaterials we consider amorphous magnetic wires of CoFe-based alloy possessing a specific helical magnetic anisotropy and microwave GMI. In amorphous materials, the magnetoelastic effects determine the magnetic behaviour via the coupling between magnetostriction and stress frozen in during the fabrication or annealing treatments. Following this, we will consider magnetoimpedance in wires with a specific magnetic anisotropy. In Section 4, the discussion on the experimental methods for measurement of magnetic field-dependent effective permittivity in the GHz frequency band on the basis of free space techniques and results on tuneable electromagnetic response in the wire composites are given. Section 5 briefly discusses the possible applications.

2. Electromagnetic spectra of magnetic wire composites

Composites containing elongated metallic inclusions can be designed to have specific frequency spectra. In particular, composites with metallic wires can have effective permittivity of plasmonic type (continuous wires), resonant type (wire length is comparable with the wavelength) or relaxation type (wire length is much smaller than the wavelength). Typically, it is considered that the wire radius is small compared to the wavelength. Then, every wire can be described in terms of effective linear current referred to the wire axis, the polarization properties of which determine the effective permittivity. It is further assumed that the wires are ideally conductive; therefore, the field distribution inside them is ignored. This is justified when the skin effect is strong. Here we are interested in realizing the conditions when the losses may become relatively large. Therefore, the approximations utilizing infinite conductivity cannot be used here. In magnetic wires, if the skin effect is essential the loss parameter is enhanced by the wire dynamic permeability. This will make it possible to change the dispersion of the effective permittivity by changing the wire magnetic properties (Makhnovskiy et al, 2003). However, the skin effect should not be too strong when the relaxation is indeed small and the internal properties of constituent wires have little effect on the permittivity spectra. For electrical systems, two main configurations are of interest as shown in Fig.1: periodic arrays of continuous wires (Brown, 1960; Rotman, 1962; Pendry et al, 1998; Belov et al, 2002) and short-cut wires arranged randomly or periodically (Lagarikov & Sarychev, 1996 ; Liu et al, 2005) . In the first case, the effective permittivity is of a plasmonic type with the plasma frequency determined by the spacing between the wires b and the wire radius a. In the second case, the permittivity is of a resonance type in the frequency band within which the half wave length condition is realized: $f_{res} = c/2l\varepsilon_d$ where l is the wire length, c is the velocity of light and ε_d is the permittivity of matrix. The volume concentration of wires should be below 0.02%. To design composite with magnetic properties, it is needed to add a magnetic subsystem with much larger concentration (~5-10%) of magnetic wires which should be placed parallel to the magnetic field of the incident electromagnetic wave and perpendicular to the electrical subsystem.

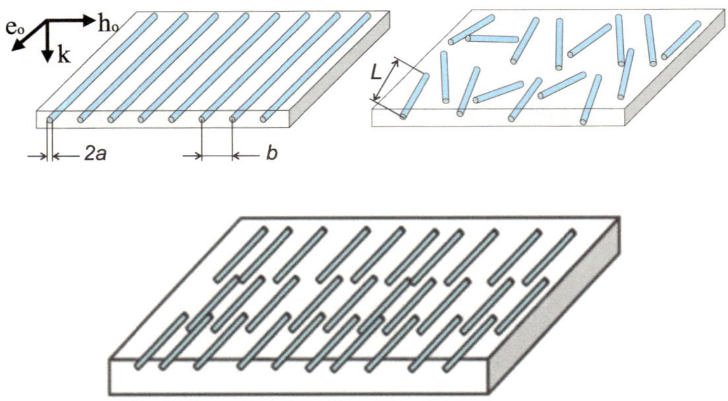

Fig. 1. Sketch of wire composites to engineer permittivity spectra. In periodic arrays, the electric field in the incident wave is parallel to the wires.

2.1 Electromagnetic field distribution in magnetic wires

Here we are interested in possible effects of the wire magnetic properties on the electromagnetic spectra of wire composites. Therefore, the electromagnetic field distribution inside magnetic wires has to be considered. The dynamic magnetic response of wires is characterized by the susceptibility tensor $\hat{\chi}$ which has the simplest form in the coordinate system with the z-axis parallel to the static magnetization $\boldsymbol{M}$ which in general has some angle θ with the wire axis as shown in Fig. 2.

$$\hat{\chi} = \begin{pmatrix} \chi_1 & -i\chi_a & 0 \\ i\chi_a & \chi_2 & 0 \\ 0 & 0 & 0 \end{pmatrix} \tag{1}$$

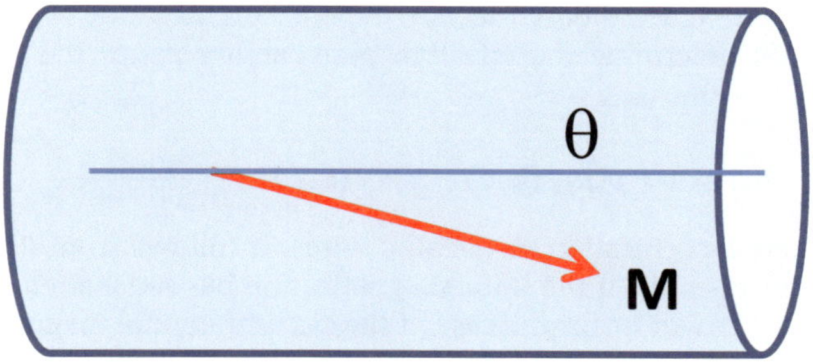

Fig. 2. Magnetic configuration in wires

In the quasi-static approximation, the Maxwell equations inside the wires can be written in terms of electric or magnetic fields. In cylindrical coordinates (r, φ, z) the magnetic induction $\boldsymbol{b}$ and magnetic field $\boldsymbol{h}$ are related as

$$b_\varphi = \mu_1 h_\varphi + \mu_3 h_z \,. \tag{2}$$

$$b_z = \mu_3 h_\varphi + \mu_2 h_z \,. \tag{3}$$

Here the parameters μ_i are related to the susceptibility tensor (1):

$$\mu_1 = 1 + 4\pi\cos^2\theta\,\chi,\;\; \mu_2 = 1 + 4\pi\sin^2\theta\,\chi,\;\; \mu_3 = -2\pi\sin 2\theta\,\chi. \tag{4}$$

$$\chi = \chi_2 - \frac{4\pi\chi_a^2}{1+4\pi\chi_1}. \tag{5}$$

The method of finding the solutions for (h_φ, h_z) for any frequencies is described in (Makhnovskiy et al, 2001). Here we are interested in the case of a moderate skin effect ($\beta = a/\delta < 1,\; \delta = c/\sqrt{2\pi\sigma\omega}, \sigma$ is the conductivity). In this case, neglecting terms proportional to the powers of β it is obtained

$$h_\varphi = h_\varphi(a)\frac{J_1(k_1 r)}{J_1(k_1 a)}, \qquad k_1^2 = \mu_1\frac{i4\pi\omega\sigma}{c^2}. \tag{6}$$

$$h_z = h_z(a)\frac{J_0(k_2 r)}{J_0(k_2 a)}, \qquad k_2^2 = \mu_2\frac{i4\pi\omega\sigma}{c^2}. \tag{7}$$

Using equations (6) and (7) it is possible to find the distribution of electric field $\boldsymbol{e}$ in magnetic wires, and hence, the surface impedance which relates the tangential components of electric $\boldsymbol{e}$ and magnetic $\boldsymbol{h}$ fields at the wire surface. For magnetic subsystems, equations (3) and (7) determine the effective permeability with respect to magnetic field parallel to the wires. In the approximation used, the longitudinal component of the surface impedance ς_{zz} is expressed as

$$\varsigma_{zz} = \frac{k_1 c}{4\pi\sigma} \frac{J_0(k_1 a)}{J_1(k_1 a)} . \tag{8}$$

It is possible to demonstrate that the surface impedance increases when the wire magnetization is rotated towards the axis (when the magnetization angle becomes zero and $\cos^2\theta = 1$). Certainly the effect of magnetic structure in wires on ς_{zz} will be essential if the dynamic permeability $\mu = 1 + 4\pi\chi$ differs from unity. Coupling between the high frequency impedance (8) and the magnetization direction depends on the value of μ.
Equations (3) and (8) determine the effective permeability μ_{ef} of the wire system in the direction parallel to the wire axis :

$$\mu_{ef} = 1 + p(\mu_2 f(k_2 a) - 1),\ f(k_2 a) = \frac{2J_1(k_2 r)}{J_0(k_2 a)(k_2 a)^2} . \tag{9}$$

Here p is the volume concentration of metallic wires. It follows from (9) that the effective permeability could be essential if the static magnetization has some angle with the wire axis. The largest values of μ_{ef} will be for the case of the circumferential magnetization ($\sin^2\theta = 1$ in equation (4)). Figure 3 shows the spectra of the permittivity parameter $\mu = 1 + 4\pi\chi$ for different values of the dc bias field H_{ex} applied along the wires. Relatively large values of H_{ex} are needed to cause substantial changes in the permeability at frequencies larger than 1 GHz ($H_{ex}/H_K > 3$, where H_K is the anisotropy field). On the other hand, the dc magnetization may be rotated from circular to axial direction by applying $H_{ex} \sim H_K$. Therefore, sensitive tuning will be realized if the external stimuli cause the reorientation of the magnetization. For circular magnetization ($H_{ex} \sim 0$), the real part of the permeability is about -20 for frequencies 1 -1.5 GHz. This suggests that it is possible to realize diluted magnetic composites ($p \sim 5\%$) with a negative effective permeability. Such wire system could be attractive for use as a component of left-handed metamaterials.

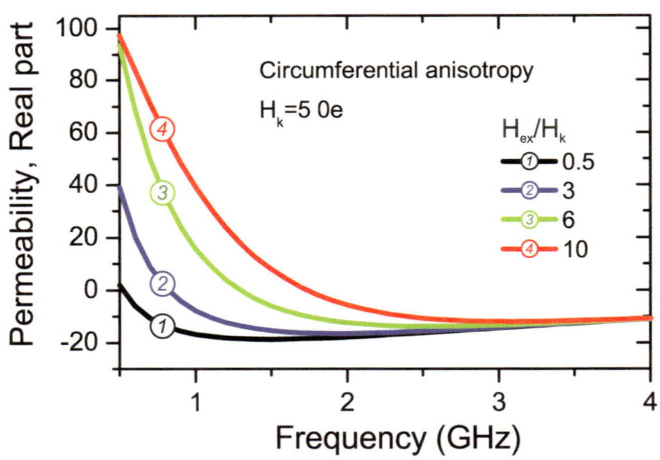

Fig. 3. Spectra of the circular permeability of a wire having a circumferential anisotropy with the axial field as a parameter. The parameters used for calculations are: $M = 500$ G, $H_K =$ 5 Oe, the anisotropy deviation from circular direction is 5 degrees.

2.1 Effective permittivity of magnetic wire systems

In a system of short-cut wires, the incident wave induces the dipole electrical moments $\mathcal{P}$ which can be found by solving the scattering problem. If the interactions between the induced dipoles are neglected (diluted system) the effective permittivity ε_{ef} is expressed as

$$\varepsilon_{ef} = \varepsilon_d + 4\pi p\alpha, \tag{10}$$

Where ε_d is the permittivity of matrix and α is the electric polarization ($\alpha = \mathcal{P}/Ve$, V is the wire volume). We are considering the case when the losses in metallic wires are important. This can be taken into account by solving the scattering problem with the impedance boundary conditions. The incident electrical field $\boldsymbol{e}$ creates a circular magnetic field $\boldsymbol{h}$ within the wire and at the surface of the wire these fields are related as

$$e_z = \varsigma_{zz}\, h_\varphi, \tag{11}$$

where the surface impedance ς_{zz} is determined by Eq.(9). For non-magnetic wires, ς_{zz} determines resistive losses and for magnetic wires this parameter also includes the relaxation due to magnetization dynamics as was discussed above. If the relaxation in the system is relatively large, the dispersion of ε_{ef} broadens and even can show a transition to a relaxation type. In composites with magnetic wires the relaxation can be controlled by the magnetic properties of wires and very large changes in the effective permittivity can be realized by applying external fields (magnetic or mechanical).

The wire polarization is expressed in a simple analytical form in the case of relatively strong skin effect when the radiation losses are neglected:

$$\alpha = \frac{1}{2\pi \ln(l/a)(\tilde{k}a)^2}\left(\frac{2}{\tilde{k}l}\tan(\frac{\tilde{k}l}{2}) - 1)\right) \tag{12}$$

Here $\tilde{k}$ is the renormalized wave number depending on the surface impedance in the following way:

$$\tilde{k} = k\left(1 + i\frac{\gamma_0}{\ln(l/a)}\right), \qquad k = \frac{\omega\sqrt{\varepsilon_d}}{c}. \tag{13}$$

$$\gamma_0 = \frac{c\varsigma_{zz}}{\omega a}\ . \tag{14}$$

It turns out that the renormalization of the wave number in (14) is essential in the case of a moderate skin effect, which is also consistent with the condition of relatively large losses.

In the case of composites with continuous wires, the dispersion of the effective permittivity corresponds to that for a diluted plasma:

$$\varepsilon_{ef} = 1 - \frac{\omega_p^2}{\omega^2(1+i\gamma)} \tag{15}$$

Different approaches to calculate ε_{ef} give slightly varying results for the plasma frequency, ω_p. Customarily, ω_p is written as

$$\omega_p^2 = \frac{2\pi c^2}{b^2\, lnb/a} \tag{16}$$

A rigorous approach allowing the determination of ε_{ef} was proposed in (Sarychev and Shalaev, 2002), which is based on the solution of the Maxwell equations in the elementary cell and the consequent homogenization procedure to find the averaged electric field and displacement. We extended the method for magnetic wires demonstrating that γ is given by (Panina, et al, 2011)

$$\gamma = \frac{\gamma_0}{lnb/a} \tag{17}$$

Therefore, for both types of magnetic wire composites the relaxation parameter in the effective permittivity spectra depends on the wire impedance. To better understand the dependence of the wire impedance on magnetic permeability, it is useful to express the surface impedance in the approximation of a strong skin effect:

$$\varsigma_{zz} = (1-i)\frac{\omega\delta}{2c}(\sqrt{\mu}cos^2\theta + sin^2\theta)\,. \tag{18}$$

Substituting (18) into (14) shows that γ_0 increases as a square root of the permeability. It is also seen that γ_0 depends on the static magnetisation angle. However, with increasing the frequency well beyond the frequency of the ferromagnetic resonance the permeability tends to be unity and the dependence on θ vanishes. Therefore, we demonstrated that the dispersion properties of permittivity in magnetic wire media depend on the wire internal magnetic structure following the magnetic behavior of the wire impedance. Combining the dispersion properties of wire media and GMI effect it is possible to actively tune the permittivity spectra of arrays of magnetic microwires by application of a small magnetic field and a stress which is now demonstrated in a number of experimental studies (as will be discussed in Section 4). However, to realize large and sensitive tuning requires the existence of very special magnetic structures in wires as that with a circular anisotropy. The methods of tailoring magnetic structure in amorphous wires will be discussed in Section 3. Here we give the results of the permittivity modeling made for amorphous Co-based wires with negative magnetostriction and nearly circular magnetic anisotropy which is presented in Fig. 4. It is seen that in the presence of the external bias field the dispersion region broadens since the losses are increased in high impedance state of the wires.

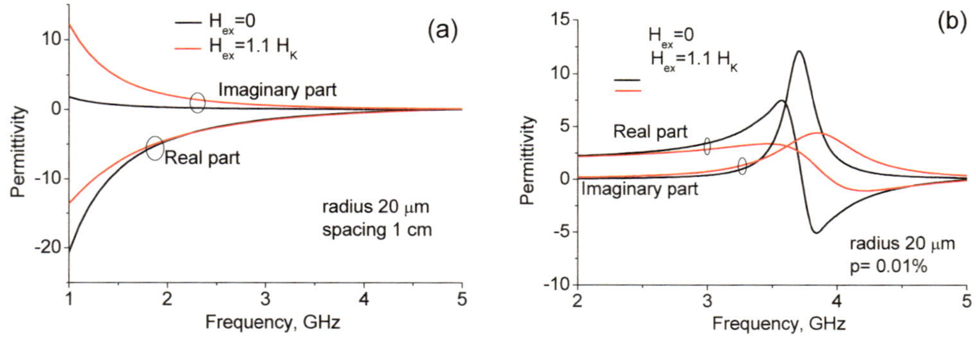

Fig. 4. Effective permittivity spectra in composites depicted in Fig.1 with the external field as a parameter. Modelling is performed for wires with a circumferential anisotropy (anisotropy field H_k=500A/m). The other parameters are: resistivity 130 μΩ·cm, magnetization 0.05T, wire radius 20 μm. For (a), *b*=1cm. For (b), *l*=4cm, *p*=0.01%.

3. Science and technology of thin magnetic wires

In this part, static and dynamic magnetic properties of amorphous magnetic wires in glass coating are reviewed in relation to their application in composites with electromagnetic functionalities.

3.1 Fabrication method of magnetic microwires

There are a number of methods of producing magnetic fibres and wires that may suit particular applications. Here we are discussing the one widely referred to as modified Taylor-Ulitovsky method or quenching-and-drawing method (Zhukov & Zhukova, 2010; Zhukov et al, 2004; Zhukov et al, 2000; Chiriac & Ovari, 1996), which is most suitable for composite technology. The method, as described elsewhere (Larin et al, 2002; Badinter et al, 1973; Ulitovski & Avernin 1964; Ulitovsky et al, 1960; Ulitovsky 1951; Taylor 1931; Taylor 1924), essentially consists of simultaneous drawing of the composite microwire (metallic nucleus inside the glass capillary) through the quenching liquid (water or oil) jet onto rotating bobbins, as schematically depicted in Fig. 5.

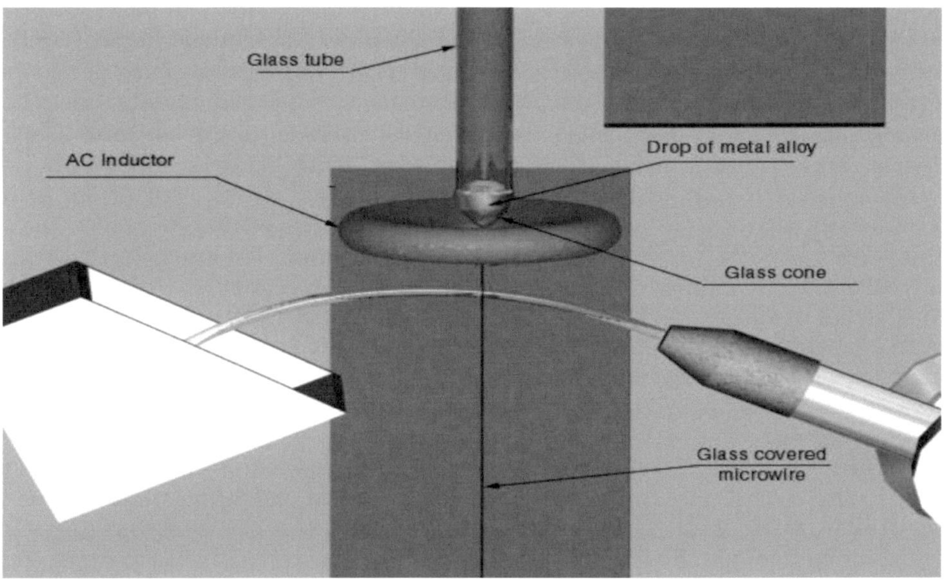

Fig. 5. Schematic drawing of microwire fabrication process by the Taylor–Ulitovsky method. Reprinted with permission from V. S. Larin et al., "Preparation and properties of glass-coated microwires", J. Magn. Magn. Mater. 249(1-2) pp.39-45 (2002). © 2002, Elsevier Science (Fig.1)

Magnetic properties of a microwire essentially determined by the structure which depends mainly on the cooling rate and the chemical composition of the alloy. Some advantages of this method of microwire fabrication important for composite technology (Larin et al, 2002) are: repeatability of microwire properties at mass-production; vast range of variation in parameters (geometrical and physical); fabrication of continuous long pieces of microwire up to 10 000 m/per bobbin; control and adjustment of geometrical parameters (inner core diameter and glass thickness) during the fabrication process. There could be some complexities related with the rapid quenching process of thin composite material subjected during the casting to the mechanical stress arising from different thermal expansion coefficients of solidifying metallic alloy and the glass coating. This potentially influences the reproducibility and homogeneity of properties along the microwire length. Another problem is a possibility of formation of the intermediate layer between the metallic nucleus and the glass coating because of the interaction between the alloy and the glass coating. The character of this interaction depends on chemical composition of the ingot as well as on type of glass used for the casting (Zhukov & Zhukova, 2010).

The other source of instability of properties of cast microwires is related with gas content inside the microwire. The sources of the gas are: the atmosphere, the gas impurities in the alloy and the glass. Some content of oxygen and/or hydrogen (in the range of 5 $cm^3/100$ g) and even nitrogen has been detected. Gaseous precipitations can cause the metallic nucleus deformation and cracks. Chemical reactions of hydrogen with the oxides of the metals can result in appearance of water bubbles inside the metallic nucleus. The electro-dynamic interaction determining the shape of the molten ingot and temperature regimes significantly affects the casting process and should be also taken into account. In general, the thermal conditions affect greatly the microwire properties, in particular, thermal phenomena inside the molten ingot under effect of electromagnetic field of the inductor and thermal conditions related with solidification of cast microwire. Much effort has been made to determine the temperature distribution inside the ingot and its dependence on the metallic bar dimensions. It was found that for Cu, increasing the diameter of the bar from 2 to 6 mm resulted in decrease in temperature inside the ingot by 10-35% (Badinter et al, 1973).

The method allows pure metals such as Ag, Au, Cu, Ni, Sn, Pb, Pt to be used. Semiconductors such as Ge, Si, or Bi can also be cast. Then, it is possible to process the alloys of these metals, but not over the whole range of compositions. For example, the content of Cr in Ni and Co based alloys can be up to 20%, the content of Mo, W, V in Ni, Co, and Fe based alloys can be up to 8 - 10 at. %. These limitations are related to the melting temperature, which must not be too high. On the other hand, rare earth metals and metals which react with the glass and atmosphere (Ti, Cr, Mo, W, Nb, Al, Na, La, Nd) are not suitable for this technology (Zhukov & Zhukova, 2010; Badinter et al, 1973).

Our prime interest here is the fabrication of magnetic microwires with soft magnetic characteristics for which Fe, Ni and Co are the main elements in alloys. Soft magnetic properties are typical for amorphous state. To prepare amorphous microwires such elements as Si, B, C, and Al are added to enable the amorphicity of the alloys. Mixed crystalline-amorphous and nanocrystalline-amorphous structure can be also obtained by thermal annealing of initially amorphous microwires (Vazquez et al, 2011; García et al 2009; Zhukov et al, 2007; Dudec et al, 2006) as well as in-prepared state using specially designed mixed compositions like (FeSiB)-Cu, (CoFeSiB)-Cu. These compositions exhibit unusual magnetic properties with irregular hysteresis loops. Additionally, recently novel magnetically soft compositions with small additions of Cr based on Ni with ferromagnetic-paramagnetic phase transitions near room temperature have been introduced (Zhukova et al, 2009; Zhukova et al, 2006).

Metastable supersaturated solid solutions can be obtained for immiscible metal systems in a solid and even liquid state, such as Cu-Co, Cu-Co–Ni, Ag-Fe, Ag-Co. As-cast microwire can exhibit a structure of a supersaturated solid solution of transition metals in Cu or Ag. After suitable thermal treatments small single domain particles of Co or Fe can precipitate. Recently giant magnetoresistance (GMR) effect attributed to the scattering of the electrons on grain boundaries between ferromagnetic grains and paramagnetic matrix has been observed in such compositions (Zhukova et al, 2009; Zhukov et al, 2004).

3.2 Magnetic properties of amorphous microwires

It is worth mentioning, that studies of magnetic properties of amorphous glass coated microwires started even in 70-th (Kraus et al, 1976), although they have been limited to Fe-Ni compositions, measurements of hysteretic properties and ferromagnetic resonance (FMR). Recently these tiny glass-coated ferromagnetic microwires newly attracted considerable attention mainly due to a number of unusual magnetic properties and their potential applications in sensors (Zhukova et al, 2009; Mohri & Honkura, 2007, Ripka, 2001, Vazquez & Hernando, 1996) and multifunctional composites (Qin et al, 2010; Panina 2009; Phan & Peng, 2008).

Generally, the magnetic properties including the shape of hysteresis loops of amorphous microwires depend on the composition of metallic nucleus as well as on the composition and thickness of glass coating. This can be illustrated by hysteresis loops in Figs. 6 & 7. The hysteresis loops of three main groups of amorphous microwires: Fe-rich, Co-rich and Co-Fe-rich with positive, negative and vanishing magnetostriction constant, respectively, show very different magnetisation reversal process. Figure 6 shows how the shape of hysteresis loops changes from rectangular typical of amorphous Fe-rich compositions to inclined typical of Co-rich compositions. The magnetisation loops of amorphous microwires with the same composition ($Co_{67}Fe_{3.85}Ni_{1.45}B_{11.5}Si_{14.5}Mo_{1.7}$ in Fig. 7) but with different ρ-ratio defined as $\rho = d/D$ (d and D are the metallic nucleus diameter and total microwire diameter, respectively) also demonstrate significant difference in anisotropy fields. Comparing the hysteresis loops in Fig.6b and Fig.6c, it can be concluded that the microwires with vanishing magnetostriction show very soft magnetic behaviour with the anisotropy field H_K being in the range of 100-400 A/m, where as the anisotropy field of Co-based wires with negative magnetostriction is about 8000 A/m. The value of H_K can be tuned by changing ρ - ratio. Therefore, magnetic microwires demonstrate large variety of magnetic behaviours which is important for sensing applications.

Strong dependence of the hysteresis behaviour on the value and sign of the magnetostriction constant, and on the glass thickness is attributed to the dominant effect of the magnetoelastic anisotropy K_{me} in the absence of the crystalline anisotropy:

$$K_{me} = \frac{3}{2}\lambda_s \sigma_i (\boldsymbol{n}_\sigma \cdot \boldsymbol{n}_m)\,. \tag{19}$$

In (19), λ_s is the saturation magnetostriction, σ_i is the internal stress, $\boldsymbol{n}_\sigma$ and $\boldsymbol{n}_m$ are unit vectors along the stress and magnetization, respectively. The magnetostriction constant depends mostly on the chemical composition and is as small as 10^{-7} in amorphous Fe-Co based alloys with Co/Fe content about 70/5 (Zhukova et al, 2009; Fujimori et al, 1976). The estimated values of the internal stresses in these glass coated microwires arising from the difference in the thermal expansion coefficients of metal and glass are of the order of 100-1000 MPa, depending strongly on the parameter ρ and generally increasing with the increase in the glass coating thickness (Vazquez et al, 2011; Zhukova et al, 2009; Chiriac & Ovari, 1996).

This allows tailoring the magnetic properties of glass-coated microwires through the change of magnetic anisotropy by tailoring the internal stresses (by using thermal treatment, chemical etching, etc.) and chemical composition of the metallic nucleus. As a confirmation, the investigations of the effect of the progressive elimination of the glass insulating coating by chemical etching on magnetic properties showed considerable improvements in soft magnetic behavior as the thickness of glass coating has been diminished (Garcia Prieto et al, 2000; Catalan et al, 1997).

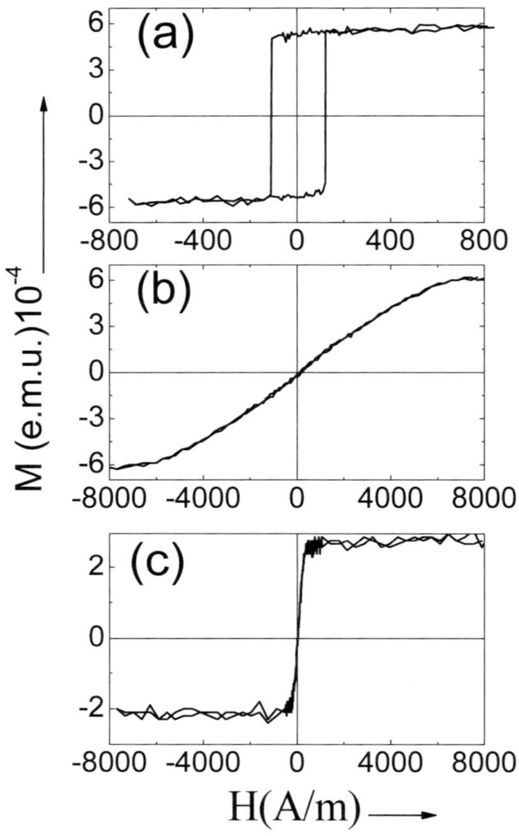

Fig. 6. Hysteresis loops of thin glass coated Fe-rich (a), C-rich (b) and Fe-Co-rich (c) microwires. Reprinted with permission from A. Zhukov et al., J. Mater. Res. 15, 2107 (2000). © 2000, Materials Research Society (Fig. 1).

Two types of microwires distinguished by the sign of the magnetostriction find quite different applications. Bistable magnetization reversal at $\lambda_s > 0$ (Fe-rich alloys) with rectangular hysteresis loops is of great interest for magnetic surveillance (Zhukova et al, 2009; Vazquez & Hernando, 1996). Possessing strongly non-linear magnetic properties, such wires generate specific high frequency harmonics, detection of which may provide information on the state of environment. Therefore, these wires can be used as local non-contact sensors of stress and temperature within composite material (Ong & Grimes, 2002). However, our prime interest here is related with wires made of nearly-zero magnetostriction alloys with good magnetic softness (Co-rich compositions). These wires exhibit almost linear and high magnetic permeability, good high frequency properties and giant magnetoimpedance (GMI) effect . Large number of magnetic sensors was proposed based on these properties alone, and in particular, based on GMI effect. Typically, for magnetic sensor applications, the interrogation frequencies are in tens MHz range. Realizing GMI at GHz frequencies, the wires can be used as embedded sensors within composite matrix. Therefore, the concept of sensory and tunable microwave composites based on magnetic microwires was put forward.

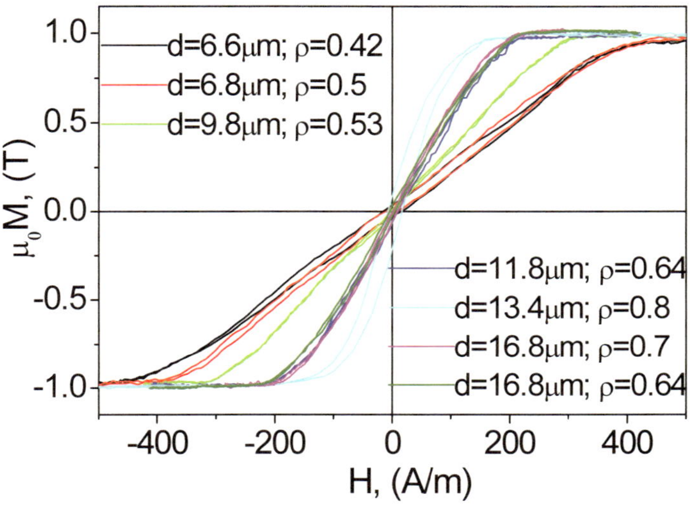

Fig. 7. Hysteresis loops of $Co_{67.1}Fe_{3.8}Ni_{1.4}Si_{14.5}B_{11.5}Mo_{1.7}$ microwires with different geometry. Reprinted with permission from A. Zhukov et al., J. Magn. Magn. Mater. 321 (2009) 822-825 Copyright Elsevier B.V (2008) (Fig. 1).

3.3 Giant magneto-impedance effect and enhanced magnetic softness

Since its discovery in 1994 (Panina &Mohri, 1994; Beach & Berkowicz, 1994) the GMI effect has become a topic of great interest in the field of applied magnetism owing to the large sensitivity of the total impedance to the applied DC field at low field magnitudes and high frequencies. Vast literature exists on this topic, and here we mention just few more relevant to this review (Zhukova et al, 2009; Mohri & Honkura, 2007; Knobel & Pirota 2002). For practical application, GMI is characterized by the relative change in impedance $\Delta Z/Z$:

$$\frac{\Delta Z}{Z} = \frac{Z(H_{ex}) - Z(H_{max})}{Z(H_{max})} \tag{20}$$

where H_{max} is the value of the external magnetic field where the impedance has a maximum which typically does not exceed few kA/m. In GMI experiments, H_{ex} is usually supplied by a long solenoid or Helholtz coils. It has been recognized that the sensitivity of the impedance change occurs due to sharp dependence of the circular permeability on axial magnetic field at a condition of a strong skin effect (see Eq.(18)). Improvements in magnetic softness allowed the GMI ratio $\Delta Z/Z$ to be increased up to 600% at few MHz in glass coated wires (Zhukova et al, 2009; Zhukova et al, 2002), as shown in Fig. 8.

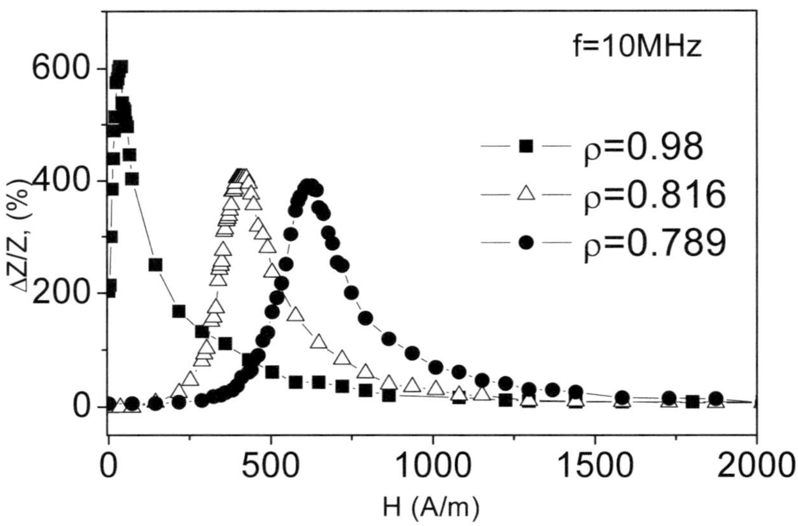

Fig. 8. Axial field dependence of $\Delta Z/Z$ measured at $f = 10$ MHz and current magnitude of 0.75 mA in microwire of composition $Co_{67}Fe_{3.85}Ni_{1.45}B_{11.5}Si_{14.5}Mo_{1.7}$ with ρ as a parameter. Reprinted with permission from V. Zhukova et al. "Optimization of Giant Magnetoimpedance in Co-Rich Amorphous Microwires" IEEE Trans. Magn., V 38, 5 part I, 2002 pp. 3090-3092. Copyright IEEE Magnetic Society (2002). (Fig. 4)

Additionally, stress-impedance (SI) and torsion impedance (TI) effects showing high sensitivity of the impedance to the applied stress with a strain gauge factor of 2000-4000 have been found in amorphous wires (Zhukova et al, 2009; Zhukov, 2006; Panina et al, 2005; Tejedor et al, 2001; Shen et al, 1997). For MI sensor applications, a CMOC IC circuitry with pulse current operation has been developed (Zhukov et al, 2008; Sandacci et al, 2004; Mohri et al, 2002). As a result, the GMI and SI sensors with the CMOC IC circuitry with advantageous features comparing with conventional magnetic sensors have been developed by different companies (Zhukova et al, 2009; Vazquez et al, 2011; Mohri et al, 2001; Honkura, 2002). Main applications of GMI effect are related with the detection of the magnetic fields, small weights, vibrations, acceleration, and recently, in microwave sensory and tunable composites.

Enhanced GMI ratios in amorphous wires were explained by a specific domain structure existing in negative magnetostrictive amorphous wires (Makhnovsky et al, 2001; Panina &Mohri, 1994; Beach & Berkowicz, 1994). Negative magnetostriction coupled with tensile stress creates alternative left and right handed circular domains in the outer sheath of wires. Such domain configuration was observed in wires produced by various method including in-rotating water developed by Unitika LTD (Takajo at el, 1993) and in glass-coated microwires (Chizhik et al, 2004; Chizhik et al, 2001). Such domain structure preserves even in wires with very small but negative magnetostriction (Vazquez et al, 2011; Zhukova et al, 2009; Chiriac & Ovari, 1996) and they demonstrate the highest sensitivity of GMI.

Strong influence of the internal stress on the magnetization behavior suggests that the magnetic structure as well as GMI can be tailored through the heat treatment using conventional annealing (CA), magnetic field annealing (MFA), or even stress annealing (SA)

(Zhukov & Zhukova, 2010; Zhukova et al, 2009; Garcia Prieto et al, 2000; Zhukov, 2006). As expected, the performed heat treatment strongly affects both the hysteresis loops and the GMI characteristics. Figure 9 shows, that both CA and MFA annealing results in considerable change in hysteresis loops. Soft magnetic behavior is improved due to stress relaxation during annealing.

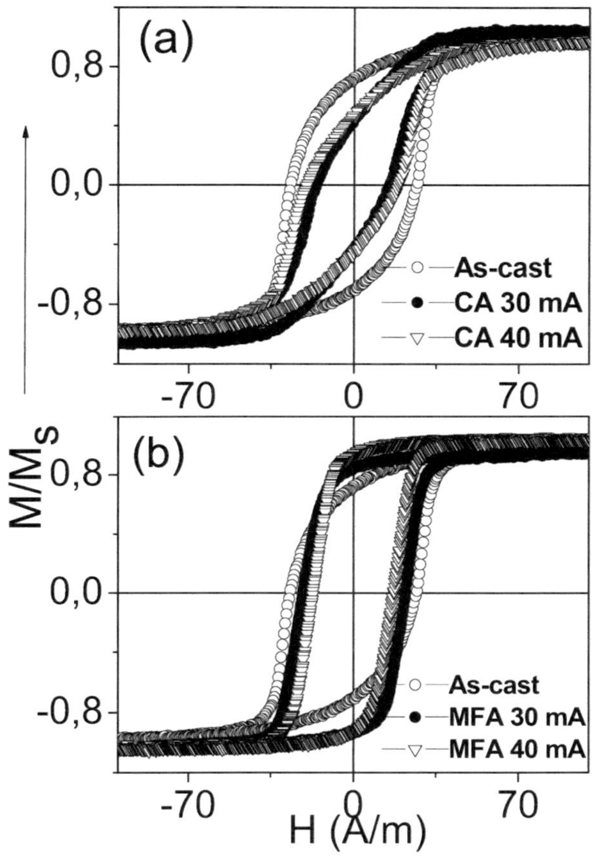

Fig. 9. Effect of CA (a) and MFA (b) treatments on bulk hysteresis loops of $Co_{67}Fe_{3.85}Ni_{1.45}B_{11.5}Si_{14.5}Mo_{1.7}$ microwires. Reprinted with permission from V.Zhukova et al., "Development of Ultra-Thin Glass-Coated Amorphous Microwires for High Frequency Magnetic Sensors Applications" Open Materials Science Reviews, 2007, 1, 1-12. Copyright 2007 Bentham Science Publishers Ltd. (Fig. 7).

Figure 10 shows the plots of relative change in impedance vs. field measured for as-prepared and annealed $Co_{67}Fe_{3.85}Ni_{1.45}B_{11.5}Si_{14.5}Mo_{1.7}$ microwires subjected to Joule heating for different annealing time (annealing current 30 mA and 40 mA). As can be seen, the value of $\Delta Z/Z$ strongly decreases at zero field resulting in much larger impedance change ratios due to annealing. However, the value of H_{max} also increases due to annealing.

One of the most important effects occurring due to stress annealing is enhancement of stress sensitivity of magnetic properties and stress- impedance (Zhukov & Zhukova, 2010;

Zhukova et al, 2009; Garcia Prieto et al, 2000; Zhukov, 2006). In particularly, it was demonstrated, that stress annealing, performed at certain annealing conditions results in induction of stress-sensitive transverse magnetic anisotropy and observation of significant (up to 60%) stress-impedance effect (Zhukov, 2006; Zhukova et al, 2003), as shown in Fig. 11. This result is of special interest for developing stress-sensitive composites with the use of magnetic microwires. The origin of this creep annealing induced anisotropy has been attributed to redistribution of the residual stresses during the stress annealing which results in drastic decrease in the longitudinal stress component and even in the appearance of the compressive longitudinal stresses (so-called "back stresses").

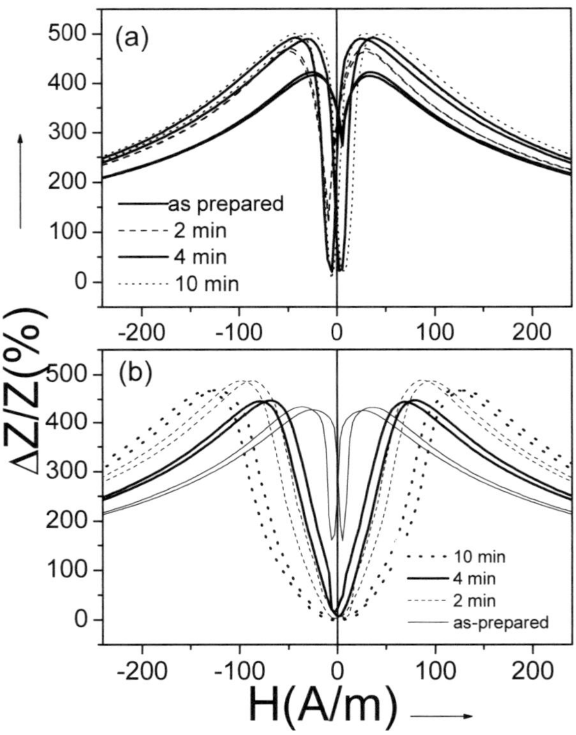

Fig. 10. Relative impedance change plots vs. external field measured at $f = 30$ MHz and current of 1 mA in microwire subjected to CA annealing at 30 mA (a) and at 40 mA (b). Reprinted with permission from A. Zhukov "Recent research on magnetic properties of glass-coated microwires" J. Magn. and Magn., Mater. 294 (2005) 182-192. Copyright Elsevier (2005). (Fig. 8).

Typically, the Curie temperature of Fe and Co-rich amorphous microwires is about 300-400 °C. Additions of Ni and Cr in the alloys make it possible to substantially decrease the Curie temperature (Trémolet de Lacheisserie, et al 2005, Zhukova et al 2006, Zhukova et al 2007). For example, the Curie temperature between 75 and 90 °C was reported for microwires with composition $Co_{60.51}Fe_{3.99}Cr_{12.13}B_{13.53}Si_{9.84}$ and $Co_{23.67}Fe_{7.14}Ni_{43.08}B_{13.85}Si_{12.26}$. This suggest a potential to develop magnetically soft microwires showing large temperature dependence of magnetization, anisotropy, magnetic permeability etc. (Fig.12) and, hence, GMI effect. Then, the microwires would be suitable for remote temperature detection in the range of moderate temperatures from room temperature to about 400 °C.

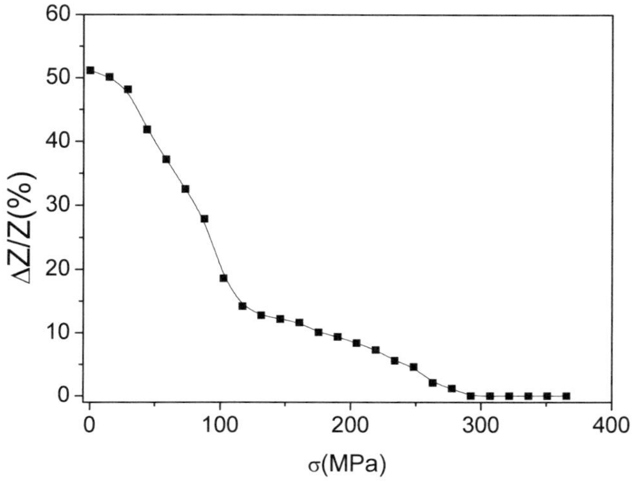

Fig. 11. Stress impedance effect in stress annealed $Fe_{74}B_{13}Si_{11}C_2$ glass-coated microwire measured at 10 MHz and driving current amplitude of 2 mA. The annealing conditions are: 468 MPa at 275°C for 0.5h for. Reprinted with permission from A. Zhukov et al. "Design of the Magnetic Properties of Fe-Rich, Glass-Coated Microwires for Technical Applications" Adv. Funct. Mater. 2006, 16, 675–680. Copyright WILEY-VCH (2006). (Fig. 6)

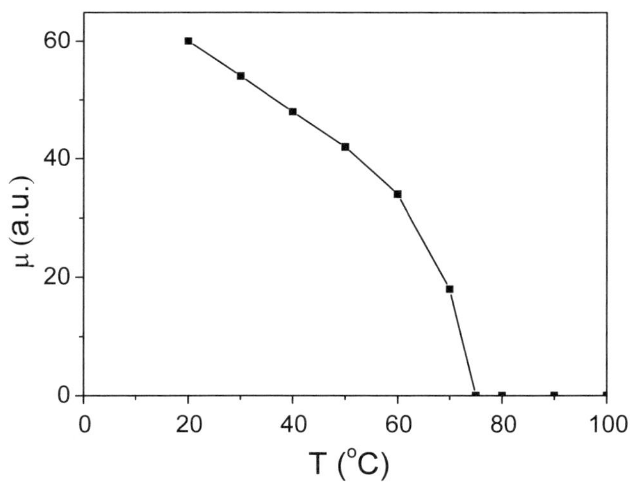

Fig. 12. Temperature dependence of permeability (given in arbitrary units) measured in $Co_{60.51}Fe_{3.99}Cr_{12.13}B_{13.53}Si_{9.84}$ microwire. Reprinted with permission from Ref. [193], V. Zhukova et al. "Development of thin microwires with low Curie temperature for temperature sensors applications" Sensors and Actuators B 126 (2007) 318–323 Copyright Elsevier (2007). (Fig. 4)

The developed magnetic field, stress, and temperature -sensitive microwires have been proposed for completely new range of applications as constituent elements of wire-composites for tunable microwave systems, and non-destructive remote control of stress, strain and temperature (Panina et al, 2011; Makhnovskiy et al, 2006; Makhnovskiy & Panina, 2005). The wires can be regarded as embedded sensors and their impedance sensitive to the wire magnetic structure will be responsible for producing a controlled microwave dielectric response, as explained in Section 2. This will be demonstrated in the next Section, where the experimental results on the microwave scattering spectra of magnetic wire composites are discussed.

4. Scattering properties of glass-coated magnetic wire composites.

4.1 Free space experimental method for composite characterization

The material parameters in microwaves frequencies usually are found from the measurement of the reflection and/or transmission coefficients from which the complex permittivity and permeability are calculated. The measurement methods can be divided in two categories: (i) transmission line methods (coaxial lines probes, rectangular waveguides, cavity resonators) and (ii) antenna techniques in free space. The methods in the first category require cutting a piece of a sample to be placed inside the transmission line or cavity making a close contact with the probe. The transmission line methods work best for homogeneous materials that can be precisely machined to fit inside the sample holder.

At microwaves, the measurement of the effective permittivity of composite materials with the inhomogeneity scale comparable with the wavelength requires large sample dimensions. In this case, the method of spot localized measurement area, such as conventional coaxial line and waveguide methods cannot be used. A free-space method is more appropriate. Generally, it is used to characterize large flat solid materials, although granular and powdered materials can also be measured in a fixture.

Free-space techniques for material property measurements have several advantages (Chen et al, 2004). Firstly, materials such as ceramics and composites are inhomogeneous due to variations in manufacturing processes. Because of inhomogeneities, the unwanted higher-order modes can be excited at an air-dielectric interface in hollow metallic waveguides, while this problem does not exist in free-space measurement. Secondly, the measurements using free-space techniques are non-destructive and contactless, so free-space methods can be used to measure samples under special conditions, such as high temperature. Thirdly, in hollow metallic waveguide methods, it is necessary to machine the sample so as to fit the waveguide cross section with negligible air gaps. This requirement limits the accuracy of measurements for materials that cannot be machined precisely; in free-space method, this problem does not exist. Finally, waveguides have a rather narrow operating frequency range. Therefore, to characterized material in a wide frequency range, a number of waveguides is required. Moreover, every waveguide, having different cross-section, will required preparation of separate sample.

Further we consider the free space method for measurement of the electromagnetic parameter of the composites as the most suitable for both laboratory investigation and *in situ* non-destructive testing and remote structural health monitoring. In free space method, materials are placed between antennas for a non-contacting measurement allowing much flexibility in studying materials under different conditions such as high temperatures and hostile environments. A key component of any free space system is antenna that is a

transition element between transmission line and free space radiating and/or receiving the electromagnetic energy into/from free space.
The experimental setup for the reflection/transmission microwave free-space measurements basically consists of vector network analyzer (VNA), a pair of broadband horn antennas and an anechoic chamber as shown in Fig. 13. A composite sample is placed in the middle of the chamber with the wire orientation along the electric-field of the incident electromagnetic wave. The desired frequency range, in which the scattering parameters will be investigated, determines the requirements to the operating frequency of the VNA, antennas and to the chamber size (distance between antennas and sample). The lens can be applied to focus the radiation pattern and minimize the effect of sample boundaries and measuring environment. It is essential to place the sample outside the reactive near-field region where the wave is not polarized and the electromagnetic interaction between the sample and the antenna can arise. The reactive near-field terminates at the distance of the order of wavelength λ from antennas.

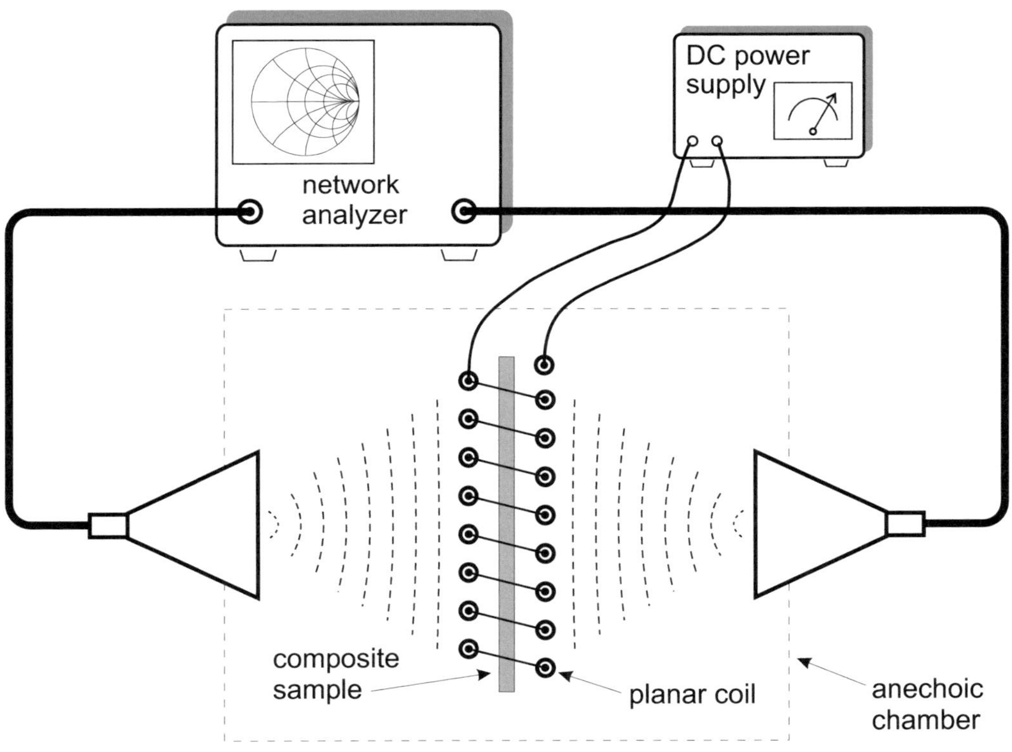

Fig. 13. Sketch of the free space microwave measurement setup.

Free space method imposes a limit on the minimal sample size. If the sample size is much smaller than the wavelength, the response of the sample to electromagnetic waves will be similar to those of a particle object. To achieve convincing results, the size of the sample should be larger than the wavelength of the electromagnetic wave. To further minimize the effects of the scatterings from the sample boundary, the sample size should be at least twice larger than the wavelength (Chen et al, 2004). Therefore if the lowest measurement frequency is 1 GHz, then the sample size should be $2\lambda = 60$ cm.

The free space technique requires precise calibration. The Thru-Rreflect-Line (TRL) and Thru-Reflect-Match (TRM) calibration techniques, that were commonly used until recently, are being widely replaced with the Gated Reflect Line (GRL) calibration (Bartley & Begley, 2005). The GRL calibration, based on the time domain gating, allows enhancing the calibration accuracy and elimination of the the need for expensive spot focusing antennas and micro positioning fixturing. Time-domain technique is important not only for calibration of the free-space measurement path but also during the measurements as it makes it possible to effectively eliminate the effects of multipath reflection to which the measurements in free-space are subjected. The main source of the reflection is the inevitable mismatch between the antenna and free space. The other error sources such as reflection from chamber's walls and noise could be also essential. The time domain procedure "gates out" these error terms and also reduces the requirements for the quality of the anechoic chamber or even allowing conducting the measurements without the chamber.

The free space setup, shown in Fig.13 is applied for the wire composite characterisation. It consists of Agilent 0.01 - 20 GHz two port VNA with time domain option, two broadband horn antennas with the operating range 0.9 - 17.0 GHz and the anechoic chamber with dimensions 80 x 80 x 80 cm^3 covered inside with a microwave absorber. The composite sheets are placed at a distance of 40 cm from each horn antenna, appearing in the radiating near-field region in the whole range of operating frequencies. 85071E Material Measurement Software (Agilent) and "Reflection/Transmission Epsilon Fast Model" can be used for calculating the complex permittivity of the composites from the experimental S-parameters.

To study the influence of magnetic field on the dispersion characteristics of the composite samples a special planar magnetic coil was constructed as described in (Makhnovskiy et all 2006). A thin planar composite sample is placed inside the magnetic coil so that the microwires were along the direction of magnetic field. The coil having the field coefficient 90 [A/m]/A creates a homogeneous along the sample surface DC magnetic field. A 35 Ampere Agilent 6674A DC power supply is used to feed the coil and permits to reach the magnetic field as high as 3000 A/m with resolution below 1 A/m. The coil turns are set perpendicular to the electrical field so there is no effect of the coil on scattering.

In a simple way, the wire composites can be prepared by gluing them on paper to form wire-lattices of needed dimensions. Firstly, 1D wire-lattices of plasmonic type with continuous wires (as shown in Fig. 1a) are arranged. Composites with short- wire pieces forming ordered electrical dipoles (seen in Fig. 1c) could be obtained by cutting the continuous wire arrays in stripes of different size to be able to change the dipole resonance frequencies. The wires used for these experiments were glass coated microwires of the composition $Co_{66}Fe_{3.5}B_{16}Si_{11}Cr_{3.5}$, with magnetic core diameter of 20 microns. These wires have small but negative magnetostriction and show GMI in the range of 100% at few GHz (Panina et al, 2011).

4.2 Tunable scattering spectra

Figure 14 shows the spectra of the reflection R and transmission T for composites with continuous wires having spacing $b = 1$cm with the magnetic field H_{ex} as a parameter. The relative change of T (Figure of Merit $FOM = (T(0) - T(H_{max})/T(0))$ is about 25% at 1.8 GHz while the phase of transmission shifts about 40 degrees with the field change from 0 to 500 A/m. Figure 15 shows the absorption parameter $A = 1 - R^2 - T^2$, which changes by 4 times from 10 % to 40 %.

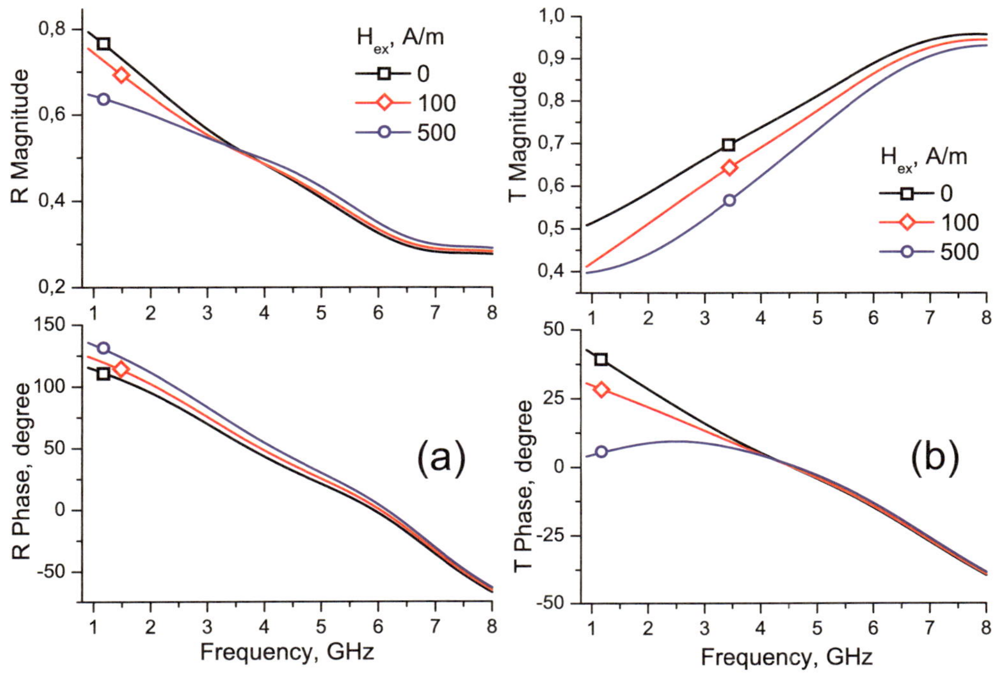

Fig. 14. Spectra of R and T for composites with continuous wire arrays with H_{ex} as a parameter.

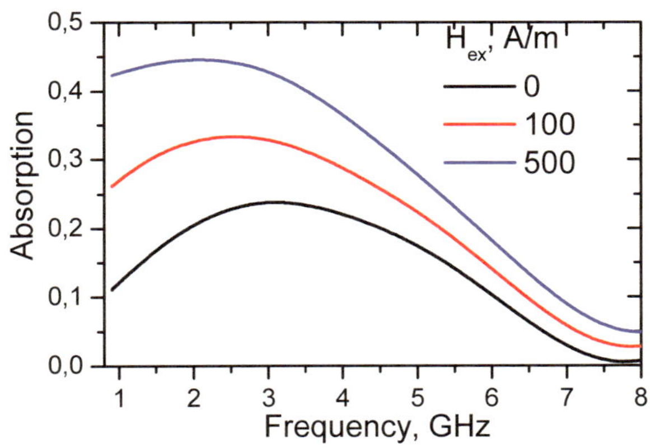

Fig. 15. Absorption spectra for composites with continuous wire arrays H_{ex} as a parameter.

The permittivity spectra shown in Fig. 16 were calculated from R and T coefficients. The effective thickness was taken equal to the lattice period b although the real composites thickness is much smaller and is defined by the host matrix thickness. The real part of ε_{ef} has a negative value below the plasma frequency which is equal to 4.7 GHz. Application of the field decreases the value of the real part (compare with theoretical plots shown in Fig. 4) as the wire impedance increases and so does the relaxation. The imaginary part of the permittivity, that directly demonstrates the losses or absorption, increases with the field. This results in reduced transmitted and reflected signals.

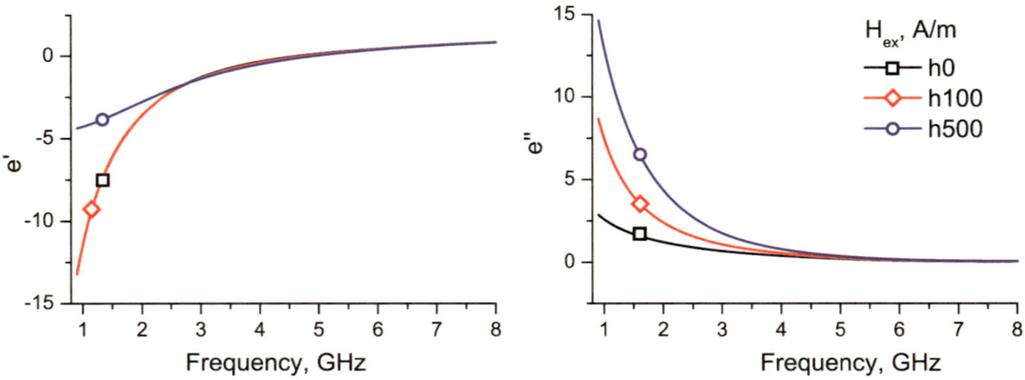

Fig. 16. Effective permittivity spectra for composites with continuous wire arrays with H_{ex} as a parameter.

The R and T spectra for the cut-wire composites with different wire length l of 40, 20 and 10 mm are shown in Fig. 17. The transmission spectra have a deep minimum near a resonance

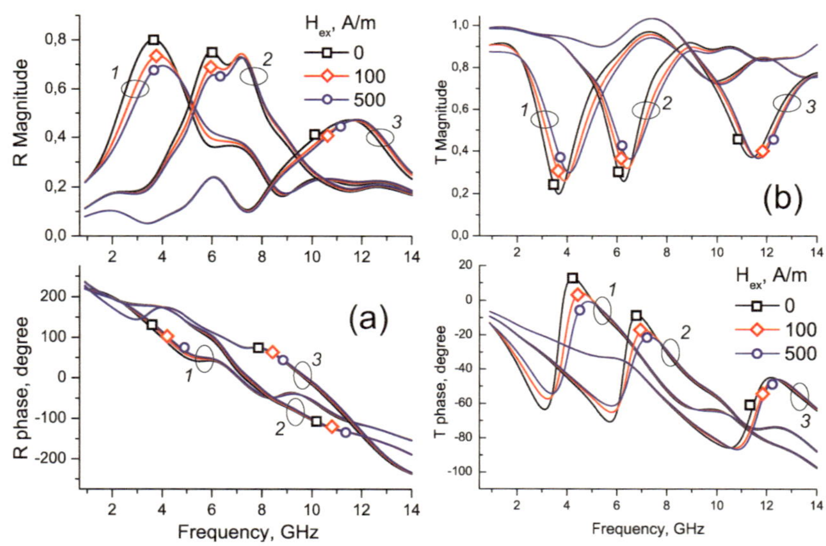

Fig. 17. Spectra of R and T of composites with cut wires of length 40 (1), 20 (2) and 10 (3)mm with the field as a parameter.

demonstrating stop-filter behaviour. The magnitude of this minimum depends strongly on the field for longer wires with lower resonance frequency. For 40 mm long-wire composites, FOM = 102 % at 3.57 GHz and for 20 mm long wires FOM=57% at 6.18 GHz. For shorter wires with the dispersion region at a higher frequency band the field dependence is not noticeable since the wire ac permeability is nearly unity and the impedance becomes insensitive to the magnetic properties, as explained in Section 2. The phase of transmission component exhibits a phase reversal and negative grope delay in the stop band region.
Fig. 18 shows the real and imaginary parts of the effective permittivity. For making comparison between the two composites, the calculation of the effective permittivity was done with the same effective thickness of 1 cm. The frequency where the real part of the permittivity is zero depends on the external magnetic field which can be useful for constructing tunable *Epsilon-Near-Zero* materials.

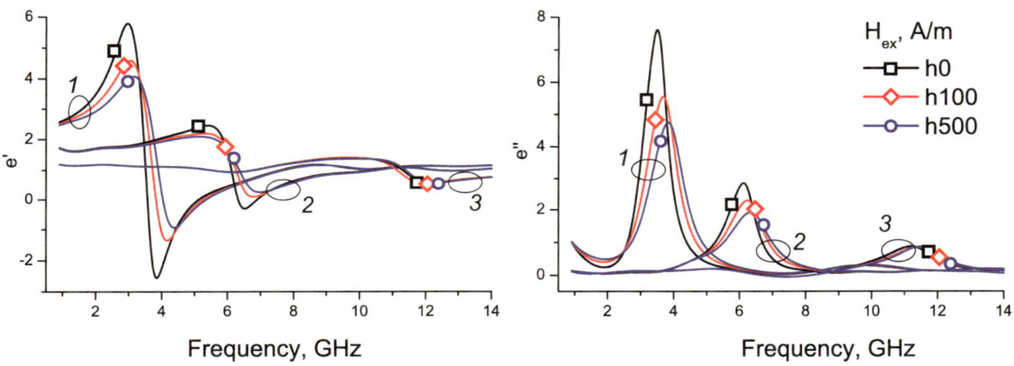

Fig. 18. Effective permittivity spectra for composites with cut wires of length 40 (1), 20 (2) and 10 (3)mm with the field as a parameter

5. Applications

Ferromagnetic microwire based composites with tuneable electromagnetic characteristics represents a new technology with potentially wide applications. These composites gain additional functionalities while their mechanical properties (mechanical strength, geometrical dimensions, weight, density and other) are defined by the host material matrix and not altered by the inclusion as they are thin and their concentration is very small, less than 0.01% for electric composites. As the wire is very inexpensive to manufacture, the overall composite cost keeps almost unchanged.
The potential applications for composite based on microwires with large and sensitive magneto-impedance effect can be divided in two categories. In the first one, the MI effect in wires is used to control the composite's electromagnetic characteristics. The application of a magnetic field or other stimuli will cause change in reflection, transmission and absorption in the composite material. For example, an "active microwave window" can be realized, the state of which can be changed from transparent (open) to opaque (close) for the microwaves. Other applications are transmission signal modulation, deferent frequency selective surfaces and reconfigurable absorbers.

The second category includes different sensing applications that use a high sensitivity of the wire impedance to external stimuli. They include non-destructive testing and structure health monitoring for detection of invisible structural damages and defects, monitoring stress concentrations and temperature distribution. The measurement can be conducted in the waveguide and in the free-space. The later being a non-contact remote method is of special interest as it allows *in-situ* health monitoring of objects such as infrastructure (bridges, buildings, etc.), pipeline and pressure vessels.

As the permittivity ε_{ef} depends on the wire surface impedance (see Eq.(14) for the relaxation parameter), which in turn is a function of the magnetic permeability of the wire (Eg.(18)), then any physical phenomenon that affects the permeability will affect ε_{ef}. Applying a stress or a torque to amorphous wires causes the change in the wire transverse magnetisation and very large variations in impedance (SI) (Garcia et al, 2005). At GHz frequencies, however, sensitive SI effect requires a special magnetic anisotropy (Panina et al, 2005; Makhnovsky et al, 2006). For example, in the case of negative magnetostrictive wires the anisotropy should be nearly axial (customarily, it is circumferential). Only in this case the applied tensile stress may produce large effect on impedance through the change in static magnetisation direction (see Eq. (18)). Such reverse anisotropy can be induced by special stress-annealing as explained in Section 3. The SI effect is promising for constructing composites with stress-sensitive dispersion of ε_{ef}. The stress applied to the composite matrix will be transmitted to each wire inclusion through matrix strain. As reported in (Sandacci et al 2005, Panina et al, 2005), the application of the tensile stress resulted in increase of the impedance up to 100% at GHz frequencies. In (Makhnovskiy et all 2006) the stress sensitivity of composite media containing amorphous microwires was demonstrated experimentally.

The magnetic structure and GMI of the ferromagnetic wires can be made highly temperature dependent for moderate temperature regions (50-200 °C) that gives a possibility to construct the composites with thermally tuneable microwave response and can find applications in remote temperature monitoring by free-space method, e.g., for composite curing control. The high sensitivity to the temperature can be achieved in the vicinity of the compensation point where the magnetostriction changes its sign. Thus, in $Fe_{73}Cu_1Nb_3Si_{16}B_6$ nanocrystalline alloy the compensation point existing due to high volume fraction of bcc-FeSi is about 170 °C (Hong Duong et al. 1996). The magnetisation of wires produced from such alloy will change from axial to circumferential direction when the temperature is increased beyond the compensation point resulting in decrease in the high frequency impedance.

The other approach to realise temperature sensitive composites is based on the ferromagnetic-paramagnetic transition at Curie temperature T_c. This transition is characterized by the drastic change of properties such as magnetization, magnetic susceptibility, anisotropy etc. It could be expected that GMI ratio will also change near T_c, constituting basis for remote temperature detection. The addition of Ni and Cr in Co/Fe amorphous alloy systems results in a decrease in T_c down to the room temperature (See Section 3). In this way a wide variety of temperature-sensitive composites based on microwires with the T_c ranging between the room temperature and 400 °C can be realized. At approaching T_c, the magnetisation saturation M_s scales with $(1 - T/T_c)^{\beta}$ and the magnetostriction scales as ${M_s}^3$. It will result in increase in the initial rotational permeability proportional to the ratio of the magnetisation and magnetostrictive anisotropy field. However, high frequency properties will deteriorate. Then, the high frequency impedance is expected to decrease near the Curie temperature.

In both proposed approaches, the decrease in the wire impedance as the temperature increases produces substantial changes in the frequency dispersion of the effective permittivity and scattering parameters from such composites. For example, in the case of cut wire composites, the resonance type dispersion of the permittivity and band gap propagation regimes will become more pronounce with increasing temperature. These investigations require modification of the microwave setup such as a construction of special thermo chamber which is quite realistic to be realised.

6. Conclusion

Here we have discussed the electromagnetic properties of composites with magnetic wires showing giant magnetoimpedance effect. A striking property of these materials is that the spectra of the effective electromagnetic parameters (permittivity and permeability) can be actively tuned by applying an external magnetic field. Similar effect could be achieved with proper wire systems by applying a stress or changing a temperature. To realize large tunability, the magnetic wires need to possess a rather unique magnetic structure. Technology of glass coated amorphous and nanocrystalline microwires allows the fabrication of continuous wires having various helical magnetic configurations sensitive to a particular magnetic or mechanical stimuli. It is also possible to vary the Curie temperature of magnetic microwires for temperature sensitive behavior.
The magnetic wire composites may either change the electromagnetic response in a desirable way or provide information about the material's properties (stress, strain, temperature). Then, they can be used as free space filters for secure wireless systems to obtain the desired band-pass or band-gap result or as sensory materials for non-destructive tests. The overall technology of magnetic wire composites is cost-effective and is suitable for large-scale applications.

7. Acknowledgment

This work was supported by EU ERA-NET programme under projects "DEVMAGMIWIRTEC" (MANUNET-2007-Basque-3) and "SoMaMicSens" (MANUNET-2010-Basque-3), by EU under FP7 "EM-safety" project and by Spanish Ministry of Science and Innovation, MICINN under Project MAT2010-18914.

8. References

Adenot, A.L.; Acher, O.; Taffary, T.; Queffelec, P. & Tanne, G. (2000). Tuneable microstrip device controlled by a weak magnetic field using ferromagnetic laminations. J.Appl. Phys, Vol. 87, No. 9, May 2000, pp. 6914-6916, ISSN 0021-8979

Agilent 85071E Materials Measurement Software Available from www.agilent.com/find/materials.

Badinter, E.Ya.; Berman, N.R.; Drabenko, I.F.; Zaborovsky, V.I.; Zelikovsky, Z.I. & Cheban, V.G. (1973) Cast micwories and its properties, Shtinica, Kishinev, USSR.

Barlevy, A. & Rahmat-Samii, Y. (1999). Control of Resonant bandwidth in frequency selective surfaces by tilting the periodic elements. Microwave and Optical Technology Letters, Vol. 21, No. 2, April 1999, pp. 114-117, ISSN 0895-2477

Bartley P. & Begley S., (2005). Improved Free-Space S-Parameter Calibration, Proceedings of the IEEE Instrumentation and Measurement Technology Conference, ISBN 0-7803-8879-8, Ottawa, Canada, May 2005.

Beach, R.S. & Berkowicz, A.E. (1994). Giant magnetic field dependent impedance of amorphous FeCoSiB wire. Appl.Phys.Lett., Vol. 64, No. 26, Jun 1994, pp. 3652-3654, ISSN 0003-6951.

Belov, P. A.; Tretyakov, S. A. & Viitanen, A. J. (2002). Dispersion and Reflection Properties of Artificial Media Formed By Regular Lattices of Ideally Conducting Wires J. Electromagn. Waves Applic. Vol. 16, No. 8, January 2002, pp 1153-1170, ISSN 0920-5071

Brown, J. (1960). Artificial dielectrics. Progress in Dielectr., Vol. 2, 1960, pp. 195-225, ISSN.

Catalan, C.F.; Prida, V.M.; Alonso, J.; Vazquez, M.; Zhukov, A.; Hernando, B. & Velazquez (1997). Effect of glass coating on magnetic properties of amorphous microwires, Rapidly Quenched and Metastable Materials J. Materials Science and Engineering A, Supplement 1997, pp. 438-441. ISSN 0921-5093

Chen, L. F.; Ong C. K.; Neo, C. P.; Varadan, V. V. & Varadan, V. K. (2004). Microwave Electronics Measurement and Materials Characterization, John Wiley & Sons, ISBN 0-470-84492-2, Chichester, England

Chiriac, H. & Ovari, T.A. (1996). Amorphous glass-covered magnetic wires: preparation properties, applications. Progress in Material Science, Vol. 40, No. 5, February 1999, pp. 333-407, ISSN: 0079-6425

Chizhik, A.; Gonzalez, J.; Zhukov, A. &Blanco, J.M. (2004). Vortex-type domain structure in Co-rich wires. J. Appl. Phys. Vol. 95, No. 5, Mart 2004 pp. 2933-2935, ISSN 0021-8979

Chizhik, A.; Zhukov, A.; Balnco, J.M; & Gonzalez J. (2001). Magneto-optical investigation of the magnetization reversal in Co-rich wires, Physica B: Condensed Matter Vol. 299, No. 3-4, June 2001, pp. 314-321, ISSN 0921-4526

Dudek, C.; Adenot-Engelvin, A . L. & Acher, O. (2006). Static and Dynamic Properties of Fe-Si-B-Nb-Cu Nanocrystallized Ferromagnetic Glass-Coated Microwires From 20 C to 350 C. IEEE Trans. Magn., Vol. 42, No. 10, October 2006, pp. 2787-2789, ISSN 0018-9464

Fujimori, H.; Arai, K.I.; Shirae, H.; Saito, H.; Masumoto, T. & Tsuya, N., (1976). Magnetostriction of Fe-Co Amorphous Alloys. Jpn. J. Appl. Phys. Vol. 15, No. 4, April 1976, pp. 705-706, ISSN 0021-4922

Garcia, C.; Zhukov, A.; Zhukova, V.; Ipatov, M.; Blanco, J. M. & Gonzalez, J. (2005). Effect of tensile stresses on GMI of Co-rich amorphous microwires. IEEE Trans Magn, Vol. 41, No. 10, October 2005, pp. 3688 - 3690, ISSN 0018-9464

Garcia, C.; Zhukov, A.; Gonzalez, J.; del Val, J.J.; Blanco, J.M.; Knobel, M. & Zhukova, V. (2009) Fabrication, structural and magnetic characterization of thin microwires with novel composition Cu70(Co70Fe5Si10B15)30. Journal of Alloys and Compounds Vol. 483, No. 1-2, August 2009, pp. 566-569, ISSN 0925- 8388

Garcia Prieto, M.J.; Pina, E.; Zhukov, A.P.; Larin, V.; Marin, P.; Vazquez, M. & Hernando, A. (2000). Glass-coated Co-rich amorphous microwires with enhanced permeability. Sensors and Actuators A: Physical, Vol. 81, No. 1, April 2000, pp. 227-231, ISSN 0924-4247

Garcia, C.; Zhukov, A.; Zhukova, V.; Ipatov,M.; Blanco, J.M. & Gonzalez, J. (2005). Effect of Tensile Stresses on GMI of Co-rich Amorphous Microwires. IEEE Trans Magn. Vol. 41, No. 10, pp. 3688-3690, October 2005, ISSN 0018-9464

Hong Duong, V.; Sato Turtelli, R.; Grossinger, R.; Hiebl, K. & Rogl, P. (1996). Magnetostriction in Fe76.5-xCu1NbxSi13.5B9. IEEE Transactions on Magnetics, Vol. 32, No. 5, September 1996, pp. 4821 - 4823, ISSN 0018-9464

Honkura, Y. (2002) Development of amorphous wire type MI sensors for automobile use. J. Magn. Magn. Mater. Vol. 249, No. 1-2, August 2002, pp. 375-381, ISSN 0304-8853

Knobel, K. & Pirota K. R. (2002) Giant magnetoimpedance: concepts and recent progress, J Magn Magn Mater, Vol. 242-245, No. 1, April 2002, pp. 33-40, ISSN 0304-8853

Kraus, L.; Schneider, J.; & Wiesner H. (1976) Ferromagnetic resonance in amorphous alloys prepared by rapid quenching from the melt. Czech. J. Phys. Vol. 26, No. 5. May 1976, pp. 601-602, ISSN: 0011-4626

Kraus, L.; Knobel, M.; Kane, S.N.; & Chiriac, H. (1999). Influence of joule heating on magnetostriction and giant magnetoimpedance effect in a glass covered CoFeSiB microwire, J. Appl. Phys., Vol. 85, No. 8, April 1999, pp. 5435–5437, ISSN 0021-8979

Lagarkov, A. N. & Sarychev, A. K. (1996). Electromagnetic properties of composites containing elongated conducting inclusions. Phys. Rev. B, Vol. 53, No. 10, March 1996, pp. 6318-6336, ISSN 1098-0121

Larin, V. S.; Torcunov, A. V.; Zhukov, A. P.; Gonzalez, J.; Vazquez, M. & Panina, L. V. Preparation and properties of glass-coated microwires. J. Magn. Magn. Mater., Vol. 249, No.1-2, August 2002, pp 39-45, ISSN 0304-8853

Liu, L.; Matitsine, S. M.; Gan, Y. B. & Rozanov, K. N. (2005). Effective permittivity of planar composites with randomly or periodically distributed conducting fibers. J. App. Phys. Vol. 98, No. 6, September 2005, pp. 063512-7, ISSN 0021-8979

Lofland, S. E.; Bhagat, S. M.; Dominguez, M.; Garcia-Beneytez, J. M.; Guerrero, F. & Vazquez ,M. Low-field microwave magnetoimpedance in amorphous microwires, J. Appl. Phys., Vol. 85, No. 8, April 1999, pp. 4442-4444, ISSN 0021-8979

Makhnovskiy, D.P.; Panina, L. V. & Mapps, D. J. (2001) Field-dependent surface impedance tensor in amorphous wires with two types of magnetic anisotropy: helical and circumferential, Phys. Rev. B, Vol. 63, No. 14, April 2001, pp. 144424-17, 2001, ISSN 1098-0121

Makhnovskiy, D.P; Panina, L.V. & Maps D. J. (2001). Field-dependent surface impedance tensor in amorphous wires with two types of magnetic anisotropy: Helical and circumferential. Phys. Rev. B Vol. 63, No. 14, March 2001, pp.144424-17, ISSN 1098-0121

Makhnovskiy, D.P & Panina, L.V. (2003). Field Dependent Permittivity of Composite Materials Containing Ferromagnetic Wires. J. Appl. Phys., Vol. 93, No. 7, April 2003, pp. 4120-4129, ISSN 0021-8979

Makhnovskiy, D. P. & Panina L. V. (2005) Field and stress-tunable microwave composite materials based on ferromagnetic wires, in Progress in Ferromagnetism Research, Murray V. N. (Ed.) pp. 255-292, Nova Science Publishers Inc., ISBN 1-59454-335-6 Hauppauge, NY, USA

Makhnovskiy, D.P.; Panina, L. V.; Garcia, C.; Zhukov, A. P. & Gonzalez, J. (2006). Experimental demonstration of tunable scattering spectra at microwave frequencies in composite media containing CoFeCrSiB glass-coated amorphous ferromagnetic wires and comparison with theory, Phys. Rev. B, Vol. 74 No. 6, August 2006, pp. 064205-11, ISSN 1098-0121

Matitsine, S. M.; Hock, K. M.; Liu, L. ; Gan, Y. B.; Lagarikov, A.N. & Rozanov, K.N. (2003) Shift of resonance frequency of long conducting fibers embedded in a composite, J. Appl. Phys., Vol. 94No. 2, July 2003, pp. 1146-1154, ISSN 0021-8979

Mohri, K.; Uchiyama, T.; Shen, L.P.; Cai, C.M. & Panina, L.V. (2001). Sensitive micro magnetic sensor family utilizing magneto-impedance (MI) and stress-impedance (SI) effects for intelligent measurements and controls. Sensors and Actuators, A: Physical, Vol. 91 No. 1-2, June 2001, pp. 85-90, ISSN: 0924-4247

Mohri, K.; Uchiyama, T.; Shen, L. P.; Cai, C. M.; Panina, L. V. & Honkura Y. (2002). Amorphous wire and CMOS IC-based sensitive micromagnetic sensors utilizing magnetoimpedance (MI) and stress-impedance (SI) effects, IEEE Trans. Magn., Vol. 38, No.5, September 2002, pp. 3063–3068, ISSN 0018-9464

Mohri K. & Honkura Y. (2007). Amorphous wire and CMOS IC based magneto-impedance sensors --- Orign, topics, and future. Sensor Letters, Vol. 5, No. 2, March 2007 pp. 267-270, ISSN 1546-198X

Ong, K .G. & Grimes, C. A. (2002). Tracking the harmonic response of magnetically-soft sensors for wireless temperature, stress, and corrosive monitoring. Sensors and Actuators A: Physical, Vol. 101, No. 1-2, September 2002, pp. 49-61, ISSN 0924-4247

Panina, L.V. & Mohri, K. (1994). Magneto-impedance effect in amorphous wires. Appl.Phys.Lett, Vol. 65, No. 9, August 1994, pp. 1189-1191, ISSN 0003-6951

Panina, L. V.; Sandacci, S.I. & Makhnovskiy, D.P. (2005). Stress effect on magnetoimpedance in amorphous wires at gigahertz frequencies and application to stress-tunable microwave composite materials. J. Appl. Phys, Vol. 97, No. 1, January 2005, pp. 013701-07, ISSN 0021-8979

Panina, L.V. (2009). Magnetoimpedance (MI) in amorphous wires: new materials and applications. Phys. Status Solidi A, Vol. 206, No. 4, April 2009, pp. 656–662, ISSN 1862-6300

Panina, L. V.; Ipatov, M.; Zhukova, V.; Zhukov, A. & Gonzalez J. (2011). Magnetic field effects in artificial dielectrics with arrays of magnetic wires at microwaves, J. Appl. Phys., Vol. 109, No. 5, March 2011, pp. 053901-8, ISSN 0021-8979

Qin, F. X. ; Peng, H. X.; Pankratov, N.; Phan, M. H. ; Panina, L. V. ; Ipatov, M.; Zhukova, V.; Zhukov, A.; & Gonzalez J. (2010). Exceptional electromagnetic interference shielding properties of ferromagnetic microwires enabled polymer composites. J. Appl. Phys., Vol. 108, No. 4, August 2010, pp. 044510-044515, ISSN 0021-8979

Ripka P. (Ed). (2001). Magnetic sensors and magnetometers, Artech House Publishers, ISBN 1-58053-057-5, Norwood, USA

Pendry, J. B.; Holden, A. J.; Robbins, D. J. & Stewart, W. J. (1998). Low frequency plasmons in thin-wire structures. J. Phys.:Condens. Matter., Vol. 10, No. 22, (June 1998), pp. 4785-4809, ISSN 0953-8984

Peng, H.X.; Qin, M.H.; Phan, F.X.; Tang Jie, Panina, L.V.; Ipatov, M.; Zhukova, V.; Zhukov, A. & Gonzalez J. (2009). Co-based magnetic microwire and field-tunable multifunctional macro-composites. J. Non-Crystalline Solids, Vol. 355, No. 24-27, August 2009, pp. 1380-1386, ISSN: 0022-3093

Reynet, O.; Adent, A.-L.; Deprot, S.; Acher, O. & Latrach, M. (2002). Effect of the magnetic properties of the inclusions on the high-frequency dielectric response of diluted composites. Phys. Rev. B, Vol. 66, No. 9, (Septembertember 2002), pp. 94412-9, ISSN 1098-0121

Phan, M. & Peng, H. (2008). Giant Magnetoimpedance Materials: Fundumentals and Applications. Progress in Materials Science, Vo.53, No 2, February 2008, pp. 323-420, ISSN 0079-6425

Rotman, W. (1962). Plasma simulation by artificial dielectrics and parallel-plate media. IRE Trans. Antennas Propag., Vol. 10, No. 1, January 1962, pp. 82-95 1962. ISSN 0096-1973

Sandacci, S. I.; Makhnovskiy, D. P.; Panina, L. V.; Mohri, K. & Honkura, Y. (2004a) Off-diagonal impedance in amorphous wires and its application to linear magnetic sensors. IEEE Trans Magn., Vol. 40, No. 6, November 2004, pp. 3505 - 3511, ISSN: 0018-9464

Sandacci, S.; Makhnovskiy, D.P.; Panina, L. V. & Larin, V. (2004b) Valve-like behavior of the magnetoimpedance in the GHz range. J. Magn. Magn. Mat., Vol. 272-276, No. 3, May 2004, pp. 1855-1857, ISSN 0304-8853

Sandacci, S.; Makhnovskiy, D.; Panina, L. & Larin, V. (2005). Stress-dependent magnetoimpedance in Co-based amorphous wires with induced axial anisotropy for tunable microwave composites. IEEE Transactions on Magnetics, Vol. 41, No. 10, October 2005, pp. 3553 - 3555, ISSN 0018-9464

Sarychev, A.K. & Shalaev, V.M. (2000) Electromagnetic field fluctuations and optical nonlinearities in metal-dielectric composites. Physics Reports, Vol. 335, No. 6, September 2000, pp. 275-371, ISSN 0370-1573, (see equation 7.67)

Shen, L.P.; Uchiyama, T.; Mohri, K.; Kita, E. & Bushida, K. (1997). Sensitive Stress-Impedance Micro Sensor Using Amorphous Magnetostrictive Wire. IEEE Transactions on Magnetics, Vol. 33, No. 5, September1997, pp. 3355 - 3357, ISSN 0018-9464

Sievenpiper, D.; Schaffner, J.; Loo, R.; Tangonan, G.; Ontiveros, S. & Harold R. (2002). A tunable impedance surface performing as a reconfigurable beam steering reflector. IEEE Trans. Antennas Propagat., Vol. 50, No. 3, March 2002, pp. 384-390, ISSN 0018-926X

Sievenpiper,D. F.; Sickmiller, M. E. & Yablonovitch, E. (1996), 3D wire mesh photonic crystals. Phys. Rev. Lett., Vol. 76, No. 14, April 1996, pp 2480-2483, ISSN 0031-9007

Smith, D. R.; Padilla, W. J.; Vier, D. C.; Nemat-Nasser, S. C. & Schultz, S. (2000). Composite Medium with Simultaneously Negative Permeability and Permittivity. Phys. Rev. Lett., Vol. 84, No. 18, May 2000, pp. 4184-4187, ISSN 0031-9007

Snoek, J.L. (1948). Dispersion and Absorption in Magnetic Ferrites at Frequencies above 1 Mc/s. Physica, Vol. 14, No. 4, May 1948, pp. 207-217, ISSN

Takajo, M.; Yamasaki, J.; Humphrey, F.B.;(1993). Domain observations of Fe and Co based amorphous wires, IEEE Trans. Magn., Vol. 29, No 6, pp. 3484-3486, ISSN 0018-9464

Taylor G.F. (1924) A Method of Drawing Metallic Filaments and a Discussion of their Properties and Uses. Phys. Rev., Vol. 23, No. 5, May 1924, pp. 655-660, ISSN 0031-899X

Taylor, G.F. Process and apparatus for making Filaments, Patented Feb. 24, 1931, United States Patent Office, 1, 793, 529.

du Trémolet de Lacheisserie, E.; Gignous, D. & Schlenker, M. (2005). Magnetism: Materials and Applications, Vol. 2 Springer, ISBN 0-387-23000-9, New York

Tejedor, M.; Hernando, B.; Sanchez, M. L.; Prida V. M.; & Vazquez, M. (2001). Magneto-impedance effect in amorphous ribbons for stress sensor application, Sensors and Actuators A: Physical ,Vol. 81, No. 1-3, 1 April 2000, pp. 98-101, ISSN 0924-4247

Ulitovsky, A.V. & Avernin, N.M. Method of fabrication of metallic microwire Patent No161325 (USSR), 19.03.64. Bulletin No7, p.14.

Ulitovsky, A.V. Micro-technology in design of electric devices, Leningrad, 1951, No.7, p.6.

Ulitovsky, A.V.; Maianski, I.M. & Avramenco, A.I. Method of continuous casting of glass coated microwire Patent No128427 (USSR), 15.05.60. Bulletin. No10, p.14.

Vazquez, M.; Chiriac, H.; Zhukov, A; Panina, L. & Uchiyama T., On the state-of-the-art in magnetic microwires and expected trends for scientific and technological studies. Phys. Status Solidi A, Vol. 208, No. 3, March 2011, pp. 493–501, ISSN 1862-6300

Vazquez, M. & Hernando, A. (1996). A soft magnetic wire for sensor applications, J Phys D: Appl Phys, Vol. 29, No. 4, April 1996, pp. 939–949

Yashchyshyn Y. & Modelski, J. (2002). The Leaky-Wave Antenna with Ferroelectric Substrate., Proceedings of XIV International Conference on Microwaves, Radar and Wireless Communications (MIKON-2002), Vol.1, p.21, ISBN 83-906662-5-1, Poland, Gdansk, May 2002

Zhukov, A.; Gonzalez, J.; Blanco, J.M.; Vazquez, M. & Larin, V. (2000). Microwires coated by glass: A new family of soft and hard magnetic materials. J. Mat. Res., Vol. 15, No. 10, October 2000, pp. 2107-2113, ISSN: 0884-2914

Zhukov, A.; Gonzalez, J.; Vazquez, M.; Larin, V. & Torcunov, A. (2004) Nanocrystalline and Amorphous Magnetic Microwires, In: Encyclopedia of Nanoscience and Nanotechnology, ed. Nalwa H. S., pp. 365-387), American Scientific Publishers, ISBN: 1-58883-001-2, Valencia, USA

Zhukov, A. (2006). Design of magnetic properties of Fe-rich glass - coated magnetic microwires for technical applications. Adv. Func. Mat., Vol. 16, No. 5, January 2006, pp.675-680, ISBN 1616-3028

Zhukov, A.; Garcia, C.; Zhukova, V.; Larin, V.; Gonzalez J.; del Val, J.J.; Knobel, M. & Blanco, J.M. (2007). Fabrication and magnetic properties of Cu50(Fe69Si10B16C5)50 thin microwires. J. Non-Cryst. Solids, Vol. 353, No. 8-10, April 2007, pp. 922-924, ISBN 0022- 3093

Zhukov, A.; Ipatov, M.; Zhukova, V.; Garcia, C.; Gonzalez, J. & Blanco, J. M. (2008). Development of ultra-thin glass-coated amorphous microwires for HF magnetic sensor applications. Phys. Stat. Sol. (A), Vol. 205, No. 6, June 2008, pp. 1367-1372, ISSN 1862-6300

Zhukov A. & Zhukova V. (2009). Magnetic properties and applications of ferromagnetic microwires with amorphous and nanocrystalline structure, Nova Science Publishers, ISBN: 978-1-60741-770-400, Hauppauge, NY, USA

Zhukova, V.; Chizhik, A.; Zhukov, A.; Torcunov, A.; Larin, V. & Gonzalez, J. (2002), Optimization of giant magnetoimpedance in Co-rich amorphous microwires. IEEE Trans. Magn., Vol. 38 No. 5, September2002, pp. 3090-3092, ISSN: 0018-9464

Zhukova, V.; Larin. V.S. & Zhukov A. (2003). Stress induced magnetic anisotropy and giant magnetoimpedance in Fe-rich glass/coated magnetic microwires. J.Appl. Phys. Vol. 94, No. 2 July 2003, pp. 1115-1118, ISSN 0021-8979

Zhukova, V.; Ipatov, M.; Zhukov, A.; Varga, R.; Torcunov, A.; Gonzalez, J. & Blanco J.M. (2006). Studies of magnetic properties of thin microwires with low Curie temperatures. J. Magn. Magn. Mater., Vol. 300, No.1, May 2006, pp. 16-23, ISSN 0304-8853

Zhukova, V.; Blanco, J.M.; Ipatov, M.; Zhukov, A.; Garcıa, C.; Gonzalez, J.; Varga, R. & Torcunov, A. (2007) Development of thin microwires with low Curie temperature for temperature sensors applications Sensors and Actuators B: Chemical, Vol. 126, No. 1, September 200), pp. 318-323, ISSN 0925-4005

Zhukova, V; Ipatov, M. & Zhukov A., (2009) Thin Magnetically Soft Wires for Magnetic Microsensors. Sensors Vol. 9, No. 11, November 2009, pp. 9216-9240, ISSN 1424-8220

23

3D Network SiC-Metals Composites for Heavy Duty Brake Applications

Hongqiang Ru, Jingyang Li and Wei Wang
School of Materials and Metallurgy, North Eastern University, Shenyang
China

1. Introduction

A critical component of high-speed trains is their braking systems. A key technology of a braking system is the braking material. Its primary function is to dissipate mechanical energy by converting it into heat during the braking process without experiencing damages in structure and deterioration in property.

Currently, the most commonly used brake materials for high-speed trains of 200~300 km/hour are copper matrix powder metallurgy materials composed by base metal Cu, Sn and Zn and additives Fe, Ni, Mo and graphite (Desplanques et al., 2001). These materials have various advantages such as stable friction coefficient, lower noise compared to iron shoe materials, high thermal conductivity and well running-in characteristic (Zhang et al., 2010) and (Wahlström et al., 2010). However, with increasing the speed to 380 km/hour, these materials suffer from several problems including adhesion, over abrasion, fatigue fracture and flake spalling. These materials become inadequate to withstand 30 MJ of braking energy during the braking process in emergency situations (Guérin et al., 1997).

Therefore, it is necessary to develop new brake materials of improved performances. The development of these materials is critical for the success of the technology of high-speed trains. The success of the project is of high technical significance.

To meet the request for brake material application for speeds above 350 km/hour, main properties of the brake material need to satisfy the following requirements (Tirovic et al., 2001):

1. Heat capacity above 500 J/kg.K;
2. Thermal conductivity above 45 W/m·K;
3. Crushing strength above 280 MPa;
4. Bending strength above 70 MPa;
5. Average friction coefficient 0.35 and the vibration below 20%;
6. Abrasion below 2×10^{-3} $cm^3/MJ \cdot cm^2$;
7. Withstand heat energy above 30 MJ during braking.

The current copper matrix braking materials are able to meet all the above except requirements (1) and (7). Their heat capacities are 390 J/kg.K. Moreover, they cannot dissipate 30 MJ. Generally, at 22 MJ, their surface temperature may reach 1100°C, above the melting point of copper (Bauer&Li, 2005). To solve the problem, an option is to introduce second phase materials of high melting points and high heat capacities into the copper matrix. These second phase materials with higher heat capacities will absorb more heat energy than copper, thus lower the surface temperature of the composite.

Silicon carbide (SiC) is widely studied for its high temperature stability, high thermal conductivity, high hardness and high heat capacity. Its melting point is above 2600°C, and the thermal conductivity is 16.7 W/m·K. Moreover, its heat capacity is 712 J/kg.K at 373 K (Lide, 2009), which increases with increasing temperature. These properties present SiC the possibility as potential additive to copper based braking material.

The concept of the new braking material is a copper matrix-SiC 3D network composite, of which the volume fraction of SiC is 40%. It is to be formed by infiltrating Cu into SiC performs. The interpenetrating composites are expected not only to preserve the properties of stable friction coefficient, low noise, high thermal conductivity and good running-in characteristics of copper, but also to ensure the mechanical and thermal properties of bending strength, crushing strength, heat capacity, as well as thermal conductivityso to overcome the problems of flaking, over abrasion and yielding. It is also intend to introduce SiC into other metal matrices, including pure iron, cast steel and stainless steel.

This chapter presents the 3D SiC ceramics network-metals composites brake materials, including the preparation procedure, the discovery of SiC-metals interfaces, the experimental evidences and discussions of reaction, as well as the brake properties of this new concept material. This chapter is divided into Three sections. The first section introduces the 3D SiC-copper composites. The second section introduces the 3D-SiC-5120 steel composites. The third section introduces the brake properties of these 3D SiC-metals composites materials.

2. SiC-Cu composites

2.1 Composites preparation

The experimental work involved the preparation of SiC skeleton. In the present study, the 6H-SiC and SiC powder have been used for preparation of SiC skeleton and sintered to SiC skeleton preform. The experiment involved the preparation of SiC-Cu composites. The method, which patented in China (Ru et al., 2006), is to cast Cu into the SiC skeleton in a special furnace at 1160 by pressure, caused by pressure difference between pressured gas and vacuum. The Cu interpenetrated the SiC skeleton and formed the composites. After the preparation, the SiC-Cu was sliced into 10mm×10mm×5mm pieces and polished metallographically. Microstructural characterization of the reaction zone was carried out by the optical microscope and the scanning electron microscopy (SEM), and the production of the solid reaction was identified by means of X-ray diffraction (XRD).

2.2 Microstructure

It is observed that the Cu in the present investigation reacted with the SiCn and develop a variety of phases close to the reaction zone (Kang&Kang, 2006). The secondary electron image and the composition profile obtained by the SEM are shown in Figure 1. The micrograph shows a nice contact between SiC and reaction zone. However, there is a gap in the interface between the SiC and the reaction zone. The gap may be due to the different thermo expansion coefficient. Meanwhile, the dark phase, which is shown in the secondary electron image, is essentially SiC particle picked up during the polish procedure from the polish paper. The sequence of reaction layers formed in the SiC-Cu system consists of SiC reaction zone, random carbon precipitation zone, regulative carbon precipitation zone and metal reaction zone, shown as Figure 1. (b), from section I to section IV. In section I, SiC decomposed to Carbon and Silicon. And the Silicon, which has good affinity with Cu, formed the compound as Cu_4Si, identified by XRD and EDS (Figure 2 and Figure 3). Meanwhile, the carbon failed to form the solid solution with Cu, according to the binary phase diagram, diffused against the SiC, and precipitated in the section II and section III. With the increasing distance from the SiC terminal to the precipitation location, the carbon scale increased and the shape transformed from the anomaly in the section II to the regulative round in the section III.

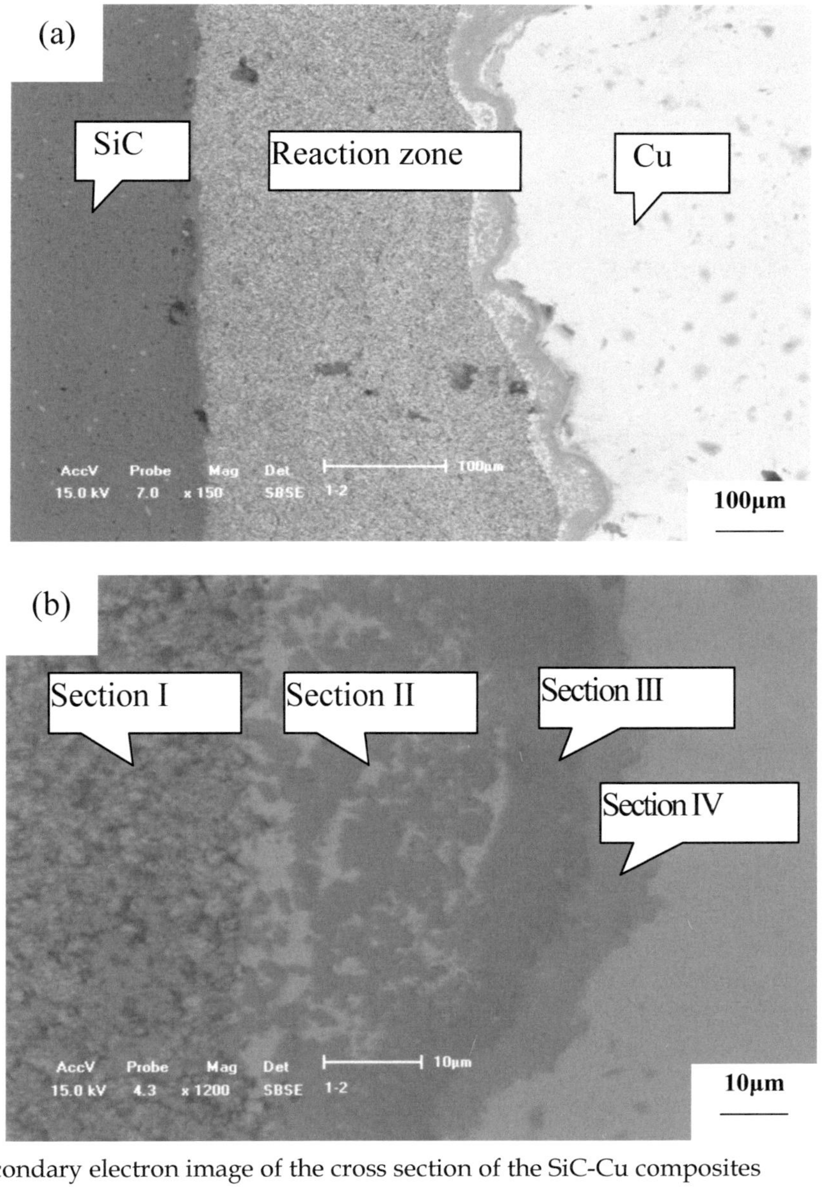

Fig. 1. Secondary electron image of the cross section of the SiC-Cu composites

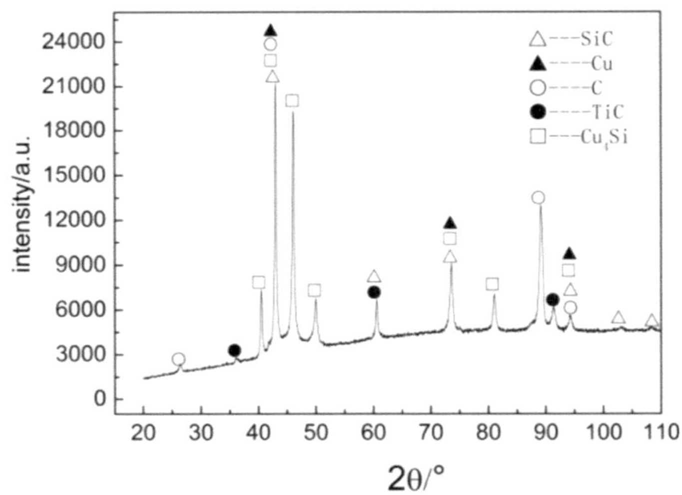

Fig. 2. The XRD patern of SiC-Cu composites

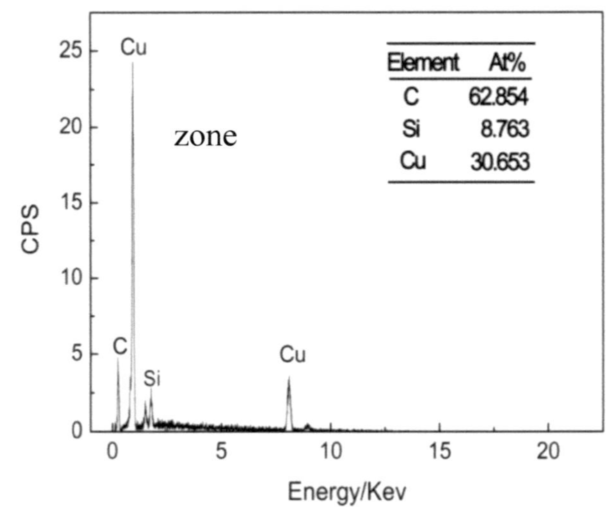

Fig. 3. The EDS patern of SiC-Cu reaction interface

The formation of the interface of SiC-Cu is co-operated by the liquid reaction during the casting process, the solid reaction during the cooling process and the diffusion of the elements. The equation 1 and 2 shows the main reaction:

$$SiO_2(s)+Cu(l)\rightarrow Cu4Si(s)+Cu_2O(s) \tag{1}$$

$$SiC(s)+Cu(l)\rightarrow Cu_4Si(s)+C(s) \tag{2}$$

At the beginning of the whole process, the silicon carbide skeleton, pre-oxidized, was eroded by the copper liquid. The SiO_2 film was crushed, decomposed into Si, formed Cu_4Si

with copper, and Oxygen, formed copper oxides and diffused against the SiC. As the copper liquid contacted with the SiC, the reaction started cosmically. Since the copper was superfluous in this certain reaction, the products and the degree of the reaction were controlled by the speed of the silicon carbide decomposition. Basically, the SiC could not decompose in the environment of 1160°C, but the presence of the copper supply the dynamic force to the SiC system (Rado&Drevet, 2000). The Si had the great affinity with the copper, forming Cu-Si compound, while the carbon failed to solute in the copper matrix and precipitated as the graphite phase. According to the previous research (Zhang, 2008), the precipitation of the carbon was oblige to meet the requests as follow: (1) the metal has greater affinity with the silicon more than the carbon; (2) the concentration of the metal atoms has to be heavy so as to supply the force to decompose the SiC; (3) the carbon has the light solubility in the metal matrix. In this paper, the SiC-Cu system happened to meet the requests above.

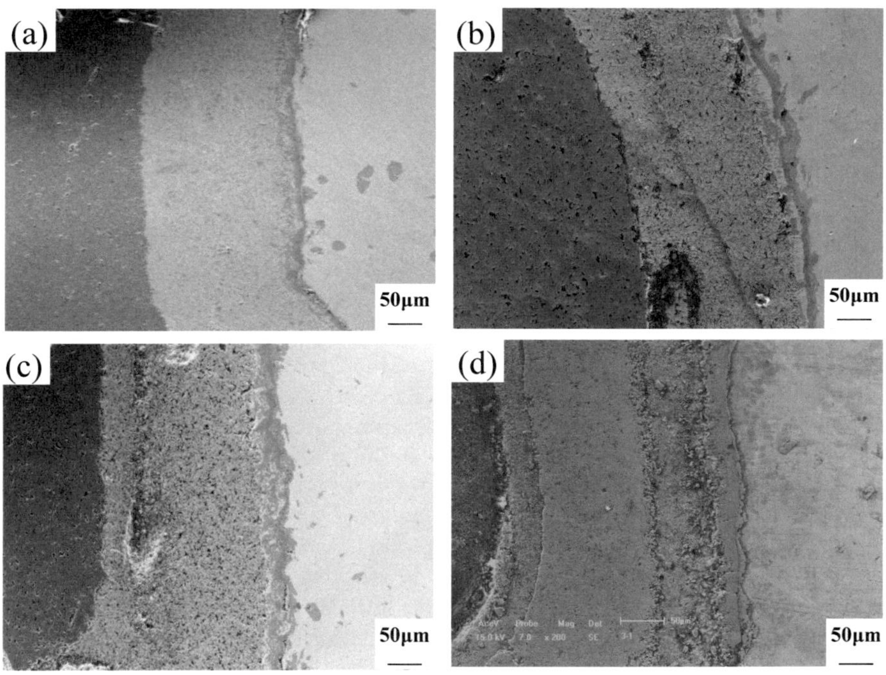

Fig. 4. The microstructure of interface region after heat treatment
(a)30min; (b)60min; (c)90min; (d)120min

In order to make clear the mechanism of the solid reaction around the interface in the SiC system, this paper has research the diffusion behavior and calculated the diffusion coefficient as well. As the Figure 4 shows, the section II of the interface was heat treated, lasting 30, 60, 90 and 120 minutes respectively. For the reason that the band of the section II was narrow, this paper chose this section so as to control the precision and to observe the change of the morphology. The distance from the copper matrix to the section II was 30μm, measured. And the content of the copper in the section II in the original specimen and the heat treated ones was shown in the table.1.

0 min	30min	60min	90min	120min
52.7%	57.4%	62.1%	66.0%	70.3%

Table 1. Content of Cu in the section II after different heat treatment time

Based on the Fick's law and the equation of semi-infinite objects (Equation 3), this paper substituted the initial condition: $C_S=1$, $C_1=0$, and drew the equation 4.

$$C=C_S-(C_S-C_1)\mathrm{erf}\left(\frac{x}{2\sqrt{Dt}}\right) \tag{3}$$

$$C=1-\mathrm{erf}\left(\frac{x}{2\sqrt{Dt}}\right) \tag{4}$$

After substituted the value as $x=30\mu m$, $C=52.7\%$ while $t=t$, $C_1=57.4\%$ while $t_1=t+1800$, $C_2=62.1\%$ while $t_2=t+3600$, $C_3=66.0\%$ while $t_3=t+5400$, and $C_4=70.3\%$ while $t_4=t+7200$ respectively, the diffusion coefficient was calculated: $D_1=2.85\times10^{-14}$ $m^2 \cdot s^{-1}$, $D_2=2.31\times10^{-14}$ $m^2 \cdot s^{-1}$, $D_3=2.79\times10^{-14}$ $m^2 \cdot s^{-1}$, and $D_4=3.09\times10^{-14}$ $m^2 \cdot s^{-1}$. The average one was $D=2.76\times10^{-14}$ $m^2 \cdot s^{-1}$.

3. SiC-5120 Steel composites

3.1 Composites preparation

A SiC-steel composite was created using a porous SiC network preform and a low alloy 5120 steel of the composition as given in Table 2. To create the composite, SiC powder was first prepared into aqueous slurry. A porous polyurethane block of 10×50×100 mm in external dimension was used as a template to produce SiC preform. The template was submerged into the SiC slurry before taken out to dry at room temperature for 24 hours to form the SiC green pattern, which was then sintered in vacuum at 2000 °C to create the porous SiC preform. The SiC preform was preheated to 450 °C in vacuum prior to steel infiltration. Ingot of 5120 steel was melted in a vacuum furnace at 1650 °C and the molten steel was then poured into the pre-heated SiC preform for infiltration. The infiltrated composite was allowed to solidify and cool down naturally in the vacuum furnace.

	C	Cr	Si	Mn	Fe
5120 steel	0.17~0.24	0.70~1.00	0.17~0.37	0.50~0.80	Balance

Table 2. 5120 steel composition in weight percentage

3.2 Microstructure of composites

Figure 5 shows optical micrographs of the samples of the experiment. Micrograph (a) shows the SiC preform before steel infiltration. The pores are spherical, interconnected and uniform in size and distribution, The average pore diameter is 3 mm and the average pore wall thickness is ~1 mm. The porosity of the SiC preform created was 15 vol%. Micrograph (b) shows the composite sample after the molten steel infiltration. The dark contrast is the SiC preform and the lighter contrast is the infiltrated 5120 steel.

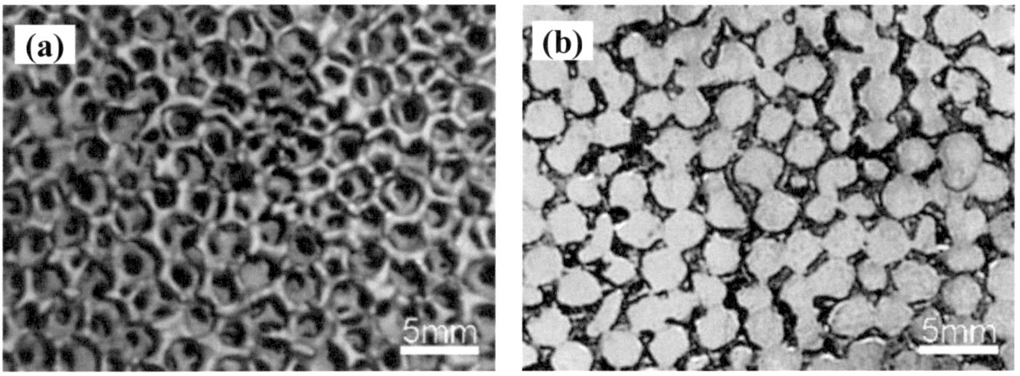

Fig. 5. Optical micrographs of (a) SiC skeleton preform and (b) SiC/20Cr composites sample.

Figure 6 shows detailed microstructures of the SiC/5120 steel composite. Micrograph (a) is an optical image showing different microstructural sections of the composite. The

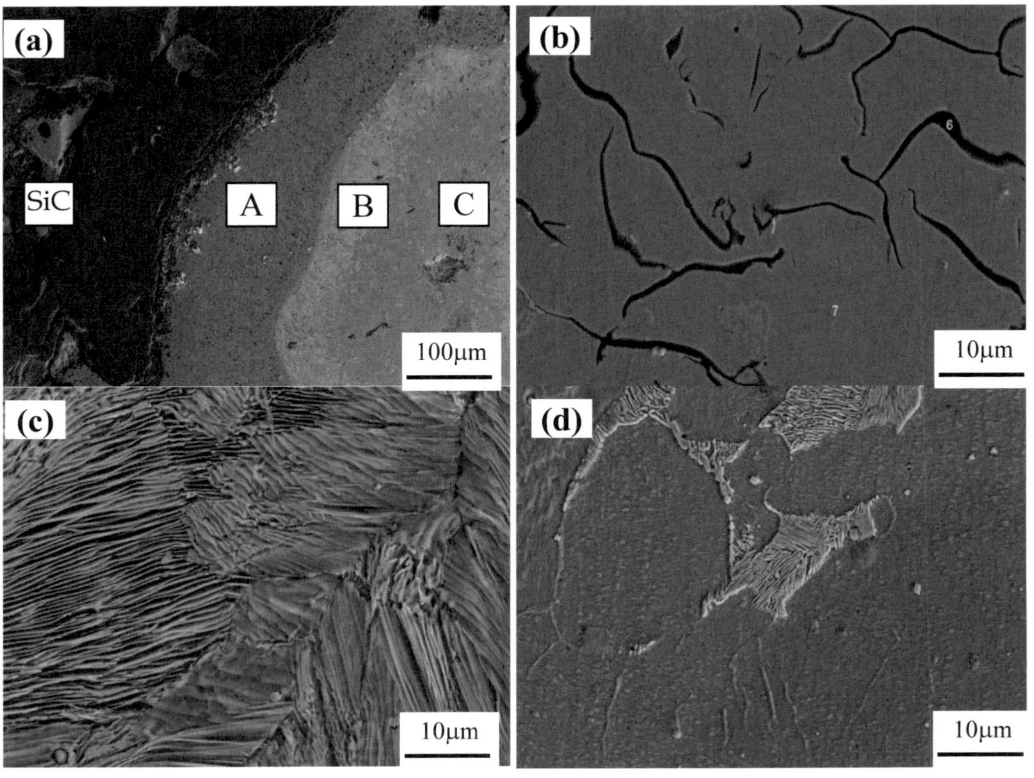

Fig. 6. Microstructures of SiC-5120 steel composite showing different section

microstructure is seen to contain four distinctive sections, labelled A, B, C and SiC. The average width of section A was 230 μm and the average width of section B was 150 μm. Micrograph (b) shows an SEM image of the structure in section A. The structure contains ferrite Fe(Si) solid solution matrix and graphite rosette flakes. EDS analysis of the matrix phase showed Si content of 13 at%, corresponding to the α1-Fe(Si) ferrite phase in the bindary Fe-Si system, according to the Fe-Si phase diagram. The volume fraction of graphite is ~8 vol%, estimated by image analysis using ImageJ software. Micrograph (c) is an SEM image of the structure in section B. The structure is pure pearlite. Microhgraph (d) is an SEM image of the structure in section C. It contains a continuous ferrite matrix and scattered colonies of pearlite. The volume fractions of the two structures are determined by means of image analysis to be 75 vol.% ferrite and 25 vol.% pearlite. The total C content in this region may be estimated to be 1 at% based on the volume fractions. It is also evident that a narrow gap of ~40 μm in width existed between SiC and region A. This is caused by the solidification shrinkage of the steel and the difference in thermal expansion coefficients between SiC and the steel, which are $2.98\times10^{-6}/°C$ and $1.5\times10^{-5}/°C$, respectively.
Figure 7 shows XRD spectrum of the SiC/5120 steel composite. The peaks are fully indexed to SiC and α-Fe(C) ferrite and/or α1-Fe(Si) ferrite. The α-Fe(C) ferrite has a bcc structure and the α1-Fe(Si) ferrite has a DO3 structure, with very similar lattice parameters [21]. Their fundamental diffraction peaks practically overlap on the spectrum. Superlattice diffraction peaks for the DO3 phase are not observed. The lattice constant of the bcc α-Fe(C) ferrite is determined to be a=0.286 nm, or a=0.567 nm for the DO3 α1-Fe(Si). Graphite and cementite (in pearlite) were not detected by XRD, because of their low volume fractions, which are about 0.8 vol.% and 0.9 vol.% in the sample.

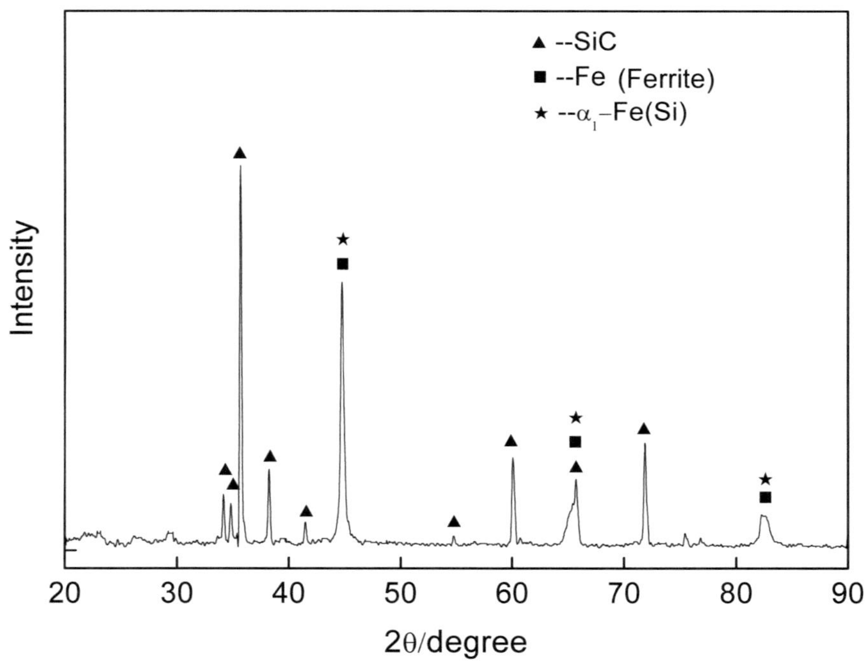

Fig. 7. X-ray diffraction pattern at room temperature of the SiC/5120 composite

4. Brake properties of SiC-5120 steel and SiC-copper composites

Stable friction coefficient guarantees the security of the brake material application. 3D SiC network-metals composite brake material could afford stable friction coefficient. For example, SiC-steel and SiC-copper sets (SiC-steel brake disc with SiC-copper pad) could keep the friction coefficient above 0.3 stably. Abrasion is another key element in evaluating the brake materials. Lower abrasion could guarantees the longer duration of the brake materials application. In SiC-steel and SiC-copper sets (SiC-steel brake disc with SiC-copper pad), the abrasion of SiC-steel composite is no more than 1.2μm every brake action, while the abrasion of SiC-copper composite is no more than 1.8μm.

Figure 8(a) shows the variation of acceleration with same brake pressure and various initial rolling speed, whereas Figure 8(b) shows the variation of acceleration with various brake pressure and same initial rolling speed.

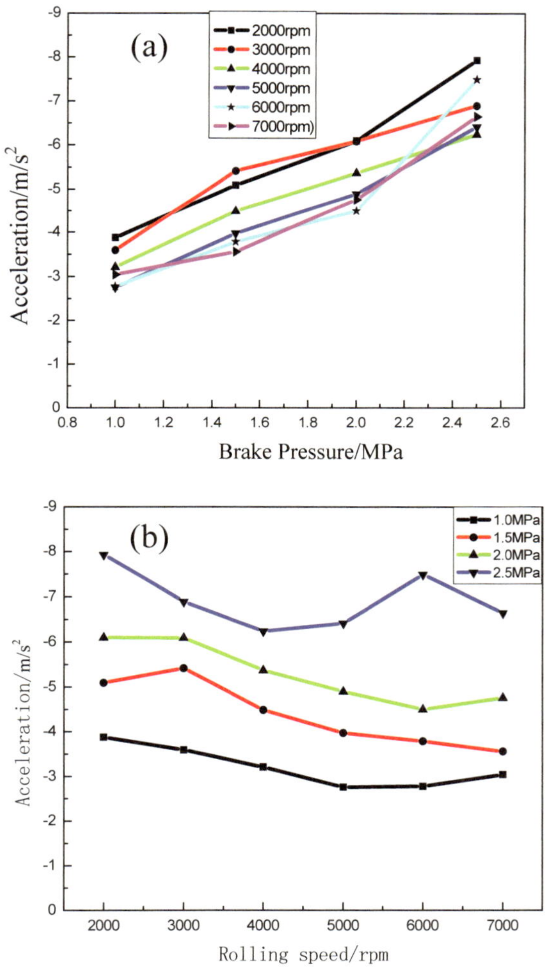

Fig. 8. The relation between the braking acceleration and the brake pressure as well as the rolling speed

It can be seen that under the same rolling speed, the with the increasing brake pressure, the acceleration is ascending, while under the same brake pressure, the acceleration trends stable. For that the more brake pressure, the shorter brake duration, leads to the more acceleration. On the contrary, the more rolling speed, the more brake duration, results in the lower acceleration.

Figure 9(a) shows the friction coefficient variation with the various pressures, whereas Figure 9(b) shows the friction coefficient variation with the same pressure. In lower rolling speed situations (2000 rpm, 3000 rpm and 4000 rpm), the average friction coefficient goes down with the increasing brake pressure, while in higher rolling speed situations (5000 rpm, 6000 rpm, and 7000 rpm), the average friction coefficient ascends after the short descending. Generally, friction coefficient is inversely proportional to brake pressure, so that friction coefficient decrease with the increasing pressure.

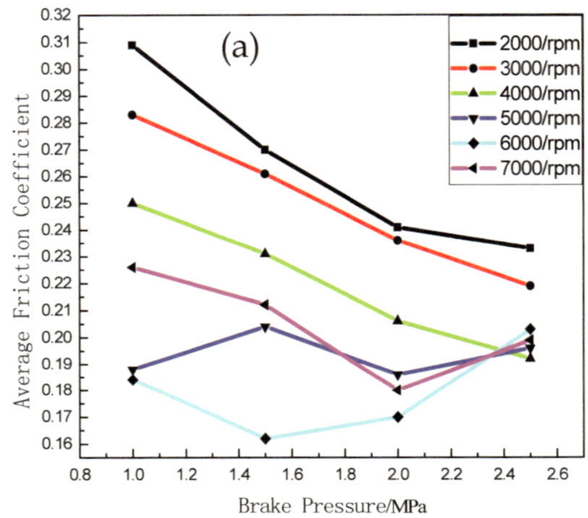

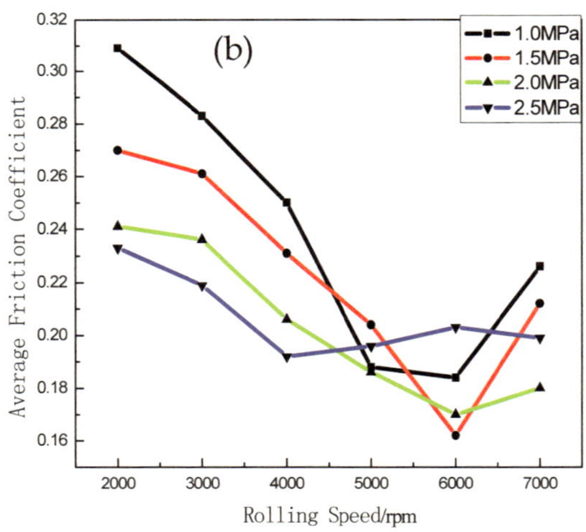

Fig. 9. The relation between the average friction coefficient and the brake pressure as well as the rolling speed

However, in higher rolling speed situations, the mechanism of abrasion is adhesive wear as the main, instead of abrasive grain abrasion as the main. The more brake pressure, the more adhesive wear leading to the more surface damage on the brake material, the less actual friction area. Moreover, the friction coefficient is inversely proportional to actual friction area. The average friction coefficient is increased by the factors described above.

The friction coefficient stability, also known as fluctuation range of friction coefficient, is the value after average friction coefficient subtracted by the maximum friction coefficient.

$$\Delta\mu = \mu_{\max} - \overline{\mu} \tag{5}$$

It is obvious that the less fluctuation range of friction coefficient, the more stable of friction coefficient.

Figure 10(a) shows the variation of friction coefficient stability with various rolling speed while Figure 10(b) shows the variation of friction coefficient stability with various brake pressure.

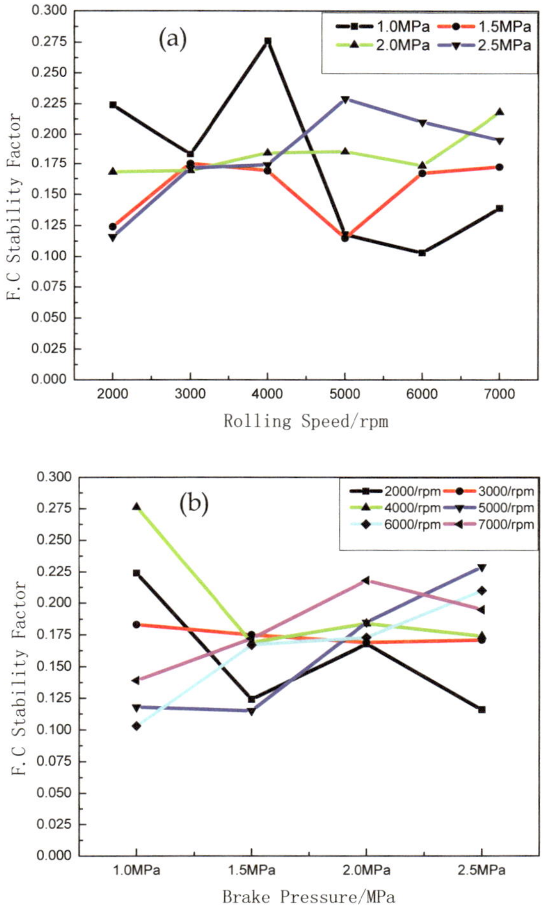

Fig. 10. The relation between the stability factor of friction coefficient and the brake pressure as well as the rolling speed

From the Figure 10(a), it can be seen that friction coefficient stability decreased with the increasing rolling speed under 1.0MPa brake pressure, but slightly increased under 1.5MPa, 2.0MPa, and 2.5MPa brake pressure condition. The reason is that the abrasion in low rolling speed and low pressure is abrasive grain wear as the main. The spalled pieces were squeezed and rolled on the friction surface, leading to the weak friction coefficient stability. However, with the increasing rolling speed, the adhesive wear became the main abrasion instead of the grain abrasion as the main. The phenomenon of friction coefficient decreasing emerged, resulted in the friction coefficient fluctuation decrease.

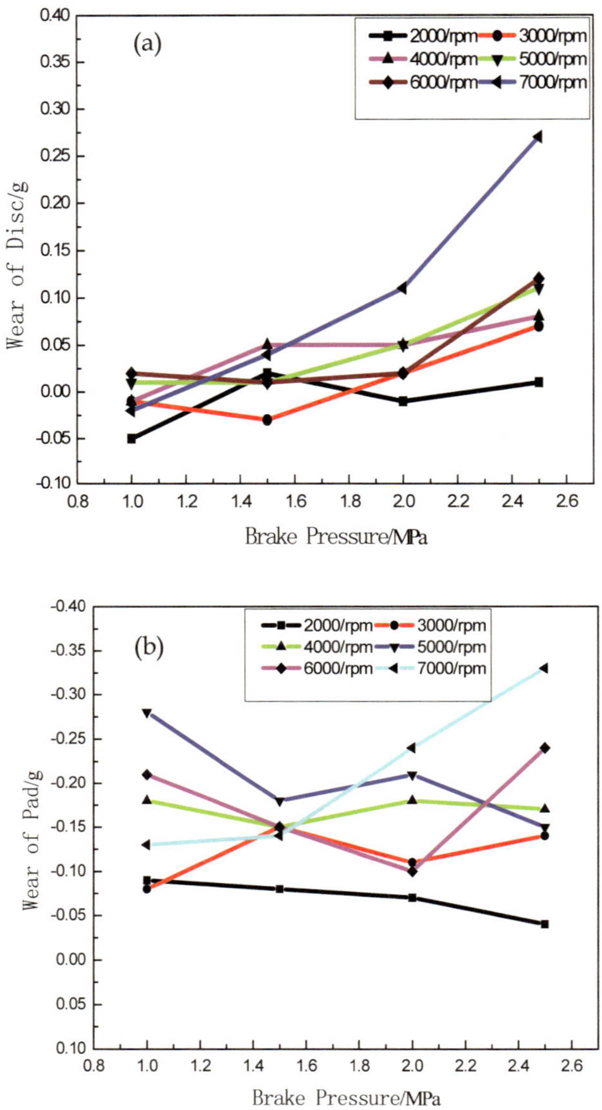

Fig. 11. The relation between the wear and the brake pressure in both disc and pad

Figure 10(b) indicated that the abrasion in lower rolling speed (2000rpm, 3000rpm, and 4000rpm) situation, is abrasive grain abrasion as the main. While the brake pressure increased, the rolling spalled pieces were easy to be squeezed, leading to the decrease of friction coefficient fluctuation. However, while the brake pressure increased further, the great amount of adhesive wear emerged, leading to the stability of the friction coefficient. In higher rolling speed situation, the friction surfaces were damaged severely with the increasing brake pressure and further adhesive abrasion. Therefore, the friction coefficient fluctuation was amplified.

Figure 11(a) shows the wear variation of SiC-5120 steel with various rolling speed, as brake disc. In lower rolling speed situation, with the increasing brake pressure, the mass of the encountered peer brake material adhered to the friction surface increased, but the adhering process kept slow. However, the adhesion mass increased aggressively in 7000 rpm situation.

Under the same rolling speed, the more brake pressure, the more damage in the friction surface and the more wear emerged. In lower rolling speed situation, the abrasion is grain abrasion and adhesive wear as the main, and the adhesive wear compensated the loss led by grain abrasion. Therefore, the mass increasing on friction surface is not much. However, the adhesion dominated the abrasion in 7000rpm situation, the great amount of friction energy welted the friction surfaces of disc (SiC-5120 steel) and pad (SiC-copper). The more brake pressure, the more friction energy was generated. Since the melting point of copper is far lower than that of steel, more pad material were welted to the disc.

5. References

Bauer H., Li L. (2005). *Foreign Locomotive & Rolling Stock Technology*, China Railway Publishing House, Beijing.

Desplanques Y., Degallaix G., Copin R., Berthier Y. (2001). A tribometer for the study of materials under railway braking conditions. *A Tribometer for the Study of Materials under Railway Braking Conditions*, Vol. 39, pp. 389-91.

Guérin J. D., Bricout J. P., Laden K., Watremez M. (1997). High thermal diffusivity materials for railway brake discs. *Tribology Letter*, Vol. 3, pp. 257-67

Kang, H.K., Kang, S. B. (2006). Thermal decomposition of silicon carbide in a plasma-sprayed Cu/SiC composite deposit. *Materials Science and Engineering: A* Vol.428, pp. 336-345

Lide D.R. (2009) *Handbook of Chemistry and Physics*, CRC Press/Taylor and Francis, Boca Taton.

Rado, C., Drevet B. (2000). The role of compound formation in reactive wetting: the Cu/SiC system. *Acta Materialia* Vol.48, pp. 4483-4491.

Tirovic M., Ali G. (2001) Design synthesis of non-symmetrically loaded high-performance disc brakes Part 1. *Proceedings of the Institution of Mechanical Engineers, Part F: Journal of Rail and Rapid Transit.* Vol. 215, pp. 101-9

Wahlström J., Olander L., Olofsson U. (2010). Size, Shape, and Elemental Composition of Airborne Wear Particles from Disc Brake Materials. *Tribology Letter.* Vol. 38, pp. 15-24.

Zhang S., Li Y., Qu S., Chen W. (2010) Friction and Wear Behaviour of Brake Pads Dry Sliding Against Semi-Interpenetrating Network Ceramics/Al-alloy Composites. *Tribology Letter*.Vol. 38, pp. 135-46.

Zhang L., Qu X., Duan B., He X., Ren S. (2008) Microstructure and thermo-mechanical properties of pressureless infiltrated SiC_p/Cu composites. *Composites Science and Technology*. Vol.68, pp.2731-2738.

24

Smart Magnetic Composites

Jerzy Kaleta, Daniel Lewandowski Rafał Mech, and Piotr Zając
Institute of Materials Science and Applied Mechanics, Wroclaw University of Technology
Poland

1. Introduction

1.1 Smart Magnetic Composites

Smart Magnetic Composites (SMC) belong to the wider group named Smart Magnetic Materials (SMM). Various properties of the SMM, involving viscosity, shape, rigidity, temperature or resistance – could be stimulated with a magnetic field. SMMs in turn, belong to the still wider group of materials called Smart Materials, SMART in short, where stimulation of properties is possible with the use of electrical or magnetic field, or heat. Nowadays, we assume that the level of dissemination of the Smart Materials is one of the measures of economy innovativeness in countries and regions. Manufacturing of Smart Materials is also stimulating development of basic research in the various cross effects. The key significance, both civil and military, have the already existing and forecasted applications of the Smart Magnetic Materials. We could specify here examples of an „intelligent" vibration damping in such stationary objects as buildings, bridges, pipelines or power networks. Equally significant field of the SMM utilisation is transport (automotive vehicles, trains, airplanes). The number of applications in medicine is growing, which could be exemplified by intelligent prostheses, remote surgery, new methods of neoplasm therapy or the magnetic "markers" of medicines. Very promising are the SMM group materials providing safety for cable transmitted data. Among the examples one could also indicate the development in new methods of non-destructive testing such as magnetovision. Within the last several years, a great interest in the use of SMMs in energy harvesting might be noticed. The listed advantages of SMMs are the reason for intensive research in many scientific institutes and industrial centers.

Smart Magnetic Materials could be divided according to various criteria. One of the possible classifications distinguishes the following SMM types:

- Materials of variable internal structure:
  - MagnetoRheological Fluids - MRF,
  - FeRroFluids – FRF,
  - Porous materials saturated with magnetorheological fluids - MagnetoRheological Composite - MRC, gels/greases/... filled with ferromagnetic material powder,
  - fluids with powdered magnetocaloric materials.
- Materials of fixed internal structure:
  - solid magnetostrictive materials, including those with giant magnetostriction - Giant Magnetostrictive Materials – GMMs,

- elastomers filled with ferromagnetic material powders (e.g. carbonyl iron, GMM or their combination), polymers on the epoxy resin base containing powdered ferromagnetic materials,
- solid magnetocaloric materials.

Below, the following Smart Magnetic Composites have been discussed in detail:

- Composites of porous matrix filled with magnetorheological fluid (MagnetoRheological Composites - MRC),
- MagnetoRheological Elastomers – MRE, known also as Magneto-Active Elastomers - MAE),
- Composites containing powdered material of giant magnetostriction (Giant Magnetostrictive Materials composites - GMMc).

In each of the above cases the manufacturing technologies, ways of stimulating with magnetic field, methodology of research and properties identification as well as application examples, have been discussed. The material has been enriched with literature overview and results of the Authors' own research.

2. Magnetorheological composites – porous materials saturated with magnetorheological fluids

2.1 Introduction

The MagnetoRheological Fluid (MRF) in the gravity conditions requires external barriers, or a vessel, which keeps it in a defined place with maintaining geometry. This inconvenience could be overcome in several ways. The first one consists in substituting the carrier fluid with a material of definitely greater viscosity (e.g. a gel or grease Malcolm et al. (2002); Rankin et al. (1999)). That way the material is being created, similarly to MRF, of the variable internal structure, but more easily kept in the required place. Disadvantage of that solution is too high viscosity in the off-state – without the magnetic field, which could be an obstacle in some applications. The effect of shaping the external dimensions may also be obtained by saturating a porous material with MR fluid. That way the material with open cellular structure creates the matrix maintaining the magnetorheological fluid within the limits determined with its dimensions, thus enabling the relatively free interaction of the magnetic and mechanical field inside the structure. Under the cellular structure the material composed mainly of the internal spaces, or open pores connected amongst others, is understood. The curing matrix of a magnetorheological composite may be fabricated from a sponge, fabric, felt or any other elastic porous material Carlson & Jolly (2000). By saturating a matrix we achieve material characterised with dependence of its mechanical parameters on the magnetic field, similarly as in case of the MRFs. The new type of material shaped that way has no single, formalised name. In the subject literature one could find such definitions as: the field-responsive fluid-impregnated cellular solids Deshmukh & McKinley (2006), the magnetorheological foams Carlson (1999); Carlson & Jolly (2000), the magnetically responsive foams Purizhansky (2004). Because of the complex structure of the matrix and a filling, the Authors of this work considered as just to name it in the simpler form as Magneto–Rheological Composite, or shortly the MRC, used from here in the work. The MRC type composites belong to the materials, which despite of the completely different build are frequently treated as one of the science fields on magnetorheological fluids. The growing interest in these materials and the attempts of their wider application, especially in the areas of active suppressions, cause that they are more and more frequently the objects of research.

Among the still scarce literature concerning the magnetorheological composites, the first information on their properties and applications may be found in the works Carlson & Jolly (2000); Deshmukh & McKinley (2006); Purizhansky (2004).
As the key issues within the magnetorheological composite (MRC) field the following were recognised:

- composite manufacturing engineering,
- strain and stress measurement methodology in the conditions of variable mechanical and magnetic parameters,
- constitutive model and its parameters identification.

The above topics have been discussed further in the work.

2.2 Magnetorheological composites manufacturing procedure

The test specimens were manufactured independently Kaleta & Lewandowski (2007). The magnetorheological composite was obtained as combination of the two basic components, i.e. the magnetorheological fluid and the porous material. The magnetorheological fluid used in the tests was prepared according to the recipe similar to the first fluids used by Rabinow (1948). A base, i.e. the magnetically active component filling the fluid, was carbonyl iron powder type CC, from BASF. As a carrier fluid the silicon oil was used. Content of the iron powder in the fluid was 80% by weight. Interactions displayed by the ferromagnetic component in the fluid are shown in Figure 1. Various structures in the form of chains being created in the magnetic field are shown. Their breaking requires applying additional external force, which is understood as the magnetorheological effect.

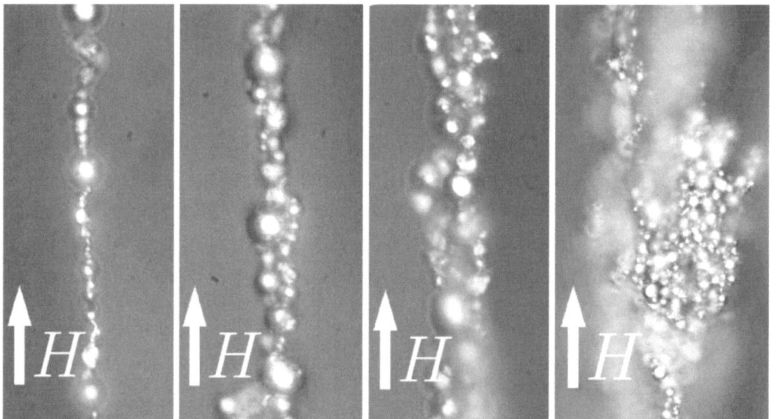

(a) (b) (c) (d)

Fig. 1. The idea of magnetorheological fluid operation: ordering of the carbonyl iron powder and creation of structures in the magnetorheological fluid Lewandowski (2005).

Schematic diagram of the composite manufacturing has been shown in Figure 2. Composite matrices are made of polyurethane foam, characterised with open pores. This enabled free penetration of MRF to its interior. The phase terminating the composite manufacturing was saturation of the matrix with magnetorheological fluid using the injection method. The process of filling was conducted so, that part of the internal foam pores was filled with air and, at the same time, the fluid was uniformly covering the internal surfaces of matrix. Composites, for which the tests were performed, had the same filling with the magnetorheological fluid. The filling amounted to 70% of the matrix volume. That way, composites of the required features were achieved.

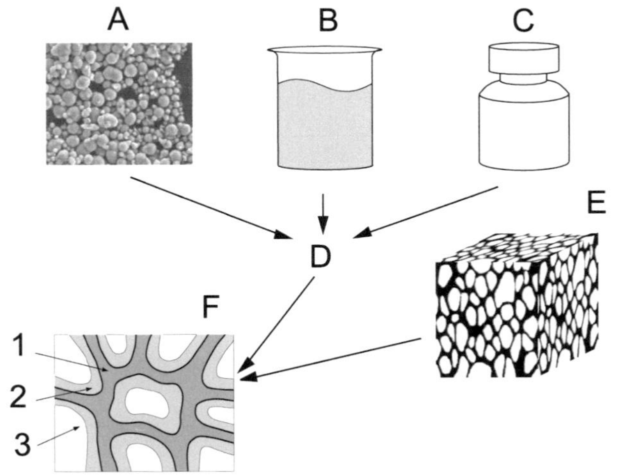

Fig. 2. Schematic diagram of the magnetorheological composite manufacturing process. Main components of the magnetorheological fluid : A – carbonyl iron powder, B – carrier fluid, C – additives. Carrier matrix of the composite – E. Schematic diagram of the matrix cell filled with fluid: 1 – matrix, 2 – the covering MR fluid 3 – remained volume filled with air Lewandowski (2005).

The spongy, elastic structure of matrix enabled maintaining of the MRF in one place defined with its external geometry and, at the same time, the free interaction of magnetic field with the fluid. Open pores enabled free flow of fluid between cells. Size of the pores was small enough to prevent the fluid getting out under the influence of gravity. Real structure of the matrix before saturation has been shown in Figure 3a and after saturation with fluid in Figure 3b.

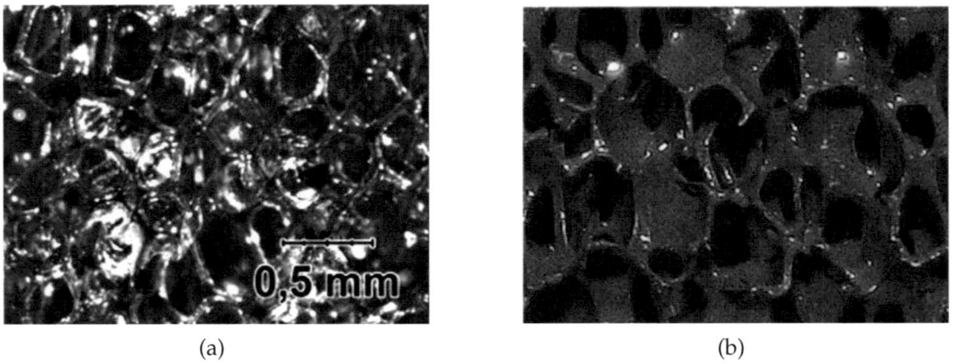

Fig. 3. (a) – internal structure of matrix before filling it with the MR fluid and (b) – after the filling.

2.3 The test stand

Schematic diagram of the test stand has been presented in Figure 4. The electromagnetic system was responsible for creating and directing the magnetic field interacting with the tested material specimen. A gap with two parallel surfaces was made in the steel core, inside which a tested material was placed. The magnetic field vector was perpendicular to the sample shear direction. The force straining the tested material was generated by an external exciter. Measurement of the magnetic field intensity was performed by the hall sensor placed beside the tested composite sample in the core gap. Samples for the tests were shaped so, as to obtain the homogeneous shear in the tested material. They were made from two identical pieces of cuboid composite attached symmetrically on both sides of the supporting plate. Fixing of a specimen in the magnetic core gap of the measurement stand has been shown in Figure 4a. That way of fixing enabled the state close to the pure shearing.

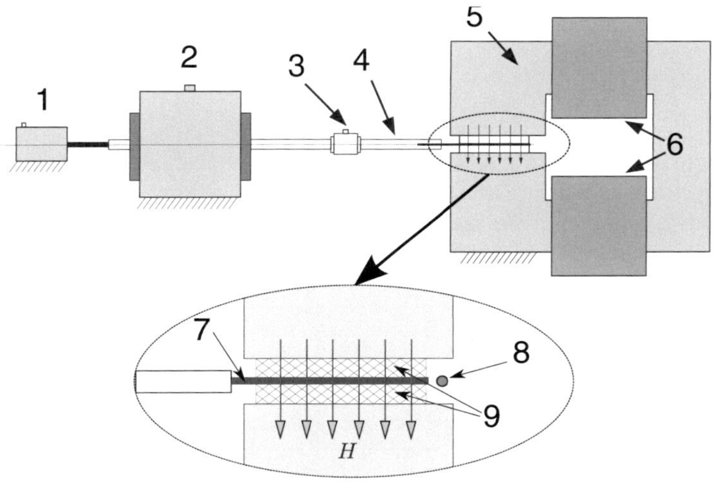

Fig. 4. The schematic diagram of the mechanical part of the test stand and mounted specimen: 1 – displacement sensor, 2 – shaker, 3 – piezoelectric force sensor, 4 – connector, 5 – magnetic core, 6 – coils, 7 – carrier plate, 8 – magnetic field sensor, 9 – magnetorheological composite material, H – direction of the magnetic field vector Kaleta & Lewandowski (2007).

2.4 Test results. Static and cyclic measurements.

Magnetorheological composite tests consist in seeking for relations between magnetic field interacting with the tested material, and mechanical properties of composite. Figure 5 presents two basic tests performed for the magnetorheological composites. The first of them is stroke load of the specimen with constant force for different values of magnetic field (Figure 5a). The changes of shearing stress τ in time were recorded. The influence of magnetic field is clearly visible. The higher values of the magnetic field intensity H, the smaller the shearing strain γ, which indicates for increase in the material stiffness. Figure 5b in turn, shows the cyclic test results. The specimen was strained with constant amplitude of the shearing strain γ_a. Increase in the magnetic field intensity H causes increase in the strain, and by that the increase in hysteresis loop area ΔW. Creation of the hysteresis loop suggests the existence of irreversible strains, which could depend on time scale, or on the strain trajectory length. Clear differences between loading and unloading suggest creation of plastic strains (independent of the time scale). Static shear tests confirm existence of both, viscous and plastic strains in the material. That is, the material is a viscoelastic body up to the yield point (dependent on the strain rate) and the viscoplastic above it Lewandowski & Ziętek (2010). Thus, above the yield point the irreversible strains appear dependent on the time scale and the plastic strain trajectory length. Therefore it is elastic/viscoplastic body.

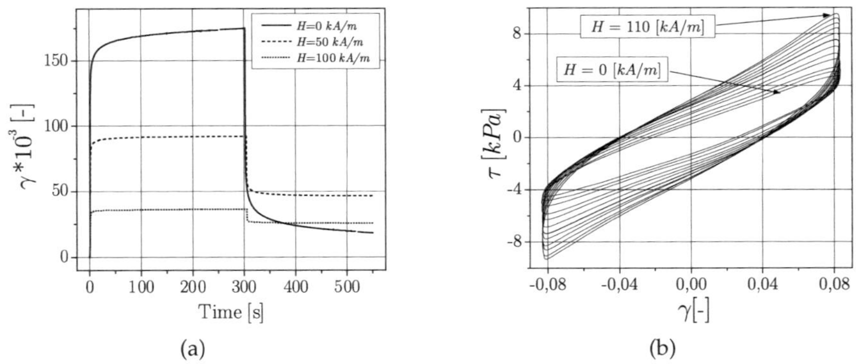

Fig. 5. (a) Change in the shearing strain of magnetorheological composite in time, under the influence of the constant loading and unloading, for three different values of the magnetic field intensity H, (b) change in the shear stress τ and the hysteresis loop area in the stress – strain coordinate system ($\tau - \gamma$), for the same amplitude value of shearing strain γ_a and the increasing values of magnetic field strength H, from 0 to 110 kA/m (with a step equal to 10 kA/m) Kaleta et al. (2007).

Next, the simultaneous influence of the shearing strain amplitude γ_a and of the magnetic field intensity H, on the character of the $\tau - \gamma$ relationship and the size of the hysteresis loop area ΔW, was tested. The tests were performed with controlled stress, yet in such a way, as to maintain the assumed value of shearing strain amplitude γ_a, independent of the applied magnetic field. The set of loops obtained for the subsequently raising values of magnetic field intensity and for three fixed values of strain amplitude has been presented in Figure 6a. It was found that the magnetic field intensity H, has significant influence on mechanical properties of the composite material. The change in magnetic field intensity H from 0 to 110 kA/m causes more than double increase in the stress amplitude τ_a, while maintaining the same amplitude of strain γ_a. Also, the increase in hysteresis loop area Δ_W and the change in its shape may be observed. The test results unequivocally confirm that value of the shearing strain, as well as the size of hysteresis loop area, depend both, on the magnetic field intensity H and on the strain amplitude value γ_a.
As a measure of the inelastic behavior of magnetorheological composite the energy dissipated by the material and expressed by the area of the hysteresis loop area ΔW (in the $\tau - \gamma$ coordinate system) was assumed. The ΔW value, acquired from the experiment, was calculated using the algorithm based at the signal analysis and synthesis by means of the Fourier transform. Figure 6b shows the ΔW values obtained for the selected frequency f = 5 Hz, at three different values of the strain amplitude γ_a. As can be seen, changes in ΔW as a function of the magnetic field intensity H, are not linear in their character.

2.5 Modeling mechanical properties of magnetorheological composites.

The basic model known from the subject literature, concerning mainly the magnetorheological fluid properties, is the so called Bingham body Seval (2002). It belongs to the "segment-linear" model group created by combining the elastic, viscous and rigid-plastic elements (Fig. 7).

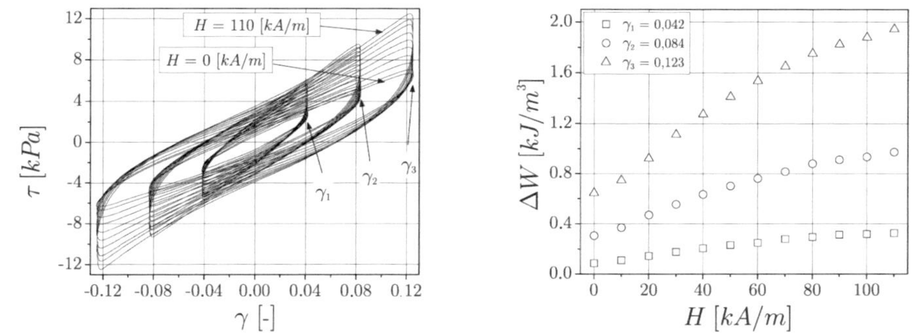

(a) (b)

Fig. 6. (a) - hysteresis loops obtained for growing values of magnetic field strength, from 0 to 110 kA/m (with a step equal to 10 kA/m), and three constant strain amplitudes at excitation frequency f = 10 Hz, (b) - changes of hysteresis loops area ΔW in function of magnetic field intensity H, at frequency excitations $f = 5\ Hz$ and three strain amplitudes γ_a Kaleta & Lewandowski (2007).

Viscous effects appearing in the material – the same as for the typical Newton liquid – are described by means of the viscosity parameter. As shown by the experimental tests Weihua (2000), the linear dependence of shear stress on the shearing rate appears only in the limited range. The additional complication results from the influence of magnetic field, which is capable of causing changes in the model parameters, also in the non-linear way. Magnetorheological fluids show, for example, the non-linear dependence of apparent viscosity on the magnetic field intensity and the strain rate Yamamoto & Nakano (1999). Also, the other models used for description of magnetorheological fluids Butz & von Stryk (1999), e.g. the models of: Cross, Casson, or Herschel–Bulkley, the so called structural and numerical simulations, have limited application.

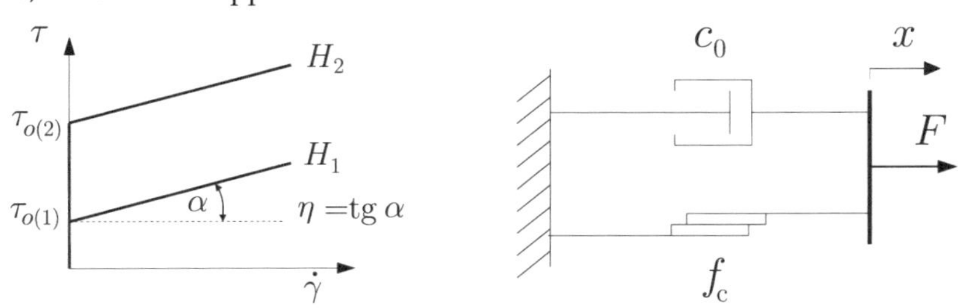

(a) (b)

Fig. 7. Bingham model. (a) – dependence of the strain rate on plasticising stress as a function of the magnetic field H, (b) – mechanical form of the Bingham model - a parallel combination of two elements - the viscous and the plastic one.

While observing the magnetorheological composite behavior it could be noted (Fig. 5a), that above the yield point the irreversible strains appear dependable on the time scale and the plastic strain trajectory length. Thus, it is the elastic/viscoplastic body Kaleta et al. (2007). One of the description methods for such a material is the parallel and series combination of the elastic, viscous and plastic elements Hart (1976). They were used in the soil rheology Kisiel (1967), for describing polymeric materials, in the small and large strain range (finite deformation) Bardenhagen et al. (1997). The viscous element could be the non-Newtonian type liquid. In that case it is assumed that the viscosity coefficient is a non-linear function of strain rate. Further, the plastic member takes into account various types of amplification both, kinematic and isotropic, as well as the mixed one. Non-linear equations and differences in the way of describing the loading and unloading process are a reason for difficulties in determining the material parameters appearing in the model. This is particularly significant when the application area involves both, the static and cyclic loads. The essential challenge here is also building a model useful also in the complex load conditions. Considering the results achieved in the previous experimental trials Kaleta & Lewandowski (2007), the possibility was analysed of describing properties of the discussed magnetorheological composite with the four-parameter rheological model Drescher (1967) of the following form:

$$\tau + t_\gamma \dot{\tau} = 2G_2(\gamma + t_\tau \dot{\gamma}), \tag{1}$$

for the first cycle, and

$$\tau + t_\gamma \dot{\tau} = 2G_2(\gamma + t_\tau \dot{\gamma}) - 2G_2\gamma_k + \tau_{o2}\mathrm{sgn}\dot{\tau}, \tag{2}$$

if the yield point in the second element has not been exceeded. Otherwise,

$$\tau + t_\gamma \dot{\tau} = 2\eta\dot{\gamma} + \tau_{o2}\mathrm{sgn}\dot{\tau} \qquad \text{for} \qquad \tau_2 = \tau_{o2}, \tag{3}$$

where

$$t_\gamma(H) = \frac{\eta(H)}{G_1(H)}, \qquad t_\tau(H) = \eta(H)\frac{G_1(H)+G_2(H)}{G_1(H)G_2(H)} = t_\gamma(H)\left(1+\frac{G_1(H)}{G_2(H)}\right). \tag{4}$$

The model diagram and comparison of the loops acquired from the experiment and from the model have been presented in Figure 8.

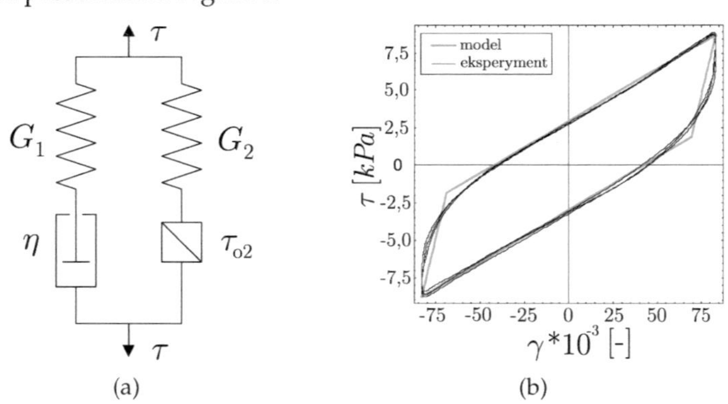

Fig. 8. a) – the elastic-viscoplastic material model of the magnetorheological composite, b) comparison of hysteresis loops obtained from the elastic-plastic model with linear amplification and the experimental ones for the selected values of magnetic field intensities H=90 kA/m Kaleta et al. (2007).

The model is well reflecting the real course during the material loading. However, for unloading phase the straight line crosses the convex part of the loop. Comparison of the experimental and the model data for the selected value of γ_a has been shown in Figure 8b .

2.6 MRC applications

Possibility of controlling the magnetorheological composite properties enables building of active devices, applicable in various structures of the SMART. As an example, a structure of semi-active damper has been presented and its damping capacities checked. Within the task the following stages have been accomplished:

- building the damper prototype with magnetorheological composite use,
- application of the damper in the selected mechanical system and analysis of its influence on vibrations.

Structure of the damper has been presented in Figure 9 Kaleta et al. (2009). Its capacities of the mechanical vibration damping were verified at the steel beam with two supports. In the subject literature, which discusses applications of similar dampers with magnetorheological fluids, usually the attempts of applying them in the discrete vibrating systems with one or two degrees of freedom could be found Carlson (1999); Müller (2000). However, the selected beam could be included to the continuous systems group, characterised with multiple forms of vibrations. For that reason the damping problem requires radically different approach, ready to take into account the nature of vibration and the frequency range appearing in a given object. Schematic diagram of the utilised measurement stand has been presented in Figure 10.

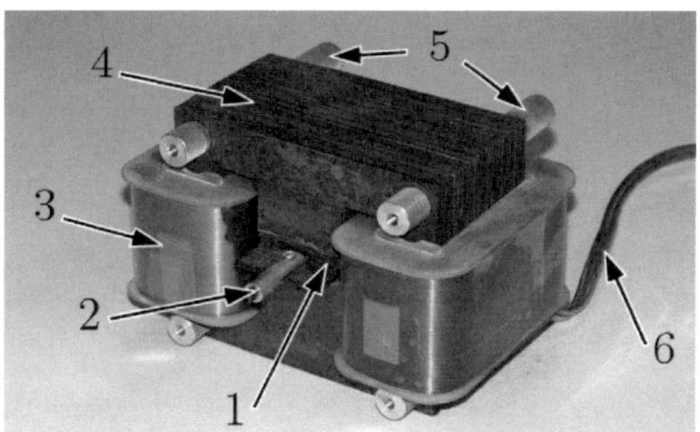

Fig. 9. General view of the damper with magnetorheological composite: 1 – MR composite, 2 – fastener, place of attachment to the vibrating structure, 3 – coils, 4 – core, 5 – damper fixing to the base, 6 – wires conducting current signal to the damper Kaleta et al. (2009).

The damper test results have been presented in Figure 11. A change in the control current intensity has essential influence on the beam vibration (Figure 11a). A drop in amplitude is the most noticeable effect here, and that is why as the estimation of damping influence just the time was assumed (marked as t_w), needed for damping the amplitude to the 20% level of its maximum value. For a beam, as the object simplified to the single-dimension scale, a full series of measurements was performed in sixteen points uniformly distributed over the beam length L. Current intensity value I was changed within the 0 to 4 Amps range for each measurement point. The collected results in the form of three-dimensional graph have been presented in Figure 11b.

2.7 Conclusions and further research directions

The research described above enabled formulation of the following conclusions:

- A composite was manufactured using the magnetorheological fluid and porous elastic matrices. Strong susceptibility of mechanical damping to the magnetic field intensity value was shown.

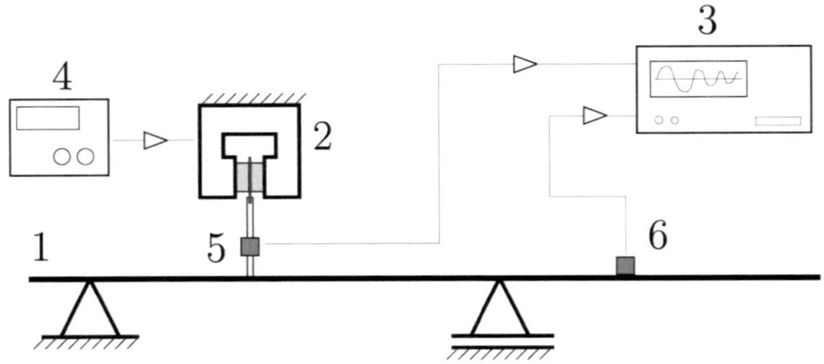

Fig. 10. Test stand: 1 – test object, a beam at two supports, 2 – the tested damper with MR composite, 3 – data collection and processing system, the HP35639A spectrum analyser, 4 – regulated power supply, 5 – force sensor, 6 – acceleration sensor Kaleta et al. (2009).

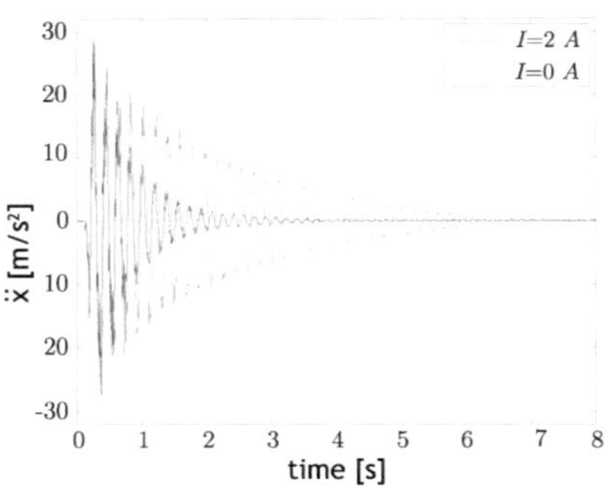

(a)

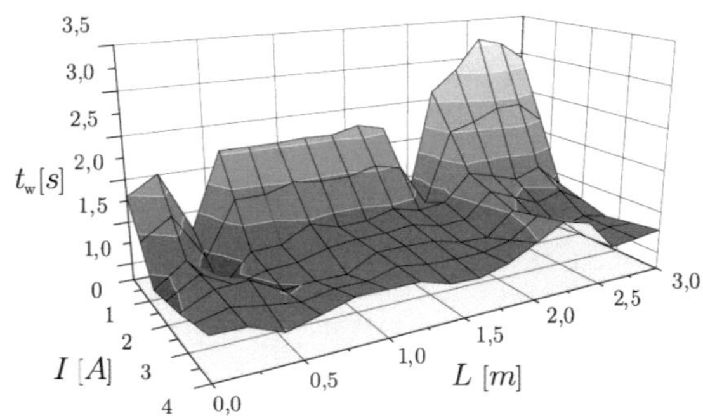

(b)

Fig. 11. a) – free vibration observed at the beam end (L = 3 m) for two intensity values of the current controlling the damper equal to I = 0 and I = 2 A, correspondingly. b) – time of the free vibration amplitude damping t_w in the individual beam locations depending on the control current intensity I Kaleta et al. (2009).

- Experimental conditions were created for determining the damping in composites influenced by the variable parameters of mechanical and magnetic field.
- A description of the tested composite properties was proposed, in the form of four-parameter model of elastic-viscoplastic body. Individual coefficients of the model were assumed as functionally depending on the magnetic field intensity H. The model identification procedure was performed.
- The own prototype structure of the so called semi-active damper with composite containing magnetorheological fluid was presented. As the test vibrating object a beam on two supports was selected. Efficiency of the vibration damping was verified with recording of the free vibration damping time.

It could be forecasted that the main research directions in the nearest future will concern:

- Improvement in the manufacturing engineering of the „wet", porous magnetorheological composites saturated with MRF. The new grades of ferromagnetic micro and nano–powders for magnetorheological fluids and ferrorheological fluids will be manufactured, in which the sol-gel technology will be used.

- The research will be undertaken on the fatigue life of MRC, as well as on the ageing processes of matrices and fluids.
- Key significance will be assigned to simplification and efficiency increase of the magnetic stimulation. As particularly promising in this area the discrete systems for magnetic field control are considered.
- A development in the constitutive MRC models and the ways of their identification will follow.
- In the MRC application domain, first of all they are going to be used for building of cheap, the so called semi-active dampers with wide application potential and the scope of loads born. The main application areas are going to be: transport, buildings and structures for the seismic or quasi-seismic conditions, large industrial structures and home appliances.

3. Magnetorheological Elastomers

3.1 Introduction

Magnetorheological Elastomers, MREs or MAEs in short (Magneto-Active Elastomers), are the controllable materials built of the magnetically polarised particles placed in the non-polarised elastomer matrix. They exhibit variation in mechanical properties under the influence of external magnetic field, which has been named as the magnetorheological effect. As opposed to the magnetorheological fluids (MRF) already known for several tens of years, the MRE are the new materials. Optimal selection of the MRE components, the technology of their manufacture, or the mathematical models of their mechanical properties, are very valid scientific issues. MREs are the investigation objects in the top research centers, among others in the USA, China, Finland, France, Canada, Switzerland, or Sweden.

3.2 Magnetorheological Elastomers build

The two basic components constituting the MRE structure are the elastomer creating the matrix, and the magnetically active particles distributed in it (Fig. 12a). As opposed to the magnetorheological fluids (MRF), the working area of which is located below the flow limit, the magnetorheological elastomers work above that limit. Thus, it could be stated, that the materials do not pose the competition for the magnetorheological fluids, but are rather their supplementation. The schematic Bingham diagram represents this at the Fig. 12b).

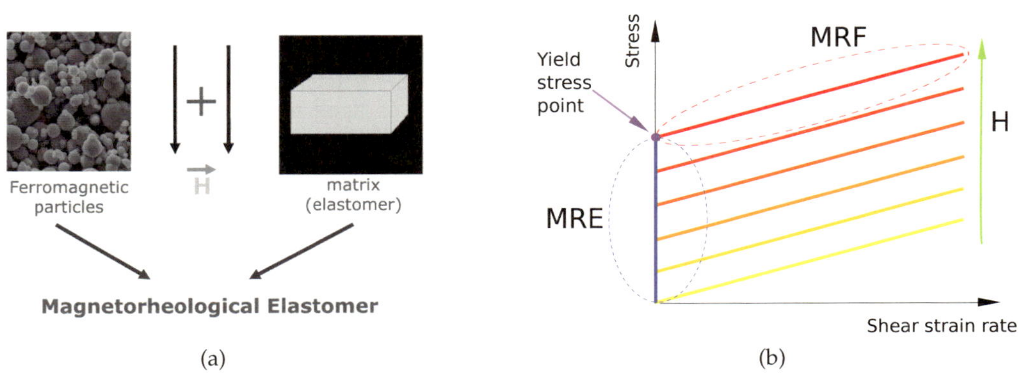

Fig. 12. (a) - the MRE build diagram and (b) - comparison of the MRE and MRF operation areas in the Bingham model Zając (2011).

Among the materials for the MRE matrix, the rubbers predominate: the natural ones Davis (1999), the silicon ones Blom & Kari (2005), as well as the butadieneacrylonitrile ones with various content of the acrylic acid Lokander & Stenberg (2003b). Natural rubbers were also modified by adding polybutadiene (Buna CB55) Ginder et al. (2002), or the zinc oxide, stearic acid and the sulphur Shen et al. (2004). The silicone rubbers, in turn, were mixed with the silicon oil Gong et al. (2005a), vinyl and toluene Wang et al. (2006), as well as polyurethanes and the silicone oil Hu et al. (2005). Also the magnetorheological elastomers with a whole range of other matrices, among others containing synthetic rubber Davis (1999) , polyurethane Shen et al. (2004) , silicone Bose & Roder (2009a) and others Kamińska et al. (2006), were created.

The ferromagnetic particles must be the soft magnetic materials with the low value of residual magnetisation. Moreover, the material of which they are made should be featured with the high permeability and high magnetic saturation value. This provides for the maximum intermolecular attraction producing the large magnetorheological effect. The most frequently selected material for production of MRE is iron, with the saturation induction equal to B_S=2,15 T, whereas the maximum value of the relative magnetic permeability μ_r=9000 – for the iron containing 0,2 % of impurities, and reaches even μ_r=200000 for the pure iron Rawa (2001). While selecting the particles, one has to consider that they should be big enough to contains several magnetic domains, as in the other case the magnetorheological effect could be unnoticeable Lokander & Stenberg (2003b).

However, most frequently the carbonyl iron is used, the particles of which have spherical shape. Usually the particles have rather small diameter, beginning with some 2 μm Bellan & Bossis (2002), through the range of 3 to 5 μm Deng & Gong (2008), and up to 10 μm Kallio (2005). The tests were also performed at materials filled with decidedly larger particles of the pure iron of diameter 400 μm and 800 μm Zhang et al. (2008). In case of the isotropic magnetorheological elastomers the particles of iron with porous surface and much larger in size, from some 60 μm Kari et al. (2002), or even up to 200 μm Lokander (2004), are usually used. The materials filled with a mixture of small particles (with diameters of 3 – 5 μm) and large (with diameters of 70 – 80 μm), in the identical proportions by volume, where all iron particles had 36% share by volume in the material were also tested Stepanov et al. (2009). The composites filled with the needle-shaped iron particles Lokander (2004), as well as the nano-wires of some 300 nm in diameter and 15 μm length Song et al. (2009) were also manufactured.

Two fundamental types of MRE could be distinguished: anisotropic and isotropic ones. In the subject literature the term of isotropic magnetorheological elastomers, the Elastomer Ferromagnet Composite (EFC) Zhou & Jiang (2004) could also be found. As the first ones, the anisotropic composites became the subject of research. In those materials the particles create the chain structures in the matrix. During hardening, the composite is being placed in the magnetic field, which induces the dipole momentum in each of the particles. The particles tend to the energy minimum, which, in this case corresponds to the chain packing with the dipole momentums in parallel to the magnetic field vector line. When later the composite is being strained in the presence of magnetic field, the particles are being led out of the energy minimum state, which requires additional work. Value of that work grows with the increase of the applied magnetic field intensity. Anisotropic MREs are the extremely curious materials from the scientific point of view, yet the technology of their manufacture may prove to be too complicated in case of the large-series production. This results from the necessity of placing the composite in magnetic field during the matrix hardening. That obstacle may be eliminated by applying the isotropic magnetorheological elastomers, in which the particles are distributed homogeneously in the whole material volume. Besides the particle size, the second important parameter is their quantity. In case of the isotropic MREs it has been found that the maximum magnetorheological effect may be obtained if the percent share of particles by volume will be close to the Critical Particle Volume Concentration (CPVC) value Lokander (2004). This corresponds to such particle packing, as appears in the container after their pouring in. The particles contact the others and gaps between them are filled with air. In the composite, air is being substituted with elastomer. If the number of particles is lower than the CPVC value, then the distances between them are larger and the result is a drop in the mutual magnetic interaction. However, if the particles will have too big percent share in relation to the matrix, then the quantity of elastomer will be insufficient for filling in all spaces between the particles. The air voids left decrease the material strength. For iron particles of about 60 μm diameter the CPVC value reaches 36,5%. Part of the isotropic MREs presented in the subject literature contains the number of particles close to the CPVC value: 37% Lokander et al. (2004), 36,5% Kari et al. (2002). MREs of lower particle content were also investigated, among others: 28% Lokander & Stenberg (2003b), 27% Farshad & Benine (2004), while they contained the carbonyl iron of much lower diameter than 60 μm, for which the CPVC value was calculated. In case of the anisotropic MREs the subject literature does not present the relationship enabling determination of the optimum number of ferromagnetic particles. The anisotropic composites of the very high range of particle share by volume from 20,1% Shen et al. (2004), through the intermittent values, among others: 25% Shen et al. (2004), 27 % Farshad & Benine (2004), or about 30% Kallio (2005), up to 50 % of the volume share were investigated Zhou & Jiang (2004). In order to improve the MRE properties, similarly as in the case of magnetorheological fluids, various additives are being applied:

- In order to increase the adhesion between the particles and the matrix, silanes are being used, which improve wettability of particle surfaces. It has been observed that in composites with silane the particles are distributed uniformly in the whole volume, and the elastomer surrounds precisely each of them. Its lack causes that particles glue up and create agglomerates, leaving free spaces between particles and the matrix. Not all types of silanes increase the magnetorheological effect – some of them even decrease it in relation to the base material, despite the improvement of adhesion between particles and the matrix Wang et al. (2006).
- Addition of carbon black to composite matrix improves binding of the matrix with iron particles, thanks to what the magnetorheological effect and tensile strength increases, though the composite suppression drops Chen et al. (2008).
- Iron particles are covered with thin layer of ferric oxides, so the large number of oxides penetrates the material. In addition, the iron ions accelerate the elastomer oxidation, and therefore the antioxidants are also being added Lokander et al. (2004).
- Plasticisers are lowering the modulus of stiffness in the absence of the external magnetic field. This, in turn, leads to among others, the greater relative magnetorheological effect Lokander & Stenberg (2003a). It has been observed that magnetic particles in the composite with silicon oil additive have greater freedom of displacement, thanks to what more regular structures are being created in the magnetic field, increasing that way the magnetorheological effect. Tests have confirmed that increase in the oil content in the composite improves the relative magnetorheological effect, but only to a certain value. On its exceeding the value of stiffness modulus drops in the absence of the external magnetic field, but the magnetorheological effect also drops.

3.3 Manufacturing of Magnetorheological Elastomers. Own research

For manufacturing the MRE the thermoplastic polymer - elastomer type TPE-S (mixture of polypropylene and the G SEBS craton) was used. While selecting the elastomer type for matrix the strength properties, large elongation at rapture and the low hardness were considered, as it has been found that the larger relative magnetorheological effect is obtained for MREs with softer matrix Wang et al. (2006). From the offered products of various hardnesses the material of 30 Shore hardness was selected, and as the ferromagnetic filling the ACS 300 iron powder (from Höganäs AB) was used. As opposed to the carbonyl iron particles, the use attempts of which in the specimens manufacturing were also made, they have rather irregular shape and porous surface as seen at the Figure 13. Most of the ASC 300 powder is constituted by particles of up to 60 μm size.

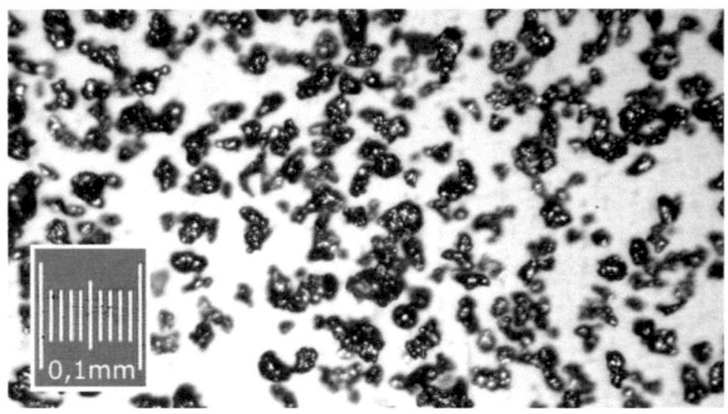

Fig. 13. Ferromagnetic material – filling of the magnetorheological elastomer Zając (2011).

Two groups of MREs were manufactured. The first one involved elastomers filled with the ASC300 iron particles only. Volume share of the iron particles in the composite was established at the base of CPVC value, which amounts to about 36,5 %for the ASC300 iron. The maximum magnetorheological effect should be obtained for specimens containing iron in quantity close to the calculated CPVC value. However, the scientific research has shown, that maximum changes in the mechanical properties induced by the magnetic field activity, were obtained for the material containing 30% by volume of iron particles Lokander (2004). Finally, the specimens containing 35% by volume of the ASC300 particles were manufactured for the tests. The production process of magnetorheological elastomer specimens consisted of the two stages: hot mixing of iron particles with elastomer and shaping the specimens in press. For mixing the two ingredients the Brabender type mixer with chamber heated to the temperature of $190^{o}C$ was used. Simultaneously with granulate, the iron particles were poured, and in case of some of the manufactured composites, also additives such as oil or silane. The whole was mixed for achieving the homogeneous constitution. In the subsequent production stage the plastic mix was placed in the mould and then pressed hot as well as cold. On removing from the mould, two cubicoids were cut out of each of the composite plates. The material sample has been shown in Figure 14 a), and Figure 14b [1] presents the internal structure of the MRE. The cut samples were glued in between the three supporting plates, as has been schematically shown in Figure 15 . This enabled the assumed state of load and strain of the sample. During cyclic tests the external plates remained fixed and the central plate was moving, enabling that way symmetric shear of both material parts. Magnetic field was applied perpendicularly to the strain direction. Figure 16 presents general view of the measurement stand – its magnetic part and the elastomer sample placed in the core center are visible.

[1] `http://www.sgml.pwr.wroc.pl/`

(a) (b)

Fig. 14. a) – macro view of the sample , b) – SEM image of the magnetorheological elastomer Zając (2011).

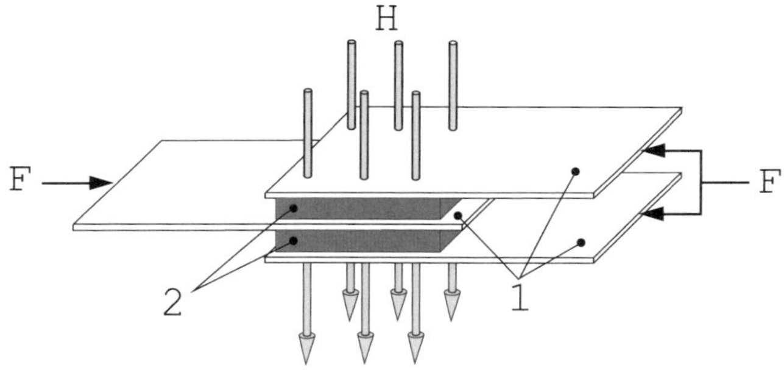

Fig. 15. Schematic diagram of the sample prepared for tests: 1 – laminate plates, 2 – magnetorheological composite Zając (2011); Zając et al. (2010).

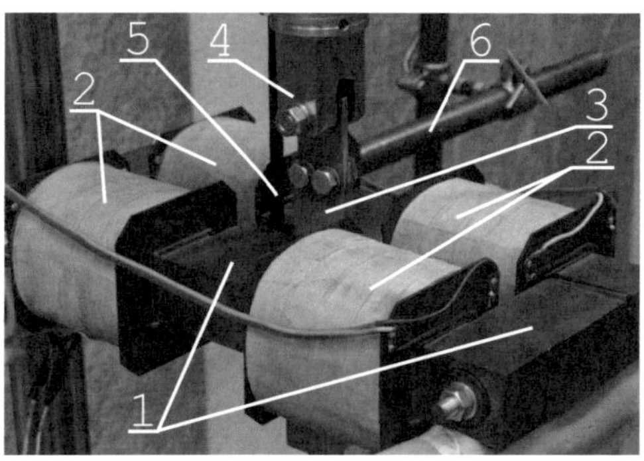

Fig. 16. View of the measurement stand: 1 – magnetic core, 2 – coils, 3 – laminate plates with the glued up magnetorheological composite samples, 4 – stainless steel holder, 5 –Hall probe sensor, 6 – support element Zając (2011); Zając et al. (2010).

3.4 Magnetomechanical properties of MRE

The magnetorheological elastomer investigations described in the subject literature involve, first of all, the sample shearing Lokander (2004); Zając et al. (2010) also their compression Farshad & LeRoux (2004); Popp et al. (2009), as well as tension Bellan & Bossis

(2002); Gong et al. (2005b) in the magnetic field of various intensity, in order to determine their magnetomechanical properties. In addition, the research has been conducted, aimed at finding the other features, such as the electrical parameters of MRE Bica (2010) and their magnetic permeability Bellan & Bossis (2002).
In order to show the changes happening in the matrix material as a result of its doping with powder of the ferromagnetic material, at the first stage the tests for samples of pure thermoplastic elastomer were performed. Next, the tests of MRE filled with the ASC 300 particles, constituting 30% of the volume were performed. The samples were tested in the pure cyclic shear conditions with various amplitude and frequency of the displacement signal values. The hysteresis loops were analysed in the stress – shearing strain ($\tau - \gamma$) co-ordinates, for all the tested materials in the absence of the magnetic field (Figure 17). The differences in the composite and pure elastomer matrix properties are clearly visible.

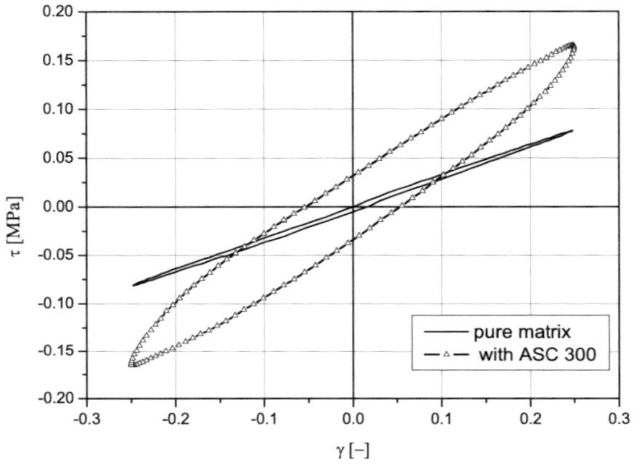

Fig. 17. Influence of the matrix filling with the magnetic particles. Hysteresis loop obtained for cyclic shearing tests Zając (2011).

The subsequent graphs (Figure 18) present the hysteresis loops within the $\tau - \gamma$ co-ordinates for the magnetorheological composite filled with the ASC 300 particles. The loops for the two extreme values of the magnetic field intensity equal to 5 kA/m and 130 kA/m correspondingly. The change in shape and area of the loop with the increase in the magnetic field intensity is visible.
Influence of the magnetic field on mechanical properties of a material is designated as the magnetorheological effect. Hysteresis loops increase their area and change the size. The stress amplitude value τ_a increases for the higher values of the magnetic field intensity, while keeping up to the constant amplitude of the shearing strain γ_a. Creation of the mechanical hysteresis loop at the harmonic loads testifies for the irreversible energy dissipation. The constitutive equations, i.e. the relations between the strain and stress tensors, should consider this phenomenon. The most frequently used models of the body in the subject literature are the linear equations for the viscoelastic body Kallio (2005). In that case, for the sinusoidal loads, the hysteresis loop is an ellipse. At Figures 18a , and 18 b, it could be observed that for describing the loops the more complex body models should be used: the non-linear viscoelastic relationships or the viscoelastic/viscoplastic body models. Selection of the model depends on the type of the irreversible strains being created in the loading process. For that

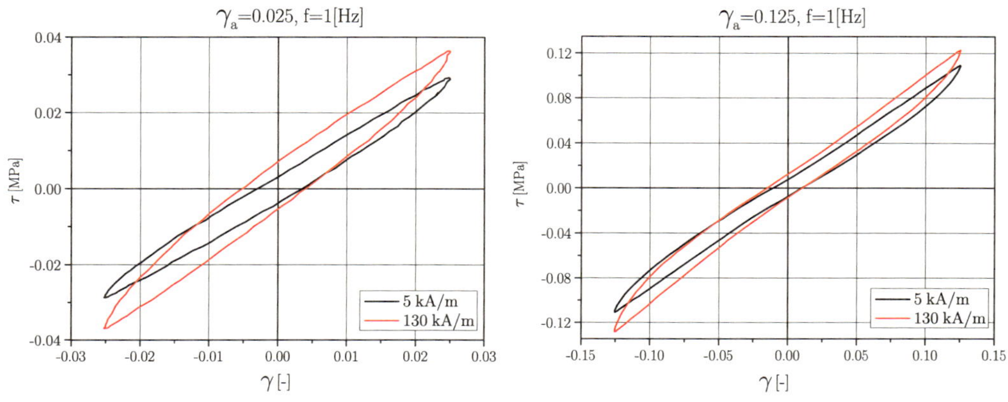

(a) (b)

Fig. 18. Hysteresis loops for the extreme enforcement values: (a) γ_a =0,025 and f=1 Hz, (b) γ_a=0,125 and f=1 Hz. Zając (2011); Zając et al. (2010).

aim it is necessary to perform subsequent experiments giving answer to the question whether the existing irreversible strains depend on the straining trajectory (the plastic strain), or these are the strains depending on the time scale (the viscous strains), as well as determining the area (the yield point), in which they appear. At present the problems are the subject of intensive research Zając (2011).

3.5 Applicability of the MRE

The magnetorheological elastomers have much shorter history than, for example, the magnetorheological fluids, and therefore they yield in the number of industrial applications. Patent applications in that domain began to appear in the 90. of the 20th century, yet their number began to grow only after the year 2000. Both, the material itself Bose & Roder (2009b), and its applications using the unique properties of the MRE have been patented. It could be noted that mainly the solutions with anisotropic magnetorheological elastomers were patented so far. On the fact that magnetorheological composites have high potential of wider existence on the market testifies the establishment in 1999 the Nevada (USA) located the Advanced Materials and Devices (AMAD) enterprise specialising just in the magnetorheological elastomers. An example of the research conducted for the US Navy was development of the suppressing inserts of variable stiffness for the missile launched system on the submarines. The inserts are located between the external missile surface and the internal surface of barrel. From the automotive branch comes large number of the patent applications for the use of magnetorheological composites. One of the first patents for the magnetorheological elastomers use developed just in the Ford Motor Company Watson (n.d.), concerns regulation of the vehicle suspension element stiffness through the use of a sleeve with the controlled stiffness. In that solution the magnetorheological elastomer is located between two sleeves, where the internal one is attached to the movable suspension element and the external one is attached to the body. Ford Global Technologies developed also a device for measuring the displacement and force in the automotive suspension on the run Elie et al. (1998). In the German ThyssenKrupp AG company a steering gear column with the adaptive energy absorption system during a car accident Klukowski & Meier (2006) was developed. Other examples of the MRE application have been described in Brei et al. (2006); Browne & Johnson (2006); Lerner & Cunefar (2006).

3.6 Further research justification and directions

Composites of the MRE type give evidence to their numerous advantages, including the operating range above the yield point (in comparison for example to the magnetorheological fluids), their recyclability in case of matrices of the plastic elastomer, as well as the relatively simple manufacturing process engineering. This allows for predicting the further numerous applications of MREs and, at the same time, justifies the need for further research.

It could be assumed that in the nearest period the activity of scientific and engineering teams will focus mainly on:

- Improvement of the magnetorheological composites manufacturing engineering. The works in that area are to be related with searching for the optimum matrices, active ferromagnetic fillings (types and sizes of powders), as well as the manufacturing parameters such as the magnetic field intensity and temperature.
- Development of the constitutive models enabling description of the MREs' magnetomechanical properties. The models identification will enable foreseeing the MREs' behaviour and will lower the costs of experimental works.
- Optimising the systems for applying magnetic field with the stress on the role of the strong permanent magnets. That would diminish the size and costs of magnetic field generation.
- Presentation of the wide spectrum of applications, mainly in the suppression of mechanical vibration, leading to increase in production of MREs and lowering their production costs.

4. Composites with giant magnetostrictive particles

4.1 Introduction

Below, the composites containing selected powder of magnetostrictive materials have been described. In relation to that, it has been considered as just to clarify the notion of magnetostriction, and then characterize the magnetostrictive materials, including the materials with the so called giant magnetostriction. The main representative of the last group is Terfenol-D. Next, the study methodology for those materials, their properties and the justification for creating composites containing the materials with giant magnetostriction, has been described. Attention was also paid to the current applications of these composites and possible directions of their further development.

4.2 Magnetostriction

Magnetostriction (from Greek 'magnet' - magnet and Latin 'strictus' - compressed, tight, tense) is a domain of magnetism dealing with "phenomena related to interaction between magnetic quantities and the stresses and mechanical deformations" Bomba (2009); Kaleta (2004). Magnetostriction as the phenomenon could also be defined as a change in material dimensions caused by a change in its magnetic state Buschow & de Boer (2004). Most often, the magnetostrictive materials change their dimensions as a result of a change in magnetic field applied to them, as well as change their magnetic properties under the force applied to them. The most commonly used is more general definition, which says that: "magnetostriction consists in changing shape and dimensions, as well as mechanical properties, as a result of magnetic field influence, or - the opposite - in changing magnetic properties, e.g. induction (or magnetisation) and permeability (or susceptibility), under the influence of stresses and strains" Bomba (2009); Kaleta (2004). The most commonly known phenomenon related to magnetostriction is the Joule effect, otherwise called as plain magnetostriction phenomenon, or the linear magnetostriction Buschow & de Boer (2004). It consists in change in length of the material in one of the directions, under the influence of the applied magnetic field with simultaneous change in the proper transverse section, while the material volume is being kept constant all the time. This is caused by the fact, that in the remaining directions the magnetic domains are being set so, as to mutually neutralise their interaction (the compensation of transverse and longitudinal magnetostriction takes place), due to which the internal energy of the system is lowered. We talk of longitudinal magnetostriction Bomba (2009); Kaleta (2004) when the strain in material, to which the magnetic field H was applied, is consistent with the field lines. The longitudinal magnetostriction coefficient is then defined as:

$$\lambda_{\parallel} = \lambda_1 = \frac{\lambda_1 - \lambda_0}{\lambda_0} = \frac{\Delta\lambda}{\lambda}, \tag{5}$$

where:

- $\Delta\lambda$ - change in length as a function of magnetic field intensity,
- λ_0 - initial specimen length,
- λ - specimen length in the magnetic field.

The transverse magnetostriction Bomba (2009); Kaleta (2004) in turn, appears when strain in material takes place in the direction perpendicular to the applied magnetic field. The coefficient of transverse magnetostriction $\lambda_{\perp}$ (or λ_t) is defined the same way as the longitudinal one. For the saturation field H_s the magnetostriction coefficients are marked as λ_s , λ_{ls} and λ_{ts} correspondingly.

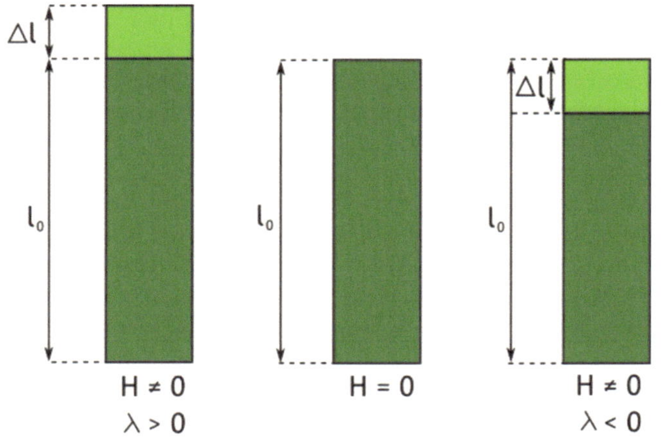

Fig. 19. The essence of positive and negative magnetostriction.

Magnetostriction could be positive or negative. The positive magnetostriction ($\lambda > 0$) occurs in magnetite, permendur, permalloys – usually with content of 45 - 65 % of Ni – alfers, or weakly magnetised iron. The negative magnetostriction ($\lambda < 0$) is characteristic for such materials as nickel and nickel ferrites. Figure 19, presents both types of magnetostriction. As seen in case of the positive magnetostriction the material increases its length, and in case of the negative one, the length is decreasing.

Features, which characterise magnetostriction are:

- even phenomenon, which signifies that a change in magnetic polarization sign (of the field or induction intensity) is not accompanied by a change in the magnetostriction sign,
- magnetic and thermal hysteresis,
- anisotropy, i.e. dependence on the shape and temperature.

Other phenomena related to magnetostriction are:

- Barnett effect,
- Villari effect or magnetoelasticity,
- Guillemin and Wiedemann effects (derivatives of the Joule effect),
- Barnett and Einstein-de Haas effects,
- Barkhausen effect.

The phenomena are widely discussed in the subject literature and are not to be analysed here.

4.3 Giant Magnetostrictive Materials

Materials, which possess the strong magnetic properties involve materials of the so called giant magnetostriction (GMM - Giant Magnetostrictive Materials). These are the alloys composed mainly of terbium (Tb), dysprosium (Dy) and pure iron Kaleta (2004). Elements such as Tb, Dy, as well as gadolinium (Gd), holmium (Ho), erbium (Er) possess the unfilled subshell 4f, due to which, similarly as it was in case of Fe, Ni, or Co, they have uncompensated spins, which characterizes their ferromagnetic properties Mech (2008). Materials with giant magnetostriction may change the magnetic energy into mechanical one and vice versa. Due to such properties the materials could be used as sensors or actuators. The GMMs obtain much larger strains (even up to 70 times) than the traditional magnetostrictive materials, and for obtaining that effect, the relatively low magnetic field intensity H is required. Very important feature of those materials is their wide working temperature range, as well as their low inertia (small area of hysteresis loop), which allows for their application in various conditions. The Curie temperature for these materials amounts to 653-693 K, and the working temperature for these materials may reach 473 K. Stresses created by the GMMs may amount up to 30 MPa Claeyssen & Lhermet (2002). Possibility of predicting the mechanical and magnetic properties of such materials as the Terfenol-D caused that they find all the wider application in various engineering domains Monaco et al. (2008). Typical examples of their application are magnetostrictive relays, which could be controlled by conventional amplifiers, much below the resonance frequencies, which in effect enables for much lower control voltage. This advantage has special meaning in case of the medical applications and enables simplification of the control systems.

4.4 Composites - new approach to GMMs

The solid magnetostrictive materials, despite their numerous advantages, have several disadvantages making their wider industrial application difficult. First of all, their significant drawback is high brittleness, resulting in the low tensile strength. The other limitations are the eddy currents of significant value, which restrict the effective working frequency of devices to several kilohertz. Important parameter is also the price of Terfenol-D, which is kept at the level of 2 $/1 g. The above limitations are generating the need for finding new solutions. One of them is magnetostrictive composites. The main object of the research are composites based on epoxy resins, where the magnetically active filling usually have the form of powder Lo et al. (2006); McKnight (2002), but also the filaments Lo et al. (2006); Or & Carman (2005), or flakes of the magnetostrictive material Liu et al. (2006). Aim of the research is determining properties of the created composites and their potential applications. Usually, properties of the composites and the monolithic magnetostrictive material, which, because of its good properties in the room temperature is Terfenol-D, are also being compared. Having in mind the above, it has been assumed that this is particularly important topic as scientific and engineering issue, and at the same time the aim of the chapter, is presentation of:

- production engineering for the new composite types with the GMM powders ,
- composite tests methodology in the conditions of simultaneous stimulation with the mechanical and magnetic field,
- magnetomechanical properties (magnetostriction) for the created composite and comparison with the solid GMM material.

4.5 Magnetomechanical properties of the GMMc. Manufacturing, research methodology, properties

Magnetostrictive composite (from now on called the GMM composite, or shortly the GMMc), has been manufactured by combining the epoxy resin and the powder of GMM type material (Terfenol-D) of grain size ranging from 5-300 μm - the shape and grain sizes are shown in Figure 20. It is clearly visible that powder particles have diversified shape and size, and their edges are sharp.

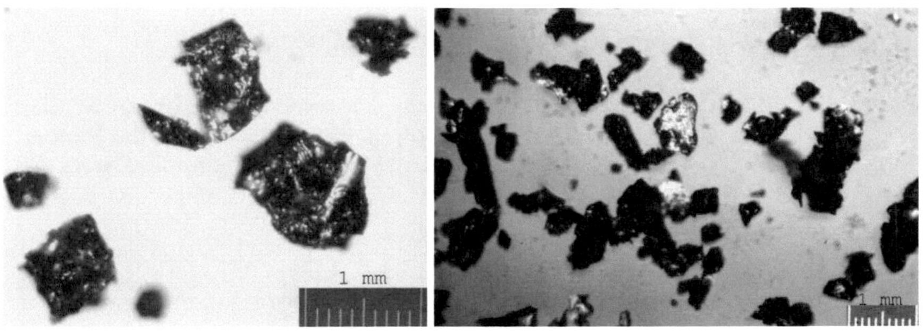

Fig. 20. Microscopic images of the Terfenol-D powder grains.

Specimens of diversified parameters have been manufactured. The manufacturing procedure is presented in Figure 21. At the first stage (A), a hardening agent was introduced into the Epolam 2015 epoxy resin (from Axons Technologies), and next (D), the measured quantity of the Terfenol-D powder (from Gansu Tianxing Rare Earth Functional Materials Co.,Ltd.). Then, (F) the whole was mixed until homogenization of all ingredients was reached. The mixture was subjected to deaeration process (G) and poured over to containers of cylindrical shape (H). The containers were subjected to initial polarization (I) and again deaerated. The specimens were subjected to subsequent polarization (J), which prevented sedimentation of the powder particles during the resin binding process.
Subsequently, the specimens were subjected to tests in order to determine their magnetomechanical properties. Magnetostriction measurement required specialist test stand. Its main part was electromagnetic circuitry composed of coil and steel casing. Current flowing through the coil created magnetic field inside it and the external casing had the task of limiting propagation of the magnetic flux outside and increasing the field inside the coil. Diameter of a hole passing through the coil axis was small enough to cause the maximum impact of the

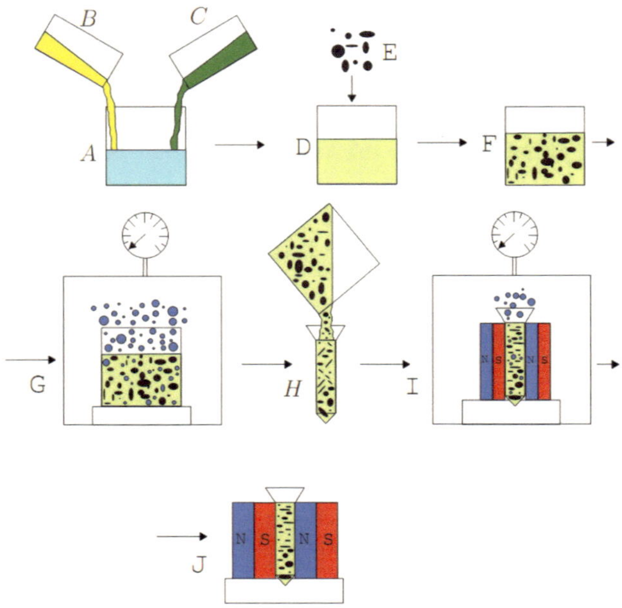

Fig. 21. Manufacturing procedure of composite specimen.

created magnetic field on a specimen located in it. Moreover, the fibre-optic displacement sensor was placed in the hole (the Fiber Bragg Gratings - FBG), as well as the temperature and magnetic field sensors. Localization of strain sensors at the specimen has been presented in Figure 22a and b.

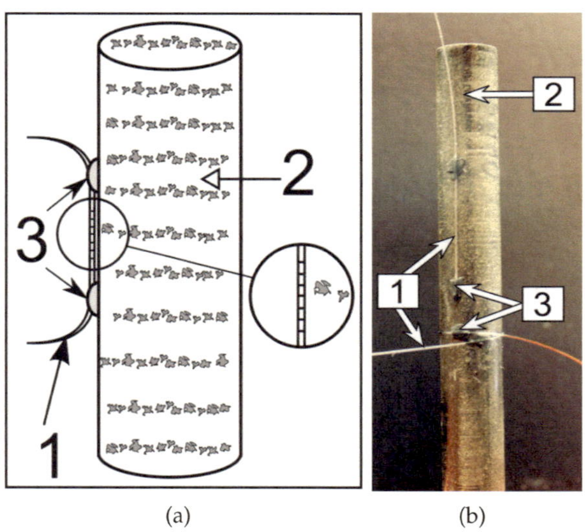

Fig. 22. The way of strain sensors distribution at a specimen: a) schematic diagram: 1 - sensor, 2 - composite, 3 - glue b) photo of the real object.

The reason for using the FBG sensors for strain measurement was their small geometric size and resistance to electromagnetic disturbance. This has a key meaning in case of tests in the conditions of stimulation with strong magnetic field (the results from other sensors, such as strain gauges and others, in which the electric quantities are recorded, are not reliable because of the high disturbances). The measurement method with the FBG use has been presented in the work Blazejewski et al. (2011).
Next, the magnetostriction value $\Delta\lambda_{max}$ was determined for the composite at the maximum intensity of the magnetic field H=158 kA/m. Figure 23a and b presents the comparison of the obtained results for the composite materials and the solid Terfenol-D, at the initial load equal to σ_0=1 MPa and σ_0=7 MPa correspondingly. Among the manufactured composite

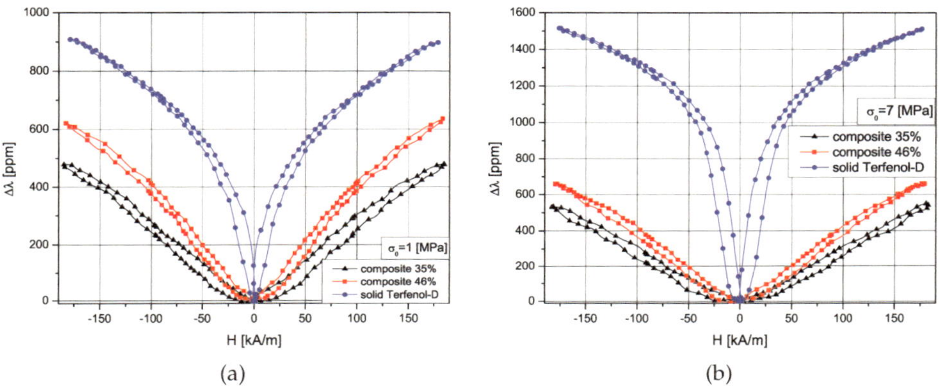

Fig. 23. Comparison of the magnetostriction results for the composite materials and the solid Terfenol-D, at the initial loads of: a) σ_0=1 MPa and b) σ_0=7 MPa correspondingly.

materials the best results were obtained for the specimen containing the highest content by volume (46%) of the Terefenol-D powder. Then, for the initial stress of σ_0=1 MPa, the magnetostriction of composite material is lower by only 280 ppm from the magnetostriction of the solid material, that is by 30%. Whereas, at the higher value of the initial load σ_0=7 MPa, it is lower from that for the solid material by 700 ppm, that is by 45%. However, it has to be noted, that by applying the composite the eddy currents phenomenon was almost eliminated and the tensile strength increased.
Figure 24a and b present measurement results for the composite material and the solid Terfenol-D, both, along the specimen axis, and in the circumferential direction. The FBG sensors glued up in both directions were used in the measurements (see Figure 22b). That way, it has been shown that both, in the solid material and in the composite, the volume magnetostriction does not appear. Moreover, the influence of the initial stress (prestress) at magnetostriction of both materials has been shown. It could be noted that increase in the σ_0 (prestress) to the value of 7 MPa significantly increases magnetostriction of the solid Terfenol-D, and it has no practical influence at the magnetostriction of composite.
Subsequent graphs in Figure 25a and b present the change in volume, which happened in case of both tested specimens depending on the magnetic field intensity and for the different values of prestress, to which the specimens were subjected during the tests, and which amounted to 0,1 MPa and 4 MPa correspondingly. From the obtained results it could be unequivocally stated that in case of the solid material, as well as in case of the manufactured composite materials, the Berrett effect is marginally small, hence one could attempt the

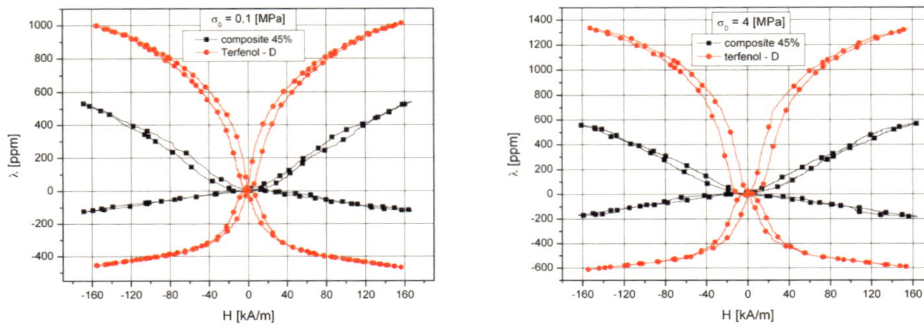

(a) (b)

Fig. 24. Comparison of magnetostriction results in the longitudinal and circumferential directions for the composite material with content of 46% by volume of the Terfenol-D powder and the solid material. The prestress amounted to: a) σ_0=0,1 MPa and b) σ_0=4 MPa, correspondingly.

statement that the effect in these materials could be neglected during the tests. Large magnetostriction of the Terfenol-D has caused that the application works concern mainly the actuators Shaffer & DeChurch (2004). It has been known however, that in the GMM materials, the reverse magnetostriction (the Villari effect, i.e. the alteration in material magnetisation influenced by the applied mechanical load) assumes large values. This allows using the GMMs for a construction of sensors Quattrone et al. (2000). Therefore, an attempt to study the Villari effect in the composites manufactured with the GMM powder was undertaken. For that aim, the proper preparation of specimen surfaces was necessary so, as to be able to make measurements of the magnetic field intensity at a constant distance from a specimen. The composite containing 46% volume fraction of Terfenol-D particles was selected for the tests.

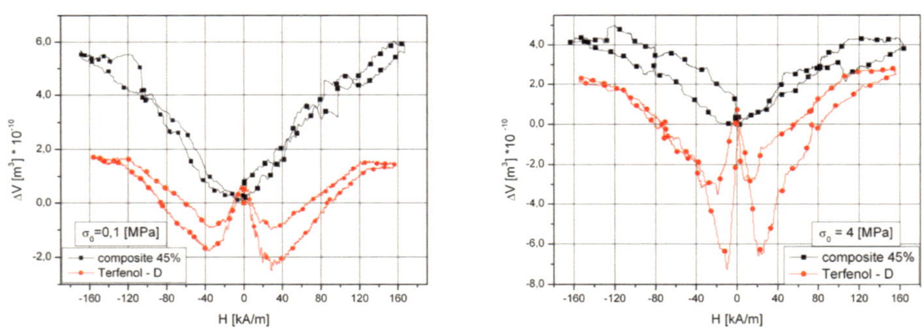

(a) (b)

Fig. 25. Change in the volume of specimen made of the composite material with 46% by volume content of the Terfenol-D powder and the solid material, depending on the magnetic field intensity value, for the prestress amounting to: a) σ_0=0,1 MPa and b) σ_0=4 MPa correspondingly.

The composite was divided into two parts, and cut crosswise and in parallel to the initial specimen polarisation line. The sample obtained that way was placed in the testing machine.

A magnetic field sensor was applied to its surface in three different directions, as shown in Figure 26a - c. The measurements were performed for cyclically loaded sample, with various force amplitudes.

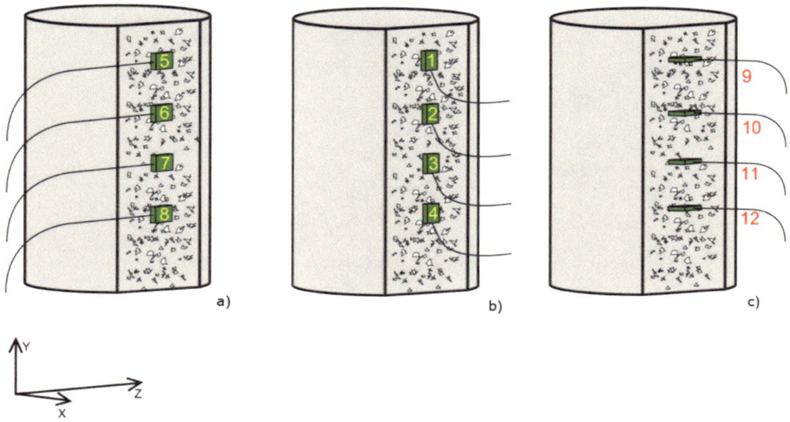

Fig. 26. The way of positioning the magnetic field sensor along the tested sample, where: a) Z component, b) Y component, and c) X component of the magnetic field.

Figure 27 presents the results obtained by means of the magnetic field sensor, for the selected sensor positioning (position 1 at Figure 26b). It could be noted that with the increase in force amplitude the magnetic response of the material measured with the magnetoresistor was growing.

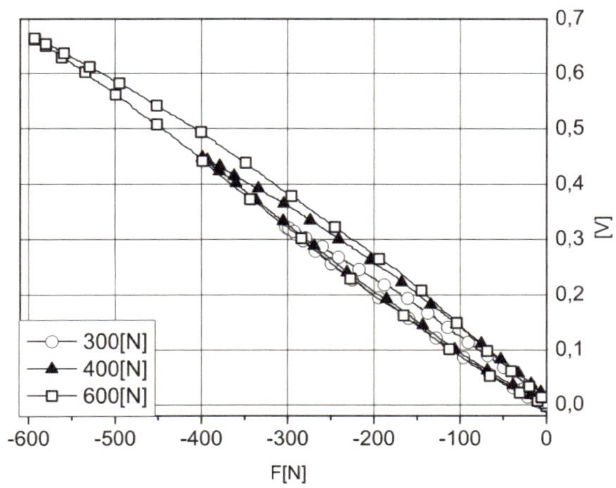

Fig. 27. Change in the magnetic field value for the sensor in position 1 (see Figure 26b).

The change in magnetic field around the sample was causing a change in the voltage induced in the sensor. It is worth of underlining that with growing load of the sample, the shape of the hysteresis loop was not changing significantly. In addition, the changes in the material response to the mechanical enforcement depending on a sensor positioning along the sample were recorded (Figure 28). The presented results concern all sensor localizations shown in Figure 26b. The changes in the hysteresis loop shape, magnetic field sign, and what is equally important, in the value of that field for the tested sample, have been shown. This indicates that the field distribution over the tested samples length is diversified.

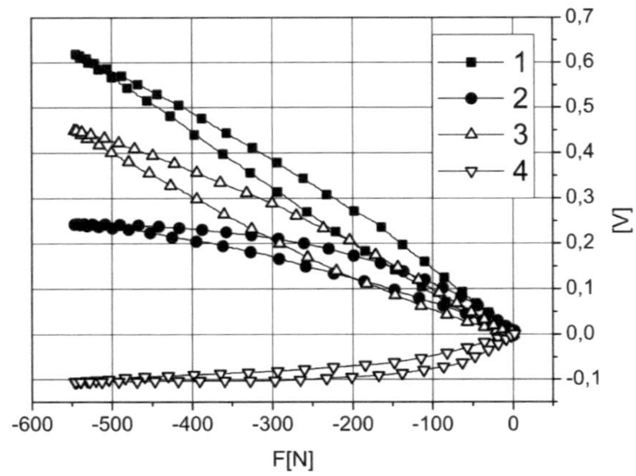

Fig. 28. Change in the magnetic field value depending on the sensor localization (for the situation as in Figure 26b).

In sum, we were able to show, that the manufactured composite materials, similarly as the solid materials, show considerable effect of the reverse magnetostriction, as well as to determine distribution of the magnetic field acting along each of the samples. That means, that composites containing powdered GMM may be used for creating the new generation sensors. The new fields of application, such as: hearing aids, extensometer systems, accelerometers, proximity detectors, momentum sensors, magnetometers and many others, could be indicated.

Study of the Villari effect in composites enabled also indicating the use of those materials in the so called harvester, which converts the mechanical energy into the electric current (known as the energy harvesting) Kaleta et al. (2010).

4.6 Justification and directions of further research

Significant advantages of the composites containing powdered GMM (elimination of the eddy currents, rise in the tensile strength), despite the lower values of magnetostriction and the Villari effect (in comparison to the solid GMM materials of the Terfenol-D type) provide for a series of applications and justify further research. As particularly valid, the following issues should be considered:

- Investigation of how magnetostriction is influenced by the powder fraction change, i.e. the use of larger grains than those used in the presented work, as well as application of grains of the up to 100 μm size. Additional parameter, that could change is the form of the powder (flakes, fibers, or the powder of oval grain shapes). The subject literature in that field is scarce and the results disputable.
- Estimation of the composite matrix influence and verification, whether there are any interactions between the matrix and the filling, as well as checking what influence has the density and hardness of the resin on the results.
- Verification of how the manufactured composite behaves during cyclic, random and pulse changes in the magnetic field intensity. Moreover, the investigation whether the use of

stronger magnetic field, over the value of 1 T, will cause saturation of the material and more non-linear distribution of the hysteresis loop field is justified.

- Verification of how the change in positioning of the magnetic field polarisation vector in relation to the main axis of a sample influences the composite magnetostriction. This could enable obtaining the optimum angle of the sample polarisation, in order to reach the highest values of magnetostriction.
- Investigation of the GMM composites magnetostriction for the case of static tension, which could enable better material characteristics and widen its application area.
- Continuation of works aimed at utilization of that composite class in the energy harvesting.

5. References

Bardenhagen, S. G., Stout, M. G. & Gray, G. T. (1997). Three–dimensional, finite deformation, viscoplastic constitutive models for polymeric materials, *Mechanics of Materials* 25: 235–253.

Bellan, C. & Bossis, G. (2002). Field dependence of viscoelastic properties of mr elastomers, *International Journal of Modern Physics B* 16: 2447–2453.

Bica, I. (2010). Magnetorheological elastomer-based quadrupolar element of electric circuits, *Materials Science and Engineering B* 166: 94–98.

Blazejewski, W., Gasior, P. & Kaleta, J. (2011). *Application of Optical Fibre Sensors to Measuring the Mechanical Properties of Composite Materials and Structures, Advances in Composite Materials - Ecodesign and Analysis*, 978-953-307-150-3, InTech.

Blom, P. & Kari, L. (2005). Amplitude and frequency dependence of magneto-sensitive rubber in a wide frequency range, *Polymer Testing* 24: 656–662.

Bomba, J. (2009). *Damping in giant magnetostrictive material. Experiment, modeling, identification (in Polish).*, PhD thesis, Wroclaw University of Technology, Institute of Materials Science and Applied Mechanics.

Bose, H. & Roder, R. (2009a). Magnetorheological Elastomers with High Variability of Their Mechanical Properties, *Journal of Physics: Conference Series* 149.

Bose, H. & Roder, R. (2009b). US Patent 7,608,197 B2. Magnetorheological Elastomers and use thereof.

Brei, D., Redmond, J., Wilmot, N., Browne, A., Johnson, N. & Jones, G. (2006). US patent application US 2006/0012191 A1. Hood latch assemblies utilizing active materials and methods of use.

Browne, A. & Johnson, N. (2006). US Patent 7,029,056 B2. Closure lockdown assemblies and methods of utilizing active materials.

Buschow, K. & de Boer, F. (2004). *Physics of Magnetism and Magnetic Materials*, 0-306-47421-2, Plenum.

Butz, T. & von Stryk, O. (1999). Modeling and simulation of rheological fluid devices, *Technical Report SFB-438-9911*, TU-München, Universität Augsburg.

Carlson, J. D. (1999). Low–cost MR fluid sponge devices, *in* R. Tao (ed.), *Proceedings of the 7th International Conference on Electro-Rheological Fluids and Magneto-Rheological Suspensions*, World Scientific, pp. 621–628.

Carlson, J. D. & Jolly, M. R. (2000). MR fluid, foam and elastomer devices, *Mechatronics* 10: 555–569.

Chen, L., Gong, X. & Li, W. (2008). Effect of carbon black on the mechanical performances of magnetorheological elastomers, *Polymer Testing* 27: 340–345.

Claeyssen, F. & Lhermet, N. (2002). Actuators based on giant magnetostrictive materials, *ACTUATOR 2002, Proceedings of 8th International Conference on New Actuators* A: 148–153.

Davis, L. (1999). Model of magnetorheological elastomers, *Journal of Applied Physics* 85(6): 3348–3351.

Deng, H. & Gong, X. (2008). Application of magnetorheological elastomer to vibration absorber, *Communications in Nonlinear Science and Numerical Simulation* 13: 1938–1947.

Deshmukh, S. S. & McKinley, G. H. (2006). Adaptive energy-absorbing materials using field-responsive fluid-impregnated cellular solids, *Smart Materials and Structures* .

Drescher, A. (1967). Analysis of some elasto-visco-plastic body model (in Polish), *Rozprawy Inżynierskie* 15: 713–729.

Elie, L., Ginder, J., Mark, J. & Nichols, M. (1998). Us patent 5,814,999. method and apparatus for measuring displacement and force.

Farshad, M. & Benine, A. (2004). Magnetoactive elastomer composites, *Polymer Testing* 23: 347–353.

Farshad, M. & LeRoux, M. (2004). A new active noise abatement barrier system, *Polymer Testing* 23: 855–860.

Ginder, J., Clark, S., Schlotter, W. & Nichols, M. (2002). Magnetostrictive phenomena in magnetorheological elastomers, *International Journal of Modern Physics B* 16: 2412–2418.

Gong, X., Zhang, X. & Zhang, P. (2005a). Fabrication and characterization of isotropic magnetorheological elastomers, *Polymer Testing* 24: 669–676.

Gong, X., Zhang, X. & Zhang, P. (2005b). Study of mechanical behavior and microstructure of magnetorheological elastomers, *International Journal of Modern Physics B* 19: 321–327.

Hart, E. W. (1976). Constitutive relations for the nonelastic deformation of metals, *Journal of Engineering Materials and Technology* 98: 193–202.

Hu, Y., Wang, Y., Gong, X., Gong, X., Zhang, X., Jiang, W., Zhang, P. & Chen, Z. (2005). New magnetorheological elastomers based on polyurethane/si-rubber hybrid, *Polymer Testing* 24: 324–329.

Kaleta, J. B. J. (2004). An initial investigation into the change in magnetomechanical properties of terfenol-d rod due to prestress and temperature, *Issue of Material Testers* 1.

Kaleta, J. & Lewandowski, D. (2007). Inelastic properties of magnetorheological composites: I. fabrication, experimental tests, cyclic shear properties, *Smart Materials and Structures* 16(5): 1948–1953.
URL: *http://stacks.iop.org/0964-1726/16/1948*

Kaleta, J., Lewandowski, D., Liberda, M., Mech, R. & Wiewiorski, P. (2010). Power generating by high pulse mechanical stimulation of magnetic coupled NdFeB and Terfenol-D, *Behavior and Mechanics of Multifunctional Materials and Composites 2010, SPIE* 7644OU: 1–12.

Kaleta, J., Lewandowski, D., Zając, P. & Kustroń, P. (2009). Magnetorheological composites as semi-active elements of dampers, *Journal of Physics: Conference Series* 149: 1.

Kaleta, J., Lewandowski, D. & Ziętek, G. (2007). Inelastic properties of magnetorheological composites: II. Model, identification of parameters, *Smart Materials and Structures* 16(5): 1954–1960.
URL: *http://stacks.iop.org/0964-1726/16/1954*

Kallio, M. (2005). *The elastic and damping properties of magnetorheological elastomers*, PhD thesis, VTT, Finlandia.

Kamińska, J., Pigłowski, J. & Zając, P. (2006). Magnetorheological properties of rubber compositions with carbonyl iron powders, *Conference Polymerwerkstoffe 2006, Halle/Salle* .

Kari, L., Lokander, M. & Stenberg, B. (2002). Structure-borne sound properties of isotropic magneto-rheological rubber, *Kautschuk Gummi Kunstoffe* 55(12): 1–5.

Kisiel, I. (1967). *Application of the rheological model body M/V in soil mechanics (in Polish)*, Ossolineum.

Klukowski, C. & Meier, R. (2006). Adaptive linear crash systems – intelligent materials for passenger car steering column, *ThyssenKrupp techforum* 1: 41–47.

Lerner, A. & Cunefar, K. (2006). US Patent 7,102,474 B2. Adaptable vibration absorber employing a magnetorheological elastomer with a variable gap length and methods and systems therefor.

Lewandowski, D. (2005). *Damping properties of magnetorheological composites. Tests, models, identification (in Polish)*, PhD thesis, Wroclaw University of Technology, Institute of Materials Science and Applied Mechanics.

Lewandowski, D. & Ziętek, G. (2010). *Advances in the mechanics of Inhomogeneous media*, University of Zielona Góra Press.

Liu, J., Or, S. W., Lo, C. Y. & Zhang, Z. (2006). Magnetic and Magnetostrictive Properties of $Tb_xDy_{0.7-x}Pr_{0.3}(Fe_{0.9}B_{0.1})_{1.93}$ Compounds and Their Composites, *IEEE TRANSACTIONS ON MAGNETICS* 42.

Lo, C. Y., Or, S. W. & Lai Wa Chan, H. (2006). Large Magnetostriction in Epoxy-Bonded Terfenol-D Continuous-Fiber Composite With [112] Crystallographic Orientation, *IEEE TRANSACTIONS ON MAGNETICS* 42.

Lokander, M. (2004). *Performance of Magnetorheological Rubber Materials*, PhD thesis, KTH, Sztokholm.

Lokander, M., Reitberger, T. & Stenberg, B. (2004). Oxidation of natural rubber-based magnetorheological elastomers, *Polymer Degradation and Stability* 86: 467–471.

Lokander, M. & Stenberg, B. (2003a). Improving the magnetorheological effect in isotropic magnetorheological rubber materials, *Polymer Testing* 22: 677–680.

Lokander, M. & Stenberg, B. (2003b). Performance of isotropic magnetorheological rubber materials, *Polymer Testing* 22: 245–251.

Malcolm, J. W., Fuchs, A. & Gordaninejad, F. (2002). Development and characterization of magnetorheological polymer gels, *Journal of Applied Polymer Science* 84(14): 2733–2742.

McKnight, G. (2002). *[112] Oriented Terfenol-D Composites*, PhD thesis, University of California.

Mech, R. (2008). *Investigation of magnetomechanical properties of composites based on powders of giant magnetostrictive materials (in polish)*, Master's thesis, Wroclaw University of Technology.

Monaco, E., Lecce, L., Natale, C., Pirozzi, S. & May, C. (2008). Active noise control in turbofan aircrafts: theory and experiments, *Acoustics 08* pp. 4629–4634.

Müller, S. (2000). *Zum Einsatz von Semi-Aktiv Aktoren zur Optimale Schwingungsreduktion in Tragwerken*, PhD thesis, Ruhr–Universität Bochum, Institut für Mechanik.

Or, S. W. & Carman, G. P. (2005). Dynamic Magnetoelastic Properties of Epoxy-Bonded Terfenol-D Particulate Composite With a Preferred [112] Crystallographic Orientation, *IEEE TRANSACTIONS ON MAGNETICS* 41.

Popp, K., Zhang, X., Li, W. & Kosasih, P. (2009). MRE properties under shear and squeeze modes and applications, *Journal of Physics: Conference Series* 149.

Purizhansky, E. M. (2004). *US Patent Nr. 6,673,258 B2. Magnetically Responsive Foam and Manufacturing Process Therefor.*

Quattrone, R., Berman, J., Trovillion, J., Feickert, C., Kamphaus, J., White, S., Giurgiutiu, V. & Cohen, G. (2000). Investigation of Terfenol-D for Magnetostrictive Tagging of Fiber-Reinforced Polymer Composites, *Technical report*, Engineering research nad Development Center.

Rabinow, J. (1948). The magnetic fluid clutch, *AIEE Trans* 67: 1308–1315.

Rabinow, J. (n.d.). *US Patent Nr. 2,575,360. Magnetic Fluid Torque and Force Transmitting Device.*

Rankin, P. J., Horvath, A. T. & Klingenberg, D. J. (1999). Magnetorheology in viscoplastic media, *Rheology Acta* 38: 417–477.

Rawa, H. (2001). *Electricity and Magnetism in Technique (in Polish)*, PWN, Warsaw, Poland.

Seval, G. (2002). *Synthesis and Properties of Magnetorheological (MR) Fluids*, PhD thesis, University of Pittsburgh, School of Engineering.

Shaffer, N. & DeChurch, D. A. (eds) (2004). *Proceedings of the 40th Forum on the Geology of Industrial Minerals*, Indiana Geological Survey.

Shen, Y., Golnaraghi, M. F. & Heppler, G. R. (2004). Experimental research and modeling of magnetorheological elastomers, *Journal of Intelligent Material Systems and Structures* 15: 27–35.

Song, H., Wereley, N., Bell, R., Planinsek, J. & Filer II, J. (2009). Field Dependent Response of Magnetorheological Elastomers Utilizing Spherical Fe Particles Versus Fe Nanowires, *Journal of Physics: Conference Series* 149.

Stepanov, G., Borin, D. & Odenbach, S. (2009). Magnetorheological effect of magneto-active elastomers containing large particles, *Journal of Physics: Conference Series* 149.

Wang, Y., Hu, Y., Chen, L., Gong, X., Jiang, W., Zhang, P. & Chen, Z. (2006). Effects of rubber/magnetic particle interactions on the performance of magnetorheological elastomers, *Polymer Testing* 25: 262–267.

Watson, J. (n.d.). US Patent 5,609,353. Method and apparatus for varying the stiffness of a suspension bushing, 1997.

Weihua, L. (2000). *Rheology of MR Fluids and MR Damper Dynamic Response: Experimental and Modeling Approaches*, PhD thesis, School of Mechanical and Production Engineering.

Yamamoto, H. & Nakano, H. (1999). Dynamic viscoelasticity and its mechanical model of an MR suspension in oscillatory flow mode, *in* R. Tao (ed.), *Proceedings of the 7th International Conference on Electro-Rheological Fluids and Magneto-Rheological Suspensions*, World Scientific, pp. 665–673.

Zając, P. (2011). *Damping properties of magnetorheological elastomers (In Polish, in preparation)*, PhD thesis, Wroclaw University of Technology, Institute of Materials Science and Applied Mechanics.

Zając, P., Kaleta, J., Lewandowski, D. & Gasperowicz, A. (2010). Isotropic magnetorheological elastomers with thermoplastic matrices: structure, damping properties and testing, *Smart Materials and Structures* 19.

Zhang, X., Peng, S., Wen, W. & Li, W. (2008). Analysis and fabrication of patterned magnetorheological elastomers, *Smart Materials and Structures* 17: 1–5.

Zhou, G. & Jiang, Z. (2004). Deformation in magnetorheological elastomer and elastomer-ferromagnet composite driven by a magnetic field, *Smart Materials and Structures* 13: 309–316.

25

Magnetic and Dynamic Mechanical Properties of Nd-Fe-B Composite Materials with Polymer Matrix

Aleksandar Grujić[1], Jasna Stajić-Trošić[1], Mirko Stijepović[2,3]
Jasmina Stevanović[1] and Radoslav Aleksić[2]
[1]Institute of Chemistry, Technology and Metallurgy, University of Belgrade
[2]Faculty of Technology and Metallurgy, University of Belgrade
[3]Department of Chemical Engineering, Texas A&M University at Qatar
[1,2]Serbia
[3]Qatar

1. Introduction

At the end of the last century whole variety of technical-technological achievements occurred. Cars overflow the streets, air-conditions coated the facades, everyday life became unimaginable without computers and cell phones; kitchens are filled with assorted appliances, and industry has been improved and enforced by high technology in order to achieve higher manufacturing products, with less energy and manpower consumption. If a deeper look is taken, behind modern design exteriors, inside all of these appliances, it is noticeable that one of the key roles in their performance is engaged by permanent magnets. These materials have very important role as functional components within the wider spectra of contemporary devices in different industrial branches, as well as in the wider consumption. One of the most important applications of the permanent magnets are: spare parts in AC and DC engines production (Slusarek & Dudzikowski, 2002), as well as synchronized motors, transformers, actuators, magnetic buffers, stationary fields, etc. (Gutfleisch et al., 2011). Information storage (hard discs), communications, medical appliances and scanners, automobile and aircrafts industries, could also be emphasized as important applications (Matsuura, 2006; Brown et al., 2002).
The nanocrystalline Nd-Fe-B alloys are one of the most superior magnetic materials with high value of maximum energy production (app 50 MGOe) (Herbst, 1991). Besides their high values of the remanence and coercivity, as well as relatively high Curie temperatures (app 312 °C) (Sagawa et al., 1984) this type of magnetic alloys are identified suitable for research and further development of magnetic composite materials with polymer matrix, so called bonded magnets (Brown et al., 2006; Ma et al., 2002). Contemporary research in the field of magnetic composite materials on the basis of Nd-Fe-B alloys are directed into four basic directions: increase of magnetic energy, meaning optimisation of magnetic capacities; improving corrosion resistance; optimisation production process of process parameters; and reduction of the subtle rare earth content (Nd), targeting decreasing the price of the final magnetic material, keeping high values of the maximum magnetic energy. Application of

various process techniques in the production process of bonded magnets, gives the possibility for utilisation of various magnetic powder in combination with different polymeric materials as binding agent. Development of bonded technology, exploring the possibility of applications of various types of magnetic powder and polymeric matrices, testing of their influence, as well as the influence of the process parameters, to achieve optimal mechanical and magnetic capacities are in research focus during the last few years (Garrell et al., 2003; Lahelin et al., 2009). Research trend is reflected in development of bonded hybrid magnetic composite materials with improved dynamic mechanical capacities and noticeably lower cost due to the replacement of the expensive Nd-Fe-B magnetic powder with cheaper ferrite magnetic materials, achieving satisfying values of the maximum magnetic energy.

The objective of this chapter is to give more insight on the role of Nd-Fe-B particles on dynamic mechanical, thermal and magnetic properties of Nd-Fe-B/epoxy resin composite materials. Replacing one fraction of Nd-Fe-B with barium ferrite the hybrid composite materials with upgraded dynamic mechanical properties is produced. Interactions between employed magnetic powders and interactions between magnetic powders and polymer binder are considered. The advantage of DMA technique compared to the standard mechanical test methods is demonstrated. In addition, predictive mathematical models are employed to evaluate behaviour of composite. Results obtained with proposed mathematical models are in very good agreement with experimental values.

2. Synthesis of Nd-Fe-B / epoxy resin composite materials

There are several different process routes for bonded magnet production (Hamano, 1995; Gronefeld, 2002). The compression moulding is the most common technique utilized for materials with thermosetting polymer matrices. Composites with varied content of Nd-Fe-B particles in epoxy matrix from 15 to 95 wt% are produced by compression moulding under a pressure of 4MPa at room temperature, using a lab scale compression moulding press. Also, Nd-Fe-B/barium ferrite/epoxy resin composites as well as pure epoxy samples are obtained. The moulded samples are then allowed to cure under a moulding pressure for about 24 hours. The synthesis is carried out under conditions that avoid air bubbles in the mixture. No external magnetic field is used during the cure.

2.1 Polymer binder

As polymer matrix thermosetting epoxy system that is a combination of liquid mixture of Bisphenol A and Bisphenol F resins and cross linking agent (hardener), which cures fully at room temperature, is used. In terms of manufacture, the curing time of the matrix must be long enough for the polymer and magnetic alloy to be properly mixed, but shorter than required for the gravitational settling of the Nd–Fe–B particles. During curing, shrinkage should be minimal, and the thermal expansion of both polymer matrix and Nd–Fe–B should be comparable. Applying these criteria, the medium hard epoxy resin has the following properties: tensile strength ~ 58 MPa, elongation ~ 2.8%, compression strength ~ 96 MPa, flexural strength ~ 78 MPa and density ~ 1.2 g/cm^3, is selected.

2.2 Magnetic materials

The rapid quenched $Nd_{11.7}(Fe,Co)_{80}B_{8.3}$ magnetic powder obtained by melt spinning method with particle size from 74 to 177 μm is employed as a magnetic filler for polymer composite magnets manufacturing. Hybrid magnetic composites are produced by replacing the part of Nd-(Fe,Co)-B particles with spherical barium ferrite ($BaFe_{12}O_{19}$) agglomerates. The magnetic properties of started magnetic materials are presented in Table 1.

Material	Chemical Formula	B_r [kG]	H_{cb} [kOe]	H_{cj} [kOe]	$(BH)_{max}$ [MGOe]
Neodymium Iron Boron	$Nd_{11.7}(Fe,Co)_{80}B_{8.3}$	8.2	6.0	8.7	13.1
Barium Ferrite	$BaFe_{12}O_{19}$	2.3	1.9	3.6	1.3

Table 1. Magnetic properties of started magnetic materials

The chemical composition of the starting Nd-Fe-B alloy is Nd: 21-25 wt%, Co: 3-5 wt%, B<1.5 wt%, Zr: 3-5 wt%, Fe: balance.

2.3 Measurements

The structure and morphology of fracture surfaces of synthesized composite materials are observed by JEOL JSM-5800 Scanning Electron Microscope (SEM), with an accelerating voltage of 20 kV. After tensile tests at room temperature, fracture sample surfaces are sputtered with gold using a POLARON SC 502 sputter coater for enhanced conductivity.
Thermal analysis of investigated magnetic composites, as well as pure epoxy material are performed by SDT Q600, TA Instruments equipment for simultaneous DSC/DTA/TGA measurements. The samples are heated at a rate of 5 °C/min in a nitrogen environment with a temperature range of ambient temperature up to 600 °C.
A TA Instruments DMA Q800 is used to obtain dynamic mechanical data of investigated magnetic composites and pure epoxy sample. Storage modulus, loss modulus, and tan delta are recorded as a function of temperature. The samples are tested using three-point bend clamp with a 20 mm span width and rectangular-edge probe, at a frequency of 1 Hz and displacement amplitude of 20 μm. A displacement control mode is used to maintain deflection through glass transition temperature *Tg*; a Force Track setting of 125% is used, i.e. an applied static force was 25% higher than the dynamic force. Samples are placed on a platform located on top of the three-point bend fixture so that no sample deflection could occur during annealing. Testing is done over a temperature range from 25 °C to 100 °C with a temperature ramp of 3 °C/min.
The Nd-Fe-B/epoxy composites and pure epoxy polymer samples are machined into tensile and flexural specimens, which are then tested at ambient temperature. The tensile specimen is dumbbell shaped, as required by ASTM D 3039-00 (American Society for Testing and Materials [ASTM], 2000), with a cross-section of 40 x 5 mm in the gauge length. The specimens for flexural tests are rectangular plates 100 x 10 x 2 mm as required by ASTM D 790-03 (ASTM, 2003). At least five specimens are tested in tests consisting of both composite and pure epoxy specimen. The dumbbell-shaped tensile specimens of pure epoxy resin are fabricated by casting the resin into the rubber moulds and following the recommended curing cycle from the material manufacturer. Flexural tests are performed using three point bend kit. A universal material testing machine (Schenck TREBEL RM 100) is used for mechanical tests. The moduli of elasticity are derived from the linear portion of the stress–strain curves obtained by both tensile and flexural tests.
The examination of macroscopic magnetic properties is tested using Superconducting Quantum Interference Device (SQUID) magnetometer. During the measurements at ambient temperature (300 K), magnetic field strength $\mu_0 H$ is varied from -5 to 5 T. Sample preparation

and experimental procedures have been conditioned such that the demagnetization factor can be neglected. While the SQUID magnetometer is a very sensitive device, a magnetic field strength from 10^{-12} up to 10^3 A/m^2 can be measured with accuracy of 0.1 %.

3. Nd-Fe-B / epoxy composite magnetic materials – bonded magnets

The Nd-Fe-B bonded magnets have been commonly used in various fields, such as electric appliances, automobile parts (Brown et al., 2002; Li et al., 2006), sensing elements (Radojević et al., 2007), electronic, communication and micro-electro-mechanical system (MEMS) applications (Chin, 2000; Hono & Ping, 2001). Advantages of the using bonded composite materials include their simple technology, possibility of forming their final properties, lowering manufacturing costs because of no costly finishing and lowering of material losses resulting from the possibility of forming any shape (Dobrzanski et al. 2007). The amount of Nd-Fe-B powder in the bonded magnet plays a crucial role in determining magnetic properties. A higher content of Nd-Fe-B powder usually results in a higher remanence magnetization (Br) and maximum energy product $(BH)_{max}$ and therefore, it is desirable from the magnetic perspectives. However, a higher content of magnetic filler may change the rheology of polymer melt during the process, subsequently, impact the mechanical strength of bonded magnets. Nevertheless, the balance between magnetic properties and corresponding dynamic mechanical behaviour is an important issue for bonded magnet applications (Garrell et al., 2003b). The presented study is undertaken with the intention to understand the effect of different filler contents on the thermal, dynamic mechanical and magnetic properties of the Nd-Fe-B/epoxy magnetic composite materials.

3.1 Structure and morphology

Uniform particle distribution and good adhesion between Nd-Fe-B particles and a polymer matrix are essential for the quality of composites, especially at temperatures above the glass transition temperature (Tg) of the polymer. The particle size of magnetic powder plays an important role in determining powder to binder ratio, degree of particle alignment and, magnetic and mechanical properties (Kokabi et al., 2005). Generally speaking, the plate-like particles would result in higher packing density under the optimal compression conditions (Rodrigues et al., 1996; Zhang et al., 2009). SEM micrographs of fracture surface morphology of Nd-Fe-B/epoxy composites are presented in Fig.1.

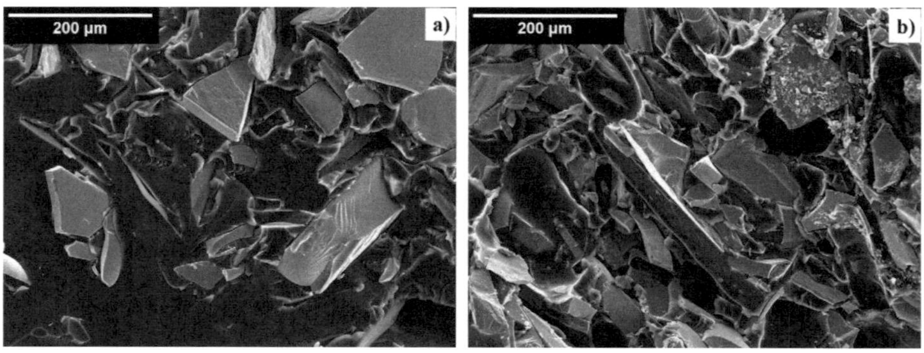

Fig. 1. SEM micrographs of fracture surface of composite material with a) 50 wt% and b) 95 wt% of Nd-Fe-B filler observed by JEOL JSM-5800 Scanning Electron Microscope

The Nd–Fe–B particles are shown as light grey and the epoxy matrix is shown as dark. The darkest gray parts represent the holes ensued by dragging the Nd-Fe-B particles during the mechanical breaking (tension). Although Nd–Fe–B particles are of variable size and shape, they seem to be attached rather well to the matrix.

3.2 Thermal properties

Using thermogravimetric analysis (TGA) and differential scanning calorimetry (DSC), different phase transformations caused by decomposition, oxidation, dehydration, etc. can be detected. Also, materials can be analysed in different experimental environments. Increasing the heating rate temperature of transformation can be moved to higher values because a temperature lag between the heater and the specimen occurs (Withaeger, 2001).

The DSC and TG curves for Nd-Fe-B/epoxy composite are shown in Fig. 2. Thermogravimetric experiments (Fig. 2a) were carried out to determine the degradation temperatures (T_d) of magnetic composite materials i.e. the temperature corresponding to the highest rate of weight loss. The results of thermal degradation studies clearly show that the samples with higher amounts of magnetic filler have a lower temperature of thermal degradation and a lower degradation rate in regard to the pure epoxy resin. The results obtained by DSC are in accordance with TG measurements.

DSC curves show an exothermal effect which appears between 300 °C and 380 °C refers to the reaction of thermal degradation or decomposition of the composite materials. The change of enthalpy which occurred during this process is decreased with a decrease in the amount of epoxy matrix in the composites. Small endothermic effects in temperature range between 47 °C and 50 °C are referring to glass transitions in the investigated composite materials (Maity et al., 2007). The quantitative values of degradation temperatures and the corresponding changes in enthalpy are presented in Table 2.

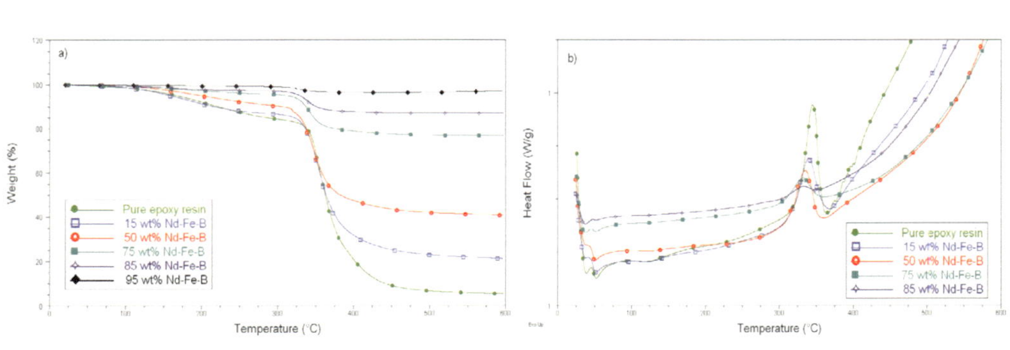

Fig. 2. TG and DSC curves for the pure epoxy resin and Nd-Fe-B composites

The quantitative values of glass transition temperatures *Tg* obtained by DSC are presented with DMA results in section 3.3 (Table 3.). Due to negligible thermal effects i.e. enthalpy change, for the sample with 95 wt% Nd-Fe-B, the corresponding DSC curve is not included in the Fig. 2b. All investigated samples have very close values of glass transition temperatures which is confirmed by DMA measurements (Fig. 3).

Sample (Composite)	TG	DSC	
	Td, °C	*Td*, °C	*ΔH*, J/g
Pure epoxy resin	354.7	344.6	267.8
15 wt% Nd-Fe-B	352.5	338.8	180.1
50 wt% Nd-Fe-B	349.7	333.5	156.3
75 wt% Nd-Fe-B	342.2	330.3	72.4
85 wt% Nd-Fe-B	341.3	329.2	18.9
95 wt% Nd-Fe-B	324.4		

Table 2. Degradation temperatures and corresponding change of enthalpy obtained from TG and DSC curves

3.3 Dynamic mechanical properties

Dynamic mechanical analysis (DMA) is probably the most sensitive single technique available for characterizing and interpreting the mechanical and curing behaviour of polymers and polymer composites. This sensitivity allows the DMA to detect the glass transition temperature, *Tg* of highly crosslinked thermosetting polymers or of thin coatings. (Menard, 1999). DMA can be simply described as applying an oscillating force to a sample and analyzing the material's response to that force. The technique separates the viscoelasticity of a material into the two components of complex modulus (E^*), a real part which is the elastic modulus (E') and an imaginary part which is the damping or viscous component (E''), and is expressed $E^* = E' + iE''$ (Ferry, 1980). These properties are often described as the ability to lose energy as heat (damping) and the ability to recover from deformation (elasticity). Due to the viscoelastic nature of polymer composites, their dynamic and thermal behaviour significantly depends on strain, frequency and temperature. For composite materials, particle shape and size, uniform particle distribution, and good adhesion between Nd-Fe-B and polymer, are important parameters which have significant impact on microstructure and stiffness of the final magnetic material (Guschl et al., 2002).
The dynamic mechanical properties of the pure epoxy polymer and the Nd-Fe-B/epoxy magnetic composite materials are examined as a function of temperature, from the glassy to the rubbery state. Applied three-point bending oscillatory testing is considered as a "pure" mode of deformation and is recommended for stiff materials. The results presented in Fig. 3 show a considerable improvement in the storage modulus (elastic component) caused by the presence of the Nd-Fe-B magnetic filler.

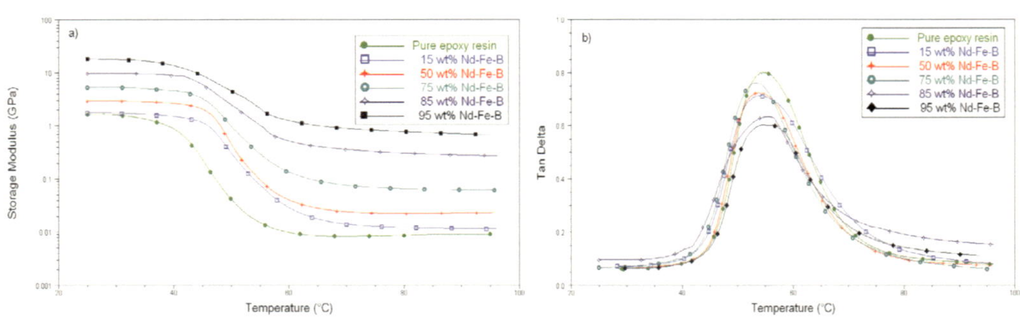

Fig. 3. DMA curves of storage modulus E' and $Tan\delta$, for the pure epoxy resin and the composites with different Nd-Fe-B filler content versus temperature

In the glassy region (around 25 °C), modulus of the pure polymer, modulus of the filler, and concentrations of both, as well as the adhesion factor between the filler and polymer, have direct influence on the total dynamic modulus of composites (Payne, 1965; Wang, 1999; Alves et al., 2004). The addition of 15 wt% and 50 wt% Nd-Fe-B powder into epoxy matrix induce an increase of roughly 6 % and 75 % in the storage modulus of pure epoxy polymer, respectively. Thus, magnetic composites with 75 wt% and 85 wt% of magnetic filler were found to have 3.2 and 5.8 times higher storage modulus than the pure epoxy sample, respectively. At the other end of temperature range, the storage modulus decreases with the temperature to the lower values in the rubbery state (Deng et al., 2007). In this region, approximately above 75 °C, the dynamic storage modulus is a function of the hydrodynamic effect, well-known to be dependent on the shape of the filler particles or agglomerates, on the concentration of the filler, and on the fillers interactions with polymer. Comparing to pure epoxy sample values for storage modulus these are 1.7, 2.9, 9.1 and 40 times higher for composites with 15 wt%, 50 wt%, 75 wt% and 85 wt% of Nd-Fe-B, respectively, which is a significant improvement of storage modulus (Grujić et al., 2010a). DMA tests for highly filled Nd-Fe-B/epoxy polymer composite (95 wt% of Nd-Fe-B) showed an 11 times higher storage modulus values at ambient temperature and up to 87 times higher at 75 °C (in rubbery state), compared with the polymer matrix. The significant enhancement of the storage modulus in the rubbery region could be explained by the variable size and shape of Nd–Fe–B particles resulting in higher packing density, good particle to particle interaction and attachment to the epoxy matrix (Payne, 1965). When comparing material properties, a material with a higher storage modulus would be stiffer and harder to deform than one with a lower E'. Besides the elastic component, a material also has a viscous component called the loss modulus (E''). This viscous component relates to the materials ability to lose energy. The material's $tan\delta$ designates the material's ratio of viscous to elastic components (E''/ E') and it is sometimes called the materials damping ability (Menard, 1999). A composite material with a higher $tan\delta$ (≈ 0.7 for composite with 15 wt% of Nd-Fe-B filler) has a higher viscous percentage than one with a lower $tan\delta$ (≈ 0.6 for composite with 95 wt% of Nd-Fe-B filler). Therefore the material would be more likely to absorb a vibration or impact, and disperse it throughout the material without failure. DMA results presented in Fig. 3b show that glass transition temperatures (Tg) obtained from $tan\delta$ curves for all investigated magnetic composites were found to lay in the same temperature region (around 54 °C). This could be a consequence of the use of Nd-Fe-B powders with similar particle size distribution and without particle surface modification (uncoated) (Otaigbe et al., 1999). The quantitative values of glass transition temperatures obtained by different methods are presented in Table 3.

Sample (Composite)	Glass Transition Temperature (T_g), [°C]						
	DSC			**DMA**	DMA (E')		
	onset	**midpoint**	**offset**	($tan\delta$)	**onset**	**midpoint**	**offset**
Pure epoxy resin	46.8	48.3	50.6	54.5	39.4	43.7	48.0
15 wt% Nd-Fe-B	45.3	47.2	49.8	53.9	43.4	47.1	50.8
50 wt% Nd-Fe-B	44.4	47.0	49.3	53.8	43.8	47.6	50.5
75 wt% Nd-Fe-B	46.7	48.4	50.2	53.3	40.7	46.9	51.6
85 wt% Nd-Fe-B	47.0	49.1	51.1	55.0	41.8	45.6	51.2
95 wt% Nd-Fe-B				55.3	39.4	45.0	52.2

Table 3. Glass transition temperatures obtained by DSC and DMA

The difference in *Tg* value for each composite obtained from DSC and DMA arises due to the difference in the method of measurement and the variations in the definition of transition temperature used (Maity et al., 2007).

3.4 Tensile and flexural properties

In order to determine the in-plane tensile and flexural properties of the polymer matrix composite materials reinforced by the Nd-Fe-B magnetic particles, standard test tensile and flexural methods are used (ASTM, 2000, 2003). The ultimate tensile stress σ_m of the investigated magnetic composite materials is calculated using:

$$\sigma_m = \frac{F_{\max}}{b \cdot d} \tag{1}$$

Where: σ_m – ultimate tensile stress, MPa; F_{max} – maximal load before failure, N; b – sample width, mm; and d – sample thickness, mm.

The values of ultimate tensile stress, elongation and modulus of elasticity have been taken from obtained stress-strain diagrams for all investigated composites, and presented as a function of the Nd-Fe-B content in the epoxy matrix respectively (Fig.4. & Fig.6.). The elastic modulus E of investigated magnetic composite materials was calculated using:

$$E = \frac{\Delta\sigma}{\Delta\varepsilon} = \frac{\Delta F}{\Delta\varepsilon} \cdot \frac{1}{b \cdot d} \tag{2}$$

Where the ratio $\Delta\sigma/\Delta\varepsilon$ is determined by linear regression method. From the linear portion of stress-strain curves i.e. in the Hookean region.

With a decreasing quantity of the thermosetting epoxy polymer in the composites i.e. with an increasing content of the Nd-Fe-B filler, the values of tensile stress and elongation are decreasing (Fig. 4.). In addition, composite materials become more brittle, and less ductile (Hemrick et al., 2004; Grujić et al., 2010b).

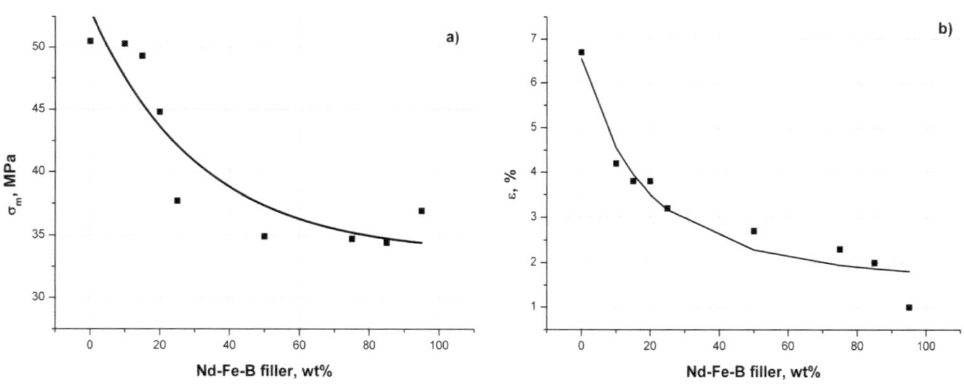

Fig. 4. a) Ultimate tensile stress, and b) strain in function of Nd-Fe-B content

The test methods according to ASTM D 790-03 (ASTM, 2003) cover the determination of flexural properties of unreinforced and reinforced plastic materials, including high-modulus composites. These test methods are generally applicable to both rigid and semi-rigid materials. However, flexural strength cannot be determined for those materials that do not break or that do not fail in the outer surface of the test specimen within a 5.0 % strain limit of these test methods. When a homogeneous elastic material is tested in flexure as a simple beam supported at two points and loaded at the midpoint, the maximum stress in the outer surface of the test specimen occurs at the midpoint. This stress may be calculated for any point on the load-deflection curve by means of the following equation:

$$\sigma_f = \frac{3F \cdot L}{2b \cdot d^2} \tag{3}$$

where: σ_f - stress in the outer surface at midpoint, MPa; F - load at a given point on the load-deflection curve, N; L - support span, mm; b - width of beam tested, mm, and d - depth of beam tested, mm.

Deformation caused by flexural test ε_f is fractional change in the length of an element of the outer surface of specimen at middle point, where stress is maximal, and it can be calculated using the following equation:

$$\varepsilon_f = \frac{6D \cdot d}{L^2} \tag{4}$$

where: ε_f - strain at outer (opposite) surface, mm/mm; D – maximal deflection in middle point of specimen, mm, L –support span, mm; and d – thickness of the specimen, mm.

The tangent flexural modulus of elasticity represents the ratio of stress to corresponding strain in the range of elastic behaviour of materials. The flexural modulus of elasticity is calculated by drawing a tangent to the steepest initial straight-line portion of the load - deflection curve. Similar to tensile test results, the flexural stress and elongation are decrease with increasing content of Nd-Fe-B filler (Fig. 5).

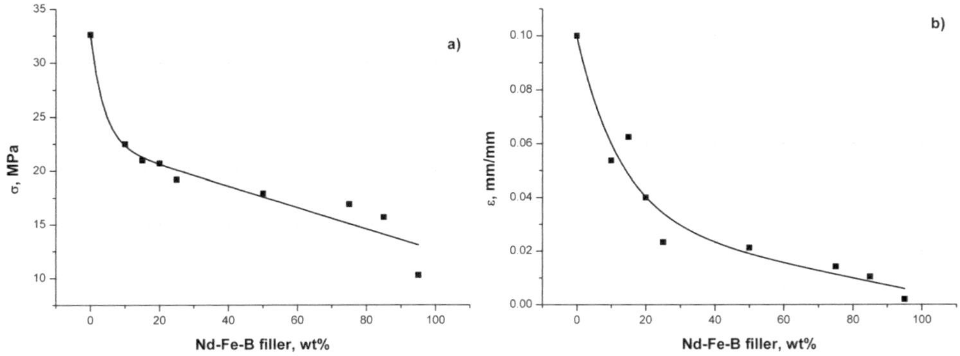

Fig. 5. Ultimate flexural stress and strain as the function of Nd-Fe-B content

Moduli of elasticity obtained by both tensile and flexural tests are increase with higher quantities of magnetic filler. This is crucial in analysis of possible use of the investigated magnetic composite materials as functional material (Fig. 6.). This means that material with higher amounts of Nd-Fe-B filler, subject to equal stress levels (ballast), tolerate 2 to 3.5 times lower deformation.

The modulus of elasticity is a very important parameter for analysis of the composite materials behaviour under discontinuous load conditions. The values of elastic modulus, obtained by tensile and flexural tests, upswing with an increasing amount of Nd-Fe-B powder from 50 wt% achieve 9.2 and 7.1 GPa, respectively. Within the narrow region, up to 20 wt% content of Nd-Fe-B, where the modulus of elasticity is practically constant according to tensile and flexural tests, dynamic-mechanical analysis could be applied to acquire additional information's related to mechanical behaviour about transitions in polymer composites (Maity et al., 2007; Almagableh et al., 2008).

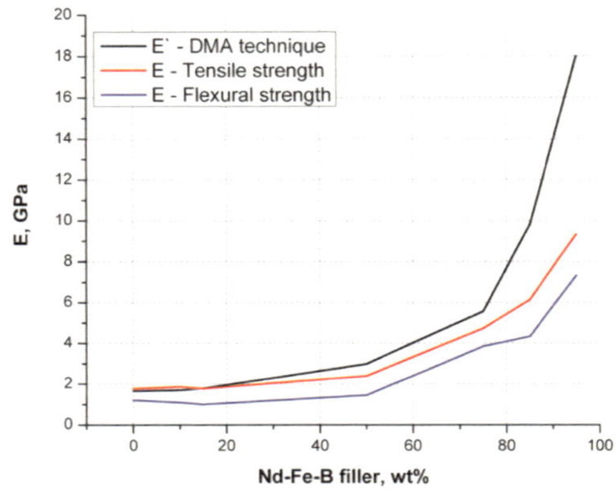

Fig. 6. Comparative view of the changes in the modulus of elasticity at 25 °C

The values of storage modulus observed by DMA were compared with elastic modulus obtained by tensile and flexural tests (Fig. 6.). In contrast to Deng, S. et al. (Deng et al., 2007) mechanical properties at temperatures higher than ambient are not compared with DMA results observed using two different clamps. It seems that observing the elastic modulus of composites by tensile, flexural and DMA tests at room temperature in the present study gives a better look at the increasing trend of elastic component of materials with increasing Nd-Fe-B filler content in the polymer matrix.

3.5 Mathematical prediction of Nd-Fe-B / epoxy composite behaviour

The strong influence of relatively small amounts of filler particles on the dynamic mechanical properties of polymers has significantly contributed to increased use of polymer materials in many commercial applications (Bergstrom & Boyce, 1999). The incorporation of filler particles is known to increase the stiffness of the material and alter time dependent aspects of material behaviour such as hysteresis and stress relaxation. Even at strains sufficiently large for the structure to have been eliminated, the storage modulus is greater than that of the pure polymer and, greater than the amount which can be predicted due to hydrodynamic interaction of the filler particles.

Ideally, in an attempt to reduce laboratory cost, one would like to make a prediction of a new material's behaviour by numerical simulation, with the primary goal being to accelerate trial and error experimental testing. The recent dramatic increase in computational power available for mathematical modelling and simulation raises the possibility that modern numerical methods can play a significant role in the analysis of heterogeneous microstructures. This section is devoted to the mathematical prediction of storage modulus of observed composite. The focus is on analytical models, due to fact that they are simple to use and required input data are only properties of individual constituents of composite and their fraction. The several proposed analytical models are tested versus experimental data as it is illustrated in following text. Some of applied models agree very well with experimental data, whilst others deviate significantly.

There have been several attempts to derive formulas giving the apparent modulus due to a dispersion of particles in polymer (Donnet, 1998). The earliest of these attempts was by Smallwood using the analogy to Einstain's viscosity equation:

$$E_{comp} = E_{matrix}(1 + 2.5c) \tag{5}$$

where: E_{comp} - storage modulus of composite (Pa), E_{matrix} - storage modulus of matrix (Pa), and c - volume faction of particle in composite

Smallwood's estimate is only good at very low filler concentrations. A number of attempts have been made to incorporate interactions between neighbouring particles to allow prediction at higher volume fractions. Most of these models add one or more terms to a polynomial series expansion. One of the most cited model of this class is the Guth-Gold (Guth & Gold, 1938):

$$E_{comp} = E_{matrix}(1 + 2.5c + 14.1c^2) \tag{6}$$

Later Guth extended the Guth-Gold model to include impact of particle shape on properties. Guth introduced a shape factor f (ratio of diameter to width of particle) and proposed a new equation (Guth, 1945):

$$E_{comp} = E_{matrix}(1 + 0.67f \cdot c + 1.62f^2 \cdot c^2) \tag{7}$$

Budiansky developed a model, for the special case of rigid particles in an incompressible matrix written as (Budiansky, 1965):

$$E_{comp} = \frac{E_{matrix}}{1 - 2.5c} \tag{8}$$

Ponte Castaneda has proposed a different self-consistent estimate for rigid particles in a neo-Hookean matrix (Ponte Castaneda, 1989):

$$E_{comp} = \frac{E_{matrix}}{1 - 3c} \tag{9}$$

Later Govindjee and Simo proposed the novel model, for the case of rigid particles in a neo-Hookean matrix written (Govindjee & Simo, 1991):

$$E_{comp} = E_{matrix}\frac{1 - c/2}{1 - c} \tag{10}$$

In addition, it is worth to mention the empirical formula suggested by Brinkmann (Brinkman, 1952):

$$E_{comp} = E_{matrix}(1 - c)^{-\frac{5}{2}} \tag{11}$$

Major characteristics of all aforementioned theoretical models are: they neglect the impact of filler properties and assume that the medium wets the filler particles, but they do not chemically react with the filler surface. Mori-Tanaka proposed a model, which takes into consideration the impact of filler properties on overall composite properties (Mori & Tanaka, 1973). The Mori-Tanaka model for spherical particles isotropically dispersed in an elastic matrix can be written as:

$$K_{comp} = K_{matrix} + \frac{c\left(K_{particle} - K_{matrix}\right)K_{matrix}}{\dfrac{3K_{matrix}(1-c)\left(K_{particle} - K_{matrix}\right)}{3K_{matrix} + 4G_{matrix}} + K_{matrix}} \tag{12a}$$

$$G_{comp} = G_{matrix} + \frac{c\left(G_{particle} - G_{matrix}\right)G_{matrix}}{(1-c)\left(G_{particle} - G_{matrix}\right)\dfrac{6\left(K_{matrix} + 2G_{matrix}\right)}{5\left(3K_{matrix} + 4G_{matrix}\right)} + G_{matrix}} \tag{12b}$$

$$E_{comp} = \frac{9K_{comp}G_{comp}}{3K_{comp} + G_{comp}} \tag{12c}$$

where: K_{comp}, K_{matrix}, $K_{particle}$ are bulk modulus of composite, matrix and particle, respectively (Pa), and G_{comp}, G_{matrix}, $G_{particle}$ are shear modulus of composite, matrix and particle, respectively.

The experimentally obtained values of storage modulus are compared with analytical models discussed above and presented in Fig. 7. Predictions of models proposed by Budiansky, Ponte Castaneda and Govindjee-Simo give inadequate estimation so they are not included in Fig. 7.

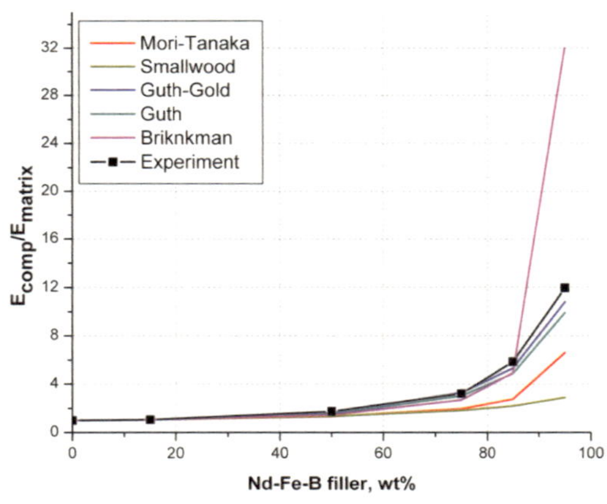

Fig. 7. Models predictions against experimental data

From Fig. 7 one may notice that all models included in analysis give very good predictions of storage modulus at lower particles concentrations (till 50%). This suggests that in this concentration range, the interactions between neighbouring particles have very low intensity. At higher concentrations, interactions become high intensity, this is why the Smallwood's model starts to show significant deviation from experimental results. Mori-Tanaka's model follows the trend of experimental results, but it gives poor predictions. Brinkman's model gives good predictions at high concentrations, but at very high concentrations of particles, it extensively overpredicts the storage modulus. The Guth and Guth-Gold models are in very good agreement with experimental results. The explanation for this behaviour lies in the fact that both models take into consideration interactions between neighbouring particles.

3.6 Magnetic properties

Characteristic of all magnetic materials is a manifestation of the hysteresis phenomena. The hard magnetic materials have the greater values of hysteresis (Goll & Kronmüller, 2000). A word of Greek derivation, hysteresis describes magnetic materials as highly nonlinear, meaning that their response to a stimulus lags behind in a repeatable manner. The stimulus in this case is an applied magnetic field and the material's response is the magnetization or induction (Trout, 2000). Magnetic properties of magnetic composite materials (bonded magnets) are affected by the magnetic properties of the magnetic powder and weight (volume) ratio of the powder. It is known that bonded magnets have inferior magnetic characteristics compared to magnetic material obtained by convectional methods (sintering for example), because in bonded technology maximal density of magnetic powder can not be achieved (Zhang et al., 2009). One of the most important characteristics of the used type of Nd-Fe-B rare-earth magnetic material is high values of remanence and coercivity, which have a direct influence on high values of maximum energy product (Gutfleisch 2000; Chen et al., 2004). The results of magnetic measurements i.e. complete hysteresis loops for bonded Nd-Fe-B/epoxy type magnets with different content of functional magnetic particles are presented in Fig. 8a. It is obvious that the largest hysteresis loop correspond to the magnetic composite with the highest amount of magnetic component.

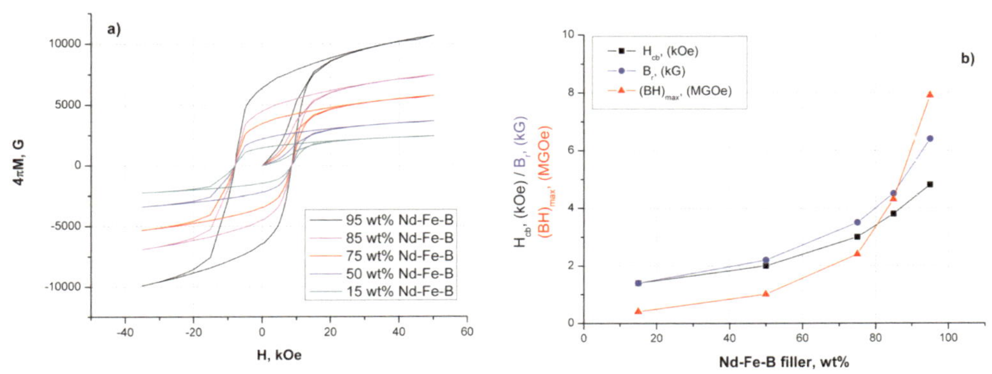

Fig. 8. a) Hysteresis loops of magnetic composites, b) magnetic properties of the composites as the function of the Nd-Fe-B filler content

On the basis of these results corresponded B-H diagrams are constructed and the changes of remanence (*Br*), coercivity (*Hcb*) and maximum energy product (BH_{max}) with an increasing content of Nd-Fe-B in the epoxy matrix are taken and presented in Fig. 8b. The presented graph illustrates the upswing of three magnetic parameters of composite materials with increasing amounts of Nd-Fe-B particles in the epoxy matrix. For example, the maximum energy product for composite with 95 wt% Nd-Fe-B is around 8 MGOe, which is two times higher than for composite with the 85 wt% Nd-Fe-B case. For composites with Nd-Fe-B content higher than 75 wt%, $(BH)_{max}$ rapidly increase i.e. for the highly filled composites even a small addition of magnetic medium have a strong influence on magnetic properties of bonded magnets. Also, the maximum energy product $(BH)_{max}$ of Nd–Fe–B bonded magnets can be simulated using a mathematical model, and choosing appropriate parameters for the magnetic texture and the magnetic coupling micro-grains, one can increase the value of $(BH)_{max}$ (Vuong et al., 2003, Xiao et al., 2000).

4. Hybrid magnetic materials

One way of improving some physical properties of the bonded magnets is to produce the hybrid magnets prepared from a mixture of two powders with different properties, e.g. the Nd–Fe–B powders with barium ferrite (Plusa et al., 2006). The addition of ferrite to the Nd–Fe–B powder decreases the temperature coefficient of coercivity (commonly known as β [%/°C]), which means that this type of bonded magnet can work under elevated temperatures. Also, for example, the bonded magnet has improved the mechanical properties with addition of iron powder (Dobrzanski & Drak, 2008). Further benefits of adding ferrite or iron would be cost reduction and ease of magnetization.

The synthesized hybrid magnetic composite materials correspond to a mixture of Nd-Fe-B and barium ferrite in different ratio. More precisely, this type of composite can be observed as a substitution of a part of Nd-Fe-B particles with barium ferrite. A better insight into the effect of added barium ferrite to the final characteristics of hybrid composite materials are examined for a constant quantity of the polymer matrix. SEM micrographs of hybrid magnetic composite materials are presented in Fig. 9.

Fig. 9. SEM micrographs of hybrid magnetic composite materials

Since the crumbled ferrite agglomerates are incorporated between bigger particles of ferrite and Nd-Fe-B, they contribute to the improved dynamic mechanical properties of composite. The results of DMA are presented in Fig. 10. For different types of composites, the values of storage modulus and $tan\delta$ are compared. All composites have improved storage modulus compared to pure epoxy resin, while the hybrid Nd-Fe-B/barium ferrite/epoxy resin type composite has the highest value (Fig. 10a). The peaks of $tan\delta$ curves lie in a temperature region between 45 -50 °C and indicate glass transition temperatures.

This phenomenon is continuously investigated using composites with different ratios of magnetic particles and for various types of polymer matrices. For this purpose, injection bonded magnets are produced using thermoplastic poly (methyl methacrylate) (PMMA) as a matrix and Nd-Fe-B and barium ferrite as a functional magnetic component, in order to investigate the dynamic mechanical properties of hybrid composite materials. The results presented in Fig. 11 show the same tendency of increasing the storage modulus as for composites with epoxy matrix. Hybrid magnetic composite has improved elastic properties compared to the pure PMMA matrix, Nd-Fe-B/PMMA and barium ferrite/PMMA composite materials. *Tanδ* curves for composites with PMMA matrix (Fig. 11b) show that the glass transition temperature is around 125 °C and strongly depends on used polymer.

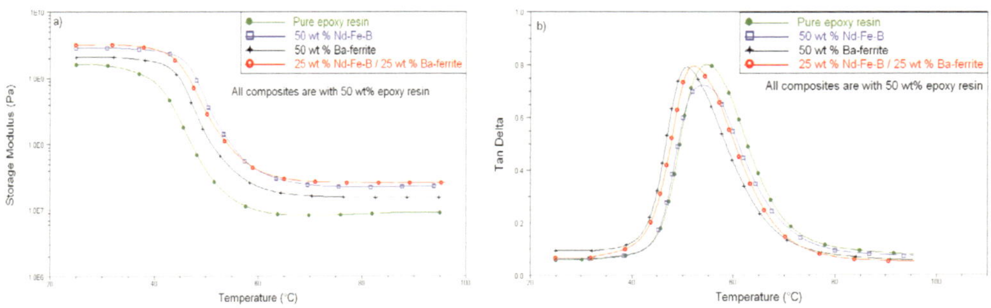

Fig. 10. Nd-Fe-B/epoxy composites: a) storage modulus, and b) *tan*δ as the function of temperature

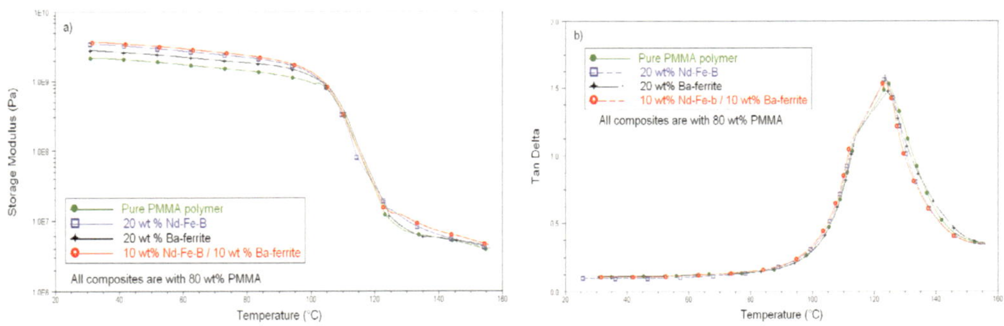

Fig. 11. Nd-Fe-B/PMMA composites: a) storage modulus, and b) *tan*δ as the function of temperature

Magnetic properties of started magnetic powders and hybrid composite materials are investigated and corresponding SQUID hysteresis loops are presented in Fig. 12a. The shape of barium ferrites hysteresis loop illustrates that magnetic saturation (*Ms*) is achieved at 70 emu/g. The coercivity, remanence and consecutively maximum energy product values are less significant as for Nd-Fe-B powder (Goll & Kronmüller, 2002). The saturation magnetization Nd-Fe-B alloy is about 3 times higher than barium ferrite. The values presented in Fig. 12a should be taken as approximate because the field strength of SQUID magnetometer (5 T) is not sufficient for full saturation of Nd-Fe-B powders. It can be seen in the first quadrant, the horizontal end of the hysteresis loop of barium ferrite indicates that this magnetic powder achieves complete saturation, as opposed to Nd-Fe-B alloy.

It is obvious from Fig. 12a that the hysteresis loops becomes more constricted as the quantity of barium ferrite increases, which is consistent with the investigations of D. Plusa et al. (Plusa et al., 2006). With an increasing content of barium ferrite in the hybrid composites, the value of magnetic remanence, coercive force, and the maximum magnetic energy decreases. Change of magnetic properties, is practically linear in relation to the magnetic powders ratio (Nd-Fe-B to barium ferrite) in hybrid composites. The magnetic properties of hybrid magnetic composites are taken from obtained hysteresis loops and used for constructing the diagram presented on Fig. 12b.

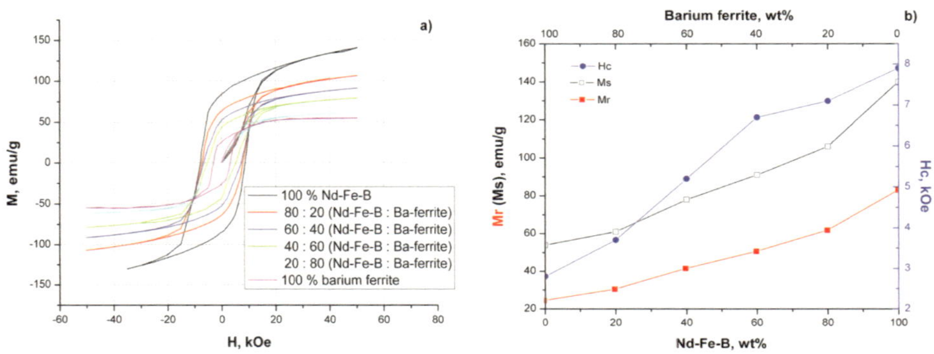

Fig. 12. a) SQUID hysteresis loops of starting magnetic powders and hybrid magnetic materials and b) magnetic properties as the function of Nd-Fe-B and barium ferrite content

5. Conclusion

The results of this study show that addition of plate shape Nd-Fe-B particles to the polymer affect the rheological properties the polymer matrix via internal structural changes and, subsequently, the impact the mechanical strength of bonded magnets. Moreover, the stability of composite materials at elevated temperatures is a crucial property has a major impact on their utilization.

There is a wide range of polymers and magnetic powders that may be used for production of magnetic composites. It seems that the mixture of Nd-Fe-B particles and epoxy resin is suitable for bonded magnet applications due to good adhesion, homogeneity, mechanical and magnetic properties. One of the characteristics of this type of materials is that the degradation temperature rises as the content of epoxy increases. The thermal experiments DSC/TGA reveals that the addition of a high quantity of Nd-Fe-B particles results in a reduction of the composite thermal decomposition enthalpy. On the other side, glass transition temperatures evaluated by DSC and DMA disclose that concentration of particles have no impact on glass transition temperature. This result could be a consequence of applied micro size particles without coupling agents or additives. Moreover, DMA data show that the value storage modulus amplifies in glassy, as well as in rubbery state, as the concentration of filler in composite rises. The tensile and flexural tests at ambient temperature show enhancement of modulus of elasticity with quantity of magnetic filler, which is a crucial parameter for analysis of composite materials behaviour. Information extracted from tensile and flexural tests are consistent with results evaluated by DMA. As expected, magnetic properties are drastically improved with higher quantities of Nd-Fe-B magnetic powder, especially for highly filled composites. These results provide information about the Nd-Fe-B/epoxy composites which could be of importance in cases where the relatively brittle metallic permanent magnets are not useable.

Hybrid materials development and utilization are economically motivated, due to fact that these materials can be produced at low cost. For example, replacing one fraction of Nd-Fe-B with less expensive barium ferrite creates a new hybrid composite. This hybrid composite shows lower intensity of magnetic property comparing to the original composite, but on the other side shows improved dynamic mechanical properties. In view of aforementioned facts on the subject of hybrid materials, it could be said that hybrid materials could replace existing composite materials in numerous applications.

For better insight into viscoelastic behaviour of composites, beside experiments, a theory that explicitly takes the particle distribution, shape factor, particle-particle interactions as well as particle-polymer matrix interactions into account is required. Considering the increasing interest in polymer composites and advanced analytical tools, the present study provides a useful basis and motivation for future experiments and theory development for the multifunctional components and commercially important polymer bonded magnets.

Further work has been focused on researching and developing bonded Nd-Fe-B composite materials with thermoplastic or rubber matrix using the injection or extrusion methods. Design of the composites with improved dynamic mechanical and optical properties can be realized by assembling magnetic powders and fibres blends. Also, corrosion resistance is one of the potential goals in further research activities.

6. Acknowledgement

The authors are grateful to colleagues from the Department for Materials and Metallurgy, ICTM and from the Department for Special Materials TMF, Belgrade for cooperation during research. Also, the authors are indebted to Nada Lazić, Zijah Burzić and Vojislav Spasojević for the measurements and results interpretation. A great appreciation for the assistance with preparation of this chapter authors owe to Ms Sabla Alnouri from TAMUQ. This work has been supported by the Ministry of Science and Technological Development of the Republic of Serbia under Projects: TR 34011 and III 45019.

7. References

Almagableh, A.; Gupta, S.; Mantena, P.R & Al-Ostaz, A. (2008). Dynamic Mechanical Analysis of Graphite Platelet and Nanoclay Reinforced Vinil Ester, and MWCNT Reinforced Nylon 6,6 Nanocomposites, Proceedings of 40th ISTC, Memphis, Tennessee, September, 2008,

Alves, N.M.; Gomez Ribelles J.L.; Gomes Tejedor J.A. & Mano,J.F. (2004). Viscoelastic Behavior of Poly(methyl methacrylate) Networks with Different Cross-Linking Degrees. Macromolecules, Vol. 37, No. 10, pp. 3735-3744, ISSN 0024-9297

ASTM standard D 3039/D 3039M–00, (2000). Standard Test Method for Tensile Properties of Polymer Matrix Composite Materials, American Society for Testing and Materials

ASTM standard D 790–03, (2003). Standard Test Method for Flexural Properties of Unreinforced and Reinforced Plastics and Electrical Insulating Materials, American Society for Testing and Materials

Bergstrom J.S. & Boyce, M.C. (1999). Mechanical Behavior of Particle Filled Elastomers. Rubber Chemistry and Technology, Vol. 72, pp. 633–656, ISSN 0035-9475

Brinkman, H.C. (1952). The viscosity of concentrated suspensions and solutions, Journal of Chemical Physics, Vol. 20, No. 4, pp. 571, ISSN 0021-9606

Brown, D.; Ma, B.M. & Chen, Z. (2002). Developments in the processing and properties of NdFeB-type permanent magnets. Journal of Magnetism and Magnetic Materials, Vol. 248, No. 3, pp. 432-436, ISSN 0304-8853

Brown, D.N.; Chen Z.; Guschl P. & Campbell P. (2006). Developments with melt spun RE–Fe–B powder for bonded magnets. Journal of Magnetism and Magnetic Materials, Vol. 303, No. 2, pp. e371–e374, ISSN 0304-8853

Budiansky, B. (1965). On the elastic moduli of some heterogeneous materials. Journal of the Mechanics and Physics of Solids, Vol.13, pp.223–227, ISSN 0022-5096

Chen, Z.; Wu, Y.Q.; Kramer, M.J.; Smith, B.R.; Ma B.M. & Huang, M.Q. (2004). A study on the role of Nb in melt-spun nanocrystalline Nd–Fe–B magnets, Journal of Magnetism and Magnetic Materials, Vol. 268, pp. 105–113, ISSN 0304-8853

Chin, T.S. (2000). Permanent magnet films for applications in microelectromechanical systems. Journal of Magnetism and Magnetic Materials, Vol. 209, No. 1-3, pp. 75-79, ISSN 0304-8853

Deng, S.; Hou, M. & Ye, L. (2007). Temperature-dependent elastic moduli of epoxies measured by DMA and their correlations to mechanical testing data, Polymer Testing, Vol. 26, No. 6, pp. 803-813, ISSN 0142-9418

Dobrzanski, L.A.; Drak, M. & Ziebowicz, B. (2007). New possibilities of composite materials application—Materials of specific magnetic properties. Journal of Materials Processing Technology, Vol.191, pp. 352–355, ISSN 0924-0136

Dobrzanski, L.A. & Drak, M. (2008). Hard magnetic composite materials Nd-Fe-B with additions of iron and X2CrNiMo-17-12-2 steel, Journal of Alloys and Compounds, Vol. 449, No. 1-2, pp. 88-92, ISSN 0925-8388.

Donnet, J.P. (1998). Black and White Fillers and Tire Compound, Rubber Chemistry and Technology, Vol. 71, pp. 323-341, ISSN 0035-9475

Ferry, J.D. (1980). Viscoelastic Properties of Polymers (3rd edition), John Wiley & Sons, Inc., ISBN 0471048941, New York

Garrell, M.G.; Shih, A.J.; Ma, B.M.; Lara-Curzio, E. & Scattergood, R.O. (2003a). Mechanical properties of Nylon bonded Nd–Fe–B permanent magnets. Journal of Magnetism and Magnetic Materials, Vol. 257, No. 1, pp. 32–43 ISSN 0304-8853

Garrell, M.G.; Ma, B.M.; Shih, A.J.; Lara-Curzio, E. & Scattergood, R.O. (2003b). Mechanical properties of polyphenylene-sulfide (PPS) bonded Nd-Fe-B permanent magnets. Materials Science and Engineering: A, Vol. 359, pp. 375-383, ISSN 0921-5093

Goll. D. & Kronmüller, H. (2000). Coercivity Mechanism in a Nanocrystalline and Bonded Magnets, In: Bonded Magnets, Series II: Mathematics, Physics and Chemistry - Vol.118, Hadjipanayis, G.C., pp.115-127, Kluwer Academic Publishers, NATO Science Series, ISBN 1-4020-1535-6, Netherlands

Goll. D. & Kronmüller, H. (2002). High-performance permanent magnets. Naturwissenschaften, Vol. 87, pp. 423-438, ISSN 0028-1042

Govindjee, S. & Simo, J. C., (1991). A micro-mechanically based continuum damage model for carbon black-filled rubbers incorporating Mullin's effect, Journal of the Mechanics and Physics of Solids, Vol. 39, No. 1, pp. 87–112, ISSN 0022-5096

Gronefeld, M. (2002). Review on Bonded Magnets, In: Bonded Magnets, Series II: Mathematics, Physics and Chemistry - Vol.118, Hadjipanayis, G.C., pp.1-12, Kluwer Academic Publishers, NATO Science Series, ISBN 1-4020-1535-6, Netherlands

Grujić, A.; Lazić, N.L.; Talijan, N.M.; Spasojević, V.; Stajić-Trošić, J., Ćosović, V. & Aleksić, R. (2010a). Polymer — Bonded Magnetic Materials with Various Nd–Fe–B Filler Content, Acta Physica Polonica A, Vol. 117, No. 5, pp. 859-863, ISSN 0587-4246

Grujić, A.; Talijan, N.; Stojanović, D.; Stajić-Trošić, J.; Burzić, Z.; Balanović, Lj. & Aleksić, R. (2010b). Mechanical and Magnetic Properties of Composite Materials with Polymer Matrix. Journal of mining and metallurgy, Section B: Metallurgy, Vol. 46, No. 1, pp. 25-32, ISSN 1450-5339

Guschl, P.C; Kim, H.S. & Otaigbe, J.U. (2002). Effects of a Nd-Fe-B Magnetic Filler on the Crystallization of Poly(phenylene sulfide). Journal of Applied Polymer Science, Vol. 83, pp. 1091–1102, ISSN 0021-8995

Gutfleisch, O. (2000). Controlling the properties of high energy density permanent magnetic materials by different processing routes. Journal of Physics. D: Applied Physics, Vol. 33, pp. R157–R172, ISSN 0022-3727

Gutfleisch, O.; Willard, M.A.; Brück, E.; Chen, C.H.; Sankar, S.G. & Liu, J.P. (2011). Magnetic Materials and Devices for the 21st Century: Stronger, Lighter, and More Energy Efficient. Advanced Materials, Vol, 23, No. 7, pp. 821–842, ISSN 1521-4095

Guth, E. & Gold, O. (1938). On the hydrodynamical theory of the viscosity of suspensions. Physical Review, Vol. 53, pp. 322, ISSN 0031-899X

Guth, E. (1945). Theory of filler reinforcement. Journal of Applied Physics, Vol.16, pp. 20–25, ISSN 0021-8979

Hamano M. (1995). Overview and outlook of bonded magnets in Japan. Journal of Alloys and Compounds, Vol. 222, No. 1-2, pp. 8-12 ISSN 0925-8388

Hemrick, J., Lara-Curzio, E. & Liu, K. (2004). Mechanical properties of thermally cycled nylon bonded Nd-Fe-B permanent magnets. Journal of Materials Science, Vol. 39, pp. 6509 - 6522, ISSN 0022-2461

Herbst, J.F. (1991.) R2Fe14B materials: Intristic properties and technological aspects. Reviews of Modern Physics, Vol. 63, No. 4, pp. 819-898, ISSN 0034-6861

Hono K. & Ping, D.H. (2001). Atom probe studies of microstructural evolution in Nd–Fe–B-based nanocomposite magnets, Materials Science and Engineering: A, Vol. 304-306, pp. 81-87, ISSN 0921-5093

Kokabi, M.; Arabgol, F. & Manteghian, M. (2005). Nd2Fe14B Permanent Polymeric Composite Magnets, IranianPolymer Journal, Vol. 14, No. 1, pp. 71-79, ISSN 1026-1265

Lahelin, M.; Aaltio, I.; Heczko, O.; Söderberg, O.; Ge, Y.; Löfgren, B.; Hannula, S.P. & Seppälä, J. (2009). DMA testing of Ni–Mn–Ga/polymer composites. Composites. Part A: Applied Science and Manufacturing, Vol. 40, No. 2, pp. 125–129, ISSN 1359-835X

Li, J.; Liu, Y.; Gao, S.J.; Li, M.; Wang, Y.Q. & Tu, M.J. (2006). Effect of process on the magnetic properties of bonded NdFeB magnet. Journal of Magnetism and Magnetic Materials, Vol. 299, No. 1, pp. 195-204, ISSN 0304-8853

Ma, B.M.; Herchenroeder, J.W.; Smith, B.; Suda, M.; Brown, D. & Chen, Z. (2002). Recent development in bonded NdFeB magnets. Journal of Magnetism and Magnetic Materials, Vol. 239, No. 1-3, pp. 418-423, ISSN 0304-8853

Maity, T.; Samanta, B.C.; Dalai, S. & Banthia, A.K. (2007). Curing study of epoxy resin by new aromatic amine functional curing agents along with mechanical and thermal evaluation. Materials Science and Engineering A, Vol. 464, No. 1-2, pp. 38–46, ISSN 0921-5093

Makled, M.H.; Matsui, T.; Tsuda, H.; Mabuchi, H., El-Mansy, M.K. & Morii, K. (2005). Magnetic and dynamic mechanical properties of barium ferrite-natural rubber composites. Journal of Materials Processing Technology, Vol. 160, pp. 229–233, ISSN 0924-0136

Matsuura, Y. (2006). Recent development of Nd–Fe–B sintered magnets and their applications. Journal of Magnetism and Magnetic Materials, Vol. 303, No. 2, pp. 344–347, ISSN 0304-8853

Menard K.P. (1999). Dynamic Mechanical Analysis: A Practical Introduction, CRC Press LLC, ISBN 0-8439-8688-8, Florida

Mori, T. & Tanaka. K, (1973). Average stress in matrix and average elastic energy of materials with missfitting inclusions, Acta Metallurgica, Vol. 21, No. 5, pp. 571–574

Otaigbe, J.U., Kim, H.S. & Xiao, J. (1999). Effect of Coupling Agent and Filler Particle Size on Melt Rheology of Polymer-Bonded Nd-Fe-B Magnets. Polymer Composites, Vol. 20, No. 5, pp. 697-704, ISSN 0272-8397

Payne A.R. (1965). Dynamic Properties of Filler-Loaded Rubbers, In: Reinforcement of Elastomers, Kraus G., pp.69-123, Interscience Publishers, a division of John Wiley & Sons, USA

Plusa, D.; Dospial, M.; Slusarek, B. & Kotlarczyk U. (2006). Magnetization reversal mechanisms in hybrid resin-bonded Nd–Fe–B magnets. Journal of Magnetism and Magnetic Materials, Vol. 306, No. 2, pp. 302–308, ISSN 0304-8853

Ponte Castaneda P. (1989). The overall constitutive behaviour of nonlinear elastic composites, Proceedings of the Royal Soiety A, Vol. 422, pp. 147–171,

Radojević, V.; Talijan, N.; Grujić, A.; Ćosović, V. & Aleksić R. (2007). Influence of composition of the magnetic composite coating on the performance of the optical fiber magnetic field sensing element. Journal of Optoelectronics and Advanced Materials, Vol.9, No. 8, pp. 2424 -2429, ISSN 1454-4164

Rodrigues, D.; Concilio, G.V. & Landgraf, F.G. (1996). Process and Magnetic Properties of Cold Pressed NdFeB Bonded Magnets, Proceedings of the 14th International Workshop Rare-Earth Magnets and Their Applications, Vol. 1, ISBN 981-02-2762-0, Sao Paulo, Brazil, September, 1996

Sagawa, M.; Fujimura, S.; Togawa, N.; Yamamoto, H. & Matsuura, Y. (1984). New material for permanent magnets on a base of Nd and Fe. Journal of Appied Physics, Vol. 55, pp. 2083-2087, ISSN 0021-8979

Slusarek, B. & Dudzikowski, I. (2002). Application of permanent magnets made from NdFeB powder and from mixtures of powders in DC motors. Journal of Magnetism and Magnetic Materials, Vol.239, No.1-3, pp. 597-599, ISSN 0304-8853

Trout, S. R. (2000). Understanding Permanent Magnet Materials; an Attempt at Universal Magnetic Literacy, Proceedings of EMCW 2000, Cincinnati, Ohio, October, 2000, http://spontaneousmaterials.com/Papers/CW2000.pdf

Vuong, N.V.; Khanh, N.V. & Thuy, D.M.; (2003). Simulation of the energy product (BH)max of Nd–Fe–B anisotropic bonded magnets, Physica B: Condensed Matter, Vol. 327, No. 2-4, pp. 349–351, ISSN 0921-4526

Wang, M.J. (1999). The Role of Filler Networking in Dynamic Properties of Filled Rubber. Rubber Chemistry and Technology, Vol. 72, pp. 430-448, ISSN 0035-9475

Withaeger, G. (2001). Techniques Utilized in the Characterization of Existing Materials for Improved Material Development, MSc Thesis, University of North Texas

Xiao, J.; Otaigbe, J.U. & Jiles, D.C. (2000). Modeling of magnetic properties of polymer bonded Nd-Fe-B magnets with surface modifications. Journal of Magnetism and Magnetic Materials, Vol. 218, pp. 60-66, ISSN 0304-8853

Zhang, X.H.; Xiong, W.H.; Li, Y.F. & Song N. (2009). Effect of process on the magnetic and mechanical properties of Nd–Fe–B bonded magnets, Materials and Design, Vol. 30, No. 4, pp. 1386-1390, ISSN 0261-3069

26

Composite Materials for Some Radiophysics Applications

S.B.Bibikov[1] and M.V. Prokof'ev[2]
[1]*Emanuel Institute of Biochemical Physics (Russ.Acad.Sci.)*
[2]*Moscow Aviation Institute (Technical State University)*
[1,2]*SPE "Radiostrim" Ltd*
Russian Federation

1. Introduction

The development of modern radio engineering can be characterized by an increase in carrier frequencies in the gigahertz range, as well as the introduction of new systems of ultrafast and ultrawideband signals. These applications require the use of the ultrawideband microwave absorbers and radio shielding materials for the equipment functionality, radioelectronic equipment compatibility, electromagnetic safety, test of the equipment. These materials must possess not only specific radio physical, but also necessary exploitational properties.

Depending on the tasks and requirements there are different approaches to design of the radioabsorbing materials. First of all, the materials which interact with non-ionization electromagnetic irradiation (EMI) can be sorted in two wide classes: the one is the shielding materials and the second is the materials absorbing EMI.

As far as materials for electromagnetic shielding, the principle of their work is simple enough. High conductivity (sometimes - magnetic permeability) provides high screening coefficient. Quit different situation is in the problem of absorbance of the electromagnetic irradiation. Finally, it is necessary that energy of electromagnetic wave should be converted to heat of the material. But substances able to absorb electromagnetic irradiation and process highest conductivity (e.g. metals) provide too high reflections of electromagnetic waves due to induced currents. To solve this contradiction, some special techniques are applied.

It is necessary to keep in mind that there are many requirements to the material. Namely the composite materials are convenient for satisfaction of a number of the necessary parameters. Commonly, a composite material consists of a matrix, usually - polymer, which determines mainly mechanical properties of a material, and a filler - dispersed phase, which provides specific electrical and magnetic properties and therefore radiotechnical parameters of a composite in whole. Different media are using as the matrix or a substance: - Si-O- based polymers, thermoplastics, porous polymer materials, etc.

To reduce the reflection from the interface between the resistance medium and the free space, the constructions with the volume resistance that is decreasing in the direction to inside are used. That's why it is necessary to know the rules which determine the resistivity of a composite material and to use the technology which guarantees homogeneous material with exactly predicted volumetric resistance. Besides, it is necessary to take in account a frequency

dispersion of a electrical properties. Electrical properties of the composite materials in DC regime are determined by a percolation theory, and the properties in radiofrequency region strongly depend on geometry of the filler and are described by a polarization theory.
Another way to avoid unnecessary reflections from the materials placed on a reflective substrate is the using of a ferromagnetic phase. In this case so-called "wave resistance" of the media is match better to the free space wave resistance. To expand frequency region of effective absorbance down, the higher magnetic permeability of the composite materials should be achieved. For example, interesting results have been received, when there was a crystallization of an amorphous magnetic alloy.

2. Resistive electroconducting composite

The development of modern radioelectronics implies both the development of new and modernization of existing measuring instruments, and new challenges associated with the problems of electromagnetic compatibility. These problems cannot be solved without application of ultrawideband radio shielding and radio absorbing materials in radio engineering systems, the equipment and test benches.
At present theoretical principles are in detail developed and practical technologies of reception of microwave absorbers for various regions of a radio spectrum are mastered. The problem of wideband absorbers for frequencies $10^6 ... 10^{11}$ Hz is still very relevant in spite of recently conducted studies of materials based on nanostructured composite thin film structures (see, e.g., [1]). For example, the non-reflective material for the walls of an anechoic chamber sometimes can provide the frequency range from ~100 MHz up to 100 GHz. It is not impossible to use ferrite-like material in this region because of magnetic permeability dispersion. But it is real to build the effective absorber with sizes of the order of a highest wavelength λ based on generally-dielectric losses.
Practice shows that the approach involving the use of resistive materials for ultra-wideband applications (Gorbatenko & Bibikov, 2006; Titov et al., 2008) will be very useful because of its adaptability, flexibility of implementation and cost effectiveness. Also, the development of radio frequency absorbers for low-frequency range-based approach uses of high magnetic permeability, has not yet been exhausted, and requires further study.
Let's consider one of the widely used approaches to obtaining microwave absorbers based on materials with Joule losses, or the so-called bulk-resistive materials. One of the advantages of this approach is the adaptability to obtain electrically conductive composite materials and the possibility of building structures on the basis of gradient materials with different concentrations of the conducting phase. Of course, one of the tasks, which must be solved with the construction of the radio absorbing materials, is the account of the dispersion of dielectric permittivity and magnetic permeability.
Commonly, the radio physical properties of the composite materials used in a practice are determined by the next:
- Conductivity of the filler;
- Concentration and conditions of the dispersion of the filler;
- Shape of the filler;
- Permittivity and electrical disruptive strength of the matrix.

It is known that the range of the bulk conductivity of composite material effectively absorbing microwave irradiation lies in the range $10^{-1} \div 10^{1}$ $\Omega^{-1}m^{-1}$ (Gorshenev et al, 1997). To reach these values it is necessary to use the fillers possessing high conductivity, e.g. of metal or carbon type.

Material of the matrix mainly provides physical and mechanical properties. Usually, polymeric materials are used as the matrices, these are: thermoplastics (for example, polyethylene, polyvinyl chloride, polyurethane, etc.), elastoplastics (silicon materials, etc.).
For the good formability of the material is sometimes necessary to use a plasticizer. For example, to improve the mechanical properties of PVC, the dioctylphthalate is used in an amount from 20 to 60%weight. Different components can be used for an improvement in the quality of composite materials, but in contrast to the functional filler and the basic component of matrix, they do not substantially influence the radio physical properties of the materials.
It is important to remember that the composite materials, in contrast to the solutions and the plasticized polymers, are not mixtures at the molecular level. The sizes of the particles of the filler always exceed the molecular dimensions. Even in the case of graphene or nanotubes used as the filler, the longitudinal particle sizes are much larger than the distance C-C bond.
According to the nature of the distribution of components, the composite materials can be divided into the following classes:

- Materials constitute the layered structure
- Materials containing fiberfill;
- Matrix mixtures;
- Statistical mixtures.

The first two are clear from the material structure determination.
In the matrix polymer mixture forms a continuous medium in which discretely distributed filler particles do not come into contact with each other. In this case, the phases are unequal mixture.
Statistical mixtures are characterized by the disordered distribution of polymer and filler, and both phases are equal.

2.1 Basis of the percolation theory

One of key parameters determining radiophysical properties of the composite is its effective conductivity, which depends on properties of the conductive filler and conditions of its distribution in the matrix. Determination of the value of effective conductivity - is the global problem in the theory of heterogeneous media, in particular, percolation theory.
To calculate the effective conductivity without the frequency dependence of the inhomogeneous system is necessary to solve the system of equations

$$\operatorname{div} \mathbf{j} = 0, \quad \operatorname{rot} \mathbf{E} = 0 \tag{1}$$

under appropriate boundary conditions. Here **E** - electric field vector, **j** - current density vector. Effective conductivity is determined as

$$\langle \mathbf{j} \rangle = \sigma_e \langle \mathbf{E} \rangle . \tag{2}$$

There < ... > means averaging over the volume V of the sample. Then we assume $V \to \infty$. Following the theory, the effective conductivity of the system can be written as

$$\sigma_e(p;\sigma_1,\sigma_2) = \sigma_1 \cdot f(p,h) \tag{3}$$

where $h = \sigma 2 / \sigma 1$ - the ratio of conductivities of the phases, p – volume concentration of the first component. The function f (p, h) – "dimensionless conductivity" – plays a fundamental role in the whole theory of transport in isotropic two-component media. It is possible to find effective conductivity for any concentration of filler with $h \sim 1$ or for arbitrary h with $p \ll 1$. In general, the problem of calculating the conductivity of disordered composites with arbitrary values of p and h has no analytical solution. To solve this problem, in the theory of heterogeneous structures, so-called "approximation of effective medium" has been proposed, or EMA (Kirkpatrick, 1973). Analysis shows that the system with the very different conductivities of components undergoes phase transition of the type metal-dielectric with the specific relationship of concentrations. According to percolation theory, the conductivity of the system in the region of the phase transition is described by the following asymptotics (Efros & Shklovskii, 1976):

$$\begin{aligned} &\tau > 0,\ \Delta_0 \ll \tau \ll 1: && f = \tau^t \left\{ A_0 + A_1\left(\frac{h}{\tau^{t/s}}\right) + A_2\left(\frac{h}{\tau^{t/s}}\right)^2 + \ldots \right\} \\ &|\tau| \ll \Delta_0: && f = h^s \left\{ a_0 + a_1\left(\frac{h}{\tau^{t/s}}\right) + a_2\left(\frac{h}{\tau^{t/s}}\right)^2 + \ldots \right\} \\ &\tau < 0,\ \Delta_0 \ll |\tau| \ll 1: && f = \frac{h}{(-\tau)^q} \left\{ B_1 + B_2\left(\frac{h}{(-\tau)^{t/s}}\right) + \ldots \right\} \end{aligned} \quad (4)$$

where $\tau = (p - p_c)/p_c$, p – critical concentration; $\Delta_0 = h^{s/t}$ - the size of the "smearing", and t, s and q – critical indexes, which are connected by the relation:

$$t / s = t + q \quad (5)$$

The values of indices depend on the dimensionality of the system:
- for the two-dimensional case $t = q \approx 1.3$ and $s = 0.5$ (exactly);
- for the three-dimensional case $t \approx 2$; $q \approx 0.8$ and $s \approx 0.7$ (exactly).

The problem of conductivity of composites with periodic structure is simpler. Here it is sufficient to find the potential within a single cell. Following (Stauffer, 2003), we assume for the calculations that the percolation threshold $p_c = 0.31$ for the three-dimensional case of a cubic lattice. In this case we can write a direct expression for the conductivity of the composite on the concentration of inclusions in the case of 3d:

$$\sigma = \sigma_1 (p - p_c)^{2\nu} \quad (6)$$

where $\nu = 0.9$ for three-dimensional case.

2.2 Type of conduction fillers

If the composite material is used as microwave absorber or shielding, the basic functional characteristics are determined by the radiophysical properties of the filler, which should ensure effective action presents the interaction with the applied field and the dissipation of electromagnetic energy. It is necessary that the filler has at least one of the following qualities:
- Ensure that the macroscopic bulk or surface currents in the material;
- Creation of a generalized dielectric loss;
- Increasing the magnetic losses;
- Increasing the effective permeability of the composite.

For this purposes, the next materials are used as the fillers:
- Materials on the basis of graphite-like structures - carbon black, carbon fibers, nanotubes, nanodisperse forms of carbon,
- Powders of metals and alloys, both in polycrystalline and amorphous state, providing electrical composite and sometimes necessary magnetic properties,
- Ferromagnetic materials the basis of iron powders,
- Conducting non-metallic materials - conductive polymers, systems with conjugated bonds (polyaniline, polypyrrole, polydiacetylene, etc.).

Commonly, the fillers based on various forms of graphite, are used. This is due to the fact that firstly, the graphite has a conductivity that lower by about 2 orders of magnitude then for metals, secondly, a more chemically inert and therefore less prone to corrosion than the metals, and in the third, it can easily be used in a variety of geometric shapes - spherical or disk-shaped particles and fibers.

Valuable feature of some kinds of carbon blacks is their ability to structured into chains. Structure and properties of different carbon blacks have a decisive importance on the properties of electromagnetic radiation shielding and absorbing materials, so you should always take into account the types of technology and preparation of various carbon blacks and graphite.

2.3 The influence of the geometry of the filler

The electrical and radiophysics properties of composite substantially depend on the geometry of filler. First of all, the discussion deals with the ratio of length to the transverse size of the inclusions. The task of determining the electrophysical parameters of the composite on the basis of the data about electrical conductivity and form of filler is no way trivial, to this the significant number of works is devoted.

Detail theoretical study of the electrophysical properties of composites containing elongated conducting inclusions – sticks – is presented in a work (Lagarkov & Sarychev, 1996). The effective-medium approximation was developed in this work and the equation was derived to calculate an effective dielectric constant of the composites in the quasistatic case and for the high frequency. The theory predicts very large values of the effective dielectric constant in a wide range of the stick concentration. There was found that the dielectric constant can exhibit various dispersive behaviors. It can has relaxation behavior, power-law scaling behavior, or resonance dependence on the frequency. The resonance dependence occurs when the skin effect is strong and wavelength is comparable to the stick length. The diffraction of electromagnetic waves on a conducting stick – is a problem of classical electrodynamics, so one has to solve Maxwell's equations to find the polarizability for a particle in the composite illuminated by an electromagnetic wave. The particle is supposed to be embedded in the "effective medium" with conductivity. Then the effective conductivity σ_e is determined by the condition that the averaged polarizability of all particles shall vanish. In the condition of for the aspect ratio of the filler: a/b ≪ 1 (a - transverse, b - longitudinal dimension), there was derived the following expression for the effective complex dielectric permittivity εe of the composite:

$$\varepsilon_e = \varepsilon_d \cdot \frac{p}{p_c} \cdot \frac{a}{b \cdot \ln(a/b)} \cdot \left[\cos\left(\sqrt{\varepsilon_d} ak\right) - i\varepsilon_d (ak)^2 \cdot \frac{\ln(a/b)}{2} \right] \quad (7)$$

where ε_d – the complex permittivity of the conducting inclusions, k - wave vector.

If the composite is a statistical mixture of interacting components, there are dependences for the effective permittivity of the material (with a limited difference of complex values ε). General formulas for calculating the ε_e have the form

$$\varepsilon_e = \varepsilon_2 + (\varepsilon_1 - \varepsilon_2) p_1 K_1 \quad (8)$$

$$(\varepsilon - \varepsilon_1) p_1 K_1 + (\varepsilon - \varepsilon_2) p_2 K_2 = 0 \quad (9)$$

where K_1 and K_2 - ratio of the average electric field strengths in the components to the average field strength in the composite. Exact solution of equation (1) is possible only for two cases: for a layered dielectric when the electric field vector parallel to 1 - the interfaces between the planes and 2 - the axes of the cylinders, uniaxially oriented environment. In these cases $K_1 = 1$, hence $\varepsilon_e = p_1\varepsilon_1 + p_2\varepsilon_2$. In other cases, for calculation of K_1 and K_2 need to make assumptions about the permeability of the medium ε^*. The assumption that $\varepsilon^* = \varepsilon_2$ corresponds to the matrix mixture, and the assumption that $\varepsilon^* = \varepsilon_e$ means equality of the components and corresponds to the condition of the statistical mixture. Thus, for the mixtures of n components, the following approximate expressions, or the so-called formulas of mixing are exist:

$$\varepsilon_e^x = \sum_{i=1}^{n} p_i \cdot \varepsilon_i^x \quad ,(10)$$

where p_i - volume fraction of the i-th component, x – a value characterizing the distribution of the components. In particular, for a system in which the components are connected in parallel (layered plastics), we put x = +1; similarly, for a system in which the components are connected in series, we put x = -1. Often the case when the components are distributed randomly. Then, after substituting x = 0 and differentiating expression (8), we obtain a formula, often used in engineering practice:

$$\begin{gathered} \ln(\varepsilon_e) = p_1 \cdot \ln(\varepsilon_1) + p_2 \cdot \ln(\varepsilon_2) + \ldots + p_n \cdot \ln(\varepsilon_n), \\ (p_1 + p_2 + \ldots + p_n = 1) \end{gathered} \quad (11)$$

Note that these expressions are used for both real and imaginary part of permittivity.
For the special case of matrix mixtures with spherical inclusions, we can write the next equation:

$$\frac{\varepsilon_e - \varepsilon_m}{\varepsilon_e + \varepsilon_m} = \sum_i p_i \frac{\varepsilon_i - \varepsilon_m}{\varepsilon_i + 2\varepsilon_m} \quad (12)$$

where ε_m - permeability of the matrix, ε_i - permeability of the components. Similar, for the statistical multicomponent mixtures:

$$\sum_i p_i \frac{\varepsilon_i - \varepsilon_e}{\varepsilon_i + 2\varepsilon_e} = 0 \quad (13)$$

From expression (7) it is possible to obtain the frequency dispersion of dielectric constant for the elongated inclusions. Analysis shows that in the high-frequency region in this case the composite is characterized by the resonance type of dispersion ε.

Many composite materials, both traditional and with nanoscale inclusions are characterized by the frequency dispersion of permeability, which includes the relaxation part, which is similar to the dipole-dipole relaxation of Debye:

$$\varepsilon_e = \varepsilon_\infty + \frac{\varepsilon_c - \varepsilon_\infty}{1 + (i\omega\tau)^{1-\alpha}} \tag{14}$$

where εc and ε∞ - the dielectric constant at low and high frequency limits, respectively, τ - time of relaxation, and a dimensionless exponent (1 - α) characterizes the distribution of relaxation times, at α = 0 there is no smearing of the relaxation times. Let us note that the same in the form expression (with [f]=0) is correct for the structure of heterogeneous layered composite, this so-called polarization of Maxwell-Wagner type. The next section shows an example of analyzing the frequency dispersion of dielectric composite material used in practice.

2.4 Porous electroconducting materials

It is possible to consider porous material with a large specific surface area, covered with the thin electroconducting film, as a variety of the composite material, which is characterized by high effective electrical conductivity with the insignificant portion of filler. Such material can be, for example, polyurethane foam, whose pores are covered with graphite film.
Fig.1 presents the material consists of polyurethane foam (PUF) with the ultra-dispersed carbonic filler, introduced into the system in an amount providing the dc conductivity. This composite material based on the polyurethane foam is a foamed structure with open pores coated with ultrafine graphite composition. Structure of the polyurethane can be characterized by a number of parameters that affect the electrical properties of the material: density, linear dimensions of cells, their degree of elongation and orientation relative to the direction of foaming, the degree of connectedness, wall thickness, etc.

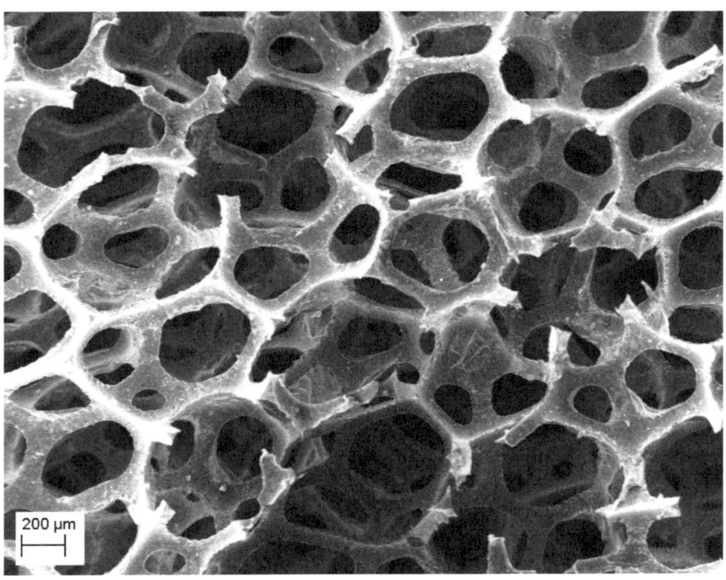

Fig. 1. Structure of resistive material on the basis of polyurethane foam.

Amount of graphite should be enough to ensure the percolation threshold. To give the properties of non-flammable material is processed with a special compound does not influence the radio-technical characteristics. Carbon particles have sizes considerably less than the wavelength and skin layer thickness, and form a continuous quasi-graphite conductive mesh. The final DC conductivity and all the features in the microwave range depend on the amount of carbon filler, the structure, thickness and length of the conducting clusters. Such material can be considered as uniform on the electrical properties (although this may be the subject of a separate study), and isotropic.

Figure 2 presents the electron microscopy of conductive graphite coating of various types. It was determined that smaller particles of graphite form more smooth continuous film on the surface of the pores of the polymer: top left - "rough" film formed by the coarse fraction of graphite particles, bottom left - "smooth" film formed by fines. A "slice" of a graphite particle is presented there, it is seen that thickness *h* is about 26 interlayer distances, hence $h = 26 \times 0.335 \approx 9$ nm. However, the average transverse size of the graphite particles is much greater than the thickness - size distribution is shown in Figure 3. Consequently, the particles are disk-shaped with a large aspect ratio. This is confirmed by data of X-ray diffraction analysis, namely, according to the size of the region of the coherent scattering, determined smaller of the sizes (Prokof'ev et al., 2010). This form with the condition of surface area contributes to shaping of the steady conductive coating.

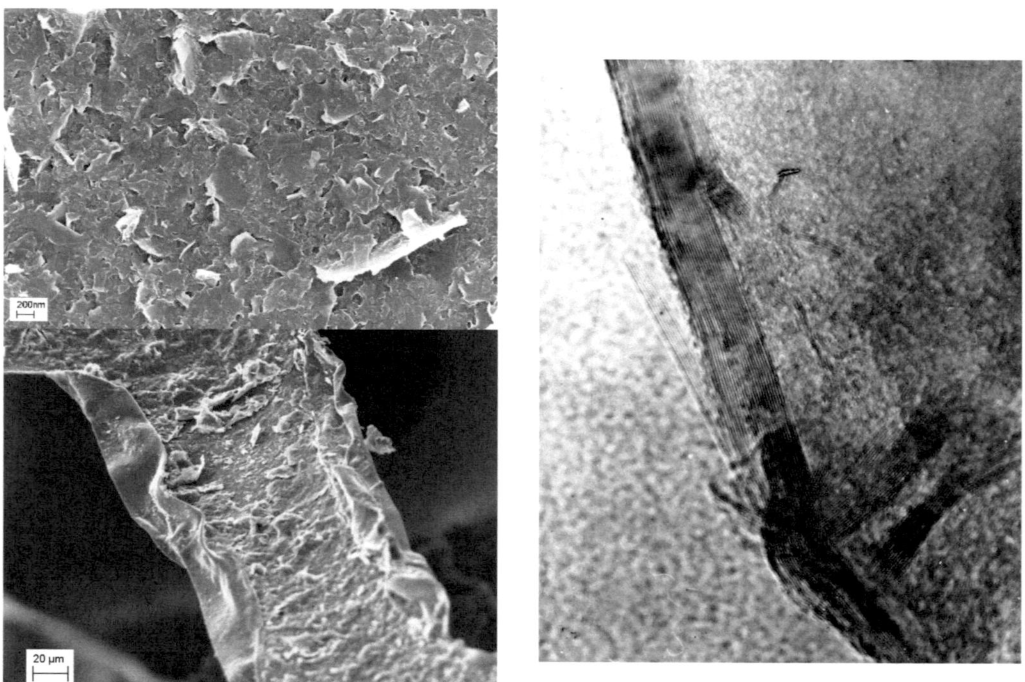

Fig. 2. Electron microscopy of conductive graphite coating of various types. Top left - "rough" film formed by the coarse fraction of graphite particles. Bottom left - "smooth" film formed by fines. Right - the "slice" of the graphite particles with thickness of about 26 interlayer distances.

With the creation of composite materials for absorbing or shielding electromagnetic radiation of radiofrequency range, are effective the composites, which possess a certain resistivity in DC regime, i.e., with the concentration of the filler $p \geq p_c$ where p_c is the percolation threshold. When optimizing the parameters of such materials is often considering that the material is completely characterized at macro scope level in two parameters: scalar dielectric permittivity $\varepsilon' = \mathrm{Re}(\varepsilon)$ and the specific volume electrical conductivity σ_V, independent of frequency in the studied range of frequencies. In some cases these assumptions are justified, because the conductivity of the medium in the radio frequency $\sigma(\omega)$ negligible differs from that of the DC conductivity σ_0, while the imaginary part of dielectric constant depends on frequency according to $\varepsilon'' \sim \omega^{-1}$ (Landau & Lifshitz, 1984). However, in some cases the condition $\sigma(\omega) \approx \sigma_0$ is not observed exactly. At the same time, during the construction of the flat multilayer constructions, where the resonance wave effects are completely important, it is necessary to know the frequency dependence (dispersion) of the generalized dielectric constant: $\varepsilon^*(\omega) = \varepsilon'(\omega) - i \cdot \varepsilon''(\omega)$.

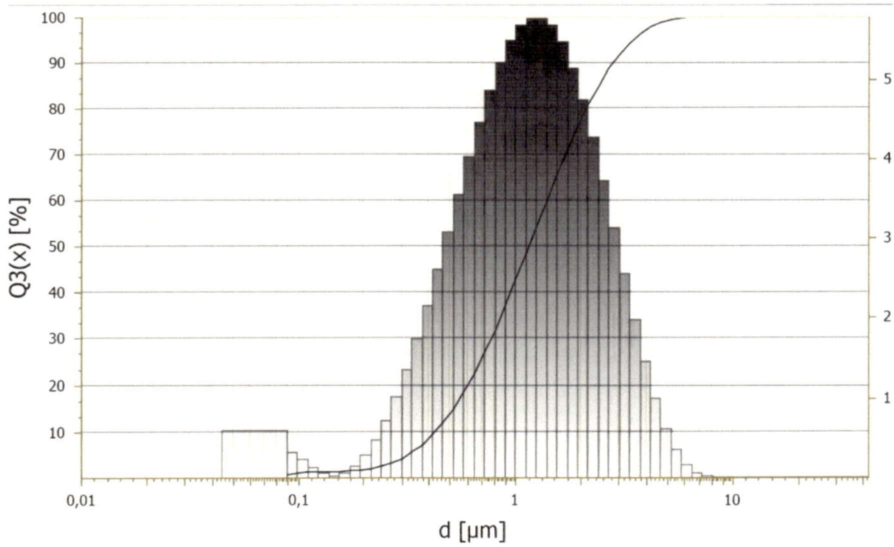

Fig. 3. Histogram of the transverse size of the colloidal graphite particles (according to photodynamic scattering).

Furthermore, often it is necessary to determine the function of the frequency dependence of dielectric constant for the correct extrapolation of experimental data into the extended frequency domain. Thus, for instance, the frequency spectrum of many UWB devices, in particular, geo-radars, stretches into region $10^7 \ldots 10^{10}$ Hz. Characterization of conductive foam is convenient to produce contactless method in the microwave range. With the measurements of such materials in the sub-gigahertz region noncontact method requires the large sizes of irradiators and models, and contact method is characterized by the poor reproducibility of results of measurements and requires special conditions for guaranteeing the electrical contacts. Determination of the dielectric permittivity values at low frequencies is possible if you know the dispersion in the microwave range.

The similar material was the base of the wideband microwave absorbers of "Mokh" type ("Radiostrim" Ltd). The analysis of the dielectric permeability dispersion of the microwave absorber, made on the basis of electrically conductive polyurethane foam, is resulted below.
In order to obtain the data about the frequency dependence and of materials, there were prepared flat models and in the free space the amplitude and phase responses of radiophysics coefficients were determined (for example, transmission and reflection coefficients). Then, after solving the inverse problem, which connects the experimental parameters with the dielectric constant of material at each frequency point, the frequency dispersion ε (complex value) was determined.
Fig. 4 (solid curves) shows the experimental plots ε' and ε", obtained from the measured values of the complex transmission coefficient for a flat sheet in an free space. Samples differ each from another by the conditions of formation of graphite coating on the surface of the pores of the polymer (a series of 3 specimens with smooth and rough coated). In each series, the fraction of graphite phase in the composite decreases from sample 1 to sample 3 ($C_1 > C_2 > C_3$).
It is logical to describe The structure by a polarization model like modified Maxwell-Wagner model of an inhomogeneous medium:

$$\varepsilon^* = \varepsilon_\infty + \frac{i \cdot \sigma_V}{\omega} + \frac{\varepsilon_c - \varepsilon_\infty}{1 + (j\omega\tau)^{1-\alpha}}, \tag{15}$$

where σ_v is the conductivity in DC regime, while ε_c , ε_∞ τ and α - the same parameters as for formula (14). The term that considers through conductivity is here added, since the threshold of percolation is passed. After decomposing (15) to the real and imaginary parts, it is possible to find the parameters of the model. Fig.4 shows the curves of the real ε' and imaginary ε" parts of the dielectric permittivity, obtained both experimentally (solid line) and from the calculation by formula (15) with the parameters: τ, α, σ_V, ε_c and ε_∞ (dotted line). Values of the parameters are given in Table 1.
At $\alpha \approx 0.15 \div 0.17$ theoretically constructed curves simultaneously well coincide with experimental values ε' and ε".
The analysis of the frequency dependences (see Table 1) shows that for the investigated samples the last term in expression (15) is more significant for the "rough" graphite covering, in contrast to the case of "smooth" covering, where value of σ_v correlates with the concentration of the graphite phase and influence of the DC-conductivity is more.
To obtain the effective microwave absorber with resistive losses is necessary to ensure optimal complex conductivity of the material in a given frequency range: it is necessary, on the one hand, efficiently convert electromagnetic energy into heat, but on the other hand - to avoid unwanted reflections from conductive surfaces (Bibikov et al., 2008). Usually for this purpose are used the structures of gradient-type material, or profiles, minimizing reflection from the front edge of the absorber (Fig.5).

3. Composite materials based on ferromagnetic fillers

Besides the microwave absorber possessing dielectric losses, the composite materials, which contain ferromagnetic phase, are used. Special features and conditions of using such materials it will be discussed further.

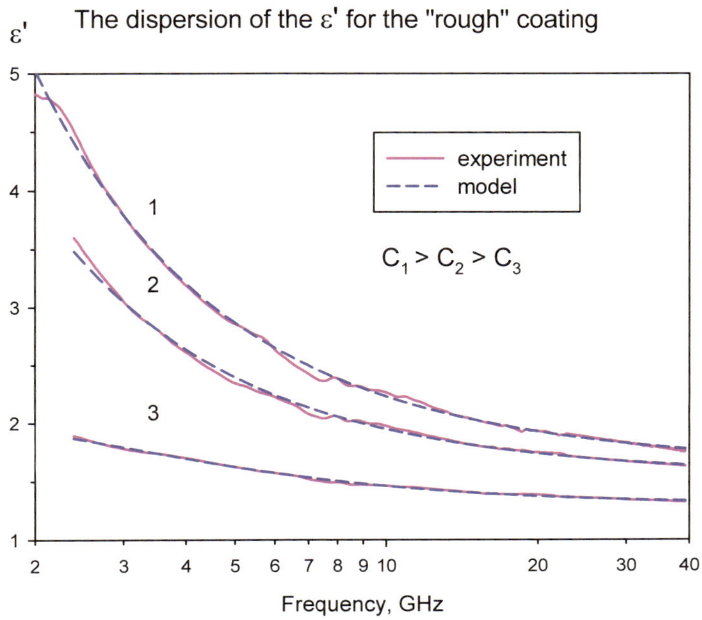

(a)

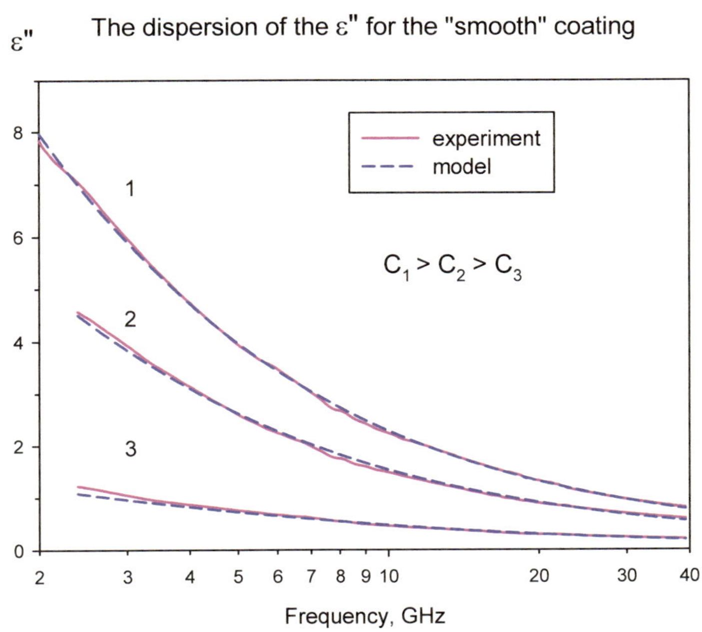

(b)

Fig. 4. Frequency dependence of dielectric permittivity of the conducting foamed PPU with the "smooth" covering: (a) – ε′, (b) – ε″.

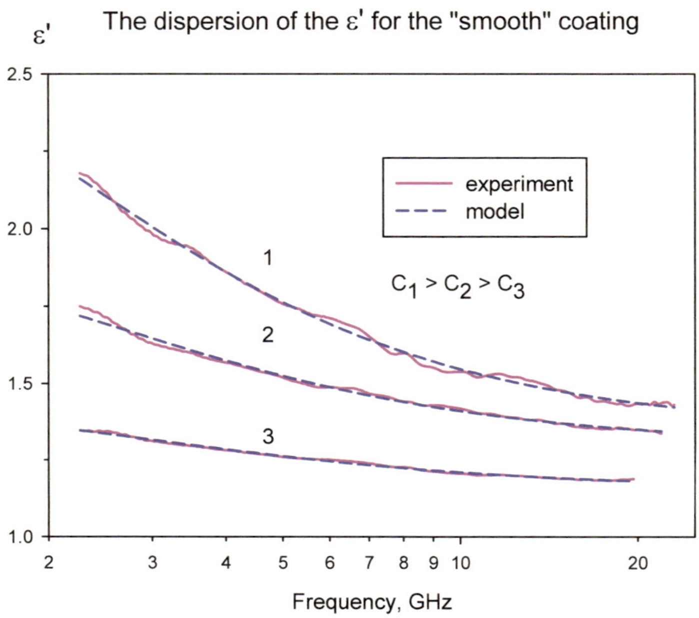

(a)

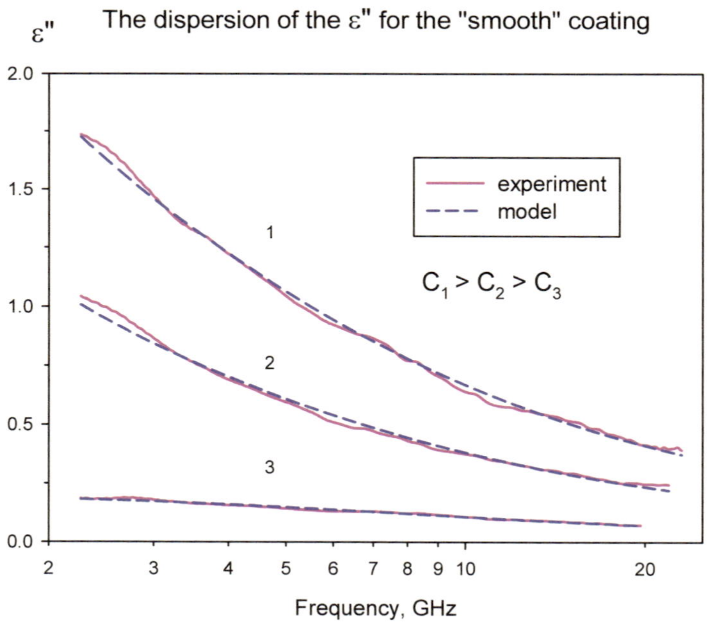

(b)

Fig. 5. Frequency dependence of dielectric permittivity of the conducting foamed PPU with the "rough" covering: (a) – ε′, (b) – ε″.

Type	Sample	τ, ns	σ_V , $\Omega^{-1}m^{-1}$	$\Delta\varepsilon$	ε_∞	1-α
"smooth"	1	0.47	0.126	1.73	1.33	0.83
	2	0.39	0.084	0.79	1.29	0.83
	3	0.36	0.006	0.33	1.15	0.83
"rough"	1	2.91	0.07	38.5	1.61	0.85
	2	2.26	0.05	20.3	1.52	0.85
	3	0.33	0.09	0.99	1.29	0.85

Table 1. Dispersion parameters of composites based on polyurethane foam with the graphite covering.

3.1 The motivation of the using of ferromagnetic phase in composite materials

One problem is the unwanted reflections from the boundary of the absorbing material. A way to reduce the reflection coefficient of the interface is the impedance matching material and free space. It is due to the mismatching of μ/ε ratio at the interface between free space and the edge of the material. Since the mismatch is due to the high complex permittivity, proportional increasing of the complex permeability can compensate it.

To ensure electromagnetic compatibility and testing a number of devices, for example, communication, UWB devices (georadar etc.), and for equipment of proper anechoic chambers, it is necessary to use radio shielding and absorbing materials of a range of frequencies from MHz up to GHz. Because of the geometric limitations, absorbers of resistive type are unacceptable here; therefore it is necessary to use materials with the high magnetic losses. For example, it possible to use the materials on the basis of ferrites or magnetic metals, which are characterized by an increase in the losses with the resonance of domain walls in the MHz region, particularly due to the special structure of the ferrite grains (Bibikov et al., 2010).

3.2 Ferrite-based composites

Analysis of the properties of ferrites at microwave frequencies shows that as an absorber of irradiation in the long wavelength part of the microwave range is preferable to use Mn-Zn ferrites, for UHF region - Ni-Zn ferrites, in the centimeter wavelength region - hexagonal ferrites. Using the high-anisotropic hexagonal ferrites as fillers of composite materials allows realizing the frequency-selective absorption of electromagnetic radiation by controlling the frequency of natural ferromagnetic resonance of the ferrites.

But the use of ferrite ceramics is often inconvenient due to specific technological and operational requirements for many electronics products. The composite material consisting of an insulating matrix and ferromagnetic filler called magnetodielectrics. As the fillers of magnetic dielectrics frequently are used carbonyl iron, Alsifer, Permalloy, etc. Magnetic permeability of magnetic dielectrics is less than in monolithic ferromagnetic materials for two reasons: in first, the presence of the significant demagnetizing factor because of the dissociation of particles and, in second, the formation of the poly-domain structure is energetically disadvantageous. In spite of it, the magnetic composite materials possess sufficient magnetic losses for the using as microwave absorbers. For example, Fig. 6 shows the reflection coefficient of the composite material based on granular Ni-Zn ferrite at a mass filling of 89%.

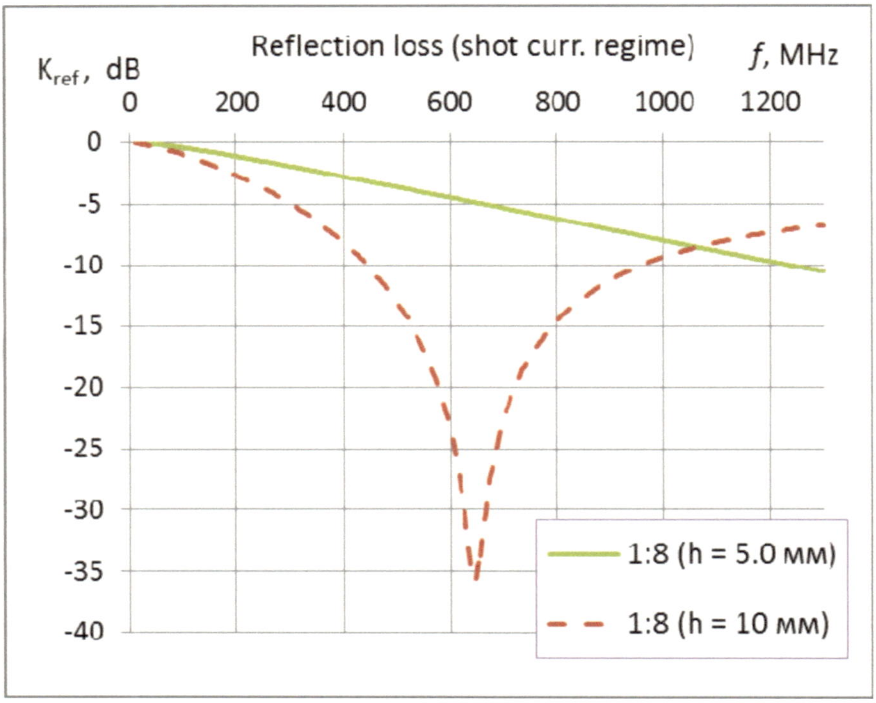

Fig. 6. The reflection coefficient of the composite materials on the basis of Ni-Zn ferrite. (Two thicknesses of the samples).

3.3 The composites based on amorphous magnetic alloys.

Ferrite–polymer composites present narrow-band electromagnetic wave absorbers for use in several hundred megahertz to several gigahertz frequency range. The merits of ferrite–polymer composite absorbers are their adequate reflection loss due to large magnetic loss. However, the major limitation for expanding the application of ferrite–polymer composites is that possesses quite different frequency dispersion of the dielectric permittivity ε and magnetic permeability μ. This limits the working frequency range of the material. The development of new thinner electromagnetic wave absorbers using amorphous alloys has been attempted in order to overcome the ferrite–polymer composites problem.

Since the permeability of amorphous alloys rapidly decreases with increasing frequency due to eddy current loss, composites which are fabricated by dispersing amorphous alloy particles in a polymer matrix are generally used in high frequency applications. However, the application of amorphous alloy composites as absorbers is restricted by large surface reflection of electromagnetic wave due to their relatively large complex permittivity values compared to complex permeability values.

Then, it is possible to smoothly change magnetic permeability, controlling by the restructuring of magnetic alloy from the amorphous into the nano-crystalline state.

The next sample of composite material consists of organosilicon thermo-stable matrix and mixed nanocrystal-amorphous magnetic filler with grain size near 50 μm (Kuznetsov et al., 2005; Bibikov et al., 2007).

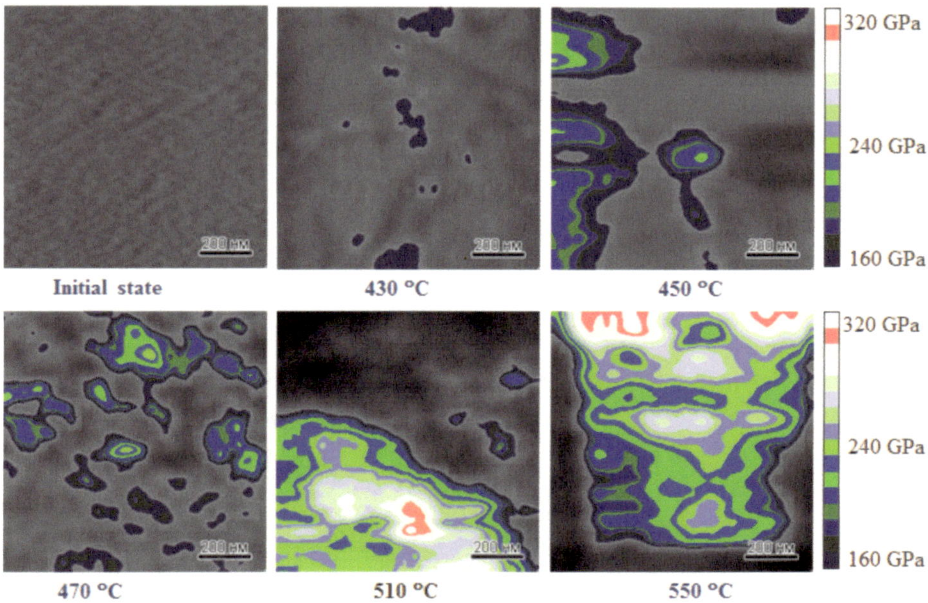

Fig. 7. Atomic force microscopy:Fe-Cu-Si-Nb-B alloy elastic modulus via temperature of the alloy treatment

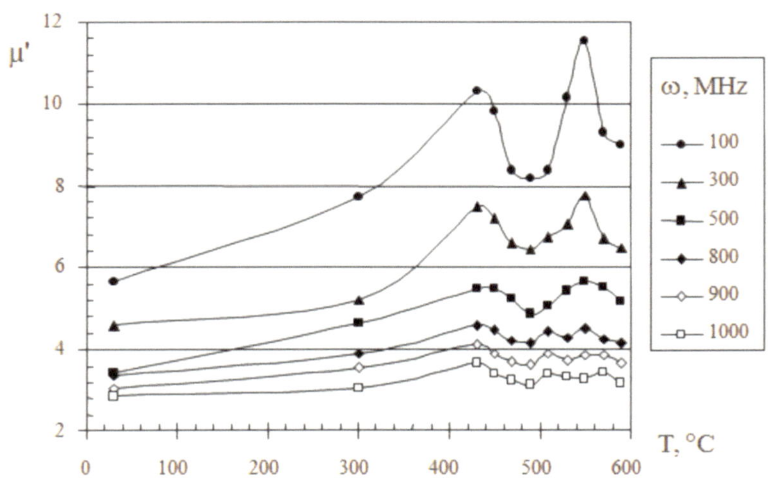

Fig. 8. Dependence of the permeability μ' at $f = 100 \ldots 1000$ MHz on the temperature treatment of the amorphous magnetic material. (Content of the magnetic powder ~50% vol.)

Using the microscope of atomic forces, it was shown (Fig.7) that a process of formation of nanocrystalline areas in the amorphous matrix takes place during mechanical and thermo treatment of the alloy. This is confirmed by the pictures of the micro-distribution of the modulus of elasticity. At the same time, it was observed the correlation between growth of the nano-crystalline regions in the filler and the increasing of the effective magnetic permeability of composite (Fig.8). This leads to noticeable changes integral radiophysical characteristics (Fig. 9).

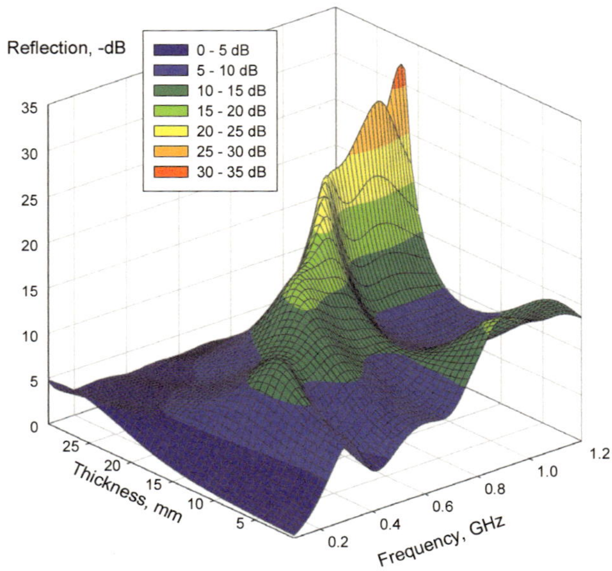

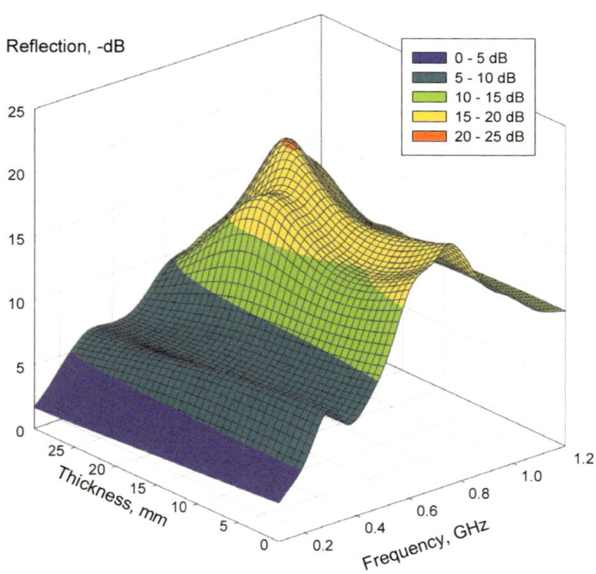

Fig. 9. The dependence of the reflection from the thick-layer coatings based on experimental data of the composite filled with amorphous crystalline alloy. Top – the filler in the amorphous state, bottom – in nanocrystalline state.

4. The possibility of the using of nanostructured composites for radiophysics applications

A number of theoretical works according to the electromagnetic and of carbonic nano-tubes shows that for further progress in the field of creating the thin-film effective microwave absorbers it is necessary to investigate the influence of magnetic nanoparticles in the insulating and semiconductor tape matrices. Also, the granular structures containing ferromagnetic nanoparticles of 3-d metals in such matrices can have a high degree of microwave absorption. It is defined by a set of factors: a large magnetization and the rapid relaxation of the spin of nanoparticles are in a superparamagnetic state, as well as a large value of dielectric constant and dielectric loss, depending on the localized electronic states in the matrix. Thus, several studies have investigated the film hydrogenated amorphous carbon nanoparticles Cm deposited on policar and Kevlar substrate (Rozanov et al., 2005).

5. Microwave absorbers based on multilayered, gradient and profiled composite materials

The absorbers of resistance type microwave emission can effectively work in entire wavelength range of radio-frequency range, but, taking into account reduction in the effectiveness with an increase in the relation of λ/h, where λ - wavelength and h - thickness of the material, they are frequently used at frequencies higher 1 GHz. . Of course, it is possible to use this material at frequencies below 1 GHz, but the size of it should be much greater (proportionally to the wavelength λ).

An example of pyramidal microwave absorber is on Fig.10. The black color is due to the carbon-graphite filler, and white color - after processing by incombustible agent. Frequency dependence of the reflectivity of the samples of composite materials such as moss of various heights are shown in Fig. 11 (Bibikov et al., 2010). The graphs in Fig.11 show that the pyramidal structure based on the PPU with fillers discussed above, provides a significant reduction of electromagnetic radiation incident along the normal. So, for the absorber, "Mokh P-330" with a height of 330 mm reflected microwave power is less than 50 dB, which is equivalent to the attenuation of the power at 10,000 times, and with the decreasing in the height of the material the reflection coefficient increases at low frequencies - for most thin absorber "Moss P-70" the interference is observed.

Fig. 10. View of microwave absorber "Mokh P-70".

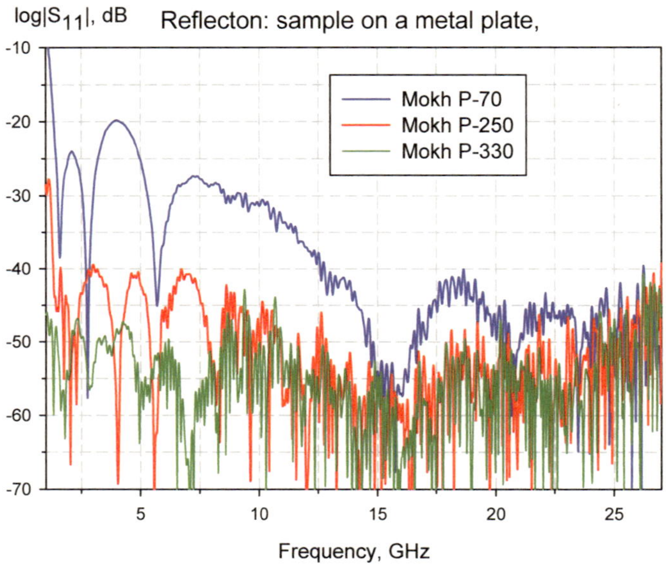

Fig. 11. Coefficient of reflection of the electromagnetic irradiation from the microwave absorber "Mokh P-70" at normal incidence in free space.

This type of material is useful both for the equipment of anechoic chambers, and for electromagnetic compatibility of radio systems (communication, radar, medical and search equipment, etc.).

6. Conclusion

The using of composite materials with desired radiophysical properties in some cases is optimal solution for the complex engineering problems in modern electronics. In particular, these are:

- Ensuring the electromagnetic compatibility of electronic equipment;
  - Information security,
- Electromagnetic security and the environment safety,
- Functionality of radio systems: communications equipment, radars, control system, ultra-wideband devices, etc.
  - Tasks of Military and special techniques.

Special feature of the composite materials for radiotechniques is the correspondence simultaneously electrophysical, mechanical, and sometimes also to the chemical parameters.
The complexity of creating a composite broadband microwave absorbers determined by the fact that the values of complex permittivity, permeability and, correspondingly, the coefficient of reflection of radio waves depends on the frequency of the electromagnetic field.
The ideal absorber of the electromagnetic radiation of radio-frequency s-band has not been created. That's why design of the similar materials is always defined by a number of requirements both for radiophysics and operational parameters of the material. Therefore there are different types of absorbers, including on the basis of composite materials: of resistive type, ferromagnetic, gradient type, shaped, etc.

7. References

Bibikov S.B., Kulikovsky Ed.I., Sharafiev R.S. & Kuznetsov P.A. (2007). New absorbing materials for applications in decimeter and millimeter wavelengths ranges, *Proceedings of the International Conference on Antenna Theory and Techniques*, pp. 295-297, ISSN 978-1-4244-1578-7, Sevastopol, Ukraine, September, 2007.

Bibikov S.B., Gorshenev V.N., Sharafiev R.S. & Kuznetsov A.M. (2008). Electrophysical properties of electroconducting papers and cardboards treated with colloid-graphite solutions *Materials Chemistry and Physics*. Vol. 108, No.1, pp. 39 - 44, ISSN 0254-0584.

Bibikov S.B., Smolnikova O.N., Prokof'ev M.V., Orlov V.V. & Men'shova S.B. (2010). Some trends in design of microwave absorbers for anechoic chambers and ultrawideband systems. *Proceedings of the 5th International Conference on Ultrawideband and Ultrashort Impulse Signals*, pp. 262-265, ISBN 978-1-4244-7468-4, Sevastopol, Ukraine, September, 2010.

Bibikov S.B., Smolnikova O.N. & Prokof'ev M.V. (2011). Dielectric properties and the microwave - conductivity of porous radioabsorbing materials. *Radiotekhnika (Radioengineering)*, No.3, Will be published (March 2011), ISSN 0033-8486.

Efros A.L. & Shklovskii B.I. (1976). Critical Behavior of Conductivity and Dielectric Constant near the Metal-Non-Metal Transition Threshold, *Phys.Stat.Sol.(B)*, Vol. 76, No2, pp. 475-785.

Gorbatenko O.N. & Bibikov S.B. (2006). The using of radar absorbing - radio scaterring for protection against electromagnetic interference of georadar. *Special Technics*, No. 3, pp.26. ISSN 1996-0506.

Gorshenev V.N., Bibikov S.B., Spector V.N. (1997). Simulation, synthesis and investigation of microwave absorbing composite materials, *Synthetic Metals*, Vol. 86, pp. 2255-2256, ISSN 0379-6779.

Kirkpatrick S. (1973). Percolation and Conduction. *Rev.Mod.Phys.*Vol. 45, No. 4, pp. 574-588.

Kuznetsov P.A., Zvorygin R.G. & Bibikov S.B. (2005). Investigation of atomic-force microscope crystallization kinetics of the nanocrystalline alloy Fe-Cu-Nb-Si-B and the creation of systems based on its electromagnetic protection. *Metally (Metals)*, 2005, No 6, pp. 25-31, ISSN 0869-5733.

Lutsev L.V., Nikolaychuk G.A., Petrov V.V., Yakovlev S.V. (2008). Multi-purpose radio-absorbing materials based on magnetic nanostructures: preparation, properties and applications. *Nanotechnics*, No.14, pp.37-43, ISSN 1816-4498.

Lagarkov A.N. & Sarychev A.K. (1996). *Phys. Rev. B*, Vol.53, No.7, (March 1996), pp. 6318-6336, ISSN 0163-1829.

Landau L.D. & Lifshitz E.M. (1984). Electrodynamics of Continuous Media, 2th ed., Pergamon, Oxford.

Maksimenko S.A., Slepian G.Ya. (2002). Electrodynamics of carbonic nanotubes. Radioelectronics and Communications Systems Vol. 47, No. 3, pp. 261-280, ISSN 0033-8494.

Prokof'ev M.V., Smolnikova O.N., Bibikov S.B. & Kuznetsov A.M. (2010). Influence of conditions of heat treatment on the structure and properties of the coatings from colloidal-graphite dispersions. *Aerospace MAI Journal*, Vol. 17, No.2, pp.78-86, ISSN 0869-6101.

Rozanov K.N., Li Z.W., Chen L.F. & Koledintseva M.Y. (2005). Microwave permeability of Co_2Z composites, *Journal of Applied Physics*, Vol.97, No.1. pp. 013905-1-7, ISSN 0021-8979.

Stauffer D. & Aharony A. (2003). *Introduction to Percolation Theory* (2003), London: Taylor & Francis.

Titov A.N., Bibikov S.B. & Kulikovskij Ed.I. (2008). *Proceedings of the 4th International Conference on Ultrawideband and Ultrashort Impulse Signals*, pp. 248-250, ISBN 978-1-4244-2738-3, Sevastopol, Ukraine, September, 2008.

27

The Composite Materials for Localization of Volatile Radioactive Iodine Forms from Steam-Air Phase during Severe Accidents at NPPs

Sergey A. Kulyukhin, Lubov' V. Mizina,
Igor' A. Rumer and Nikolay B. Mikheev
A.N.Frumkin Institute of Physical Chemistry and Electrochemistry
Russian Academy of Sciences
Russia

1. Introduction

The safety of NPPs is currently a problem of particular concern in all industrially developed countries throughout the world. The various science and technology projects being developed in this area seek to minimize the possibility of accidents. During severe accidents the formation of large quantities of gaseous products, which leads to an abrupt increase in pressure inside the containment (protective casing) take place. Eventually, there is a risk of destroying the containment as such. Besides, such accidents may cause contamination of the nuclear power plant closed areas, such as the intercontainment area, technical and service corridors, reactor area, et cetera, due to leakage through non-airtight sectors of the containment. It was established that such a leak of the air-and-steam mix out of the containment can reach a rate of 0.3 to 1.0 % per 24 h. Today, in order to prevent environmental pollution in the event of severe accidents, the operational nuclear power plants are equipped with emergency valves that can be activated to reduce the pressure inside the containment, in which case the air-and-steam flow, carrying radioactive aerosols and volatile fission products, is forced through special decontamination systems (Asmolov, 1994; Particulate Filtration in Nuclear Facilities, 1991; Containment Systems and Components for LOCA and Severe Accident Mitigation, 1993; Rogers, 1990). The all systems, excepting sand bed and gravel filter, have a high efficiency of localization of aerosols, and some of them have a high efficiency relative to molecular form of iodine. However, these systems, excepting Germany system, including the filter based on molecular sieve, practically do not localize the organic form of iodine, such as methyl iodide.
WWER-440/V-230 NPPs are first-generation Russian NPPs that have been operated for more than 25 years now. As the engineering of the designs of these stations was realized long before severe accidents at Three Mile Island and Chernobyl NPPs, all system of localization of accidents was plotted on the concept of impossibility of severe accidents at NPPs. However now point of view varies and the concept of improvement of a protection system of Russian WWER-440/V-230 NPPs with allowance for possibilities of severe accidents is now considered. All modifications of Russian WWER-440/V-230 NPPs belong to this first generation (Kovalevich et al., 2000; Dragunov et al., 2006).

Modern designs of new generation NPPs (EPR /Framatome ANP/, SWR-1000 /Framatome ANP/, AP-1000 /Westinghouse/, ESBWR /General Electric/, APWR /Mitsubishi/, ABWR /Toshiba and Hitachi/) do not provide for pressure-drop systems, assuming that the entire peak load will be withheld by the containment (Reactor Design, 2002; Krebs, 1999; Tomorrow's reactors, 2003). Calculations show that during 122 h, pressure under the containment can raise up to 0.7 MPa, with the steam/air ratio being 2.7:1 (Dillman, 1991). This time is enough for the steam-air joint containing volatile radioactive fission products to penetrate (0.3 to 3 % per day) into the intercontainment area and then into the environment. To prevent radioactive contamination of the environment, modern designs propose using an active air decontamination system, which using high-capacity ventilators, will forward the air from the intercontainment area to filters filled with impregnated activated charcoal. This system requires regular electric power supply, which is a significant disadvantage of this technological approach.

However, in accordance with the general requirements for safety systems, alongside active ventilation systems for intercontainment space, nuclear power plants should include a passive ventilation system, which can be used during hypothetical severe accident with a complete loss of power supply.

Now a passive-mode environmental protection system for decontamination of intercontainment area, which can be used in the event of severe accidents, was developed (Bronnikov, 2004; Zeyen et al., 2006; Friesen et al., 2001; Asmolov, 1994; Kulyukhin et al., 2008; State of the art report on iodine chemistry, 2007; State of the art report on nuclear aerosols, 2009; Kulyukhin et al., 2011). The basic purpose of developed passive filtration system is creation of a rarefaction in an intercontainment space due to the organized removal of the radioactive steam-air mixture from an intercontainment space at full loss of all power sources. Steam-air mixture containing radioactive aerosols and flying fission products removal from an intercontainment space is passed through the special filtering module. Efficiency of cleaning of a radioactive steam-air mixture is determined by filtration and sorption properties of used filtering elements.

One of the nuclear fuel fission products of high environmental emergency is radioactive iodine. The International Atomic Energy Agency (IAEA) classifies accidents and other events occurring at nuclear industry enterprises in terms of the amount of radioiodine release into the environment (Bukrinskii et al., 1991). Currently we know about the existence of 14 iodine nuclides, of which only ^{127}I is stable. Given an average nuclear fuel burn up depth of 45 GWd/t, one ton of irradiated nuclear fuel of PWR type (in Russia, WWER) contains 200 g to 300 g of iodine radionuclides or about 12 kg per the entire amount of fuel in the reactor (Radioiodine Removal in Nuclear Facilities, 1980; Ustinov et al., 2005). During a long exposure of fuel elements, all short-lived iodine radionuclides decompose, and as a result the irradiated nuclear fuel that goes to reprocessing plants includes either two (^{127}I, ^{129}I) or three (^{127}I, ^{129}I, ^{131}I) iodine nuclides, depending on the time of storage.

On the other hand, a totally different situation may occur during accidents at nuclear power plants and especially during non-design accidents associated with nuclear fuel melting. Depending on the length of the event, the amount of radioiodine in the fuel that fully penetrates into the primary containment environment and further, possibly, into the open environment may vary from hundreds of grams to tens of kilograms. To predict the chemical forms of radioiodine and their concentrations in the containment environment during non-design accidents, the Library of Iodine Reactions in Containment (LIRIC) was created. It includes over 150 chemical reactions of iodine, including those with organic species (Evans et al., 1990). The variety of the chemical forms of radioactive iodine in gaseous media needs sophisticated protective system to localize it during irradiated nuclear fuel reprocessing, as well as in the protective systems at nuclear power plants (Particulate Filtration in Nuclear Facilities, 1991). Despite a great number of volatile forms of radioiodine, the focus of attention in modern irradiated nuclear fuel reprocessing systems in both Russia and abroad, as well as at all nuclear power plants, is the localization of CsI radioactive aerosols and gaseous I_2, HI, and CH_3I.

The study of the sorption of volatile compounds of radioactive iodine from a gas phase on a wide spectrum of sorbents has shown that the most effective are the sorbents containing 8-12 wt% of Ag (Kulyukhin et al., 2008; State of the art report on iodine chemistry, 2007; Kulyukhin et al., 2011). Because of high cost of Ag, which is the one of the basic components connecting radioactive iodine, there is a problem of decreasing of its quantity in sorbents by their modifying, or makes topical its replacement. Apparently, the most suitable for this purpose are *d*-elements whose physicochemical properties are close to those of Ag, namely, Cu, Ni and Zn.

The aim of this study was to develop new composites based on Ag-, Ni-, Zn- and Cu-containing silica gel for localization of volatile compounds of radioactive iodine from steam-air phases. These composites will be used during modernization of the environmental protection systems during severe accidents at the different types of NPPs.

2. Experimental

In our study we used the radionuclide ^{131}I supplied by Izotop Joint-Stock Company in the form of carrier free $Na^{131}I$ solution. The radioactivity of the nuclides was determined by γ-ray spectrometry with a semiconductor Ge–Li detector on a multichannel analyzer. The radionuclide ^{131}I was used as this study as a spike for weighable amount of inactive iodine. Therefore, designations like $^{131}I_2$ and $CH_3{}^{131}I$ refer to labeled species and not to species of pure ^{131}I radionuclide.

We used $^{131}I_2$ (10-100 mg) and also $CH_3{}^{131}I$ (1-10 mg) with the specific activity from 3×10^2 to 3×10^3 Bq/mg. The $^{131}I_2$ generated in a gas phase by bubbling of water solution $K^{131}I$-$^{131}I_2$ by a steam-air flow. As a collector of CH_3I labeled with ^{131}I we used a quartz vessel into which, prior to experiment, we condensed with liquid nitrogen from a helium flow 1-100 mg of $CH_3{}^{131}I$ prepared by the reaction of dimethyl sulfate with $K^{131}I$ (Murata et al., 1968).

In the experiment, methyl iodide was introduced into the system by passing air at a definite rate through the quartz vessel with $CH_3{}^{131}I$.

To prepare composites, we used macroporous silica gel with a particle size of 2.00–3.00 mm as matrices.

Nitrates of Zn^{2+}, Cu^{2+}, Ni^{2+}, and Ag^{+}, and also all the acids, alkalis and hydrazine hydrate (HH) used in the study were of chemically pure grade. Pure grade hydroxylamine sulfate (HA) was used without additional purification.

Mechanical strength of granules of developed composites was measured on MP-2C device.

Electron micrographs of developed composites were obtained on the electron microscope TEM-301 ("Philips").

X-ray electron probe microanalysis (EPM) of developed composites was made on the scanning electron microscope JSM-U3 (Japan) with energy dispersive spectrometer "Eumex" (Germany).

Thermal gravimetric analysis of developed composites was performed with a Q-1500 derivatograph (MOM, Budapest, Hungary). Samples were heated in platinum crucibles in air at a rate of 10 deg/min.

2.1 Ag-containing composites

2.1.1 Synthesis of Ag-containing composites

The composites containing 1-8 wt% of $AgNO_3$ were prepared by impregnation with an aqueous solution of $AgNO_3$ with the following treatment by different N-containing compounds with the following drying on air at increasing of temperature from 20 to 300°C and next air-conditioning at 300°C.

Composites were synthesized in two stages:

1. Impregnation of macroporous silica gel with a solution of $AgNO_3$ with the following drying on air at 110°C;
2. Treatment of precursors with solutions of N-containing compounds with the following drying on air at increasing of temperature from 20 to 300°C and next air-conditioning at 300°C during 4 h.

The quantity of a solution of the $AgNO_3$, which is necessary for impregnation of silica gel, was taken in an amount ensuring 50% filling of the volume of the silica gels. The volume of silica gel was calculated from its bulk weight, which equal 0.576 kg/m^3.

The quantity of $AgNO_3$ for synthesis of composites was calculated on the basis of data about necessary Ag concentration in precursor.

The first stage of synthesis of precursors of composites on a basis of macroporous silica gel containing Ag was carried out as follows. At intensive mixing a 434 cm^3 of a solution of $AgNO_3$ with necessary concentration was added to 500 g of macroporous silica gel. The obtained mixture was mixed during 60 min for obtaining of uniformity of impregnation of total joint of silica gel with solution of $AgNO_3$.

After obtaining of proportional impregnation of total joint of silica gel with solution of $AgNO_3$ the wet sample was placed in BINDER drier. Precursors were dried on air at temperature 110°C to an air-dry state.

As a result of these operations there was a partial destruction of granules of precursor. Therefore before carrying out of the second stage of synthesis of composites it has been made the fractioning. As result of the fractioning of precursor granules with the size less than 1 mm has been separated. The quantity of the given fraction did not exceed 2-3% from total synthesized joint of precursor.

As a result the precursors, containing 1, 2, 4 and 8 wt% of Ag on the basis of macroporous silica gel, were synthesized.

At the second stage of synthesis of composites the precursors were treated by different N-containing compounds with the following drying on air at increasing of temperature from 20 to 300°C and next air-conditioning at 300°C during 4 h.

In our study 0.01, 0.05 and 0.1 M water solutions of HH and HA, and 0.5, 1.0 and 2.0 M water solutions of NH_4OH were used as N-containing compounds.

The second stage consists on treatment of precursor by N-containing compounds for obtaining of composites with nanometric particles of different compounds of Ag, was carried out as follows. At intensive mixing the 80 cm^3 of solution of N-containing compound with necessary concentration were added to 100 g air-dry precursor, containing 1, 2, 4 or 8 wt% of $AgNO_3$. The obtained mixture was mixed during 60 min for obtaining of uniformity of impregnation of joint of precursor with solution of N-containing compounds. During impregnation the colour of precursor changed from white to grey.

After obtaining of proportionality of impregnation of total joint of precursor that proved to be true of uniformity of colour, the wet samples were placed in exiccator on 3 h. Through 3 h wet samples were placed in BINDER drier. The wet samples of composites were dried on air

at increasing of temperature from 20 to 275-300°C with the following air-conditioning during 4 h at temperature 275-300°C.
As a result of these operations the composites, containing 1, 2, 4 and 8 wt% of nanometric particles of different compounds of Ag, on a basis of macroporous silica gel were obtained.
As above mentioned, during treatment of precursors by N-containing compound its colours changed. Besides, the change of precursor's colour, containing $AgNO_3$, was observed at long-term exposure on them of sunlight. The change of colour is connected with photo reduction of ions of Ag to a metal state on the surface of material.

2.1.2 The content of Ag in synthesized composites

Percentage content of Ag in synthesized precursors and composites is specified by quantity of $AgNO_3$, which was used for impregnation of macroporous silica gel. Whereas during synthesis of precursors and its modifying it is possible the loss of Ag with dust fraction, the final content of Ag in precursors and composites is determined by an analytical method by the following technique. The joint of precursor or composite (~1 g) carefully triturate in a porcelain mortar in the presence of 10 cm³ of water. Suspension quantitatively transfers in 30-cm³ centrifuge tube and 10 cm³ of concentrated HNO_3 was added to suspension. Than centrifuge tube with suspension was heated up during 1 h on a boiling water bath. After heating centrifuge tube was cooled and a solution was separated from solid phase by centrifugation. The solid phase was washed 3 times by 25 cm³ of distilled water. The solution and all washings were merged in 100-cm³ volumetric flask for titration. The solution in volumetric flask was reduced to 100 cm³ and than the determination of Ag by titration on Volgard's method from 0.01 N solution of NH_4SCN at presence of iron ammonium alum indicator was carried out (Charlout, 1961). The content of Ag was calculated by the formula:

$$\%Ag = (N_1 x V_1 x M)/(V_2 x P), \quad (1)$$

where N_1 - normality of NH_4SCN solution; V_1 - volume of NH_4SCN solution, which was used for titration, cm³; V_2 - aliquot of Ag solution for analysis, cm³; M - gramm-equivalent of Ag; P - joint of composite, g.
The analysis of data on the Ag content in synthesized precursors and composites shows that the content of Ag in synthesized composites is in the range from 1.0 to 8.0 wt% that is from 95 to 99.9% from Ag quantity taken for synthesis.

2.1.3 Desorption of Ag from synthesized composites into water

During treatment of precursors by N-containing compounds the poorly soluble compounds of Ag in the composites are formed. At treatment of precursors by HH and HA the Ag^0 particles can be formed. At treatment of precursors by NH_4OH both Ag_2O and Ag^0 particles can be formed. For study of degree of conversion of $AgNO_3$ to poorly soluble compounds studies of Ag desorption from synthesized precursors and composites into water were carried out.
The study of Ag desorption from composites into water was carried out as follows. The joint of composite (~1 g) carefully triturate in a porcelain mortar in the presence of 10 cm³ of water. Suspension quantitatively transfers in 30-cm³ centrifuge tube. The suspension was conditioned on 24 h at temperature 18-20°C. After that a solution was separated from solid phase by centrifugation. The solid phase was washed 3 times by 25 cm³ of distilled water.

The solution and all washings were merged in 100-cm^3 volumetric flask for titration. The solution in volumetric flask was reduced to 100 cm^3 and than the determination of Ag by titration on Volgard's method from 0.01 N solution of NH_4SCN at presence indicator iron ammonium alum was carried out (Charlout, 1961). The content of Ag was calculated by the formula (1).
Analysis of data on the desorption of Ag from synthesized composite into water for 24 h at temperature 18-20°C shows that the quantity of Ag which has passed into water from synthesized composites does not exceed 18% from total quantity of Ag in composites.

2.1.4 Fractional content of synthesized composites.

Determination of fractional content of synthesized composites was carried out using a nest of sieves (hole diameters were from 7.0 to 0.3 mm) by shaking of 200-400 g of composite for 15 min in according with technique, presented in (State Standard 16187-70, 1970). The resultant separated fractions were weighted. The percentage of each fraction was calculated by the formula:

$$\% = (m/M) \times 100, \quad (2)$$

where m - weight of separated fraction, g; M - total weight of composite taken for the analysis, g.
Analysis of data on the fractional content of averaged sample of synthesized composites shows that during synthesis the composites practically have not changed its initial fractional content. The changes of basis fraction of macroporous silica gel with the size of granules from 2.0 to 3.0 mm were not more than 2%.

2.1.5 Free volume of synthesized composites

The free volume, which is a sum of pore volumes of composite and intergranule space, was determined in the following way. 10 cm^3 of hexane was added to 5 g of air-dry composite. Suspension carefully was mixed before full removal of air bubbles and than was conditioned during 24 h at temperature 20°C. Through 24 h summary volume of composite and hexane was measured. Knowing volume of air-dry composite, the free volume of composite V_{free} was calculated by the formula:

$$V_{free} = \{[(V_1+V_2)-V_3]/V_1\} \times 100\%, \quad (3)$$

where V_1 - volume of a joint of air-dry composite, cm^3; V_2 - volume of hexane, cm^3; V_3 - final volume of composite and hexane through 24 h, cm^3.
Analysis of data on free volume of composites shows that the free volumes for synthesized composites were in the range from 72 to 80%. The obtained data on free volume of material allow concluding that synthesized composites have large specific surface and total volume of pores. The extended surface and large porosity of composites suggests both high rate of sorption of volatile compounds, and low aerodynamic resistance to a gas flow.

2.1.6 Bulk weight of synthesized composites

Bulk weight of composites was determined as the ratio of mass of an air-dry material and the volume taken by it. Analysis of data on bulk weight of synthesized composites shows that synthesized composites have bulk weights from 0.5 to 0.7 g/cm^3 or 500-700 kg/m^3. The sufficiently low density of synthesized composites, as well as high values of free volume, indicates to large porosity of composites and, hence, a well developed specific surface.

2.1.7 Mechanical strength of synthesized precursors and composites

Mechanical strength of granules of precursors and composite was measured on MT-2C device which scheme is presented on Fig. 1. The device was developed and designed in Institute of Physical Chemistry and Electrochemistry of Russian Academy of Sciences. The principle of operation of this device consists in loading the sample at constant speed of a working table movement and fixing the breaking force on the indicating gage. Mechanical strength σ_p was calculated by the formula:

$$\sigma_p = 4P/\pi d^2, (4)$$

where P - the load (kg), d - the diameter of granules of materials under study (cm).

Samples of composites were a mixture of different particles (mall, hemispherical and spherical). Only spherical granules (12 samples) with an average diameter 3.0 mm were used for measurement of mechanical strength σ_p. The granules were tested on squeeze in static conditions on MT-2C device at loading rate 1 mm/min. During measurements mechanical strength of granules σ_p (kg/cm^2) and a roof-mean-square deviation $S_{n\bar{x}}$ from arithmetical mean S_n for 36 samples were determined.

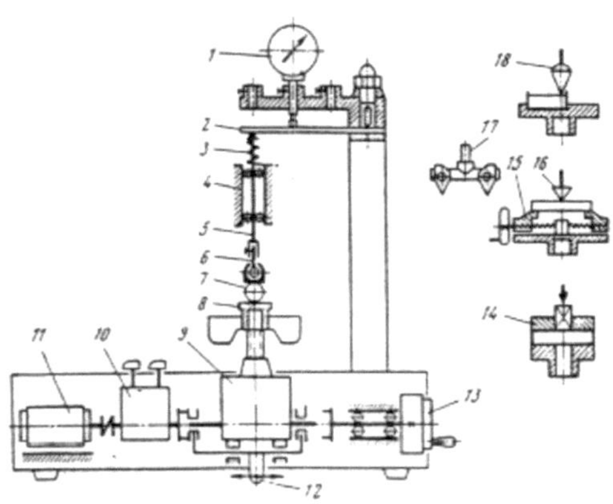

Fig. 1. The scheme of MT-2C device (1 - indicator of displacements, 2 - a flat spring of weighting device, 3 - a spring for rod weighing, 4 - guide ways, 5 - a rod, 6 - the top support, 7 - the sample, 8 - a working little table, 9 - a reduction unit, 10 - a speed gear box, 11 - electric motor, 12 - the shifter of the electric motor or handcrank to a reduction unit, 13 - a handcrank, 14 - the adaptation for test of samples for a cut, 15 - a movable support for test of samples for a bending under 3th and 4th dot scheme, 16 - a prism, 17 - a movable wedge for test of samples for a bending under 4th dot scheme, 18 - a cone with various corners for test of pastes on a method conical plastometer)

Analysis of results of tests of 36 samples of granules of composites under static loading shows that strength of granules increases with an increase in the quantity of Ag in material, i.e., the structure of the materials strengthens. On the other hand, strength of granules of composites was lower than strength of analogous precursors. Decreasing of mechanical strength occurs as a result of changing of structure of macroporous silica gel during treatment of precursors by N-containing compounds. It is known, that during treatment of silica gel by alkaline solutions dissolution of gel silica acid, which are in a pores of silica gel, takes place. The similarly process of dissolution of gel silica acid in pores of silica gel occurred during treatment of precursors by N-containing compounds. Besides, during treatment of precursors by N-containing compounds there was a change of the chemical forms of Ag in the materials under study. As a result of the various physico-chemical processes, which occurred in granules of material, there was an increasing of pore volumes of granules that leads to decrease in mechanical strength of granules. Mechanical strength of composites was more than 85 kg/cm^2.

2.1.8 Electron microscopy study of synthesized composites

Figure 2 show the electron micrographs of the synthesized composites SiO_2-n%Ag (n = 1, 8), obtained using a Philips TEM-301 transmission electron microscope. As shown in Fig. 2, the synthesized materials contain Ag^0 nanometer particles not larger than 10 nm. The amount of the nanometer particles increases with an increase in the Ag concentration in the materials. Note that the amount of Ag^0 nanometer particles depends not only on the Ag concentration in the material, but also on the nature of the nitrogen-containing compound used for the modification of the precursor.

A study of the microphotographs showed that the largest particles were formed during treatment of the precursor with ammonia solutions. The particles form aggregates (Fig. 2B). When HH or HA is used, highly dispersed particles stabilized in the matrix pores are formed (Fig. 2A). On the hand, decreasing the Ag concentration during treatment of the precursor with ammonia solutions leads to the formation of highly dispersed particles in the silica gel matrix (Fig. 2B).

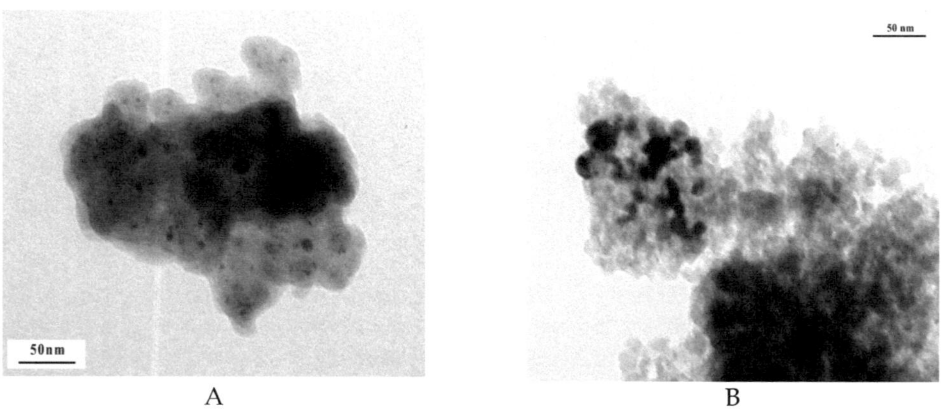

Fig. 2. The electron microphotograph of the composite SiO_2-8%Ag obtained by treatment of the precursor with solutions of 0.1 M HH (A) and 0.5 M NH_4OH (B)

As follows from Fig. 2, the amount of nanometer Ag^0 particles is very small, which, as we believe, indicates the formation in the matrix of Ag particles smaller of a size smaller than the equipment resolution power, i.e., <1 nm. In addition, it is necessary to take into account a low concentration of Ag in the matrix of the composite with a strongly developed surface.

The Ag distribution in the matrix of the composites was studied using their electron probe X-ray microanalysis. Figure 3 shows a typical microphotograph of the surface of the chip of the granule of the composite, and fig. 4 shows the X-ray diffraction spectrum of the sample surface. As follows from Figure 3, particles not larger than 1 μm are evenly distributed on the granule surface and that the amount of the aggregates larger than 10 μm is relatively small. An elemental composition study of the surface showed that the composites contained O, Si, and Ag (Fig. 4).

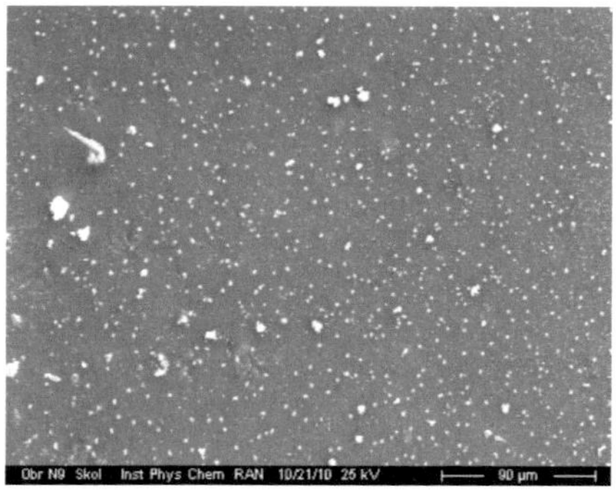

Fig. 3. The microphotograph of the surface of the SiO_2-4%Ag composite

Figure 5 shows the X-ray spectroscopy data on the Ag distribution on the surface of granules of SiO_2-n%Ag (n = 1, 2, 4, 8) composites after treatment of the precursors with solutions of HH and HA. The surface of the granules was scanned at a radiation power only characteristic of Ag.

As follows from fig. 5, the Ag particles are virtually evenly distributed on the surface of the composite granules. The amount of the particles on the surface decreases with a decrease in the Ag concentration.

The obtained results are good agreement with the conclusions drawn from the microphotographs obtained using transmission electron microscopy. Actually, the well-developed surface of the synthesized composites stimulates a strong distribution of Ag in the volume of the granules of the material. As a result, when the Ag concentrations are low, the number of Ag particles per surface unit of the material proves to be very low.

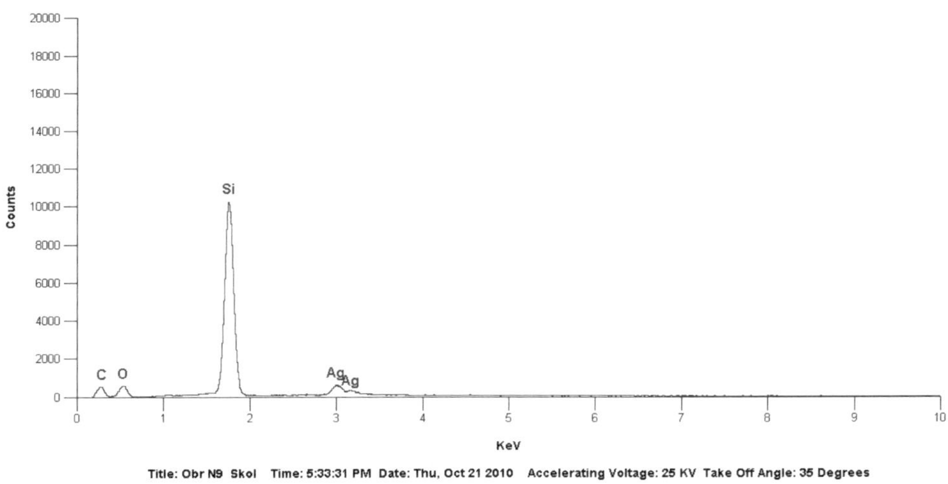

Fig. 4. The X-ray diffraction spectrum of the SiO_2-4%Ag composite after treatment of the precursors with solutions of HH

2.1.8 Localization of molecular iodine and methyl iodide from a steam-air flow

The efficiency of molecular iodine and methyl iodide localization by composites under analogous experimental conditions depends on many factors, the main of which are as follows: active metal concentration, preparation conditions, quantity of methyl iodide or molecular iodine, steam-air flow rate, and time of contact between the steam-air flow and the composite.

2.1.8.1 A study of the localization of molecular radioiodine from a steam-air flow

To study the localization of $^{131}I_2$ from a steam-air flow, we used the procedure and installation described in (Kulyukhin et al., 2007).

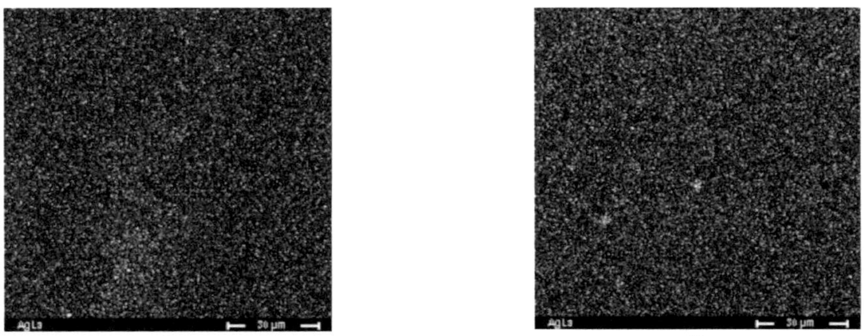

A B

Fig. 5. The distribution of Ag particles on the surface of the granule of the composite based on silica gel containing 4 wt% Ag (A) and 8 wt% Ag (b) after treatment of the precursor with a 0.1 M solution of HH and conditioning in the air at 300°C for 4 hours

It is necessary note that the efficiency of molecular iodine absorption is higher than 99% virtually in all cases. After the localization of molecular iodine, the composite changed its color from grey to lettuce green. The appearance of the lettuce green color is due to the formation of $Ag^{131}I$ in the composition of the composite.

Note that irrespective of the Ag concentration in the composite, the absorption efficiency is the highest in the case of using HA and is the lowest in the case of using the ammonia solution (Fig. 6). The composites obtained by treatment with HH take an intermediate position. The observed difference is most likely due to the formation of metal Ag particles in the case of HH and HA versus the formation of Ag oxide in the case of using ammonia solutions. Note that treatment of HA leads to the formation of smaller particles versus HH.

As follows from Fig. 6, when the Ag concentration in the material is increased from 1 to 8 wt%, the total molecular iodine sorption by the total layer of the sorbent remains virtually unchanged and higher than 99%. Despite the virtually equal sorption efficiencies, different distributions of molecular iodine in the layers of the column are observed. Analysis of the layer-by-layer distribution of radioactive iodine in the column as a function of the Ag concentration in the material shows that when the Ag concentration in the material is increased, the sorption front becomes steeper, i.e., the major part of molecular iodine is absorbed in the first layer of the material. The formation of the nanometer Ag particles in the case of HH and HA results in a high sorption efficiency of the synthesized materials. On the other hand, the formation of nanometer Ag oxide particles in the case of treatment of the precursors with an ammonia solution leads to a lower-angle front of the sorption curves along the columns. The difference in the slopes of the sorption curves allows us to draw conclusions regarding not only the possible Ag form in the composite, but also regarding the chemosorption speed for its different forms. For Ag oxide, the molecular iodine chemosorption rate is about 1.5 times slower than that on the metal Ag particles obtained by treatment of the precursors with HH or HA.

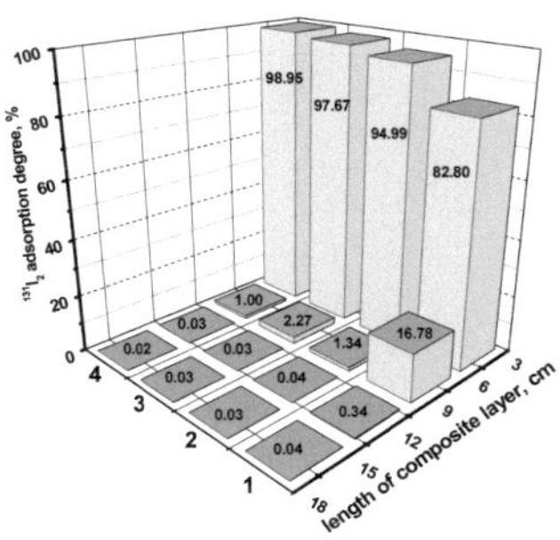

A

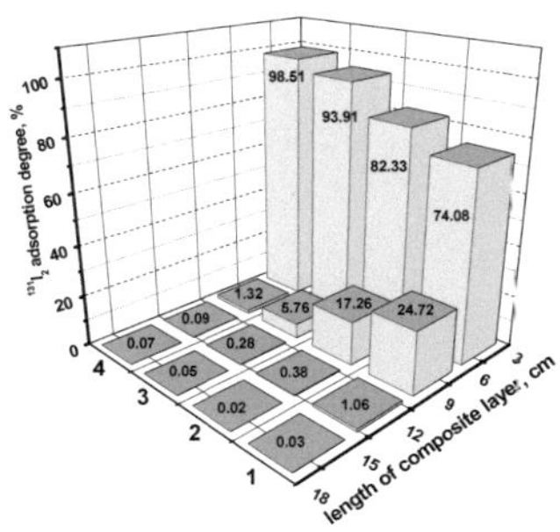

B

Fig. 6. The ^{131}I layer-by-layer distribution on the column with the composite materials obtained by treatment of the precursor with a 0.1 M solution of HA (A) and NH_4OH (B) as a function of the Ag concentration in the material (1 - wt% Ag, 2 - 2 wt% Ag, 3 - 4 wt% Ag, 4 - 8 wt% Ag)

All synthesized composites display high molecular iodine sorption efficiency. Note that the sorption efficiency of each of the studied materials is virtually independent of the nature of the N-containing compound used for treatment of the precursor.

Analysis of the layer-by-layer distribution of radioactive iodine on the column containing the SiO_2-2%Ag composite obtained by treatment of the precursor by different N-containing compounds shows the layer-by-layer distribution of radioactive iodine along the column filled with the precursor of the given composite (Fig. 7). The lowest-angle molecular iodine sorption front in the layers of the column is observed for the composite obtained by treatment of the precursor with the ammonia solution; the rate of the chemosortion by Ag oxide particles is the lowest as compared with other Ag forms. The chemosorption speed for the non-modified Ag form ($AgNO_3$) is virtually the same as that on the metal Ag particles obtained by treatment of the precursor with a 0.1 M solution of HA. The highest value is observed for the composite obtained during treatment of the precursor with a 0.1 M solution of HH. During treatment of the precursor with HH, metal Ag particles are formed in the near-to-surface layer of the granules, whereas during treatment with HA, the metal Ag particles are formed in the entire volume. As a result, gaseous molecular iodine takes a longer time to enter the chemosorption reaction with the Ag particles in the volume of the granules versus its reaction with the Ag particle in the near-to-surface layer or on the surface of the granules. Note that because of the high porosity of the material the difference in the molecular iodine sorption efficiencies in the 1st layer of the sorbents obtained by treatment of the precursor with HH and HA is very small.

Figure 8 shows the data on the molecular iodine sorption efficiency on the SiO_2-2%Ag composite obtained by treatment of the precursor with a 0.01 M solution of HH as a function of the molecular iodine quantity supplied to the column. As follows from Fig. 8, when the molecular iodine quantity is increased tenfold (from 10 to 100 mg), the sorption efficiency remains virtually unchanged and is very high (higher than 99.9%). On the other hand, increasing the molecular iodine quantity leads to its redistribution in the layers of the column.

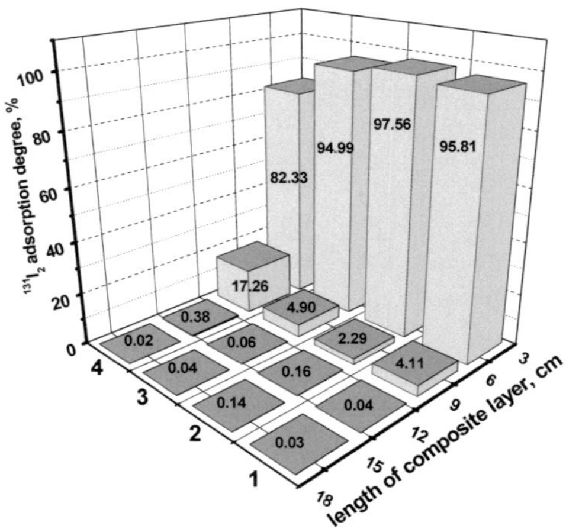

Fig. 7. The ^{131}I layer-by-layer distribution on the column containing the SiO_2-2%Ag composite material as a function of the nature of the N-containing compound used for treatment of the precursor (1 - without modification, 2 - 0.1 M hydrazine hydrate, 3 - 0.1 M hydroxylamine, 4 - 0.5 M NH_4OH)

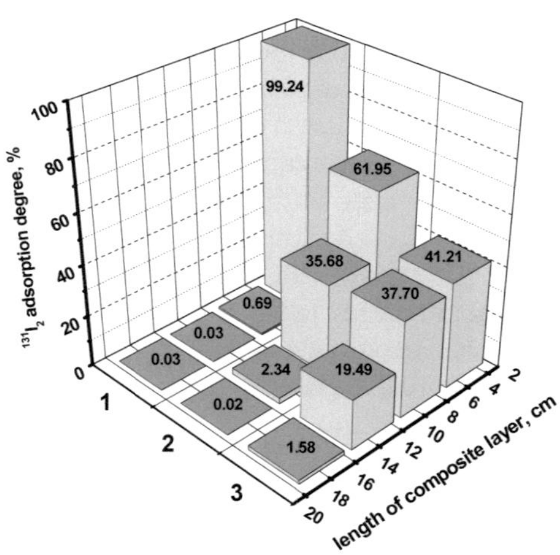

Fig. 8. The ^{131}I layer-by-layer distribution on the column containing the SiO_2-2%Ag composite material obtained by treatment of the precursor with 0.01 M solution of HH as a function of the $^{131}I_2$ quantity (1 - 10 mg, 2 - 50 mg, 3 - 100 mg)

One of the major factors influencing the degree of the molecular iodine sorption from the gas flow is the time of contact between the composites and the gas flow. The time of the contact is determined by the linear gas flow rate and the height of the sorbent in the column. Figure 9 shows the data on the molecular iodine sorption from the gas flow as a function of the time of contact between the steam-air flow and the SiO_2-2%Ag composite obtained by treatment of the precursor with a 2.0 M ammonia solution. As follows from Fig. 9, when the time of the contact between gas phase and composite is decreased from 4.5 to 1.1 s, the molecular iodine sorption remains virtually unchanged and is higher than 99%. An increase the linear gas flow rate and hence, a decreases in the time of contact between the gas flow and the composite lowers the sorption front angle. Note that it takes two or three layers of the composite to localize radioactive iodine.

Another factor influencing the sorption efficiency is the composition of steam-air phase. The presence of water molecules in the gas phase has a competitive influence during the interaction of molecular iodine with active centers of the sorbent.

Figure 10 show the data on the molecular iodine sorption on the SiO_2-8%Ag composite obtained by treatment of precursors with 0.1 M solution of HH from steam-air flows with different steam quantities. As follows from Fig. 10, when the steam content in the gas phase is increased about twofold, the total degree of the molecular iodine sorption remains virtually unchanged and equal to more than 99%. On the other hand, it is necessary note that an increase in the steam quantity in the gas phase causes the radioactive iodine to move along the column. Thus, even a small change in the steam quantity impacts the sorption capacity of the composites.

In conclusion note that all synthesized composites have high sorption efficiency for molecular iodine in the gas flow.

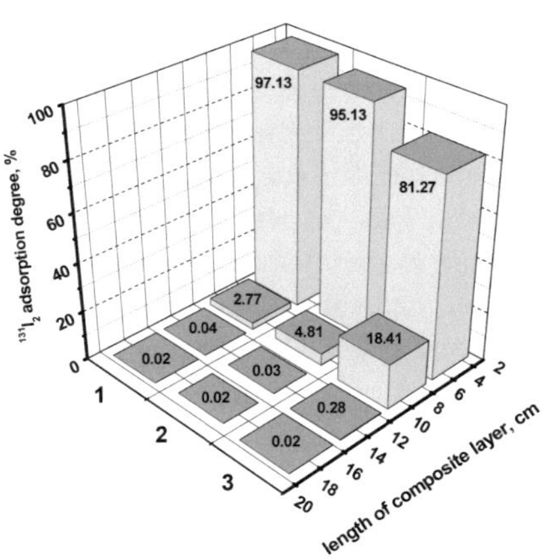

Fig. 9. The ^{131}I layer-by-layer distribution on the column containing the SiO_2-2%Ag composite material obtained by treatment of the precursor with 2.0 M solution of NH_4OH as a function of the time of contact between the steam-air flow and composite (1 - 4.53 s, 2 - 2.06 s, 3 - 1.11 s)

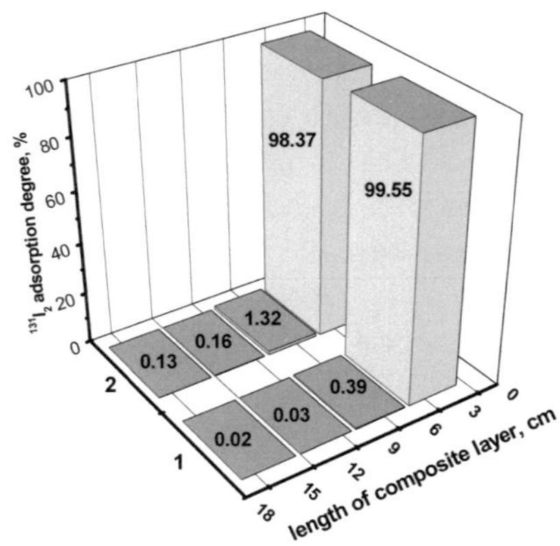

Fig. 10. The ^{131}I layer-by-layer distribution on the column containing the SiO_2-8%Ag composite material obtained by treatment of the precursor with 0.1 M solution of HH as a function of the steam quantity in the steam-air flow (1 - 5-6 vol%, 2 -10-11 vol%)

2.1.8.2 A study of the $CH_3{}^{131}I$ sorption from the steam-air phase by the synthesized composites

To study the localization of $CH_3{}^{131}I$ from a steam-air flow, we used the procedure and installation described in (Kulyukhin et al., 2007).

In contrast to the molecular iodine sorption, for which the sorption degree was higher than 99% in all cases, the $CH_3{}^{131}I$ sorption efficiencies of the synthesized composites differ a lot. An increase in the Ag quantity increases the total $CH_3{}^{131}I$ localization degree. Thus, increasing the Ag quantity from 1 to 8 wt% for the composites prepared by treatment of the precursor with a 0.1 M solution of HH, the localization degree increases from ~1.8% to ~60.2% (Fig. 11A). An analogous picture is also observed for the composites obtained as a result of treatment of the precursors with solutions of HA or ammonia (Fig. 11B).

Analysis of the data on the total $CH_3{}^{131}I$ sorption on the composites obtained by treatment with 0.1 M solutions of HH and HA, as well as with a 0.5 M solutions of ammonia as a function of the Ag content in the material shows that increasing the Ag concentration in the material increases its sorption capacity. The highest $CH_3{}^{131}I$ sorption capacity of the composites is observed for the materials obtained by treatment of the precursors with a 0.1 M solution of HA, and the lowest sorption capacity is observed for the samples treated with a 0.5 M solution of ammonia.

For the composites containing 1, 2 and 4 wt% of Ag, the sorption capacity of the precursor is higher (Fig. 12A). For the composite containing 8 wt% of Ag the sorption efficiency of the material obtained by treatment of the precursor with a 0.1 M solution of HA is higher than that of the precursor (Fig. 12B).

Note that for all composites containing 1, 2, and 4 wt.% of Ag, the quantity of the sorbed $CH_3{}^{131}I$ increases from layer to layer (Fig. 13). The only exception are composites containing 8 wt% of Ag, which display standard dependences; i.e., a decrease in the sorbed $CH_3{}^{131}I$ along the column length (Fig. 13). The observed picture is likely to be due to a drop in the flow rate in the composite layer because of a rise in its turbulence, which leads to an increase in the time of contact between the gas phase and the composite.

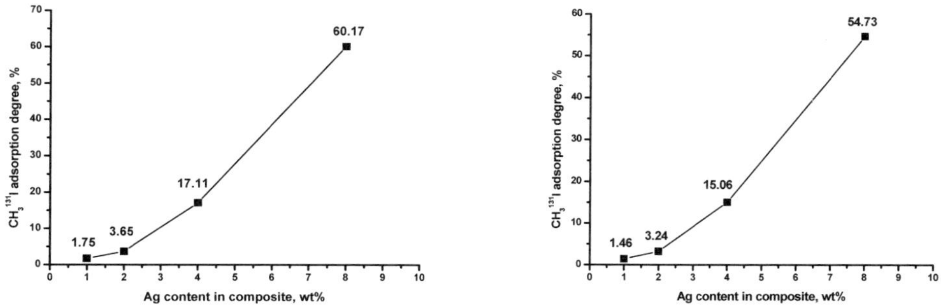

A B

Fig. 11. The efficiency of $CH_3{}^{131}I$ sorption on the composite materials obtained by treatment of the precursor with a 0.1 M solution of HH (A) and 0.5 M solution of NH_4OH (B) as a function of the Ag concentration in the material

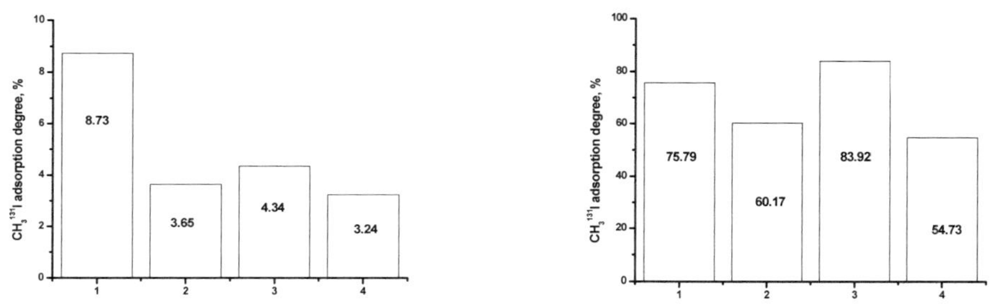

A B

Fig. 12. The efficiency of the $CH_3{}^{131}I$ sorption on the SiO_2-2%Ag (A) and SiO_2-8%Ag (B) composite materials as a function of the nature of the N-containing compound used for treatment of the precursor (1 - without the modification, 2 - 0.1 M HH, 3 - 0.1 M HA, and 4 - 0.5 M NH_4OH)

The obtained data on the $CH_3{}^{131}I$ sorption are very curious. Ag^0 is known not to react with $CH_3{}^{131}I$, whereas Ag_2O can react with the compound in the presence of water steam by the reaction:

$$2CH_3{}^{131}I + Ag_2O + H_2O \Rightarrow 2Ag^{131}I + 2CH_3OH. \quad (5)$$

Treatment of the precursors with HH and HA solutions leads to the formation of nanometer Ag^0 particles in the material, and treatment with NH_4OH yields nanometer Ag_2O particles, which should suggest that the highest sorption efficiency for $CH_3{}^{131}I$ would be characteristic of the composites obtained by treatment of the precursors with solutions NH_4OH. However in reality, the sorption efficiency of these composites proved to be the lowest. The nanometer Ag^0 particles are likely to act as catalysts of the $CH_3{}^{131}I$ decomposition into $CH_3^{\bullet}$

and $^{131}I^{\bullet}$ radicals. In this case, the $CH_3^{\bullet}$ can react with the components of the gas phase (oxygen and water steam) to form various volatile organic compounds. In turn, the $^{131}I^{\bullet}$ can form $^{131}I_2$ in the gas phase, which will be absorbed at a high speed by the silver compounds in the composition of the composites. The smaller the Ag^0 particles are the higher their catalytic activity is.

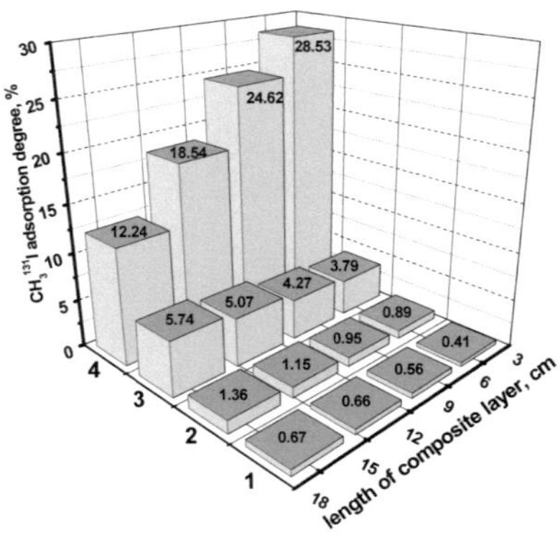

Fig. 13. The ^{131}I layer-by-layer distribution on the column with the composite materials obtained by treatment of the precursor with a 0.1 M solution of HA as a function of the Ag concentration in the material (1 - 1 wt% Ag, 2 - 2 wt% Ag, 3 - 4 wt% Ag, 4 - 8 wt% Ag)

Increasing the concentration of the HH or HA should lead to both an increase in the quantity of nanometer Ag^0 or Ag_2O particles and their further aggregation. The aggregation will result in the formation of larger particles with a smaller specific surface. In this case, the catalytic activity of the composites should decrease, which can influence their sorption capacity.

Figure 14 shows the data on the total sorption degree for the synthesized composite materials as a function of the concentration of the N-containing compounds and used for treatment of the precursor. As was expected for the SiO_2-8%Ag composite obtained by treatment of the precursor with solutions of HH and HA, the sorption degree decreased with an increase in the concentration of the N-containing compound (Fig. 14A). On the other hand, the sorption degree of these materials is higher than that of the precursor. In the case of using NH_4OH solutions, the sorption efficiency also decreases with an increase in the concentration of the N-containing compound; however, it is not higher than the sorption capacity of the precursor (Fig. 14B).

Figure 15 shows the data on the ^{131}I layer-by-layer distribution in the columns containing the SiO_2-8%Ag composite obtained by treatment of the precursor with solutions of N-containing compounds of different concentrations. Figure 15A shows that increasing the HA (or HH) concentration leads to a noticeable decrease in the $CH_3{}^{131}I$ sorption in the 1st layer of the materials. Simultaneously, the angle of the sorption front becomes lower. In the case of NH_4OH (Fig. 15B), ^{131}I is virtually evenly distributed over the layers, and as in the case of HH and HA, the highest sorption is observed for the lowest NH_4OH concentration. Note that in the case of the composites obtained by treatment of the precursor with 1.0 and 2.0 M solutions of NH_4OH, the sorption degree increases from layer to layer. The effect is again likely to be due to an increase in the turbulence of the gas flow as it passes through the column, which leads to an increase in the time of contact between the gas phase and the composite.

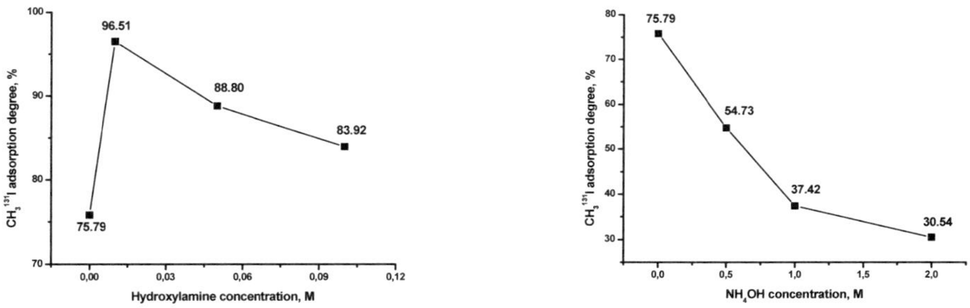

A B

Fig. 14. The efficiency of the $CH_3{}^{131}I$ sorption on the SiO_2-8%Ag composite, obtained by treatment of the precursor with a solution of HA (A) and NH_4OH (B) as a function of the N-containing compounds concentration

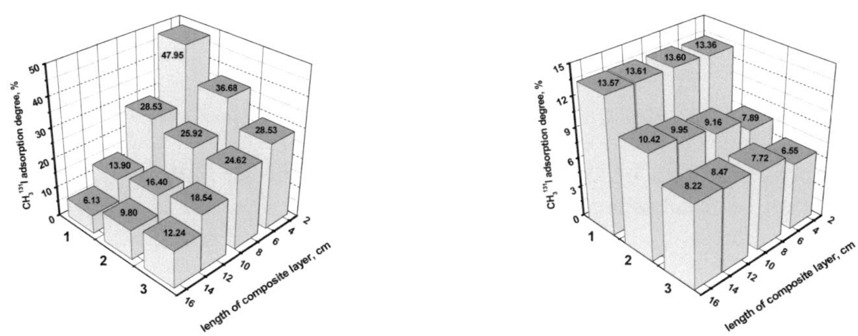

A B

Fig. 15. The ^{131}I layer-by-layer distribution on the column with the SiO_2-8%Ag composite obtained by treatment of the precursor with a solution of HA (A, 1 - 0.01 M, 2 - 0.05 M, 3 - 0.1 M) and NH_4OH (B, 1 - 0.5 M, 2 - 1.0 M, 3 - 2.0 M) as a function of N-containing compound concentration

As established earlier, the degree of the $CH_3{}^{131}I$ and $^{131}I_2$ sorption depends on the quantity of radioactive compound supplied to the column.
Figure 16 shows the data on the $CH_3{}^{131}I$ sorption from the steam-air flow on the mixture (1:1) of composites SiO_2-8%Ag obtained by treatment of the precursor with a 0.01 M solution of HH and HA as a function of $CH_3{}^{131}I$ quantity. As follows from Fig. 16, a tenfold increase in the $CH_3{}^{131}I$ quantity leads to a decrease in its sorption degree from 99.3% to 95.6%. On the other hand, increasing the $CH_3{}^{131}I$ quantity from 5 to 10 mg has hardly any effect on the degree of its sorption on the composites. The ^{131}I distribution in the layers changes accordingly. Increasing the $CH_3{}^{131}I$ quantity lowers the angle of the sorption front. On the

other hand, increasing the $CH_3{}^{131}I$ quantity from 5 to 10 mg gives us virtually identical ^{131}I distribution patterns in the layers of the composites in the column.

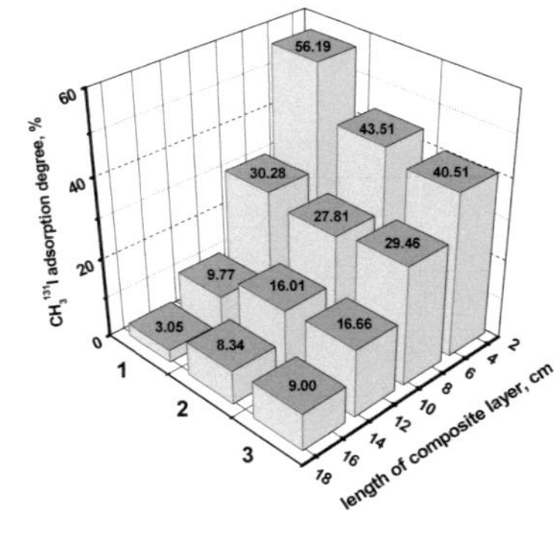

Fig. 16. The $CH_3{}^{131}I$ layer-by-layer distribution on the column with the mixture (1:1) of composites SiO_2-8%Ag obtained by treatment of the precursor with a 0.01 M solution of HH and HA as a function of $CH_3{}^{131}I$ quantity supplied to the column (1 - 1 mg, 2 - 5 mg, 3 - 10 mg)

The linear gas flow rate and the related time of contact between the gas phase and the composite also impact the sorption degree. Increasing the linear gas flow rate decreases the time of contact of $CH_3{}^{131}I$ present in the gas phase with the elements of the composites responsible for its chemosorption (in the synthesized materials, these are nanometer Ag particles).

Figure 17 shows the data on the $CH_3{}^{131}I$ sorption from the air flow on the composite SiO_2-8%Ag obtained by treatment of the precursor with a 0.01 M solution of HA as a function of the linear gas flow rate. A threefold increase in the linear gas flow rate decreases the

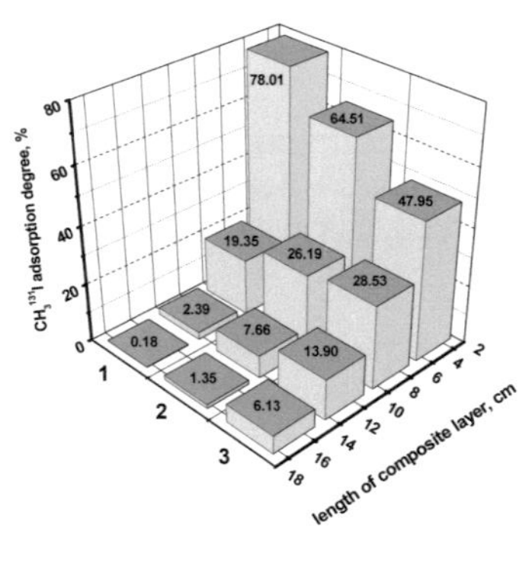

Fig. 17. The $CH_3{}^{131}I$ sorption efficiency on the composite SiO_2-8%Ag obtained by treatment of the precursor with a 0.01 M solution of HA as a function of the linear gas flow rate (1 - 4.4 cm/s, 2 - 8.1 cm/s, 3 - 14.8 cm/s)

sorption degree of $CH_3{}^{131}I$ from 99.9% to 96.5%. The ^{131}I distribution in the layers of the composites in the columns also changed. As follows from Fig. 17, decreasing the linear gas flow rate increases the angle of the $CH_3{}^{131}I$ sorption front. Thus, increasing the linear gas flow rate from 4.4 to 14.8 cm/s decreases the quantity of ^{131}I in the 1st layer of the composite in the column from ~78 to ~48%. At the same time, the ^{131}I quantity in the 4th layer increased from ~0.2% to ~6.1%.

The findings of the investigations allow us to conclude that the most suitable composite material for $CH_3{}^{131}I$ localization from the gas flow is SiO_2-8%Ag obtained by treatment of the precursor with a 0.01 M solution of HA.

2.2 Composites, containing Ag, Ni, Cu, and Zn

Presently, there is carrying out investigations aiming to produce a new granulated sorbent based on macroporous silica gel containing simultaneously nanometric particles of Ag and Ni (or Zn, or/and Cu) compounds for use in the filtering units of the emergency filters at WWER-1000 nuclear power plants (NPP-2006 project).

2.2.1 Synthesis of composites

The sorbents containing 10 wt% of $AgNO_3$ and $Ni(NO_3)_2$ [or $Zn(NO_3)_2$, or/and $Cu(NO_3)_2$] were prepared by impregnation with an aqueous solution of d-element nitrates with the following treatment by ammonia with the following drying on air at increasing of temperature from 20 to 300°C and next air-conditioning at 300°C.

Composites were synthesized in two stages:

1. Impregnation of macroporous silica gel with a solution of d-element nitrates in different ratio with the following drying on air at 110°C;
2. Treatment of precursors with solutions of ammonia with the following drying on air at increasing of temperature from 20 to 300°C and next air-conditioning at 300°C during 4 h.

The quantity of a solution of d-element nitrates, which is necessary for impregnation of silica gel, was taken in an amount ensuring 50% filling of the volume of the silica gels. The volume of silica gel was calculated from its bulk weight, which equal 0.576 kg/m^3.

The quantity of d-elements nitrates for synthesis of composites was calculated on the basis of data about necessary d-element concentration in precursor.

The first stage of synthesis of precursors of composites on a basis of macroporous silica gel containing Ag and Ni (or Zn, or/and Cu) was carried out as follows. At intensive mixing a 434 cm^3 of a solution of d-element nitrates with necessary concentration was added to 500 g of macroporous silica gel. The obtained mixture was mixed during 60 min for obtaining of uniformity of impregnation of total joint of silica gel with solution of d-element nitrates.

After obtaining of proportional impregnation of total joint of silica gel with solution of d-element nitrates the wet sample was placed in BINDER drier. Precursors were dried on air at temperature 110°C to an air-dry state.

As a result of these operations there was a partial destruction of granules of precursor. Therefore before carrying out of the second stage of synthesis of composites it has been made the fractioning. As result of the fractioning of precursor granules with the size less than 1 mm has been separated. The quantity of the given fraction did not exceed 2-3% from total synthesized joint of precursor.

As a result the precursors, containing 10 wt% of Ag and Ni (or Zn, or/and Cu), on the basis of macroporous silica gel were synthesized.

At the second stage of synthesis of composites the precursors were treated by solution of NH_4OH with the following drying on air at increasing of temperature from 20 to 300°C and next air-conditioning at 300°C during 4 h.
In our study 2.0 M water solutions of ammonia were used for treatment of precursors.
The second stage consist on treatment of precursor by solution of 2.0 M solution of NH_4OH for obtaining of composites with nanometric particles of different compounds of d-element, was carried out as follows. At intensive mixing the 80 cm³ of 2.0 M solution of NH_4OH were added to 100 g air-dry precursor, containing 10 wt% of d-element nitrates. The obtained mixture was mixed during 60 min for obtaining of uniformity of impregnation of joint of precursor with solution of ammonia. During impregnation the colour of precursor changed from green to black for composites containing Ag, Ni or/and Cu, and from light grey to grey for composites containing Ag and Zn.
After obtaining of proportionality of impregnation of total joint of precursor that proved to be true of uniformity of colour, the wet samples were placed in exiccator on 3 h. Through 3 h wet samples were placed in BINDER drier. The wet samples of composites were dried on air at increasing of temperature from 20 to 275-300°C with the following air-conditioning during 4 h at temperature 275-300°C.
As a result of these operations the composites, containing 10 wt% of nanometric particles of different compounds of Ag and Ni (or Zn, or/and Cu), on a basis of macroporous silica gel were obtained (Table 1). The figure in the brackets shows the molar ratio of the metal.

Composites	Precursor		Composite	
	wet	dried	wet	dried
SiO_2-10%AgNi (1 : 4)	green	light green	black	black
SiO_2-10%AgCuNi (1.6 : 0.4 : 4)	green	light green	black	black
SiO_2-10%AgCuNi (1 : 1 : 4)	green	light green	black	black
SiO_2-10%AgCuNi (0.4 : 1.6 : 4)	green	light green	black	black
SiO_2-10%AgCu (1 : 4)	green	light green	black	black
SiO_2-10%AgZn (1 : 4)	white	light grey	grey	grey

Table 1. The colour parameters of the synthesized composites

2.2.2 Analysis of synthesized composites

To determine the total metal content, the composite was treated with concentrated nitric acid for 2 hours. The mother solution was separated from it. The composite was washed with water, and the metal content was then determined. The Ag content in the solutions was measured by titration by the Volgard method. The nickel, copper and zinc contents were determined by direct titration with murexide (Schwarzenbach, 1958).
The amount of the metals in the form of nitrates in the composites was determined by the above methods in the solution which was obtained as a result of treatment of the composite with distilled water for 24 hours. The amount of d-elements in poorly soluble forms in the composite was calculated as the difference between the total concentrations of each metal in the composite and the amounts of the metals in nitrate form.
As a result, we obtained a granulated composites with a Ag content of 0.4 ÷ 2.0 wt% and a d-element (Ni, Zn, Cu) content equal to its molar ratio to Ag from 14:1 to 4:1 (Table 1).

2.2.3 Electron microscopy study of synthesized composites

Figure 18 shows the electron micrographs of the composite SiO_2-10%AgNi(1:4), obtained using a Philips TEM-301 transmission electron microscope. As shown in Fig. 18, the synthesized materials contain nanometer particles of Ag^0 and possibly NiO not larger than 10 nm. As follows from Fig. 18, the amount of nanometer Ag^0 and NiO particles is very high. In addition, it is necessary to take into account a low concentration of Ag in the matrix of the composite with a strongly developed surface.

The Ag distribution in the matrix of the composites was studied using their electron probe X-ray microanalysis.

Figure 19 shows a typical microphotograph of the surface of the chip of the granule of the composite, and figure 20 shows the X-ray diffraction spectrum of the sample surface.

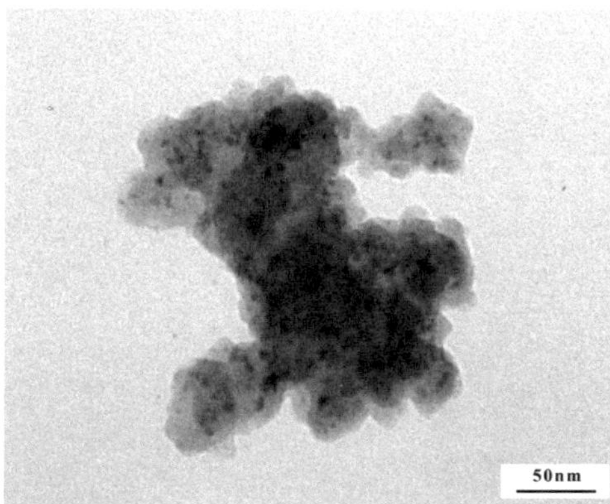

Fig. 18. The electron microphotograph of the SiO_2-10%AgNi(1:4) composite

Fig. 19. The microphotograph of the surface of the SiO_2-10%AgNi(1:4) composite

As follows from Fig. 19, particles not larger than 1 μm are evenly distributed on the granule surface and that the amount of the aggregates larger than 10 μm is relatively small.

An elemental composition study of the surface showed that the composites contained O, Si, Ni, Ag and trace quantity of Na (Fig. 20).

Figure 21 shows the X-ray spectroscopy data on the Ag (A) and Ni (B) distribution on the surface of granules of SiO_2-10%AgNi(1:4) composite. The surface of the granules was scanned separately at a radiation power characteristic of Ag or Ni. As follows from Fig. 21, the Ag and Ni are virtually evenly distributed on the surface of the composite granules.

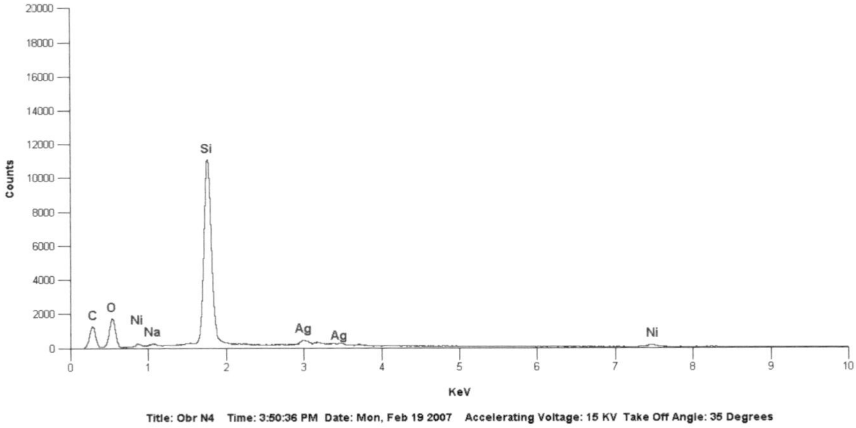

Fig. 20. The X-ray diffraction spectrum of the SiO_2-10%AgNi(1:4) composite

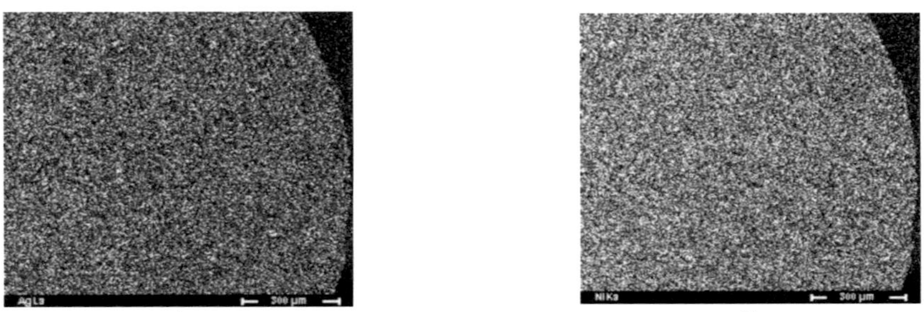

A B

Fig. 21. The distribution of Ag (A) and Ni (B) particles on the surface of the granule of the SiO_2-10%AgNi(1:4) composite

2.2.4 Thermal gravimetric analysis of synthesized composite

The results of thermal gravimetric analysis of SiO_2-10%AgNi(1:4) and SiO_2-10%AgCu(1:4) composites are shown in Figs. 22-23. It can be seen that thermal decomposition of the composites is a multistep process accompanied by both evolution and absorption of heat.

As seen from Fig. 22, the patterns of thermal decomposition in air in a platinum crucible of SiO_2-10%AgNi(1:4) composite without conditioning during 4 h at 300°C and with it differ essentially. The composite obtained without conditioning during 4 h at 300°C (Fig. 22A) decomposes in two steps, the first of which is apparently associated with endothermic loss of water (65-175°C). The second step is apparently related to partial exothermic decomposition of d-elements and ammonium nitrate (265°C). The total loss of mass was about 30 wt%. At the same time, as follows from Fig. 22B, the composite obtained with conditioning during 4 h at 300°C decomposes without thermal effects. The total loss of mass was about 11 wt%.

As follows from Fig. 23, the character of decomposition of SiO_2-10%AgCu(1:4) composite is similar to that for SiO_2-10%AgNi(1:4) composite. The SiO_2-10%AgCu(1:4) composite also decomposes without thermal effects. The total loss of mass was about 9 wt%.

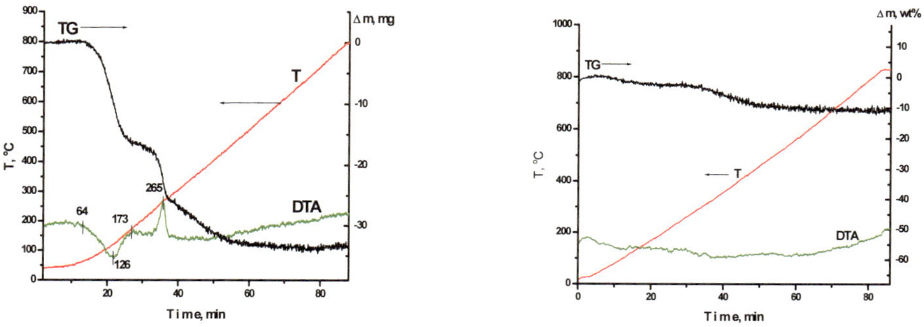

A B

Fig. 22. Results of thermal gravimetric analysis of SiO_2-10%AgNi(1:4) composite without conditioning during 4 h at 300°C (A, sample weight 102.4 mg) and with it (B, sample weight 115.7 mg)

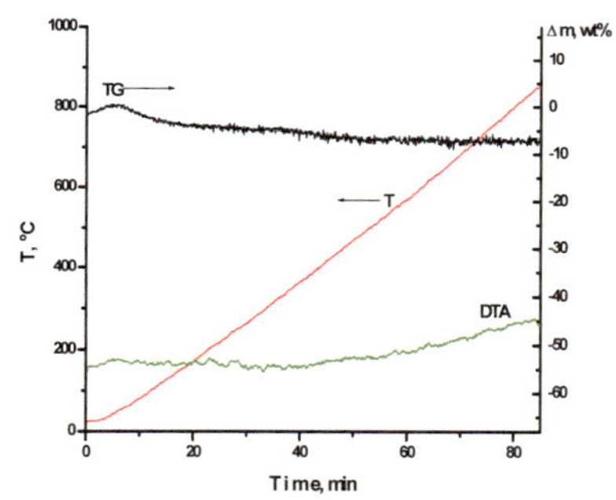

Fig. 23. Results of thermal gravimetric analysis of SiO_2-10%AgCu(1:4) composite (sample weight 89.9 mg)

2.2.5 Localization of molecular iodine and methyl iodide from a steam-air flow

To study the localization of $CH_3{}^{131}I$ and ${}^{131}I_2$ from a steam-air flow, we used the procedure and installation described in (Kulyukhin et al., 2007).

Results of the studies of the sorption of $CH_3{}^{131}I$ and ${}^{131}I_2$ from a steam-air flow on developed composites are presented in Table 2. As follows from the Table 2, all composites have high sorption efficiency for $CH_3{}^{131}I$ and ${}^{131}I_2$ in the gas flow. The highest $CH_3{}^{131}I$ sorption capacity is observed for the SiO_2-10AgNi(1:4) composites, and the lowest sorption capacity is observed for the SiO_2-10AgCuNi(0.4:1.6:4) composites.

It is necessary note that with increasing of time of the contact between gas flow and composite the sorption efficiency appreciably increases. For example, at increasing of time of the contact between gas flow and SiO_2-10%AgNi(1:4) composite from 1 to 5 s the $CH_3{}^{131}I$ adsorption degree increases from 88.74% to 99.99%.

In conclusion note that synthesized composites containing 10 wt% of Ag and Ni have high sorption efficiency for $^{131}I_2$ and $CH_3{}^{131}I$ in the gas flow.

3. Conclusion

The developed composites do not have analogs in the world and can compete with other sorption materials designed to prevent the release of volatile radioiodine compounds into the environment. In addition, the developed composites can be used for effective localization of volatile chlorine, bromine, and arsenic, as well as polonium radionuclide into the environment.

Composites	$^{131}I_2$, %	$CH_3{}^{131}I$, %		
	20°C	20°C	110°C	170°C
SiO_2-10%AgNi (1 : 4)	99.95	88.74	97.00	98.28
SiO_2-10%AgCuNi (1.6 : 0.4 : 4)	99.98	74.28	96.67	98.58
SiO_2-10%AgCuNi (1 : 1 : 4)	99.98	69.04	95.69	96.02
SiO_2-10%AgCuNi (0.4 : 1.6 : 4)	99.99	33.71	86.30	86.34
SiO_2-10%AgCu (1 : 4)	99.99	64.89	95.38	97.15
SiO_2-10%AgZn (1 : 4)	99.97	52.34	90.52	95.32

Table 2. The sorption of $^{131}I_2$ (100 mg) and $CH_3{}^{131}I$ (20 mg) from the air flow on the developed composites containing 10 wt% of d-elements (the surface area of the cross-section of the column - 3.3 cm^2; the time of the experiment, including the time of the $^{131}I_2$ or $CH_3{}^{131}I$ supply - 4.0 h; the composite particle size - 3.0-6.0 mm, the linear steam-air flow speed in the column - 4-6 cm/s, the time of contact between gas phase and composite - 1 s).

4. Acknowledgment

Authors thank Department of Chemistry and Material Science of Russian Academy of Sciences (program of fundamental research № 8 /academician A.Yu.Tsivadze/) for financial support.

5. References

Asmolov, V. G. (1994). Test data of severe accidents at WWER, *Atomnaya Energiya (Russia)*, Vol.76, No.4, (April 1994), pp. 282-302, ISSN 0004-7163

Bronnikov, V. A. (2004). Eksperimentalnye Issledovaniya Protsessov Razrusheniya Aktivnoi Zony Reaktorov LWR Pri Tyazheloi Avarii (Proekt LACOMERA), *Atomnaya Tekhnika za Rubezhom (Russia)*, No.11, (November 2004), pp. 16-20, ISSN 0320-9326

Bukrinskii, A. M. & Fedulov, A. F. (1991). International scale of estimation of event risk at NPPs, *Atomnaya Energiya (Russia)*, Vol.70, No.1, (January 1991), pp. 3-8, ISSN 0004-7163

Charlot, G. (1961). *Les methods de la chimie analytique. Analyse quantitative minerale*, Paris: Masson, 1961, 4th ed. Translated under the title: *Metody analiticheskoi khimii.*

Kolichestvennyi analiz neorganicheskikh soedinenii, Vol.2, 1108, Khimiya, Moscow, Russia.

Containment Systems and Components for LOCA and Severe Accident Mitigation (1993). Promotion Materials from "Siemens AG. Power Generation Group (KWU)".

Dillman, y. G. & Bunz H. (1991). Design of a Filter System for PWR Containment Venting, *Journal of Aerosol Science*, Vol.22, Supplement 1, pp. S701-S704, ISSN 0021-8502

Dragunov, Yu. G. & Denisov, V. P. (2006). Support of safe operation of WWER unit at NPPs, *Atomnaya Energiya (Russia)*, Vol.101, No.2, (August 2006), pp. 87-93, ISSN 0004-7163

Evans, G. J.; Melnyk, A. et al. (1990). The LIRIC database/model, *Proceeding of 2nd International Conference on Containment Design and Operation*, 7.1-7.20, Toronto, Canada, October 1990.

Friesen, E.; Meseth J. et al. (2001). Containment behaviour in the event of core melt with gaseous and aerosol releases (CONGA), *Nuclear Engineering and Design*, Vol.209, No.1-3, (November 2001), pp. 253-262, ISSN 0029-5493

Kovalevich, O. (2000). What is preferentially: reconstruction or construction?, *Journal of Russian Nuclear Society*, No.1, (March 2000), pp. 34-38.

Krebs W.-D. (1999). The new generation, *Siemens Power Journal*, No.4, (April 1999), pp. 10-13

Kulyukhin, S. A.; Mizina, L. V. et al. (2007). Sorption of $^{131}I_2$ and $CH_3{}^{131}I$ from steam-air phase on porous inorganic sorbents containing d-elements, *Radiokhimiya (Russia)*, Vol.49, No.2, (April 2007), pp. 156–158, ISSN 0033-8311

Kulyukhin, S. A.; Kamenskaya, A. N. et al. (2008). Chemistry of radioactive iodine in gas phase: fundamental and applied aspects, *Radiokhimiya (Russia)*, Vol.50, No.1, (January 2008), pp. 3–21, ISSN 0033-8311

Kulyukhin, S. A.; Kamenskaya, A. N. & Konovalova, N. A. (2011). Chemistry of radioactive iodine in aqueous phases: fundamental and applied aspects, *Radiokhimiya (Russia)*, Vol.53, No.2, (April 2011), pp. 107–123, ISSN 0033-8311

Particulate Filtration in Nuclear Facilities (1991). *Technical Report Series*, No.241, 101, ISBN 92-0-125491-1, IAEA, Vienna, Austria.

Radioiodine Removal in Nuclear Facilities. Methods and Techniques for Normal and Emergencies Situations (1980). *Technical Report Series*, No.201, 98, ISBN 92-0-125280-3, IAEA, Vienna, Austria.

Reactor Design (2002). *Nuclear Engineering International*, Advert. Supplement.

Rogers, J. T. (1990). *Fission Product Transport Processes in Reactor Accidents*, ISBN 978-0891168768, Himespere Publishing, New York, USA.

Schwarzenbach, H. (1958). Complexometric titration, In: *Complexonometry*, N.I. Stupnikova, (Ed.), 104, Goskhimizdat, Moscow, Russia.

State of the art report on iodine chemistry (2007). *Report NEA/CSNI*. No.R1.

State of the art report on nuclear aerosols (2009). *Report NEA/CSNI*. No.R5.

State Standard 16187-70, Sorbents. A method of determination of fractional composition (1970). Committee of standards, measures and measuring devices at Ministerial council of the USSR, Moscow, Russia.

Tomorrow's reactors (2003). *Alternatives*, No. 2, (March 2003), pp. 4-5

Murata, T. & Kamiya, K. (1968). Removal of iodine and methyl iodide by a full-sized charcoal filter, *Proceedings of a symposium on operating and developmental experience in the treatment of airborne radioactive wastes*, 451-463, New York, USA, August 26-30, 1968.

Ustinov, O. A.; Sukhanov L.P. et al. (2005). Incidents at nuclear power plants, ways of their prevention and the localization of the incidents consequences, *Russian Chemical Journal*, Vol.49, No.4, (August 2005), pp. 54-60, ISSN 0373-0247

Zeyen, R. & Clement, B. (2006). The PHEBUS Fission Product And Source Term International Programmes, *Proceedings of Annual Meeting on Nuclear Technology "AMNT 2006"*, 234-239, Aachen, Germany, May 16-18, 2006.

28

Composite Materials under Extreme Radiation and Temperature Environments of the Next Generation Nuclear Reactors

Nicholas Simos
Brookhaven National Laboratory
USA

1. Introduction

In the nuclear energy renaissance, driven by fission reactor concepts utilizing very high temperatures and fast neutron spectra, materials with enhanced performance that exceeds are expected to play a central role. With the operating temperatures of the Generation III reactors bringing the classical reactor materials close to their performance limits there is an urgent need to develop and qualify new alloys and composites. Efforts have been focused on the intricate relations and the high demands placed on materials at the anticipated extreme states within the next generation fusion and fission reactors which combine high radiation fluxes, elevated temperatures and aggressive environments. While nuclear reactors have been in operation for several decades, the structural materials associated with the next generation options need to endure much higher temperatures (1200°C), higher neutron doses (tens of displacements per atom, dpa), and extremely corrosive environments, which are beyond the experience on materials accumulated to-date. The most important consideration is the performance and reliability of structural materials for both in-core and out-of-core functions.

While there exists a great body of nuclear materials research and operating experience/performance from fission reactors where epithermal and thermal neutrons interact with materials and alter their physio-mechanical properties, a process that is well understood by now, there are no operating or even experimental facilities that will facilitate the extreme conditions of flux and temperature anticipated and thus provide insights into the behaviour of these well understood materials. Materials, however, still need to be developed and their interaction and damage potential or lifetime to be quantified for the next generation nuclear energy. Based on material development advances, composites, and in particular ceramic composites, seem to inherently possess properties suitable for key functions within the operating envelope of both fission and fusion reactors. In advanced fission reactors composite materials are being designed in an effort to extend the life and improve the reliability of fuel rod cladding as well as structural materials. Composites are being considered for use as core internals in the next generation of gas-cooled reactors. Further, next-generation plasma-fusion reactors, such as the International Thermonuclear Experimental Reactor (ITER) will rely on the capabilities of advanced composites to safely withstand extremely high neutron fluxes while providing superior thermal shock resistance.

In addition it will be required by the composite to possess and maintain under severe neutron irradiation extremely high thermal conductivity to enable the flow of the anticipated extreme thermal heat loads generated in the core. The first wall and blanket surrounding the core in the fusion reactor are the two elements where composites are considered leading candidates.

Composites of special interest to both fission and fusion next generation nuclear reactors are carbon-fiber (C/C) and silicon carbide fiber (SiC_f/SiC), and more recently, C/SiC composites. These are continuous fiber-reinforced materials of either carbon or silicon carbide fibers infiltrated with a similar matrix. During the last two decades a number of studies have been conducted to address the feasibility and response of the two composites to different radiation environments of fission and fusion reactors and identify their limitations. While these composite structures have a significant advantage over materials used in the same reactor applications (i.e. nuclear graphite, BeO and metal alloys) because of the physical and mechanical properties they possess, they also experience limitations that require quantification. Carbon-fiber composites for example while they can have customized architecture to enhance desired properties, such as thermal conductivity, they too may experience anisotropic dimensional changes and be susceptible to irradiation-induced degradation. SiC_f/SiC composites, on the other hand exhibit good fracture resistance and low induced activity due to the irradiation stability of the SiC crystal but their technology is less mature. Critical issues such as cost, fabrication and joining as well as uncertainties due to lack of experience data in performance/survivability and lifetime in the combined extremes of high temperature and high fast neutron fluxes require further evaluation to qualify and quantify their performance.

During the last three decades and driven primarily by the fusion reactor needs (i.e. ITER) a extensive array of neutron-irradiation experiments at high temperatures have been conducted using available test reactors while the technology in composites was maturing. Facilities such as the High Flux Isotope Reactor (HFIR) at Oak Ridge National Laboratory, Japan Materials Testing Reactor (JTMR) as well as test reactors in Europe have been used to irradiate newly developed alloys and composites to anticipated fluence levels in the next generation fusion and fission reactors. As mentioned above, the understanding of the behaviour of reactor materials through studies and experience from the operating low energy neutron reactors may not necessarily transfer to the new materials under a different neutron spectrum. While the peak of the energy spectrum in the fission reactors to-date is ~1 MeV, neutrons in fast-spectrum reactors of the next generation are of several MeV while neutrons from fusion reactions will have energies of ~14 MeV. Based purely on modelling it is anticipated that the induced damage in the microstructure of materials will be similar to the one corresponding to that induced by less than 1 MeV neutrons. While the assumption may be generally correct, there is greater uncertainty with composites which do not form a perfectly oriented structure and are subject to the effects of dissimilar response between the fibre structure and the matrix. In addition, interaction with higher energy neutrons will generate more hydrogen and helium (as a result of nuclear interactions and transmutation products). Helium trapped within basal graphite planes and irradiation-induced defects will form bubbles and degrade the microstructure of the constituents. Therefore understanding how these composite structures behave in the fast neutron environment and what the degradation rate of their key properties (thermal conductivity, expansion, strength, etc.) is important and assessment by means of extrapolation from available data will be risky.

To observe the effects of high energy irradiating particles on composite structures, irradiation studies have been launched using the Brookhaven National Laboratory (BNL) proton linear accelerator and the target station at the isotope production facility (BLIP). The tuneable, ~25 kW (~110 μA peak) accelerator can accelerate protons to energies up to 200 MeV. Irradiation damage studies on graphite, carbon composites and new alloys have been performed in different phases using the proton beam directly on the materials and reaching fluences of ~10^{21} protons/cm^2. Results on the finding following proton irradiations on carbon composites, graphite and composite-like structures are presented in subsequent sections. In a different irradiation mode where the isotope targets completely absorb the 116 MeV protons, a mostly isotropic fast neutron is generated downstream of the isotope targets from the spallation process. This neutron spectrum with mean energy of ~9 MeV is utilized for the irradiation of composites and new alloys. While peak fluences are of the order of 10^{19} - 10^{20} neutron/cm^2 for each yearly proton beam run (much lower than the ~10^{21} -10^{22} n/cm^2) expected in the fusion reactor, still the interaction of these new composites with predominantly fast neutrons is expected to provide useful indicators of their stability and resilience to damage. Neutron irradiation studies at BNL BLIP have been completed recently for a number of super alloys and nano-structured coatings (Al_2O_3, TiO_2) on various substrates. Nanostructured coatings, along with AlBeMet, an aluminium-beryllium metal matrix composite, and structures made of fusion-bonded dissimilar materials constitute a special class of composites which are discussed in a subsequent section.

2. Composites and extreme environments

The development of advanced materials to be used in next generation reactors (Zinkle, 2004) has been driven by the need to endure the extreme environment consisting of prolonged, highly damaging radiation fluxes (tens of dpa), extreme temperatures (above 1000°C and up to 1200°C) and high stress conditions, which together can push well-understood and widely used materials beyond their limit. In fusion and fission reactor applications there are certain key properties that the materials which are playing a pivotal role (i.e. first wall and blanket in a fusion reactor) must maintain. These include low activation, structural integrity, dimensional stability, thermal conductivity and inherent ability to absorb thermal shock.

2.1 C/C composites

Carbon fibre reinforced composites (C_f/C) are an attractive choice for use in the extreme environment of the next generation reactors because of some key properties they inherently possess, namely enhanced strength as compared to nuclear graphite, thermal shock resistance because of their unique structure, extremely low thermal expansion, enhanced thermal conductivity due to the presence and directionality of fibres and low neutron activation. Due to their attractive properties carbon-carbon composites have enjoyed widespread use in advanced technologies which have led to the maturity of their technology and fabrication. A wide variety of architectures of the fibre/matrix have been developed as well as fabrication techniques. Most widely used architectures are the two-dimensional (2D C_f/C) and three-dimensions (3D C_f/C) forms. Shown in Figures 1 and 2 are sections of the three-dimensional architecture of the composite (FSI 3D C_f/C) indicating the orderly fibre bundle (thickness of ~265μm) and matrix arrangement.

While carbon composites exhibit enhanced properties when compared to graphite, radiation-induced damage from neutrons or other energetic particles such as protons is far less well understood. To the contrary, nuclear graphite has been extensively studied for radiation-induced degradation for almost sixty (60) years and so the degradation of the key properties as a function of the neutron fluence such as thermal conductivity, dimensional stability and strength has been established thus leading to limitation thresholds for its use in more extreme environments (Gittus, 1975; Maruyama & Harayama 1992; Nikolaenko et al. 1999). Key findings from these studies on graphite are the anisotropic dimensional changes that take place at higher radiation doses and most importantly the degradation of the thermal conductivity. Within the last two decades a body of experimental research work on irradiation damage of carbon-carbon composites has been reported prompted primarily by the need to identify higher performance, low neutron activation for the first wall of fusion reactors such as the International Thermonuclear Experimental Reactor (ITER). Of primary interest in these reported studies (Burchell, 1992, 1994; Burchell et al. 1996; Barabash, et al., 1998) are neutron irradiation induced dimensional changes, thermal conductivity and mechanical properties.

Fig. 1. Image of a 3D C_f/C composite structure using an optical microscope. Shown is the orderly arrangement of the fibre bundles and the matrix

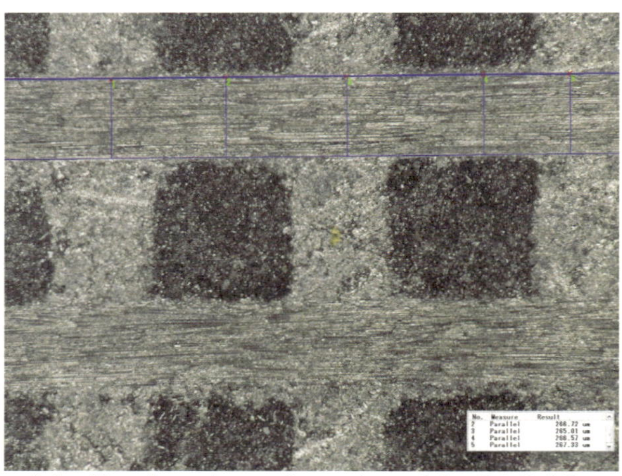

Fig. 2. A closer view of the fibre/matrix configuration of the 3D C_f/C composite and measured fibre bundle thicknesses

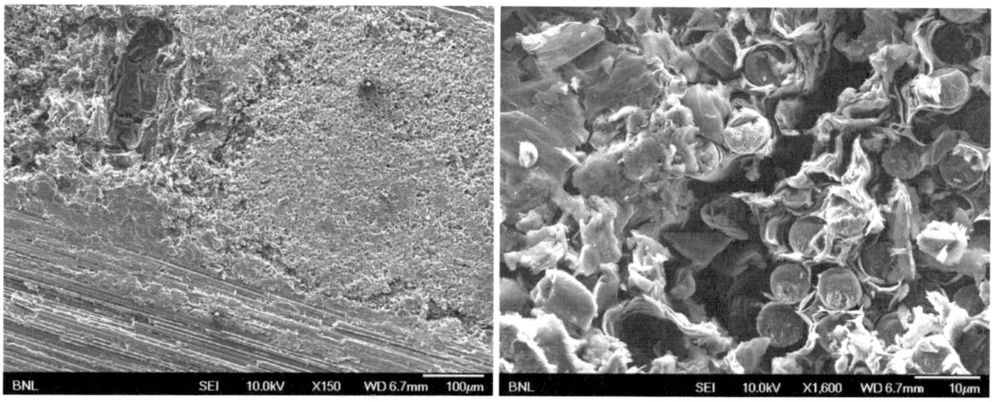

Fig. 3. SEM image of the matrix/fibre bundle interface in 3D C/C composite including the presence of voids (left) and closer depiction (right) of one of the pre-existing voids, individual fibres and interface

2.1.1 Shock resistance

One of the important attributes of C/C composites in the fusion reactor environment is their inherent ability to absorb thermal shock. In an effort to quantify the ability of the C/C composite to absorb thermal shock and so be used as the material of choice in a number of high power accelerator applications including accelerator targets for the Long Baseline Neutrino Experiment, Neutrino Factory (LBNE), beam collimating elements for the Large Hadron Collider (LHC) or energetic beam absorbers, experiments have been performed using intense pulses of energetic protons. In these experiments (Simos et al., 2005) performed using the 24 GeV proton beam at the Accelerating Gradient Synchrotron (AGS) at BNL the shock performance of FMI 3D C/C composite targets (16-cm long, 1-cm diameter rods) was measured and compared to that of ATJ graphite. Shown in Figure 4 is the shock test arrangement where 3D C/C composite and ATG graphite targets are instrumented with fibre-optic strain gauges mounted on the surface of the target and measuring extremely fast axial strain transients in the target resulting from its interaction with the 24 GeV proton beam. The response from the intense (3.0 e+12 protons) and focussed (0.3mm x 0.7mm) proton pulses on the ATJ graphite and 3D C/C targets is shown in Figure 5 where the two are compared. It is evident from the comparison that the carbon-carbon composite shows a much lower response to the shock induced by the beam while radial reverberations indicated by the high frequency cycles within each axial cycle are damped out as a result of the impedance interfaces (fibre/matrix) and the voids that are present as shown in Figure 3. Due the potential implications and applications of high shock resistance in the carbon-carbon composites which stem from the "effective" low thermal expansion coefficient specific studies (Hereil, 1997) have focussed on experimentally verifying the compressive wave velocities in a plate-impact configuration by wave decomposition. As observed in (Simos et al., 2006; Hereil et al, 1997) the problem of shock in materials such as C/C remains very complex due to the anisotropy and the fibre-matrix interfaces as well as the response to dynamic loads of wave propagation in the individual components. In such materials, understanding the behaviour at the mesoscale is important for modelling and implementation of these composites in large-scale designs.

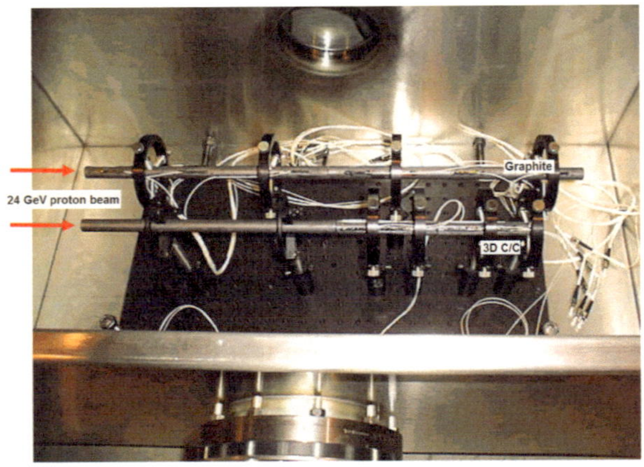

Fig. 4. C/C composite and ATJ graphite target shock test arrangement for the BNL AGS proton beam tests

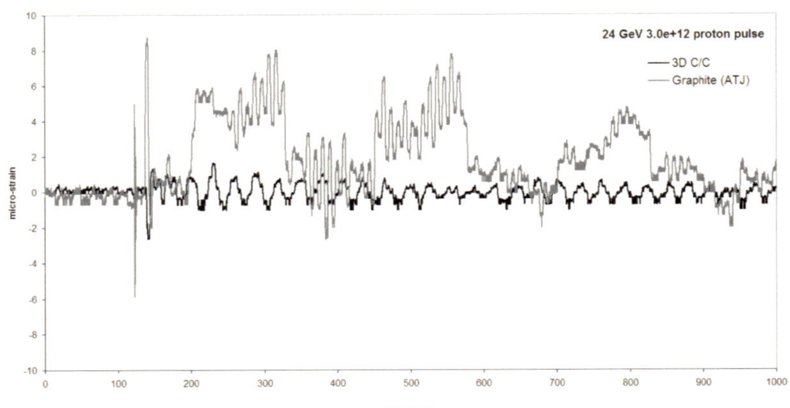

Fig. 5. Comparison of shock response in terms of axial strain recorded on 3D C/C and ATJ graphite targets

2.1.2 Radiation damage

While radiation damage in the carbon-carbon composites have been studied using neutrons from various reactors such as the High Flux Isotope Reactor (HFIR) at Oak Ridge National Laboratory or the Japan Materials Testing Reactor (JMTR), the radiation damage from very energetic accelerator protons is very limited. Reported in the neutron-induced damage studies (Burchell, 1992, 1996; Bonal et al., 2009) is that the composite undergoes dimensional changes as a function of the fluence (or dpa) with the 3D architecture to exhibit an isotropic behaviour. Under neutron irradiation and for a fluence of 1.0 dpa the thermal conductivity reduces by ~50% (Burchell, 1992) while other studies (Maruyama & Harayama 1992) on CX-2002U carbon-carbon composite suggest (see Figure 6) that there is a dramatic drop of the thermal conductivity even at very low fluences (0.01 dpa) when the irradiation temperature is 200°C and below the threshold where induced vacancies and interstitials become mobile.

Initiation of the degradation of the 3D C/C structure following neutron irradiation have been observed (Snead, 2004) and (Bonal et al, 2009) at the 2 dpa dose level leading to serious structural disintegration at about 10 dpa. These reported levels of irradiation damage onset pose a serious limitation on the desired lifetime of C/C plasma facing components in the fusion reactors.

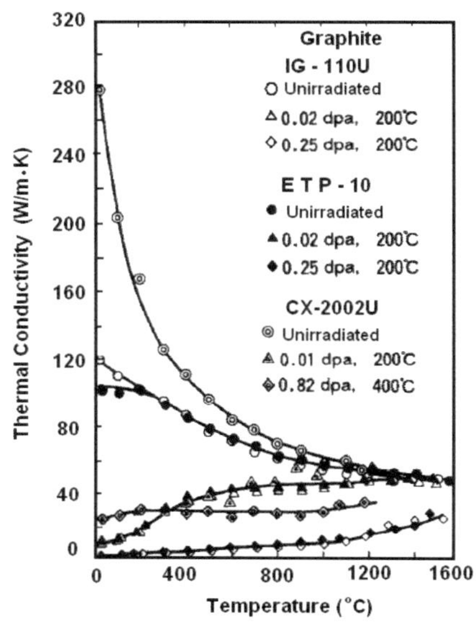

Fig. 6. Thermal conductivity of neutron-irradiated graphite and carbon-carbon composite (Maruyama, 1992)

Fig. 7. C/C composite following irradiation (10 dpa at 800 °C) showing serious degradations and anisotropic dimensional change in the form of swelling and shrinkage of fibre bundles (Snead, 2004)

For high power accelerators, while the thermal shock resistance is superior to graphite and so is the retention of thermal conductivity, degradation as a result of energetic proton irradiation can be a serious limiting factor along with the dimensional stability required of critical elements such as primary beam intercepting accelerator target and collimators. Irradiation damage studies have been conducted in recent years (Simos et al, 2006a, 2006b, 2008) using the BNL 200 MeV Linac beam at the isotope production facility (BLIP). The main objective was to assess the proton-induced damage at energies higher than the thermal and fast neutrons these composite structures have been exposed in test reactors like HFIR and JMTR and qualify the differences stemming from the irradiating species (protons vs. neutrons) and energies (neutron energies a few MEV and proton energies up to 200 MeV).

Two architectural types of the composite were proton-irradiated and studied. A 2D C/C structure (AC-200) made by Toyo Tanso to be used as a primary beam collimating material at the Large Hadron Collider at CERN where 3.5 TeV protons at the beam halo (>6σ) will be intercepted and diverted away from the circulating beam and a 3D C/C architecture made by FMI as a target candidate material for the high power accelerators (LBNE and Neutrino Factory). In the case of the 2-D C/C, dimensional stability in the direction along the fibres was extremely important and so was the thermal conductivity and structural degradation resistance. The first in series of long exposures (>10^{20} p/cm^2 or >0.2 dpa) for these two composites revealed that both architectures experience serious structural degradation. Shown in Figure 8 is the structural degradation of the AC-200 2D C/C of specimens formed normal to the fibre planes and along the fibres. Also shown in Figure 8 (right) is the structural degradation of the FME 3D C/C composite. The peak proton fluence within the damage area is ~5.0 - 7.0 10^{20} p/cm^2 (~0.6-0.8 dpa) a level significantly lower than the one associated with neutron-based irradiation of 2 dpa (Bonal et al, 2009). Subsequent irradiations to similar fluence thresholds verified the damage initiation onset at these low proton fluences. Recent studies of the FMI 3D C/C composite irradiated in an inert gas (argon) environment showed that there exists an environmental factor associated with the damage given that the apparent fluence threshold for structural degradation has increased slightly. Further investigation on this finding is currently under way.

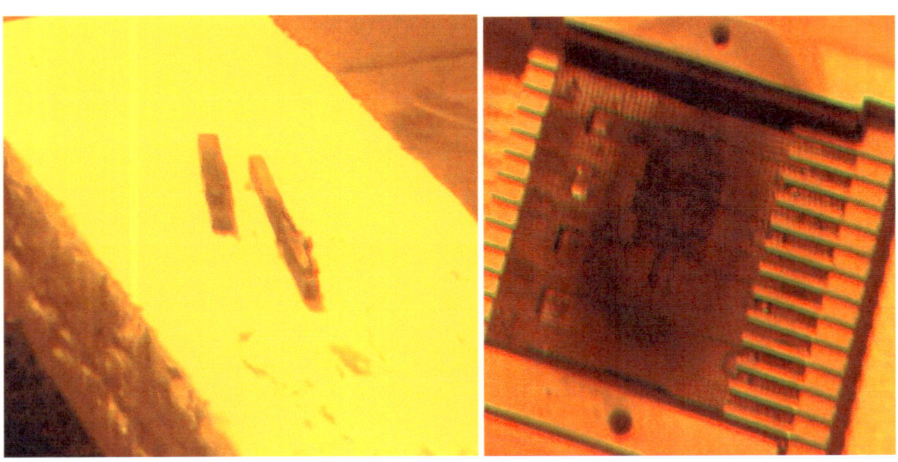

Fig. 8. Irradiation damage induced by 200 MeV (left) protons on AC-200 2D C/C (left) and damage induced by 160 MeV protons on FMI 3D C/C composite

The dimensional stability of the proton-irradiated 2D and 3D carbon composites was measured using a LINSEIS high precision dilatometer in the BNL hot cell facility. Of primary interest were the irradiation effects on the thermal expansion coefficient which represents a crucial parameter in the accelerator applications discussed above. From the precise measurements made it was observed that the fibres in both architectures undergo shrinkage as a result of the irradiation. This is evident in dimensional change data depicted in Figure 9 (left). The 2D C/C in its un-irradiated state will shrink along the directions of the fibres (negative thermal expansion) and expand in the direction normal to the fibre planes acting more like typical graphite. Following irradiation, however, and during the first thermal cycle that was applied to the 2-D specimen made along the fibres there is an accelerated expansion beyond the irradiation temperature (irradiation temperature is depicted as the inflection point in the curve). This is the result of the expansion of the fibres which experienced shrinking during irradiation. Interesting to note is that the composite with progressive upper temperature limit in a series of thermal cycles restores the uni-irradiated behaviour up to that temperature. Observing the dimensional change along the normal direction shown in Figure 9 as a function of fluence one notes the growth that has occurred along the direction due to irradiation indicated by the apparent shrinkage during the thermal cycle when the composite is being restored towards the original configuration rather that expand with increasing temperature. This is evident in Figure 10 (left) which depicts a subsequent thermal cycle where the dimensional behaviour is aiming towards that of the un-irradiated composite. Comparing this restoration along the normal direction with that of irradiated IG-43 graphite, Figure 10 (right), the similarities can be seen which should be anticipated along this normal direction except that for the 2D C/C the growth in the lattice between the planes has occurred at lower irradiation temperatures.

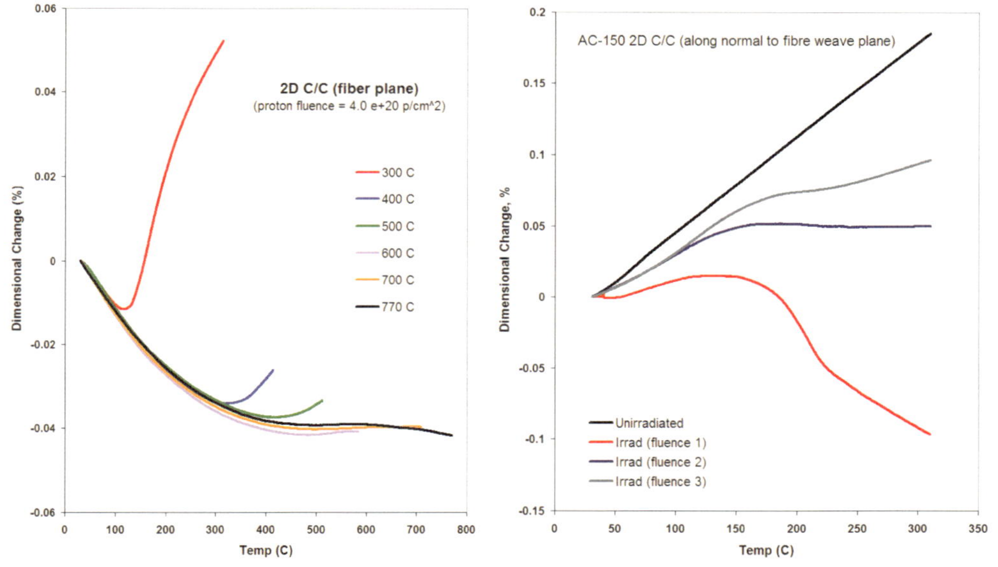

Fig. 9. Dimensional change in 2D C/C composite following proton irradiation and removal of fibre shrinkage (left) with thermal annealing in the fibre direction and growth (right) induced in lattice along the normal to fibre direction as a function of fluence (1 = $6.0 \bullet 10^{20}$ p/cm^2; 2 = $3.0 \bullet 10^{20}$ p/cm^2; 3 = $1.0 \bullet 10^{20}$ p/cm^2)

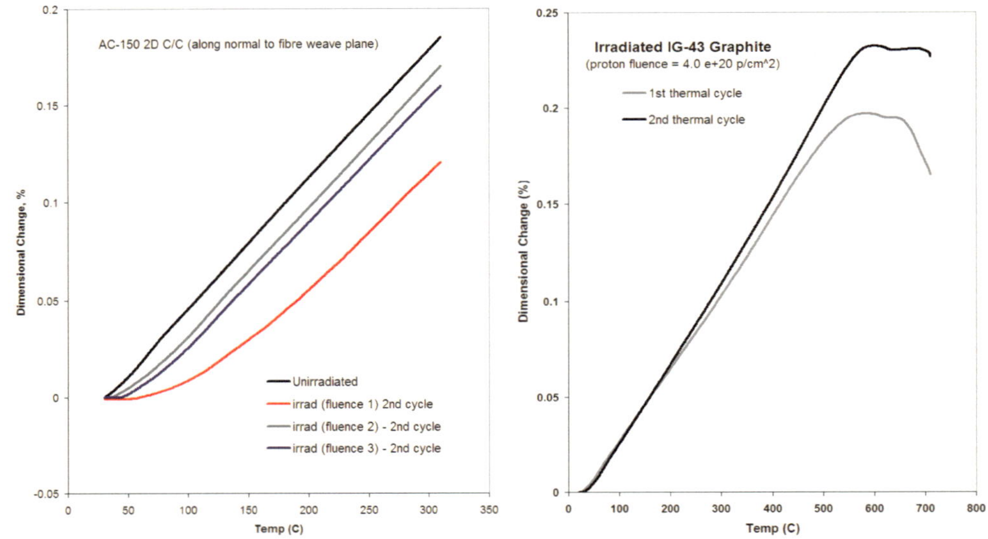

Fig. 10. Restoration of growth along the normal direction of irradiated 2D C/C fibres through thermal cycling and as a function of proton fluence (left) and of irradiated IG-43 graphite irradiated with fluence 4.0 10^{20} p/cm^2

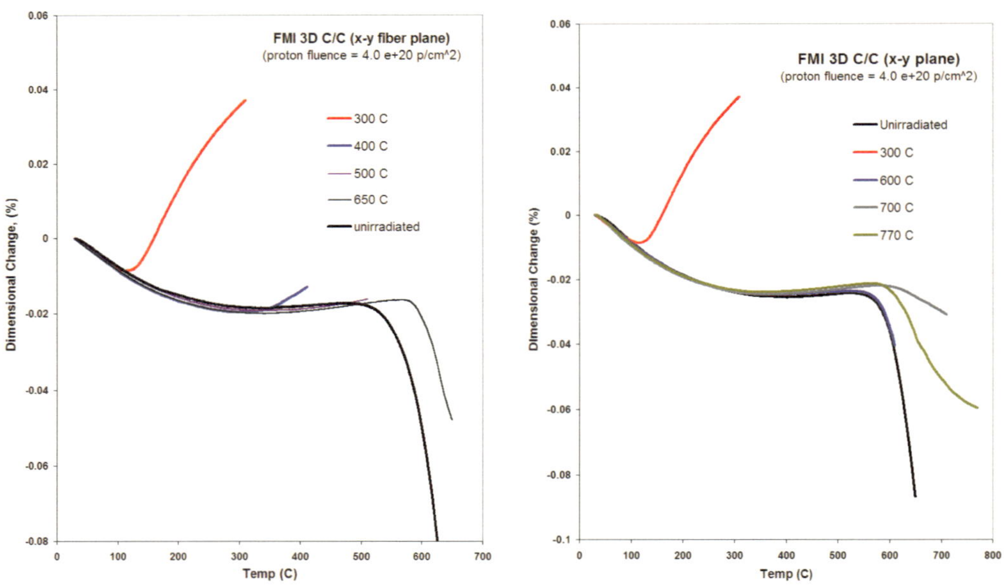

Fig. 11. Dimensional change and thermal annealing of 3D C/C composite structures following proton irradiation with fluence 4.0 10^{20} p/cm^2

The dimensional changes of irradiated 3D C/C composite are shown in Figure 11. For the 3D architecture the material exhibits a negative CTE in all directions. As seen in Figure 11 for the un-irradiated samples, there is accelerated shrinkage >600°C attributed to the influence of the matrix within the fibre structure. Reversal from shrinkage to growth at these temperatures was observed in previous studies (Burchell, 1994). Measurements of thermal conductivity of the irradiated 3D C/C and IG-43 graphite samples following proton irradiation revealed that thermal conductivity reduced by a factor of three (3) for the 3D C/C for 0.25 dpa fluence and by a factor of six (6) for IG-43 and similar fluence. To address the significantly higher irradiation damage from energetic protons a new irradiation experiment has been initiated where the carbon composite will be exposed to a neutron flux at the BNL BLIP facility which results from the spallation of protons with upstream isotope targets. The goal is to compare the damage at similar fluences of energetic protons and energetic neutrons.

2.2 SiC_f/SiC

Silicon carbide fibre reinforced composites (SiC_f/SiC), along with the C/C composites, considered as prime candidates for the first wall and blanket structural material in fusion reactors. Due to their low activation and irradiation stability these composites have a clear advantage over the C/C composites especially when the neutron doses are expected to be high as it is the case of fusion reactors where tens of dpa over the lifetime will be accumulated. Its inherent stability stems from the isotropic dimensional change of the cubic SiC crystal (Bonal et al. 2009) which tends to saturate at modest irradiation levels. It also exhibits good fracture resistance and excellent mechanical properties at high temperatures. While the carbon composite technology and manufacturing is more mature than that of the SiC composites, which currently have limited structural applicability outside nuclear reactors. Significant progress has been made in recent years to both eliminate issues of early grades of the composite associated with poor irradiation performance (Snead et al., 1992) and reduce cost through adoption of novel fabrication techniques (Katoh et al., 2010) such as the nano-infiltrated transient-eutectic (NITE) process (Bonal et al., 2009). For nuclear grade SiC_f/SiC composites the costly chemical vapour infiltration technique (CVI) is used.
To assess the neutron irradiation damage of SiC_f/SiC composites grades produced by a variety of approaches a number of irradiation experiments have been launched using the HFIR and JTMR reactors (Katoh et al., 2010). Irradiation of the various grades at HFIR to levels up to 10 dpa and at elevated temperatures of 800°C has also been conducted. The key objective is to assess the strength and stability of the improved composites and the thermal conductivity degradation. In contrast to the early SiC_f/SiC grades which suffered significant irradiation damage from de-bonding of fibre matrix interface driven by fibre densification the new grades show minimal degradation of strength and stability of mechanical properties up to 10 dpa (Katoh et al., 2007). Thermal conductivity, however, remains an issue for low irradiation temperatures.
An experimental study on the effects of high temperatures and proton or fast neutron irradiation has been initiated at Brookhaven National Laboratory. The goal is to subject SiC_f/SiC composite to similar irradiation fields that the carbon composites have been exposed to and make comparison in the irradiation-induced damage. As discussed in the previous section, significant damage was observed in C/C composites at levels far below the observed limits in neutron irradiation environments. Shown in Figure 12 is an optical image of SiC_f/SiC composite section showing the fibre bundle thickness (~128μm) and the SiC matrix thickness (~343μm). Also shown in Figure 12 is the distribution of voids within the architecture. Figures 13 and 14 are SEM images of the fibre/matrix interfaces and of the individual fibres.

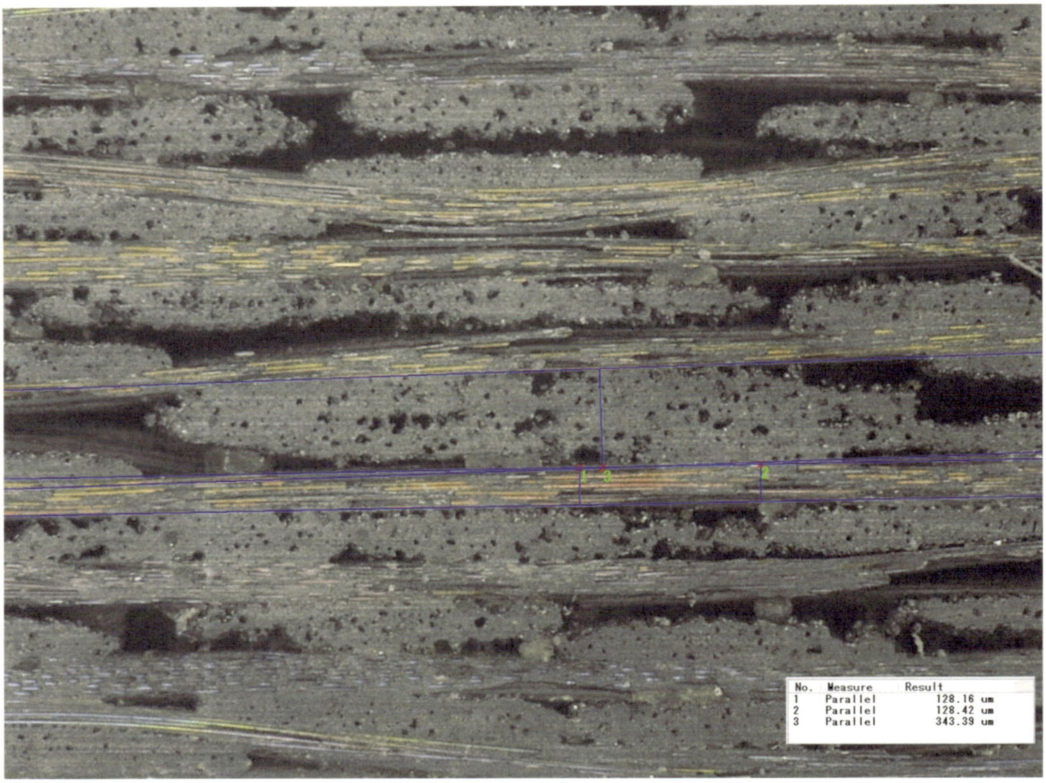

Fig. 12. Optical microscope image of a SiC_f/SiC structure showing the arrangement of the weaved fibre bundles and the matrix

To assess the effect of high temperature on the SiC_f/SiC in terms of dimensional changes, structure and density composite samples were brought to 1000°C in atmosphere and the changes were made with precise instruments at the BNL isotope facility. Dimensional changes were more pronounced along the fibres (shrinkage of ~1%) while in the direction normal to the fibres they were of the order of 0.09%. Density reduction of ~0.8% was also observed (ρ_{rt} = 2.4324 g/cc) following the annealing of the sample for one hr at 1000°C. Shown if Figure 17 is dimensional changes obtained for temperatures up to 610°C and are compared with those of 3D C/C. As seen in Figure 17 during the first thermal cycle there is an adjustment in the both structures except more pronounced in SiC_f/SiC which expands with increasing temperature in contrast to C/C which shrinks for the selected temperature range. Based on the stabilized thermal expansion, the thermal expansion coefficient (CTE) in the range of 200-600°C was estimated as 3.7 10^{-6}/K. Experimental results (Zhang, 2006) on carbon fibre reinforced SiC (via CVI method) up to 1400°C showed values similar average values in the 200-600°C but with dramatic fluctuations above 800°C. Following the planned proton irradiation of the SiC_f/SiC and the fast neutron irradiation using the spallation process at the isotope production facility at BNL the effects on the physio-mechanical properties will be studied and compared with the C/C composites.

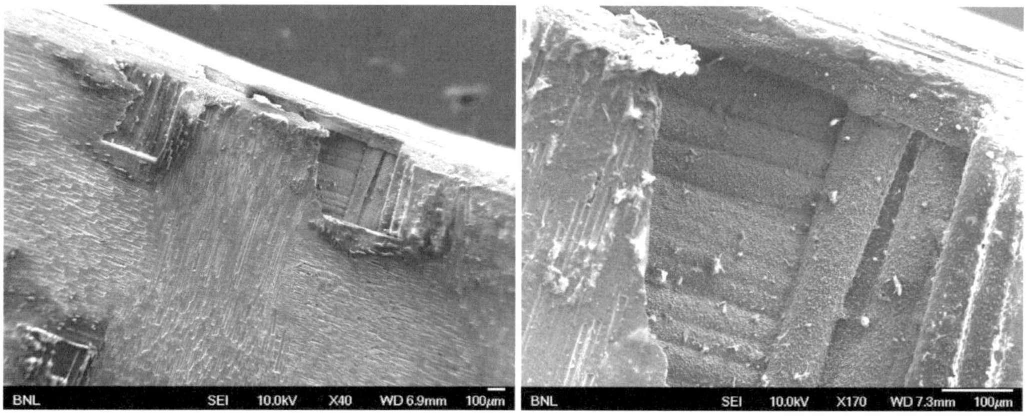

Fig. 13. SEM photograph of a SiC_f/SiC cut section (left) and of fibre arrangement (right)

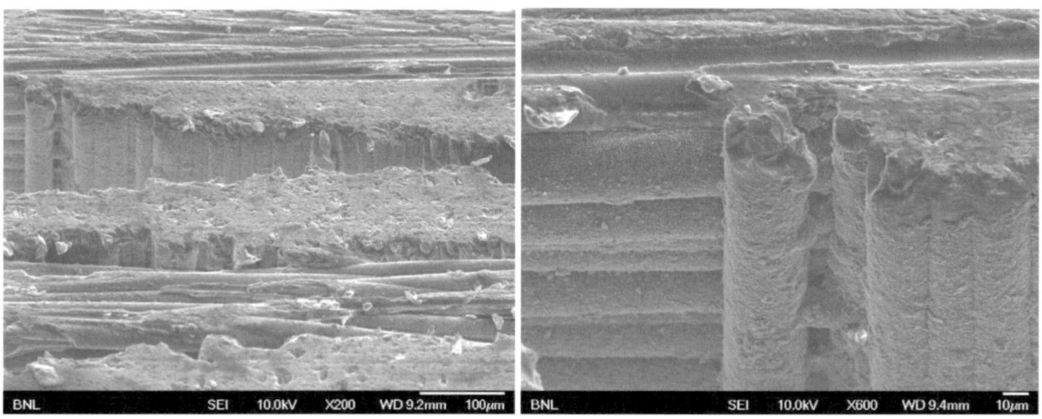

Fig. 14. SEM photograph of a fibre bundle (left) and fibre detail (right) of SiC_f/SiC

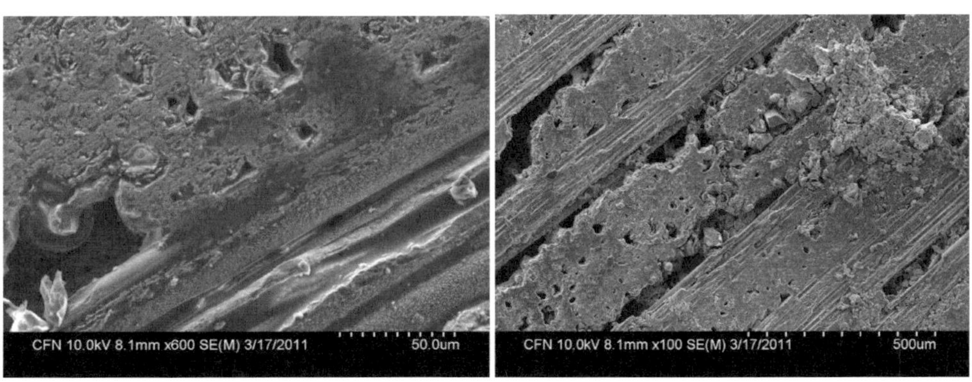

Fig. 15. SEM of a SiC_f/SiC section at room temperature (left) and following annealing at 1000°C for one hour (right)

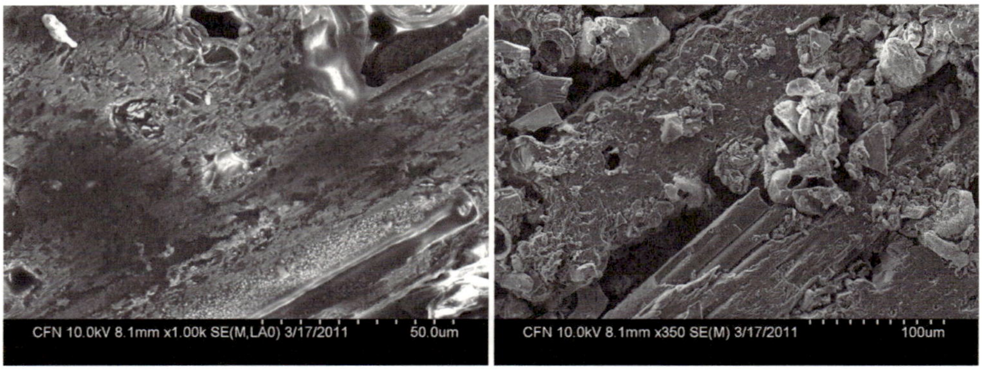

Fig. 16. Close SEM view of a SiC/SiC section at room temperature (left) and following annealing at 1000°C for one hour (right)

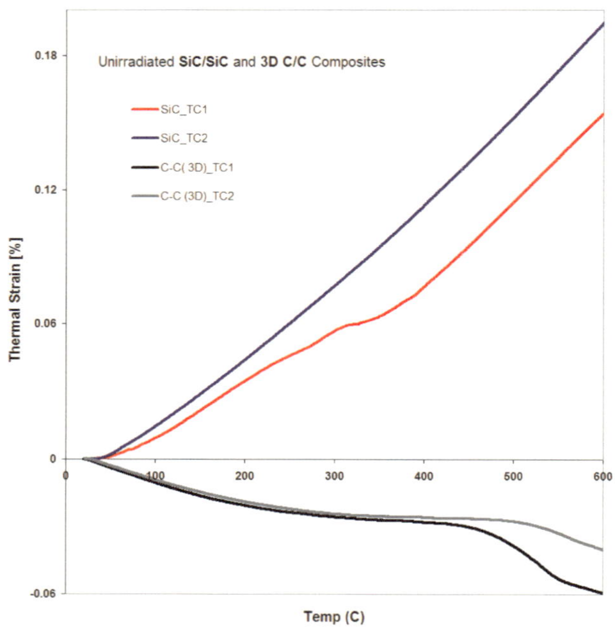

Fig. 17. Dimensional change of un-irradiated SiC/SiC composite over two thermal cycles (noted as TC1 and TC2) and comparison with 3D C/C

2.3 Composite-like structures

A number of material structures not adhering to the classical definition of composites involving a matrix with fibre-reinforcement, i.e. C/C, C/SiC, SiC_f/SiC, etc., can still be considered composites, or more appropriately composite-like with potential applications in the next generation fusion and fission nuclear reactors. These can be based on (a) the embedment of particles of one material into the lattice of another thus maintaining the individual characteristics, (b) the bonding of dissimilar materials using solid state reaction of chemical vapour deposition with the help of an interface layer, and (c) on deposition of nanostructured coatings on substrates to either enhance the properties of the combined structure or protect the substrate. Because of their potential for use in nuclear reactors, some of these composite-like structures have been studied for radiation damage and extreme temperatures.

2.3.1 AlBeMet

AlBeMet, while by metallurgical definition may be considered an alloy, is in fact a composite of aluminium and beryllium consisting typically of ~62% commercially pure beryllium and 38% of commercially pure aluminium by weight. The two metals involved in forming AlBeMet do not fully mix but instead the beryllium particles are embedded in a pure aluminium lattice. The powder is produced by a gas atomization process yielding a fine beryllium structure. The two granular forms are mixed at temperatures just below the melting points of the two metals and a pressure that prompt the particles to form a stable bond. The result is a non-typical composite with some very appealing thermo-physical properties since it combines the workability of aluminium and, for the most part, the hardness of beryllium. Interest in this special composite has been increased in recent years primarily for use in special components of particle accelerator systems and in particular in the accelerator target envelope characterized by high-radiation, high temperature and thermal shock conditions. The combination of low-Z, good thermal conductivity (210 W/m-K) and low electrical resistivity (3.5e-6 Ohm-cm), combined with its workability and hardness, make it very attractive for special components such as magnetic horns and targets. Unknown was its radiation resistance and dimensional stability which are key parameters for potential applications in nuclear reactor systems. The effects of radiation on the physical and mechanical properties of this unique composite have been studied using direct energetic protons and secondary fast neutrons of the BNL accelerator complex through a series of irradiation experiments. Peak proton fluences of 3.0 10^{20} p/cm^2 at 140 MeV using the 200 MeV proton beam at BNL BLIP and, through a different study, fast neutron fluences of ~10^{19} n/cm^2 were achieved in these accelerator-based irradiation experiments. The arrangement of specially designed test samples of beryllium (similar to AlBeMet samples) in the irradiation space intercepting the proton beam is shown in Figure 18. The numbered tensile specimens (dog-bone) are 42mm long and 1.5mm thick and have a strain gauge length of 6mm. The matching specimens are used for post-irradiation analysis of thermal expansion, electrical and thermal conductivity.

Post-irradiation studies revealed that AlBeMet is dimensionally stable following irradiation and that it resists embrittlement and degradation even at high proton fluences where materials such as graphite and carbon composites have shown to undergo serious degradation. As indicated above, the proton fluences received may be representative of a much more severe irradiation condition when correlated to either thermal or fast neutrons. Depicted in Figure 19 is the thermal expansion of both AlBeMet and beryllium for peak fluence of 1.2 10^{20} p/cm^2 and the measured CTE of AlBeMet as a function of average proton fluence. The effects of proton irradiation on the stress-strain relation of AlBeMet and its comparison with beryllium are shown in Figure 20. While it is confirmed that the ultimate tensile strength of beryllium is higher than that of AlBeMet, the AlBeMet appears to increase its strength following irradiation (as expected in all metals due to the pinning of dislocations) but without loss of the ductility anticipated to accompany the induced hardening. Further tests are planned where AlBeMet will be exposed to higher proton and accelerator-produced fast neutron fluences to explore its mechanical behaviour.

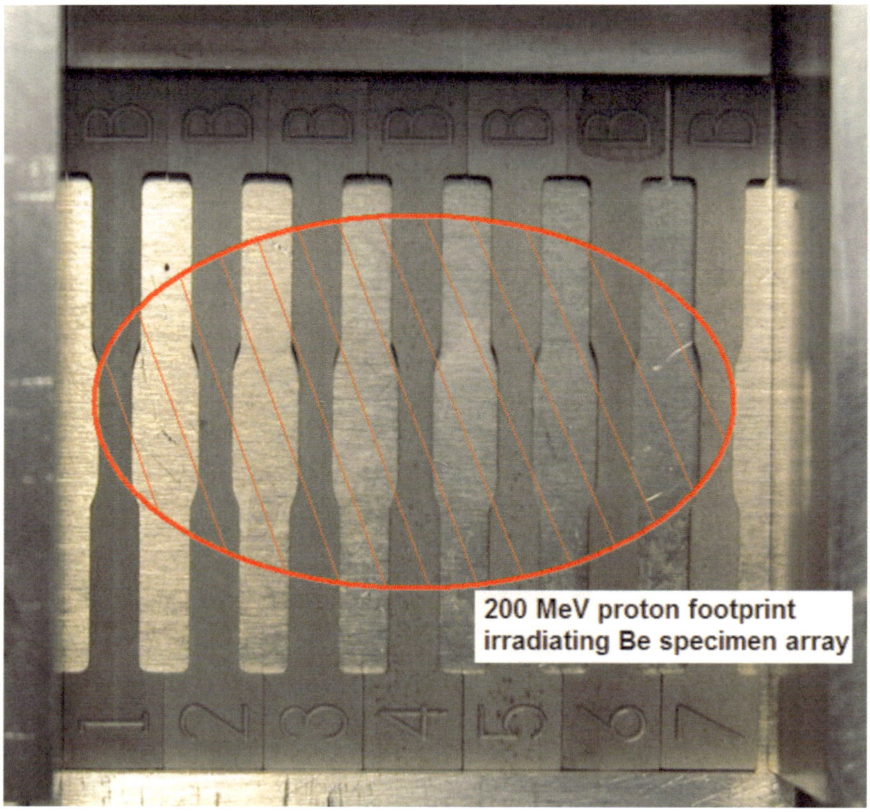

Fig. 18. Arrangement of irradiated test samples of beryllium intercepting the 200 MeV proton beam at BNL isotope production facility.

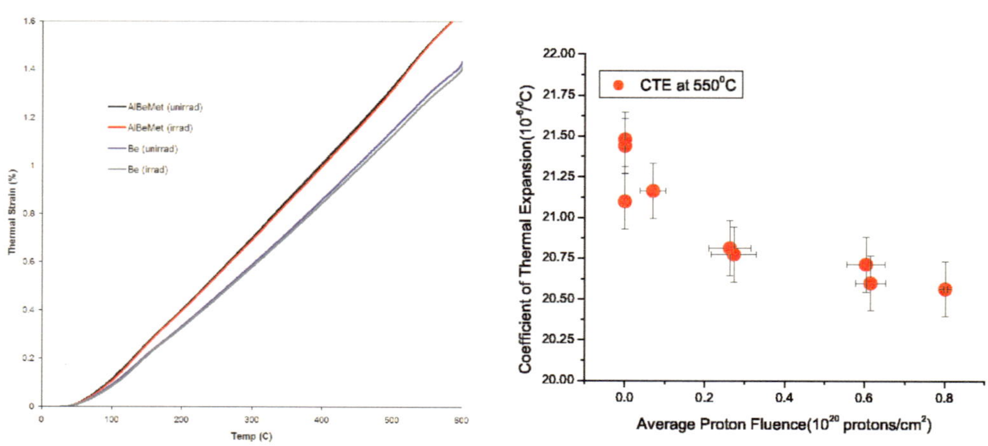

Fig. 19. Thermal expansion (left) comparison of AlBeMet and beryllium irradiated by 200 MeV protons and peak fluence of 1.2 10^{20} p/cm^2, and coefficient of thermal expansion of AlBeMet (right) at 550°C as a function of proton fluence

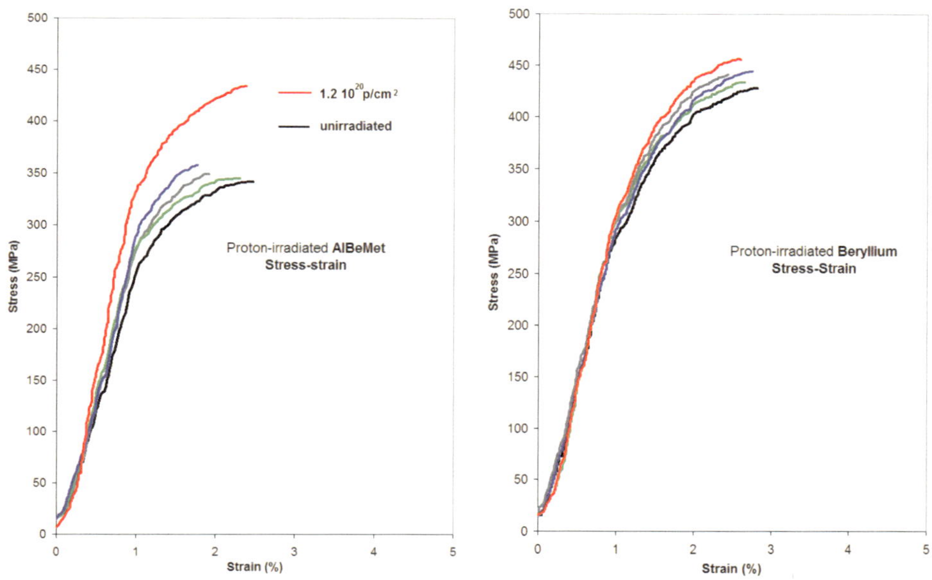

Fig. 20. Effects of 140 MeV proton irradiation on AlBeMet (left) and beryllium (right) on the stress-strain relations (range of fluence of irradiated samples 0.4 - 1.2 10^{20} p/cm^2)

2.3.2 Bonded dissimilar materials

Bonding of dissimilar materials to create a "composite-like" structure and ensuring its ability to maintain the integrity of the interfaces under extreme temperature conditions and high radiation fluences is an important challenge. Applications of such composites can be seen in nuclear fuel elements, fusion reactor plasma facing components where high-Z materials such as tungsten with higher erosion resistance will protect low Z materials like carbon or beryllium and particle accelerator targets where the variation of the material atomic number Z from the centre of the intercepted beam is important for both particle-mass interaction and heat removal from the hot central part (Simos et al., 2006b). In all applications and due to the dissimilarity of the thermal expansion in the bonded material structure, high stresses can develop at the interfaces leading to micro-cracking or even separation. Such condition can dramatically reduce a key property of the composite layer that controls the primary function such as heat transfer across the interfaces through thermal conduction. Shown in Figure 21 is a schematic of a TRISO-coated particle and pebble bed fuel sphere for Generation-IV Very High Temperature Reactor (VHTR). Maintaining the integrity of the interfaces between the various layers around the fuel kernel under high temperatures and extreme radiation fluxes is crucial. The dimensional changes occurring as a result of the two simultaneous effects do not necessarily coincide in terms of direction (growth or shrinkage) and thus a better understanding of the interface mechanics under such conditions is required.

There is a significant need in the nuclear industry and in particular in the first wall of fusion reactors, of graphite/metals bonding to form a coating or cladding on the low Z materials (i.e. graphite) and reduce the erosion rate. Direct graphite-metal junctions (Brossa et al., 1992) for use in the first wall of fusion reactors are extremely sensitive to thermal cycling due to differences in thermal expansion so techniques have been developed and applied with the introduction of interface agents (such as silver) that will prevent the formation of fragile components and metal dissolution. Thus brazing between stainless steel and graphite, Mo and graphite as well as W and graphite has been produced using vacuum plasma spray (VPS) and chemical vapour deposition (CVD) techniques. To study the resilience to thermal shock, graphite/Mo or graphite/W bonding was achieved by using an intermediate layer of Mo, V, or Mo-Ti and applying a solid state reaction bonding technique (Fukatomi et al., 1985).

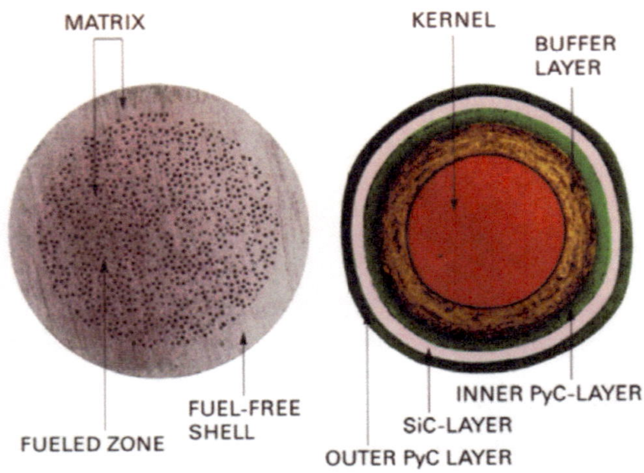

Fig. 21. TRISO-coated particle and fuel element for a pebble bed reactor (US DOE, 2006)

To assess the effect of proton irradiation bombarding the muon target at J-PARC facility where a 3 GeV, 333 μA proton beam is intercepted by a graphite target at a rate of 25 Hz an experimental study was initiated at Brookhaven National Laboratory. The study consisted of an irradiation phase using the 200 MeV proton beam of the BNL Linac and of a post-irradiation analysis to observe the degradation of the target-like composite structure that was made for the study. Shown in Figure 22 is a test specimen consisting of three materials (copper and titanium alloy Ti-6Al-4V) and graphite (IG-43) and two interfaces (graphite to titanium and copper to titanium alloy). To form the two interfaces the silver brazing technique in vacuum was applied. Two types of geometry in the 42mm long (4mm x 4mm cross sectional area) specimens was used, one at 45^0 (as shown) and one with normal or 90^0 interfaces. Shown in the SEM image of Figure 22 and prior to irradiation is the achieved bonding/interface (extremely faint) between copper and titanium alloy for the 90^0 interface. The composite specimens were placed in the proton beam and received peak fluence over the two interfaces of ~8.0 10^{20} p/cm^2. Post-irradiation examination revealed the loss of integrity of the graphite/copper interface and the complete separation. Shown in Figure 23 is the graphite-to-titanium interface zone prior to irradiation and the condition of the composite structure after irradiation. As it has been assessed by focussing on the graphite susceptibility to high energy proton irradiation (elevated irradiation damage due to the nuclear interactions and the accelerated formation of He bubbles trapped in the lattice) as part of the same irradiation study, the degradation of the interface is attributed to graphite radiation damage. The composite dimensional changes, which will inevitably result in interface stresses due to thermal expansion coefficient differences, were studied by analyzing the thermal strain as a function of temperature using the high-sensitivity dilatometer in the hot cell. Shown in Figure 24 are the thermal expansion of the composite structure (graphite/copper/titanium) for both 45 and 90-degree interface orientations and the three components of the composite when all have received the same proton fluence. The dissimilar rates of expansion between the three constituents and the dimensional change reversal of graphite above 600°C (discussed previously) introduce a non-liner thermal expansion trace. Important to note is that subsequent thermal cycles above the 600°C threshold, resulted in a complete failure and fracturing of the irradiated graphite in the composite. This was observed for both interface orientations and attributed to the growth reversal in the graphite leading to fracture of an already compromised graphite strength resulting from radiation damage.

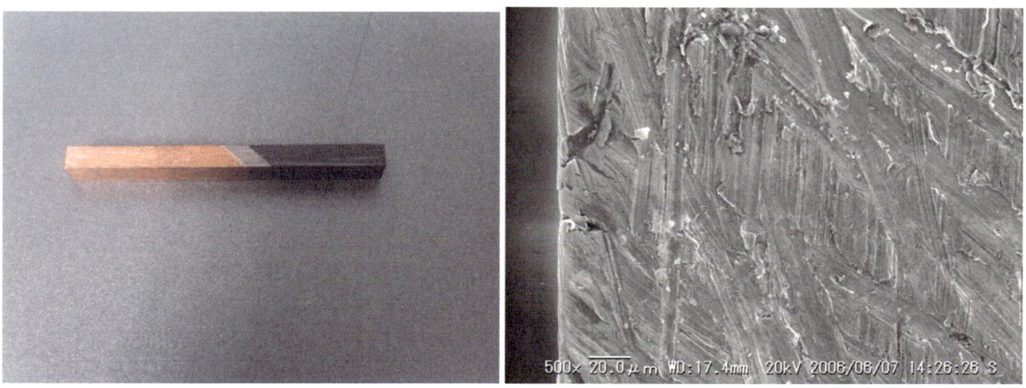

Fig. 22. Graphite-titanium-copper composite sample (left) used for the BNL irradiation studies and SEM image of the copper-titanium interface prior to irradiation

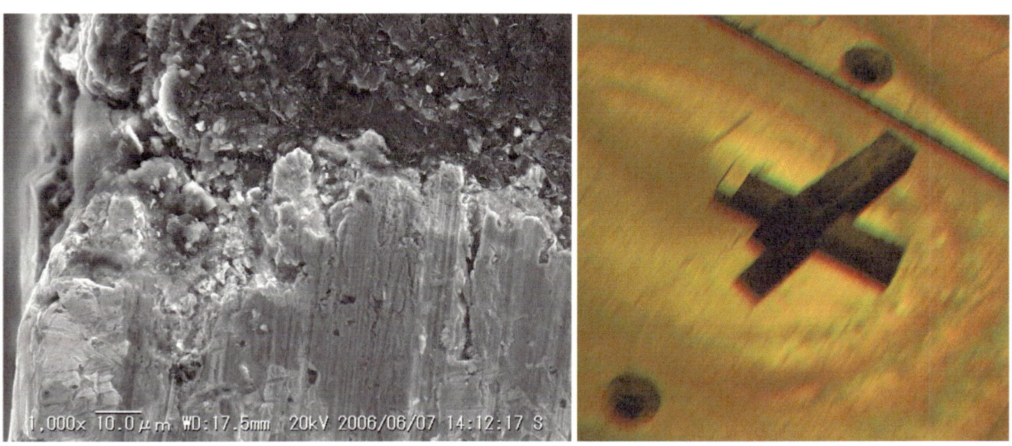

Fig. 23. SEM image (left) of the graphite (top) and titanium interface prior to the 200 MeV proton irradiation and image of the degraded interface following irradiation (right)

The study demonstrated the serious effects that energetic protons at fluences above ~5.0 10^{20} protons/cm^2 have on graphite and thus its interface or bonding with metals. In the fusion reactor, however, composite structures that involve graphite with metal cladding or coating will be exposed to higher fluences of 14 MeV fast neutrons. While for such neutron energies it is anticipated that the cross section of nuclear interaction is similar to that of the 200 MeV protons, confirmatory investigations are necessary. It should be emphasized that nuclear graphite exposed to higher fluences of thermal neutrons (< 1 MeV) than the ones achieved with the 200 MeV protons has shown much greater resilience to irradiation damage. Therefore, to address the potentially different response of such graphite to metal bonding when fast neutrons are the irradiating species (as in the fusion reactor) a radiation damage experiment using the spallation-produced fast neutrons at the BNL isotope facility has been launched. During this irradiation phase fast neutron fluences of ~2.0e+19 n/cm^2 and dominated by energies between 1 MeV and 30 MeV will be achieved. While these levels are far below the fusion reactor fluences anticipated for the graphite/metal composite structures, a comparison between proton and neutron irradiation at similar exposure levels can be made.

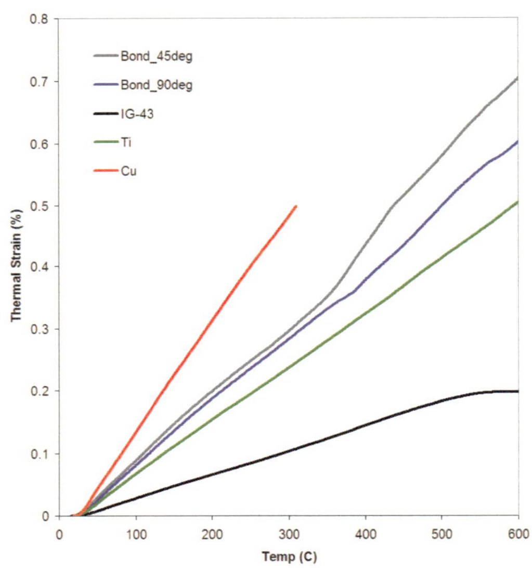

Fig. 24. Thermal strain of the composite structures and of the three constituents all exposed to the same level of proton fluence (~2.0 10^{20} protons/cm^2).

2.3.3 Nanostructured coatings

While nano-structured coatings on metal substrates form a unique class of materials with a wide range of applications, the combined coating/substrate structure can be also characterized as a composite. These structures exhibit similar behaviour at the interface between the substrate and the coating as fibre-reinforced matrices stemming from the mismatch of thermal expansion coefficient which leads to elevated interface stress fields at high temperatures. With recent advances in the techniques and application of nanocoatings on base materials such as thermal spray deposition (Tsakalakos, 2009), interest has increased for their potential use in nuclear reactor systems and in particular in plasma-facing components of fusion reactors.

In fusion reactor environments where the low Z materials of the plasma-facing wall (carbon or beryllium) require protection from erosion, coatings based on tungsten and its alloys have been explored (Koch, 2007). In such setting, however, extreme radiation damage presents an additional challenge for these relatively untested structures which alters the physio-mechanics of the interaction between the substrate and the coating due to the fact that the rate of change of the thermal expansion (CTE) as a function of the radiation fluence may differ significantly between the distinct materials. Therefore, the potential for micro-cracking and even separation between the substrate and the coating under a combination of extreme temperatures and radiation fluxes requires experimental investigation.
Experimental studies focussing on the radiation and extreme temperature effects on alumina (Al_2O_3) and titania (TiO_2) nano-structured coatings applied on Ti-6Al-4V and 4130 steel alloy substrates were launched to assess their susceptibility. Specifically, 200 μm-thick coatings consisting of 87% Al_2O_3 and 13% TiO_2 (grid-blasted) and 600 μm-thick Al_2O_3 (thermally sprayed) on Ti-6Al-4V substrates were used along with 600μm-thick Al_2O_3 and 600 μm-thick amorphous Fe coating on alloy steel 4130 substrates. The behaviour of the interface of the between the substrate and the coating was evaluated for temperatures reaching 1200°C. In addition, the radiation damage from the spallation-based radiation field at BNL BLIP using 116 MeV protons. During irradiation, nano-coated samples received a neutron fluence of ~2.0e+19 n/cm^2 with mean energy of 9 MeV. Combined with the neutron fluence, the coatings received a secondary proton fluence of ~3.2e+15 p/cm^2 of 23 MeV mean energy, a photon fluence of ~3.0 e+19 γ/cm^2 of 1 MeV mean energy and ~2.4e+16 e/cm^2 of 1 MeV mean energy. Of primary interest was the effect of irradiation on the thermo-mechanical behaviour of the structures. Shown in Figure 25 are changes that occur at the interface of 600μm-thick Al_2O_3 on Ti-6Al-4V substrate from room temperature, to 900°C and 1200°C. Demonstrated is the resilience of the composite structure despite the high interface stresses which result in shear failure planes in the substrate (middle image). At higher temperatures (1200°C) the substrate material begins to re-arrange across the shear failure plane. The effect of extreme temperatures on the 600μm-thick Al_2O_3 layer deposited on alloy steel 4130 substrate is quite different as shown in Figure 26. At elevated temperatures (>600°C) an inter-metallic layer begins to form at the interface eventually leading to complete separation of the nano-structured alumina layer at 1200°C.

Fig. 25. Effect of temperature on the interface of nano-structured alumina coating on Ti-6Al-4V substrate from room temperature (left), to 900°C (middle) and 1200°C (right)

The effects of irradiation on the thermal expansion of the coated samples were studied and are depicted in Figures 27 and 28 where they are compared with their un-irradiated counterparts. Observed in the un-irradiated case is that the coating and substrate adjust at certain temperatures to accommodate for the dissimilarities in thermal expansion coefficient. The adjustment which first occurs at a low temperature (~150°C) re-occurs at a higher temperature in a subsequent thermal cycle. This behaviour in which the two dissimilar materials adjust at a particular temperature is very similar to what has been observed near phase transitions in bcc metals such as tungsten. The radiation effects on the stress-strain behaviour based on specially-designed and irradiated specimens with the fluences mentioned above are currently being evaluated.

Fig. 26. Effect of temperature on the interface of nano-structured alumina coating on 4130 steel alloy substrate from room temperature (left), to 900°C (middle) and 1200°C (right)

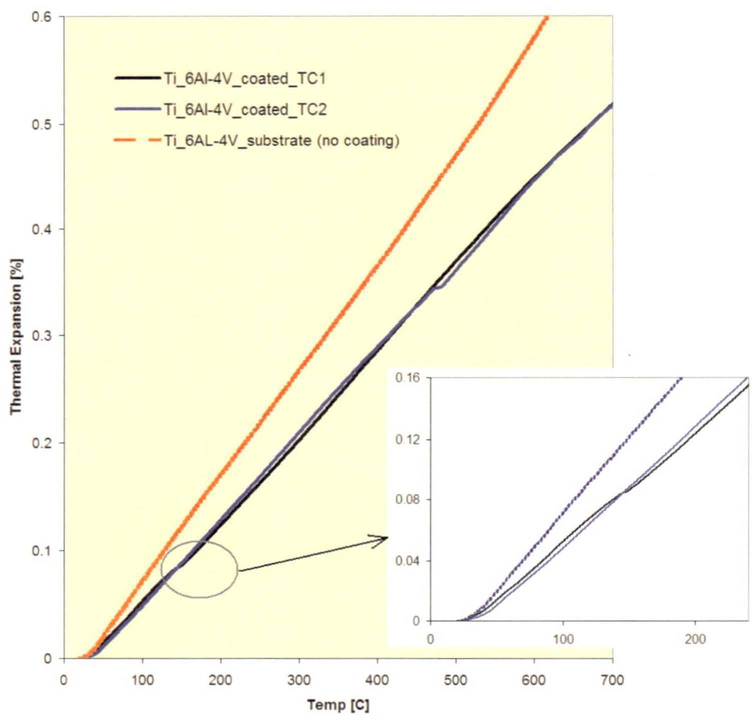

Fig. 27. Thermal expansion of Ti-6Al-4V substrate coated with 200μm-thick Al_2O_3 -TiO_2 layer compared with the thermal expansion of the substrate alone for the same temperature range. Shown is the re-adjustment that takes place in the coated sample at different temperatures in the thermal cycles (TC1=1st thermal cycle, TC2=2nd thermal cycle).

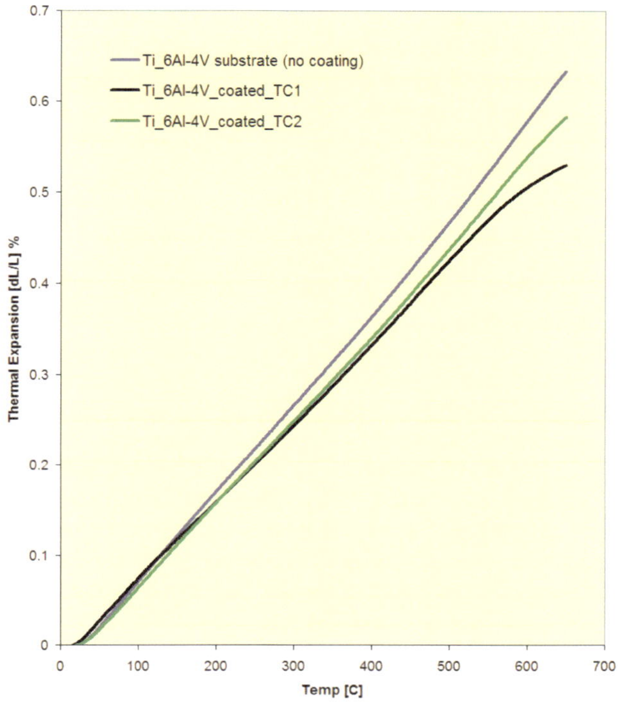

Fig. 28. Thermal expansion of Ti-6Al-4V substrate coated with 200 μm-thick Al_2O_3 -TiO_2 layer compared with the thermal expansion of the substrate alone for the same temperature range following irradiation with neutrons of fluence 2.0 e+19 n/cm^2.

3. Conclusion

With the development of the new generation composites such as C/C, and SiC/SiC as well as special bonds and coatings rapidly advancing and, in the process, performance in extreme environments is better understood and quantified, there is high degree of confidence that these material structures will be able to support the needs of the next generation reactors. A multitude of efforts world-wide have been aiding in closing the knowledge gap on these very promising materials during the last three decades while, with the adoption of novel processing techniques, have made their fabrication at a large-scale feasible. However, due to the harshness of the nuclear environment of the future reactors, consisting of a combination of extremes in temperature and radiation flux, further work is necessary to qualify these composites since the available data are the results of small-scale experimental efforts. The extensive experience on the constituents of these composites from fission reactors may not necessarily provide a good basis to assess the performance of the integrated composite in the elevated nuclear environments. As some of the irradiation experiments using more energetic particles than the thermal neutrons from fission reactors on materials with well understood behaviour (i.e. graphite) showed is that irradiation-induced damage may occur at a faster rate at much lower thresholds. This emphasizes the need to understand and quantify the performance of both the constituents and the final composite structure under prolonged exposure to higher energy neutrons that make up the flux in the next generation fusion and fission reactors.

Irradiation damage studies to-date focussing on the next generation composites and using the available facilities have shown that it is feasible with these new material forms to achieve the performance required through extrapolation. However, the actual conditions in the fusion and next generation fission reactors are expected to be more severe in terms of flux, fluence and temperature. These may result in a much greater spectrum of changes in the physio-mechanical properties of these materials especially in hydrogen and helium formation. The knowledge of the behaviour of these promising composite materials at these levels, either extrapolated or acquired through tests simulating the anticipated conditions, will still be at a small scale. For application in the large-scale of the fusion or fission reactor environment the small-scale must be extrapolated to the realistic size of the components. Therefore a better handle of the scaling must be achieved with the development and implementation of numerical codes. Because of the variability within their structure the numerical models need to consider spatial variation of the properties. Code benchmarking efforts focussed in the specifically on the prediction of the response of these composites will be necessary. Important attributes that make these composites attractive, such as shock resistance, need special attention and further experimental work due to the enormous complexity of the problem associated with fibre-reinforced composites. The effect of irradiation on the degradation of the physical and mechanical properties that control the response to shock absorption, for example, down to the interface between the fibre and the matrix need to be understood so the performance of the bulk composite can be assessed.

4. Acknowledgment

The author gratefully acknowledges the input of Dr. H. Ludewig and Dr. H. Kirk for discussions on the subject and for reviewing and commenting on the manuscript, Dr. L. Snead for providing SiC/SiC samples for the study and Prof. T. Tsakalakos for providing the nanostructured coating samples and for discussions on the subject. The help of A. Kandasamy, Dr. A. Stein and Dr. J. Warren for facilitating the optical microscopy and SEM images is much appreciated and acknowledged.

5. References

Barabash, V., Akiba M., Bonai J., Federici, G., Matera, R., Nakamura, K., Pacher, D., Rodig, M., Vieder, G., & Wu, H. (1998). Carbon fibre composites application in ITER plasma facing components. Journal of Nuclear Materials, Vol. 258-263, P, pp. 149-159

Bonal, J-P., Kohyama K., van der Laan J & Snead L. (2009). Graphite, Ceramics and Ceramic Composites for High-Temperature Nuclear Power Systems. MRS Bulletin, Vol. 34, pp. 28-34

Brossa, F., Franconi E. & Schiller P. (1992). Development of graphite/metals bondings for fusion reactors applications. Journal of Nuclear Materials, Vol. 191-194, pp. 469-472

Burchell, T. (1994). Irradiation-induced structure and property changes in tokamak plasma-facing, carbon-carbon composites. Proceedings of the 39th International Symposium of the Society for the Advancement of Material and Process Engineering and Exhibition, Anaheim, CA, April 1994

Burchell, T. (1996). Radiation Damage in Carbon-Carbon Composites: Structure and Property Effects. Physica Scripta, Vol. T64, pp. 17-25

Burchell, T., Eatherly, P., Robbins, M. & Strizak P. (1992). The effect of neutron irradiation on the structure and properties of carbon-carbon composite materials. Journal of Nuclear Materials, Vol. 191-194 Part 1, pp. 295-299

Fukatomi, M., Fujitsuka, M. & Okada, M. (1985) The development of graphite/metal bonds for fusion reactor applications, Journal of Nuclear Materials, Vol. 133-134, pp. 769-772

Gittus, J.H., (1975). Creep, Viscoelasticity and Creep Fracture in Solids, Halsted Press, ISBN 0-470-30265-8, New York, USA

Goto, K. & Kagawa Y. (1996). Fracture behaviour and toughness of a plane-woven SiC fibre-reinforced SiC matrix composite. Materials Science and Engineering A211, pp. 72-81

Hereil, P.-L., Allix, O. & Gratton, M. (1997). Shock Behaviour of 3D Carbon-Carbon Composite. Journal Phys IV France, Vol. 7, pp. 529-534

Jones, H., Steiner D., Heinisch H.L., Newsome A. & Kerch M. (1997). Radiation resistant ceramic matrix composites. Journal of Nuclear Materials, Vol. 245, pp. 87-107

Katoh, Y., Snead, L., Nozawa, T., Kondo, S. & Busby, T. (2010a). Thermophysical and mechanical of near-stoichiometric fiber CVI SiC/SiC composites after neutron irradiation at elevated temperatures. Journal of Nuclear Materials, Vol. 403, Issues 1-3, pp. 48-61

Katoh, Y. Nozawa, T., Snead, L. & Hinoki T. (2010b). Effect of neutron irradiation on tensile properties of unidirectional silicon carbide composites. Journal of Nuclear Materials, Vol. 367-370, pp. 774-779

Koch, F. & Bolt, H. (2007). Self passivating W-based alloys as plasma facing materials for nuclear fusion. Physica Scripta, Vol. T128, pp. 100-105

Kohyama, A., Hasegawa, A., Noda, T. & Katoh, Y. (n.d.). Present status of SiCf/SiC composites as low-activation structural materials of fusion reactors in Japan. IAEA-F1_CN-69/FTP/37, pp. 2-5

Nikolaenko, A., Karpukhin, I., Kuznetsov, N., Platonov A., Alekseev, M., Chugunov, K., Stronmbakh, I., Baldin, D., Rodchenkov, S., Smirnov, I, Subbotin, V., Khadomirov, E. & Lebedev, G. (1999). Effect of the composition of radiation on the radiation damage to graphite. Atomic Energy, Vol. 87, No. 1, pp. 480-484

Maruyama, N. & Harayama M. (1992). Neutron irradiation effect of thermal conductivity and dimensional change of graphite materials. Journal of Nuclear Materials, Vol. 195, pp. 44-50

Simos, N., Kirk, H., Thieberger, P., Ludewig, H., Weng, W.-T., Trung P.-T., McDonald, K., Sheppard, J., Yoshimura, K. & Hayato, Y. (2006a). Solid Target Studies for Muon Colliders and Neutrino Beams. Nuclear Physics B - Proceedings Supplements, Vol. 155, Issue 1, pp. 288-290

Simos, N., Kirk, H., Thieberger, P., Ludewig, H., Weng, W.-T., Trung P.-T., McDonald, K., Sheppard, J., Evangelakis, G., & Yoshimura, (2005). Target material irradiation studies for high-intensity accelerator beams. Nuclear Physics B - Proceedings Supplements, Vol. 149, pp. 259-261

Simos, N., Kirk, H., Thieberger, P., Ludewig, H., Weng, W.-T., Trung P.-T., McDonald, K., Sheppard, J., Yoshimura, K. & Hayato, Y. (2006a). Solid Target Studies for Muon Colliders and Neutrino Beams. Nuclear Physics B - Proceedings Supplements, Vol. 155, Issue 1, pp. 288-290

Simos, N., Kirk, H., Ludewig, H., Mausner, L., O Conor J., Makimura, S. Yoshimura K., , McDonald, K.,& Trung, P-T. (2006b). Material irradiation damage studies for high power accelerators. Proceedings of EPAC'06, Edinburg, Scotland, June 2006, pp. 1816-1818

Simos, N., Kirk, H.G., Thieberger, P., Ludewig, H., O Conor, J., Mausner, L. Trung P-T., McDonald, K., Yoshimura, K. & Bennett, R. (2008). Irradiation damage studies of high power accelerator materials. Journal of Nuclear Materials, Vol. 377, Part 1, pp. 41-51

Simos, N., Kirk, H., Ludewig, H., Mausner, L., O Conor J., Makimura, S. Yoshimura K., , McDonald, K.,& Trung, P-T. (2006b). Material irradiation damage studies for high power accelerators. Proceedings of EPAC'06, Edinburg, Scotland, June 2006, pp. 1816-1818

Snead, L., Steiner D. & Zinkle S. (1992). Radiation resistant ceramic matrix composites. Journal of Nuclear Materials, Vol. 191-194, pp. 566-570

Snead, L. (2004). Ceramic structural composites. The most advanced structural material. Presentation at the International School on Fusion Reactor Technology, Erice, Italy, July 26 - August 1

Tsakalakos, T. (2009). Nanostructured Coatings & their Characterization under Nuclear Energy System Extremes: Energy Dispersive X-ray Diffraction (EDXRD) Synchrotron Probe. Presentation at Characterization of Advanced Materials under Extreme Environment of the Next Generation Energy Systems Workshop, Brookhaven National Laboratory, USA, September 25-26

Ullmaier, H. & Carsughi F. (1995). Radiation damage problems in high power spallation sources. Nuclear Instruments and Methods in Physics Research B, Vol. 101, pp. 406-421

US Department of Energy (2006), Basic Research Needs for Advanced Nuclear Energy Systems, Available from: www.sc.doe/gov/bes/reports/files/ANES_rpt.pdf

Zhang, Q., Cheng L., Zhang L. & Xu Y (2006). Thermal expansion behaviour of carbon fibre reinforced chemical-vapor-infiltrated silicon carbide composites from room temperature to 1400oC. Materials Letters, Vol. 60, pp. 3245-3247

Zinkel, S. & Ghoniem N.M. (2000). Operating temperature windows for fusion reactor structural materials. Fusion Engineering and Design, Vol. 51-52, pp. 55-71

Zinkle, J. (2004). Advanced Materials for Fusion Technology. Proceedings of the 23rd Symposium on Fusion Technology, Venice, Italy, September 2004

29

Graphite-Composites Alternatives for Electrochemical Biosensor

Ninoska Bojorge[1] and Eliana Alhadeff[2]
[1]*Chemical and Petroleum Engineering Department / Fluminense Federal University*
[2]*Chemical School / Federal University of Rio de Janeiro*
Brazil

1. Introduction

The basic principle of detection of a biosensor is based on the specific interaction between the analyte of interest and the recognition element. As a result of this specific interaction, changes are produced one or several physical-chemical properties (pH, electron transference, heat transfer, change of potential or mass, variation of optical properties, etc.). These changes are detected and can be measured by a specific transductor (Thévenot et al., 2001; Wang, 2004). Electrochemical biosensors are based on the electrochemical species consumed and / or generated during a biochemical interaction process of a biological active substance and analyte. Electrochemical biosensors, especially the amperometrics, have an important position among the biosensors. Since 2000s until the completion of this review, intensive research activity has been devoted to the development of amperometric biosensors. The statistic in ScienceDirect search showed 6950 items found for publication made with pub-date > 1999 and keywords Electrochemical biosensor. Of which about 38% with application to detect ethanol, 47 % for glucose and 15% to phenolic compound.

The electrochemical biosensor usually consists of three phases: receptors phase, transducer phase and a signal amplifier phase. The **receptor phase** incorporates a biological or biomimetic recognition element (nucleic acid, enzyme, anti-body, tissue, organelles or whole cells). The most important phase in an amperometry biosensor is the receptor biomolecule by the selectivity of the device to a given analyte or condition. The **transducer** is the conductive phase, which converts a biochemical signal into a reading or measurement. And the **amplifier** is the computerized apparatus/software using to boosts/readout of signal.

The role of the transducer in a biosensor is to generate a measurable signal of the analyte interacts with the biological molecule associated with the transducer surface. So, in the case of the optical transducer, it generates a signal measured as a light intensity proportional to the concentration of analyte in the sample; this may be an inverse relationship. The composite-films of the surface optical biosensor has an important role in the process in which changes in optical characteristics such as absorption, rotation, refractive index, bio/chemiluminescence, and fluorescence are related to the analyte concentration (Koncki et al., 2001). Martin (2002) showed how the fiber optical immunosensors based on long-period gratings that have limited sensitivity at the refractive index of ordinary aqueous solutions (~1.33). And using composite such as films of titanium dioxide, for example, can raise the local refractive index of the sensor (~1.42), thus increasing sensitivity. Titanium dioxide is commercially available and has been extensively used in the paint industry because of its ability to scatter visible light efficiently.

Electrochemical transducers generate a current or voltage in proportion to the analyte being measured as result of the electron transference; again this may be an inverse relationship. There are numerous components to any biosensor configuration. Through the years, a great many combinations have been proposed and demonstrated, although far fewer have been commercial successes. Figure 1 presents the two most popular platforms for electrochemical biosensor including: carbon paste electrode for discrete or continuous measurements and disposable screen-printed electrode (SPE) for low-cost and single-use field applications.

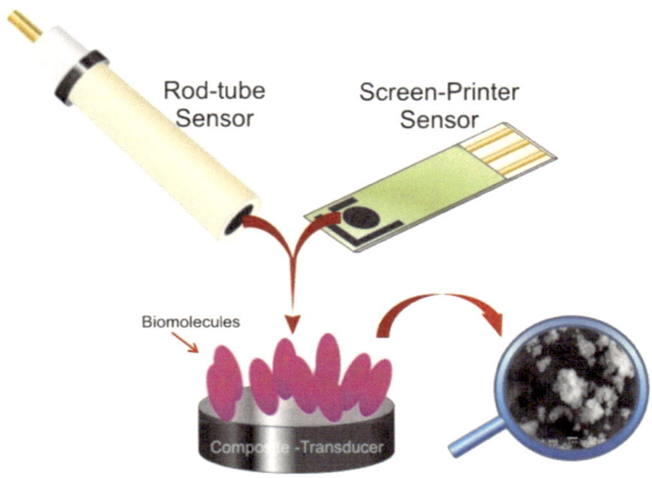

Fig. 1. Electrochemical composite biosensor for analyte detection

In the literature different procedures have been developed to manufacture electrochemical transducers. These conducting composite materials could be considered to be random assemblies of minielectrodes, in which "edge effects" contribute significantly to the Faradic current. Several transducer composite show electrochemical advantages over those built using a single conductive material (platinum, gold, silver, carbon, mercury, graphite, etc.). The conductive parts of the arrays have different sizes and shapes, and it is this randomness that prevents theoretical models based on homogeneously distributed discs, to be used to describe the composite electrode behaviour (Barsan et al., 2009). Using different types of carbon, entrapped in a suitable, normally polymeric, binder, various composite electrodes have been extensively used in electroanalytical measurements.

The desirable characteristics of composite-biosensors usually cited are: i) high versatility in shape and size; ii) lower cost; iii) easy fabrication; iv) higher signal-to-noise ratio; v) surface regeneration; vi) provide suitable mechanic and conducting characteristics, vii) possibility to incorporate other components in the bulk of composite (mediators, cofactors or other bio-molecules) to enhance selectivity or sensitivity, and viii) a long-term stability and lifetime. Conducting composites are interesting alternatives for the construction of electrochemical biosensors. The capability of integrating various materials is one of their main advantages. Thus, the composites used for the construction of biosensors, that have been proposed by several researchers and are used efficiently in different types of analytical determinations of various electroactive species, showing the robustness and sensitivity (Kress-Rogers & Brimelow, 2003; Ahammad et al., 2009; Lojou & Bianco,2006; Bojorge et al., 2009). Several materials have been used to construct electrochemical composite biosensor. Among them, carbon-based matrices are the most applied due to costs and their electronic conductivity.

Aspects related with the protocols for bio-molecules immobilization on the composite support have been widely proposed for many researchers (Caramori & Fernandes, 2004; Ikeda et al., 2002; Shan et al., 2007; Mateo et al., 2000). Difficulties still pose problems such as being reproducible, simple to use, and stable immobilization of the biological component. For example, to obtain a fast response time and a reliable reading a thin layer of immobilized bio-molecule is desirable and shelf life and operational stability demand a high value of immobilized enzymatic activity. Adsorption to adequate surfaces including a metal electrode layer yields relatively unstable systems (Mateo et al., 2006).

The immobilization of bio-molecules on composite surface, the composite supports should be designed to permit a minimum desorption of native proteins, which it is possible through the preparation and selection of arrays that generate for a very strong ionic adsorption of enzymes. In this aspect, Montes et al., 2006 shows as the surface of the enzyme penicillin G acylase was chemically modified under controlled conditions: chemical amination of the protein surface of carboxylic groups (using soluble carbodiimide and ethylendiamine) and chemical succinylation (using succinic anhydride) of amino groups. The full chemical modification produced some negative effects on enzyme stability and activity, although partial modification (mainly succinylation) presented negligible effects on both enzyme features. The chemical amination of the protein surface permitted the immobilization of the enzyme on carboxymethyl and dextran sulphate - coated support, while the chemical succinylation permitted the enzyme immobilization on DEAE-agarose and polyethylenimine coated supports. Immobilization was very strong on these supports, mainly in the polymeric ones, and dependent of the degree of modification, although the enzymes still can be desorbed or inactivated after incubation under drastic conditions.

The immobilization on ionic polymeric beds allows a significant increase in enzyme stability against the inactivation and inhibitory effects of organic solvents, very likely by the promotion of a certain partition of the organic solvent out of the enzyme environment. These results suggest that the enrichment of the surface of proteins with ionic groups may be a good strategy to take advantage of the immobilization of enzymes via ionic exchange on ionic polymeric beds. For example, one should mention the work of Hentze & Antonietti (2002) that describe conventional and modern techniques of porous organic polymers synthesis. A great variety of polymer architectures and functions can be gained by foaming, phase separation, imprinting or templating approaches. Several applications of porous polymers are discussed, focusing on biotechnological and biomedical applications, such as chromatography, protein synthesis, drug carrier systems, tissue engineering and others.

This work describes several approaches for the construction of device and rigid-composites application for detection health and environmental target analyte, such as ethanol and phenol. The first approach is based on a matrix of graphite-epoxy as electrochemical transducer where the immobilization of HRP enzyme for detection of phenol. The second approach is based on matrixes of graphite-epoxy and graphite-Teflon, which are incorporated with HRP/TYR enzymes for the detection of phenol. These bio-composites offer several potential advantages over more traditional to the electrodes based on a modified surface phase driving. The ability to integrate different materials into one is its main advantage, besides the improved electrochemical properties. The different properties of these materials are discussed and compared. The third approach is based on composite Graphite / Epoxy / Pani. Here the polyaniline polymer (PANI) was selected because of its ease of preparation and economic good environmental stability, and conductivity that occur after doping. Several arrays of composite were prepared with different fractions of PANI and were characterized by their morphological, electrical and mechanical. Ahuja et al. (2007) reviewed the state of the art of the use conducting polymers to construct biosensors, studying the different aspects of biomolecules immobilization techniques. Electrochemical biosensors based on polyaniline immobilized with different recognizing biomolecules like as oxide-reduction enzymes, nucleic acids, antibodies, were just constructed and reported (Wei & Ivaska, 2006; Dhand et al., 2011). The term polyaniline is employed to refer a class of polymers with repeated units of 'ring-N' and the base form is [Ping et al., 1997]:

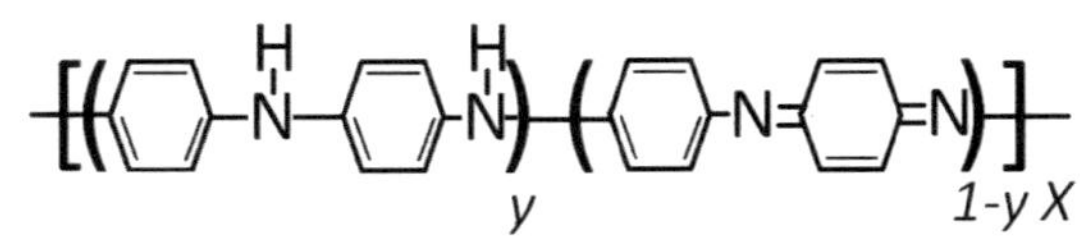

(1)

The value of y in (I) can varied from unity, the completely reduced polymer with amine form for all nitrogen atoms, to zero, which is the completely oxidized form with imines form for all nitrogen atoms. The degree of protonation of the polymeric base depends on its oxidation state and on the pH of the environment. The terms leucoemeraldine (LE), emeraldine (EB) and perniganiline refer to the different oxidation states of the polymer where y = 1, 0.5 and 0, respectively. The emeraldine half-oxidised and protonated form is the conductive polymer (Abdiryim et al., 2005).

2. Materials composite

Composite biosensors are made from two or more constituent materials, conductive material and nonconducting-binding material, with significantly different physical or chemical properties which remain separate and distinct at the macroscopic or microscopic scale within the finished structure. The conductive materials involve the use of nature different of carbon-based matrices, such as: glassy carbon, graphite and nano-carbon and or also by mix of a polymer with electrical conductivity property. They are of highest purity grade and used as delivered from commercial sources (Aldrich, Sigma-Aldrich, Fluka). The main purpose of the conducting phase in the composite is to supply the electrical conductivity needed for the conduction of the electrical biosignal. Another material consists of agglutinating agents (insulator materials), such as: epoxy resins, silicone, polyurethane, metacrylate resin, Teflon, etc. The main purpose of agglutinating agents is mechanical consistence assuring the durability and resistance to the electrolytic working medium.
The conductivity is defined by the connectivity of the conducting particles in the midst of the polymer. When a biocomposite is prepared, it is imperative that its calibration curve is studied to define the composition that guarantees a proper electrical conductivity without losing the mechanical and physical rigidity of the biocomposite. Electrochemical composites can be classified according to how the phases are distributed within the composite matrix. For example, a conducting composite electrode surface can be prepared as an ordered array or as a random arrangement (ensemble) of conducting regions separated by an insulator. The random composite mixtures are classified according to the distribution of the conductor within the composite matrix. If the conductor particles are distributed randomly within the

composite matrix the composite is said to be of the dispersed type. If the conductor extends throughout the composite in a random fashion with regions of pure insulator and pure conductor that do not intermix, the composite is said to be of the consolidated type (Ates & Sarac, 2009).

2.1 Carbon

Carbon is an ideal electrode substrate due to its wide anodic potential range, low residual current, chemical inertness, low cost, fast response time, and ease for fabrication in different configuration and size and suitability for various sensing and detection applications (Wang, 2001). Carbon electrodes allow scans to more negative potentials than platinum or gold, as well as good anodic potential windows. Carbon is observed in several variants such as, glassy carbon, graphite, fullerene, graphene and carbon nanotubes (Ates & Sarac, 2009) .

2.2 Graphite

Graphite is a dark gray, soft and porous material with adsorption capacity and is naturally abundant and highly conductive fillers (with an electrical conductivity of 10^4 S/cm at ambient temperature) for conducting polymer composites (Du et al., 2004) and electrical resistivity is 50 $\mu.\Omega$.m (Sengupta et al. 2010) and its density is 1.95-2.3 g/cm^3. In graphite, the carbon atoms are only bonded in two dimensions. Bond angle in graphite is 120°. Each carbon atom in graphite is sp2 hybridized and with a distance between basal planes of 3.35 Å. Three out of four valence electrons of each carbon atom are used in bond formation with three other carbon atoms while the fourth electron of each carbon atom forms delocalized π-bonds which spreads uniformly over all carbon atoms. The carbon atoms are arranged hexagonally in a planar condensed ring. Also, the layers are stacked parallel to each other, with the atoms within the rings bonded covalently, whereas the layers are loosely bonded together by Van der Waal's forces. The anisotropic nature of graphite is the result of the two types of bonding acting in different crystallographic directions. The ability of graphite to form a solid film lubricant may be attributed to these two contrasting chemical bonds. The weak Van der Waal's forces govern the bonding between the individual layers and also there are no bonds between the layers, the layers can easily slip off one to another, making it an ideal lubricant, and resulting in a reduced coefficient of friction and, hence, wear, and by this same reason, the graphite is a good material for pencils - layers come off and get left on the paper as you write.

On the other hand, given the good electro-catalytic properties, the graphite has been used as an electrode material in the construction of disposable amperometric biosensors for the detection of various analytes such as ethanol, phenol and glucose, and many others (Kirgöz et al., 2006; Llopis et al., 2005; Mailley et al., 2003). By this, the aim of this study was to show the potential use of graphite in biosensor, based on the direct electronic transfers between the enzyme and mediator conducting salt, which are contained in a polymeric matrix of epoxy resin and graphite powder or Teflon and graphite powder. These devices combine the advantages of the biosensors based on solid composites and the electro-catalytic properties of an organic conducting salt, such as PANI.

2.3 Glassy carbon

Another common form of carbon electrode material is the glassy carbon (GC), which is relatively expensive and difficult to build. Glassy carbon also called vitreous carbon, is an advanced material of pure carbon combining glassy and ceramic properties with these of graphite. GC is mechanically hard material, negligible porosity. The higher electrical resistance (0.48 and 0.09 vs. 0.28 and 0.18 Ω-cm at 325 and 725 °C, respectively) and its linear temperature dependence in the annealed sample are attributed to formation of graphite crystalline nuclei in the previously homogeneous and amorphous glassy carbon (Barykin et al., 1976). GC is a class of non-graphitizing carbon that is widely used as an electrode material in electrochemistry and for high-temperature crucibles. It is prepared by subjecting the organic precursors to a series of heat treatments at temperatures up to 3000°C (Kinoshita et al., 1988; Jenkins et al., 1972, 1976). Unlike many non-graphitizing carbons, it is impermeable to gases and chemically extremely inert, especially when prepared at very high temperatures. It has been demonstrated that the rates of oxidation of glassy carbon in oxygen, carbon dioxide, or water vapour are lower than those of any other carbon (Harris, 2004). It is also highly resistant to attack by acids. Thus, whereas normal graphite is reduced to a powder by a mixture of concentrated sulphuric and nitric acids at room temperature, glassy carbon is unaffected by such treatment, even after several months. The structure of glassy carbon has been the subject of research since it was first produced in the early 1960s. However, the properties which make glassy carbon so valuable in these applications are poorly understood, since its detailed atomic structure is not known. A model for the structure of glassy carbon put forward many years ago has gained wide acceptance, but appears to suffer from serious shortcomings. In particular, it fails to account for the chemical inertness of the carbon, and for its high proportion of closed porosity (Harris, 2004).

2.4 Carbon paste

The carbon paste (CP) is used for construction of carbon paste electrodes (CPE´s) for voltammetric determination, which are made usually of high purity graphite powder or another type of carbon, as carbon nanotubes, dispersed in a non-conductive mineral oil such as Nujol or silicone oil to form a paste. Common types of carbon pastes are soft and non-compact, and have to be kept in special bodies. A holder for carbon pastes can be realized as a well drilled into a short Teflon rod (Peng et al., 1993) a glass tube (Pei et al., 1991) or a polyethylene syringe (Švancara et al., 2005) filled with a paste, which is electrically contacted via a conducting wire. Exhaustive reviews on CP-based sensors have been published the last two decades, where various types of biosensors (e.g., those for amino acids, ethanol, fructose, galactose, glucose, glycerol, lactate, xanthine, etc.) based on related oxidases and dehydrogenases, whole cells, and plant tissues are reviewed. A major advantage of CP-based biosensors are very low background current and are the feasibility of bulk modification of the electrode material with biocatalyst as well as with other components essential for their effective functioning. Renewable or disposable surfaces so that each measurement can be performed on the new surface and not be affected by the residuals from the previous measurement (Bard & Rubinstein, 1996).

The constructions of CP biosensor are very simple; however, there is one aspect which makes them not very convenient for practical use and this is the necessity of refilling the carbon paste in experiments requiring a regular removal of the electrode surface layer. Another smart construction circumventing this time-consuming procedure were proposed by Švancara et al., 2005 and Kalcher et al., 2009 who proposed piston-driven electrode holders where the desired amount of the used paste could simply be extruded from the electrode body and smoothed away or cut off.

Another advantage of the CP is that has no need for the sample to have a high electrical conductivity because the conduction is mainly accomplished by the graphite. The currents obtained are within a suitable range due to the small sample amounts. This means that the signal resolution is improved. The use of CP composite biosensor helps in investigating substances in small sample amount and the electrochemical reactions proceed at the surface of the paste electrode. At which, an ion transfers between the solid sample and the electrolyte solution is possible.

The properties of the CP depend on the specific components employed the manner of preparation and maintenance. The properties of CP-based biosensor can be partially improved by the incorporation of additives into the paste, e.g., polyethylenimine, acetylenic polymers, polyaniline, chitosan, glutamate, cationic antibiotics, silica and carbon nanotubes to prevent the leakage of mediator, covalent binding of the mediator to a polymer matrix (Mailley et al., 2003; Anik & Çevik , 2009; Tingry et al., 2006). Recently a novel carbon paste ion selective electrode for determination of trace amount of holmium was prepared by Ganjali M. R. et al., 2009. The authors used multi-walled carbon nanotubes and nanosilica for improvement of a holmium carbon paste sensor response. The approaches to improve the properties of CP mentioned above have certain limitations and deficiencies. On one hand, it is still unclear how additives might affect the biosensor performance at the molecular level, thus making the search for suitable additives for each specific biosensor rather difficult and more random than systematic. On the other hand, covalently bound mediators exhibit modified electrochemical properties and reduced mobility, which affect the reaction rate with enzymes.

The main disadvantages of the CP are the fragile surface or weak mechanical properties due to their creamy texture, which can easily lead to disintegration of the system; and the solubility of the pasting liquid in organic solvents, which often show voltammograms with a higher irreversibility than in cases where no organic binders are used. Therefore, several carbon composite electrodes based on carbon nanotubes and a solid matrix have been proposed. Dues to its insulating nature efficiency in the presence of a solid matrix decreases electrode reaction rates more than a pasting liquid does unless an active carbon surface is exposed (Bard,1996; Kalcher et al., 2006).

2.5 Carbon nanotubes

Carbon nanotubes (CNTs) represent one of the best examples of novel nanostructures derived by bottom-up chemical synthesis approaches. CNTs are molecular-scale tubes of graphitic carbon with outstanding properties. They are among the stiffest and strongest fibres known, and have remarkable electronic properties and many other unique characteristics. CNTs has received a great deal of attention as an electrode material, because these have good electrocatalytic properties (Wang, 2005). Merkoçi, 2006 showed an interesting review paper on CNTs in analytical sciences covers the full calendar from their discovery in 1991 until 2005 and treat analytical aspects of interest in the coupling of CNTs to enzymes, DNA, proteins and, finally, the use of CNTs for several applications such as in chromatography, sensors and biosensors, nanoprobes, etc. Commercial applications have been rather slow to develop, however, primarily because of the high production costs of the best quality nanotubes. For these reasons they have attracted huge academic and industrial interest, with several articles on nanotubes being published the lasts years (Ruoff & Lorents, 1995; Gouveia-Caridade et al., 2008; Yadav et al., 2011).

Carbon nanotubes have received considerable attention in the field of electrochemical sensing, due to their unique structural 1D nature (Javey, 2009), for instance, unique tubular nanostructure, due to their superior mechanical (Schadler et al., 1998) and electronic and chemical properties (Saito et al.,1998), large specific surface, excellent conductivity, superior to 100 Scm^{-1} (Yao et al., 2000), modifiable sidewall, good biocompatibility, and so on. Solubilization and biological functionalization of carbon nanotubes have greatly increased the usage of carbon nanotubes in biomedical applications such as biosensors and nanoprobes (Teker, 2008).

The structure of CNTs is like a sheet of graphite rolled up into a tube of diameter 1-10 nm, and hence forms hollow tubules of a single layer of carbon atoms (Wang, 2005). CNTs include both singlewalled (SWNT or SWCNT) and multiwalled (MWNTS) structures (Baughman et al., 2002). CNTs have very high surface-to-volume ratios and, therefore, promise depending on the direction of hexagons, nanotubes can be classified as either zigzag, armchair or chiral. When scientists make nanotubes, they tend to get a mixture of several types due to different types of nanotubes have different properties. Today, a major challenge in nanoscience is finding a way to make just one type of nanotube. In this aspect, an interesting study made by Safarova et al., 2007, who shows how parameters of SWCNTs, specially a diameter and length of one nanotube or a bundle of nanotubes and a number of nanotubes in the bundle, can be determinated using the techniques as Transmission Electron Microscopy (TEM), Scanning Electron Microscopy (SEM) and Atomic Force Microscopy (AFM).

2.6 Rigid carbon composites

According to Céspedes & Alegret (2000), rigid carbon composites are ideal for the construction of electrochemical sensors. The plastic nature of these materials makes them modifiable, permitting the incorporation of fillers before they are cured. A large number of biological materials can be immobilised by blending them with these composites to form new biocomposite materials, showing robustness and sensitivity. These biocomposites not only act as reservoirs of the biological material but may also contain catalysts, mediators and cofactors that improve the response of the resulting electrochemical biosensors. The carbon-based matrices are the most applied due to their high conductivity, whereas epoxy resins, silicone, polyurethane, metacrylate resin, Teflon, etc, can be employed as agglutinating agents (insulator materials). The main characteristic of these composites is their rigidity, resulting in a high mechanical stability over time. This type of composite biosensor offers additional important advantages: the sensing surface can be renewed by a simple polishing procedure. The proximity of the redox centers of the biological material and the conducting sites on the sensing surface favours the transfer of electrons between electroactive species. So, for example, a universal affinity platform for electrochemical genosensing can be easily achieved by modifying the graphite-epoxy composite with avidin to obtain an avidin biocomposite , where biotinylated DNA can be rapidly single-point attached (Pividori et al., 2005). Thus, this often allows the regeneration of the biological component without using co-substrates and mediators, it can be customized easily according to biocatalytical and/or electrochemical requirements; it presents low background currents, favoring a high signal-to-noise ratio, and lower detection limits and it offers a new active layer after removal of the outer surface, extending the lifetime of such electrodes.

2.6.1 Graphite-epoxy composite

Graphite and epoxy resins are employed to construct rigids composites that can be used in aqueous or nonaqueous media. Serra et al., 2005 showed a comparison study of the behavior of different rigid composite matrices for the construction of amperometric tyrosinase biosensors, widely used for the detection of phenolics compounds. So, they showed that the great advantage of Graphite-epoxy composite (GEC) composite over carbon paste composite is their tolerance to organic solvents, due to a better reproducibility of the amperometric measurements both with and without regeneration of the electrode surface by polishing.
Kırgöz et al., 2006 showed the modification of a GEC electrode with bacterial cells to detection of xenobiotic in waste water samples. Pseudomonas putida DSM 50026 was used as a biological component and the measurement was based on the respiratory activity of the cells. This study the combination of microorganisms with GEC composite electrodes provided economic and practical disposable biosensors.
The physical and chemical effects of water on graphite/epoxy composite were investigated in an interesting work of Zhou and Luke (1995). This paper demonstrates by diffusion data that the time for the onset of non-Fickian behavior is inversely related to the increase in exposure temperature. So, if a relatively low temperature of exposure there is no dissolution of the surface of the material or physical damage and the behaviour profile of weight gain is Fickian. Pointing out that Fick's laws are differential equations that describe the flow of a substance and the concentration versus time and position. And with the temperature increase, cracks, voids, surface peeling and dissolution occur. An explanation for this effect is that the cracks retain water which contributes to the behavior of absorption higher than the theoretical Fickian diffusion curve or the epoxy resins used in manufacturing GEC are capable of undergoing a significant and irreversible thermal oxidative degradation at high temperatures after submission or within a narrow range of its maximum temperature of superior service. When exposed to temperatures high enough to cause degradation of resin, these materials experience a drop in glass transition temperature which effectively increases service temperature and significantly reduces the room-temperature mechanical strength properties of the composite. Below a certain threshold of exposure, these compounds are visually and microscopically damaged, embrittlement and cracking of the surface causes a loss in resistance the impact of the material. Therefore, composite GEC exposed to conditions of overheating may suffer irreversible damage and catastrophic in a very short time period.
From the viewpoint of the effects of electrochemical of the GEC electrode have an anion response close to the Nernstian behavior, 1-3 potential salt and are more sensitive to OH^{-}. This response of the transducer is due to an electroreduction of the adsorbed oxygen on the graphite surface in such way that an increase in the electrode potential takes place due to a change in the oxidation state (Rodriguez-Huerta et al., 2006).

2.7 Carbon/PANI or derivates composites

Li & Shi (2011), published a review about the electrochemical applications of composites prepared with chemically converted graphenes (CCGs) and conducting polymers as polyaniline (PANI), polypirrole (PPy), polythiophene (PTh), to construct sensors and, consequently, biosensors. The electrochemical activity and sensing of the composites prepared with SPANI (acid doped PANI) and CGC were improved and tested with cyclic voltammetric studies. The electropolymerization of polyaniline on the carbon surface was discussed and described by Ates and Sarac (2009) in order to improve the proprieties of the conducting polymers and the application to construct sensors, biosensors, capacitors and batteries. Gómez et al., 2011, reported a chemical precipitation technique to synthesized graphene-polyaniline nanocomposite for electrodes and electroanalysis, and the morphology was characterized by using scanning electron microscopy, transmission electron microscopy, Fourier transform infrared spectroscopy and Raman and cyclic voltammetry. Tung at al., 2011, prepared a nanocomposite with selenious acid, doped polyaniline and graphite nanoplatelet (GNP), and characterized by using SEM, FTIR, X-ray diffraction, and electrical conductivity measurement. A crystalline structure was shown and the electrical conductivity increases with the increase of the GNP content although lower than the HCl doped one. Kim and Park (2011), used multi-walled carbon nanotube and grapheme to prepare a nanocomposite with polyaniline (M-GR/PANI) and higher sensitive CVs were observed when compared with a graphite-polyaniline composite CV. A bienzymatic biosensor was designed for glucose oxidase and horseradish peroxidase based on covalent attachment onto carboxylic-derived multi-walled carbon nanotube for the deposition electroactive polyaniline by Sheng and Zheng (2009), and the linear range for glucose detection was 0.05 mM - 8.0 mM. Zhan et al., 2009, proposed a new strategy for a highly sensitive amperometric biosensor immobilizing tyrosinase on the surface of the polyaniline-ionic liquid-carbon nanofiber composite for phenols detection. A linear range from 4.0×10^{-10} to 2.1×10^{-6} mM was obtained for catechol detection.

3. Methodology for preparation of composite transducers

In this section, the protocols for construction to make the maximum use of composites-transducers are described. Especially the behavior of graphite to transfer electronic device promoted the electrochemical sensor. Rigid matrices based on epoxy/graphite, and the use of thermoplastic resins such as Teflon are very useful tools in building composite electrode for its flexibility in shape and size, allowing easy adaptation to a variety of electrode configurations (conventional flow-through, screenprinted, etc.).

3.1 Preparation of the carbon paste electrodes

Carbon paste electrodes are the most popular electrodes in electroanalytical chemistry and bioelectrochemistry due to that exhibit low background current and are easily prepared. The types of carbon pastes are soft and non-compact, and have to be kept in special bodies. A holder for carbon pastes can be realized as a well drilled into a short Teflon rod, a glass tube or a polyethylene syringe filled with a paste, which is electrically contacted via a conducting wire. Carbon pastes usually employed for the fabrication of CPE are usually prepared as follow: (i) Carbon paste are prepared by mixing graphite powder with Nujol oil or paraffin oil in a mortar until it was uniformly wetted using a graphite/Nujol. The ratio varies for each biosensor proposed in the literature: of 4/1 w/w; 50:50 % m/m; 75:25% m/m. These ratios are employed as it provides convenient analytical properties. (ii) The enzymes are incorporated in the carbon paste using an enzymatic ratio up to 10% w/w. (iii) Sometimes, redoxmediators substances such as quinones and ferrocenes, are incorporated to facilitate electron transfer between the electrode and enzyme employed with ratios of mediators ranging from 1 to 6% w/w. (iv) After blending, the pastes were packed into a Teflon electrode holder (geometric surface area of about 0.07 to 3 cm^2) with electrical wire. (v) The electrode surface was smoothed on a paper to produce a reproducible working surface.

The use of CNTs also has recently been reported to build nanotubes paste electrode (MWCNTPE). Patrascu et al., 2011 showed the use of MWCNTPEs as voltammetric sensors to selectively detect dopamine in the presence of serotonin. Which, it was prepared by thoroughly hand-mixing the multi-wall carbon nanotubes powder with paraffin oil (60:40 w/w) in a mortar. The paste is packed into 1.0 mL polyethylene disposable syringes with a copper wire being used for electrical contact. The surfaces of the electrodes were smoothed by polishing with emery paper.

3.2 Preparation of GEC composites

The epoxy composites filled with graphite are prepared by solution intercalation method, whereby graphite is added to the mixtures of epoxy resin and curing agent. The materials were mechanically stirred, in order for the epoxy resin to intercalate inside the conductive fillers, especially into the graphite interlayers and pores. After the conductive fillers were mixed with epoxy resins, solvents presents in the mixture are evaporated with continuous stirring. The polymer matrix system consists of epoxy resin and a cure agent, and generally all the components are commercial products, and are used as received, without purification.
Epoxy resins, aromatic or aliphatic, have similar ether linkages as that of the Nafion® membranes commonly used in fuel cells. So it is reasonable to select epoxy resins as the polymer matrix in the composite bipolar plates. One of the most popularly used in composites is formed by curing diglycidal ether of bisphenol A (DGEBA). It was used under its commercial designation (Dow Chemical Company; DER 332) whit triethyltetramine (TETA). The basic reaction involves an amine reacting with the epoxy ring. Resulting in simple chain extension and cross-linking. The former reaction generates a new secondary amine that can also react with DGEBA. This is often referred to as branching. The reaction also produces a hydroxyl group that has long been considered as being ideally located to aid (catalyze) the branching reaction.
The criterion of the selection between the type epoxy resin and the cure agent observed in the majority of the publishing paper is in order to adjust the viscosity of the solution during mixing and provide flexibility to the cured products. However, usually the curing processes of these materials need high temperatures, curing at 150°C, 80% of epoxide groups were converted (Merad et al.,2007; Laza et al., 2005), that would be fatal for the biological component. So that, should be reduced the curing temperature and prolong the curing time. These precautions can be extended to other chemical modifiers added to the biocomposite such as redoxmediators, catalysts, etc. that are quite sensitive to wide temperature variations.
Rigid phenyl rings are particularly useful to provide resins with improved heat and thermal resistance and char formation. Undesirably, too rigid backbone structures also drastically reduce the processability of a resin due to viscosity increase, and produce a relatively still but brittle material.
The table 1 show different ratios of the composition of rigid composites based in GEC resins, which isn´t standard. So, when to same resin is used, the curing conditions can vary and should be optimised depending on the nature of the fillers present in the biocomposite. For example, a study proposed by Fatibello et al., 2007 show the effect of graphite powder varying from 30 to 80% (m/m), epoxy resin from 5 to 60% (m/m), and tissue from 5 to 20% (m/m) on the biosensor response using 0.025 mol L^{-1} catechol solution at pH 7. The best composition obtained was in function of the remarkable robustness and sensitivity of the sensor.

Graphite	epoxy	Hardener	Biomolecule	Other	Time cure	Ref
1 part wt	1 part wt	a part wt	-	-	55 °C for 2 h	Wang J. et al., 1989
4 part wt	16 part wt	-	15 mg GOD/g composite	one part wt	35 - 40°C, for a week	Céspede et al., 1993
1 part wt	4 parts wt	-	20mg HRP /g graphite	-	40°C for 4 days	Morales et al., 1996
18%	71%	-	2%	9%	-	Martorell et al., 1997
1 part wt	4 parts wt	-	g 0.9% (w/w) RIgG	-	40 °C for 1 week	Santandreu et al, 1997
10%	40%	40%	5%	5%	72 h at 28 °C	Dutra et al., 2000
1 : 1 (graphite: epoxy-hardener)	1 part	0.4 part	0.15 g	-	60°C for 24 h	Puig-Lleixà et al., 2001
1 part wt	4 part wt	-	150 µl	-	40 °C for 1 week	Pividori et al., 2003
9.5%	76%	-	5% GOD	9.5%	40 °C for 1 week	Llopis et al., 2005
75%	10%	-	15%	-	-	Lupetti et al., 2006
100 mg	60%	40%	10mg IgG-HRP	-	25°C for 96 h	Bojorge et al., 2007
100 mg	70%	30%	-	-	-	Pauliukaite et al. 2009

Table 1. Compositions of Graphite/Epoxy composites Biosensor.

3.3 Preparation of graphite/teflon composites

The same as epoxy resin, Teflon (polytetrafluoroethylene, PTFE) is other the nonconducting binder used to construct rigid composite transducers with electrochemical sensing applications (Peña et al., 2001; Wang & Musameh, 2003]. The main advantage of the incorporation of enzymes into graphite-Teflon composite matrixes is the compatibility of these electrodes with organic or predominantly organic solvents.

The most popular procedure is summarized as: Teflon powder is dissolved in hexane by gentle mixing. Then, the hexane is evaporated under vacuum. Hexane is used to precipitate any asphaltenes presents. Then, an appropriate amount of graphite powder is added to the Teflon and mixed with the enzyme in a mortar for a time to incubate. Finally, the mixture is pressed in into pellets, by means of a Carver pellet press at 10,000 kg cm^{-2} for 10 min. From this mother pellet, several cylindrical portions of each pellet are bored, and each portion is press-fitted into a Teflon holder. Electrical contact is made through a stainless steel screw (Carralero et al., 2006).

4. Experimental section

One of the most important steps of building a biosensor is to immobilize the biomolecules. A successful matrix-transducer-composite should immobilize or integrate biomolecules stably at a transducer surface and efficiently maintain the functionality of the biomolecules, while providing accessibility towards the target analyte and an intimate contact with the transducer surface.

The bases immobilization proposed here are: Enzymes are hydrophobically adsorbed on the supports of high ionic strength. There is a covalent "intermolecular" reaction between the adsorbed absorbed protein and the media. The immobilized protein is incubated at alkaline pH to increase multipoint covalent attachment to stabilize the enzyme. The hydrophobic surface of hydrophylized support is the reaction of other groups of amino acids in order to reduce unfavourable interactions enzyme-support hydrophobic. This strategy produced a significant increase in stability, for example, immobilization of enzymes such as HRP, TYR and AOX, which has been extensively used in determination of analyte-target different such as ethanol and phenol and, in comparison with the stability achieved using conventional protocols. The development of a good biocompatible matrix for the immobilization of biomolecules is very crucial to improve the analytical performance of biosensor. A total of four electrochemical sensors were made using different composites: carbon paste, GEC and Teflon/Graphite, to which different enzymes were added to the bulk mixture (see Figure 2). The composition of the composites were selected among successful combinations in the literature (Bojorge et al., 2007; Carralero et al., 2006), and according to preliminary exploratory experiments in our laboratory.

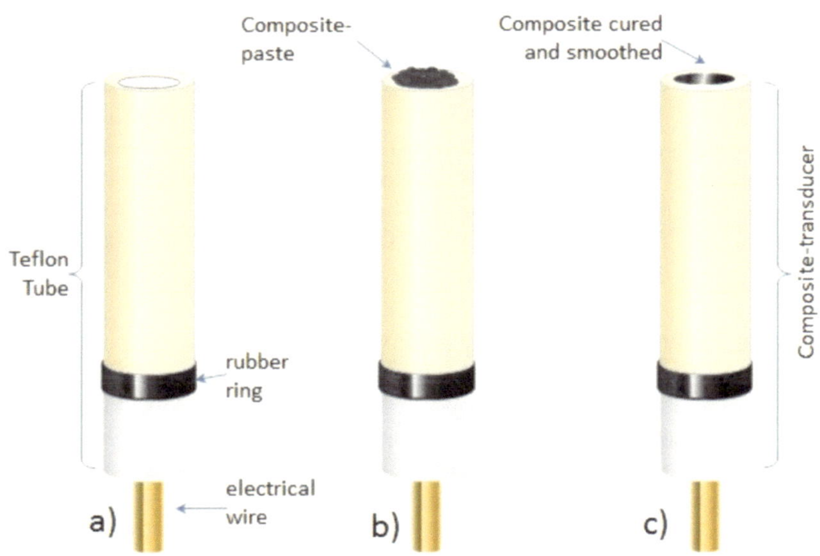

Fig. 2. Rigid-Composite electrodes based on graphite used in the array. a) Initial assembly with electrical contact in a Teflon-tube; b) preparation of the composite paste mixture: an excess of the mixture is placed on the tip of the electrode body during the curing time; c) final aspect after curing and polishing, which is carefully performed to ensure a smooth and flat surface area of 0.015 ± 0.005 cm^2.

4.1 Preparations of graphite composites biosensors

4.1.1 Graphite-epoxy/peroxidase-tyrosinase composite

The bi-enzymatic composite was initially prepared by homogenizing the mass from 0.150 to 0.200 g of graphite powder (Fluka, Cat. No. 50870), previously treated with H_2O_2 [Bojorge et al., 2007] and a specific mass of 0.050g of Tyrosinase (Sigma) and HRP (Toyobo-Brazil) in a mortar for at least 20 min. After that, epoxy resin at a ratio of 1: 4 wt/wt was subsequently added to this mixture in a mortar and mixed for at least 20 min to produce the final biocomposite paste, that was packed (1000 mg) into the tip of depth of 3mm of a cylindrical Teflon sleeve body (1.5 mm I.D) and a copper wire was used to provide the external electric contact. All composites were prepared at room temperature.

4.1.2 Graphite/teflon-peroxidase-tyrosinase composite electrode

Composite enzyme electrodes were fabricated in the form of cylindrical pellets, as follows. Graphite (Fluka, Cat. No. 50870), 0.10 g horseradish peroxidase (E.C. 1.11.1.7, 270 I.U. mg^{-1}; Toyobo-Brazil).), 0.014 g and mushroom Tyrosinase (EC.1.14.18.1, Sigma T3824, Sigma-Aldrich Inc.), were accurately weighed and thoroughly mixed by mechanic stirring for 1 h in a 0.4mL suspension of a 0.1 mol L^{-1} phosphate buffer solution of pH 7.2 at 4 °C. Next, put it desiccator at room temperature to evaporate the water. Next, Teflon powder (Aldrich) was added in at proportion of 70% wt. and mixed thoroughly by hand. The mixture was pressed into pellets by using a Carver pellets press at 10000kg cm^{-2} for 10min. The diameter of these pellets was 1.3 cm and their thickness 0.4 cm, approximately. Several 3.0-mm diameter cylindrical portions of the pellet were bored, and each portion was press-fitted into a Teflon tube. Electrical contact was made through a stainless steel flat-tip screw.

4.1.3 Graphite – PANI /epoxy –peroxidase composite electrode

The powder graphite received a previous treatment with oxygen peroxide and dried in desiccators. The chemical modified graphite was mixed with the emeraldine polyaniline (Sigma) in a proportion of 30:70 (percentage, w/w) and epoxy resin added (60:40, w/w). The paste was used to fill the cylindrical Teflon sleeve body (1.5 mm ID) electrode and rest 2 hours at 60°C to dry. After that, the composite was well polished to immobilize the horseradish peroxidase (Toyobo-Brazil) with glutaraldehyde (2.5%, v/v) as the agent for the covalent linkage.

4.1.4 Graphite-Polyaniline/epoxy-Bi-enzymatic composite electrode

The chemically modified graphite-emeraldine polyaniline composite (30:70, w/w) was mixed with horseradish peroxidase (Toyobo of Brazil) and alcohol oxidase (E.C.1.1.3.13, Sigma-Aldrich Inc.) adjusting the corrects UI/mg composite values. After that, the epoxy resin was added to the composite and the moisture rested for a few minutes at environment temperature for aggregation. The bi-enzymatic immobilized paste was used to fill the cylindrical Teflon electrode (Serra et al., 2003).

4.1.5 Graphite paste - peroxidase electrode

0.3 g graphite powder was added to the equivalent of 0.2 g of Nujol mineral oil (density = 0.838 g/mL) and then was added 0.01 g of HRP dispersing it slowly for about 10 minutes until a homogeneous paste. The resulting paste is inserted into a pipe end polymer (made from the body of a commercial syringe) into the tip of depth of 3mm of a cylindrical Teflon sleeve body (1.5 mm I.D) and a copper wire was used as electrical contact. The surface is gently polished with A4 paper or tissue paper to achieve a smooth flat surface. Rinsed the surface of the electrode with deionized water.

4.2 Electrical resistivity measurements

Electrical resistivity measurements of the all composites pellets based on graphite and were performed according to the ASTM D 257 standard testing method with a Keithley 6517A electrometer (Cleveland, OH, USA) as the source. From these measurements and the geometric dimensions of the sample composites electrodes, both the volume resistivity and surface electrical insulating materials and can be calculated as well as the corresponding conductivities.

4.3 Morphology

Morphology studies of all composites pellets were carried out with SEM. All the composites pellets were mounted directly onto aluminum specimen stubs with two-sided adhesive carbon tabs (Pelco, Redding, CA) and coated with gold for approximately 45 s at 20 lA and 75 mTorr. Samples were analyzed in a JEOL JSM-6460 emission scanning electron microscope dotted with detector electron back-scattered diffraction pattern and system integrated analysis station software used to interface with the SEM and EDS detector operating at an accelerating voltage of 30 KV. Representative micrographs were chosen to illustrate each composite´s pellet.

4.4 Electrochemical technique and procedure

The electrodes were cycled for 3–5 times in distilled water in order to get stable voltammetric responses before performing the measurements with analyte samples. Cyclic Voltammetry measurements were taken using a potentiostat Autolab/PGSTAT12 (Ecochemie, Netherlands, http://www.ecochemie.nl/). The following parameters where fixed: First potential: −1.0 V; Second Potential: 1.0 V; Scan rate: 0.1 V s^{-1} and Step potential: 0.00244 V. All experiments were carried out without any oxygen removal from the sample and with no physical surface regeneration of the working electrodes after each measure.
In order to obtain reliable measurements, and to prevent the accumulative effect of impurities on the working electrode surfaces, an electrochemical cleaning stage was considered between measures. This stage was performed by applying a conditioning potential of +1.5 V for 30 s after each experiment, in a cell containing 5 ml of distilled water.

4.5 Results and discussion

4.5.1 Conductivity characteristics

Chemical compositions of the materials of different composites obtained, the variation of the electrical conductivities are presented in Figure 3. To single filler epoxy composite as a function of the graphite loading concentration the electrical conductivity of graphite/epoxy resin increase gradually with increasing graphite filler. This phenomenon shows that the graphite acts as the transfer medium for electrons hence the electrical conductivity of the composite would be increased. However, the electrical conductivity of the graphite composite is still low, reaching only 0.0078 S/cm at loading concentration of 80 w/w %. This suggests that composites filled with graphite only are not suitable for the achievement of high electrical conductivity. A low resistivity is desirable when high currents are flowing through the electrode body in order to avoid the iR drop of the applied potential. However, when the composite are constituted of graphite/Teflon the performance of conductivity is different. The electrode resistivity was lower when working with high ratios of graphite, but an inverse tendency for mechanical strength was observed. This means that when graphite is fully intercalated with Teflon, the material is a rather poor electrical conductor , so, the change in Teflon composition fraction could create variations in conductivity probably due to an induces defects in the graphite structure with the intercalation with Teflon. In relation to the conductivity of the PANI- Graphite/epoxy composite (GEC-PANI), it can be see that the conductivity of composites increased with increasing graphite content. The conductivity found for neat polyaniline was 0.00249 Scm^{-1}. However after addition of 1, 2 and 3 w/w % of graphite the electrical conductivity of composites increased slightly and for pellets prepared using 10-30 w/w % of graphite the electrical conductivity of composites increased significantly showing a maximum rate conductivity of 28 μScm^{-1} for GEC-PANI 30 w/w % sample as shown in Figure 3, indicating that, for the electrical properties, the PANI-Graphite/epoxy 30 w/w% present the better compromise between filler and polymeric matrix.

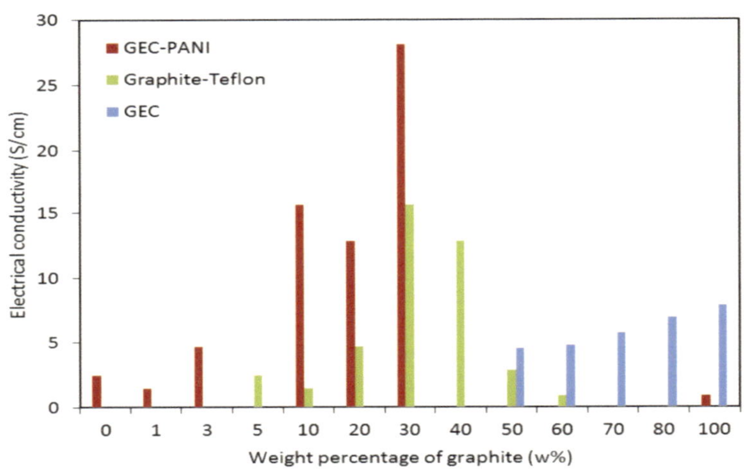

Fig. 3. Electrical conductivity of composites in variation graphite composition

4.5.2 SEM analysis

The SEM is a very good technique for investigating the morphology of composites based on graphite, due to provide information about the structure, size and distribution of the graphite particles in the matrix of the composite. Figure 4 shows scanning electron micrographs of carbon paste, GEC, Graphite-PTFE and GCE-PANI composites pre-treated as described in the experimental section. As can be seen, carbon paste composite made of graphite powder mixed and Nujol oil (Figure 4a) the electrode surface has a rough surface morphology and is heterogeneous with lighter areas, associated with the Nujol oil, and darker areas corresponding to the graphite conducting micro-structures. Figure 4b shows the SEM of GEC surface, which is less rough than CP. However, cluster appears also, which it appears agglomerated in random areas. This is due to the graphite particles randomly distributed and randomly oriented in the epoxy resin. Another important characteristic of the graphite-epoxy composite is the rather flat fracture surface indicating the nature of brittle fracture. Figure 4c show SEM image of the worn surface for Graphite-Teflon, SEM examinations of Graphite-Teflon show slightly ribbed i.e., less creation and development of the cracks occurred on the surfaces. This can be seen with the naked eye during the preparation, whose surface is brighter and sharper with Teflon than epoxy, because developing a uniform transfer layer on the composite-surface. Figure 4d shows SEM image for graphite-PANI-epoxy that revealed also the uniform morphology of the structures in which the outer layers were PANI and the inner layer consisted of graphite. Such a uniform morphology is desirable because it enables a material with high ionic conductivity to achieve fast charge/discharge rates. In the present study, however, the morphology of PANI was strictly granular and it coalesced, making the surface rough with no uniformity.

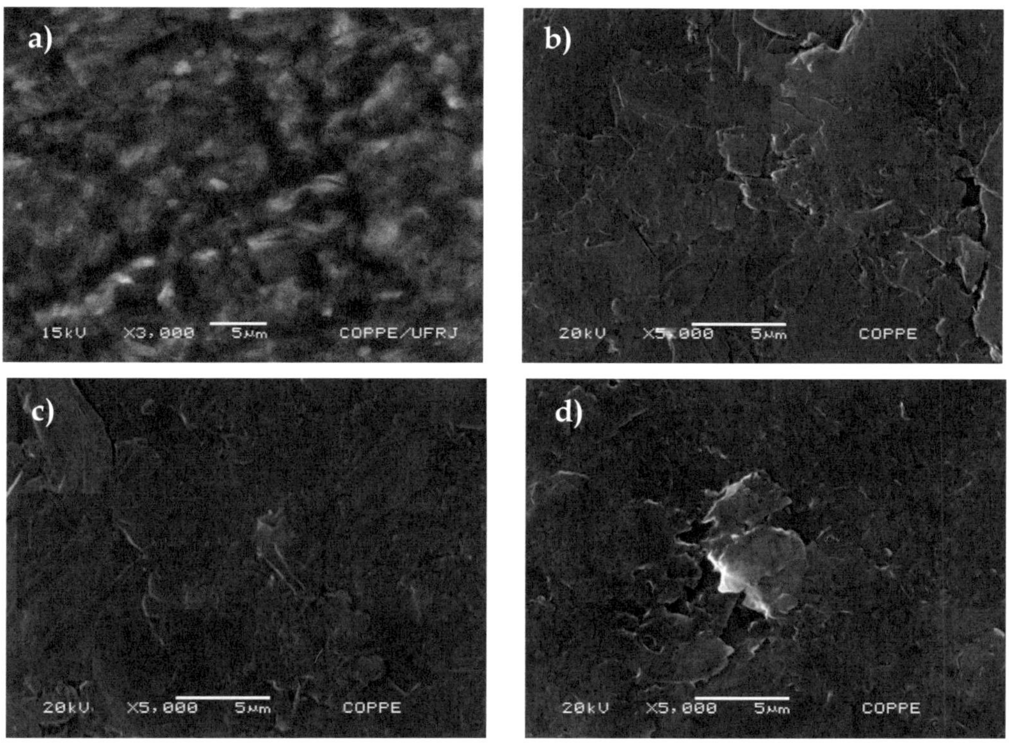

Fig. 4. Electronic microscopic images of composites: a) GP (Graphite Paste) electrode; b) GEC (graphite epoxy composite) electrode; c) GEC (Graphite Teflon composite) electrode and d) GEC-PANI. All electrode surfaces were polished as explained in the text. The same accelerated voltage (20 KV) and resolution (5μm) were used.

4.5.3 Electrochemical properties

Figure 5 shows the cyclic voltammetric behaviors of different composites based on graphite with potential sweep rates of 100 and 50 mV/s between -0.1 and 1 V in solution containing 1 mM $K_4[Fe(CN)_6]$ and were recorded on the five cycle of a repetitive potential sweep program. As it can be seen, the voltammograms that show the well-defined redox peaks is the GEC, which it is related to the intrinsic characterization of graphite. However, for the voltammogram of the PANI-graphite composites show a nearly rectangular-shaped voltammogram, typical of capacitive systems. This characteristic is attributed mainly to the enhancement of the conductivity due to the graphite particles added to the polyaniline chains. The anodic peak, occurred at potential of ~ 0.2 V, is connected to doping of chloride anions indicating transition of leucoemeraldine form of PANI to emeraldine salt. Further increase in potential, above ~ 0.50 V, refers to transition of emeraldine salt to perningraniline salt. Therefore, it can be observed that the electrochemical kinetics of redox transition of PANi-graphite composites has been performed.

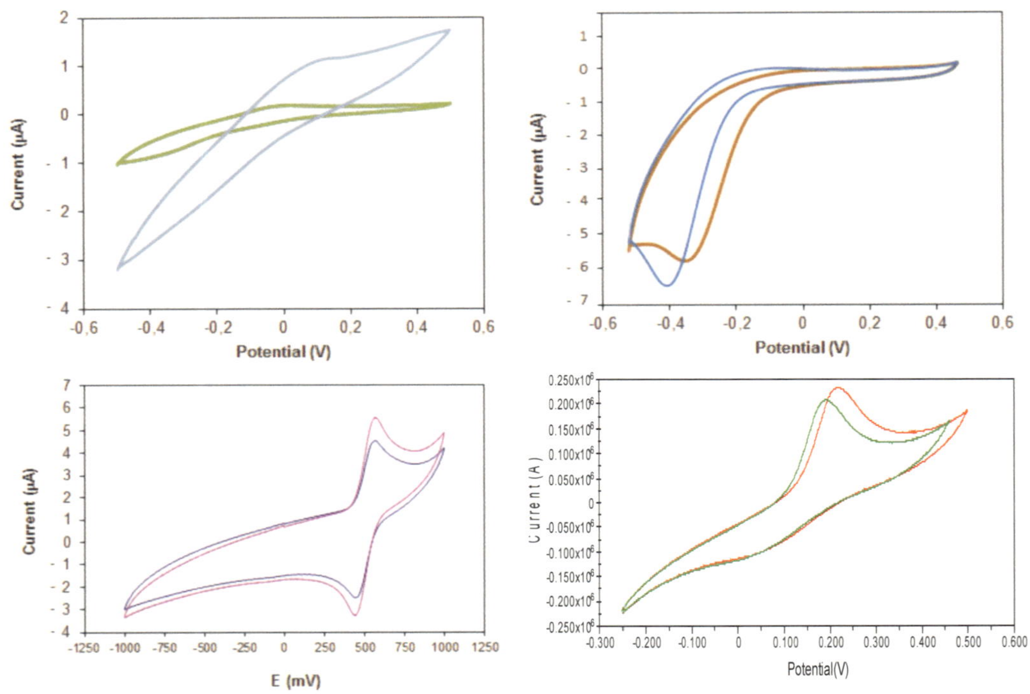

Fig. 5. Cyclic voltammograms for 10 mM potassium ferricyanide employing the composites sensors: a) using GP composite, b) using Graphite-Teflon composite, c) Graphite-Epoxy composite, d) Polyaniline Graphite-Epoxy composite. Scan rate 100 mV/s

In agreement with this, the reaction kinetics of the electroactive species in concentrated solutions may be slowed due to the slow mobility of the molecules. In the Figure 5d, the degradation of the PANI-perningraniline salt electrode almost was not observed for that potential region. The possible explanation could be to the fact that some forms of PANI are potential dependent and pH dependent [Tawde et al., 2001] or the fractal dimensions dependent [Ghanbari et al., 2006] evaluated from the cyclic voltammetry. Hence, at such low pH, extent of the degradation products was negligible and practically had no influence on charge/discharge characteristics of the PANI electrode during initial cyclization. However, this work is still in progress, in further studies it would be beneficial to investigate the influence of pH and anions on degradation of PANI during the cyclization. Though, attention must be paid to each measurement made with a composite film as accurate reproducibility of the arrays and polymeric composites is very difficult. The results indicated that the electrochemical methods could be used as a simple tool for analyzing the structure of conducting polymers and their composites.

5. Conclusion

The electrochemical performance of the composite electrodes is strongly affected by the presence of the insulator matrix of the polymeric binder, which coats the graphite particles, at the same time increasing electrode resistivity which is influenced by the distance between the particles. Since it is known that a low rate of evaporation of the solvents provides the smoothest surfaces, the material was dried at low temperatures of ≈20 to 30°C.
Different morphologies are expected for different composite compositions. We can conclude that the nature of the surface film plays a key role for the electrode stability or to capacity decrease always related to an increase of the electrode itself impedance

6. Acknowledgment

The Authors thanks for the resources received from Fundação de Amparo à Pesquisa do Estado do Rio de Janeiro (Proc. No. E26/150.997/2006) and Conselho Nacional de Pesquisa e Desenvolvimento - CNPq (Edital Universal, Process number 476654/2008-4). and in special to Professor Belkis Valdman, our inspiration, and to Students João Paulo Barros Guimarães Mendes, Erica Ferreira Southgate, Leonardo Ivar Gomes Jaldin, Gleice Santos Lima Magalhães by theirs participations in the LABSENS by the Scientific initiation program - PIBIC - UFRJ, to which we encourage to continue in this way.

7. References

Abdiryim T., Xiao-Gang Z., Jamal R., (2005). Comparative studies of solid-state synthesized polyaniline doped with inorganic acids, *Materials Chemistry and Physics*, 90, 367-372.

Ahammad S. A. J. , Lee J-J and Aminur Rahman Md., (2009). Electrochemical Sensors Based on Carbon Nanotubes, *Sensors*, 9, 2289-2319.

Ahuja T., Mir A. I., Kumar D., Rajesh, (2007). Biomolecular immobilization on conducting polymers for biosensing applications, *Biomaterials*, 28, 791-805.

Anik Ü., Çevik S., (2009). Double-walled carbon nanotube based carbon paste electrode as xanthine biosensor, *Microchim Acta*, 166: 209–213.

Arai, T. & Kragic, D. (1999). Variability of Wind and Wind Power, In: Wind Power, S.M. Muyeen, (Ed.), 289-321, Scyio, ISBN 978-953-7619-81-7, Vukovar, Croatia.

Ates M., Sarac A. S., (2009). Conducting polymer coated carbon surfaces and biosensor applications, *Progress in Organic Coatings*, 66, 337-358.

Bard AJ, Rubinstein I. (eds.), *Electroanalytical chemistry: a series of advances,* (1996). 19. Marcel Dekker Inc.; New York, 5.

Barsan M.M, Pinto E. M., Florescu M., Brett C.M.A., (2009). Development and characterization of a new conducting carbon composite electrode, *Analytica Chimica Acta* , 635, 71–78.

Barykin, B. M.; Spiridonov, E. G.; Tarabanov, A. S., (1976). Electrical conductivity of glass-carbon, In NASA STI/Recon Technical Report N, - adsabs.harvard.edu, 116-119.

Baughman, R. H., Zakhidov, A. A., de Heer, W. A. (2002); Carbon Nanotubes—The Route Toward Applications. *Science*, 297, 787–792.

Bojorge Ramírez N.I., Salgado A.M e Valdman B., (2007). Amperometric Immunosensor for Detecting Schistosoma mansoni Antibody, *ASSAY and Drug Development Technologies*. October 2007, 5(5): 673-682.

Bojorge Ramírez N.I., Salgado A.M e Valdman B., (2009). The Evolution and Developments of Immunosensors for Health and Environmental Monitoring: Problems and Perspectives, Vol. 26, No. 02, pp. 227 - 249.

Caramori S. S., Fernandes K. F., (2004). Covalent immobilisation of horseradish peroxidase onto poly (ethylene terephthalate)-poly (aniline) composite, *Process Biochemistry*, 39, 883–888.

Carralero V., Mena M.L., Gonzalez-Cortés A. , Yáñez-Sedeño, Pingarrón J.M., (2006). Development of a high analytical performance-tyrosinase biosensor based on a composite graphite–Teflon electrode modified with gold nanoparticles, *Biosensors and Bioelectronics*, 22, 730–736.

Céspedes F. and Alegret S., (2000). New materials for electrochemical sensing II. Rigid carbon polymer biocomposites, *Trends in Analytical Chemistry*, 19, 4, 276-285.

Céspedes F., Martinez-Fabregas E., Alegret S., (1993). Amperometric glucose biosensor based on an electrocatalytically bulk-modified epoxy-graphite biocomposite, *Analytica Chimica Acta*, Volume 284, Issue 1, Pages 21-26.

Dhand C., Das M ., Datta M., Malhotra B.D., (2011). Recent advances in polyaniline based biosensors, *Biosensors and Bioelectronics*, 26, 2811-2821.

Du, X. S., Xiao, M. and Meng, Y. Z. (2004). Synthesis and characterization of polyaniline/graphite conducting nanocomposites. *Journal of Polymer Science Part B: Polymer Physics*, 42: 1972–1978. doi: 10.1002/polb.20102

Dutra R.F., Coelho G.D., Silva V.L., Ledingham W.M. & Lima Filho J.L., (2000). A reusable amperometric biosensor based on a novel silver-epoxy electrode for immunoglobulin detection, *Biotechnology Letters* 22: 579–583,

Kress-Rogers E., Brimelow C. J. B. (2003). *The rapid detection of pesticides in food. In Instrumentation and sensors for the food industry;* Ibtizam E. Tothill , Woodhead Publishing Limited, Cambridge, pp. 55-73.

Fatibello Filho, O. ; Lupetti, K. O. ; Leite, O. D. ; Vieira, I C . (2007). Electrochemical biosensors based on vegetable tissues and crude extracts for environmental, food and pharmaceutical analysis. Part 1-Fundamentals and Applications. In: Salvador Alegret & Arben Merkoçi. (Org.). *Electrochemical Sensor Analysis*. 1 ed. Amsterdam: Elsevier, 49, 355-375.

Ganjali R., Motakef-Kazemi N., Norouzi P. and Kho S., A Modified Ho3+ (2009). Carbon Paste Electrode Based on Multi-walled Carbon Nanotubes (MWCNTs) and Nanosilica Mohammad, *International Journal Electrochemical Science*., 4 , 906 - 913.

Ghanbari Kh., Mousavi M.F., Shamsipur M., Rahmanifar M.S., Heli H., (2006). Change in morphology of polyaniline/graphite composite: A fractal dimension approach, Synthetic Metals, 156 (14-15), pp.911-916, ISSN 0379-6779, DOI: 10.1016.

Gómez H., Ram M. K., Alvi F., Villalba P., Stefanakos E., Kumar A., (2011). Graphene-condunting polymer nanocomposite as novel electrode for supercapacitores, *Journal of Power Sources*, 196, 4102-4108.

Gouveia-Caridade C., Pauliukaite R., Brett C.M.A., (2008). Development of electrochemical oxidase biosensors based on carbon nanotube-modified carbon film electrodes for glucose and ethanol, *Electrochimica Acta* 53 6732–6739.

Harris, P. J. F., (2004). Fullerene-related structure of commercial glassy carbons, *Philosophical Magazine*, 84 (29), 3159–3167.

Hentze H-P.and Antonietti M., (2002). Porous polymers and resins for biotechnological and biomedical applications, *Reviews in Molecular Biotechnology* , 90 (1), 27-53.

Ikeda Y., Kurokawa Y., Nakane K. and Ogata N., (2002). Entrap-immobilization of biocatalysts on cellulose acetate-inorganic composite gel fiber using a gel formation ofcellulose acetate–metal (Ti, Zr) alkoxide, *Cellulose*, 9, 369–379.

Javey A., Carbon nanotube Field-Effect Transistors in: Carbon nanotube electronics, Series on Integrated Circuits and Systems, Javey A., Kong J., (Eds), Springer, 2009 –ISBN 978-0-387-36833-7, p.82

Jenkins, G. M. and Kawamura, K., (1976). *Polymeric carbons – Carbon Fibre, Glass and Char*, Cambridge University Press.

Jenkins, G. M., Kawamura, K. and Ban, L. L., (1972). Formation and structure of polymeric carbons. *Proc. Roy. Soc.*, A327, 501.

Kalcher K., Švancara I., Metelka R., Vytřas K., Walcarius A., (2006). *Heterogeneous Carbon Electrochemical Sensors, in Encyclopedia of Sensors*, C. A. Grimes, E. C. Dickey, M. V. Pishko, Eds., American Scientific Publishers, Stevenson Ranch, 4, 283.

Kim K., Park S., (2011). Influence of multi-walled carbon nanotubes on the electrochemical performance of grapheme nanocomposites for supercapacitors electrodes, *Electrochimica Acta*, 56, 1629-1635.

Kinoshita, K., *Carbon: Electrochemical and Physicochemical Properties*, 1988, Wiley : New York

Kirgöz Ü. A., Odaci D., Timur S., Merkoc A., Alegret S., Besün N., Telefoncu A., (2006). A biosensor based on graphite/epoxy composite electrode for aspartame and ethanol detection,. *Analytica Chimica Acta* , 570, 165–169

Koncki, R. Lenarczuk T., Radomska A. and Glab S., (2001). Optical biosensors based on Prussian Blue films, *Analyst*, 126, 1080–1085.

Li C., Shi G., (2011). Synthesis and electrochemical applications of the composites of conducting polymers polymers and chemically converted graphene, *Electrochimica Acta,* in Press.

Laza J. M., Vilas J. L., Garay M. T. ,Rodriguez M. , León L. M., (2005). Dynamic Mechanical Properties of Epoxy-Phenolic Mixtures, *Journal of Polymer Science Part B: Polymer Physics,*Volume 43, Issue 12, Article first published online: 29 Apr

Lee H.S., Kim H.J., Kim S.G. , Ahn S.H., (2007) Evaluation of graphite composite bipolar plate for PEM (proton exchange membrane) fuel cell: Electrical, mechanical, and molding properties, *Journal of Materials Processing Technology*, 187–188, 425–428.

Li C., Shi G., Synthesis and electrochemical applications of the composites of conducting polymers and chemically converted graphene, *Electrochimica Acta*, 2011, in Press.

Li, B.; Xu, Y. & Choi, J. (1996). Applying Machine Learning Techniques, Proceedings of ASME 2010 4th International Conference on Energy Sustainability, pp. 14-17, ISBN 842-6508-23-3, Phoenix, Arizona, USA, May 17-22, 2010

Lima, P.; Bonarini, A. & Mataric, M. (2004). Application of Machine Learning, InTech, ISBN 978-953-7619-34-3, Vienna, Austria

Llopis X., Merkoci A., del Valle M., Alegret S., (2005). Integration of a glucose biosensor based on an epoxy-graphite-TTF[middle dot]TCNQ-GOD biocomposite into a FIA system, *Sensors and Actuators B: Chemical*, Volume 107, Issue 2, 29, Pages 742-748

Lojou É , Bianco P. , (2006). Application of the electrochemical concepts and techniques to amperometric biosensor devices, *J Electroceram*, 16, 79–91.(DOI 10.1007/s10832-006-2365-9)

Lupetti K. O., Zanotto-Neto G. and Fatibello-Filho O., (2006). Sweet Potato Tissue-Epoxy Resin Composite Biosensor for Hydroquinone Determination in Photographic Process Wastewater, *Journal of Brazilian Chemical Society*, Vol. 17, No. 7, 1329-1333,

Mailley P., Cummings E. A., Mailley S. C., Eggins B. R., McAdams E., Cosnier S., (2003). Composite Carbon Paste Biosensor for Phenolic Derivatives Based on in Situ Electrogenerated Polypyrrole Binder, *Analytical Chemistry*, 75, 5422-5428.

Martin J. E., (2002). Composite films for modifying evanescent wave characteristics in long-period grating biosensors, Thesis (M.S.) -Virginia Polytechnic Institute and State University, (http://scholar.lib.vt.edu/theses/available/etd-02172001-014116/unrestricted/JEMartinETD.pdf).

Martorell D., Cespedes F., Martinez-Fabregas E., Alegret S., (1997). Determination of organophosphorus and carbamate pesticides using a biosensor based on a polishable, 7,7,8,%tetracyanoquinodimethane-modified, graphite-epoxy biocomposite, *Analytica Chimica Acta*, 337, 305-3 I3

Mateo C., Abian O., Fernandez-Lafuente R., Guisan J. M., (2000). Reversible enzyme immobilization via a very strong and nondistorting ionic adsorption on support-polyethylenimine composite, *Biotechnology and Bioengineering*., 68, (1), 98 - 105.

Mateo C., B. Pessela C. C., Fuentes M., Torres R., Ortiz C., López-Gallego F., Betancor L., Alonso-Morales N., Guisan J. M., R. Fernandez-Lafuente, (2006).Very Strong But Reversible Immobilization of Enzymes on Supports Coated With Ionic Polymers, In: Immobilization of Enzymes and Cells - Methods in Biotechnology, José M. Guisán ,2nd Ed., Humana Press,

Mehrvar M. and Abdi M., (2004). Recent Development, Characteristics and Potential Applications of Electrochemical Biosensors, *Analytical Sciences*, 20, 1113 - 1126.

Merkoçi A. , Carbon Nanotubes in Analytical Sciences, *Microchim Acta*, (2006). 152, 157–174.

Merad L., CochezM. , Margueron S., JauchemF. , Ferriol M. , Benyoucef B. and Bourson P., In-Situ Monitoring of the Curing of Epoxy Resins by Raman Spectroscopy, DOI 10.1051/names2007029 155. Article available at http://names.edpsciences.org or http://dx.doi.org/10.1051/names2007029

Montes T., Grazu V., López-Gallego F., Hermoso J.A., Guisán J. M. and Fernández-Lafuente R., (2006) Chemical modification of protein surfaces to improve their reversible enzyme immobilization on ionic exchangers, *Biomacromolecules*, 7 (11), 3052–3058.

Morales A., Cespedes F., Munoz J., Martinez-Fabregas E., Alegret S., Hydrogen peroxide amperometric biosensor based on a peroxidase-graphite-epoxy biocomposite, *Analytica Chimica Acta*, Volume 332, Issues 2-3, 21 October 1996, Pages 131-138

Pauliukaite R., Ghica M. E, Fatibello-Filho O.,. Brett C. M. A, (2009). Comparative Study of Different Cross-Linking Agents for the Immobilization of Functionalized Carbon Nanotubes within a Chitosan Film Supported on a Graphite−Epoxy Composite Electrode, *Analytical Chemistry*, 81 (13), 5364-5372.

Patrascu D., David I., David V., Mihailciuc C., Stamatin I., Ciurea J., Nagy L., Nagy G., Ciucu A.A., (2011). Selective voltammetric determination of electroactive, neuromodulating species in biological samples using iron(II) phthalocyanine modified muli-wall carbon nanotubes paste electrode., In Press, Accepted Manuscript, *Sensors and Actuators B*.

Pei J. Jin Q., Zhong J., (1991). Potentiometric determination of trace silver based on the use of a carbon paste electrode, *Talanta*, 38, 1185-1189.

Peng T. Z., Li H. P., S. Wang W., (1993). Selective extraction and voltammetric determination of gold at a chemically modified carbon paste electrode, *Analyst* , 118, 1321.

Peña N., Ruiz G., A. Reviejo J. and Pingarrón J.M., (2001). Graphite-Teflon Composite Bienzyme Electrodes for the Determination of Cholesterol in Reversed Micelles. Application to Food Samples, *Analytical Chemistry*, 73, 1190-1195.

Pierson, Hugh O., (1993) Graphite structure and Properties In: Handbook of Carbon, Graphite, Diamonds and Fullerenes, Noyes Publications, 43-68.

Ping, Z., Nauer, G., Neugebauer, H., Theiner, J., Neckel, A., (1997). Protonation and electrochemical redox doping process of polyaniline in aqueous solutions: Investigations using in situ FTIR-ATR spectroscopy and a new doping system, *Journal of Chemical Society, Faraday Transform*, 93, 1, 121-129.

Pividori M. I., Alegret S., (2005). Electrochemical Genosensing Based on Rigid Carbon Composites. A Review , *Analytical Letters*, 1532-236X, 38, 15, 2541 - 2565

Pividori M. I., Merkoci A., Alegret S., Graphite-epoxy composites as a new transducing material for electrochemical genosensing, *Biosensors and Bioelectronics*, Volume 19, Issue 5, 30 December 2003, Pages 473-484,

Puig-Lleixà C. , Jiménez C. and Bartrolí, J., (2001). Acrylated polyurethane — photopolymeric membrane for amperometric glucose biosensor construction, *Sensors and Actuators B: Chemical*, Volume 72, Issue 1, 5, Pages 56-62

Rodriguez-Huerta L.A., Galán-Vidal C.A.; Álvarez Romero G.A, and Páez-Hernández M.E., (2006). Potentiometric behavior of graphite-epoxy electrochemical transducers towards anions, cations and pH in aqueous media, *Rev. Mex. Fís.* S, 52, (2) 17–19.

Ruoff R. S., Lorents D. C., (1995). Mechanical and Thermal Properties of Carbon Nanotubes, *Carbon*, 33, 925.

Safarova K., Dvorak A., Kubinek R. , Vujtek M., Rek A., (2007). Usage of AFM, SEM and TEM for the research of carbon nanotubes, Modern Research and Educational Topics in Microscopy, A. Méndez-Vilas and J. Díaz (Eds.), 513-519.

Saito R., Dresselhaus G., Dresselhaus M. S., *Physical Properties of Carbon Nanotubes*, Imperial College Press, 1998.

Santandreu M. , Céspedes F., Alegret S., and Martínez-Fábregas E., (1997). Amperometric Immunosensors Based on Rigid Conducting Immunocomposites , *Analytical Chemistry*, 69, 2080-2085.

Sengupta R., Bhattacharya M., Bandyopadhyay S., Bhowmick A. K., A review on the mechanical and electrical properties of graphite and modified graphite reinforced polymer composites, Progress in Polymer Science, In Press, Corrected Proof, Available online 19 November 2010, ISSN 0079-6700, DOI: 10.1016/j.progpolymsci. 2010.11.003

Schadler L. S., Giannaris S. C., and Ajayana P. M., (1998). Load transfer in carbon nanotube epoxy composites, *Applied Physics Letters*, Volume 73, Number 26

Serra, B., Reviejo, A.J., Pingarrón, J.M., (2003). Composite Multienzyme Amperometric Biosensors for an Improved Detection of Phenolic Compounds. *Electroanalysis*, 15, 22, 1737-1744.

Shan D., Yao W. and Xue H., (2007). Electrochemical study of ferrocenemethanol-modified layered double hydroxides composite matrix: Application to glucose amperometric biosensor, *Biosensors and Bioelectr*onics., 23, (3), 432-437.

Sheng Q., Zheng J., (2009). Bienzyme system for the biocatalyzed deposition of polyaniline templated by multiwalled carbon nanotubes : A biosensor design, *Biosensors and Bioelectronics*, 24, 1621-1628.

Siegwart, R. (2001). Indirect Manipulation of a Sphere on a Flat Disk Using Force Information. *International Journal of Advanced Robotic Systems*, Vol.6, No.4, (December 2009), pp. 12-16, ISSN 1729-8806

Švancara I., Metelka R. and Vytřas K., (2005). Piston-Driven Carbon Paste Electrode Holders for Electrochemical Measurements, in Sensing in Electroanalysis, K. Vytřas, K. Kalcher, Eds., University of Pardubice, Pardubice, , pp. 7.

Tawde S., Mukesh D., Yakhmi J. V., (2001). Redox behavior of polyaniline as influenced by aromatic sulphonate anions: cyclic voltammetry and molecular modeling, Synthetic Metals, 125,(3), 401-413, ISSN 0379-6779, ISSN 0379-6779, DOI: 10.1016/S0379-6779(01)00483-0.(http://www.sciencedirect.com/science/article/B6TY7-44KM0ST-P/2/0326413824dc09086b82037c2197231a).

Teker K., (2008). Confocal Microscopy of Bioconjugated Carbon Nanotubes for Biosensor Applications, *Sensors & Transducers Journal* , 88, (2), ISSN 1726-5479

Yadav S, Devi R., Kumari S., Pundir C.S., (2011). An amperometric oxalate biosensor based on sorghum oxalate oxidase bound carboxylated multiwalled carbon nanotubes–polyaniline composite film, *Journal of Biotechnology*, Volume 151, Issue 2, Pages 212-217

Thévenot D. R.; Toth K.; R. Durst A., Wilson G S., (2001). Electrochemical biosensors: recommended definitions and classification, *Biosensors and Bioelectronics*, 16, 1-2, 121-131.

Tingry S., Innocent C., Touil S., Deratani A., Seta P., (2006) Carbon paste biosensor for phenol detection of impregnated tissue: modification of selectivity by using [beta]-cyclodextrin-containing PVA membrane, *Materials Science and Engineering: C*, 26, 2-3, 222-226.

Tung N. T., Khai T. V., Lee H., Sohn D., (2011). The effects of dopant on morphology formation on polyaniline graphyte nanoplatelet composite, *Synthetic Metals*, 161, 177-182.

Van der Linden, S. (June 2010). Integrating Wind Turbine Generators (WTG's) with Energy Storage, In: Wind Power, 17.06.2010, Available from http://sciyo.com/articles/show/title/wind-power-integrating-wind-turbine-generators-wtg-s-with-energy-storage. Zhou J. and Lucas J. P., The effects of a water environment on anomalous absorption behavior in graphite/epoxy composites, Composites Science and Technology 53 (1995) 57-64.

Vytřas K., Švancara I. and Metelka R., (2009). Carbon paste electrodes in electroanalytical chemistry, *J. Serb. Chem. Soc.*, 74 (10) 1021–1033.

Wang J. *Analytical Electrochemistry*, 2nd Ed.; Wiley-VCH, New York, US, 2001, 113-114

Wang J. & Musameh M., (2003). Carbon Nanotube/Teflon Composite Electrochemical *Sensors and Biosensors, Analytical Chemistry*, 75, 2075-2079.

Wang J., Electrochemical Sensors for environmental monitoring: A review of recent technology, A Note from the U.S. Environmental Protection Agency (EPA), 2004.

Wang J., (2005). Carbon-nanotube Based Electrochemical Biosensors: Review, *Electroanalysis*, 17, 1, 7-14.

Wang J., Golden T., Varughese K. I, and El-Rayes I., (1989). Polishable and Robust Modified Graphite Epoxy Electrodes, *Analytical Chemistry*, 6 1 , 508-512

Wei, D., Ivaska, A., (2006). Electrochemical Biosensors Based on Polyaniline, *Chemical Analitycal* (Warsaw), 51, 839-852.

Yao Z., Kane C. L., Dekker C., (2000). High-Field Electrical Transport in Single-Wall Carbon Nanotubes, *Physics Review Lett*ers , 84, 2941-2944.

Zhang J., Lei J., Liu Y., Zhao J., Ju H., (2009). Highly sensitive amperometric biosensors for phenols based on polyaniline-ionic liquid-carbon nanofiber composite, *Biosensors and Bioelectronics*, 24, 1858-1863.

30

Composite Cathode Material for Li-Ion Batteries Based on $LiFePO_4$ System

Janina Molenda[1] and Marcin Molenda[2]
[1]*AGH University of Science and Technology*
Faculty of Energy and Fuels, Krakow
[2]*Jagiellonian University, Faculty of Chemistry,Krakow*
Poland

1. Introduction

Huge development electric vehicles market and storage energy in renewable energy systems forces usage of batteries characterized by high volumetric and gravimetric density of stored energy, exhibiting large number of charge/discharge cycles, being safe for user as well as of low environmental impact and low cost. The widespread technology of reversible Ni-Cd and NiMH cells reached the theoretical limit of available improvements. Promising for the future and still undergoing rapid development is the technology of reversible lithium cells, commonly known as *Li-ion batteries*. The required capacity of a Li-ion battery for vehicle applications and renewable energy systems is much bigger than for portable electronics – 20-100 kWh. The security of usage of that kind of batteries is an important issue. The bigger the capacity, the more energy is accumulated – therefore, more strict security measures must apply. This raises numerous challenges to develop new material technologies – cell components with a better chemical and thermal stability, as well as to solve problems such as heat dissipation dependent on the cell housing system. The lack of experimental data on real lifetimes of Li-ion batteries in changeable climatic conditions – from +40°C to -40°C (the required lifetime of batteries is ten years for vehicles and 1-4 years for laptops) constitutes also a big problem.

2. How does a Li-ion battery function and how can it be improved?

Li-ion batteries are based on capability of transition metals compounds M_aX_b (M- transition metal; X = O, S) with layered or tunnel structure to reversibly insert lithium (one or more mol Li per mol M_aX_b) at room temperature without significant changes in their crystallographic structure (*intercalation process*) (Whittingham, 1978; Tarascon & Armand, 2001; Ohzuku, 1993). Fig.1 presents different type structures capable to intercalation process. In this process, the basic elements of the structure do not undergo any changes, except for minor reversible variations of lattice parameters. Stability of structure during the whole process is due to strong ionic and covalent bonds between M and X atoms.
The intercalation of lithium (which is always a combined ionic- electronic transport process, involving insertion of Li^+ ions and an equivalent number of electrons) to transition metal compounds M_aX_b can be put down as follows:

$$xLi^{+} + x\,e^{-} + M_aX_b = Li_xM_aX_b \qquad (1)$$

This reaction makes use of deep *d-type* energy levels in transition metal compounds, with the energy of several eV/atom, which can accumulate the energy of several kWh/kg; thus enabling construction of a power supply with significant volumetric and gravimetric energy density.

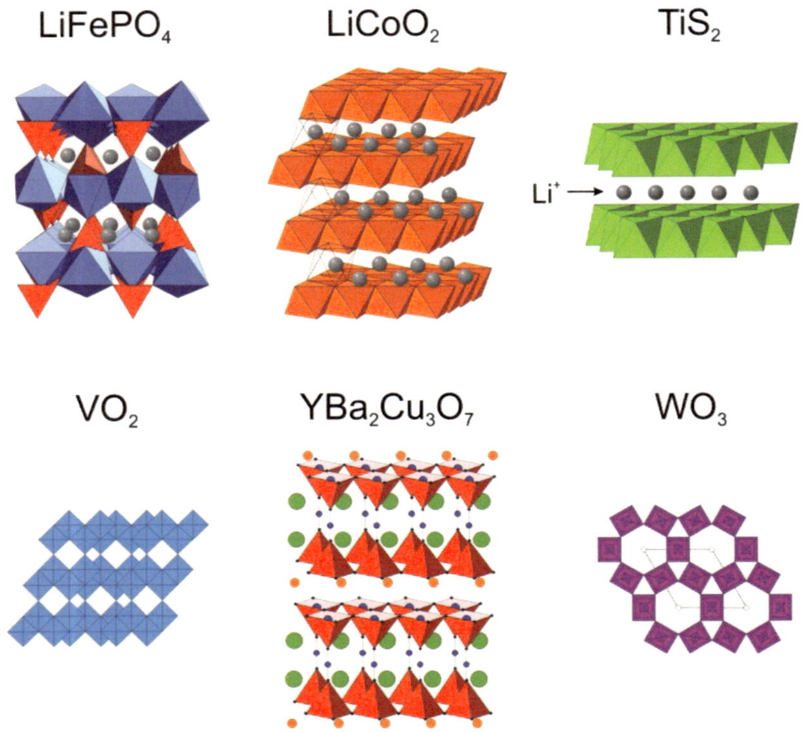

Fig. 1. Different type structures capable to lithium intercalation.

Numerous works on cathode materials by J. Molenda: Li_xTiS_2 (Than et al., 1991), Li_xVO_2 (J. Molenda & Kubik, 1993), Li_xCoO_2 (J. Molenda & Stokłosa, 1989), Li_xNiO_2 (J. Molenda et al., 2002), Li_xWO_3 (J. Molenda & Kubik, 1989), $Li_xYBa_2Cu_3O_7$ (J. Molenda et al., 1993), $Li_xMn_2O_4$ (J. Molenda et al., 2000; J. Molenda et al., 2004), Li-graphite (J. Molenda, 1997), Li_xFePO_4 (J. Molenda et al., 2006) prove the key role of electronic structure on the intercalation process. Variations of the EMF of the $Li/Li^{+}/Li_xM_aX_b$ cell corresponds to those of electrochemical potential of electrons (Fermi level) of the cathode material brought about by lithium intercalation (J. Molenda et al., 2005). The electrons introduced during intercalation process (with equivalent number of Li^{+} ions) take available electronic levels and rise the position of the Fermi level in a way dependant of the density of states function (DOS). A high density of states at the Fermi level results in a weak composition dependence of the electromotive force, what is advantageous from the application point of view. The lithium chemical diffusion coefficient, which determines the current density is a function of the mobility of lithium ions and electrons in the cathode materials. In the layered and skeleton structures (Fig.1), paths of rapid diffusion exist and they ensure sufficiently high mobility of lithium ions, which does not limit the efficiency of the intercalation process. However, the localization of electronic states at the Fermi level, often observed in transition metal compounds, leads to a kinetic limitation of the intercalation process.

The first lithium battery technology i.e. $Li/Li^+/Li_xTiS_2$ cell technology was rapidly withdrawn from the market in the beginning of 70-ties of the XX-th century due to formation of lithium dendrites which short-circuit the cell. Twenty years later, in 1991, a new generation of lithium batteries, i.e. Li-ion batteries ($Li_xC_6/Li^+/Li_{1-x}CoO_2$), was commercialized by Sony Cor. The metallic lithium anode was replaced with graphite, which has the ability to reversibly intercalate lithium and has a reasonably low potential versus lithium. Charging and discharging is related to a reversible „pumping" of lithium ions from one electrode to another (subsequent, reversible intercalation and deintercalation processes). Fig.2 presents the working mechanism of Li-ion batteries.

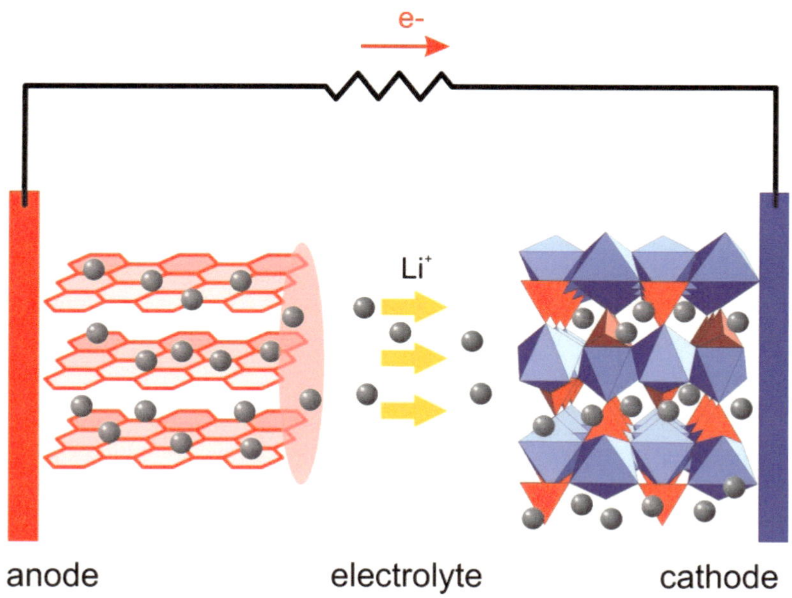

Fig. 2. Working mechanism of Li-ion batteries.

The effectiveness of the intercalation process in $Li_xC_6/Li^+/Li_{1-x}M_aX_b$ type cell, which at the graphite anode can be written as:

$$C_6 + xLi^+ + xe^- \leftrightarrow Li_xC_6 \tag{2}$$

and analogously at the cathode

$$Li_1M_aX_b \leftrightarrow Li_{1-x}M_aX_b + xLi^+ + xe^- \tag{3}$$

is determined by ionic-electronic transport properties of both electrode materials, number of sites available for lithium ions Li^+ and density of available electronic states around the Fermi level in both electrode materials. Li-ion batteries $Li_xC_6/Li^+/Li_{1-x}M_aX_b$ principle parameter, i.e. energy density – per unit mass or volume, dependent on the cell electromotive force and its capacity, is defined by electronic and crystallographic structure of both electrode materials in relation to lithium intercalation reaction. Current density of the cell depends on ionic-electronic transport properties in both electrode materials (ambipolar diffusion). Consequently, the cell voltage, capacity, energy density and current density are defined by properties of the cathode and anode materials. The number of charge and discharge cycles and cell lifetime are significantly conditioned by processes taking place on electrode material/electrolyte interfaces. Cell safety depends on thermal and chemical stabilities of electrode materials and electrolyte. It has been shown (J. Molenda, 1997), that graphite anode does not limit Li-ion batteries operational parameters such as current density or voltage. Favourable structure properties of graphite (layered structure) and its high electrical conductivity together with delocalisation of electronic states at the Fermi level ensure a high efficiency of ionic-electronic processes in the graphite electrode. **Therefore the only way for substantial improvement of Li-batteries parameters is to improve and upgrade the cathode materials.**

Presently in the Li-ion batteries technology, beside $LiCoO_2$, there are also used as cathode materials $LiCo_{1-y}Ni_yO_2$ and $LiMn_2O_4$. However these cathode materials exhibit some practical faults. In case of $LiCoO_2$ only half of its theoretical capacity can be used. This is due to reversible intercalation/deintercalation process only within Li_1CoO_2 - $Li_{0.5}CoO_2$ range, what yields this low reversible capacity of 130 mAh g^{-1} (J. Molenda & Stokłosa, 1989). Moreover the $LiCoO_2$ is not environmental friendly and expensive. $LiNiO_2$ reveals higher reversible capacity (190 mAh g^{-1}) than $LiCoO_2$, however there are some basic difficulties in obtaining ordered structure of $LiNiO_2$ due to strong cation mixing (Li- Ni) effect in this compound, what substantially worsens its transport and electrochemical properties (J. Molenda et al., 2002). Moreover at high deintercalation degree an exothermic reaction with liquid organic electrolyte takes place. Manganese spinel $LiMn_2O_4$ exhibits phase transition at room temperature and limited stability, what decrease cyclability of the battery (J. Molenda et al., 2000; J. Molenda et al., 2004).

3. Cathode materials based on iron oxides

A cathode material for Li-ion technology based on iron oxides was always a desired object. The interest in iron compounds arises from the fact that iron is cheap, abundant in the earth crust and friendlier for the environment than cobalt, nickel or manganese. Unfortunately the first taken-intoconsideration iron-containing, layered compound $LiFeO_2$, isostructural with $LiCoO_2$ and $LiNiO_2$, is found to be metastable. Generally, stability of ABO_2 oxides with the layered structures of α-$NaFeO_2$ (R3m) type can be estimated from Pauling's rule, according to which the r_B/r_A ratio should be less than 0.86. In the case of $LiCoO_2$ and $LiNiO_2$, this ratio is 0.77 and 0.78, respectively. For $LiFeO_2$ $rFe^{3+}/rLi^+ = 0.88$, and the structure is unstable. Another important reason, which excludes $LiFeO_2$ from the application in batteries, is disadvantageous position of redox potentials of iron in relation to those of lithium (Fig.3).

The Fe^{3+}/Fe^{4+} potential is too distant from that of Li/Li^+ and located beyond the electrochemical window of the electrolyte, which cannot guarantee the neutrality of the electrolyte versus the cathode. On the other hand, the Fe^{3+}/Fe^{2+} potential is too close to that of Li/Li^+, which results in a too low voltage of the cell. Such behavior is related to the high spin configuration of Fe^{3+} and strong interactions between the d electrons. The problems of structural instability and unfavorable position of redox potentials of iron versus lithium can be overcome by using new series of iron compounds, e.g. $LiFeXO_4$, as proposed by Goodenough (Goodenough, 1998) with large polyanions $(XO_4)^{y-}$ (X =S, P, As, Mo, W; y = 2 or 3) that might stabilize the structure. The presence of such $(XO_4)^{y-}$polyanions with a strong covalent bond X–O stabilizes the anti-bonding state of Fe^{3+}/Fe^{2+}. The Fe–O bond becomes less covalent due to the induction effect in the Fe–O–X system, which raises the electrode potential in $LiFePO_4$ (Fig. 3).

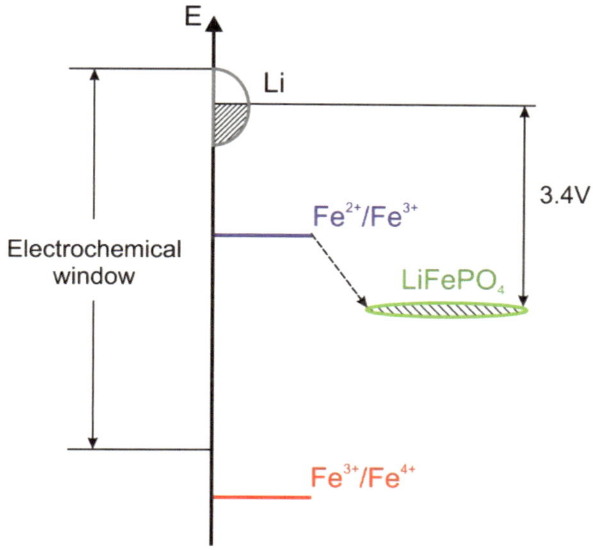

Fig. 3. Energy diagram showing the Fe^{3+}/Fe^{4+} and Fe^{2+}/Fe^{3+} potentials in the cathode materials based on iron in octahedral coordination.

4. Phospho-olivine $LiFePO_4$

The olivine structured $LiFePO_4$ belongs to a family of super ionic conductors called NaSICON, known as fast ionic conductors and used as solid electrolytes in electrochemical cells. In $LiFePO_4$, the hexagonal close-packed lattice of oxygen has one dimensional channels which act as potential fast diffusion paths for the lithium ions (Fig.4).
However, due to the specific crystal structure (FeO_6 octahedra linked via corners lead to a significant, over 4Å, Fe-Fe distance, while the M-M distance for conductive oxides is below 3Å) phospho-olivine is practically an electronic insulator. Its electrical conductivity at room temperature equals to 10^{-10}S $cm^{-1}$is extremely low as for a cathode material (Fig.5).
Low electrical conductivity is a reason for which the observed $LiFePO_4$ delithiation is not really a diffusional deintercalation process described by the following equation:

$$LiFePO_4 - xLi+ - xe- \leftrightarrow Li_{1-x}FePO_4 \quad (4)$$

that yields a homogeneous material with varying lithium content. Literature reports (Yang et al., 2002) and our own studies indicate that lithium extraction from $LiFePO_4$ in the charge cycle consists of decomposition of the cathode material into two phases of which one contains lithium and the other is lithium-free:

$$LiFePO_4 - xLi+ - xe- = xFePO_4 + (1 - x)LiFePO_4 \quad (5)$$

and a similar reverse reaction (discharging cycle):

$$FePO_4 + xLi+ + xe- = xLiFePO_4 + (1 - x)FePO_4 \quad (6)$$

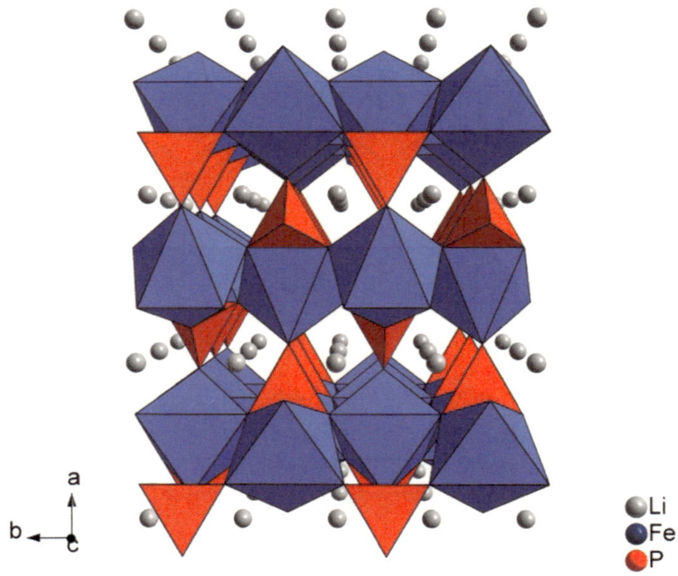

Fig. 4. Structure of $LiFePO_4$.

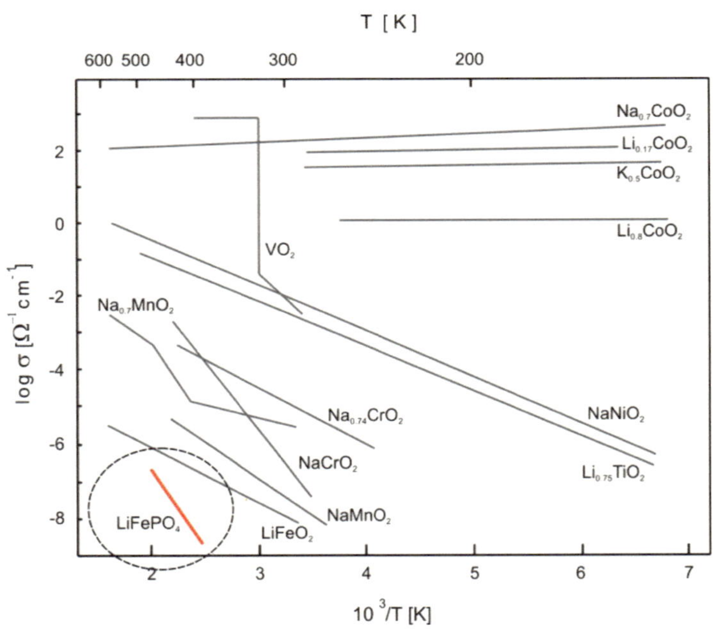

Fig. 5. Temperature dependence of electrical conductivity for different cathode materials in comparison to $LiFePO_4$.

Observed high reversibility of the charge/discharge processes is related to vast similarity of the $LiFePO_4$ and $FePO_4$ structures (the same space group and volume difference of only 6.81%). The described behavior of the cathode material is not at all beneficial – only the grain surfaces of the cathode are effectively used and current density of the cell is low. The reason for the reactions (5) and (6) to occur is low electronic conductivity.

On the other hand, $LiFePO_4$ shows some important advantages: the highest theoretical capacity of all known cathode materials (170mAh g^{-1}), the highest thermal stability, which guarantees safety of use and stable capacity after numerous work cycles. Therefore the most challenging issue in the search for a cathode material based on phospho-olivine is to get a mixed ionic–electronic conductivity, which should activate the diffusional mechanism of deintercalation/ intercalation process (i.e. proceed according to reaction (4)).

Much hope for commercial Li-ion batteries was associated with Chiang's revolutionary report on phospho-olivine doping (Chung et al., 2002), indicating a possibility of electrical conductivity increase by a factor of 10^7. However, examination of doped phospho-olivine surfaces by J. Molenda et al. (Marzec et al., 2006; Ojczyk et al. 2007) and the results published by Nazar (Herle et al., 2004) have proved that the high values of conductivity of doped phospho-olivine are not due to bulk metallic properties but due to the formation of metallic iron phosphides on the surface of phospho-olivine grains, which are the effect of partial reduction of $LiFePO_4$ to Fe_2P during the synthesis (Fig. 6). Formation of Fe_2P showing metallic type conductivity on the surface of the $LiFePO_4$ grains in the course of phospho-olivine synthesis, give rise to an effective technology for producing composite $LiFePO_4$–Fe_2P cathode materials, exhibiting remarkably improved properties.

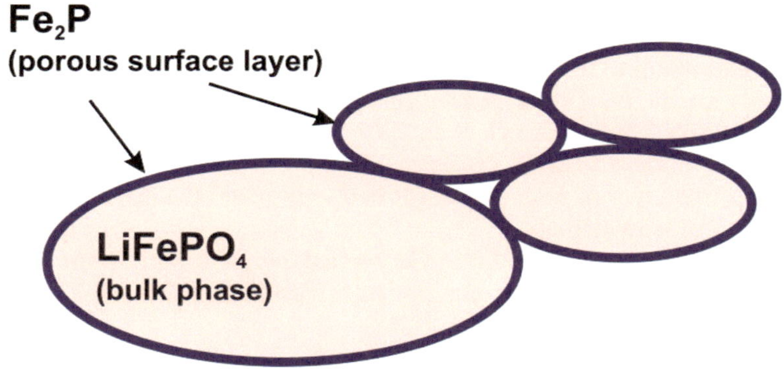

Fig. 6. Composite $LiFePO_4$–Fe_2P cathode material.

5. Preparation of a composite phase $LiFePO_4$–Fe_2P

For a successful preparation of the Fe_2P films on the phospho-olivine grains, the presence of a highly reducing agent is required. Such a reducing agent should be formed during the synthesis process, being an intermediate product of the reaction between $Fe_2C_2O_4$ xH_2O, $NH_4H_2PO_4$ and Li_2CO_3 (Fe, $Fe_x(CO)_y$, C, Co and NH_3). The analysis of chemical properties of the particular substrates indicates that the reducing agent can be possibly formed as a product of a $FeC_2O_4 \cdot 2H_2O$ thermal decomposition reaction. However, the results of a thermal analysis (MS-TGA/DTG) carried over the thermal decomposition of $FeC_2O_4 \cdot 2H_2O$ has shown that the mechanism of the decomposition reaction strongly depends on the reaction environment. Our investigations demonstrated that the maximum amount of the reducing agent, namely fine dispersed iron nanoparticles, was obtained during the decomposition carried in a dry argon flow. This condition seems to be crucial, since the presence of steam in the reaction environment causes a secondary oxidation of iron and, consequently, a disappearance of the reducing agent, according to the reaction:

$$3Fe + 4H_2O = Fe_3O_4 + 4H_2\uparrow \tag{7}$$

A formation of the lower oxides is also possible in case of the steam deficiency (playing the role of the oxidizing agent). A uniform dispersion of the reducing agent is also required for the creation of Fe_2P thin layer on the $LiFePO_4$ phospho-olivine surface. On the basis of the synthesis tests previously performed, we found that the best results were achieved for the reaction mixtures milled in a ball mill. The additional DSC measurements demonstrated that milling the reaction mixture affects the water distribution in the system (Ojczyk et al., 2007). Iron nanoparticles as a reducing agent is highly unstable in the presence of even trace amounts of oxygen. Still, when it is rapidly exposed to the environment with a high concentration of oxygen (like air), it may undergo a passivation reaction (a surface oxidation and the formation of a protective, gas-tight oxide layer). In our case it can be assumed that the oxide layer, resulting from the passivation, prevented the oxidation of the reducing agent with oxygen from air at the intermediate stage of the synthesis, i.e. milling of the mixture after the calcinations at 350°C (Ojczyk et al., 2007).

The formation of iron phosphide Fe_2P resulting from the reduction of phosphate radicals with iron nanoparticles occurs only at higher temperatures (above 600°C). Below this temperature iron can be oxidized with oxygen residues. Consequently, the exposure time of the reaction mixture at low temperatures should be as short as possible, therefore high rates of the furnace heating up to the optimum reaction temperature are essential. The results of a thermal analysis for the reaction mixtures yielded a temperature of 800°C for 10 h as the optimum synthesis conditions (Ojczyk et al., 2007).

On the basis of the obtained results of thermal analysis and synthetic route trials, it can be concluded that a thin layer of Fe_2P can be formed only when the appropriate conditions of the synthesis process are satisfied:

- The products of the reaction (H_2O) can be carried away properly from the reactor at the volume flow rate (F_v) of the carrying gas (Ar = 99.999%, O_2<0.0005%) being at least: F_v= $20V_{react}/t$ for 1 g of the reagents batch

where: F_v— volume flow rate [$cm^3\ min^{-1}$]; V_{react} — the pipe flow reactor volume [cm^3]; t — time =60 minutes, both at the initial stage of the decomposition (350°C) and the main high-temperature reaction (800°C),

- Linear rate of carrying gas over the reagents should be at least 0.5 cm s^{-1}; this would prevent secondary reactions (iron oxidation)
- Milling the substrates in a ball mill prior to the reaction ensures an appropriate dispersion of the substrates
- A high heating rate (β= 10°min^{-1}), ensuring the maximum concentration of the reductive agent at the optimal temperature of the re action (800°C).

6. Carbon composite electrode materials

The high performance lithium ion batteries need electrode (cathode and anode) materials showing high ionic and electronic conductivities as well as high chemical stability, as it was mentioned above. Currently, the standard composite cathodes used in Li-ion cells are prepared as physical mixture of powders (active material, carbon and binder) (Tarascon &

Armand, 2001). Higher performances of cathode as well as anode materials may be achieved by lowering the size of active material grains. Nanosized grains reveal better electrochemical properties and also lower chemical stability towards electrolyte (Armand &Tarascon, 2008). The latest may provide dangerous and uncontrolled self ignition of a cells. Coating of active material by conductive carbon layer (CCL) increase the chemical stability and safety of the composite cathode (M. Molenda et al., 2008). There are many reports on the improvement of electrochemical performance of electrode materials for lithium-ion batteries using carbon coatings (Armand &Tarascon, 2008; Cushing & Goodenough, 2002; Kim et al., 2008; Lin et al., 2008; Kim, et al., 2008; Fey et al., 2008; Choi et al., 2008; Guo et al., 2009; Hassoun et al., 2008). The compounds reported for the coating formation include carboxylic acids (Lin et al., 2008), poly-alcohols (Guo et al., 2009), resins (Cushing & Goodenough, 2002; He et al., 2007) and sugars (Kim et al., 2008). To form the carbon coating, the compounds were deposited on the electrode material grains and pyrolysed, from which carbon layers resulted. However, the morphology of these layers was not discussed in detail. Characteristic of the carbon coatings deposited on electrodes is that they can react with the electrolyte thereby giving rise to the formation SEI or of passive insulating layers. The carbon coatings formed as carbon matrixes are applied in silica, tin or antimony based high capacity anode composites (Ng et al., 2007; He et al., 2007; Trifonova et al., 2008; Li & Li, 2008; Hassoun et al., 2008). In such the composites the carbon layers act as conducting and stress buffering agent compensating a volume changes during lithium insertion-deinsertion from active material, avoiding lost of electrical contact in the composite. However, high capacity electrodes requires a nano-sized grains of active materials which reveal high chemical reactivity thus it makes difficult the composites preparation. A novel method of preparation of conductive carbon layers with controlled morphology on fine particle powders, based on direct polymer deposition, has been developed by M. Molenda et al. (M. Molenda et al., 2007, 2008, 2010).

7. Preparation and properties of carbon coatings

A model of carbon coatings or carbon matrixes formation process are presented on Figs. 7. In this process prepared model $C/\alpha-Al_2O_3$ composites to revealed the possibility of control of the CCL morphology and its electrical properties. Two ways of the composite precursors formation can be applied. In the first (Fig. 7A), the free-radical precipitation polymerization of freshly distilled acrylonitrile (AN) was performed in the presence of $\alpha-Al_2O_3$ grains (POCh, Poland, 99.99 %, S_{BET} = 24 m^2g^{-1}), according to procedure described by M. Molenda et al. (M. Molenda et al., 2007). In brief, $\alpha-Al_2O_3$ grains were suspended in water solution of AN (7 wt%) and the polymerization was initiated by 2,2′-azobis(isobutyramidine hydrochloride) (Aldrich), upon which the obtained $PAN/\alpha-Al_2O_3$ samples were washed, filtered and dried in the vacuum at 50°C. In the second way (Fig. 7B), the $\alpha-Al_2O_3$ grains were impregnated with polymers composition in water solutions. The polymers used, were poly-N-vinylformamide (PNVF) obtained by radical-free polymerization from N-vinylformamide (Aldrich) (M. Molenda et al., 2008) and pyromellitic acid (PMA) modified (5-10 wt%) PNVF (called MPNVF) (M. Molenda et al., 2010). To achieve the impregnation, $\alpha-Al_2O_3$ grains were suspended in the solutions of respective polymers in water (8-15 wt%).

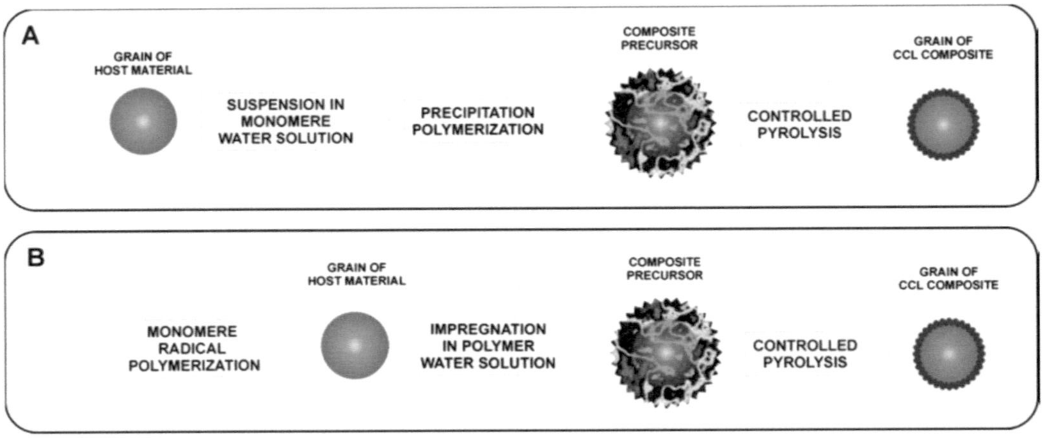

Fig. 7. Process for the preparation of conductive carbon layers (CCL) on powdered support.

The suspensions were stirred continuously until the solvent evaporated and the viscosity of the solutions became high enough to avoid sedimentation. Then the samples were dried in an air drier at 90°C overnight. To obtain C/α–Al_2O_3 composites, following the preparation, the polymer/α–Al_2O_3 composite precursors were pyrolysed in a tube furnace under the flow of 99.999 % argon (5 dm^3/h) at 550° and 600°C for 24h. The C/α–Al_2O_3 composites were deep black with foamed slag-like structure or lustrous and glassy like graphite (for PMA modified PNVF carbon precursor).

Electrical properties of the model C/α–Al_2O_3 composites, presented in Fig. 8, reveal increase of the electrical conductivity with an increase in carbon loading. For the model support, the minimal carbon loading for electrical conductivity through CCL (continuous path) is above 12 wt% of C. However, below this limit a conduction percolation path formation is observed. By a contrast, the activation energy of electrical conductivity remained nearly constant, suggesting a preservation of the conductivity mechanism through the CCL. The best electrical properties, i.e. the highest conductivity and the lowest activation energy, were revealed by the model composite based on the MPNVF precursor. This behavior was supported by the results of the Raman spectroscopy (RS) (Fig. 9).

The correlation between degree of carbon materials graphitization, characterized by intensity ratio of D/G bands, were found. The D band (defect mode; at about 1350 cm^{-1}) corresponds to sp^3 diamond-like carbon structures while the G band (about 1600 cm^{-1}) corresponds to sp^2 graphitic structures (Ferrari & Robertson, 2000; Fauteux et al., 2004; Osswald et al. 2005; Pantea et al., 2001). The graphitization degree (decrease in the D/G ratio) significantly increased after PMA modification of PNVF polymer, what may suggest formation of the highest amounts of the graphene domains in CCL. Also downshift and width decrease of the G peak of the carbonized MPNVF precursor suggest 2D ordering (1st phase of graphitization) (Fauteux et al., 2004). The improvement of the polymer carbonization upon CCL formation after PMA modification may results from the fact that planar structure of pyromellitic acid molecules serve as a nucleus of the graphene domains, and compete with the formation of the disordered structures.

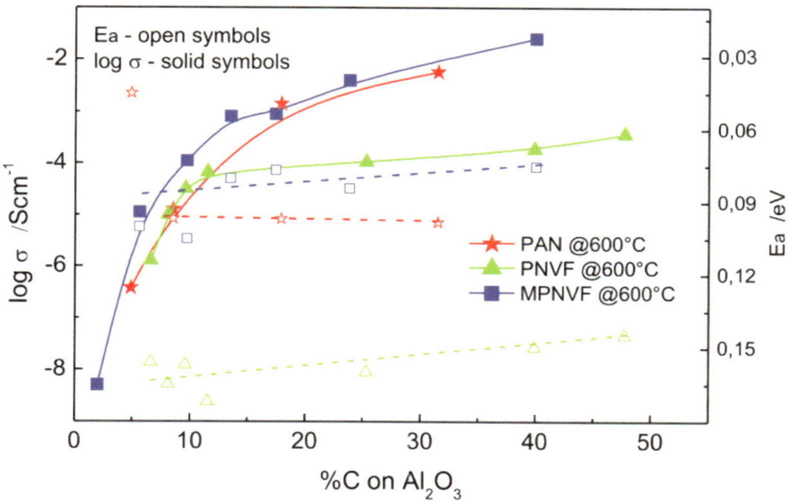

Fig. 8. Electrical properties of the model $C/\alpha-Al_2O_3$ composites.

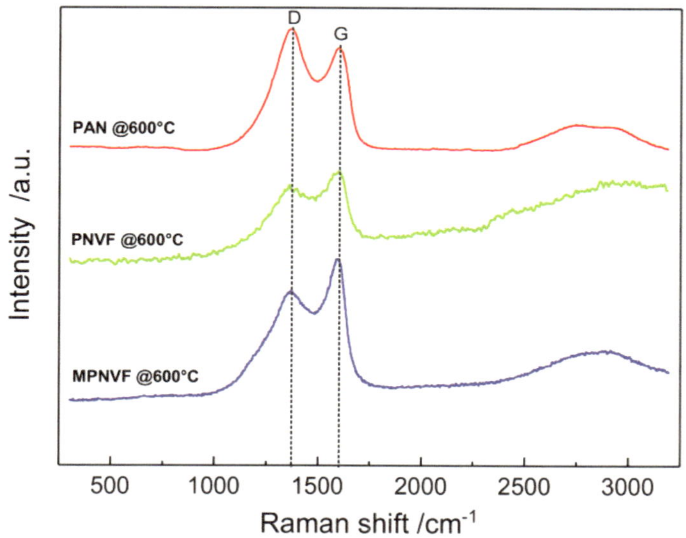

Fig. 9. Raman spectra of pyrolytic carbon coatings obtained from different polymer precursors.

In terms of SEI formation, a specific surface area of the CCL coating should be controlled as well as its pores structure. The resulting specific surface area of the composite depends on polymer precursor used (Fig. 10) and increases linearly with carbon loading up to 25-30 wt%.

This suggests that the initial porous structure of the CCLs is preserved regardless of the carbon content. The PMA modification of polymer precursor results in reducing of specific surface area of formed CCL, probably with locally arranged graphene domains what limits the formation of disordered carbon.

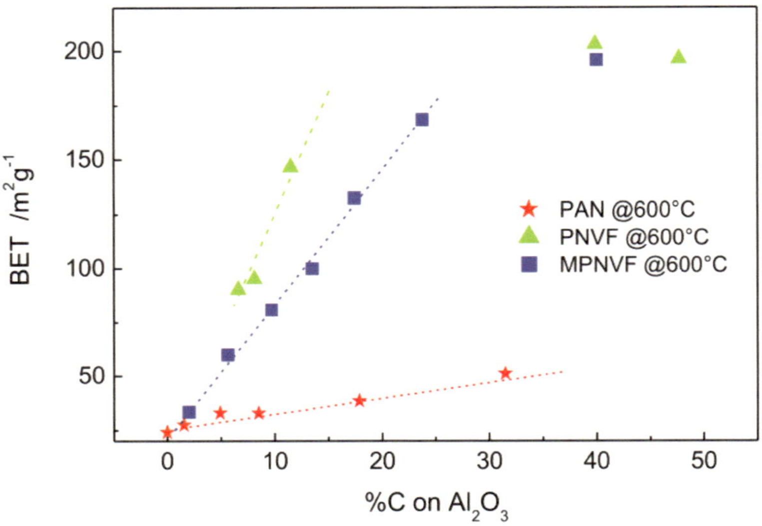

Fig. 10. Specific surface area of the composites as function of carbon loading.

The pores structure and morphology of the CCL in the model C/α–Al_2O_3 composites evaluated from N_2-adsorption-desorption isotherms (Fig. 11) reveal strong dependence on starting polymer composition (carbon precursor). The observed shapes of the isotherms (Fig. 11a-11c) correspond to the mixed I and IV types of isotherms (according to IUPAC nomenclature). Such shapes indicate the presence of micro- and mesopores within the CCLs. The hysteresis loops of the H4 type (IUPAC) suggest the slotted pores located within the intergranular spaces. The pore size and their distribution varies on carbon precursor used (Fig. 11d-11f) and may be controlled by optimization of polymer composition, e.g. modification of PNVF by PMA results in very uniform distribution of the mesopores with sizes within the range of 3-4.5 nm (Fig. 11f).

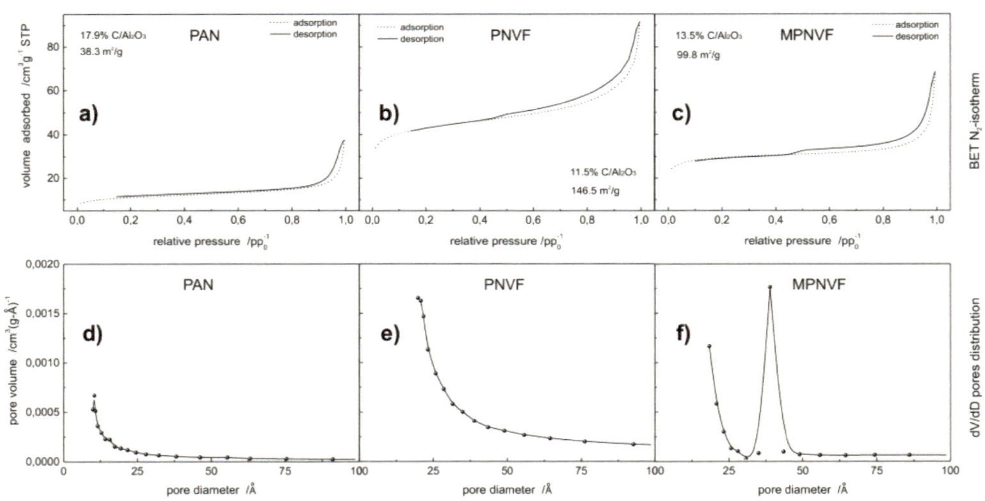

Fig. 11. BET N_2-adsorption-desorption isotherms and pore size distribution of the composites.

The morphological and electrical properties as well as observed local structure suggest the following structural model of the $C/\alpha-Al_2O_3$ composites derived from the polymer precursors (Fig. 12).

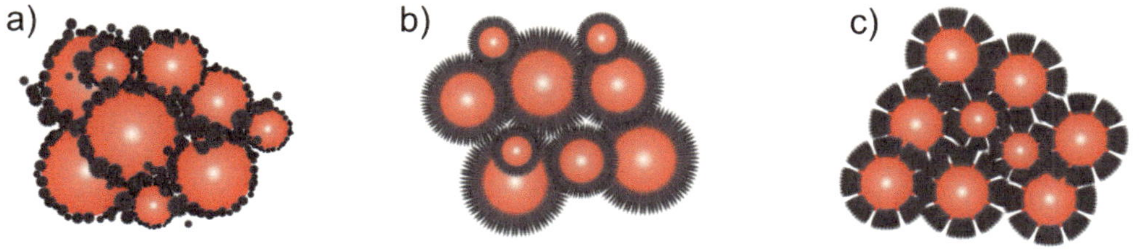

Fig. 12. Structural model of the CCL carbon coatings.

The CCL obtained from the PAN precursor (Fig. 12a) consists of tightly packed small carbon particles, while that obtained from the PNVF precursor (Fig. 12b) is built of carbon whiskers, this latter structure being reflected in a high specific surface area and a high share of the micropores in this sample. This same PNVF precursor, after modification with pyromellitic acid, strongly diminishes the specific surface area of the resulting composite, which is due to the formation of a tight, highly conductive carbon film with the defined porous structure that is dominated by mesopores with a narrow size distribution (Fig. 12c). The formed CCL coating build of amorphous carbon is thin and uniformly dispersed on the host grains what can be seen from HR-TEM picture (Fig. 13).

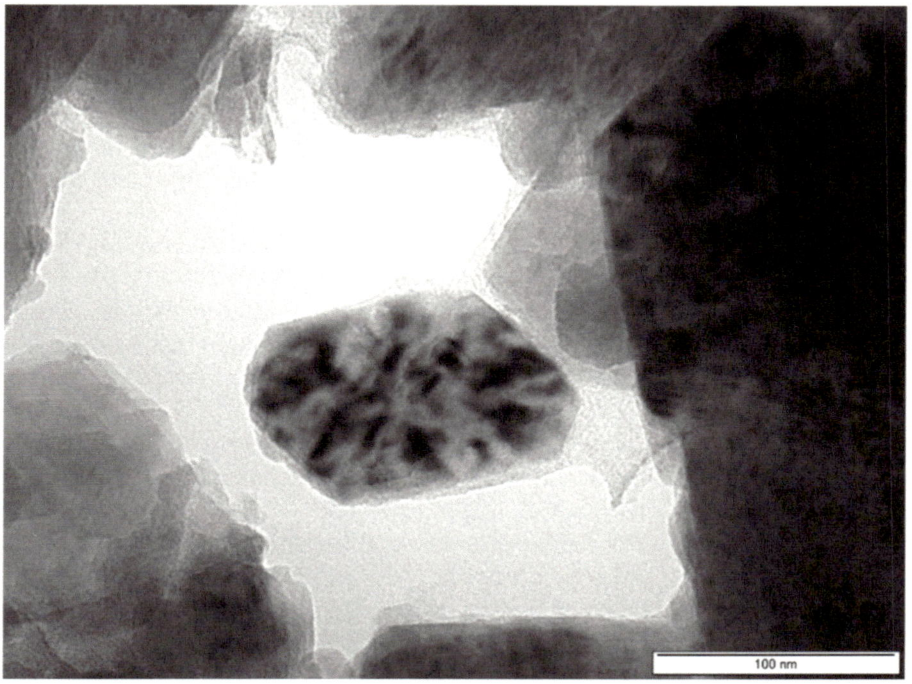

Fig. 13. HR-TEM picture of the $C/\alpha-Al_2O_3$ composite.

8. Acknowledgements

This work was financially supported by EU under the grant No. UDA-POIG 01.01.02-00-108/09-01.

9. References

Chung, S.Y., Bloking, J.T. & Chiang, Y.M., Electronically conductive phospho-olivines as lithium storage electrodes. *Nature Materials*, Vol. 1, No. 2, (October 2002), pp. 123-128.

Goodenough, J.B. (1998). *General concepts*, In: Lithium Ion Batteries. Fundamentals and Performance, Wakihara, M., Yamamoto, O., pp. 1-25, Wiley-VCH, ISBN: 9783527295692, Tokyo.

Herle, P.S., Ellis, B., Coombs, N. & Nazar, L.F. , Nano-network electronic conduction in iron and nickel olivine phosphates. *Nature Materials*, Vol. 3, No. 3, (March 2004), pp. 147-152.

Marzec, J., Ojczyk, W. & Molenda, J. , Delithiation of olivine- structured $LiFe_xMn_{1-x}PO_4$ cathode materials. Mössbauer studies. *Materials Science-Poland*, Vol. 24, No. 1, (2006), pp. 69-74.

Molenda, J., Bąk, J. Effect of tungsten on the electrical and electrochemical properties of the $Li_yW_xVO_2$ cathode. *Physica Status Sololidi (a)*, Vol. 135, No. 1, (1993), pp. 263-271.

Molenda, J., Kubik, A., Transport properties and reactivity of tungsten trioxide. *Solid State Ionics*, Vol. 117, No. 1-2 (February 1999), pp. 57-64.

Molenda, J., Stokłosa, A. Electronic structure and electrochemical properties of VO_2. *Solid State Ionics*, Vol. 36, No. 1-2, (October 1989), pp. 43-52.

Molenda, J., Bąk, T. & Stokłosa, A., Influence of lithium on the electronic structure of $YBa_2Cu_3O_{7-\delta}$. *Physica C*, Vol. 207, No. 1-2 (March 1993), pp. 147-158.

Molenda, J., Świerczek, K., Molenda, M. & Marzec, J., Electronic structure and reactivity of $Li_{1-x}Mn_2O_4$ cathode. *Solid State Ionics*, Vol. 135, No. 1-4, (November 2000), pp. 53-59.

Molenda, J., Wilk P. & Marzec, J. Structural, electrical and electrochemical properties of $LiNiO_2$. *Solid State Ionics*, Vol. 146, No. 1-2, (January 2002) 73-79.

Molenda, J., Marzec, J. ,Świerczek, K., Ojczyk, W., Ziemnicki, M., Molenda, M., Drozdek, M. & Dziembaj, R. The effect of 3d substitutions in the manganese sublattice on the charge transport mechanism and electrochemical properties of manganese spinel. *Solid State Ionics*, Vol. 171, No. 3-4, (July 2004), pp. 215-227.

Molenda, J., Ojczyk, W., Świerczek, K., Zając, W., Krok, F., Dygas, J. & Liu, R.S. Diffusional mechanism of deintercalation in $LiFe_{1-y}Mn_yPO_4$ cathode material. *Solid State Ionics*, Vol. 177, No. 26-32, (October 2006), pp. 2617-2624.

Molenda, J. Correlation between electronic and electrochemical properties of carbon. *Bulletin of Polish Academy of Sciences - Chemistry*, Vol. 45, (1997) pp. 449.

Molenda, J. Electronic limitations of lithium diffusibility. From layered and spinel toward novel olivine type cathode materials. *Solid State Ionics*, Vol. 176, No. 19-22, (June 2005), pp. 1687-1694.

Ohzuku, T. *Four-volt cathodes for lithium accumulators and the Li-ion battery concept*, In: Lithium Batteries, New Materials, Development and Perspectives, Pistoia, G., pp. 239-280, Elsevier, Amsterdam.

Ojczyk, W., Marzec, J., Świerczek, K., Zając, W., Molenda, M. , Dziembaj, R. & Molenda, J. Studies of selected synthesis procedures of the conducting $LiFePO_4$-based composite cathode materials for Li-ion batteries. *Journal Power Sources*, Vol. 173, No. 2, (November 2007), pp. 700-706.

Tarascon, J.-M. & Armand, M. Issues and challenges facing rechargeable lithium batteries. *Nature*, Vol. 414, (November 2001) pp. 359-367.

Than, D., Molenda, J. & Stokłosa A. Correlation between electronic and electrochemical properties of $Li_xTi_{1+y}S_2$. *ElectrochimicaActa*, Vol.36 No. 10, (1991), pp. 1555-1560.

Whittingham, M.S. Chemistry of intercalation compounds: Metal guests in chalcogenide hosts. *Prog. Solid State Chem.*, Vol. 12, No. 1, (1978), pp. 41-99.

Yang, S., Song, Y., Zavalij, P.Y. & Whittingham, M.S., Reactivity, stability and electrochemical behavior of lithium iron phosphates. *Electrochemistry Communications*, Vol. 4, No. 3, (March 2002) pp. 239-244.

Armand, M. &Tarascon, J.-M, Building better batteries. *Nature*, Vol. 451 (February 2008) pp. 652-657.

Molenda, M., Dziembaj, R., Piwowarska, Z. & Drozdek M., Electrochemical properties of $C/LiMn_2O_{4-y}S_y$ ($0 \leq y \leq 0.1$) composite cathode materials. *Solid State Ionics*, Vol. 179 No. 1-6 (2008) pp. 88-92.

Cushing, B.L. & Goodenough, J.B., Influence of carbon coating on the performance of a $LiMn_{0.5}Ni_{0.5}O_2$ cathode. *Solid State Science*, Vol. 4, (2002) pp. 1487-1493.

Kim, J.K., Cheruvally, G., Ahn, J.H. & Ahn, H.J., Electrochemical properties of $LiFePO_4/C$ composite cathode material: Carbon coating by the precursor method and direct addition. *Journal of Physics and Chemistry of Solids*, Vol. 69, (2008) pp. 1257-1260.

Lin, B., Wen, Z., Han J. & Wu X., Electrochemical properties of carbon-coated $Li[Ni_{1/3}Co_{1/3}Mn_{1/3}]O_2$ cathode material for lithium-ion batteries. *Solid State Ionics*, Vol. 179, (2008) pp. 1750-1753.

Kim, J.K., Cheruvally, G. & Ahn, J.H., Electrochemical properties of $LiFePO_4/C$ synthesized by mechanical activation using sucrose as carbon source. *Journal of Solid State Electrochemistry*, Vol. 12, (2008) pp. 799-805.

Fey, G.T.K., Lu, T.L. & Wu, F.Y., Carboxylic acid-assisted solid-state synthesis of $LiFePO_4/C$ composites and their electrochemical properties as cathode materials for lithium-ion batteries. *Journal of Solid State Electrochemistry*, Vol. 12, (2008) pp. 825-833.

Choi, Y.J., Chung, Y.D., Baek, C.Y., Kim, K.W., Ahn, H.J. & Ahn, J.H., Effects of carbon coating on the electrochemical properties of sulfur cathode for lithium/sulfur cell. *Journal of Power Sources*, Vol. 184, (2008) pp. 548-552.

Guo, R., Shi, P., Cheng, X. & Du, C., Synthesis and characterization of carbon-coated $LiNi_{1/3}Co_{1/3}Mn_{1/3}O_2$ cathode material prepared by polyvinyl alcohol pyrolysis route. *Journal of Alloys Compounds*, Vol. 473, (2009) pp. 53-59.

Ng, S.H., Wang, J., Konstantinov, K., Wexler, D., Chew, S.Y., Guo, Z.P. & Liu, H.K., Spray-pyrolyzed silicon/disordered carbon nanocomposites for lithium-ion battery anodes. *Journal Power Sources*, Vol. 174, (2007) pp. 823-827.

He, X., Pu, W., Wang, L., Ren, J., Jiang, C. & Wan, C., Synthesis of spherical nano tin encapsulated pyrolyticpolyacrylonitrile composite anode material for Li-ion batteries. *Solid State Ionics,* Vol. 178, (2007) pp. 833-836.

Trifonova, A., Stankulov, T. & Winter, M., Study of metal-supported carbon matrix as a high-capacity anode for Li-ion battery. *Ionics,* Vol. 14, (2008) pp. 421-425.

Li Y. & Li J., Carbon-Coated Macroporous $Sn_2P_2O_7$ as Anode Materials for Li-Ion Battery. *Journal of Physical Chemistry C,* Vol. 112, (2008) pp. 14216-14219.

Hassoun, J., Derrien, G., Panero, S. & Scrosati, B., The role of the morphology in the response of Sb–C nanocomposite electrodes in lithium cells. *Journal of Power Sources,* Vol. 183, (2008) pp. 339-343.

Molenda, M., Dziembaj, R., Piwowarska, Z. & Drozdek, M., A new method of coating powdered supports with conductive carbon films. *Journal of Thermal Analysis and Calorimetry,* Vol. 88, (2007) pp. 503–506.

Molenda, M., Dziembaj, R., Podstawka, E., Proniewicz, L.M. & Piwowarska, Z., An attempt to improve electrical conductivity of the pyrolysed carbon-$LiMn_2O_{4-y}S_y$ ($0<y<0.5$) composites, *Journal of Power Sources,* Vol. 174, (2007) pp. 613-618.

Molenda, M., Dziembaj, R., Drozdek, M., Podstawka, E. & Proniewicz, L.M., Direct preparation of conductive carbon layer (CCL) on alumina as a model system for direct preparation of carbon coated particles of the composite Li-ion electrodes. *Solid State Ionics,* Vol. 179, (2008) pp. 197–201.

Molenda, M., Dziembaj, R., Kochanowski, A., Bortel, E., Drozdek, M. & Piwowarska, Z., Process for the preparation of conductive carbon layers on powdered supports, *Patent Application,* WO 2010/021557 A2 (2010).

Ferrari, A.S. & Robertson J., Interpretation of Raman spectra of disordered and amorphous carbon. *Physical Review B,* Vol. 61, (2000) pp. 14095-14107.

Fauteux, C., Longtin, R., Pegna, J. & Boman, M., Raman characterization of laser grown carbon microfibers as a function of experimental parameters. *Thin Solid Films,* Vol. 453-454, (2004) pp. 606-610.

Osswald, S., Flahaut, E., Ye, H. & Gogotsi, Y., Elimination of D-band in Raman spectra of double-wall carbon nanotubes by oxidation. *Chemical Physics Letters,* Vol. 402, No. 4-6, (2005) pp. 422-427.

Pantea, D., Darmstadt, H., Kaliaguine, S., Summchen, L., & Roy, C., Electrical conductivity of thermal carbon blacks - Influence of surface chemistry. *Carbon,* Vol. 39, No. 8, (2001) pp. 1147-1158.

31

Multilayer Fresnel Zone Plate with High-Diffraction Efficiency: Application of Composite Layer to X-ray Optics

Shigeharu Tamura
National Institute of Advanced Science and Technology (AIST),Osaka
Japan

1. Introduction

Composite materials are widely utilized in a number of fields, such as materials science, metallurgy, polymer science, interface science, mechanical engineering and aerospace engineering. In this chapter, as an application of a composite material layer (thin film) produced by co-sputtering to the hard X-ray focusing optical technique, a multilayer Fresnel zone plate (ML-FZP) with high diffraction efficiency is described. X-rays in the energy region of 100 – 2,000 eV (2 keV) are called soft X-rays (Snigirev & Snigireva, 2008), while those with higher energy are called hard X-rays. Soft X-rays, especially in the "water window" region (Spiller, 1994), are mainly used in the field of biotechnology. On the other hand, hard X-rays are used in various research fields, including materials science, environmental science and medical science.

High brilliant hard X-ray beams with submicron- or nanometre-scale spot sizes generated by third-generation synchrotron radiation (SR) facilities such as the APS (USA), ESRF (France) or SPring-8 (Japan), especially for use in the high-energy region, have great potential for use in various fields of research. They are remarkably powerful tools. Recently, higher energy (shorter wavelength) X-ray beams above 20 keV have been utilised in a number of applications, including residual stress measurement in metal matrix composites at 40 keV (Korsunsky & Wells, 2000), local strain measurement within bulk materials at 52 and 90 keV (Lienert et al., 2000), a novel experimental scheme for high-resolution X-ray analysis of deeply buried interfaces at 71.3 keV (Reichert et al., 2003), study of the ice–SiO_2 model interface, using X-ray transmission–reflection scheme at 71.3 keV (Engemann et al., 2004), mapping of Sr in $(Ba,Sr)TiO_3$ dielectric ceramics using Kα fluorescence X-rays at 25 keV (Takeuchi et al., 2005), micro-XRF (X-ray fluorescence) analysis of heavy metals in the cells of hyperaccumulator plants at 37 and 75 keV (Terada et al., 2004; Terada et al., 2005), non-destructive imaging of integrated circuits (ICs) at 25 keV, and imaging of Au mesh by three types of X-ray microscopy at 82 keV (Awaji et al., 2003; Suzuki et al., 2006). In addition, microscopic imaging of Au mesh at 200 keV has also been reported (Kamijo et al., 2009).

Many types of focusing optics have been developed for hard X-rays and their focusing abilities have been improved over the past two decades. Especially within the last several years, there have been dramatic changes in the performances of focusing optics. The main types of focusing optics and their performances are shown in Fig. 1.

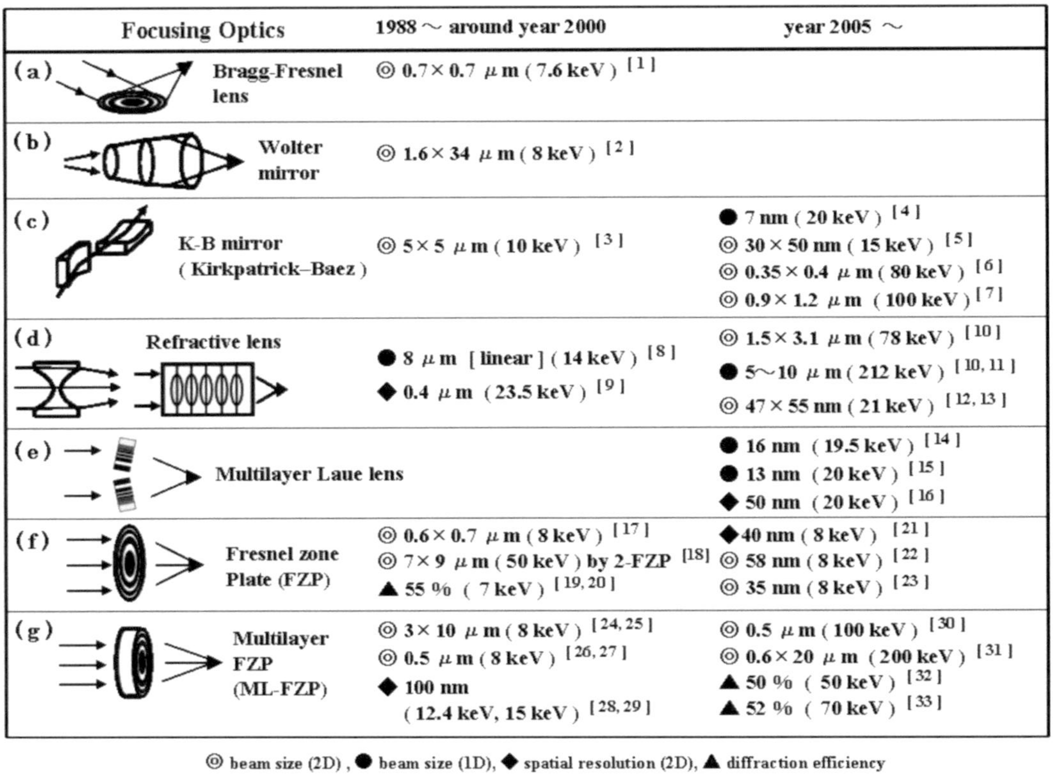

Fig. 1. Various focusing optics for synchrotron radiation (SR) X-ray.

Representative results (measured numerical datum) in Fig.1 are as follows: [1] (Snigirev et al., 1995), [2] (Hayakawa et al., 1989), [3] (Underwood et al., 1988), [4] (Mimura et al., 2010), [5] (Matsuyama et al.; 2006), [6-7] (Suzuki et al.; 2007), [8] (Snigirev et al., 1996), [9] (Lengeler et al., 1999), [10] (Snigirev et al., 2004), [11] (Nazmov et al, 2005), [12] (Schroer et al., 2006), [13] (Schroer et al., 2005), [14] (Kang et al., 2008), [15] (Koyama et al., 2010a), [16] (Koyama et al., 2010b), [17] (Lai et al., 1992), [18] (Shastri et al., 2001), [19] (Fabrizio et al., 1999a), [20] (Fabrizio et al., 1999b), [21] (Chu et al., 2008), [22] (Suzuki et al., 2005), [23] (Suzuki et al., 2010), [24] (Saitoh et al., 1988), [25] (Saitoh et al., 1989), [26] (Kamijo et al., 1997), [27] (Suzuki et al., 1997), [28] (Kamijo et al., 2002a), [29] (Kamijo et al., 2002b), [30] (Kamijo et al., 2003), [31] (Kamijo et al., 2009), [32] (Tamura et al., 2006), [33] (Tamura et al., 2008). At present, only the ML-FZP, K-B mirror and Refractive lens have achieved focusing above an X-ray energy of 100 keV. Recently, new types of optics have been proposed (Takano et al., 2010; Suzuki et al., 2010a). Among the optics shown in Fig. 1, FZP has advantages in that the optical system can be constructed easily (the tilting angle of the FZP is easily adjustable to the optical axis). In addition, ML-FZP has the important advantages that the kinoform type (theoretical diffraction efficiency, 100%) (Erko et al., 1996) can be fabricated, and that it can operate in the high-energy X-ray region (> 100 keV) because a large "aspect ratio" structure can be realised relatively easily. The most important characteristics of the FZP are its high diffraction efficiency, high spatial resolution and availability in the high-energy X-ray region. Here, the author describes (1) the process of fabrication and testing of the ML-FZP and (2) the development of ML-FZP with high diffraction efficiency using composite thin film layers.

2. Fresnel zone plate (FZP)

The normal FZP is a circular diffraction grating of alternate transparent and opaque zones. The outline of the FZP is shown in Fig. 2 (details of the schematics of various types of the FZP are shown in Fig.6). The FZPs have mainly been fabricated by a lithography-based technique or by a sputtered-sliced one (ML-FZP). As mentioned above, the ML-FZP has many advantages. The zone width decreases gradually from the centre to the outer edge.

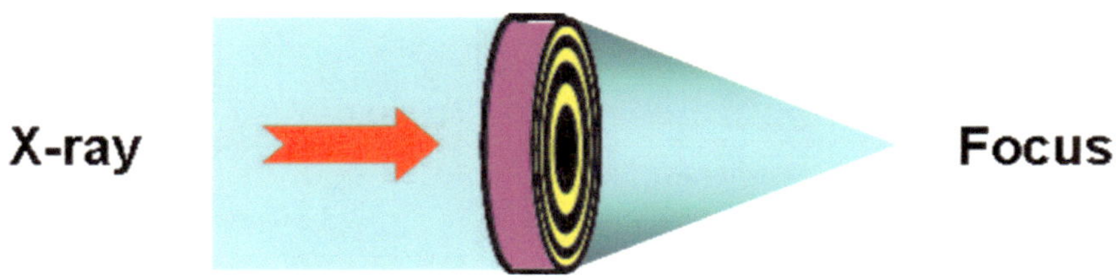

Fig. 2. Schematic of Fresnel zone plates

The radius r_n of the nth zone is given by

$$r_n^2 = r_0^2 + n\lambda f \tag{1}$$

where r_0 is the radius of the central zone (wire substrate), λ is the X-ray wavelength and f is the focal length. The spatial resolution D of the FZP for the first-order focus is determined by the outermost (minimum) zone width dr

$$D = 1.22\, dr \tag{2}$$

The focal spot size is comparable to the spatial resolution. The theoretical diffraction efficiency of the normal type (an amplitude modulation) FZP is 10 %. The efficiencies of the phase modulating type FZP are described elsewhere (Kirz, 1974; Yun et al., 1992; Bionta et al., 1994; Tamura et al., 2000; Kamijo et al., 2000).

3. Fabrication of normal multilayer Fresnel zone plate (ML-FZP)

3.1 Deposition of multilayer

The ML-FZP is or often called a "sputtered-sliced" zone plate. The conventional-type ML-FZP is composed of multiple concentric layers of alternating high-Z (*i.e.*, large atomic number) and low-Z (*i.e.*, small atomic number) materials. Hart et al. (Hart et al., 1966) and Rudolph et al. (Rudolph et al., 1981) proposed the concept of the ML-FZP. WSi_2/C FZP and Cu/Al FZP were subsequently developed and tested by Saitoh et al. (Saitoh et al., 1988; Saitoh et al., 1989) and Bionta *et al.* (Bionta et al., 1989; Bionta et al. 1990), respectively. However, these experiments did not achieve focusing with submicron spot size. Figure 3 presents an outline of the fabrication process of ML-FZP (normal type), which consisted of two parts: deposition and mechanical manufacturing. As a preliminary experiment, various concentric multilayer samples (W/C, Cr/C, Ag/C, Cu/C, Cu/Al, NiCr/ Al, *etc.*) were fabricated and the roughness of their multilayer interfaces (zone) was examined. Among these samples, the Cu/Al multilayer formed comparatively smooth zones and a polished surface (Tamura et al., 1994). Therefore, the Cu/Al multilayer coating is described in this chapter.

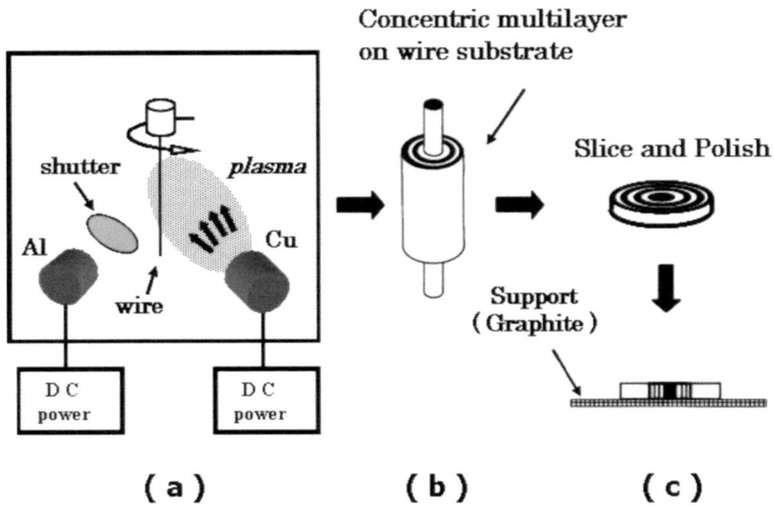

Fig. 3. Fabrication of multilayer Fresnel zone plate (ML-FZP): (a) schematic diagram of DC sputtering apparatus, (b) deposited sample, and (c) fabrication of FZP.

The Cu/Al concentric multilayer samples were deposited onto a rotating (15 – 50 rpm) Au wire substrate 50 micron in diameter by a DC magnetron sputtering apparatus with two DC-sputtering guns (3 inches in diameter) placed at the base points of a right triangle with the Au wire at the apex. The gun-to-substrate distance was 50 mm. The layer thickness was monitored and controlled by a quartz thickness monitor. To achieve smooth zones, if necessary, a cylindrical slit (linear slit fabricated on the surface of a stainless cylindrical shield) was used between the target and the substrate in the sputtering apparatus (Yasumoto et al., 2001; Tamura et al., 2002). The slit was controlled automatically such that it faced the operating sputtering gun. Figure 4 shows photographs of the sputtering apparatus. The Cu/Al samples were fabricated under conditions with various deposition parameters. Representative deposition parameters were as follows: sputtering power, 6 W cm^{-2} for Cu and 10 W cm^{-2} for Al; total gas pressure, 0.2 Pa; and coating rate, 0.3 nm s^{-1} for both Cu and Al layers. A protective overcoat layer (Cu, 3 micron) was also deposited.

3.2 Fabrication of zone plate

After the deposition process, the multilayer wire sample was embedded into low melting point alloy (melting point: 453 K), sliced into a plate 1 mm thick perpendicular to the wire axis using a band saw micro-cutting machine. One sliced surface was polished mechanically using a micro-grinding machine (wet process) and the layer structure was examined. Next, this polished surface was fixed onto a graphite plate with resin glue and polished until the desired thickness was achieved. Photographs of a sliced sample (embedded in alloy), the FZP fixed on the graphite plate and the scanning electron microscopic (SEM) image of an FZP are shown in Fig. 5.

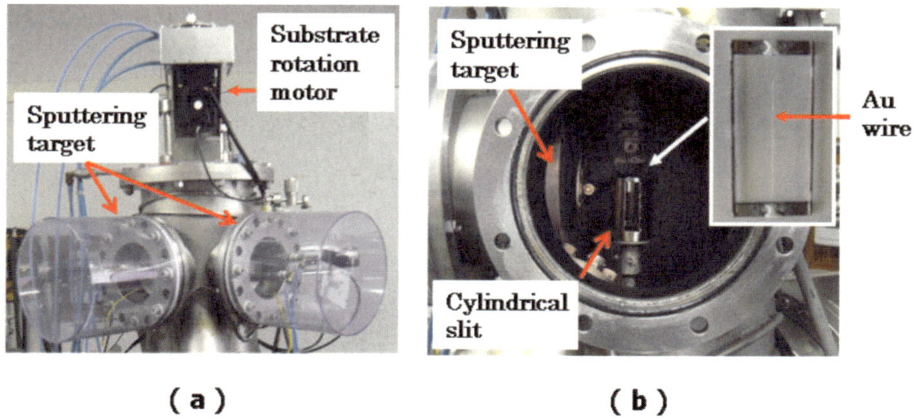

Fig. 4. Photographs of DC sputtering apparatus: (a) sputtering chamber, (b) inside of chamber.

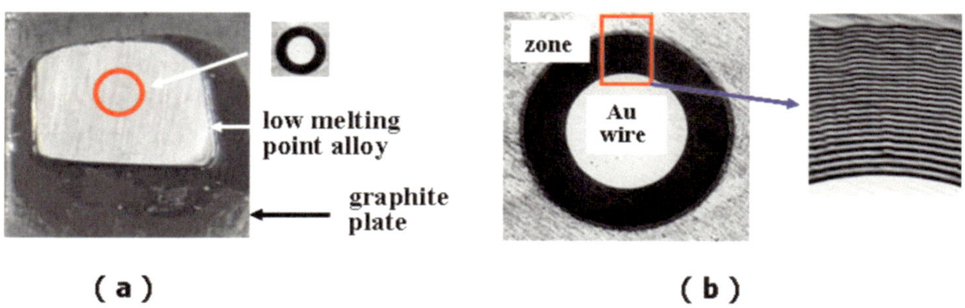

Fig. 5. Multilayer Fresnel zone plate (ML-FZP) : (a) ML-FZP on graphite, (b) Scanning electron micrographs. Black and white zones are Al and Cu layers, respectively.

4. Fabrication of multilavel-type multilayer Fresnel zone plate

4.1 Deposition of multilayer

As mentioned in Section 1, the high diffraction efficiency is one of important performance characteristics of the Fresnel zone plate. The theoretical diffraction efficiency of the normal type (amplitude modulation) FZP is 10% and that of the phase modulating type FZP is 40%. However, the actual efficiencies are 10% - 20%. This reduction in performance is due to fabrication error and absorption by the material. In X-ray microscopy optical systems using the FZP, higher diffraction efficiency is necessary to reduce radiation damage to biological specimens (Fujisaki & Nakagiri, 1990) and to simplify the optical system because this considerably suppresses unwanted diffraction orders (Fabrizio et al., 1999a). Among the various types of FZP, the kinoform FZP has the highest theoretical diffraction efficiency of 100% (Erko et al., 1996), although it is extremely difficult to fabricate such an FZP precisely. A multilevel-type FZP fabricated by the lithography-based technique has been proposed (Fabrizio et al., 1994) and developed as an approximate structure for the kinoform FZP, and it showed efficiency of 55% at 7 keV (Fabrizio et al., 1999a; Fabrizio et al., 1999b). Such an FZP, however, cannot function in the high-energy region (above 20 keV) because it is technically difficult to realise a high aspect ratio (*i.e.*, the ratio of FZP thickness to the zone width). Schematic representations of various types of FZP are shown in Fig. 6.

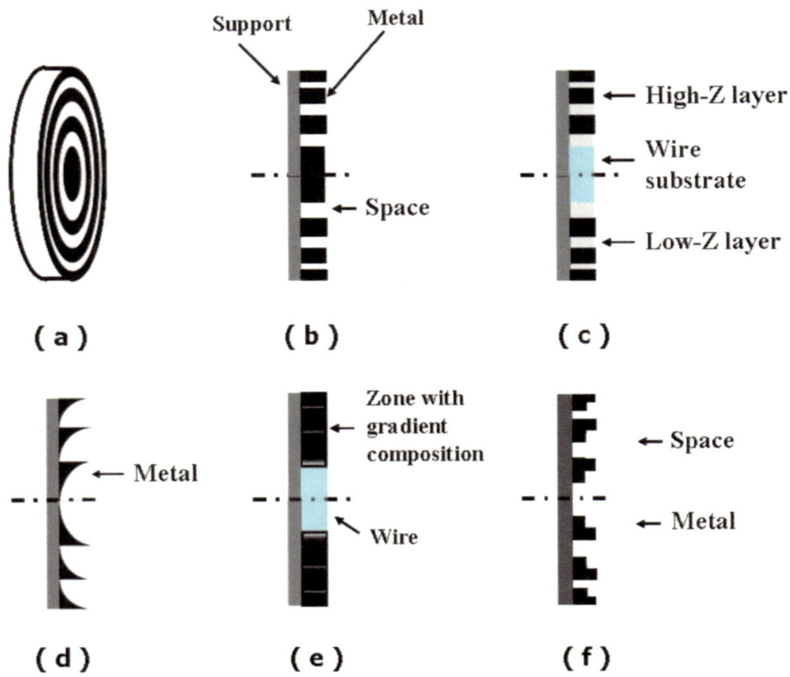

Fig. 6. Schematics of Fresnel zone plates (FZP): (a) schematic image, (b) FZP by lithography-based technique, (c) FZP by sputtered-sliced technique, (d) kinoform zone plate by lithography-based technique, (e) kinoform sputtered-sliced zone plate, and (f) 3-step multilevel zone plate by lithography-based technique.

4.2 Fabrication of zone plate

Schematic representations of the 1-period structure of various FZPs are shown in Fig. 7. As a preliminary experiment, a multilevel-type ML-FZP (5-period, total of 15 layers) with 3-step structure (Fig. 7(e)) was fabricated on the assumption that this ML-FZP composed of high-Z (Cu), low-Z (Al) and composite layers (Cu:Al = 1:1) was functionally equivalent to the FZP with the structure shown in Fig. 7(b). The composite layer was deposited by co-sputtering of high-Z and low-Z materials with the same coating rate. The focusing characteristics were evaluated at SPring-8 by comparing the 1st and –1st order diffraction at a CCD detector located downstream of the focal point. The details of the focusing test are described later. At the normal FZP, the intensity of the 1st order diffraction was the same as that of the –1st order diffraction. On the other hand, at this 3-step ML-FZP, the intensity of the 1st order diffraction was somewhat higher than that of the –1st order diffraction. These observations indicated that this composite layer fabricated by co-sputtering functions well and contributes to the improvement of diffraction efficiency.
Several types of 3-step, 4-step and 6-step ML-FZP were fabricated. As an example, the deposition of the 4-step multilayer is described. The concentric multilayer sample for the multilevel-type FZP was designed according to the theory presented by Fabrizio et al.

(Fabrizio et al., 1994). The designed focal length was 200 mm at 12.4 keV. Such a multilayer is composed of multiple concentric layers of alternating high-Z (Cu), low-Z (Al), and two types of composite material (Cu, Al) layers: zone-1 is a Cu layer, zone-2 is a composite layer (Cu:Al = 2:1), zone-3 is a composite layer (Cu:Al = 1:2) and zone-4 is an Al layer. The composite layers were deposited by co-sputtering of high-Z and low-Z materials, and the composition of the composite layers was controlled by changing the power of each target (*i.e.*, by controlling each coating rate). A total of 60 layers (15 pairs) were deposited. The inner zone (Al) was 0.2 micron in width and the outermost zone (Cu) was 0.15 micron in width. The theoretical focused beam size was 0.61 multiplied by the zone width of the outermost one pair zones (*i.e.*, the amount of 4 kinds of zones). Deposition parameters were as follows: sputtering powers were 2.7 W/cm^2 for Cu (zone-1), 5 W/cm^2 for Al (zone-4), 1.8 W/cm^2 for Cu, and 1.6 W/cm^2 for Al (zone-2), 0.9 W/cm^2 for Cu and 3.2 W/cm^2 for Al (zone-3). Ar gas pressure was 0.2 Pa. The substrate temperature was not controlled, the coating rate was 0.7 nm/s and the thickness of the FZP was 55 micron.

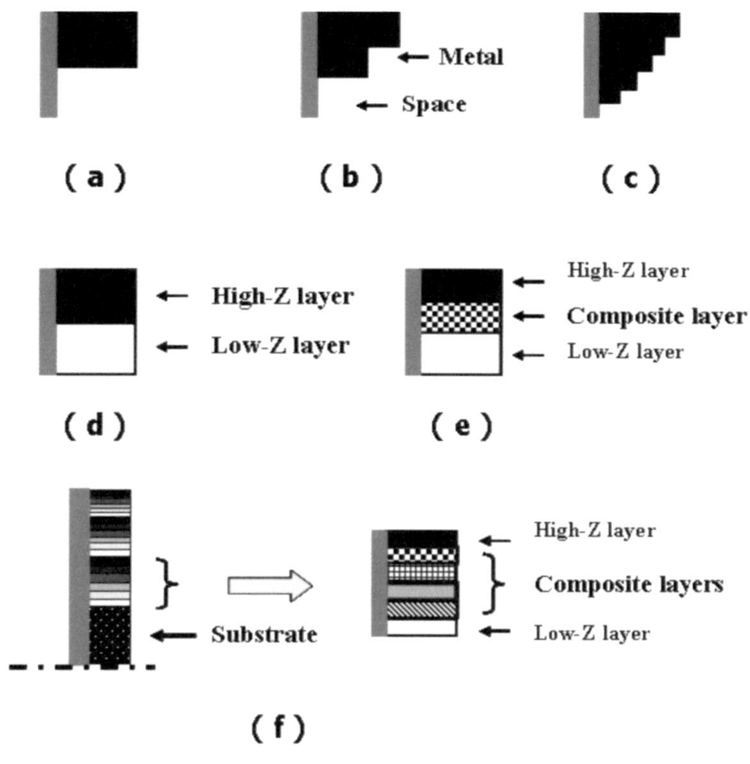

Fig. 7. Schematics of one-period structure of Fresnel zone plates (FZP): (a) conventional FZP by lithography-based technique, (b) 3-step multilevel FZP by lithography-based technique, (c) 6-step multilevel FZP by lithography-based technique, (d) conventional FZP by sputtered-sliced technique, (e) 3-step multilevel FZP by sputtered-sliced technique, and (f) 6-step multilevel FZP by sputtered-sliced technique.

In a similar manner, two 6-step ML-FZPs were fabricated with thicknesses of 35 micron and 66 micron, respectively. SEM images of the 3-step and 6-step ML-FZP are shown in Fig. 8, along with the results of Energy Dispersive X-ray (EDX) line scanning along the multilayer surface. Accurate analysis could not be done because of the resolution of the EDX equipment (0.5 micron). As shown in Fig.8, zone (multilayer interface) roughness towards the top of the layer stack is observed. Such a roughness is caused by the complex structure of the sputtered thin film (Thornton, 1974; Thornton, 1986).

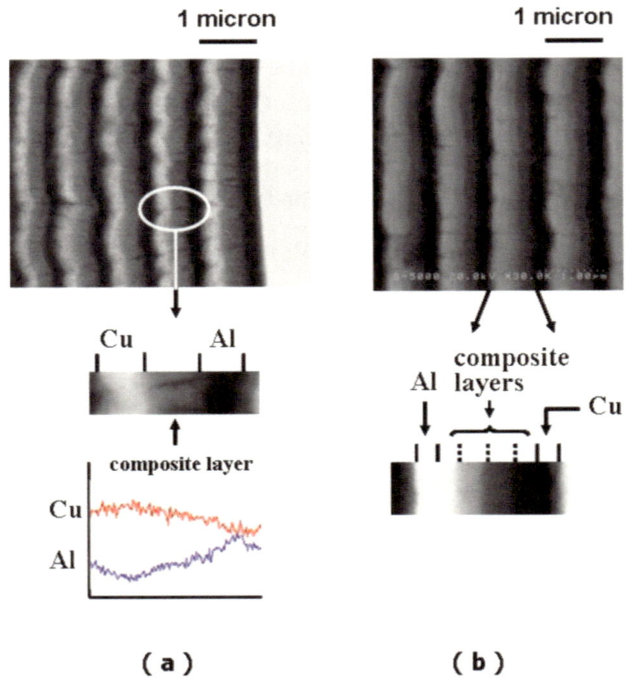

Fig. 8. Scanning electron micrographs (close-up view) of multilevel multilayer FZP: (a) 3-step FZP and its scanning of the energy dispersive X-ray (EDX) along the multilayer surface, and (b) 6-step FZP. Black and white zones are Al and Cu layers, respectively. Another gray zones are composite layers.

5. Focusing test of multilayer Fresnel zone plate

5.1 Focused beam size and spatial resokution

Focusing tests of various ML-FZPs were performed at the BL20XU or BL47XU undulator beamline of SPring-8. A representative experimental setup of the beamline is described previously (Kamijo et al., 2002a; Kamijo et al., 2003; Suzuki et al., 2001). Photographs of the experimental facilities are shown in Fig. 9. The focused beam size, the spatial resolution and the diffraction efficiency were measured as necessary. The focused beam size [full width at half maximum (FWHM)] was measured by conventional knife-edge scan. As shown in Fig. 1, representative results were a spot size of 500 nm at 100 keV and focusing at 200 keV (Kamijo et al., 2003; Kamijo et al., 2009). The spatial resolution was measured by observing an X-ray microscopic image of a resolution test pattern with a tantalum microstructure. The representative result was 100 nm at 12.4 keV and at 15 keV (Kamijo et al., 2002b; Kamijo et al., 2003).

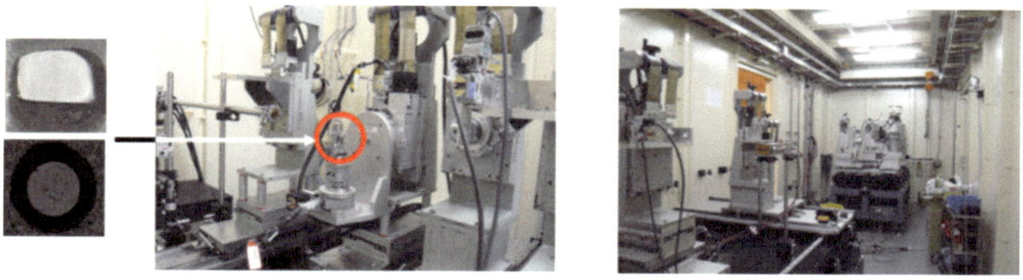

(a) (b)

Fig. 9. Photographs of the experimental facilities at Spring-8: (a) facilities around FZP, and (b) facilities in experimental hutch.

5.2 Diffraction efficiency

To examine the effectiveness of the use of the composite layer as shown in Fig. 7, the diffraction efficiencies of three types of ML-FZP discussed above were measured. The diffraction efficiency at the focal point was estimated by comparing the incident beam intensity measured by ion chambers upon the OSA (order sorting aperture: tantalum cross slit) and the total intensity of the focused beam through the OSA (Kamijo et al., 2006). The diffraction efficiency of the 1st order focus and the focused beam size were measured.

For the 4-step FZP (thickness 55 micron), the intensity the 1st order diffraction was compared with that of the –1st order diffraction using the refraction corn image at 30 keV. The results indicated that the intensity of the 1st order diffraction was higher than that of the –1st order diffraction (Fig. 10). This FZP was confirmed to function well as an FZP with high diffraction efficiency. The details of this experiment are described were described previously (Kamijo et al., 2006). The diffraction efficiency was next measured in the high energy region and an efficiency of 50% was achieved at 50 keV with a focused beam size of 1 micron.

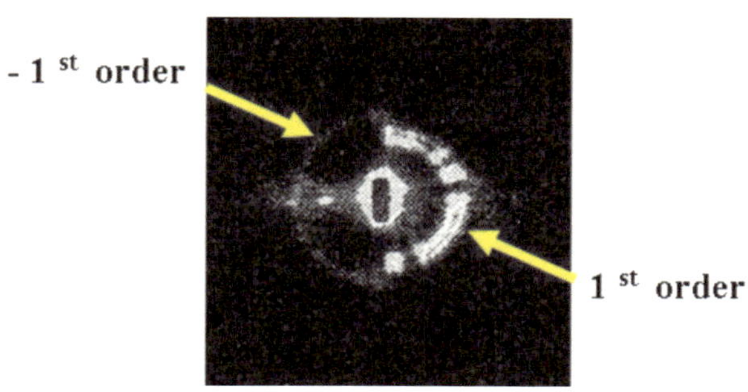

Fig. 10. Refraction corn of 4-step multilayer FZP.

For the 6-step FZP (thickness 35 micron), a 1st order diffraction efficiency of 52% was attained at 41.3 keV with a focused beam size of 0.8 micron. For the 6-step FZP (thickness 66 micron), a 1st order diffraction efficiency of 51% was attained at 70 keV with a focused beam size of 1.2 micron. The efficiency measured over a wide range is shown in Fig. 11. High diffraction efficiencies over 40%, which is the theoretical limit of the efficiency of the normal FZP, were attained. These experimental results confirmed that the multilevel type ML-FZP using composite layers functioned well like the FZP with a step structure, such as that shown in Fig. 7(c). Despite the zone roughness, these ML-FZPs worked well as X-ray focusing optics.

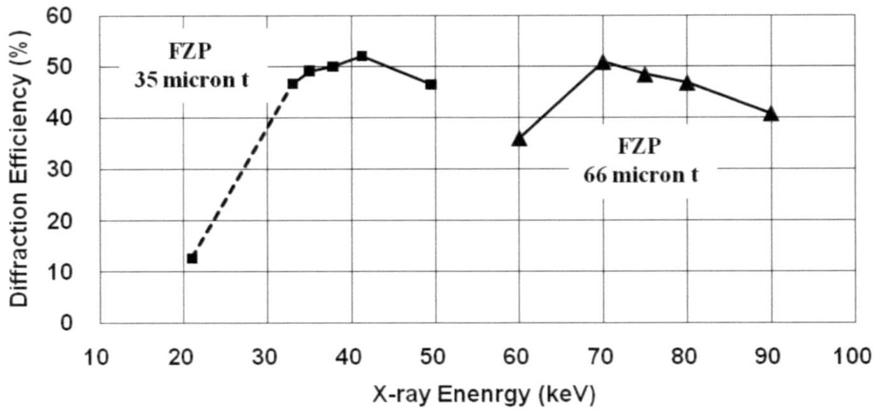

Fig. 11. Measured diffraction efficiencies for 1st-order diffraction for two types of 6-step multilayer FZP.

To achieve high diffraction efficiency in the high-energy X-ray region, multilevel-type multilayer FZPs fabricated with composite layers were developed and tested at the synchrotron radiation beamline at SPring-8. Despite the existence of zone roughness, the experimental results described here confirmed that the proposed multilevel type ML-FZP with composite layers functioned well like the FZP with a step structure fabricated by the lithography-based technique. By precise control of the coating rate in the co-sputtering deposition system, it will be possible to fabricate thin film layers with graded composition structures, which will allow the realisation of kinoform type ML-FZPs with diffraction efficiencies of greater than 80%.

7. X-ray microscopy

As mentioned in Section 1, the ML-FZP is a type of focusing optics for high-energy X-rays at SR facilities. One promising application of the ML-FZP is in X-ray microscopy. Main results of several X-ray microscopy studies using normal ML-FZPs or the 4-step ML-FZP have been reported elsewhere (Takeuchi et al., 2005; Terada et al., 2004; Terada et al., 2005; Kamijo et al., 2003; Suzuki et al., 2001; Kamijo et al., 2006; Takeuchi et al., 2008).

In this section, therefore, other results are shown. Figure 11 presents schematics of representative X-ray microscopy: (a) a scanning microscopy and (b) an imaging microscopy.

A scanning X-ray microscopic image of a Au mesh (64 micron pitch) on the PC monitor is shown in Fig.12-(a). This image was taken in the fluorescent mode (detecting Au K-fluorescent X-ray) at 200 keV (Kamijo et al., 2009). A scanning X-ray microscopic image of a Ta test pattern (thickness: 0.5 micron) is shown in Fig.12-(b). This image was taken by using the 4-step multilevel-type ML-FZP at 23 keV (2nd order light was used without the OSA). The spatial resolution was 500 nm (Kamijo et al., 2006).

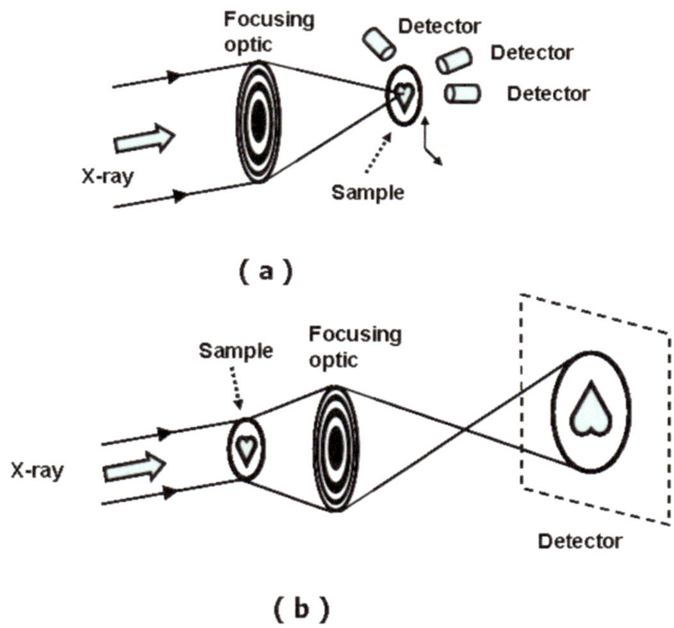

Fig. 11. Schematics of X-ray microscopy: (a) scanning microscopy, (b) imaging microscopy.

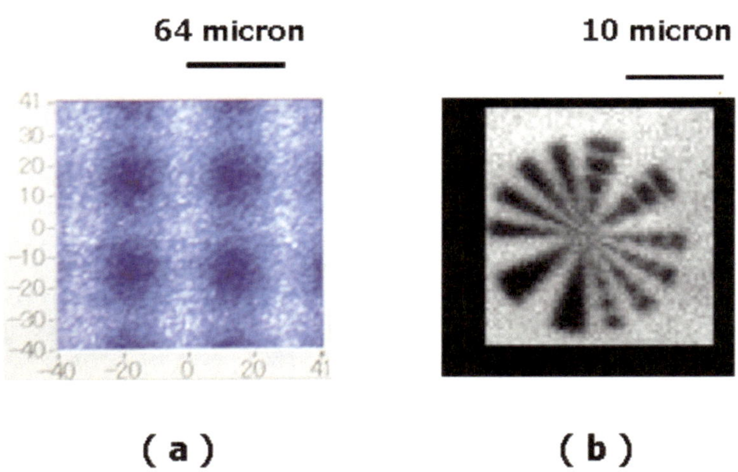

Fig. 12. Scanning X-ray microscopic images by using multilayer Fresnel zone plate: (a) image on PC monitor. (b) transmission x-ray image of Ta test pattern with 0.125 micron step, 0.2 s/pixel.

A scanning microscopic image of a GaAs FET (Gallium Arsenide Field-Effect Transistor) at 27.8 keV is shown in Fig.13 (Tamura et al., 2009).

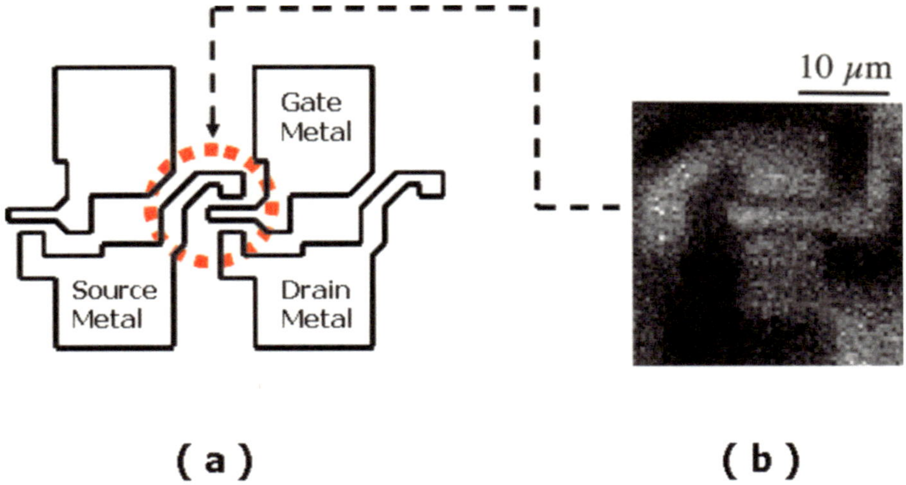

Fig. 13. Scanning X-ray microscopic image of Gallium Arsenide Field-Effect Transistor (GaAs FET) by using multilayer Fresnel zone plate: (a) schematic of GaAs FET, (b) Scanning X-ray microscopic image. Au L-fluorescent X-ray are measured.

8. Conclusions

To achieve high diffraction efficiency in the high-energy X-ray region, multilevel-type multilayer FZPs fabricated with composite layers were developed and tested at the synchrotron radiation beamline at SPring-8. Despite the existence of zone roughness, the experimental results described here confirmed that the proposed multilevel type ML-FZP with composite layers functioned well like the FZP with a step structure fabricated by the lithography-based technique. By precise control of the coating rate in the co-sputtering deposition system, it will be possible to fabricate thin film layers with graded composition structures, which will allow the realisation of kinoform type ML-FZPs with diffraction efficiencies of greater than 80%.

Recently, the microscopic imaging of uranium (L-fluorescent X-ray) by using the K-B mirror as the focusing optics is reported (Terada et al., 2010). Some detail articles have been published concerning X-ray microscopy (Howells et al., 2006) or X-ray focusing optics (Snigirev et al., 2008). Great advance, therefore, will be expected in the field of X-ray microscopy in the near future. In addition, the ML-FZP including multilevel type ML-FZP will be also powerful tools in the field of the hard x-ray three-dimensional microtomography (Takeuchi et al., 2002; Toda et al., 2006) .

9. Acknowledgment

The research described in this chapter was performed as collaborative research with SPring-8 (Japan Synchrotron Radiation Research Institute: JASRI). Thanks are due to long-time collaborators Dr. Yoshio Suzuki (SPring-8), Dr. Akihisa Takeuchi (SPring-8), Dr. Kentaro Uesugi (SPring-8), Dr. Mitsuhiro Awaji (SPring-8), Dr. Yasuko Terada (SPring-8), Dr. Masato Yasumoto (AIST) and Dr. Nagao Kamijo (SPring-8, Kansai Medical University, retired AIST). The synchrotron radiation experiments have been performed at SPring-8 with the approval of the Japan Synchrotron Radiation Research Institute (JASRI) (Proposal No.2004B0180-NM-np, 2005B0269, 2006A1153 and 2006B1747).

10. References

Awaji, M.; Suzuki, Y.; Takeuchi, A.; Takano, H.; Kamijo, N.; Yasumoto, M.; Terada, Y. & Tamura, S. (2003). Microfocusing of 82 keV X-rays with a sputtered-sliced Fresnel zone plate, *Review of Scientific Instruments*, 74, pp.4948-4949, ISSN 0034-6748

Bionta, R.M.; Ables, E.; Clamp, O.; Edwards, O.D.; Gabriele, P.C.; Makowiecki, D.; Ott, L.L.; Skulina. K.M. & Thomas, N. (1989). 8 keV x-ray zone plates, *Proc. SPIE (The International Society for Optical Engineering)*, 1160, pp.12-18, ISSN 0277-786X

Bionta, R.M.; Ables, E.; Clamp, O.; Edwards, Gabriele, P.C.; Ott, L.L., Skulina, K.M. & Viada, T. (1990). Tabletop X-Ray Microscope Using 8 keV Zone Plates, Optical Engineering, 29, pp.576-580 , ISSN: 0091-3286

Bionta, R.M.; Skulina, K.M. & Weinberg, J. (1994). Hard x-ray sputtered-sliced zone plates, *Apped.Physics Letters*, 64, pp.945-947, ISSN 0003-6951

Chu, Y.S.; Yi, J.M.; Carlo, F.De.; Shen, Q.; Lee, Wah-Keat.; Wu, H.J.; Wang, C.L.; Wang, J.Y.; Liu, C.J.; Wang, C.H.; Wu, S.R.; Chien, C.C.; Hwu, Y.; Tkachuk, A.; Yun, W.; Feser, M.; Liang, K.S.; Yang, C.S.; Je, J.H. & Margaritondo, G. (2008). Hard x-ray microscopy with Fresnel zone plates reaches 40 nm Rayleigh resolution, *Applied Physics Letters*, 92, 103119, ISSN 0003-6951

Engemann, S.; Reichert, H.; Dosch, H.; Bilgram, J.; Honkimäki, V. & Snigirev, A. (2004). Interfacial Melting of Ice in Contact with SiO_2, *Physical Review Letters*, 92, 205701, ISSN 0031-9007

Erko, A.I.; Aristov, V.V. & Vidal, B. (1996). "*Diffraction X-ray Optics*", Institute of Physics Publishing, Bristol and Philadelphia, pp.36-40, ISBN 0-7503-0359-X

Fabrizio, E. Di,; Gentili, M.; Grella, L.; Baciocchi, M.; Krasnoperova, A.; Cerrina, F.; Yun, W.; Lai, B. and Gluskin, E. (1994). High-performance multilevel blazed x-ray microscopy Fresnel zone plates: Fabricated using x-ray lithography, *Journal of Vacuum Science and Technology B*, 12, pp.3979-3985, ISSN 0734-211X

Fabrizio, E.Di.; Romannato, F.; Gentili, M.; Cabrini, S.; Kaulich, B.; Susini, J. & Barrett. B. (1999a). High-efficiency multilevel zone plates for keV X-rays, *Nature*, 401, pp.895-898, ISSN 0028-0836

Fabrizio, E.Di. & M.Gentili, M. (1999b). X-ray multilevel zone plate fabrication by means of electron-beam lithography: Toward high-efficiency performances, *Journal of Vacuum Science and Technology B*, 17, pp.3439-3443, ISSN 0734-211X

Fujisaki, H. & Nakagiri, N. (1990). Design of a gradient refractive index phase zone plate for soft x-rays, *Applied Optics*, 29, pp.483-488, ISSN 0003-6935

Hart, H.E.; Scrandis, J.B.; Mark, R. & Hatcher, R.D. (1966). Diffraction Characteristics of a Linear Zone Plate, *Jopurnal of the Optical Society of America*, 56, pp.1018-1023, ISSN 0030-3941

Hayakawa, S.; Iida, A.; Aoki, S. & Gohshi, Y. (1989). Development of a scanning x-ray microprobe with synchrotron radiation, *Review of Scientific Instruments*, 60, pp.2452-2455, ISSN 0034-6748

Howelles, M.; Jacobsen, C. And Warwick, T. (2006). Principles and Applications of Zone Plate X-Ray Microscopes, *Science of Microscopy (Springer-Verlag)*, pp.835-926, ISBN-10: 0387252967

Kamijo, N.; Suzuki, Y.; Tamura, S.; Handa, K.; Takeuchi, A.; Yamamoto, S.; Ando, M.; Ohsumi, K. & Kihara, H. (1997). Fabrication of Hard X-ray Sputtered-Sliced Fresnel Phase Zone Plate, *Review of Scientific Instruments*, 68, pp.14-16, ISSN 0034-6748

Kamijo, N.; Suzuki, Y.; Tamura, S.; Awaji, M.; Yasumoto, M.; Kohmura, Y,; Handa. K. & Takeuchi, A. (2000). Fabrication of high energy X-ray Fresnel phase zone plate, *X-Ray Microscopy (Procedings of the Sixth International Conference)*, pp.672-675, ISSN 0094-243X.

Kamijo, N.; Suzuki, Y.; Awaji, M.; Takeuchi, A.; Takano, H.; Ninomiya, T.; Tamura, S. & Yasumoto, M. (2002a). Hard X-ray microbeam experiments with a sputtered-sliced Fresnel zone plate and its applications, *Journal of Synchrotron Radiation*, 9, pp.182-,186, ISSN 0909-0495

Kamijo, N.; Suzuki, Y.; Tamura, S.; Takeuchi, A.; Takano, H.; Yasumoto, M. & Awaji, M.(2002b). High resolution microbeam experiment with sputtered-sliced zone plate (II) at 250 m beam line (BL20XU), *SPring-8 User Experiment Report*, No.9, p.134

Kamijo, N.; Suzuki, S.; Takano, H.; Tamura, S.; Takeuchi, A.; Yasumoto, M. & Awaji, M, (2003). Microbeam of 100 keV X-ray with a sputtered-sliced Fresnel zone plate, *Review of Scientific Instruments*, 74, pp.5101-5104, ISSN 0034-6748

Kamijo, N.; Suzuki, Y.; Tamura, S.; Takeuchi, A. & Yasumoto, M. (2006). Practical use ofquasi-kinoform zone plate : Towards high-efficiency microbeam for hard /high-energy x-rays, *IPAP Conference Series 7 (Institute of Pure and Applied Physics, Japan)*, pp.97-99, ISBN4-900526-21-5

Kamijo, N.; Suzuki, Y.; Takeuchi, A.; Itou, M. & Tamura, S. (2009). Microbeam of 200 keV x-ray with a sputtered-sliced zone plate, *Japanese journal of Applied Physics*, 48, 010209. ISSN 0021-4922

Kang, H.C.; Yan. H.; Winarski, R.P.; Holt, M.V.; Maser. J.; Liu, C., Conley, R.; Vogt, S.; Macrander, A.T. & Stephenson, G.B. (2008). Focusing of hard x-rays to 16 nanometers with a multilayer Laue Lens, *Applied Physcs Letters*, 92, 221114, ISSN 0003-6951

Kirz, J. (1974). Phase zone plates for x rays and the extreme uv, *Jopurnal of the Optical Society of America*, 64, pp.301-308, ISSN 0030-3941

Korsunsky, A.M. & Wells, K.E. (2000). High Energy Synchrotron X-Ray Measurements of 2D Residual Stress States in Metal Matrix Composites, *Materials Science Forum*, 321-324, p.218, ISSN 1662-9752

Koyama, T.; Takenaka, H.; Ichimaru, S.; Ohchi, T.; Tsuji, T.; Takano, H. & Kagoshima, Y. (2010a). Development of Multilayer Laue Lenses --- (1) Linear Type ---, *Abstract of 10th International Conference on X-ray Microscopy*, P.39

Koyama, T.; T.; Tsuji, Takano, H.; & Kagoshima, Y.; Ichimaru, S.; Ohchi, T. & Takenaka, H. (2010b). Development of Multilayer Laue Lenses --- (2) Circular Type ---, *Abstract of 10th International Conference on X-ray Microscopy*, P.135

Lai, B.; Yun, W. B.; Legnini, D.; Xiao, Y.; Crzas, J.; Viccaro, P.J.; White, V.; Denton, D.; Cerrina, F.; Fabrizio, E.Di; Grella, L. & Baciocchi, M. (1992). Hard x-ray phase zone plate fabricated by lithographic techniques, *Applied Physics Letters*, 61, pp.1877-1879, ISSN 0003-6951

Lengeler, B.; Schroer, C. G.; Richwin, M.; Tümmler, J.; Drakopoulos, M.; Snigirev, A. & Snigireva, I. (1999). A microscope for hard x-rays based on parabolic compound refractive lenses, *Appled Physic Letters*, 74, pp.3924-3926, ISSN 0003-6951

Lienert, U.; Poulsen, H.F.; Martins, R.V. & Kvick, Å. (2000). A High Energy Microscope for Local Strain Measurements Within Bulk Materials, *Materials Science Forum*, 347-349, P.95, ISSN 1662-9752

Matsuyama, S.; Mimura, H.; Yumoto, H.; Sano, Y.; Yamamura, K.; Yabashi, M.; Nishino, Y.; Tamasaku, K.; Ishikawa, T. & Yamauchi, K.(2006). Development of scanning X-ray fluorescence microscope with spatial resolution of 30nm using K-B mirrors optics, *Review of Scientific Instruments*, 77, 103102, ISSN 0034-6748

Mimura, H.; Handa, S.; Kimura, T.; Yumoto, H.; Yamakawa, D.; Yokoyama, H.; Matsuyama, S.; Inagaki, K.; Yamamura, K.; Sano, Y.; Tamasaku, K.; Nishino, Y.; Yabashi, M.; Ishikawa, T. & Yamauchi, K. (2010). Breaking the 10 nanometer barrier in hard X-ray focusing, Nature Physics, 6, pp.122-125, ISSN 1745-2473

Nazmov, V.; Reznikova, E.; Snigirev, A,; Snigireva, I.; DiMichiel, M.; Grigoriev, M.; Mohr, J.; Matthis, B.& V. Saile, V. (2005). LIGA fabrication of X-ray Nickel lenses *Microsystem Technologies*, 11, pp.292-297. ISSN 0946-7076

Reichert, H.; Honkimäki, V.; Snigirev, A.; Engemann, S. & Dosch, H. (2003). A new Transmission-Reflectivity Scheme using High-Energy X-ray Microbeams for the Study of Deeply buried Interfaces, *Physica*, B336, pp.46-55, ISSN 0921-4526

Rudolph, D.; Niemann, B. and Schmahl, G, (1981). Status of the Sputtered Sliced Zone Plates For X-Ray Microscopy, *Proc. SPIE (The International Society for Optical Engineering)* , 316, pp.103-105, ISSN 0277-786X

Saitoh, K.; Inagawa, K.; Kohra, K.; Hayashi, C.; Iida, A. & Kato, N. (1988). Fabrication and Characterization of Multilayer Zone Plate for Hard X-Rays, *Japanese Jourbnal of Applied Physics*, 27, pp. L2131-L2133, ISSN 0021-4922

Saitoh, K.; Inagawa, K.; Kohra, K.; Hayashi, C.; Iida, A. & Kato, N. (1989). Characterization of sliced multilayer zone plates for hard x rays, *Review of Scientific Instruments,*, 60, pp.1519-1523, ISSN 0034-6748

Schroer, C.G.; Kurapova, O.; Patommel, J.; Boye, P.; Feldkamp, J.; Lengeler, B.; Burgahammer, M.; Riekel, C.; Vincze, L.; Hart, A. van der. & Küchler, M. (2005). Hard x-ray nanoprobe based on refractive x-ray lenses, *Applied Physics Letters*, 87, 124103, ISSN 0003-6951

Schroer, C.G.; Kurapova, O.; Patommel, J.; Boye, P.; Feldkamp, J.; Lengeler, B.; Burgahammer, M.; Riekel, C.; Vincze, L.; Hart, A. van der. & Küchler, M. (2006). Hard X-Ray Nanoprobe with Refractive X-Ray Lenses, *IPAP Conference Series 7 (Institute of Pure and Applied Physics, Japan)*, pp.94-96, ISBN4-900526-21-5

Shastri, S.D.; Maser, J.M.; Lai, B. & and Tys. J. (2001). Microfocusing of 50 keV undulator radiation with two stacked Zone pates, *Optics Communications*, 197, pp.9-14. ISSN 0030-4018

Snigirev, A.; Snigireva, I.; Engström, P.; Lequien, S.; Suvorov, A.; Hartman, Ya.; Chevallier, P.; Idir, M.; Legrand, F.; Soullie, G. & Engrand, S. (1995). *Review of Scientific Instruments*, 66, pp.1461-1463, ISSN 0034-6748.

Snigirev, A.; Kohn, V.; Snigireva, I. & Lengeler, B. (1996): A compound refractive lens for focusing high-energy X-rays, *Nature*, 384, pp.49-51, ISSN 0028-0836

Snigirev, A.; Snigireva, I.; Michiel, M.Di.; Honkimaki, V.; Grigoriev, M.; Nazmov, V.; Reznikova, E.; Mohr. J. & Saile, V. (2004). Submicron focusing of high-energy x-rays with Ni refractive lenses, *Proc. SPIE (The International Society for Optical Engineering)* , 5539, pp.244-250, ISSN 0277-786X

Snigirev, A. & Snigireva, I. (2008). Hard X-ray microoptics, *Modern Development in X-ray and Neutron Optics*, p.256, ISBN 978-3-540-74560-0

Spiller, E. (1994). *Soft X-ray Optics*, P.179, SPIE (The International Society for Optical Engineering) Press, ISBN 0-8184-1654-1

Suzuki, Y.; Kamijo, N.; Tamura, S.; Handa, K.; Takeuchi, A.; Yamamoto, S.; Sugiyama, H.; Ohsumi, K. & Ando, M. (1997). Hard X-ray Microprobe Experiment at the TRISTAN MAIN Ring Test Beamline of the KEK, *Journal of Synchrotron Radiation*, 4, pp.60-63. ISSN 0909-0495

Suzuki, Y.; Awaji, M.; Kohmura, Y.; Takeuchi, A.; Takano, H.; Kamijo, N.; Tamura, S.;

Yasumoto, M. & Handa, K. (2001). X-ray Microbeam with Sputtered-Sliced Fresnel Zone Plate at SPring-8 Undulator Beamline, *Nuclear Instruments and Methods in Physics Research*, A467-468, pp.951-953, ISSN 0168-9002

Suzuki, Y.; Takeuchi, A.; Takano, H. & Takenaka, H. (2005). Performance Test of Fresnel Zone Plate with 50nm Outermost Zone Width in Hard X-ray Region, *Japanese Journal of Applied Physics*, 44, pp. 1994–1998, ISSN 0021-4922

Suzuki, Y.; Takeuchi, A.; Uesugi, K.; Awaji, M.; Yasumoto, M.; Tamura, S. & Kamijo, N. (2006). X-ray Imaging Microscopy at 82 keV with Sputtered-sliced Fresnel Zone Plate Objective, *IPAP Conference Series 7 (Institute of Pure and Applied Physics, Japan)*, pp.47-49, ISBN4-900526-21-5

Suzuki, Y.; Takeuchi, A.; & Terada, Y. (2007). High-energy x-ray microbeam with total-reflection mirror optics, *Review of Scientific Instruments*, 78,053713, ISSN 0034-6748

Suzuki, Y. & Takeuchi, A. (2010a). X-ray Microfocusing by Combination of Grazing-Incidence Spherical-Concave Mirrors, *Japanese Journal of Applied Physics*, 49, 106701, ISSN 0021-4922

Suzuki, Y.; Takeuchi. A.; Takenaka, H. & Okada, I. (2010b). Fabrication and Performance Test of Fresnel Zone Plate with 35nm Outermost Zone Width in Hard X-Ray Region, *X-Ray Optics and Instrumentation (Hindawi Publishing Corporation)*, Article ID 824387, ISSN 1687-7632

Takeuchi, A.; Uesugi, K., Takano, H. & Suzuki, Y. (2002). Submicrometer-resolution three-dimensional imaging with hard x-ray imaging microtomography, *Review of Scientific Instruments*, 73, pp.4246-4269, ISSN 0034-6748

Tamura, S.; Ohtani, K. & Kamijo, N. (1994). Materials for Multilayer Zone Plates --- Development of a focusing element for use in SR photo-excited process ---, *Applied Surface Science*, 79/80, 99.514-518, ISSN 0169-4332

Tamura, S.; Kensuke, M.; Nagao, K.; Yoshida, K.; Kihara, H. & Suzuki, Y. (2000). Focusing efficiency of a multilayer Fresnel zone plate for hard X-ray fabricated by DC sputtering deposition: *Vacuum*, 59, pp.553-558, ISSN 0042-207X

Tamura, S.; Yasumoto, M.; Mihara, T.; Kamijo, N.; Suzuki, Y.; Awaji, M.; Takeuchi, A.;Takano, H. & Handa, K. (2002). Multilayer Fresnel zone plate for high energy X-rays by DC sputtering deposition: *Vacuum*, 66, pp.495-499, ISSN 0042-207X

Tamura, S.; Yasumoto, M.; Kamijo, N.; Suzuki, Y.; Awaji, M.; Takauchi, A.; Uesugi, K.; Terada, Y & Takano, H. (2006). New approaches to fabrication of multilayer Fresnel zone plate for X-rays, *Vacuum*, 80, pp.823-827, ISSN 0042-207X

Tamura, S.; Yasumoto, M.; Kamijo, N.; Takeuchi, A.; Uesugi, K. & Suzuki, Y. (2008). Multilevel-type multilayer X-ray lens (Fresnel zone plate) by sputter deposition, *Vacuum*, pp.691-694, ISSN 0042-207X

Tamura, S.; Yasumoto, M.; Kamijo, N.; Takeuchi, A. & Suzuki, Y. (2009). X-ray Focusing Optics with Multilayer Structure and its Application, *Journal of the vacuum society of Japan*, 52, pp.212-217, ISSN 1882-2398

Takano, H.; Tsuji, T.; Hashimoto, T.; Koyama, T.; Tsusaka, T. & Kagoshima, Y. (2010). Sub-15nm Hard X-Ray Focusing with a New Total-Reflection Zone Plate, *Applied Physics Express*, 3, 076702, ISSN 1882-0778

Takeuchi, T.; Kageyama, H.; Tamura, S.; Kamijo, N.; Suzuki, Y. & Takeuchi, A. (2005): Preparation of Composite Dielectric Ceramics by Spark-Plasma-Sintering Method, *Proceedings of International Symposium on Novel Materials Processing by Advanced Electromagnetic Energy Source (MAPEES'04)*, pp. 285-288, ISBN 10-0080445047

Takeuchi, T.; Kageyama, H.; Nakazawa, H.; Atsumi, T.; Tamura, S.; Kamijo, N.; Suzuki, Y. & Takeuchi, A. (2008). Preparation of fluorine-containing indium tin oxide sputtering targets using spark plasma sintering process, *Journal of American Ceramics Society*, 91, pp.2495-2500, ISSN 0002-7820

Terada, Y.; Suzuki, Y.; Kamijo, N.; Tamura, S.; Onuma, R.; Hokura, A. & and Nakai, I. (2004). High Energy Micro-XRF Imaging of Heavy Metals in the Cells of Hyperaccumulator Plants, *Abstract of 8 th International Conference on Biology and Synchrotron Radiation (BSR2004)*

Terada, Y.; Takeuchi, A.; Suzuki, Y.; Onuma, R.; Shirai, K.; Nakai, I.; Kamijo, N. & Tamura, S. (2005). Development of high energy micro-XRF analysis using fresnel zone plate optics, *Abstract of 8th International Conference of X-ray Microscopy (XRM2005)*, PD22.

Terada, Y.; Takeda, S.Homma.; Takeuchi, A. and Suzuki, Y. (2010). High-Energy X-Ray Microprobe System with Submicron Resolution for X-Ray Fluorescence Analysis of Uranium in Biological Specimens, *X-Ray Optics and Instrumentation (Hindawi Publishing Corporation)*, Article ID 317909, ISSN 1687-7632

Thornton, J. A. (1974). Influence of apparatus geometry and deposition conditions on the structure and topography of thick sputtered coatings, *Nuclear Instruments and Methods in Physics Research* , 11, pp.666-670, ISSN 0022-5355

Thornton, J. A. (1986). The microstructure of sputter-deposited coatings, *Nuclear Instruments and Methods in Physics Research Section J. Vac. Sci. Technol.*, A4, pp.3059-3065, ISSN 0734-2101

Toda, H.; Uesugi, K.; Takeuchi, A.; Minami, K.; Kobayashi, M. & Kobayashi, T. (2006). Three-dimensional observation of nanoscopic precipitates in an aluminum alloy by microtomography with Fresnel zone plate optics, *Appled Physic Letters*, 89, 143112, ISSN 0003-6951

Underwood, J.H.; Thompson, A.C.; Wu, Y. & Giauque, R.D. (1988). X-ray microprobe using multilayer mirrors, *Nuclear Instruments and Methods in Physics Research*, A266, pp.296-302, ISSN 0168-9002

Yasumoto. M.; Tamura, S.; Kamijo, N.; Suzuki, Y.; Awaji, M.; Takeuchi, A.; Takano, H.; Kohmura, Y. & Handa, K. (2001). Suppression of corrugated boundaries in multilayer Fresnel zone plate for synchrotron radiation hard X-ray using cylindrical slit, *Japanese Journal of Applied Physics*, 40, pp.4747-4748, ISSN 0021-4922

Yun, W. B.; Viccaro, P.J.; Lai, B. & Chrzas. J. (1992). Coherent hard x-ray focusing optics and application, *Review of Scientific Instruments*, 63, pp.582-585, ISSN 0034-6748

32

New Composite Materials for Decreasing of Radioactive Molecular Iodine in the Water Coolant in Working and New Developed NPPs

Sergey A. Kulyukhin and Lubov' V. Mizina
A.N. Frumkin Institute of Physical Chemistry and Electrochemistry
Russian Academy of Science
Russia

1. Introduction

Radioactive iodine activity in the water coolant sharply grows during shutdown of the working WWER or RBMK reactors. Therefore, it is marked that radioactivity of ^{131}I in water coolant grows in 30-200 times during shutdown of the RBMK reactor (the Leningrad and Smolensk NPPs) (Kriytskii et al., 2005). Authors of the given paper explain such increase of activity in water coolant as follows. During shutdown of a reactor the temperature of fuel inside fuel elements is reduced, that results in reduction of volume of fuel and occurrence of free space inside fuel elements. At the same time, water from coolant penetrates inside fuel elements through defects in fuel cladding and dissolves radioactive iodine and caesium, which were diffused in the fuel and accumulated in a gap between fuel and cladding at a high degree of the fuel burnup. As a result of dissolution of radioactive iodine and caesium, they are passed as ions into a water coolant. Radioactive iodine can be present in a water coolant as ionic (I^- and IO_3^-) and molecular (I_2) forms due to the oxidation-reduction reactions. Furthermore, because of a radiation destruction of the ion-exchanged resins the organic form (CH_3I) can be also formed in a water coolant. At the same time the content of radioactive iodine in a gas phase grows more than by the order as molecular and organic forms of radioactive iodine make the basic contribution to pollution of a gas phase since their distribution factors between gas and water phases are higher on some orders than similar for ionic forms. The similar effect of increase in the radioactive iodine content in the water coolant during shutdown can be also observed as on working (PWR, LWR, BWR), and on new developed (EPR /Framatome ANP/, SWR-1000 /Framatome ANP/, AP-1000 /Westinghouse/, ESBWR /General Electric/, APWR /Mitsubishi/, ABWR /Toshiba and Hitachi/) NPPs (Krebs, 1999; Reactor Design, 2002; Tomorrow's reactors, 2003).

Usually for decreasing of the radionuclide content in a water coolant the cation- and anion-exchange resins are used on NPPs during normal operation.

Moreover, the 50% decreasing in the power capacity are used on the Leningrad NPP for decreasing of the radioactive iodine content in the water coolant during shutdown of a reactor, i.e. the reactor works on 50% of the power capacity during 2-5 days (Kriytskii et al., 2005). The given approach allows to lower radioactivity of iodine in the water coolant due to short-lived radionuclides decay. However, these actions do not influence the ^{131}I behavior, which makes the basic contribution to activity of the water coolant even in 5 days.

In this connection, it was suggested to continue circulation of the water coolant during 2-5 d with its simultaneous clearing from radioactive iodine after shutdown of a reactor (Kriytskii et al., 2005).

As anion-exchange resins are capable to absorb only ionic forms of the radioactive iodine, there was a problem to develop a new material, capable to absorb molecular and organic iodine from water solutions. Absorption of all forms radioactive iodine from a water phase during shutdown of a reactor, and also, probably, and at normal operation will allow not only to lower radioactivity of the water coolant, but also will sharply lower emission of radioactive iodine into a gas phase. Thus, it is desirable to use materials, which are applied on the working NPPs to clearing coolant, for example ion-exchange resins.

We have shown previously (Krasavina et al., 2008; Kulyukhin et al., 2008, 2008a) that composite materials based on KU-2 cation exchanger, containing nanometer-sized particles of Ag and Cu in an amount of 10 wt %, are capable to take up the molecular form of radioactive iodine from aqueous solutions to 95–99%. The distribution coefficients (K_d) of I_2 exceeded 10^3 cm^3/g upon 15-min contact of the solid and liquid phases at V/m = 100 cm^3/g. The high cost of silver, which is a component of the examined sorbent, makes topical its replacement. Apparently, the most suitable for this purpose are *d*-elements whose physicochemical properties are close to those of silver, namely, Cu, Ni and Zn.

The goal of this study was to develop new composite materials based on Cu-, Ni- and Zn-containing ion exchangers, capable of decreasing the concentration of the molecular form of radioactive iodine in aqueous primary coolant of NPP.

2. Experimental

In our study we used the radionuclide ^{131}I supplied by Izotop Joint-Stock Company in the form of carrier free Na^{131}I solution. The radioactivity of the nuclides was determined by γ-ray spectrometry with a semiconductor Ge–Li detector on a multichannel analyzer. The radionuclide ^{131}I was used as this study as a spike for weighable amount of inactive iodine. Therefore, designations like $^{131}I^-$ and $^{131}IO_3^-$ refer to labeled species and not to species of pure ^{131}I radionuclide. In our study we used aqueous 10^{-5} M solutions of K^{131}I or K$^{131}IO_3$ with the specific activity from 2×10^3 to 3×10^4 Bq/cm^3, and also aqueous solutions of 9×10^{-4} M I_2 containing ~5 mg/cm^3 KI.

All salts, alkalis, and acids used in this study were of chemically pure grade.

Composite materials based on KU-2 cation exchanger and containing *d*-elements were synthesized in two steps, following the procedure described in the patent (Kulyukhin, 2009). In the first step, we performed sorption of *d*-elements (Cu^{2+}, Ni^{2+}, Zn^{2+}) on KU-2 cation exchanger by the following procedure. To aqueous solutions containing appropriate amounts of *d*-element nitrates, we added a weighed portion of KU-2 cation exchanger. The suspension was shaken for 2 h, and then the mother liquor was separated from the cation exchanger impregnated with *d*-elements by centrifugation. The cation exchanger was washed two or three times with distilled water. As a result, we obtained samples of KU-2 cation exchanger containing *d*-elements (Cu^{2+}, Ni^{2+}, Zn^{2+}). The content of metals in KU-2 cation exchanger was determined from the difference between the initial and residual concentrations of metals in the mother liquor and wash waters. Analysis was performed by complexometric titration with EDTA in the presence of murexide or Eriochrome Black, following standard procedures (Charlot, 1961). The results are given in Table 1.

Composite material	Content of d-element in composite material		
	mM/g	mg/g	wt %
$KU\text{-}2\text{-}Cu^{2+}$	0.830	53.12	5.3
	0.844	54.02	5.4
$KU\text{-}2\text{-}Ni^{2+}$	0.931	54.93	5.5
	0.920	54.28	5.4
$KU\text{-}2\text{-}Zn^{2+}$	0.869	56.49	5.6
	0.872	56.68	5.7

Table 1. Content of *d*-elements (Cu^{2+}, Ni^{2+}, Zn^{2+}) in the composite material based on KU-2 cation exchanger

In this step of the synthesis we prepared precursors of the composite materials, based on KU-2 cation exchanger containing ~5.5 wt % *d*-elements. In the second step of the synthesis, the precursors of composite materials were modified. For this purpose, the materials were treated with a 1 M hydrazine hydrate (HH) solution at the metal : HH molar ratio of 1 : 3. Modification was performed at room temperature for 24 h. In all the cases we observed gas evolution and a change in the sample color: In the case of KU-2-Cu-m the color changed from light green to brown with metal lustre, in the case of KU-2-Ni-m, violet and turquoise disseminations appeared against the background of light green color, and in the case of KU-2-Zn-m the color changed from yellow to pale yellow (Fig. 1). The modified cation exchanger containing *d*-elements was separated from the mother liquor by centrifugation and then washed with distilled water until hydrazine hydrate was removed completely (to pH 6–7 of wash waters).
To study localization of ionic ($^{131}I^-$, $^{131}IO_3^-$) and molecular (I_2) iodine species, we used KU-2-M^{II}-m composites (M^{II} = Cu, Ni, Zn) and, along with distilled water, solution of the following composition: NaCl 1.17 mg/dm^3, NH_4Cl 29.96 mg/dm^3, H_3BO_3 5 g/dm^3, pH 7–8.5. The salt concentrations were chosen on the basis of data on the averaged composition of aqueous primary coolant from the first units in Russian NPPs with WWER-type reactors (Kalinin and Novovoronezh NPPs).

Fig. 1. Composite materials KU-2-Zn-m (pale yellow), KU-2-Ni-m (violet and turquoise disseminations) and KU-2-Cu-m (brown with metal lustre)

Experiments on sorption of $^{131}I^-$ or $^{131}IO_3^-$ from the aqueous phase on composite materials were performed as follows. To 25 cm^3 of an aqueous solution containing 10^{-5} M KI (or KIO_3) and the ^{131}I radionuclide, we added 25 mg of the composite material. The suspension was vigorously stirred for a definite period of time, after which the solid phase was separated from the mother liquor by centrifugation. The precipitate was washed two times with water. The wash waters and mother liquor were combined in one fraction, after which the content of ^{131}I in the solution and precipitate was determined by γ-ray spectrometry. Knowing the V/m ratio, from the distribution of ^{131}I between the solid and liquid phases we calculated the distribution coefficient K_d by the formula

$$K_d = (A_1 / A_2)\cdot(V/m) \quad (1)$$

where A_1 and A_2 - the content of radionuclide ^{131}I in solid and liquid phases accordingly; V - volume of liquid phase (cm^3); m - weight of the composite material based on ion-exchange resin (g).

Because in aqueous solution radioactive iodine can occur not only in molecular form, but also in several other forms simultaneously (I^-, IO_3^-, IO_4^-), it was impossible to use the radioactive tracer technique for studying localization of specifically I_2. Therefore, the efficiency of the I_2 sorption on AV-18 anion exchanger and on the synthesized composite materials was studied by colorimetry using inactive aqueous solutions of I_2.

Experiments were performed as follows. To 25 cm^3 of an aqueous solution of 9×10^{-4} M I_2 we added 25 mg of the composite material. The suspension was vigorously stirred for 120 min, after which the precipitate was separated by centrifugation and washed two times with water. The wash waters obtained and the mother liquor were combined. The amount of I_2 remaining in the final mother liquor and wash waters and that in the initial solution were determined colorimetrically with a fairly high accuracy from a change in the absorption of I_3^- ions in aqueous solution (Ksenzenko et al., 1979). The procedure is very sensitive (detection limit 5 μg/cm^3). The molar extinction coefficient ε for I_3^- at λ = 350 nm is 25000 dm^3/mol×cm. The content of I_2 was determined as follows. An aliquot of a neutral solution being analyzed (5 cm^3), containing from 20 to 200 μg of I_2, was placed in a volumetric flask (25 cm^3), and 1 cm^3 of 1.0 M KI was added. The solution volume was brought to 25 cm^3, and the contents were thoroughly mixed. After 1 h, the solution transparence was measured using a blue color filter at the absorption band of I_3^- ions (λ = 350 nm) in 1–3-cm-thick cells. The transparence of the reference solution was also determined after 1 h. The amount of I_2 was determined with a calibration plot constructed using standard solutions with the I_2 concentration ranging from 0 to 1.0 mg. The standard solutions for the measurements were prepared for measurements by the same procedure as the solution being analyzed.

Thermal gravimetric analysis of KU-2-M^{II}-m composites (M^{II} = Cu, Ni, Zn) and of KU-2 cation exchanger was performed with a Q-1500 derivatograph (MOM, Budapest, Hungary). Samples were heated in platinum crucibles in air at a rate of 10 deg/min.

3. Results and discussion

It is known that, in aqueous solution, molecular iodine (I_2) can form complex ions of the composition I_3^-. Therefore, prior to developing new materials for localization of the molecular form in the primary coolant, it was necessary to examine the sorption properties of the anion exchangers toward I_2 in aqueous solution. The distribution coefficients K_d of I_2 on AV-18 anion exchanger from a solution of iodine in distilled water and simulated coolant are given in Table 2.

Time of contact between phases	K_d, cm^3/g	
	water	primary coolant
5 min	100 ± 3	100 ± 3
15 min	130 ± 10	150 ± 12
30 min	415 ± 45	250 ± 25
60 min	532 ± 50	480 ± 50
120 min	540 ± 52	485 ± 52
24 h	7900 ± 200	9100 ± 250
72 h	> 10^4	> 10^4

Table 2. Distribution coefficients K_d of I_2 on AV-18 anion exchanger from solutions in water and simulated primary coolant at 20°C (V/m = 100 cm^3/g, $[I_2] = 9\times10^{-4}$ M)

As can be seen, AV-18 anion exchanger fairly efficiently takes up molecular iodine from aqueous solutions. The distribution coefficients K_d of I_2 increase with the time of contact of the solid and liquid phases, and after 24 h of contact of the solid and liquid phases more than 99.0% of iodine is recovered from its solutions in distilled water and from the simulated primary coolant.

It should be noted that data on sorption of molecular iodine, given in Table 2, were obtained under static conditions. Actually the contact time of the coolant with the ion exchangers does not exceed 5–15 min (Glanneskog et al., 2006). Therefore, low rate of sorption of molecular iodine from aqueous solutions on AV-18 anion exchanger does not allow its use for efficient removal of this form of radioiodine from primary coolant both in the course of the campaign and during the reactor outage.

Data on localization of I_2 from distilled water on wet composite materials at various contact times of the solid and liquid phases are given in Table 3. To avoid oxidation of the developed materials with atmospheric oxygen, they were stored under a layer of distilled water and, in experiments on I_2 localization, used wet. As seen from Table 3, the synthesized composite materials exhibit high ability to localize the molecular form of iodine from aqueous solutions, and virtually complete localization of I_2 (>99.0%) is attained already in 5 min. At the same time, KU-2 cation exchanger does not participate in localization of the molecular form of radioiodine to a noticeable extent.

Time of contact between phases	C/C_0			
	KU-2	KU-2-Ni-m	KU-2-Zn-m	KU-2-Cu-m
5 min	< 0.0001	0.994 ± 0.004	0.993 ± 0.005	0.991 ± 0.004
15 min	< 0.0001	0.994 ± 0.004	0.993 ± 0.003	0.992 ± 0.005
30 min	< 0.0001	0.994 ± 0.005	0.992 ± 0.004	0.991 ± 0.003
60 min	< 0.0001	0.994 ± 0.005	0.992 ± 0.004	0.992 ± 0.004
120 min	< 0.0001	0.992 ± 0.004	0.990 ± 0.002	0.991 ± 0.004
72 h	< 0.0001	0.995 ± 0.004	0.995 ± 0.004	0.992 ± 0.005

Table 3. Fraction of I_2 (C/C_0) localized on various materials at various contact times of the solid and aqueous phases at 20°C ($[I_2] = 1\times10^{-5}$ M)

It was interesting to elucidate the mechanism of localization of molecular iodine on composite materials. A study of the I_2 localization from aqueous solutions on massive Cu, Zn, and Al plates (Glanneskog et al., 2006) showed that localization of I_2 on Cu plates occurs via adsorption with the formation of CuI on the surface, whereas in the presence of Zn and

Al plates the I_2 localization involves its conversion into ionic species. It was shown in (Glanneskog et al., 2006) that, the larger the ratio of the metal surface area to the solution volume, the higher the adsorption (or conversion) rate. In addition, the reaction rates increased with an increase in the solution temperature. To elucidate the mechanism of localization of molecular iodine from aqueous solution, it was necessary to perform experiments to examine the material balance of iodine in the liquid phase before and after localization on composite materials.

However, because iodine in the aqueous phase can occur in both molecular and ionic forms, it was preliminarily necessary to examine the sorption properties of the composites synthesized toward ionic iodine species, I^- and IO_3^-. The distribution coefficients K_d of $^{131}I^-$ and $^{131}IO_3^-$ from aqueous solutions on composite materials based on KU-2 cation exchanger containing *d*-elements and on straight KU-2 resin after 120-min contact of the solid and liquid phases are given in Table 4. As can be seen, the synthesized composite materials KU-2-M^{II}-m (M^{II} = Ni, Zn), as well as KU-2 cation exchanger, do not noticeably sorb ionic radioiodine species from aqueous solutions. The degree of sorption of $^{131}I^-$ and $^{131}IO_3^-$ does not exceed 0.02%, and K_d of $^{131}I^-$ and $^{131}IO_3^-$ is below 1.5 cm^3/g. At same time, the degree of sorption of $^{131}I^-$ and $^{131}IO_3^-$ from water solutions on composite material KU-2-Cu-m changes from 12 to 45%, and K_d of $^{131}I^-$ and $^{131}IO_3^-$ is 15-85 cm^3/g.

Composite materials	K_d, cm^3/g			
	water		primary coolant	
	$^{131}I^-$	$^{131}IO_3^-$	$^{131}I^-$	$^{131}IO_3^-$
KU-2	1.1 ± 0.05	1.1 ± 0.05	1.1 ± 0.06	1.1 ± 0.06
KU-2-Ni-m	0.4 ± 0.02	0.16 ± 0.02	0.1 ± 0.01	0.5 ± 0.02
KU-2-Zn-m	0.1 ± 0.01	0.16 ± 0.02	0.1 ± 0.01	0.1 ± 0.01
KU-2-Cu-m	14.8 ± 0.7	24.3 ± 1.2	46.8 ± 2.3	81.0 ± 4.0

Table 4. Distribution coefficients (K_d) of $^{131}I^-$ and $^{131}IO_3^-$ on composite materials from solutions in distilled water and in simulated primary coolant at 20°C ([$^{131}I^-$] = [$^{131}IO_3^-$] = 1×10^{-5} M, [NaCl] =1.17 mg/dm^3, [NH_4Cl] = 29.96 mg/dm^3, [H_3BO_3] = 5 g/dm^3, contact time 120 min, V/m = 100 cm^3/g)

Experiments on studying the material balance of iodine in the liquid phase before and after localization on composite materials were performed as follows. To 100 cm^3 of an aqueous solution containing definite amounts of I_2, I^-, and IO_3^-, at 20°C, we added 250 mg of the composite material. The suspension was shaken for 120 min, and the solid phase was separated from the solution by centrifugation. The solid phase was then washed two or three times with distilled water.

The mother liquor and the wash waters were combined. In the initial and final solutions, we first determined the amount of the molecular form (I_2) and then, after the reduction of I_2 with hydrazine hydrate to I^-, the total iodine content. Because the developed composite materials do not noticeably absorb ionic iodine species, from the change of the iodine amount in solution we could evaluate the localization of I_2. Similar experiments were performed at 40 and 60°C, but first we examined the effect of temperature on the dissociation of I_2 in aqueous solution. The experiments were performed as follows. Ampules with an aqueous I_2 solution of known concentration were placed in a BINDER drier oven and kept at a definite temperature for 5, 15, 30, 60, and 120 min. The I_2 concentration in the final solution was determined, and the degree of dissociation of I_2 was calculated from difference. This correction for the change in the I_2 concentration was taken into account both when constructing the iodine balance and when performing experiments on I_2 localization from various solutions. Data on the kinetics of variation of the I_2 content in aqueous solution at 20, 40, and 60°C are given in Table 5. As can be seen, at 20 and 40°C the I_2 concentration in the solution does not change for 120 min, whereas at 60°C in 120 min the I_2 concentration in the solution decreases by approximately 20%. Because in all the experiments the contact time of the solid and liquid phases did not exceed 120 min, it was unnecessary to perform thermostating of I_2 solutions for a longer time.

t, min	20°C	40°C	60°C
5	220 ± 3	220 ± 3	220 ± 3
15	220 ± 4	220 ± 2	200 ± 2
30	220 ± 3	220 ± 2	190 ± 4
60	220 ± 2	220 ± 3	185 ± 2
120	220 ± 3	220 ± 4	180 ± 3

Table 5. Concentration of I_2 in aqueous solution (mg/dm^3) in relation to thermostating time *t*

Data on the iodine balance in the aqueous solution–composite system at 20, 40, and 60°C are given in Table 6. As can be seen, in all the cases, excepting system "water solution - KU-2-Cu-m" at 20°C, the total amount of iodine in the system before and after the interaction with the composite materials coincides. Each value of the I_2 and ΣI amount in the initial and final solutions was calculated as the mean value from three replicate runs. This fact suggests that, in all the systems, in the presence of the developed composite materials I_2 undergoes conversion to ionic species. In the case system "water solution - KU-2-Cu-m" at 20°C about 20% of I_2 absorb on the composite material and about 80% of I_2 converse to ionic forms. So, in what follows we study the localization of I_2 in the system "water solution - KU-2-Cu-m" at 20°C without isolation of each processes contribution.

Data on localization of I_2 from aqueous solutions on wet composite materials KU-2-M^{II}-m (M^{II} = Ni, Zn, Cu) at 20, 40, and 60°C are given in Table 7. As can be seen, wet composite materials KU-2-M^{II}-m (M^{II} = Ni, Zn, Cu) exhibit high localizing ability toward molecular iodine both in distilled water and in simulated primary coolant of water-cooled water-moderated reactors. Despite certain decrease in the degree of I_2 localization in the simulated primary coolant, the degree of localization exceeds 95%. The study of the kinetics of these processes shows that already after 5-min contact of the solid phase and solution the degree of localization of I_2 exceeds 95%.

Taking into account high reduction potential of the Ni^{2+}/Ni^0 and Zn^{2+}/Zn^0 couples [–0.234 and –0.763 V, respectively (Lidin et al., 2006)] and the capability of Ni^{2+} and Zn^{2+} ions to form stable complexes with hydrazine [log*K* = 2.76–11.99 for Ni^{2+} and 2.4–6.3 for Zn^{2+} (Chemist's Handbook, 1964)], we could expect that modification of the materials with hydrazine hydrate could involve partial or complete reduction of metal ions to the metallic state or formation of complex ions $[M^{II}(N_2H_4)_n]^{2+}$ (M^{II} = Ni, Zn) in the composite materials. Figure 2 shows electron micrographs of the composite materials KU-2-M^{II}-m (M^{II} = Ni, Zn), obtained with an EM-301 electron microscope (Philips). As can be seen, the materials developed contain nanometric M^0 particles (M = Ni, Zn) of the size that does not exceed 50 nm. However, their concentration is low. Therefore, we could assume formation of complex ions in the composite material on treatment of the precursor with hydrazine hydrate.

Figure 3 shows the results of thermal gravimetric analysis of the synthesized materials KU-2-M^{II}-m (M^{II} = Ni, Zn). As seen from Fig. 3a, in the DTA curve of KU-2-Ni-m at 252°C there is an exothermic effect accompanied by a sharp weight loss. This effect is parently associated with decomposition of the complex $[Ni(N_2H_4)_n]^{2+}$ in the material. The subsequent exothermic effect is due to combustion of the ion-exchange matrix. In the DTA curve of KU-2-Zn-m (Fig. 3b), there is also an exothermic effect accompanied by a sharp weight loss. It is observed at 285°C and is, apparently, associated with decomposition of the complex $[Zn(N_2H_4)_n]^{2+}$ in the material. The subsequent exothermic effect, as in the case of KU-2-Ni-m, is due to combustion of the ion-exchange matrix. The data we obtained allow a conclusion that treatment of the precursor, KU-2-M^{2+} (M^{2+} = Ni, Zn), with hydrazine hydrate mainly involves formation of complex ions $[M^{II}(N_2H_4)_n]^{2+}$ (M^{II} = Ni, Zn) in the composite material and only partial reduction of *d*-element ions to the metal. It should be noted that both diagrams of thermal decomposition of KU-2-M^{II}-m materials (M^{II} = Ni, Zn) differ from the curves of thermal decomposition of KU-2 (Fig. 4). Thus, the localizing effect of KU-2-M^{II}-m composites (M^{II} = Ni, Zn) is based on the reduction of molecular iodine to ionic species with hydrazine molecules incorporated in the complex $[M^{II}(N_2H_4)_n]^{2+}$ (M^{II} = Ni, Zn) in the structure of the materials.

Composite material	T, °C	Initial solution		Final solution		Composite material
		I_2	ΣI	I_2	ΣI	
KU-2-Ni-m	20	0.039 ± 0.002	0.064 ± 0.003	< 0.0001	0.064 ± 0.003	< 0.0001
	40	0.037 ± 0.001	0.062 ± 0.002	< 0.0001	0.062 ± 0.002	< 0.0001
	60	0.038 ± 0.002	0.063 ± 0.003	< 0.0001	0.063 ± 0.003	< 0.0001
KU-2-Zn-m	20	0.039 ± 0.002	0.064 ± 0.003	< 0.0001	0.064 ± 0.003	< 0.0001
	40	0.037 ± 0.001	0.062 ± 0.002	< 0.0001	0.062 ± 0.002	< 0.0001
	60	0.038 ± 0.002	0.063 ± 0.003	< 0.0001	0.063 ± 0.003	< 0.0001
KU-2-Cu-m	20	0.039 ± 0.002	0.064 ± 0.003	< 0.0001	0.056 ± 0.002	0.008 ± 0.001
	40	0.037 ± 0.001	0.062 ± 0.002	< 0.0001	0.062 ± 0.002	< 0.0001
	60	0.038 ± 0.002	0.063 ± 0.003	< 0.0001	0.063 ± 0.003	< 0.0001

Table 6. Iodine balance (content, mM) in the aqueous solution–composite system at various temperatures

Composite materials	T, °C	C/C_0	
		water	primary coolant
KU-2-Ni-m	20	0.992 ± 0.004	0.960 ± 0.004
	40	0.982 ± 0.005	0.958 ± 0.005
	60	0.985 ± 0.007	0.963 ± 0.006
KU-2-Zn-m	20	0.995 ± 0.004	0.968 ± 0.004
	40	0.993 ± 0.005	0.961 ± 0.005
	60	0.991 ± 0.007	0.966 ± 0.006
KU-2-Cu-m	20	0.991 ± 0.004	0.963 ± 0.004
	40	0.987 ± 0.005	0.957 ± 0.005
	60	0.984 ± 0.007	0.955 ± 0.006

Table 7. Degree of conversion of I_2 (C/C_0) into ionic species with wet composite materials KU-2-M^{II}-m (M^{II} = Ni, Zn, Cu) from a solution in distilled water and from simulated primary coolant at various temperatures ($[I_2] = 1\times10^{-5}$ M, [NaCl] = 1.17 mg/dm^3, $[NH_4Cl]$ = 29.96 mg/dm^3, $[H_3BO_3]$ = 5 g/dm^3, contact time 120 min)

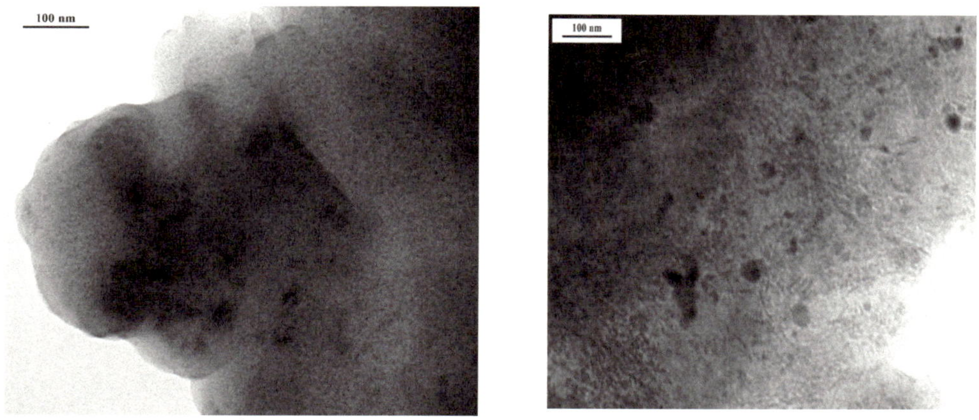

A B

Fig. 2. Electron micrographs of the composite material KU-2-Ni-m (A) and KU-2-Zn-m (B).

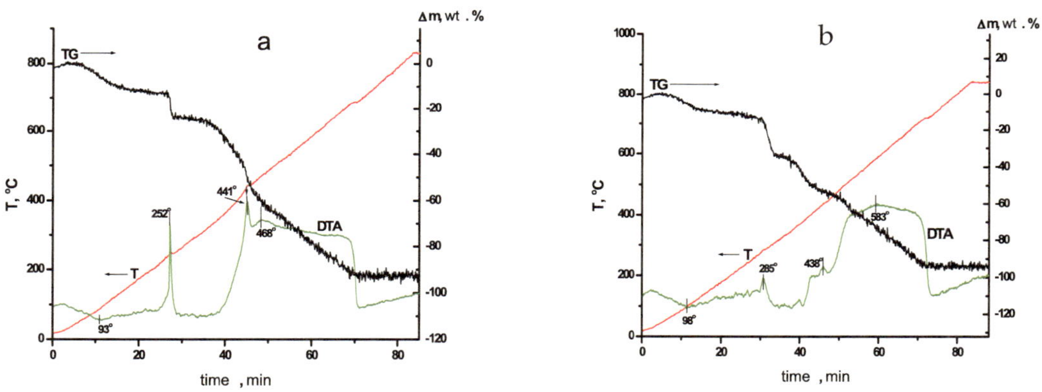

Fig. 3. Results of thermal gravimetric analysis of (a) KU-2-Ni-m (sample weight 69.0 mg) and (b) KU-2-Zn-m (sample weight 62.6 mg).

Based on the data of conversion of I_2 in the presence of copper nanometric particles (Fedushchak et al., 2008), it was interested to study the structure of developed composite material KU-2-Cu-m.

During reduction of Cu by hydrazine hydrate the formation of metal particles can take place. On the dependence from reduction conditions the size of metal particles can change more then order (Fedushchak et al., 2008). The higher copper concentration in material, the bigger particle formed during reduction.

Figure 5 shows electron micrographs of the composite material KU-2-Cu-m, obtained with an EM-301 electron microscope (Philips). As can be seen, the materials developed contain nanometric Cu^0 particles of the size that does not exceed 50 nm. Due to the formation of nanometric particle of Cu^0 the conversion of I_2 takes place in these systems.

At the same time the $[Cu(N_2H_4)_n]^{2+}$ can formed during treatment of the precursor with hydrazine hydrate (HH). Figure 6 shows the results of thermal gravimetric analysis of the synthesized materials KU-2-Cu-m. As seen from Fig. 4 and 6, the decomposition curves of KU-2 and KU-2-Cu-m are similar. The DTA curve of KU-2-Cu-m does not include additional exothermic effect in comparison with DTA curve of KU-2. These data allow us to conclude that during treatment of the precursor KU-2-Cu^{2+} with HH the reduction of Cu^{2+} to Cu^0 take place.

Figure 7 shows comparative data on localization of the I_2 from simulated primary coolant on AV-18 anion exchanger and on wet samples of the developed composite materials. As can be seen, contact for 15 min ensures virtually complete localization of I_2 (>99.0%) on the materials we developed, whereas on AV-18 anion exchanger the degree of the I_2 sorption is as low as 60%.

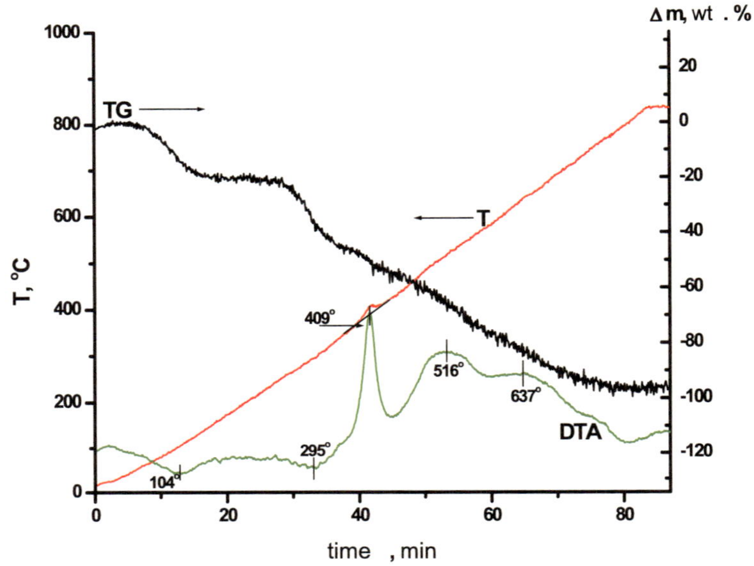

Fig. 4. Results of thermal gravimetric analysis of KU-2 (sample weight 60.4 mg).

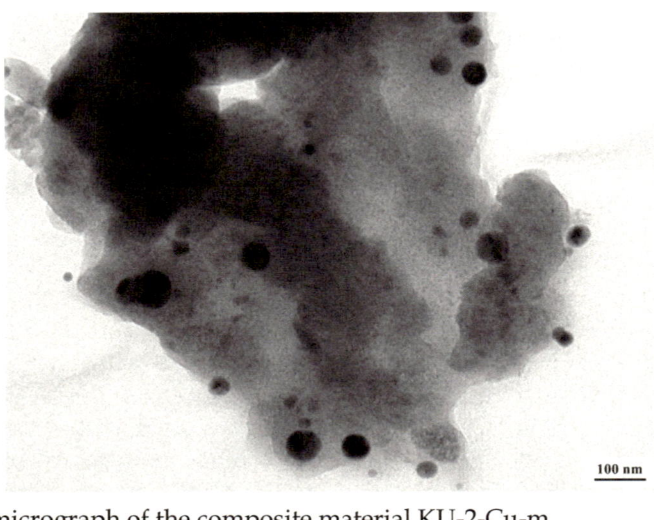

Fig. 5. Electron micrograph of the composite material KU-2-Cu-m.

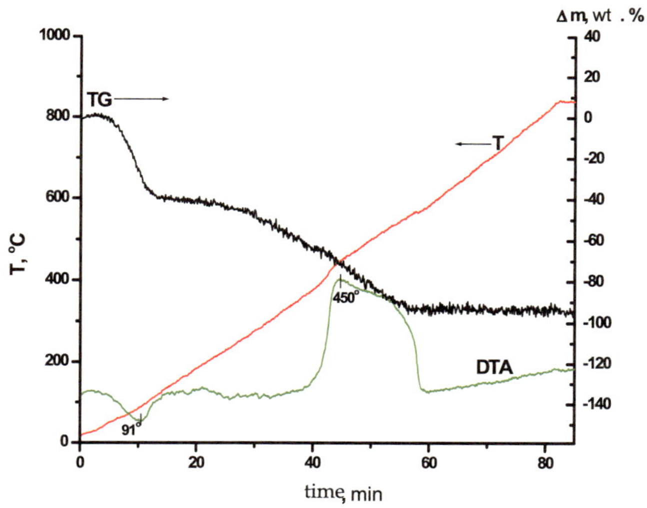

Fig. 6. Results of thermal gravimetric analysis of KU-2-Cu-m (sample weight 54.2 mg)

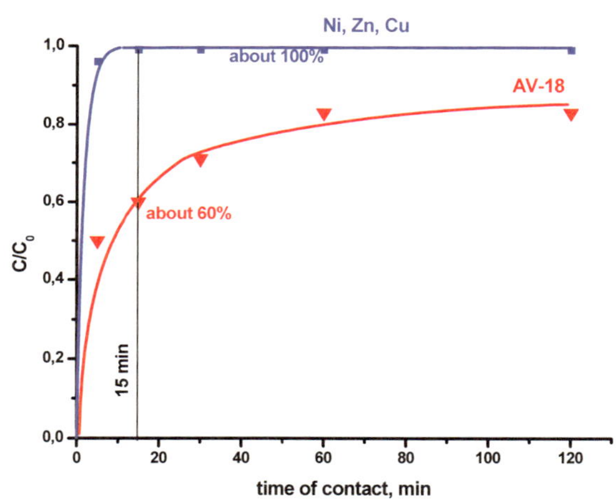

Fig. 7. Fraction of I_2 taken up by AV-18 anion exchanger and by wet samples of composite materials KU-2-M^{II}-m (M^{II} = Ni, Zn, Cu) as a function of the contact time of the solid and liquid

Thus, wet composite materials based on KU-2 cation exchanger containing *d*-elements, after modification with hydrazine hydrate, show high localizing ability toward molecular iodine in aqueous solutions. At the same time, from the viewpoint of possible application of these materials and their transportation to the site of use, it is important to evaluate the localizing properties of these materials after their drying to the friable state. Below are data on the degree of localization of molecular iodine in simulated primary coolant in the presence of dried composite materials KU-2-M^{II}-m (M^{II} = Ni, Zn, Cu) (Table 8). The analogous data for the wet samples are given for comparison.

It can be seen that the degree of I_2 localization from simulated primary coolant in the presence of dried Ni and Zn-containing samples is close to that obtained with the wet samples. At the same time, the localizing ability of the Cu-containing samples decreases on drying by a factor of approximately 5. This may be due to the fact that drying of the Cu-containing materials is accompanied by changes in the state of a part of metallic copper. Thus, KU-2-M^{II}-m composites (M^{II} = Ni, Zn) can be used in both wet and dried state.

Composite material	C/C_0	
	Wet sample	Dried sample
KU-2-Ni-m	0.960 ± 0.004	0.961 ± 0.003
KU-2-Zn-m	0.968 ± 0.004	0.970 ± 0.005
KU-2-Cu-m	0.963 ± 0.004	0.180 ± 0.001

Table 8. Degree of conversion of I_2 into ionic species (C/C_0) with the KU-2-M^{II}-m composites (M^{II} = Ni, Zn, Cu) in simulated primary coolant at 20°C ($[I_2] = 1\times10^{-5}$ M, [NaCl] = 1.17 mg/dm^3, [NH_4Cl] = 29.96 mg/dm^3, [H_3BO_3] = 5 g/dm^3, contact time 120 min)

Along with studying the localizing ability of the materials synthesized, it was interesting to estimate the maximal amount of I_2 that can be converted in aqueous solutions and to compare the data obtained with the analogous characteristics of AV-18 anion exchanger. The experiment was performed as follows. A 250-ml Erlenmeyer flask was charged with 500 mg of KU-2-M^{II}-m composite (M^{II} = Ni, Zn, Cu) or AV-18 anion exchanger. Then 100 cm^3 of a 9×10^{-3} M aqueous solution of I_2 was added. The suspension was vigorously stirred for 120 min. Then an aliquot was taken from the solution, and the I_2 amount in it was determined colorimetrically. If the sample contained no I_2, one more portion (100 cm^3) of the same aqueous solution of I_2 was added. The experiment was performed until unchanged I_2 appeared in the mother liquor. The maximal amounts of I_2 localized from various aqueous solutions with the materials developed are given in Table 9. As can be seen, for distilled water and simulated primary coolant the results are well consistent. The amount of I_2 localized with wet KU-2-M^{II}-m samples (M^{II} = Ni, Zn) is considerably higher than with AV-18 anion exchanger.

Material	Aqueous solution	Localized I_2, mg/g	
		wet sample	dried sample
KU-2-Ni-m	Water	> 3100	> 3100
	Primary coolant	> 3100	> 3100
KU-2-Zn-m	Water	> 3100	> 3100
	Primary coolant	> 3100	> 3100
KU-2-Cu-m	Water	> 3100	101 ± 5.05
	Primary coolant	> 3100	81.5 ± 4.40
AV-18	Water	-	158 ± 7.6
	Primary coolant	-	208 ± 10.0

Table 9. Maximal amount of I_2 converted into ionic species in the presence of the developed materials in distilled water and simulated primary coolant at 20°C ([NaCl] = 1.17 mg/dm^3, [NH_4Cl] = 29.96 mg/dm^3, [H_3BO_3] = 5 g/dm^3, contact time 120 min)

Thus, composite materials KU-2-M^{II}-m (M^{II} = Ni, Zn) that we developed allow more than 95.0% localization of molecular iodine from aqueous solutions. At the same time, it should be taken into account that iodine after the conversion remains in the solution phase, i.e., its back conversion to the molecular form is probable. Therefore, we developed a two-step procedure for iodine localization from aqueous solutions. The experiment on two-step localization of molecular iodine was performed as follows. To 100 cm^3 of a 1×10^{-5} M I_2 solution, we added 500 mg of the composite material KU-2-M^{II}-m (M^{II} = Ni, Zn) and vigorously stirred the suspension for 5 min. Then the mother liquor was separated by decantation, the composite material was washed with two portions of distilled water, and the wash waters and mother liquor were combined. An aliquot of the combined solution was taken for colorimetric determination of the concentration of molecular iodine. After that, the combined solution was spiked with ^{131}I and thoroughly mixed for 5 min, and 5.0 g of AV-18 anion exchanger was added. The suspension was vigorously stirred for 15 min. The anion exchanger was separated by centrifugation and washed with two portions of distilled water. The mother liquor was combined with the wash waters. The amounts of ^{131}I in the solution and resin were determined by γ-ray spectrometry. Below are results of experiments on localization of molecular iodine from aqueous solutions by the two-step scheme (Table 10). As can be seen, the use of the two-step treatment first with the composite material and then with AV-18 anion exchanger allows more than 95.0% localization of the molecular form of radioactive iodine from aqueous solutions of various compositions.

Material	C/C_0	
	water	primary coolant
KU-2-Ni-m	0.956	0.961
KU-2-Zn-m	0.959	0.966

Table 10. Degree of I_2 localization (C/C_0) from aqueous solutions by two-step treatment first with the composite material and then with AV-18 anion exchanger at 20°C ($[I_2] = 1\times10^{-5}$ M, [NaCl] = 1.17 mg/dm^3, [NH_4Cl] = 29.96 mg/dm^3, [H_3BO_3] = 5 g/dm^3, contact time 120 min)

4. Conclusion

It should be noted in conclusion that the use of the materials we developed does not require changes in the NPP design and significant changes in the units for treatment of aqueous primary coolant. At the same time, these materials ensure not only a decrease in the amount of radioactive iodine in aqueous primary coolant, but also a decrease in the radioactive iodine discharge into working rooms of an NPP. That is, the load on iodine filter is reduced both in the course of the reactor outage and spent nuclear fuel reloading and at normal NPP operation.

5. Acknowledgment

Authors thank Department of Chemistry and Material Science of Russian Academy of Sciences (program of fundamental research № 8 /academician A.Yu.Tsivadze/) for financial support.

6. References

Charlot, G. (1961). Les methods de la chimie analytique. Analyse quantitative minerale, Paris: Masson, 1961, 4th ed. Translated under the title: Metody analiticheskoi khimii. Kolichestvennyi analiz neorganicheskikh soedinenii, Vol.2, 1108, Khimiya, Moscow, Russia.

Chemist's Handbook (1964). Vol.3, 127, Khimiya, Moscow, Russia.

Fedushchak, T. A.; Ermakov, A. E. et al. (2008). Physico-chemical properties of surface of copper nanopowders produced by blasting and gas-cycle methods, Zhurnal Fizicheskoy Khimii (Russia), Vol.82, No.4, (April 2008), pp. 708–712, ISSN 0044-4537

Glanneskog et al., 2006 Glanneskog, H.; Liljenzin, J.-O. & Sihver, L. (2006). Reactions between reactive metals and iodine in aqueous solutions, Journal of Nuclear Materials, Vol. 348, No.1-2, (January 2006), pp. 87–93, ISSN 0022-3115

Krasavina, E. P.; Kulyukhin, S. A. et al. (2008). Nanocomposites for localization of radioactive iodine compounds in aqueous and steam-air phases at NPPs, Abstracts of Papers of VIII All-Russia Conference "Physical Chemistry of Ultradisperse and Nanosystems", pp. 279–280, ISBN 978-7262-0987-6, Belgorod, Russia, November 10-14, 2008

Krebs W.-D. (1999). The new generation, Siemens Power Journal, No.4, (April 1999), pp. 10-13

Kriytskii V. G.; Rodionov Yu. A. & Ampelogova N. I. (2005). Behavior of radioactive iodine isotopes in a contour of repeated compulsory circulation at RBMK, Atomnaya Energiya (Russia), Vol.99, No.2, (August 2005), pp. 103 - 108, ISSN 0004-7163

Ksenzenko, V. I. & Stashkevich, D. S. (1979). Chemistry and Technology of Bromine, Iodine, and Their Compounds, Khimiya, ISBN 5-7245-0845-1, Moscow, Russia.

Kulyukhin, S. A.; Mizina, L. V. et al. (2008). Sorption of I_2 and $CH_3{}^{131}I$ on ion-exchanhe materials from water solutions at 20°C, Radiokhimiya (Russia), Vol.50, No.3, (April 2008), pp. 258–260, ISSN 0033-8311

Kulyukhin, S. A.; Mizina, L. V. et al. (2008a). Nanocomposites for Localization of Radioactive Iodine at Nuclear Power Facilities, Abstracts of Papers of International Forum on Nanotechnologies, Vol.1, pp. 342–344, Moscow, Russia, December 03-05, 2008

Kulyukhin, S. A.; Mizina, L. V. et al. (2009). Sorbent based on silica gel for localization of volatile radioactive iodine forms, Russia Patent no. 2 345 431, Priority of January 17, 2007, Registered January 27, 2009.

Lidin, R. A.; Andreeva, L. L. & Molochko, V. A. (2006). Constants of Inorganic Substances: Handbook, 552–578, Drofa, ISBN 5-7107-8085-5, Moscow, Russia.

Reactor Design (2002). Nuclear Engineering *International*, Advert. Supplement.

Tomorrow's reactors (2003). Alternatives, No. 2, (March 2003), pp. 4-5

33

The Glass Transition Temperature in Dental Composites

J.C.S. Moraes[1], M.M.D.S. Sostena[1] and C.R. Grandini[2]

UNESP – Univ Estadual Paulista, [1]Campus de Ilha Solteira and [2]Bauru

Brazil

1. Introduction

In dentistry, two types of composites are used: light- and chemical-cured. Light-cured composites are preferred by dentists, mainly due to their longer working time before the curing process is initiated by exposure to visible light. Such materials have been indicated for various clinical procedures such as aesthetic and functional restorations of anterior and posterior teeth, preventive restorations, cementation of endodontic posts, indirect restorations with prostheses, and bonding of orthodontic appliances.

The importance of the curing efficiency to the performance of light-cured composites is well established. The physical and mechanical properties of these materials are influenced by the level of monomer to polymer conversion achieved during the polymerization process (Manffezzoli et al., 1994). Light intensity and irradiation time are important factors to achieve an appropriate degree of conversion (DC) of photoactivated composites (Rueggeberg et al., 1992). DC also depends on the medium in which the material is being cured. An important physical property of the cured matrix is indicated by the glass transition temperature (T_g). The T_g of a dental composite is only of relevance if it lies within the range of intraoral temperatures (Knox et al., 2000; Moore et al., 1999). Inadequate polymerization determines a low final T_g in material. Intraoral temperatures that exceed the T_g may result in material softening and, consequently, in failure of the clinical procedure (Rueggeberg et al., 1992).

Glass transitions are one of the most important phenomena in polymeric materials because they determine the physical state and final mechanical properties of the material. The T_g represents a point at which dramatic changes take place in polymer properties. T_g is the temperature at which amorphous polymers undergo a transition from a glassy to a rubbery state. At temperatures lower than T_g, only localized molecular movements are observed (secondary relaxations); for $T > T_g$, the flow of chains occurs. Between both (i.e., $T = T_g$), a rather complicated behavior known as the primary (or α) relaxation is observed. Moreover, crystallization is possible for a great number of polymers so that such materials are formed by two phases differing highly in their rheological properties. In the case of totally amorphous polymers, the occurrence of localized molecular motion below T_g in a glassy polymer (β relaxation) is considered a precursor of the large-scale molecular motion (α relaxation). In the case of amorphous polymers, the former is attributed to the availability of a set of configurational states through the faster, uncorrelated, rotational motions of the smallest unit of the chains within the randomly distributed sites of frozen-in density fluctuations (McCrum et al., 1991).

For composites with application in dentistry, the T_g is an especially important parameter to be taken into account. The dental composites must have a T_g higher than the maximum temperature in the oral cavity in order to preserve the material's physical and mechanical properties. Therefore, to illustrate the importance of the glass transition temperature of the material to be applied in dentistry, this chapter reports the results obtained from the authors'study that evaluated the influences of thermal shock and of factors related to photopolymerization (exposure time and power density of the light-curing unit) on the T_g of three light-cured orthodontic composites.

2. Dental composites

The composites are formed by combining two or more chemically different components, which in general, have better physical and mechanical properties than either of the original components individually. The dental composites are basically composed of an organic matrix, inorganic particles (fillers) and a coupling agent, usually an organo-silane. The resin matrix is comprised of a monomer system, an initiator system for free radical polymerization, and stabilizers for maximizing the storage stability of the uncured composite and the chemical stability of the cured composite (Peutzfeldt, 1997). The monomers are the most important components of the organic part because they become converted to solids during the process of polymerization. The bisphenol A glycol dimetacrilate (Bis-GMA) is the most common dimethacrylate monomer used in current commercial dental composites (Figure 1), which is described as an ester of an aromatic dimethacrylate. Other monomers, such as bisphenol A ethoxylated dimethacrylate (Bis-EMA), urethane dimethacrylate (UDMA), and ethyleneglycol dimethacrylate (EGDMA) can also be found in the dental composites. The Bis-GMA is the reaction product of an epoxy resin (bisphenol A ethylene glycol) and glycidyl ester methacrylate (GMA). It is a long and rigid molecule with reactive carbon-to-carbon double bonds at both ends. Its high molecular weight, relatively high viscosity, and the presence of aromatic groups contribute to the rigidity of the molecular chain.

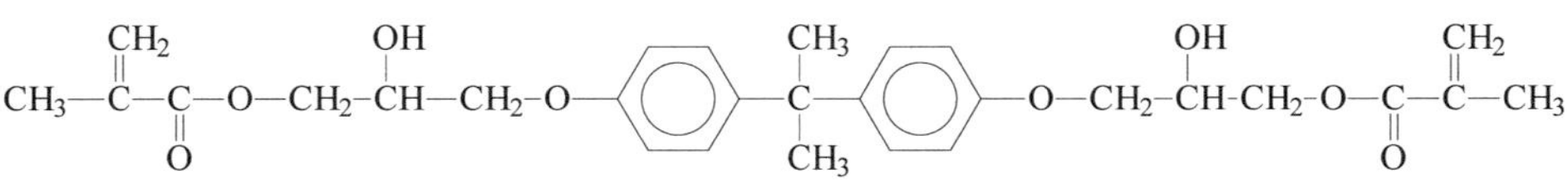

Fig. 1. Bis-GMA molecule.

The high viscosity of Bis-GMA is lowered by dilution with a monomer of lower molecular weight to facilitate the incorporation of inorganic particles and to improve the material homogenization. The most commonly used diluent is triethyleneglycol dimethacrylate (TEGDMA) (Figure 2), which is a relatively flexible linear molecule with carbon unsaturated bonds at both ends. All other components being equal, an increase of filler content will improve several physical and mechanical properties. However, the dilution of BisGMA also has negative effects such as increased polymerization shrinkage of the mixture.

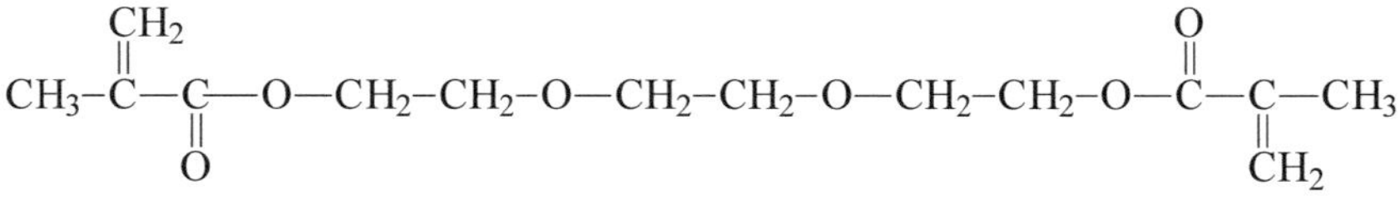

Fig. 2. TEGDMA molecule.

Different fillers have been used in formulating dental composites: barium glass, quartz, pyrolytic silica, silicate-lithium-aluminum, and colloidal silica. The incorporation of filler particles into the organic matrix improves the physical and mechanical properties of the cured material. The amount of filler added and the size and form of the particles are factors that determine the final properties of the composite such as fracture toughness, wear resistance, hardness, and thermal expansion. Furthermore, the filler particles tend to influence the stiffness of the polymer chain. The fillers are treated with silane coupling agents to reinforce the adhesion between the filler and the matrix polymer and also to increase the hydrolytic stability of the polymer (Halvorson et al., 2003).

The properties of composites are determined by the properties of the components, by the shape of the filler phase, by the morphology of the system, and by the nature of the interface between the phases (Nielsen & Landel, 1994). A great variety of properties can be obtained by changing one of the components. For example, the use of monomers that form a rigid polymer network with higher cross-linking density can lead to a composite with high T_g. Because the composition greatly influences the behavior of dental composites, better clinical results will be obtained as professionals gain more information about materials and their clinical implications.

2.1 Polymerization process

The cure reaction of the dental composite occurs through a chemical process of polymerization where the monomers join together through chemical bonds to form large-chain molecules called polymers.

The dimethacrylate undergoes a chain polymerization also called addition polymerization. This reaction begins when a reactive species and a monomer react to form an active center (Fraunhofer, 2010; Peacock & Calhoun, 1959; Tager, 1978). The chain polymerization process occurs in a sequence of three stages: initiation, propagation and termination. In initiation, the free radical (R•) reacts with the monomer molecule (M), producing another active free radical species (excited monomer molecule, MR•), which is capable of further reaction. In propagation, this process is followed by successive additions of monomer units to the polymer resulting in a macroradical (MR_n•).

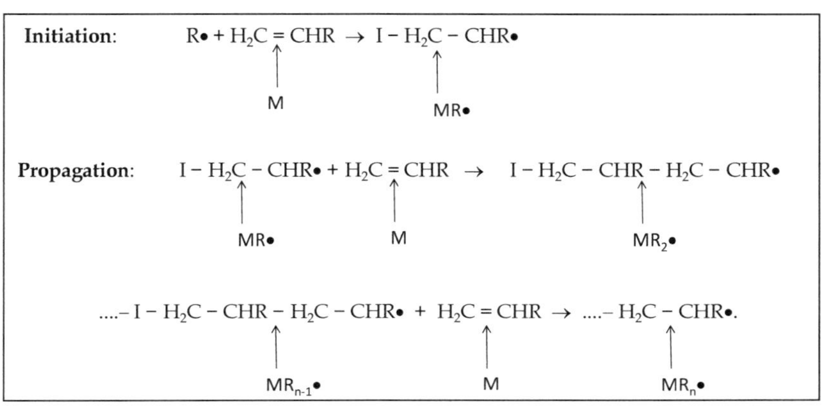

The resulting macroradical may change into an inactive polymer molecule (chain termination). The termination of polymerization occurs with the suspension of chain growth due to the disappearance of the active center through combination, disproportionation, or transfer, depending on the type of monomer and polymerization conditions. In combination, two free radicals interact and form a covalent bond. Disproportionation occurs when two growing chains interact, while a transfer occurs when the growth of the original chain stops and another shorter chain starts the propagation process (Fraunhofer, 2010). These reactions produce dead polymer chains that are not capable of further additions (McCabe & Walles, 2008). The term dead polymer represents the cessation of growth for the propagation radical.

Termination

Combination:

$$....-H_2C-CHR\bullet + \bullet RHC-CH_2-.... \rightarrow ....-H_2C-HRC-CHR-CH_2-....$$

Disproportionation:

$$....-H_2C-CHR\bullet + \bullet RHC-CH_2-.... \rightarrow ....-H_2C=CHR + RH_2C-CH_2-....$$

Transfer:

$$....-H_2C-CHR\bullet + H_2C=CHR \rightarrow ....-CH_2-CHR + H_2C-CHR\bullet$$

The polymerization of dimethacrylate results in a highly cross-linked structure. However, the monomer conversion is not complete and there will always be the presence of considerable amounts of residual monomer in the cured composite. The extent to which the monomer is converted to polymer is called the DC. T_g has been found to be well correlated to DC in polymers, based on a number of different dimethacrylates (Baran et al., 1994; Ferracane & Greener, 1986; as both cited in Peutzfeldt, 1997).

Free radicals (R•) can be generated by chemical activation or external activation energy (light or heat). The dental composites polymerized by chemical systems are named chemical-cured and those polymerized using external irradiation are named light-cured composites.

The first composites used in dentistry were chemical-cured composites. These materials are presented in the form of two pastes: one contains an initiator, generally benzoyl peroxide, and the other contains an activator, a tertiary amine (typically N, N - dimethyl-p-toluidine). When the two pastes are mixed, a reaction occurs between the initiator and the activator, resulting in the appearance of free radicals leading to the polymerization of Bis-GMA. The reaction is characterized by heat release and contraction of the polymer. In the chemical-activated system, a conversion is expected to occur uniformly throughout the volume of the material. However, the chemical-cured composites have some disadvantages: the possibility of the incorporation of air bubbles during mixing, the possibility of incomplete mixing of the pastes, and the possible difficulties resulting from the operator having to proportion the pastes. Moreover, there is a reduced working time due to rapid polymerization. These facts contributed to the replacement of the chemical-cured composites with light-cured materials.

The visible light-cured composite is available in one-paste form, packaged in syringe. This system has a photoinitiator, typically camphorquinone, which has an absorption spectrum in the range between 400 and 500 nm, with an absorption peak around 467 nm. In the initiation stage, the camphorquinone and a reducing agent represented by a tertiary amine, the dimetilaminoetilmetacrilato (DMAEMA), generate free radicals when irradiated by visible light, initiating the polymerization process (Figure 3). When light-cured composite are irradiated, the molecules of camphorquinone exhibit an excited triplet state that reacts with tertiary amines (DMAEMA), forming complex excitations (exciplex) that are converted in free radicals (Figure 3). The amounts of initiator and activator in the composite formulation may influence the DC of the material (Schneider, et al., 2008).

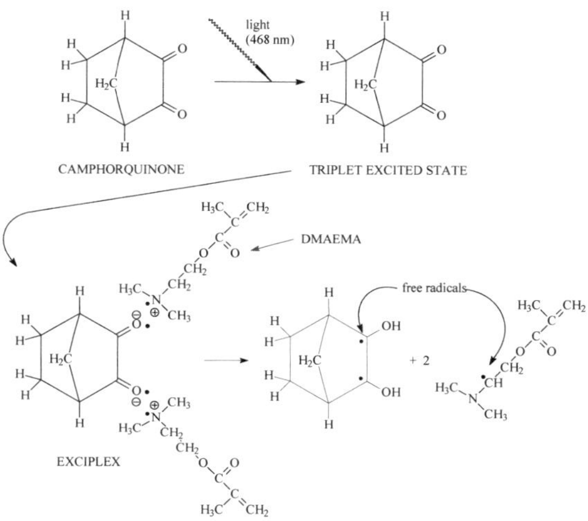

Fig. 3. Scheme shows the mechanism of the free radical formation of light-cured composites.

Halogen lamps and LEDs (Light-emitting diode) are light sources commonly used to activate the light-cured composites. In the lamp units, the light is produced when electrical energy heats a small tungsten filament to extremely high temperatures (Dunn et al, 2002). In this type of light source, a filter is used to select a particular range of wavelengths relevant to the activation of camphorquinone.

In the LED source, the light is generated through junctions of doped semiconductors, emitting light beams with wavelengths between 450 and 490nm, coinciding exactly with the absorption band of the spectrum of camphorquinone. This source presents a higher efficiency than the halogen source due to the greater coupling between its emission band and the camphorquinone absorption band (Stahl, et al., 2000). The halogen light unit has a broader emission spectrum when compared to the camphorquinone absorption spectrum.

Several concepts are relevant to the emission of light by light-curing units. The light intensity of the source is also called the power density or irradiance and it is expressed as mW/cm^2. The power density (PD) is the ratio of average power (P) and unit area (A) of the active tip, i.e., ($PD = P/A$). Moreover, the energy density (ED) represents the amount of power supplied to a given irradiated surface, which is expressed as J/cm^2. The energy density (ED) is the product of the power density (PD) of the source and the exposure time (t) from surface to light, or ($ED = PD \times t$).

In the process of the polymerization, it is necessary to expose the material to light; this process can significantly affect the number of free radicals formed, and consequently, the cure of material. A sub-polymerization can reduce the strength of the material and affect its biocompatibility. Thus, the amount of radiation energy used must be sufficient to promote the complete polymerization of the material. Factors such as light intensity, energy density, and exposure time are relevant for evaluating the efficiency of the source used for the cure of the composite. The light has an influence on the DC of materials, affecting its mechanical and physical properties.

3. Glass transition and measurement techniques

Glass transition is the temperature where the polymer goes from a hard, glass-like state to a rubber like state and it provides a method to characterize a property of a polymeric material. The best way to envision this type of transition is to put a rubber band (rubber like state, very flexible) into a container of liquid nitrogen. When removed, the rubber band is solid and inflexible (glass state), and in fact, the rubber band can be shattered. Upon standing and warming to room temperature, the rubber band will again become flexible and rubbery (rubber like state).

Routinely, there are two primary ways to measure T_g: Differential Scanning Calorimetry (DSC, the "official" method), and Mechanical Analysis (MA). DSC defines T_g as a change in the heat capacity as the polymer matrix goes from the glass state to the rubber state. This is a second order endothermic transition (requiring heat to go through the transition) so, in the DSC, the transition appears as a step transition and not as a peak, as might be seen with a melting transition. MA defines T_g in terms of the mobility of the chains as the polymer goes from glass to rubber state with the associated change in free molecular volume. Each of these techniques measures a different result of the change from glass to rubber. The DSC is measuring a heat effect, whereas the MA is measuring a physical (mechanical) effect. Both techniques assume that the effects are created over a narrow range of a few degrees in temperature. If the T_g is very broad, it may not be seen with either approach. The DSC and MA may also often give results that differ from one another by 5-10 °C when used to test a polymer. Moreover, some polymers are more amenable to DSC or to MA, because the transition is easier to observe using one technique over the other (McCrum, 1991).

Mechanical spectroscopy (MS) is a MA particularly well suited to the study of the behavior of dental composites because it studies the absorption spectra of mechanical energy under the conditions of applied periodic external stress.

In the elastic analysis of materials, it is assumed that elastic strain is a function of stress only. This is not strictly true since there is time dependence to the elasticity. In metals, the effect is very small and generally negligible, but in polymers, the effect is much more significant. The general name for this time dependence is anelasticity.

A solid is considered anelastic when there is a phase difference or delay of the effect (strain) in relation to the cause (stress), making Hooke's law with time as a variable. From a thermodynamic point of view this means that, in response to a change in the mechanical forces applied, it takes some time for the anelastic solid to achieve equilibrium. Then, the auto-adjust system, in response to a change in the external variable, is known as relaxation. When the external variable is mechanical (a strain or stress), the phenomenon is known as anelastic relaxation. One can, therefore, interpret anelastic behavior as a manifestation of internal relaxation processes.

The anelastic behavior of a material can be described in terms of a group of parameters considered to be intrinsic attributes of the material. The starting point for this approximation is a stress-strain relation in the form of a linear differential equation involving stress, strain, and their derivatives in time (Nowick & Berry, 1972). The simplest differential stress-strain equation capable of representing anelasticity involves three independent parameters and the mechanical model most appropriate to describe the behavior of an anelastic solid, where it is present only as a relaxation process, is known as the Three Parameters Model; it consists of a Voight unit composed of a spring connected in parallel with a damper, associated in series with another spring.

When a uniform stress is applied in t = 0, the spring deforms instantly, while the Voigt unit will take some time to deform completely, due to the damper. When the stress is eliminated, the spring will return immediately to the starting position while the Voigt takes some time to reach its initial position, also due to the damper. Whereas the damper consists of a viscous linear liquid, the stress-strain differential equation for the Three Parameters Model is:

$$J_R \sigma + \tau J_U \dot{\sigma} = \varepsilon + \tau \dot{\varepsilon} \quad (1)$$

where σ is the stress, ε is the strain, τ is the relaxation time, J_R is the relaxed compliance, and J_U is the unrelaxed compliance.

When a dynamical stress is applied to the system, the real and imaginary parts of the complex compliance are (Schaller et al., 2001):

$$J_1 = J_U + \frac{\delta J}{1+(\omega\tau)^2} \quad (2)$$

$$J_2 = \delta J \frac{\omega\tau}{1+(\omega\tau)^2} \quad (3)$$

where $J_1 = \varepsilon_1/\sigma_0$, $J_2 = \varepsilon_2/\sigma_0$, $J^*(\omega) = J_1 - i J_2$, $\delta J = J_R - J_u$, σ_o is the stress amplitude, ε_1 and ε_2 are, respectively, the components of the strain in phase and 90º out of phase with the stress, ω is the vibration angular frequency, and in $\omega=2\pi f$, f is the vibration frequency.

Equations (2) and (3) are called the Debye equations and are shown in Figure 4. The quantity J_1 will of J_U at high frequencies ($\omega\tau >> 1$) to J_R at low frequencies ($\omega\tau << 1$). The quantity J_2 shows a peak function, it has small values in high and low frequencies and goes to a maximum at $\omega\tau = 1$. Any function that varies with frequency $\omega\tau/(1 + \omega^2\tau^2)$ can be called a Debye peak.

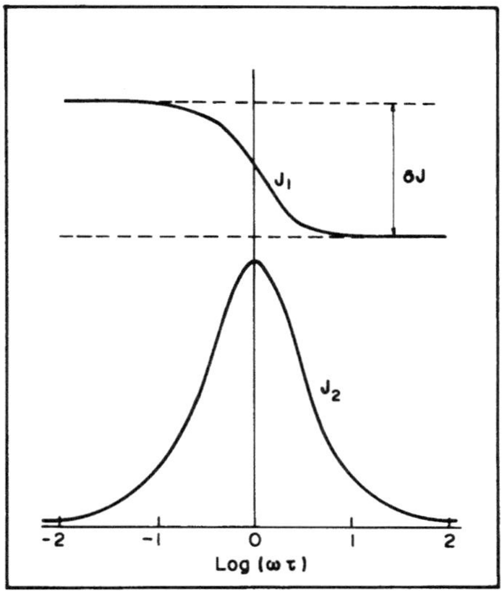

Fig. 4. Comparison of $J_1(\omega)$ and $J_2(\omega)$ as a function of log(ωt) for a single relaxation process.

Internal friction is related to the real and imaginary parts of the compliance by Nowick & Berry (1972). Then,

$$Q^{-1} = \Delta\left(\frac{\omega\tau}{1+(\omega\tau)^2}\right) \qquad \Delta << 1 \tag{4}$$

where the dimensionless quantity Δ is called relaxation strength.
Using the Arrhenius equation for the relaxation time, the internal friction can be written as a function of temperature:

$$Q^{-1} = \frac{Q_m^{-1} T_m}{T} \operatorname{sech}\left[\frac{E}{k}\left(\frac{1}{T_m} - \frac{1}{T}\right)\right] \tag{5}$$

where $\Delta=(2Q_m^{-1} T_m)/T$ for $\Delta<<1$, Q^{-1}_m is the maximum of internal friction at temperature T_m, E is the activation enthalpy, k is the Boltzmann constant, and T is the absolute temperature.
The principle of internal friction measurement is very simple; the sample is put to vibrate on its fundamental mode (free or forced vibration), which causes the dissipation of energy in the heat form, due to internal friction. Internal friction can be measured statically or dynamically. Dynamic experiments are most suitable because of the relatively short time needed for their execution. Basically, dynamic experiments are divided into three groups: wave propagation, sub-resonant, and resonant.
In the case of wave propagation, high frequency short pulses (around 10^6-10^9 Hz) propagate through the sample and the attenuation coefficient and the velocity of these waves are measured. The internal friction is related to the attenuation coefficient of the waves and the elasticity modulus is related to the velocity of the propagation of these waves. Sub-resonant experiments highlight the torsion pendulum with forced vibration and dynamic-mechanical analysis equipment (DMA). The main advantage of this equipment is the possibility of working in a large and continuous frequency range (10^{-5}-10^2 Hz), without the need to change the geometry of the sample, i.e. with the same sample. Resonant experiments constitute the largest and oldest group of methods for measuring mechanical spectroscopy. The most important techniques are the torsion pendulum, vibrant bars, and composed oscillators, with frequency from 10^{-1} to around 10^6 Hz. Requiring different sample geometries for each technique is the major inconvenience of these techniques.
In general, a resonant system must have two elements: the elastic element (which, in fact, is anelastic), and inertia. The situation is considerably simplified when this system, for example, is a wire stretched with a weight placed at one end, possessing torsional vibrations. This is the case of the torsion pendulum, in which the deflection, at any point, can be expressed in terms of a single parameter, the angular displacement, Θ, of the inertial component. The physical quantity that is responsible for this shift is the torque, Γ, which operates in the sample. In the case of an ideal elastic solid, $\Gamma = G\Theta$, where G is the rigidity modulus of the sample. For an anelastic solid $\sigma = M^* \varepsilon$, with $\sigma \equiv \Gamma$, $M^* \equiv G^*$ and $\varepsilon \equiv \Theta$, and they result in,

$$\omega^2 \approx G/I \tag{6}$$

$$\delta \approx \pi\phi \tag{7}$$

where I is the inertia momentum, ϕ is the phase difference between σ and ε, and δ is a constant.
Considering two oscillations separated by N periods:

$$\delta = \frac{1}{N}\ln\left(\frac{A_1}{A_2}\right) \quad (8)$$

The dimensionless quantity δ represents the natural logarithm of the ratio of amplitudes A_n on two successive vibrations, and is called the logarithmic decrement. Then, the internal friction is given by:

$$Q^{-1} = \tan\phi \approx \phi \approx (\delta/\pi) = \frac{1}{N\pi}\ln\left(\frac{A_1}{A_2}\right) \quad (9)$$

4. Evaluation of glass transition temperature by mechanical spectroscopy

The specifications, main ingredients and manufacturers of the orthodontic composites used are displayed in Table 1.

Composite	Manufacturer	Composite composition	Inorganic content (wt%)
Fill magic	Vigodent (Rio de Janeiro - RJ, Brazil	BisGMA, Methacrylate acid ester, and fluorosilicate glass	38.1
Ortho Lite Cure	Orthosource (Hollywood - CA, USA	BisGMA, TEGDMA, glass and silica	70.0
Transbond XT	3M Unitek (Monrovia - CA, USA)	BisGMA, N-dimethyl benzocaine, hexafluoride phosphate, silane and silica	74.6

Table 1. Orthodontic composites used in this study.

Beam-shaped specimens of each material were prepared in a brass mold with internal dimensions of 40 x 4 x 4 mm for T_g measurements. The samples were irradiated using two different LED light-curing units: Free Light II Elipar (3M ESPE, Seelfeld - Germany) with a power density 232 mW/cm^2 and LDIII (Gnatus, Ribeirão Preto - Brazil) with a power density of 52 mW/cm^2, which will be named here for convenience as F1 and F2, respectively. The power output was measured with a power meter (Model 13PEM001; Melles Griot, Irvine, CA, USA). F1 was used for an exposure time of only 40 s; the exposure times for F2 were 40, 60, 90, and 120 s due to its lower power density. Two specimens were made for each exposure time and were irradiated on the opposite faces of the lateral surface. The samples were kept in the dark at room temperature until completely cured. Half of the specimens were submitted to thermo shock (or thermocycling) before T_g measurements.
Thermal shock was performed in a humid environment in order to simulate daily meals. For this, the samples were immersed in tanks with distilled water at 5, 37, and 60 °C. The immersion time in each tank was 15 s. This process was carried out for an hour, with

approximately 42 cycles and a break of three hours at 37°C. The process was repeated four times a day for three days, totaling 504 cycles. The T_g was evaluated using the group of samples that was submitted (CT) to the thermocycling and the other group that was not (ST). Measurements of mechanical spectroscopy (internal friction) were performed in a torsion pendulum, operating at a 1.4 to 9.4 Hz frequency range.
A typical internal friction curve versus temperature is shown in Figure 5, where the maximum intensity occurs in the T_g. Table 2 shows the T_g value obtained for the investigated composites in the CT and ST conditions.

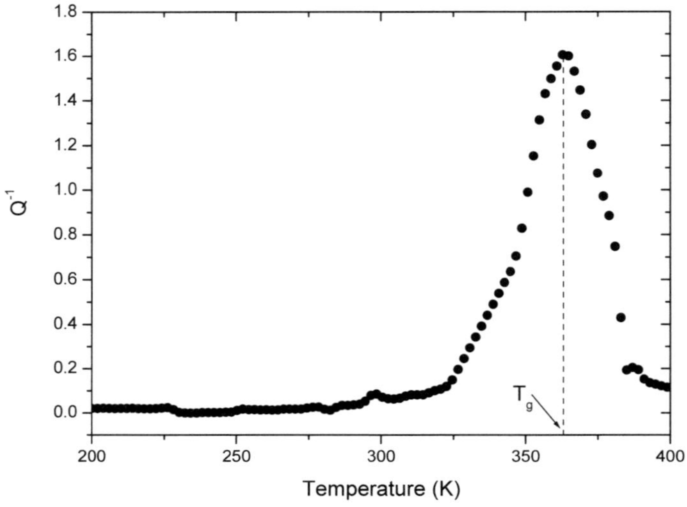

Fig. 5. Typical internal friction curve obtained by mechanical spectroscopy technique.

The higher T_g values were reached when the three composites were irradiated with the F1 source for both conditions (ST and CT). The Ortho Lite and Transbond composites presented higher T_g values in the ST condition and the lowest T_g values were observed for Fill Magic. The three composites showed T_g values within the range of oral temperature (273 to 340 K).
The effect of *exposure time* could be observed when the composites were irradiated with the source F2. For the ST condition, the T_g value of the Fill Magic and Ortho Lite composites increased when the exposure time was increased. In contrast, the T_g of the Transbond did not depend on the duration of the light exposure time. For the CT condition, when the exposure time was increased, the T_g of Fill Magic composite increased and the T_g of Ortho Lite decreased. For Transbond, there was an initial decrease between 40 and 90 s and the highest value was obtained at 120 s. For both conditions (ST and CT), the values of Tg obtained with an exposure time of 120 s reached values close to those obtained using 40 s with source F1, except for the Ortho Lite composite in the CT condition.
The thermal shock factor changed the T_g final reached by the three composites. The thermal shock prompted a rise in the T_g values of the Fill Magic composite and a reduction in T_g values for the other two composites.
The differences observed in the T_g of the studied composites can be attributed to the structural characteristics of each material. Factors such as monomer type, inorganic filler type, and content determine the physical and mechanical properties of the cured composite.

The T_g determines the physical state of a polymer and influences several properties. T_g variation has been attributed to various molecular parameters, such as molecular weight, stiffness of the cross-linked chains, and free volume entrapped in the network (Min et al., 1993). Other investigations have shown that T_g can be affected by cross-linked chains and filler content (Emami & Söderholm, 2005; Lu et al., 2001; Papadogiannis et al., 2003; Nielsen & Landel, 1994).

Composite	Source	Exposure time (s)	T_g (K)	
			ST	CT
Fill Magic	F1	40	363	377
	F2	40	308	300
	F2	60	295	338
	F2	90	323	342
	F2	120	357	362
Ortho Lite Cure	F1	40	387	372
	F2	40	370	332
	F2	60	380	318
	F2	90	368	315
	F2	120	383	307
Transbond XT	F1	40	384	376
	F2	40	378	354
	F2	60	373	340
	F2	90	378	318
	F2	120	378	372

Table 2. T_g values obtained for investigated composites in function of light source, exposure time, and treatment condition (ST and CT).

Figure 6 compares the internal friction curve of the three composites, which have different compositions (Table 1). The Fill Magic composite has in its composition the Bis-GMA monomer and 38.1 wt% of inorganic content. The Ortho Lite contains the Bis-GMA and TEGDMA monomers, and 70 wt% of glass and silica as inorganic filler. The Transbond has the Bis-GMA as monomer and 74.6 wt% of silanized silica as filler. As can be seen in the figure, the curves of the Transbond and Ortho composites are very similar, with peaks around 370 K. These two composites have practically the same amount of filler in their composition and the Ortho Lite has TEGDMA beyond Bis-GMA. Although the composites possess different organic composition, no difference is observed in the loss tangent curves. On the other hand, the Fill Magic presents a loss tangent curve more intense, with a peak around 310 K. This composite also has the Bis-GMA monomer in its composition, but has half the amount of filler that the other two composites have. Studies indicate that the increase of filler particles in the composite's composition introduces a broadening and reduction in the loss tangent value (Nielsen & Landel, 1994). The polymerization of multifunctional monomers produces network polymers with highly heterogeneous environments as they contain very highly cross-linked regions as well as less densely cross-linked regions connecting them (Kannurpatti et al., 1997). This heterogeneity results in a broad distribution of mobilities or relaxation times [Kannurpatti et al., 1997). Therefore, from comparing these three curves, one can conclude that the main cause of the observed difference in the peak position and curve form is the amount of filler. The Ortho and Transbond samples present broader curves and less intensity than the Fill Magic composite does because both have more filler particles in their compositions.

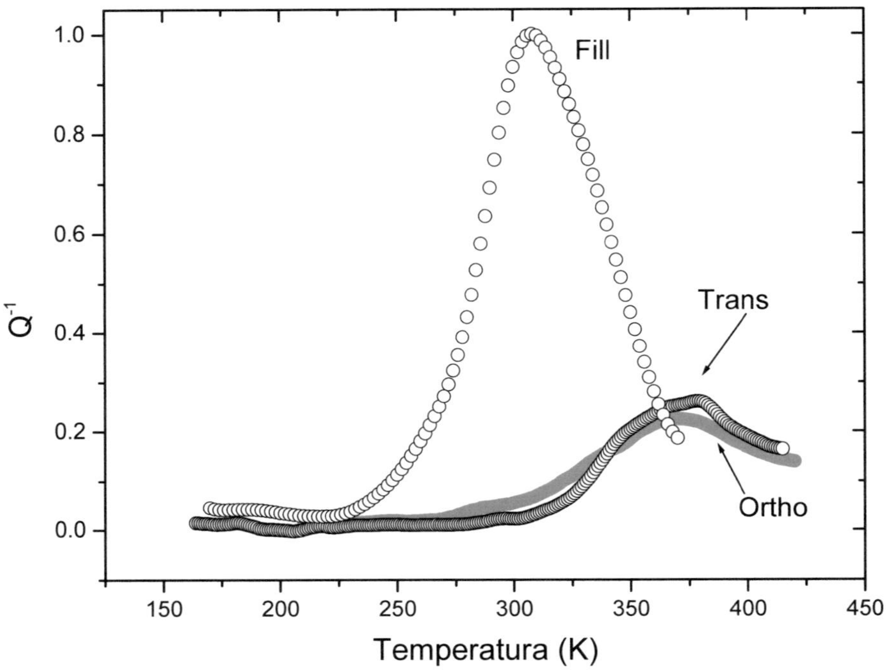

Fig. 6. Internal friction for three tested composites with the same exposure time (40 s) and thermal condition (ST).

The polymerization of dimethacrylates monomers produces densely cross-linked networks. The network is characterized by the heterogeneity caused by the formation of regions with higher and lower cross-linking. High heterogeneity is characteristic of a polymer structure with a larger number of cross-links. Hi Lu et al. attributed to broadening of the loss tangent peak to an increase in the heterogeneity of the network. Therefore, the characteristic of the relaxation peak of each composite points to differences in the final structure, suggesting that the Lite Cure and Transbond composites could have a higher cross-link density than Fill Magic, which would have a more homogeneous newtwork.

The influence of the exposure time in the T_g is shown in Figure 7A. The light source used to cure has a power density of 52 mW/cm², which is smaller than the minimum necessary to produce an adequate polymerization (233 mW/cm²) (Rueggeberg, 1994). Using this light-curing unit, it was possible to reach only an adequate polymerization and a T_g higher than the maximum oral temperature using an exposure time of 120 s. Better than power density, studies have shown that the properties of the light-cured composites are mainly influenced by the amount of energy delivered during irradiation (Dewaele et al., 2009). As mentioned earlier, the energy density is the product between the power density of the light source and the duration of irradiation (exposure time). T_g values above the oral temperature range were obtained when the three composites were irradiated for 40 s with source F1 (ED = 232 mW/cm² x 40 s = 9.28 J/cm²) in both thermal conditions. On the other hand, in the Fill Magic/source F2 combination it was possible to obtain adequate T_g value only with exposure times of 90 s (4.68 J/cm²) and 120 s (6.24 J/cm²) in the CT and ST conditions, respectively.

Figure 7B shows how thermal shock in a humid environment can influence the T_g of the material. For the Fill Magic composite, the obtained loss tangent curves indicate that the thermal treatment improved the mechanical and thermal properties of the material. The thermal treatment promoted a broadening and a decrease of intensity, indicating the formation of a more heterogeneous network with reduced chain mobility and, consequently, an elevation of the T_g. This effect could have been caused by thermal treatment in a humid environment (thermocycling). The absorbed water and temperature rise during the thermocycling contribute to molecular mobility, increasing the probability of contact between the groups that have not reacted, and therefore, of creating chemical bonds (Mesquita et al., 2006). Higher mobility and the increase of chemical reactions contribute to an increase in cross-link density, causing an increase in T_g.

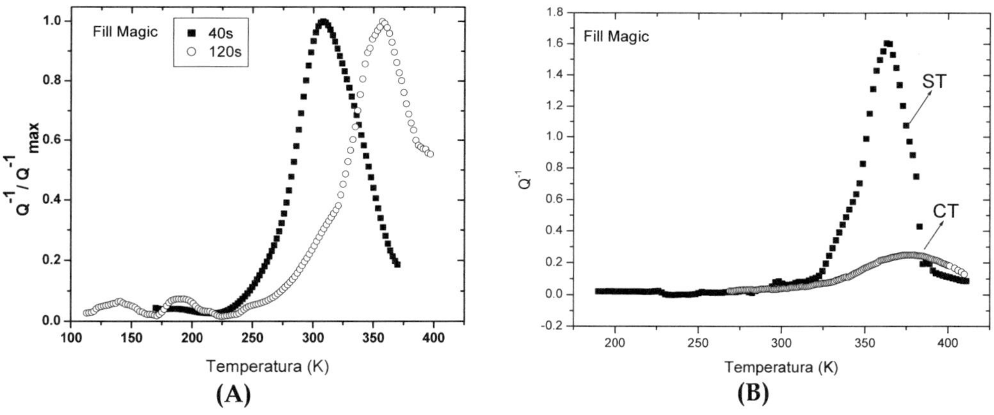

Fig. 7. Internal friction for the Fill Magic composite: (A) for 40 and 120 s exposure times in the same thermal conditions; (B) in different thermal conditions for the same exposure time

5. Conclusion

Within the limitations of in vitro study, the investigation with three orthodontic composites demonstrated the following:

To reach T_g above the oral temperature range, the energy density of light sources and the environment where the composite will be cured are relevant factors to be considered.

Each dental composite behaves differently when cured in a moist environment, subject to temperature variation.

The glass transition temperature reached by the composite depends on the amount of filler particles contained in its composition.

For application using composites with small amounts of filler in its composition and light sources with power density less than 232 mW/cm², the exposure time is a relevant factor to reach a T_g above the oral temperature range. For Fill Magic orthodontic composite, it was necessary to use 6.24 J/cm² of energy density (source F2 - 120 s) in the ST condition and 4.68 J/cm² (source F2 - 90 s) in the CT condition to reach an adequate T_g value.

For light sources with power density higher than 232 W/cm², an exposure time of 40 s is sufficient to reach satisfactory values of T_g, independent of temperature and humidity conditions.

6. Acknowledgment

The authors thank Vigodent and Orthosource for supplying the Fill Magic and Ortho Lite Cure orthodontic composites, respectively, and CAPES, CNPq, and FAPESP Brazilian research agencies for the financial support.

7. References

Baran, G.; Shin, W.; Abbas, A.; Wunders, S. (1994). Indentation cracking of composite matrix materials. *Journal of Dental Research*, Vol.73, No.8, (August 1994), pp.1450-6, ISSN 0022-0345.

Dewaele, M.; Asmussen, E.; Peutzfeldt, A.; Munksgaard, E.C.; Benetti, A.R.; Finné, G.; Leloup, G.; Devaux, J. (2009). Influence of curing protocol on selected properties of light-curing polymers: Degree of conversion, volume contraction, elastic modulus, and glass transition temperature. *Dental Materials*, Vol.25, No.12, (December 2009), pp.1576-84, ISSN 0109-5641.

Dunn, W.J.; Bush, A.C. A comparison of polymerization by light-emitting diode ad halogen-based light-curing units. *Journal of American Dental Association*, Vol.133, No.03, (March 2002), pp.335-341, ISSN 0002-8177.

Emami, N; Söderholm, K.J.M. (2005). Dynamic mechanical thermal analysis of two light-cured dental composites. *Dental Materials*, Vol.21, No.10, (October 2005), pp.977-83, ISNN 0109-5641.

Ferracane, J.L., Greener, E.H. (1986). The effect of resin formulation on the degree of conversion and mechanical properties of dental restorative resins. *Journal of Biomedical Materials Research*, Vol.20, No.1, (January 1986), pp.121-131, ISSN 0021-9304.

Fraunhofer, J.A. (2010). *Dental Materials at a Glance*. Wiley-Black Well. ISBN 978-0-8138-1614-2, Oxford, United Kingdom.

Halvorson, R.H.; Erickson R.L.; Davidson C.L. (2003). The effect of filler and silane content on conversion of resin-based composite. *Dental Material*, Vol.19, No.4, (June 2003), pp.327-33, ISSN 0109-5641.

Kannurpatti, A.R.; Anderson, K.J.; Anseth, J.W.; Bowman C.N. (1997). Use of "living"radical polymerizations to study the structural evolution and properties of highly corsslinked polymer networks. *Journal Polymer Science Part B: Polymer Physics*, Vol.35, No.14, (October 1997), pp.2297–307, ISSN 1099-0488.

Knox, J.; Jones, M.L.; Hubsch, P.; Middleton, J. (2000). The influence of orthodontic adhesive properties on the quality of orthodontic attachment. *Angle Orthodontist*, Vol.70, No. 3, (June 2000), pp.241-6, ISSN 0003-3219.

Lu, H.; Lovell, L.G.; Bowman, C.N. (2001) Exploting the heterogeneity of cross-linked photopolymers to create high Tg polymers from polymerizations performed at ambient conditions. *Macromolecules*, Vol.34, No.23, (November 2001), pp.8021-5, ISNN 0024-9297.

Maffezzoli, A.; Della Pietra, A.; Rengo, S.; Nicolais, L.; Valletta, G (1994). Photopolymerization of dental composite matrices. *Biomaterials*, Vol.15, No.15, (December 1994), pp.1221-8, ISSN 0142-9612

McCrum, N.G.; Read, B.E. and Willians, G. (1991). *Anelastic and Dielectric Effects in Polymeric Solids*. Dover Publications. ISBN-13: 978-048-6667-522, New York, USA.

McCabe, J. F; Walls, A.W.G. (2008). *Applied dental materials*. (9nd ed). Blackwell Munsksgaard., ISBN 978-1-4051-3961-8, Oxford, United Kingdom.

Mesquita, R.V.; Axmann, D.; Geis-Gerstorfer, J. (2006). Dynamic visco-elastic properties of dental composites resins. *Dental Materials*, Vol.22, No.3, (March 2006), pp.258-67, ISSN 0109-5641.

Min, B.G.; Stachurski, Z.H.; Hodgkin, J.H. (1993). Cure kinetics of elementary reactions of a DBEBA DDS epoxy-resin. 1. Glass-transition temperature versus conversion. *Polymer*, Vol.34, No.23, (December 1993), pp.4908-12, ISSN 0032-3861.

Moore, R.J.; Watts, J.T.F.; Hood, J.A.A.; Burritt, D.J (1999). Intra-oral temperature variation over 24 hours. *European Journal of Orthodontics*, Vol. 21, No. 3, (June 1999), pp.249-61, ISSN 0141-5387.

Nielsen, L.E & Landel, R.F. (1994). *Mechanical properties of polymers and composites* (2nd ed), Marcel Dakker Inc., ISBN 0-8247-8964-4 , New York, USA.

Nowick, A.S. & Berry, B.S. (1972). *Anelastic Relaxation in Crystalline Solids*. Academic Press. ISBN 978-012-5226-509, New York, USA.

Papadogiannis, Y.; Boyer, D.B.; Helvatjouglu-Antoniades, M.; Lakes, R.S.; Kapetanios, C. (2003). Dynamic viscoelastic behavior of rein cements measured by torsional resonance. *Dental Materials*, Vol.19, No.6, (September 2003), pp.510-6, ISSN 0109-5641.

Peacock, A.J; Calhoun, A.R. (1959). *Polymer chemistry: properties and applications* (1nd ed), Hanser Gardner Publication Inc, ISBN 978-1-56990-397-1, Ohio, USA.

Peutzfeldt, A. (1997). Resin composites in dentistry: the monomer systems. *European Journal of Oral Science*, Vol.105, No.2, (April 1997), pp.97-116, ISSN 0909-8836.

Rueggeberg, F.A.; Maher, F.T.; Kelly, M.T (1992). Thermal-properties of a methyl methacrylate-based orthodontic bonding adhesive. *American Journal of Orthodontics and Dentofacial Orthopedics*, Vol. 101, No. 4, (April 1992), pp.342-9, ISSN 0889-5406.

Rueggeberg, F.A.; Caughman, W.F.; Curtis Jr, J.W. (1994). Effect of light-intensity and exposure duration on cure of resin composite. *Operative Dentistry*, Vol.19, No.1, (January-February 1994), pp.26-32, ISSN 0361-7734.

Schaller, R.; Fantozzi, G. and Gremaud, G. (2001). *Mechanical Spectroscopy Q^{-1}: With Applications to Materials Science.* Trans Tech Publications. ISBN 978-087-8498-765, Stafa-Zurich, Switzerland.

Schneider, L.F.J.; Pfeifer, C.S.C.; Consani, S.; Prahl, S.A.; Ferracane, J.L. Influence of phtoinitiator type on the rate of polymerization, degree of conversion, hardness and yellowing of dental resin composites. Dental Materials, Kidlinton, Vol.24, No 09, (September 2008), pp.1169-1177, ISSN 0109-5641.

Stahl, F.; Ashworth, S.H.; Jandt, K.D.; Mills, R.W. (2000). Light-emitting diode (LED) polimerisation of dental composites: flexural properties and polimerisation potential. *Biomaterials*, Vol.21, No.13, (July 2000), pp.1379-1385, ISSN 0142-9612.

Tager, A. (1978). *Physical Chemistry of Polymers*. Mir Publishers. ISBN 978-0714712369, Moscow, USSR.